Organometallic Compounds of

Cobalt, Rhodium and Iridium

Organometallic Compounds of

Cobalt, Rhodium and Iridium

Edited by

C. White

University of Sheffield

LONDON NEW YORK

CHAPMAN AND HALL

First published 1985 by Chapman and Hall Ltd
11 New Fetter Lane, London EC4P 4EE
733 Third Avenue, New York NY 10017

Phototypeset in the United States of America by
Mack Printing Company, Easton, Pennsylvania 18042
Printed in Great Britain by J.W. Arrowsmith Ltd, Bristol

ISBN 0 412 26800 0

Library of Congress Cataloging in Publication Data

Main entry under title:

Organometallic compounds of cobalt, rhodium, and iridium.
(Chapman and Hall chemistry sourcebooks)
Includes index.
1. Organocobalt compounds—Handbooks, manuals, etc.
2. Organoiridium compounds—Handbooks, manuals, etc.
3. Organorhodium compounds—Handbooks, manuals, etc.
I. White, C., 1944- . II. Series.
QD412.C6'74 1985 547'.05 84-29351

ISBN 0-412-26800-0

British Library Cataloguing in Publication Data

Organometallic compounds of cobalt, rhodium and iridium.—(Chapman and Hall chemistry sourcebooks)
1. Organometallic compounds
2. Transition metal compounds
I. White, C.
547'.056 QD411

ISBN 0-412-26800-0

Contents

Preface

This is one of the first volumes to be published in the series of *Chapman and Hall Chemistry Sourcebooks* which provides carefully tailored information to workers in specialized areas of chemistry. The information contained in this book is derived from the *Dictionary of Organometallic Compounds*, published in November 1984.

The organometallic compounds of cobalt, rhodium and iridium constitute an important class of organometallics: not only do they merit study in their own right but they are also increasingly used both as catalysts and as laboratory reagents in organic synthesis. It is therefore anticipated that this particular compendium will reach a wide readership.

The databank on the properties of organometallic compounds, which is represented in its current form by the *Dictionary of Organometallic Compounds* and its subset publications such as this volume, will be kept continuously up-to-date. Supplements to the main *Dictionary* will appear annually and revised editions of this *Sourcebook* will be published from time to time as demands permits.

C. White

Introduction

1. Using the Sourcebook

The *Sourcebook* is divided into element sections: within each section the arrangement of entries is in order of molecular formula according to the Hill convention (i.e. C, then H, then other elements in alphabetical sequence of element symbol; where no carbon is present, the elements including H are ordered strictly alphabetically).

Every entry is numbered to assist ready location and the entry number consists of a metal element symbol followed by a five-digit number.

Indexes

There are three printed indexes: a name index which lists every compound name or synonym in alphabetical order; a molecular formula index which lists all molecular formulae, including those of derivatives, in Hill convention order; and a CAS registry number index listing all CAS numbers included in the *Sourcebook* in serial order. All indexes refer to the entry number. In the name index an entry number which follows immediately upon an index term means that the term itself is used as the entry name but an entry number which is preceded by the word 'see' means that the term is a synonym to an entry name. In all three indexes an entry number which is preceded by the word 'in' refers the reader to a specified stereoisomer or derivative which is to ɔe found embedded within the particular entry.

In addition to the three printed indexes, each element section is preceded by a graphical structure index allowing the rapid visual location of compounds of interest. The structure index reproduces all structure diagrams present in that element section in reduced size and printed in entry number order.

The following paragraphs summarize important considerations in compiling the information in this *Sourcebook*. For more detailed information, see the Introduction to the *Dictionary of Organometallic Compounds* from which this *Sourcebook* derives.

2. Compound Selection

In compiling this *Sourcebook* the aim has been to include from the primary literature up to mid 1983:

(1) Compounds representative of all important structural types (typically, the parent member of each series, where known, together with a selection of its homologues).

(2) Any compound with an established use, such as in catalysis, as a synthetic reagent or starting material.

(3) Other compounds of particular chemical, structural, biological or historical interest, especially those thought to exhibit unusual bonding characteristics.

Some compounds which are not considered sufficiently important to justify separate entries of their own have been included as derivatives in the entries of other compounds. These may include for example:

(1) Organic derivatives in the classical sense.

(2) Donor-acceptor complexes.

(3) The various salts of an anion or cation. In nearly every case, the entry for an ionic substance refers to the naked anion or cation, and the molecular formula, molecular weight and CAS registry number given for the main entry are those of the ion, in agreement with current CAS practice. Salts of the ion with various counterions are then treated as derivatives and the molecular formulae of all of these are given.

(4) Oligomeric compounds. Where a compound is known in several states of molecular aggregation these are all included in the one entry, which usually refers to the monomer. Compounds which are known only in dimeric form are entered as such, but the hypothetical monomers are included as derivatives to ensure that the names and molecular formulae of the monomeric forms occur in the indexes.

All names and molecular formulae recorded for derivatives occur in the Name and Molecular Formula Indexes respectively.

3. Chemical Names and Synonyms

The naming of organometallic compounds is frequently problematic and so in selecting the range of alternative names to present for each compound or derivative, editorial policy has been to report names which are found in the literature, including *Chemical Abstracts*, and not to attempt to impose a system of nomenclature. The editorial generation of new names has therefore been kept to a minimum required by consistency. Most names given in the *Sourcebook* are those given in the original paper(s) and in *Chemical Abstracts*.

Names corresponding to those used by CAS during

the 8th, 9th, and 10th collective index periods (1967-71, 1972-6 and 1977-81 respectively) are labelled with the suffixes 8CI, 9CI and 10CI respectively.

4. Toxicity and Hazard Information

Toxicity and hazard information is highlighted by the sign ▷ which also appears in the indexes.

All organometallic compounds should be treated as if they have dangerous properties.

The information contained in the *Sourcebook* has been compiled from sources believed to be reliable. No warranty, guarantee or representation is made by the Publisher as to the correctness or sufficiency of any information herein, and the Publisher assumes no responsibility in connection therewith.

The specific information in this publication on the hazardous and toxic properties of certain compounds is intended to alert the reader to possible dangers associated with the use of those compounds. The absence of such information should not, however, be taken as an indication of safety in use or misuse.

5. Bibliographic References

The selection of references is made with the aim of facilitating entry into the literature for the user who wishes to locate more detailed information about a particular compound. Reference contents are frequently indicated using mnemonic suffixes. In general recent references are preferred to older ones, and the number of references quoted does not necessarily indicate the relative importance of a compound.

Journal abbreviations generally follow the practice of *Chemical Abstracts Service Source Index* (CASSI). In patent references, no distinction is made between patent applications and granted patents.

6. Sources of Further Information

The following books and review series provide more information about various aspects of organometallic chemistry. Lists of reviews specific to organic compounds of particular metals may be found in the introductory sections of the metals concerned.

General

Comprehensive Organometallic Chemistry, Wilkinson, G. *et al.* Eds, Pergamon, Oxford, 1982. This book represents the most complete and up to date review of the whole subject. In addition to sections for each element there are chapters on the use of organometallics in organic synthesis and catalysis.

Comprehensive Inorganic Chemistry, Trotman-Dickenson, A.F. *et al.* Eds, Pergamon, Oxford, 1973. Contains information about organometallics as well as discussions of oxidation states, coordination chemistry and analysis of the metals.

Gmelins Handbuch der Anorganischen Chemie, 8th Edn, Springer-Verlag, Berlin. Some volumes of Gmelin covering organometallic compounds have been updated relatively recently and can therefore be consulted for comprehensive data on some types of organometallics. Some Gmelin element sections, however, are many years out of date.

Houben-Weyls Methoden der Organischen Chemie, 4th Edn, Band XIII, *Metallorganische Verbindungen*, Thieme-Verlag, Stuttgart.

The Chemistry of the Carbon-Metal Bond, Hartley, F.R. and Patai, S. Eds, Wiley, New York, 1982-. Contains sections on the synthesis, analysis and thermochemistry of various classes of organometallic compounds.

Transition-Metal Complexes of Phosphorus, Arsenic and Antimony Ligands, McAuliffe, C.A. Ed., Macmillan, London, 1973.

Methods of Elemento-Organic Chemistry, Kocheshkov, K.A. Ed., North Holland, Amsterdam, 1967.

MTP International Review of Science: Inorganic Chemistry, Series 2, Emeléus, H.J. Ed., Butterworths, London; University Park Press, Baltimore, 1974-5.

Advances in Organometallic Chemistry, Academic Press, 1964-.

Annual Surveys of Organometallic Chemistry, Elsevier, 1964-7.

Organometallic Chemistry Reviews, Elsevier, 1966-7.

Organometallic Chemistry Reviews, Section A: Subject Reviews 1968-72.

Organometallic Chemistry Reviews, Section B: Annual Surveys 1968-74.

Journal of Organometallic Chemistry: This incorporates reviews and surveys after the discontinuation of the two series of *Organometallic Chemistry Reviews*.

Organometallic Chemistry, 1972-, (Specialist Periodical Reports), RSC.

Coordination Chemistry Reviews, Elsevier, 1966-.

Progress in Inorganic Chemistry, Interscience, 1959-.

Advances in Inorganic Chemistry and Radiochemistry, Academic Press, 1959-.

Analysis

Scott's Standard Methods of Chemical Analysis, Furman, N.H. Ed., 6th Edn, Van Nostrand, New York, 1962.

Crompton, T.R., *Chemical Analysis of Organometallic Compounds*, Academic Press, London, 1973.

Spectroscopy

Nuclear Magnetic Resonance Spectroscopy of Nuclei Other than the Proton, Axenrod, T. and Webb, G.A. Eds, Wiley, London, 1974.

NMR and the Periodic Table, Harris, R.K. and Mann, B.E. Eds, Academic Press, London, 1978.

^{13}C NMR Data for Organometallic Compounds, Mann, B.E. and Taylor, B.F. Eds, Academic Press, London, 1981.

Spectroscopic Properties of Inorganic and Organometallic Compounds, 1968-, (Specialist Periodical Reports), RSC.

Handling

Shriver, D.F., *The Manipulation of Air-Sensitive Compounds*, McGraw-Hill, 1969.

Organometallic Syntheses, Academic Press, New York, 1965, Vol. 1.

Co Cobalt

C. White

Cobalt (Fr.), Kobalt (Ger.), Cobalto (Sp., Ital.), Кобальт (Kobal't) (Russ.), コバルト (Japan.)

Atomic Number. 27

Atomic Weight. 58.9332

Electronic Configuration. [Ar] $3d^7\ 4s^2$

Oxidation States. Organocobalt chemistry is particularly diverse since all the oxidation states from Co(−I) to Co(+III) show quite extensive organometallic chemistry; a few organocobalt(IV) compounds such as Co(1-norbornyl)$_4$ are also known. A large number of stable alkyl and hydrido compounds of Co(III) have been reported including the Vitamin B_{12} models. Organometallic compounds of Co(II) are less common; they tend to be paramagnetic and readily oxidised. Co(I) and Co(0) form a multitude of organocobalt compounds, most of which contain carbonyl and/or phosphine ligands. The organometallic chemistry of Co(−I) is almost exclusively restricted to that of carbonyl compounds and their derivatives. Such compounds are formed only under strongly reducing conditions and tend to be extremely reactive and air-sensitive.

Co(I) compounds usually undergo oxidative addition reactions although less readily than their rhodium or iridium analogues. Oxidative addition may also take place with Co(0) or Co(II) compounds although this occurs at two metal centres ie.

$$Co_2(CO)_8 + H_2 \rightleftharpoons 2CoH(CO)_4$$

$$2[Co(CN)_5]^{3-} + H_2 \rightleftharpoons 2[CoH(CN)_5]^{3-}$$

Coordination Number. Since cobalt has an odd number of electrons it might be expected that compounds containing the metal in an even oxidation state would be paramagnetic. This is often true for Co(II) compounds and typically such compounds are four-coordinate such as the planar mesityl complex *trans*- $Co(Mes)_2(PEt_2Ph)_2$ or five-coordinate such as the square-pyramidal $[Co(CNPh)_5]^{2+}$. In the case of Co(0), however, diamagnetic polynuclear compounds are normally formed; these contain cobalt—cobalt bonds and also usually bridging carbonyl ligands. It is common for such compounds to exist in several isomeric forms and for interconversion between such isomers to occur readily eg. $(OC)_4Co{-}Co(CO)_4$ and $[(OC)_3Co{-}Co(CO)_3](\mu\text{-}CO)_2$. Thus, the coordination number of Co(0) compounds may vary from four as in the paramagnetic tetrahedral anion $[Co(CN)_4]^{4-}$ to seven as in $Co_4(CO)_{12}$.

Organocobalt(III) compounds are normally six-coordinate and octahedral. The majority of Co(I) compounds are five-coordinate such as the trigonal-bipyramidal $Co(CF_2CF_2H)(CO)_3PPh_3$. Four-coordinate 16-electron Co(I) compounds are also known, especially among compounds containing bulky ligands as for example the planar complex $CoMe(PPh_3)_3$. Co(−I) compounds such as the anion $[Co(CO)_4]^-$ are typically tetrahedral.

Colour. Organometallic compounds of cobalt are coloured and the colours cover the whole spectrum including black.

Availability. Common starting materials are $Co(OAc)_2{\cdot}4H_2O$ and $CoCl_2{\cdot}6H_2O$. The latter is usually dehydrated by heating at ca. 160° under vacuum until it turns completely blue (without any pinkish tinge) prior to use. $Co_2(CO)_8$ is the usual precursor to cobalt carbonyl compounds, and is also available commercially. $Co_4(CO)_{12}$ is also available commercially but is expensive.

Handling. It is normally essential to carry out organocobalt chemistry under an inert atmosphere. Strong light should also be avoided if preparing or investigating the chemistry of cobalt—alkyl compounds of macrocyclic ligands since the cobalt—carbon bond is often cleaved by light.

Toxicity. Cobalt carbonyls are often relatively volatile and extremely toxic. Therefore they must only be handled in an efficient fume-cupboard by trained operators; hydridotetracarbonylcobalt $HCo(CO)_4$ requires particular care.

The toxicity of other organocobalt compounds has not been investigated in detail although a general review on the toxicity of cobalt compounds has been published: Morel, C. *et al.*, *Cah. Notes Doc.*, 1977, **87,** 259.

Isotopic Abundance. ^{59}Co, 100%.

Spectroscopy. ^{59}Co ($I = \frac{7}{2}$) has relative receptivity 0.277 (^1H = 1.000), placing it amongst the top six nuclear species for ease of detection. Its quadrupole moment of 0.4×10^{-28} cm^2 affects the line width, which is sensitive to the electric field gradient at cobalt. Thus any deviations from strict octahedral or tetrahedral symmetry will lead to line broadening. This may of course be used to advantage to detect subtle variations in structure. Texts summarizing the bulk of the published work on ^{59}Co nmr spectroscopy and ^{13}C nmr spectroscopy of organocobalt compounds have been published and recent developments in these fields are surveyed annually.

^{59}Co nqr spectroscopy may also provide detailed information about the environment around the metal in organocobalt solids.

Analysis. The analytical chemistry of cobalt is well documented with atomic absorption and emission spectroscopy being the methods of choice. Iodine oxidations have proved to be useful in cobalt carbonyl chemistry because coordinated carbon monoxide is normally liberated quantitatively by iodine from Co(0) compounds. Thus the number of carbonyl ligands per cobalt atom is often conveniently determined by measuring the volume of carbon monoxide liberated and the concentration of cobalt in solution.

References. In addition to reviews listed in the introduction to the *Sourcebook* the following provide further reading:

General

Heck, R. F. *et al.*, *Adv. Organomet. Chem.*, 1966, **4**, 243.
Nicholls, D., *The Chemistry of Iron, Cobalt and Nickel*, Pergamon, Oxford, 1973.
Pratt, J. M. *et al.*, *Adv. Organomet. Chem.*, 1973, **11**, 331.
Dodd, D. *et al.*, *J. Organomet. Chem.*, 1973, **52**, 1.
Parshall, G. W., *Catal. Rev.*, 1981, **23**, 107.

Nqr

Brown, T. L., *Acc. Chem. Res.*, 1974, **7**, 408.
Ramakrishnan, L. *et al.*, *Coord. Chem. Rev.*, 1977, **22**, 123.

Analysis

Young, R. S., *The Analytical Chemistry of Cobalt*, Pergamon, Oxford, 1966.
Young, R. S., *Chemsa*, 1976, **2**, 196.
Sternberg, H. W., *Anal. Chem.*, 1952, **24**, 174.

$BrCo(CO)_2NO^{\ominus}$

Co-00001

$ClCo(CO)_2NO^{\ominus}$

Co-00002

$ICo(CO)_2NO^{\ominus}$

Co-00003

$Co(NO)(CO)_3$

Co-00004

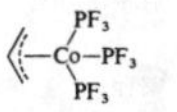

Co-00005

$Co(CO)_4^{\ominus}$

Co-00006

$HCo(CO)_4$

Co-00007

$Co(CF_3)(CO)_4$

Co-00008

$[Co(CN)_5]^{\ominus\ominus\ominus}$

Co-00009

$HCo(CN)_5^{\ominus\ominus\ominus}$

Co-00010

$MeCo(CO)_4$

Co-00011

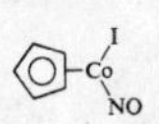

Co-00012

$F_3CCOCo(CO)_4$

Co-00013

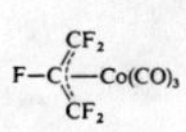

Co-00014

$MeCOCo(CO)_4$

Co-00015

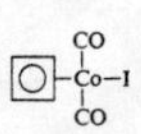

Co-00016

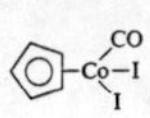

Co-00017

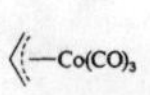

Co-00018

$CoEt(CO)_4$

Co-00019

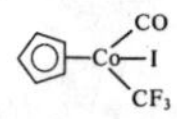

Co = 00020

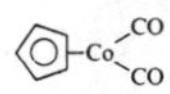

Co-00021

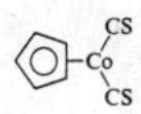

Co-00022

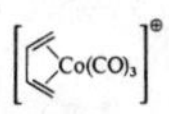

Co-00023

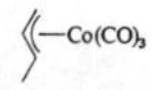

(E)-form

Co-00024

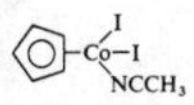

Co-00025

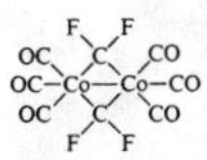

Co-00026

$(OC)_4Co—Hg—Co(CO)_4$

Co-00027

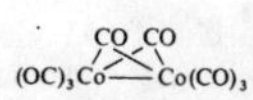

Co-00028

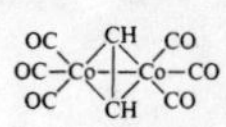

Co-00029

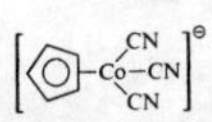

Co-00030

$[H_2C{=}CHCH_2Co(CN)_5]^{\ominus\ominus\ominus}$

Co-00031

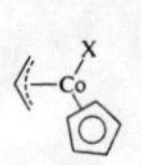

X = Br

Co-00032

As Co-00032 with X = I

Co-00033

$(H_3CCN)_4CoBr_2$

Co-00034

$(MeNC)_4CoI_2$

Co-00035

$[Co(CNMe)_4]^{\oplus\oplus}$

Co-00036

$CoMe_2(NO)(PMe_3)_2$

Co-00037

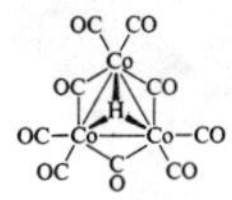

Co-00038

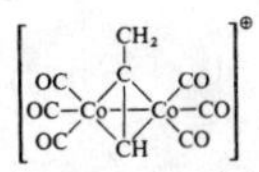

Co-00039

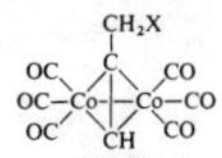

X = H

Co-00040

As Co-00040 with X = OH

Co-00041

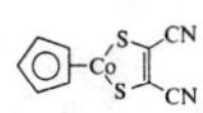

Co-00042

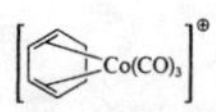

Co-00043

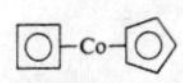

Co-00044

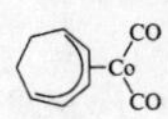

Co-00045

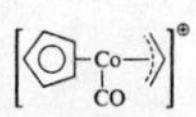

Co-00046

$Et_3PCoBr(CO)_3$

Co-00047

$Et_3PCoCl(CO)_3$

Co-00048

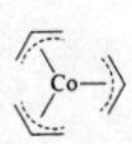

Co-00049

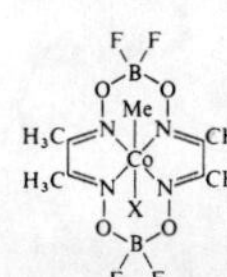

X = OH_2

Co-00050

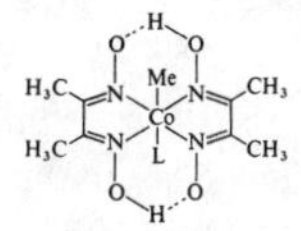

L = OH_2

Co-00051

$[(MeO)_3P]_2CoMe(CO)_2$

Co-00052

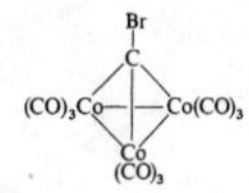

Co-00053

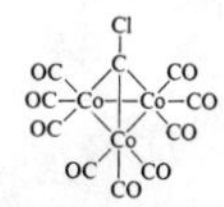

Co-00054

$(C_6F_5)Co(CO)_4$

Co-00055

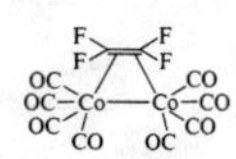

Co-00056

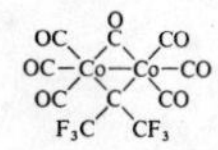

Co-00057

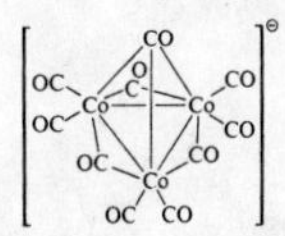

Co-00058

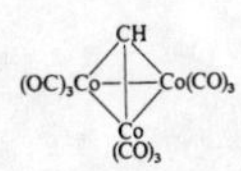

Co-00059

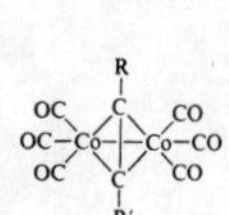

R = R′ = CH_3

Co-00060

$(OC)_4Co—Co(CO)_3P(OMe)_3$

Co-00061

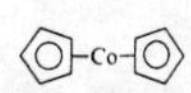

Co-00062

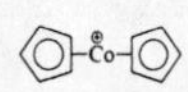

Co-00063

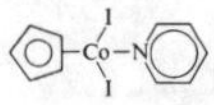

Co-00064

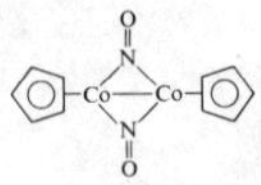

Co-00065

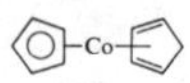

Co-00066

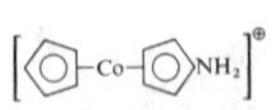

Co-00067

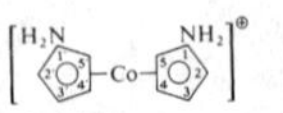

Co-00068

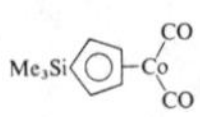

Co-00069

$(MeNC)_5Co^{\oplus}$

Co-00070

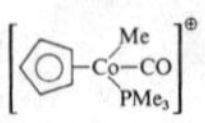

Co-00071

$[(MeO)_3P]_2Co(CO)_2Ac$

Co-00072

$(Me_3P)_3CoCl(CO)$

Co-00073

$(Me_3P)_3CoI(CO)$

Co-00074

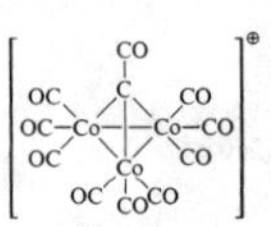

Co-00075

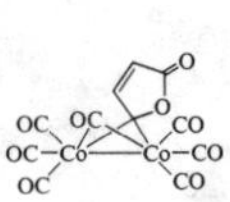

Co-00076

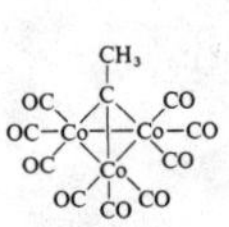

Co-00077

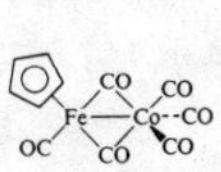

Co-00078

$[PhCo(CN)_5]^{\ominus\ominus\ominus}$

Co-00079

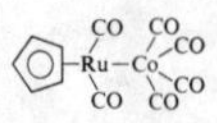

Co-00080

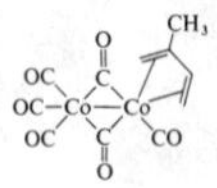

Co-00081

$(H_3C)_3CC{\equiv}PCo_2(CO)_6$

Co-00082

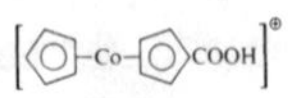

Co-00083

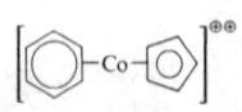

Co-00084

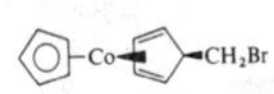

Co-00085

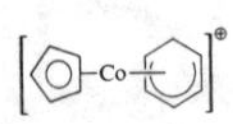

Co-00086

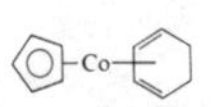

Co-00087

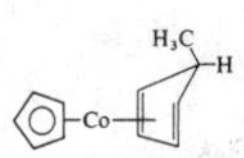

Co-00088

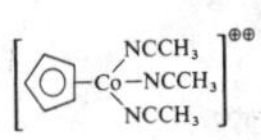

Co-00089

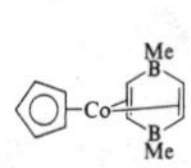

Co-00090

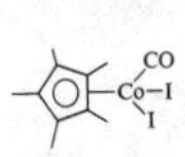

Co-00091

$[(H_3C)_3CNC]_2Co(CO)NO$

Co-00092

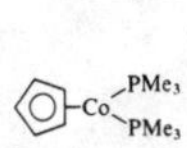

Co-00093

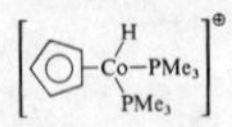

Co-00094

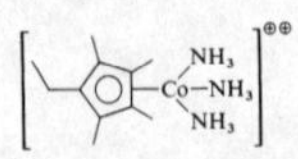

Co-00095

$[(MeO)_3P]_3Co(CO)_2^{\oplus}$

Co-00096

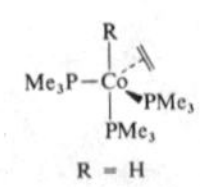

Co-00097

$CoMe_2Br(PMe_3)_3$

Co-00098

$CoMe_2I(PMe_3)_3$

Co-00099

$CoMe_2(PMe_3)_3$

Co-00100

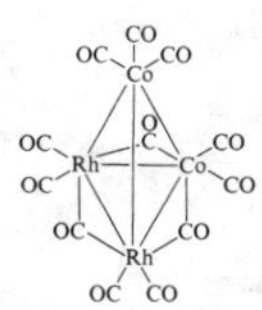

Co-00101

$Er[Co(CO)_4]_3$

Co-00102

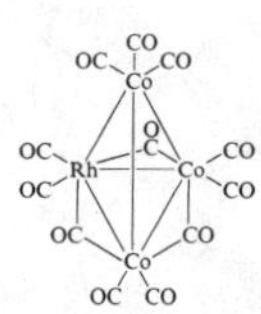

Co-00103

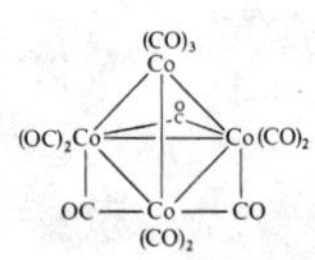

Co-00104

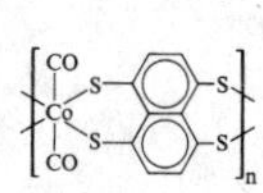

Co-00105

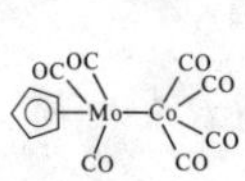

Co-00106

$[Co(C{\equiv}CH)_6]^{\ominus\ominus\ominus\ominus}$

Co-00107

$(NC)_5CoCH_2Ph^{\ominus\ominus\ominus}$

Co-00108

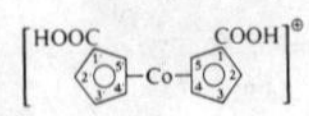

Co-00109

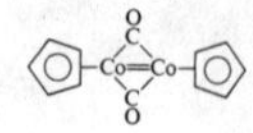

Co-00110

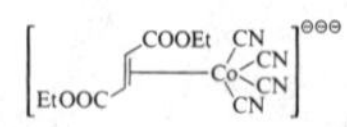

(E)-form

Co-00111

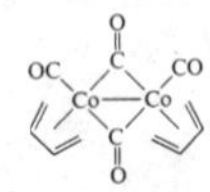

cis-form

Co-00112

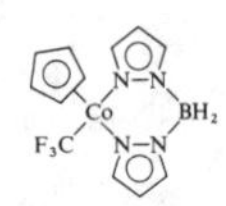

Co-00113

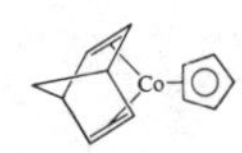

Co-00114

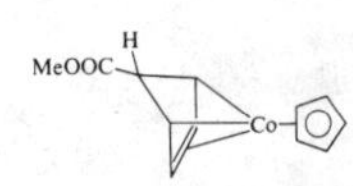

Co-00115

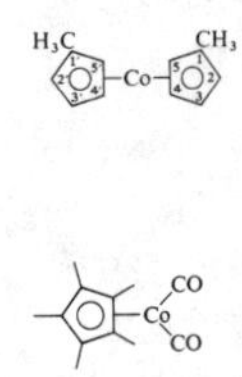

Co-00116

Co-00117

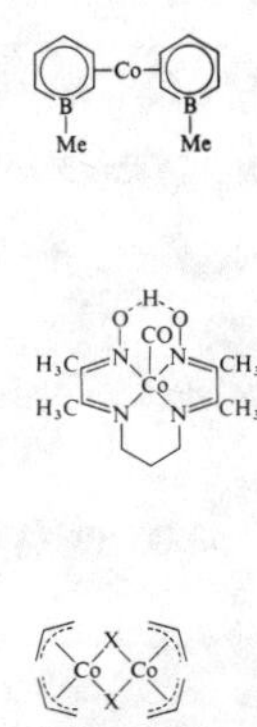

Co-00118

Co-00119

Co-00120

As Co-00120 with
X = Cl

Co-00121

As Co-00120 with
X = I

Co-00122

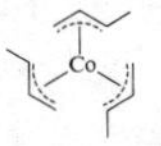

Co-00123

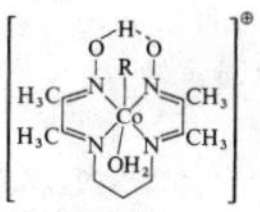

R = Me

Co-00124

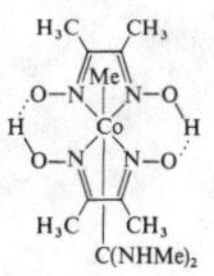

Co-00125

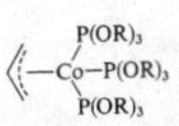

R = Me

Co-00126

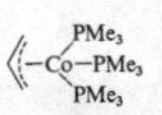

Co-00127

As Co-00097 with
R = Me

Co-00128

$Co[P(OMe)_3]_4^{\oplus}$

Co-00129

$CoMe_3(PMe_3)_3$

Co-00130

$HCo[P(OMe)_3]_4$

Co-00131

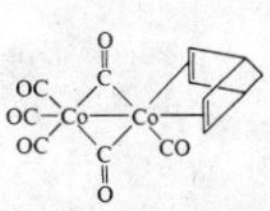

Co-00132

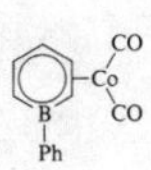

Co-00133

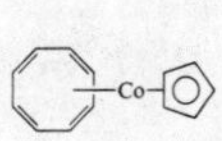

Co-00134

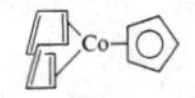

Co-00135

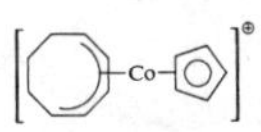

Co-00136

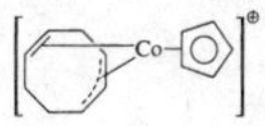

Co-00137

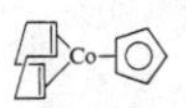

Co-00138

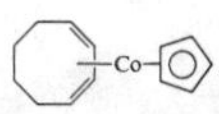

Co-00139

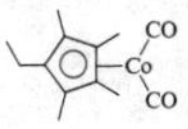

Co-00140

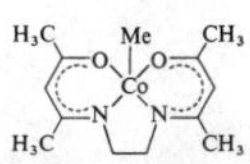

Co-00141

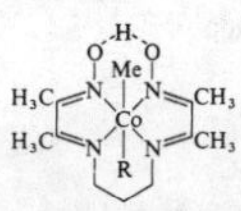

R = Me

Co-00142

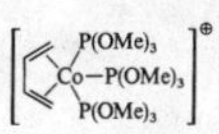

Co-00143

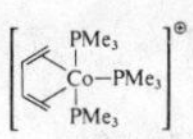

Co-00144

$[(Me_3P)_4CoCO]^{\oplus}$

Co-00145

$[(MeO)_3P]_4Co(CO)^{\oplus}$

Co-00146

$CoMe(PMe_3)_4$

Co-00147

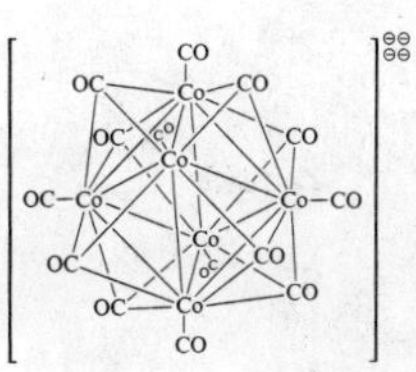

Co-00148

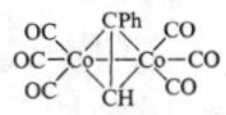

Co-00149

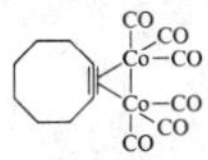

Co-00150

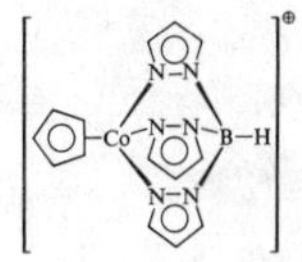

Co-00151

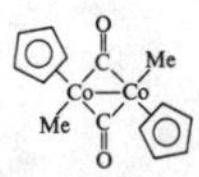

Co-00152

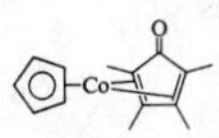

Co-00153

As Co-00050 with
X = Py

Co-00154

As Co-00051 with
L = Py

Co-00155

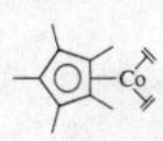

Co-00156

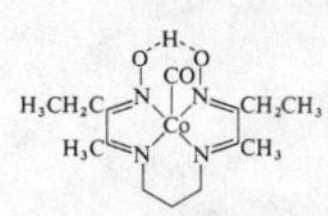

Co-00157

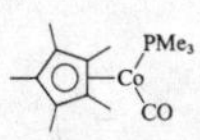

Co-00158

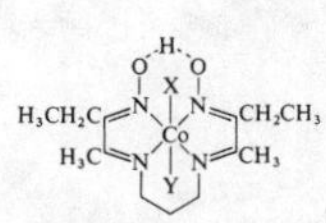

X = Me,Y = I

Co-00159

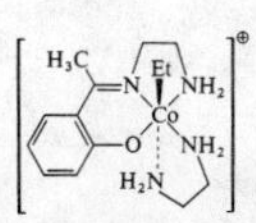

Co-00160

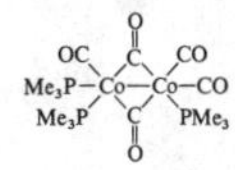

Co-00161

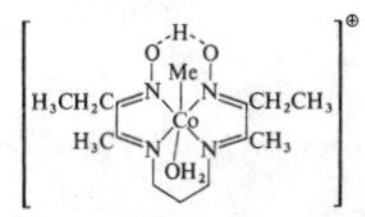

Co-00162

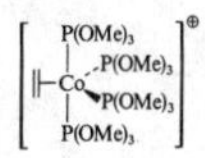

Co-00163

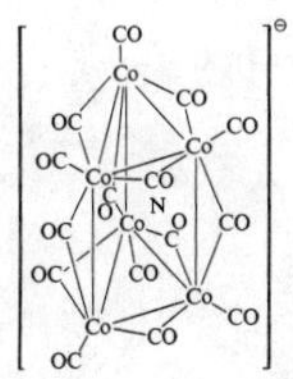

Co-00164

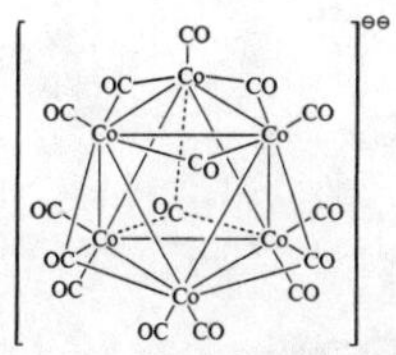

Co-00165

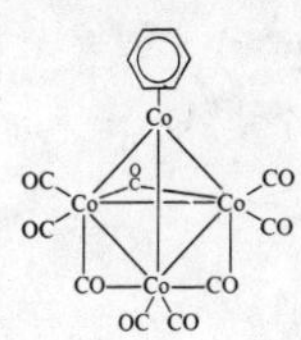

Co-00166

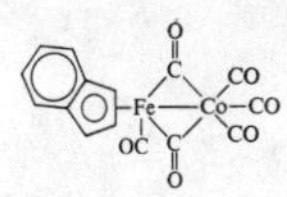

Co-00167

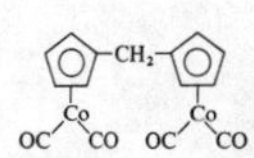

Co-00168

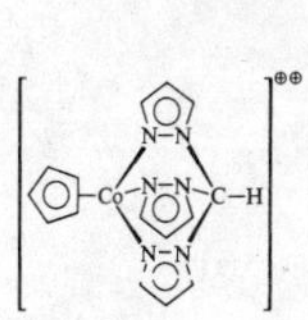

Co-00169

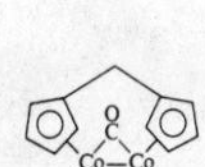

Co-00170

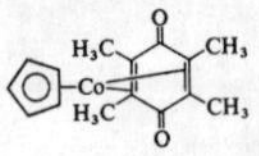

Co-00171

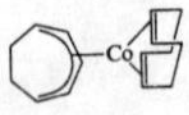

Co-00172

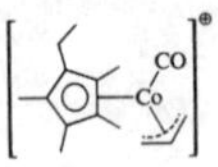

Co-00173

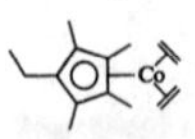

Co-00174

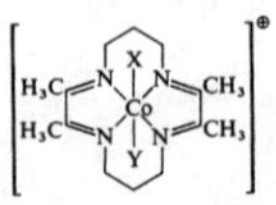

X = Me, Y = Br

Co-00175

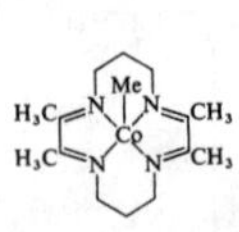

Co-00176

[$(H_3C)_3CNC]_3CoNO$

Co-00177

$(H_3CCH_2CH_2CH_2)_3PCoH(CO)_3$

Co-00178

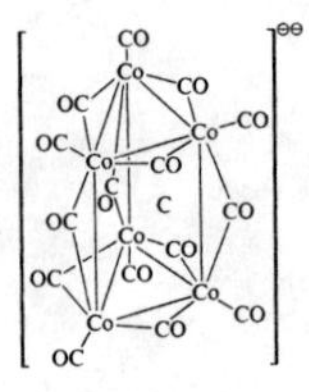

Co-00179

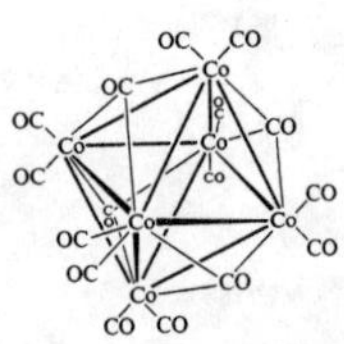

Co-00180

As Co-00183 with
n = 1

Co-00181

As Co-00183 with
n = 2

Co-00182

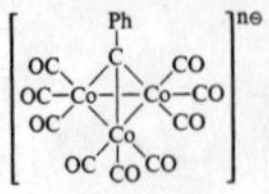

n = O

Co-00183

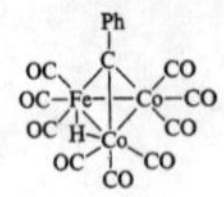

Co-00184

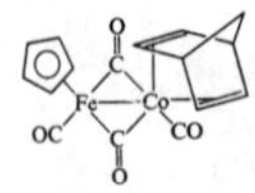

cis-form

Co-00185

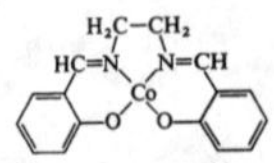

Co-00186

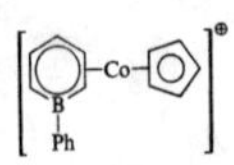

Co-00187

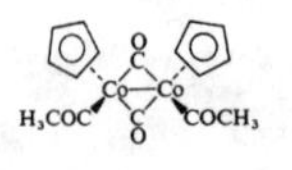

cis-form

Co-00188

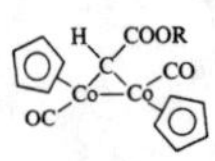

R = Et

Co-00189

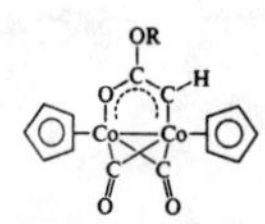

R = Et

Co-00190

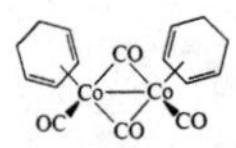

Co-00191

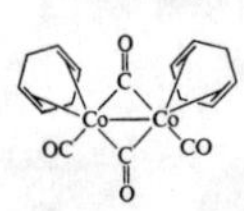

Co-00192

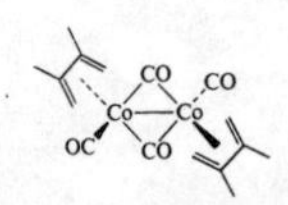

Co-00193

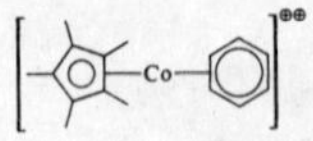

Co-00194

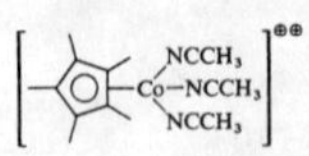

Co-00195

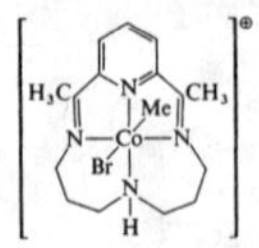

Co-00196

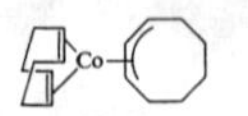

Co-00197

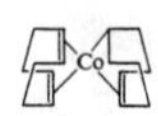

Co-00198

As Co-00175 with
X = Y = Me

Co-00199

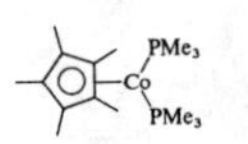

Co-00200

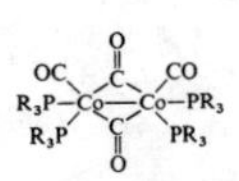

R = Me

Co-00201

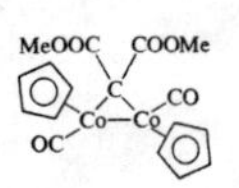

Co-00202

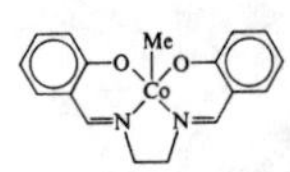

Co-00203

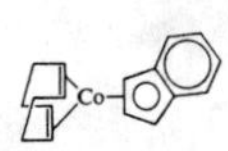

Co-00204

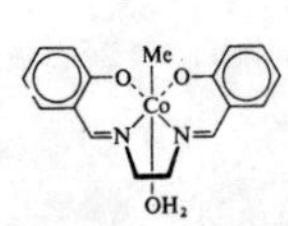

Co-00205

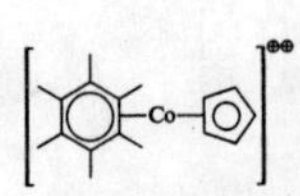

Co-00206

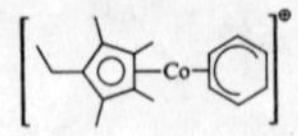

Co-00207

As Co-00124 with
R = Ph

Co-00208

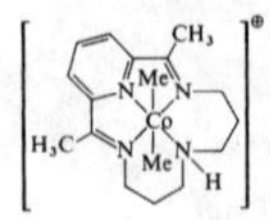

Co-00209

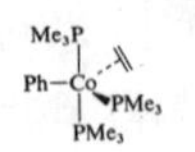

Co-00210

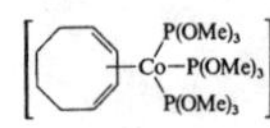

Co-00211

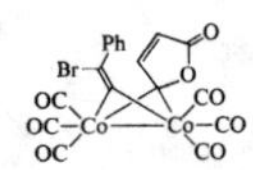

Co-00212

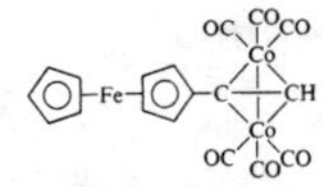

Co-00213

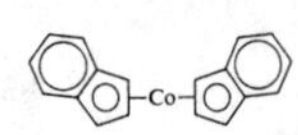

Co-00214

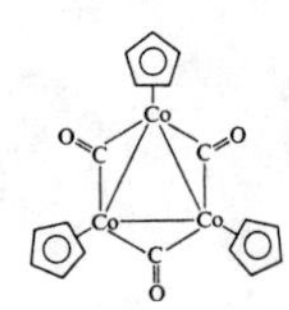

Co-00215

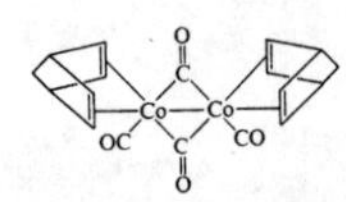

cis-form

Co-00216

Co-00217

Co-00218

As Co-00189 with
R = $(H_3C)_3C-$

Co-00219

As Co-00190 with
R = $(H_3C)_3C-$

Co-00220

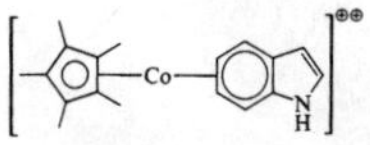

Co-00221

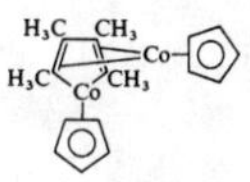

Co-00222

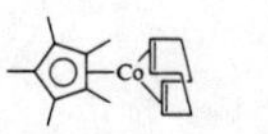

Co-00223

As Co-00142 with
R = Ph

Co-00224

As Co-00124 with
R = CH_2Ph

Co-00225

$PhCo[P(OMe)_3]_4$

Co-00226

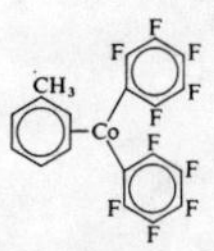

Co-00227

$(PhO)_3PCo(NO)_2CF_3$

Co-00228

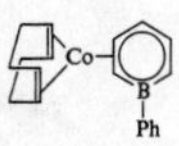

Co-00229

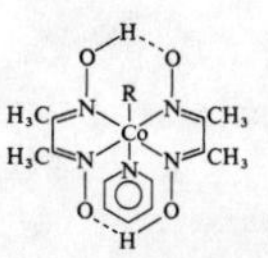

R = Ph

Co-00230

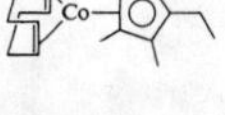

Co-00231

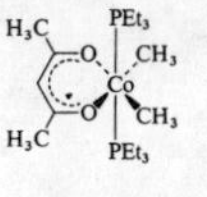

Co-00232

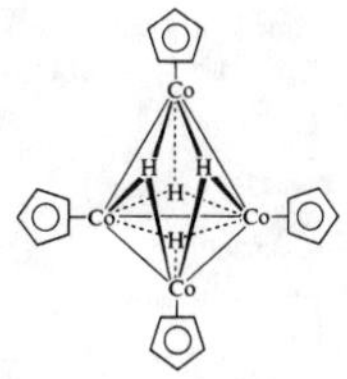

Co-00233

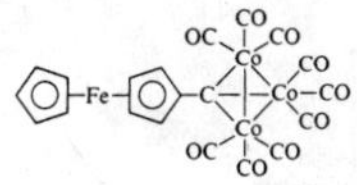

Co-00234

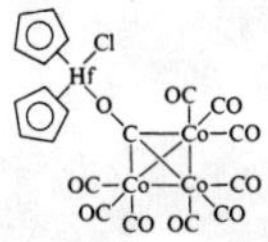

Co-00235

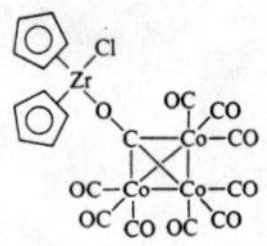

Co-00236

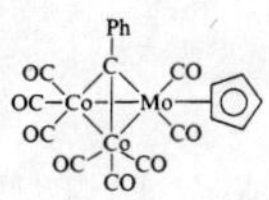

Co-00237

As Co-00060 with
R = Ph

Co-00238

$Ph_2Pb[Co(CO)_4]_2$

Co-00239

$Ph_3PCo(NO)(CO)_2$

Co-00240

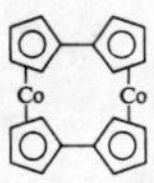

Co-00241

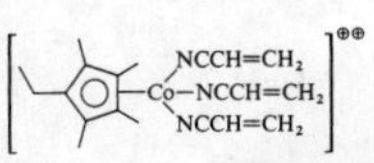

Co-00242

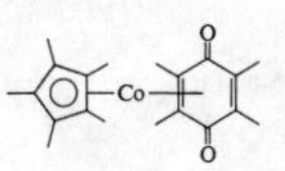

Co-00243

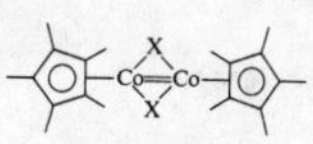

X = Br

Co-00244

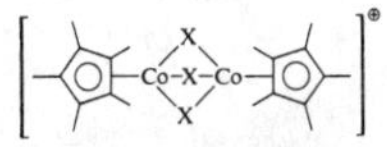

X = Br

Co-00245

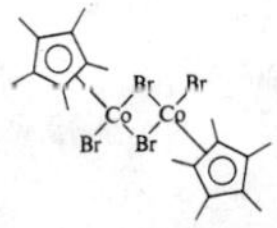

Co-00246

As Co-00159 with
X = $-CH_2Ph$, Y = Cl

Co-00247

As Co-00244 with
X = Cl

Co-00248

As Co-00245 with
X = Cl

Co-00249

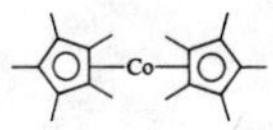

Co-00250

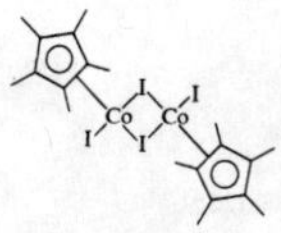

Co-00251

$Co_4(CO)_8[P(OMe)_3]_4$

Co-00252

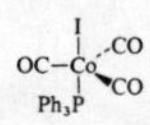

Co-00253

$HCo(CO)_3PPh_3$

Co-00254

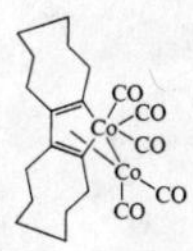

Co-00255

As Co-00230 with
R = $H_2C{=}CPh-$

Co-00256

As Co-00230 with
R = $PhCH{=}CH-$ (*E*- or *Z*-)

Co-00257

As Co-00175 with
X = $PhCH_2-$, Y = Cl

Co-00258

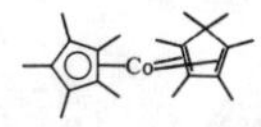

Co-00259

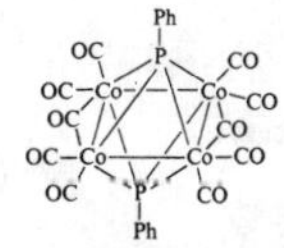

Co-00260

$Ph_3PCo(CO)_4^{\oplus}$

Co-00261

$Ph_3PbCo(CO)_4$

Co-00262

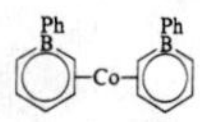

Co-00263

$Co(CO)_4 \cdot [H_3CC_5H_4]_3Sm$

Co-00264

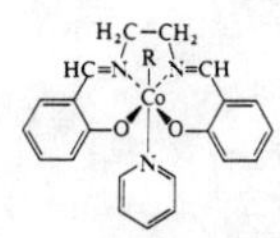

R = Me

Co-00265

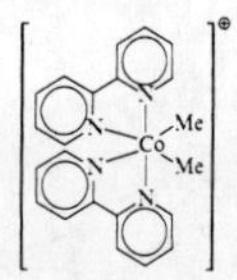

Co-00266

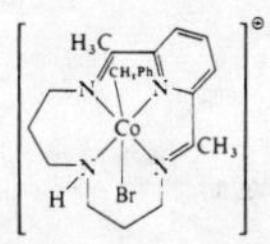

Co-00267

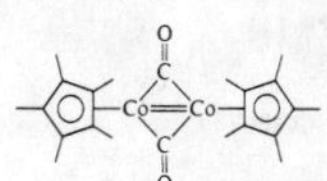

Co-00268

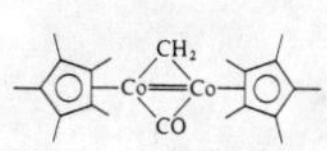

Co-00269

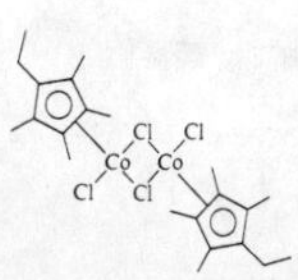

Co-00270

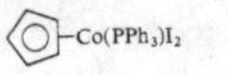

Co-00271

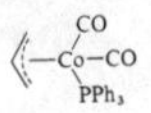

Co-00272

Co(CHNMe$_2$)(CO)$_2$Br(PPh$_3$)

Co-00273

Co(CHNMe$_2$)(CO)$_2$Cl(PPh$_3$)

Co-00274

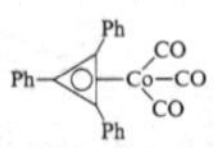

Co-00275

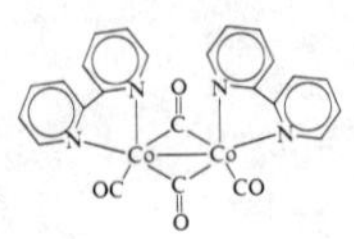

Co-00276

Co(CH$_3$CO)(CO)$_3$(PPh$_3$)

Co-00277

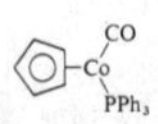

Co-00278

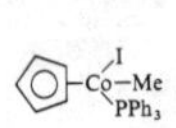

Co-00279

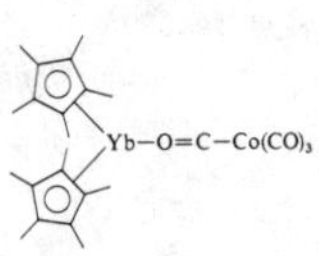

Co-00280

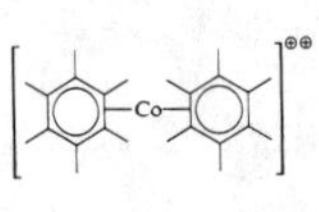

Co-00281

Co$_2$[P(OMe)$_3$]$_8$

Co-00282

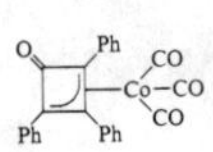

Co-00283

CO$_2$(CO)$_7$(PPh$_3$)

Co-00284

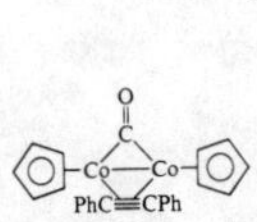

Co-00285

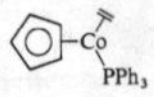

Co-00286

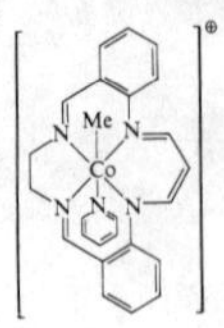

Co-00287

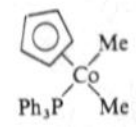

Co-00288

[(H$_3$C)$_3$CNC]$_5$Co$^{\oplus}$

Co-00289

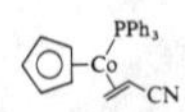

Co-00290

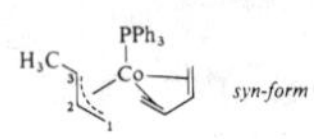

Co-00291

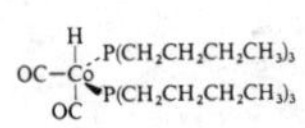

Co-00292

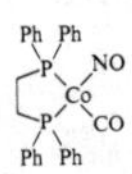

Co-00293

As Co-00265 with
R = Ph

Co-00294

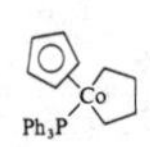

Co-00295

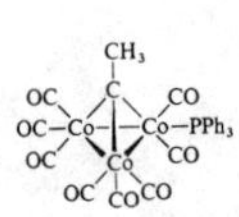

Co-00296

CoBr$_2$(CNPh)$_4$

Co-00297

(PhNC)$_4$CoI$_2$

Co-00298

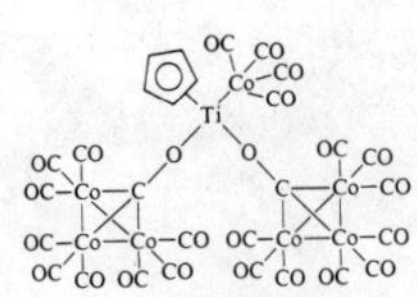

Co-00299

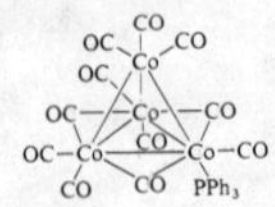

Co-00300

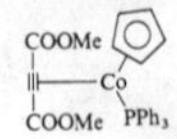

Co-00301

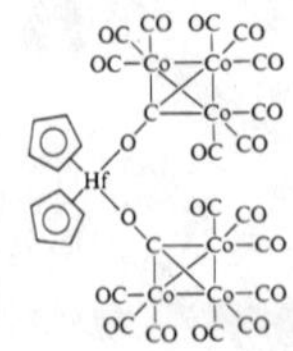

Co-00302

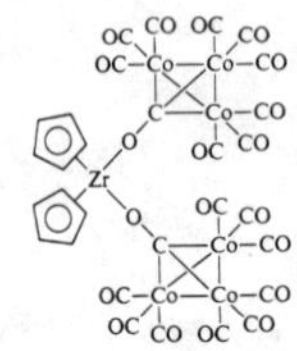

Co-00303

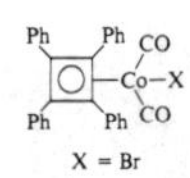

Co-00304

As Co-00304 with
X = Cl

Co-00305

As Co-00304 with
X = I

Co-00306

(H$_3$CCH$_2$CH$_2$CH$_2$)$_3$PCo(CO)$_3$–Hg–Co(CO)$_3$P(CH$_2$CH$_2$CH$_2$CH$_3$)$_3$

Co-00307

(H$_3$CCH$_2$CH$_2$CH$_2$)$_3$P–Co(CO)$_3$–Co(CO)$_3$–P(CH$_2$CH$_2$CH$_2$CH$_3$)$_3$

Co-00308

As Co-00126 with
R = (H$_3$C)$_2$CH–

Co-00309

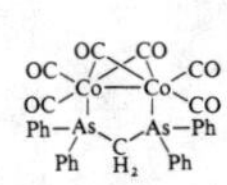

Co-00310

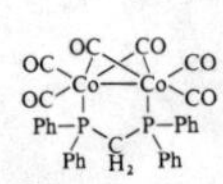

Co-00311

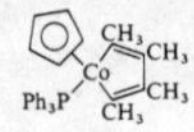

Co-00312

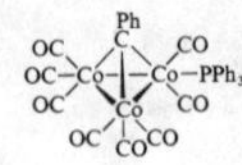

Co-00313

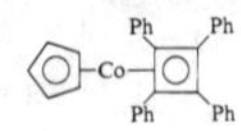

Co-00314

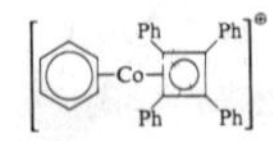

Co-00315

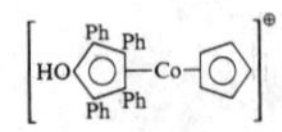

Co-00316

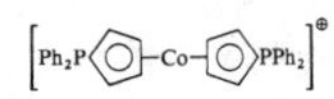

Co-00317

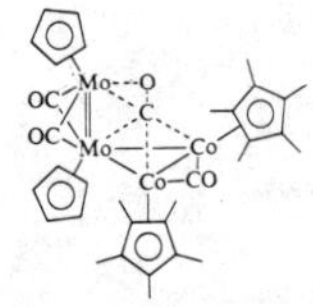

Co-00318

Co(NCPh)$_5^{\oplus}$

Co-00319

[Co(CNPh)$_5$]$^{\oplus\oplus}$

Co-00320

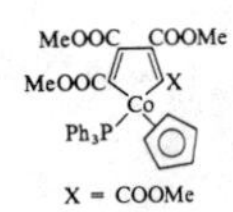

Co-00321

(Ph$_3$P)$_2$Co(NO)$_2^{\oplus}$

Co-00322

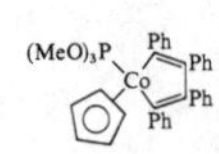

Co-00323

(Ph$_3$P)$_2$Co(NO)(CO)

Co-00324

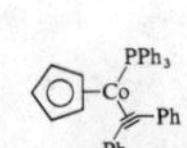

Co-00325

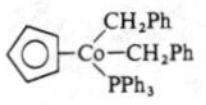

Co-00326

$[(H_3CCH_2CH_2CH_2)_3P]_3CoH(CO)$

Co-00327

$(Ph_3P)_2CoCl(CO)_2$

Co-00328

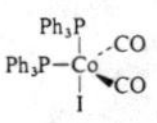

Co-00329

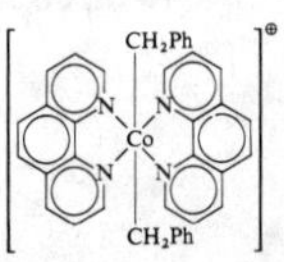

Co-00330

$[Co(CO)_3(PPh_3)_2]^{\oplus}$

Co-00331

As Co-00321 with
X = Ph

Co-00332

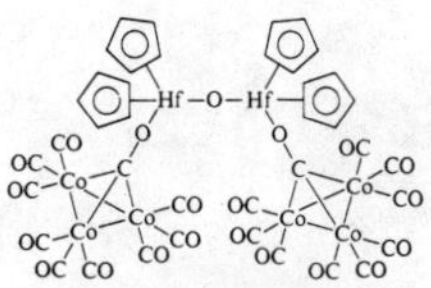

Co-00333

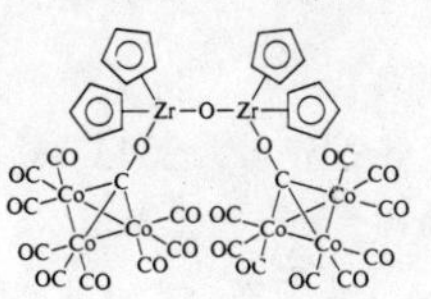

Co-00334

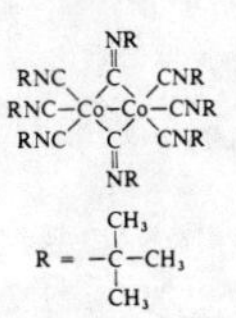

Co-00335

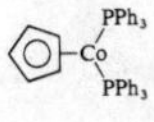

Co-00336

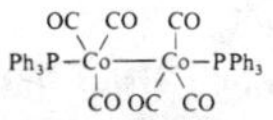

Co-00337

$Co_2(CO)_6[P(OPh)_3]_2$

Co-00338

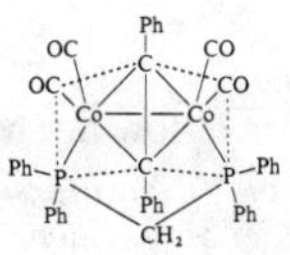

Co-00339

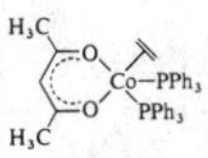

Co-00340

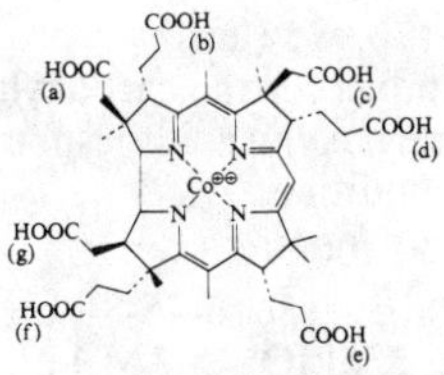

Co-00341

$[Co(C{\equiv}CPh)_6]^{\ominus\ominus\ominus\ominus}$

Co-00342

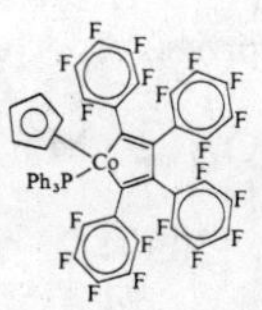

Co-00343

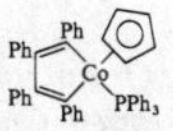

Co-00344

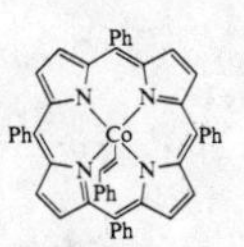

Co-00345

$CoCl(PPh_3)_3$

Co-00346

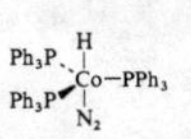

Co-00347

$H_3Co(PPh_3)_3$

Co-00348

$CoMe(PPh_3)_3$

Co-00349

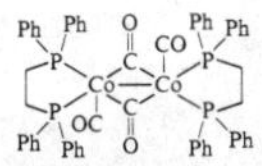

Co-00350

$(H_3CCN)CoH[P(OPh)_3]_3$

Co-00351

$Co(H_2C{=}CH_2)(PPh_3)_3$

Co-00352

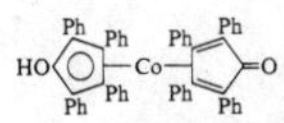

Co-00353

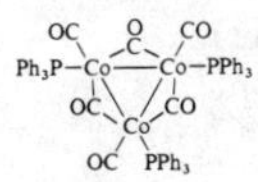

Co-00354

As Co-00359 with
R = H, R′ = $CONH_2$

Co-00355

As Co-00359 with
R = ONO, R′ = $CONH_2$

Co-00356

As Co-00359 with
R = OH, R′ = $CONH_2$

Co-00357

As Co-00359 with
R = CN, R′ = OH

Co-00358

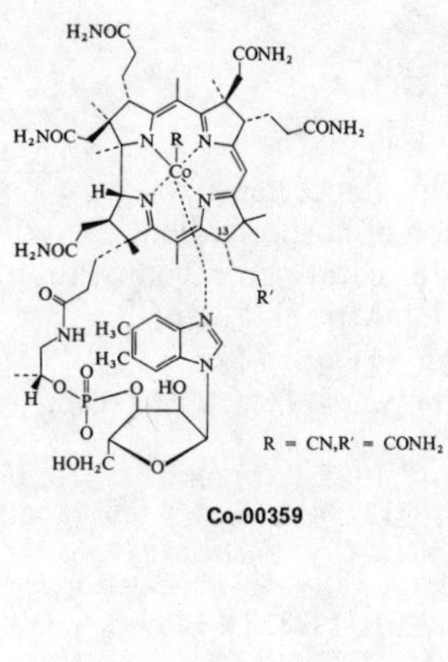

Co-00359

As Co-00359 with
R = Me, R′ = $CONH_2$

Co-00360

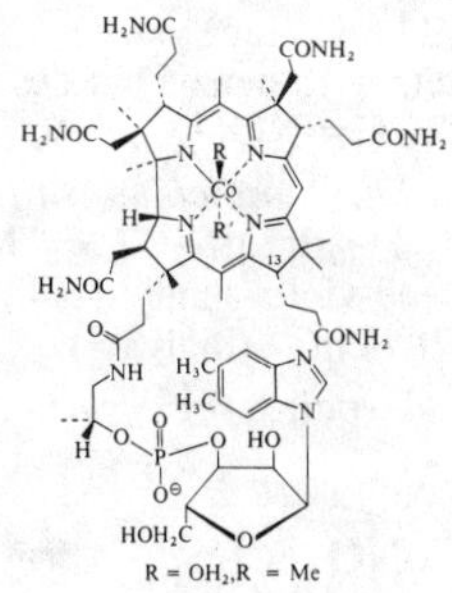

Co-00361

As Co-00361 with
R = Me, R′ = OH_2

Co-00362

As Co-00359 with
R = $-C{\equiv}CH$, R′ = $CONH_2$

Co-00363

As Co-00359 with
R = $-CH{=}CH_2$, R′ = $CONH_2$

Co-00364

$HCo[P(OPh)_3]_4$

Co-00365

As Co-00335 with
R = 2,6-dimethylphenyl

Co-00366

As Co-00359 with

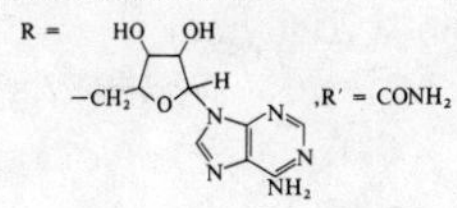

Co-00367

As Co-00201 with
R = OPh

Co-00368

$Co_2(CO)_2(PPh_3)_6$

Co-00369

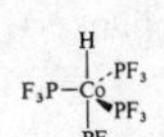

Co-00370

$C_2BrCoNO_3^{\ominus}$ Co-00001

Bromodicarbonylnitrosylcobaltate(1−), 8CI

$BrCo(CO)_2NO^{\ominus}$

M 224.864 (ion)

Reacts with benzyl halides to give benzaldoxime, PhCH=NOH.

Tetrabutylammonium salt: [33361-06-7].
$C_{18}H_{36}BrCoN_2O_3$ M 467.331
Deep-violet solid. Mp 42-3°. ir ν_{CO} 2013, 1940, ν_{NO} 1707 cm^{-1} (diglyme).

Foa, M. *et al*, *J. Organomet. Chem.*, 1971, **30**, 123 (*synth, ir*)

$C_2ClCoNO_3^{\ominus}$ Co-00002

Dicarbonylchloronitrosylcobaltate(1−), 8CI

$ClCo(CO)_2NO^{\ominus}$

M 180.413 (ion)

Reacts with benzyl halides to give benzaldoxime, PhCH=NOH.

Tetraheptylammonium salt: [33393-84-9].
$C_{30}H_{60}ClCoN_2O_3$ M 591.202
ir ν_{CO} 2012, 1938, ν_{NO} 1703 cm^{-1} (diglyme).

Foa, M. *et al*, *J. Organomet. Chem.*, 1971, **30**, 123 (*synth, ir*)

$C_2CoINO_3^{\ominus}$ Co-00003

Dicarbonyliodonitrosylcobaltate(1−), 8CI

$ICo(CO)_2NO^{\ominus}$

M 271.865 (ion)

Reacts with benzyl halides to give benzaldoxime, PhCH=NOH.

Tetrabutylammonium salt: [33361-05-6].
$C_{18}H_{36}CoIN_2O_3$ M 514.332
Violet solid. Mp 73-5°. ir ν_{CO} 2011, 1943, ν_{NO} 1712 cm^{-1} (diglyme).

Foa, M. *et al*, *J. Organomet. Chem.*, 1971, **30**, 123 (*synth, ir*)

C_3CoNO_4 Co-00004

Tricarbonylnitrosylcobalt, 10CI, 9CI, 8CI

Cobalt tricarbonylnitrosyl

[14096-82-3]

$Co(NO)(CO)_3$

M 172.971

Catalyses cyclic oligomerization of acetylenes to benzenes and the dimerization of norbornadiene. Catalyst for the phase-transfer-catalysed carbonylation of benzyl halides. Deep-red liq. Mp −11°. Bp 78.6°.

Mond, L. *et al*, *J. Chem. Soc.*, 1922, **121**, 21 (*synth*)
Blanchard, A.A. *et al*, *J. Am. Chem. Soc.*, 1934, **56**, 16 (*synth*)
Inorg. Synth., 1946, **2**, 238 (*synth*)
Bramley, R. *et al*, *J. Chem. Soc. (A)*, 1967, 861 (*cmr*)
Bor, G., *J. Organomet. Chem.*, 1976, **122**, 413 (*ir*)
Job, R. *et al*, *Synth. React. Inorg. Met.-Org. Chem.*, 1976, **6**, 367 (*synth*)
Schumann, H. *et al*, *Z. Naturforsch., B*, 1978, **33**, 1489 (*nmr, uv*)
Gambarotta, S. *et al*, *J. Organomet. Chem.*, 1981, **212**, C23 (*use*)

$C_3H_5CoF_9P_3$ Co-00005

Tris(phosphorous trifluoride)(η^3-2-propenyl)cobalt, 9CI

(1,2,3-η-Allyl)tris(trifluorophosphine)cobalt

[42043-90-3]

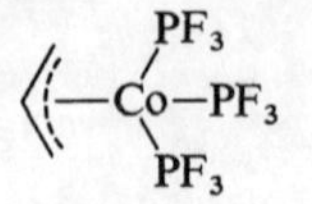

M 363.913

Air-sensitive yellow liq. Mp 22°.

Kruck, T. *et al*, *Z. Natuforsch., B*, 1973, **28**, 28 (*synth, ir*)
Cairns, M.A. *et al*, *J. Chem. Soc., Dalton Trans.*, 1974, 2001 (*synth, ir, ms, pmr*)

$C_4CoO_4^{\ominus}$ Co-00006

Tetracarbonylcobaltate(1−), 10CI

Cobalt tetracarbonyl(1−)

[14971-27-8]

$Co(CO)_4^{\ominus}$

M 170.975 (ion)

Catalyst for the carbonylation of halohydrocarbons to carboxylic acids, esp. under phase-transfer catalysis condition.

▷GF9520000.

Li salt: [15616-75-8].
C_4CoLiO_4 M 177.916
White cryst.

Na salt: [14878-28-5].
C_4CoNaO_4 M 193.965
White solid.

K salt: [14878-26-3].
C_4CoKO_4 M 210.073
No phys. props. reported.

In(3+) salt: [15820-67-4].
$C_{12}Co_3InO_{12}$ M 627.744
Red cryst.

Tl(1+) salt: [38991-21-8].
C_4CoO_4Tl M 375.358
Yellow cryst. (H_2O). Mp 120-30° dec. ir ν_{CO} 2043, 1972, 1937 cm^{-1} (CH_2Cl_2).

▷Toxic

Clark, R.J. *et al*, *J. Organomet. Chem.*, 1968, **11**, 637 (*synth*)
Edgell, W.F. *et al*, *Inorg. Chem.*, 1970, **9**, 1932 (*synth*)
Lyford, P. *et al*, *J. Chem. Phys.*, 1970, **52**, 4329 (*synth, ir*)
Edgell, W.F. *et al*, *Ions Ion Pairs Org. React.*, 1972, **1**, 153 (*ir, rev*)
Burlitch, J.M. *et al*, *J. Chem. Soc., Dalton Trans.*, 1974, 828 (*synth, ir, ms*)
Schlusser, D.P. *et al*, *Inorg. Chem.*, 1974, **13**, 153 (*synth, cryst struct*)
Gladysz, J.A. *et al*, *Inorg. Chem.*, 1979, **18**, 553 (*synth, ir*)
Habeeb, J.J. *et al*, *Can. J. Chem.*, 1979, **57**, 2196 (*synth, ir*)
Alper, H., *Adv. Organomet. Chem.*, 1981, **19**, 183 (*use, rev*)

C_4HCoO_4 Co-00007

Tetracarbonylhydrocobalt, 10CI, 9CI, 8CI

Cobalt tetracarbonyl hydride. Cobalt hydrocarbonyl

[16842-03-8]

$HCo(CO)_4$

M 171.983

Active intermed. in the cobalt-catalysed hydroformylation of alkenes. Also catalyses hydrocarboxylation, hydrogenation and isomerization of alkenes. Light-yellow liq. or gas. Mp −33°, −26.2°. Bp 10°. pK_a <0.4. Unstable, but distillable in current of CO.

▷Extremely toxic vapour. GG0900000.

Gilmont, P. *et al*, *J. Am. Chem. Soc.*, 1940, **62**, 1192 (*synth*)
Inorg. Synth., 1946, **2**, 238 (*synth*)
Sternberg, H.W. *et al*, *J. Am. Chem. Soc.*, 1953, **75**, 2717 (*synth, ir, ms*)
Inorg. Synth., 1957, **5**, 192 (*synth*)
Farrar, T.C. *et al*, *Inorg. Chem.*, 1967, **6**, 161 (*pmr*)
Gmelin's Handbuch der Anorg. Chem.; 8th Ed., 1974, Suppl. 6, 104.
Orchin, M., *Acc. Chem. Res.*, 1981, **14**, 259 (*use, rev*)

$C_5CoF_3O_4$ Co-00008

Tetracarbonyl(trifluoromethyl)cobalt, 10CI
[15892-59-8]

$Co(CF_3)(CO)_4$

M 239.981
Light-amber liq. Mp 10.5-11°. Bp 91°. ir ν_{CO} 2130, 2068, 2055, 2015w cm^{-1} (neat).

McClellan, W.R., *J. Am. Chem. Soc.*, 1961, **83**, 1598 (*synth, nmr, ir*)
Van Den Berg, G.C. *et al*, *J. Organomet. Chem.*, 1974, **69**, 169 (*ir, raman*)
Lichtenberger, D.L. *et al*, *J. Am. Chem. Soc.*, 1977, **99**, 8187 (*cmr*)

$C_5CoN_5^{\ominus\ominus\ominus}$ Co-00009

Pentakis(cyano-*C*)cobaltate(3−), 10CI
Pentacyanocobaltate(3−)
[14971-18-7]

$[Co(CN)_5]^{\ominus\ominus\ominus}$

M 189.022 (ion)
Li and K salts thought to form dimers, $M_6[Co_2(CN)_{10}]$. Catalyst for hydrogenation of alkenes, dienes, dienones, Schiff bases, cyanation of alkenes and for deuterium-water exchange.
Tri-Li salt: [24803-72-3].
$C_{10}Co_2Li_6N_{10}$ M 419.689
Violet-red cryst. Dimeric.
Tri-K salt:
$C_5CoK_3N_5$ M 306.317
Fuel-cell electrolyte, catalyst in hydrogenation of coal. Brown solid. ir ν_{CN} 2129, 2077, 2042 cm^{-1}.
Tri-K sclt, dimer:
$C_{10}Co_2K_6N_{10}$ M 612.633
Violet cryst. (H_2O). ir ν_{CN} 2128, 2120, 2110, 2079 cm^{-1}.
Tris(tetraethylammonium) salt: [38141-18-3].
$C_{29}H_{60}CoN_8$ M 579.780
Yellow cryst. (DMF).
Tris(tetrabutylammonium) salt: [38141-19-4].
$C_{53}H_{108}CoN_8$ M 916.423
Yellow cryst. (hexamethylphosphoramide).

Kwiatek, J., *Catal. Rev.*, 1967, **1**, 37 (*use*)
Pregaglia, G. *et al*, *Discuss. Faraday. Soc.*, 1968, 110 (*synth*)
Reinstein, J.G. *et al*, *Inorg. Chem.*, 1969, **8**, 2499.
Griffith, W.P. *et al*, *J. Chem. Soc., Dalton Trans.*, 1972, 158 (*struct, ir, raman*)
White, D.A. *et al*, *Inorg. Chem.*, 1972, **11**, 2160 (*synth, uv*)
Ohkubo, K. *et al*, *Bull. Chem. Soc. Jpn.*, 1975, **48**, 1114 (*use*)

$C_5HCoN_5^{\ominus\ominus\ominus}$ Co-00010

Pentakis(cyano-*C*)hydrocobaltate(3−), 10CI
Pentacyanohydrocobaltate(3−), 8CI.
Hydridopentacyanocobaltate(3−)
[17168-56-8]

$HCo(CN)_5^{\ominus\ominus\ominus}$

M 190.030 (ion)
Homogenous hydrogenation catalyst for the reduction of conjugated dienes to olefins. Useful under micellar and phase transfer conditions.
Tri-Li salt: [24803-73-4].
$C_5HCoLi_3N_5$ M 210.853
Characterised in soln. only.
Tri-K salt: [18117-30-1].
$C_5HCoK_3N_5$ M 307.325
Characterised in soln. only.

Griffith, W.P. *et al*, *J. Chem. Soc.*, 1959, 2757 (*synth, pmr, uv*)
Pregaglia, G. *et al*, *Disc. Faraday Soc.*, 1968, **46**, 110 (*synth, use*)
Lim, H.S. *et al*, *Inorg. Chem.*, 1971, **10**, 103 (*synth*)
Venerable, G.D. *et al*, *J. Am. Chem. Soc.*, 1971, **93**, 2176 (*synth, bibl*)
Funabiki, T. *et al*, *Bull. Chem. Soc. Jpn.*, 1972, **45**, 2945 (*ir*)
Clase, H.J. *et al*, *J. Chem. Soc., Dalton Trans.*, 1973, 2546 (*raman*)
Reger, D.L., *Adv. Chem. Ser.*, 1979, 173 (*use, rev*)

$C_5H_3CoO_4$ Co-00011

Tetracarbonylmethylcobalt, 10CI, 9CI, 8CI
Methyltetracarbonylcobalt
[14709-64-9]

$MeCo(CO)_4$

M 186.010
Light-yellow cryst. Mp −44° (−35° dec.).

Vohler, O. *et al*, *Z. Naturforsch., B*, 1958, **13**, 192 (*synth, ir*)
Breslow, D.S. *et al*, *Chem. Ind. (London)*, 1960, 647 (*ir*)
Beck, W. *et al*, *Chem. Ber.*, 1964, **97**, 2098 (*synth, ir*)
Bor, G., *Inorg. Chim. Acta*, 1967, **1**, 81 (*ir*)
Gmelins Handbuch Anorg. Chem., Suppl., **5**, 126.
Cser, F. *et al*, *Inorg. Chim. Acta*, 1979, **37**, L517 (*uv*)

C_5H_5CoINO Co-00012

(η^5-2,4-Cyclopentadien-1-yl)iodonitrosylcobalt, 9CI
[33114-12-4]

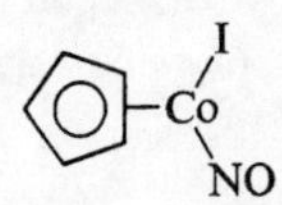

M 280.938
Black cryst. by subl. Mp 99.5-100°. ν_{NO} 1830 cm^{-1} (CH_2Cl_2). Subl. at 100-10° *in vacuo*.

Brunner, H. *et al*, *J. Organomet. Chem.*, 1968, **15**, 409 (*synth, ir, ms, pmr*)

$C_6CoF_3O_5$ Co-00013

Tetracarbonyl(trifluoroacetyl)cobalt, 9CI
[57307-02-5]

$F_3CCOCo(CO)_4$

M 267.991
Liq. Reacts readily with PF_3 to yield complexes of type $CF_3Co(CO)_{4-n}(PF_3)_{4-n}$. Decarbonylates >55°.

Hieber, W. *et al*, *Chem. Ber.*, 1961, **95**, 2042 (*synth*)
Udovich, C.A. *et al*, *Inorg. Chem.*, 1969, **8**, 938.

$C_6CoF_5O_3$ Co-00014

Tricarbonyl(η^3-1,1,2,3,3-pentafluoro-2-propenyl)cobalt, 9CI

Tricarbonyl(η^3-pentafluoroallyl)cobalt

[57665-36-8]

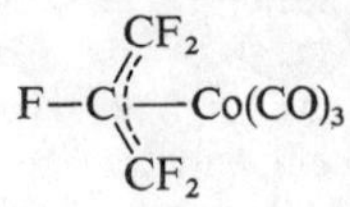

M 273.989

Yellow liq., yellow cryst. (pentane at −80°). Mp 5°. ir ν_{CO} 2065, 1998cm^{-1}.

Stanley, K. *et al*, *Can. J. Chem.*, 1975, **53**, 2537 (*synth, ir, nmr*)

$C_6H_3CoO_5$ Co-00015

Acetyltetracarbonylcobalt, 10CI

[13964-92-6]

MeCOCo(CO)$_4$

M 214.020

Unstable. Yellow cryst. Mp −30° (−20° dec.).

Breslow, D.S. *et al*, *Chem. Ind. (London)*, 1960, 467 (*synth, ir*)
Breslow, D.S. *et al*, *J. Am. Chem. Soc.*, 1962, **84**, 2499 (*synth*)
Heck, R.F., *J. Am. Chem. Soc.*, 1963, **85**, 651.
Fonnesbeck, N. *et al*, *Int. J. Quantum Chem.*, 1978, **12** (*Suppl. 2*), 95; *CA*, **89**, 214751 (*struct*)
Gmelin Suppl., **5**, 127.

$C_6H_4CoIO_2$ Co-00016

(η^4-1,3-Cyclobutadiene)dicarbonyliodocobalt, 9CI

Dicarbonyliodo(η^4-cyclobutadiene)cobalt

[12306-58-0]

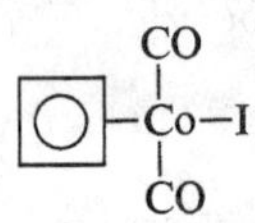

M 293.934

Red cryst. (pentane). Mp 85°. pmr δ 4.58 (C_4H_4).

Amiet, R.G. *et al*, *J. Am. Chem. Soc.*, 1968, **90**, 1059 (*synth, pmr*)

$C_6H_5CoI_2O$ Co-00017

Carbonyl(η^5-2,4-cyclopentadien-1-yl)diiodocobalt, 9CI

[12012-77-0]

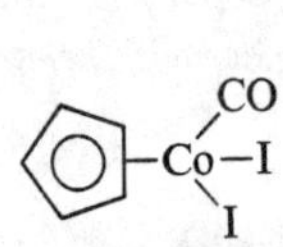

M 405.847

Starting material for synth. of metallocycles and a range of cyclopentadienyl-complexes including isocyanides, phosphines, dithioformates and metal alkyls. Dark-purple cryst. ($CHCl_3/Et_2O$/pentane). Mp 185° dec.

Heck, R.F., *Inorg. Chem.*, 1965, **4**, 855 (*synth, uv*)
King, R., *Inorg. Chem.*, 1966, **5**, 82 (*synth, ir, pmr*)
Dineen, J.A. *et al*, *J. Organomet. Chem.*, 1974, **71**, 77 (*use*)
Epley, L.J. *et al*, *J. Organomet. Chem.*, 1975, **87**, 317 (*use*)
Lee, W.-S. *et al*, *J. Organomet. Chem.*, 1977, **127**, 93 (*use*)
Spencer, A. *et al*, *J. Organomet. Chem.*, 1979, **171**, 209 (*use*)
Evill, E.R. *et al*, *J. Am. Chem. Soc.*, 1980, **102**, 7003 (*use*)

$C_6H_5CoO_3$ Co-00018

Tricarbonyl(η^3-2-propenyl)cobalt, 10CI, 9CI

η^3-Allyltricarbonylcobalt. Tricarbonylallylcobalt

[12144-85-3]

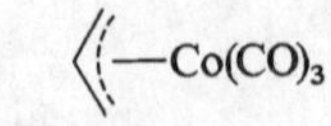

M 184.037

Red-yellow liq.; cryst. (pentane at −80°). Mp −33° to −32°. May be dist. in high vacuum.

Heck, R.F. *et al*, *J. Am. Chem. Soc.*, 1961, **83**, 1097 (*synth, props*)
McClellan, W.R. *et al*, *J. Am. Chem. Soc.*, 1961, **83**, 1601 (*synth, props*)
Siep, R., *Acta Chem. Scand.*, 1972, **26**, 1966 (*struct*)
Alper, H. *et al*, *J. Organomet. Chem.*, 1976, **121**, C31 (*synth*)
Brown, D.A. *et al*, *Inorg. Chem.*, 1977, **16**, 2497.

$C_6H_5CoO_4$ Co-00019

Tetracarbonylethylcobalt

CoEt(CO)$_4$

M 200.036

Intermediate in the cobalt-catalysed hydroformylation of ethene.

Heck, R.F. *et al*, *J. Am. Chem. Soc.*, 1961, **83**, 4023 (*synth*)
Heck, R.F., *Adv. Organomet. Chem.*, 1966, **4**, 243.

$C_7H_5CoF_3IO$ Co-00020

Carbonyl(η^5-2,4-cyclopentadien-1-yl)iodo(trifluoromethyl)-cobalt, 10CI

Iodocarbonylcyclopentadienyl(trifluoromethyl)cobalt

[12078-24-9]

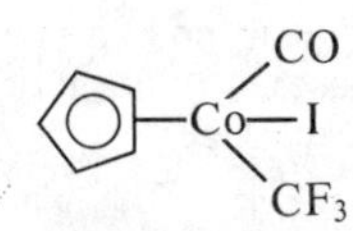

M 347.949

Black cryst. Bp$_{0.1}$ 100° subl. ir ν_{co} 2073 cm^{-1} (CS_2). Dec. without melting at ~145°.

King, R.B. *et al*, *J. Am. Chem. Soc.*, 1961, **83**, 3593 (*synth, ir, pmr*)
McCleverty, J.A. *et al*, *J. Chem. Soc.*, 1964, 4200 (*nmr*)
Gardner, S.A. *et al*, *Inorg. Chem.*, 1974, **13**, 997.
King, R.B. *et al*, *J. Am. Chem. Soc.*, 1974, **96**, 1334.
Brunner, H. *et al*, *J. Organomet. Chem.*, 1977, **139**, 223.

$C_7H_5CoO_2$ Co-00021

Dicarbonyl(η^5-2,4-cyclopentadien-1-yl)cobalt, 9CI

Cyclopentadienyldicarbonylcobalt

[12078-25-0]

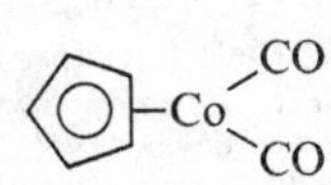

M 180.049

Catalyses cocyclization of α,ω-diynes with alkynes. Dark-red liq. Mp −22°. Bp$_{22}$ 75°. Sensitive to air and heat.

Cotton, F.A. *et al*, *J. Inorg. Nucl. Chem.*, 1955, **1**, 165 (*synth, ir*)
Fischer, E.O. *et al*, *Z. Naturforsch.*, *B*, 1955, **10**, 355 (*synth*)
Ger. Pat., 1 044 805, (*1958*); *CA*, **55**, 17648 (*synth*)
Inorg. Synth., 1963, **7**, 99 (*synth*)
Vollhardt, K.P.C., *Acc. Chem. Res.*, 1977, **10**, 1 (*use*)
Beagley, B. *et al*, *J. Mol. Struct.*, 1979, **52**, 47 (*struct*)
Fieser, M. *et al*, *Reagents for Organic Synthesis*, Wiley, 1967-83, **10**, 126 (*use*)

$C_7H_5CoS_2$ — Co-00022

Bis(carbonothioyl)(η^5-2,4-cyclopentadien-1-yl)cobalt, 10CI
η-Cyclopentadienylcobalt dithiocarbonyl
[65835-86-1]

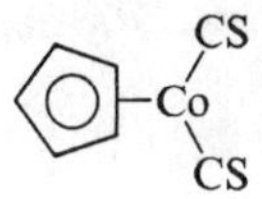

M 212.170

Rajan, S., *Indian J. Chem., Sect. A*, 1977, **15**, 920 (*synth, ir*)

$C_7H_6CoO_3^{\oplus}$ — Co-00023

(η^4-1,3-Butadiene)tricarbonylcobalt(1+), 9CI

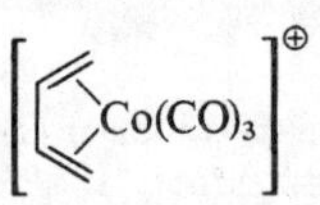

M 197.056 (ion)
Hexafluorophosphate(1−): [53382-27-7].
$C_7H_6CoF_6O_3P$ M 342.020
No phys. props. reported.

Chaudary, F.M. *et al*, *J. Organomet. Chem.*, 1974, **69**, C31.

$C_7H_7CoO_3$ — Co-00024

[(1,2,3-η)-2-Butenyl]tricarbonylcobalt, 10CI, 9CI
η^3-Crotyltricarbonylcobalt. Tricarbonylcrotylcobalt. η^{1-3}-Butenyltricarbonylcobalt
[12078-39-6]

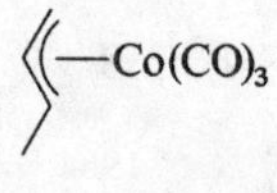

(*E*)-*form*

M 198.064
Catalyst for the alkylation of allyl chlorides, formates and acetates by dialkyl malonates. Red-orange liq. Bp_6 34-5°.

(*E*)-*form* [31627-44-8]
syn-*form*
Obt. pure by heating mixed isomers at 120°.

Heck, R.F. *et al*, *J. Am. Chem. Soc.*, 1961, **83**, 1097 (*synth, props*)
McClellan, W.R. *et al*, *J. Am. Chem. Soc.*, 1961, **83**, 1601 (*synth, props*)
Bertrand, J.A. *et al*, *Inorg. Chem.*, 1963, **2**, 601 (*pmr*)
Rupilius, W. *et al*, *J. Org. Chem.*, 1971, **36**, 3604 (*synth*)
Alper, H. *et al*, *J. Organomet. Chem.*, 1976, **121**, C31 (*synth*)
Roustan, J.L. *et al*, *Tetrahedron Lett.*, 1979, 3721 (*use*)

$C_7H_8CoI_2N$ — Co-00025

(Acetonitrile)(η^5-2,4-cyclopentadien-1-yl)diiodocobalt, 10CI
[73065-75-5]

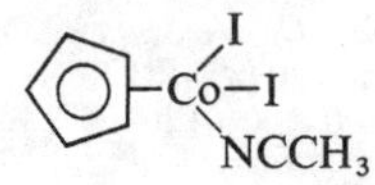

M 418.889
Formed by action of KI on $[Co(C_5H_5)(CH_3CN)_3]^{\oplus\oplus}$. Experimental details not given in reference.

Koelle, V., *J. Organomet. Chem.*, 1980, **184**, 379 (*synth, pmr, uv*)

$C_8Co_2F_4O_6$ — Co-00026

Hexacarbonylbis[μ-(difluoromethylene)]dicobalt, 10CI
Bis(μ-difluoromethylene)hexacarbonyldicobalt. Di-μ-difluoromethylenebis(tricarbonylcobalt)
[26025-17-2]

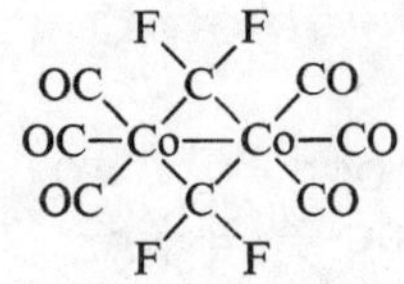

M 385.944
Yellow cryst. (MeOH). Bp 45° subl. ir ν_{co} 2128w, 2095s, 2086m, 2070vs, 2041m cm^{-1} (hexane).

Seel, F. *et al*, *Angew. Chem., Int. Ed. Engl.*, 1970, **9**, 166 (*synth, ms, ir*)
Seel, F. *et al*, *Z. Anorg. Allg. Chem.*, 1971, **386**, 297 (*synth, ir, ms*)

$C_8Co_2HgO_8$ — Co-00027

Octacarbonyl(mercury)dicobalt, 9CI
[13964-88-0]

$(OC)_4Co—Hg—Co(CO)_4$

M 542.540
Catalyst for cyclotrimerisation of alkynes and perfluoroalkynes, intermed. in $Co_2(CO)_8$-catalysed synthesis of ketones from diarylmercury compds. Orange cryst. (Me_2CO). Mp 81-2°. ir ν_{CO} 2072, 2021, 2007 cm^{-1} (heptane).

Hubel, W. *et al*, *Chem. Ber.*, 1960, **93**, 103.
Lucken, E.A. *et al*, *J. Chem. Soc. (A)*, 1967, 148 (*nmr*)
Sheldrick, G.M. *et al*, *J. Chem. Soc., Chem. Commun.*, 1967, 1015 (*cryst struct*)
Manning, A.R., *J. Chem. Soc. (A)*, 1968, 1018 (*synth, ir*)
Seyferth, D. *et al*, *J. Am. Chem. Soc.*, 1969, **91**, 6192 (*use*)
Gastinger, R.C. *et al*, *J. Org. Chem.*, 1978, **43**, 159.

$C_8Co_2O_8$ — Co-00028

Di-μ-carbonylhexacarbonyldicobalt, 10CI, 9CI
Octacarbonyldicobalt. Cobalt carbonyl. Dicobalt octacarbonyl
[10210-68-1]

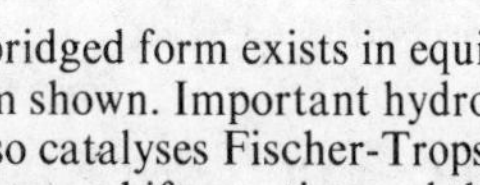

M 341.950
In soln. a nonbridged form exists in equilib. with the bridged form shown. Important hydroformylation catalyst. Also catalyses Fischer-Tropsch hydrogenation of CO, water gas shift reaction and the carboalkoxylation of alkyl halides and other compds. Orange-red cryst. Mp 51-2°. Air-sensitive.

▷Highly toxic. GG0300000.

Inorg. Synth., 1946, **2**, 238 (*synth*)
Adkins, H. *et al*, *J. Am. Chem. Soc.*, 1948, **71**, 3051 (*synth*)
Wender, I. *et al*, *J. Am. Chem. Soc.*, 1950, **72**, 4375 (*synth*)
Adkins, H. *et al*, *J. Am. Chem. Soc.*, 1954, **76**, 3951 (*ir, uv*)
Inorg. Synth., 1957, **5**, 190 (*synth*)
Noack, K., *Helv. Chim. Acta*, 1964, **47**, 1555 (*ir*)
Christahl, M. *et al*, *Z. Naturforsch., B*, 1977, **32**, 637 (*uv*)
Evans, J. *et al*, *J. Chem. Soc., Dalton Trans.*, 1978, 626 (*cmr*)
Hagen, A.P. *et al*, *Inorg. Chem.*, 1978, **17**, 1369 (*synth*)
Murai, S. *et al*, *Angew. Chem., Int. Ed. Engl.*, 1979, **18**, 837 (*use, rev*)
Fieser, M. *et al*, *Reagents for Organic Synthesis*, Wiley, 1967-83, **10**, 129.
Sax, N.I., *Dangerous Properties of Industrial Materials*, 5th Ed., Van Nostrand-Reinhold, 1979, 509.

$C_8H_2Co_2O_6$ Co-00029

Hexacarbonyl[μ-(η^2:η^2-ethyne)dicobalt, 9CI
Acetylenedicobalt hexacarbonyl.
Acetylenehexacarbonyldicobalt
[12264-05-0]

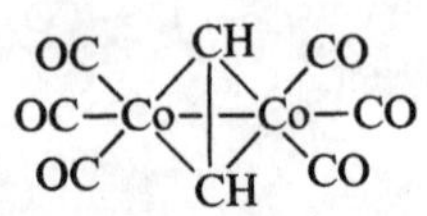

M 311.967
Red oil. Mp 13.0-13.6°. $Bp_{2.5\text{-}4}$ 64-6°. ir ν_{CO} 2098s, 2059vs, 2034vs, 2028vs, 2017m, 1985w cm^{-1} (hexane).

Greenfield, H. *et al*, *J. Am. Chem. Soc.*, 1956, **78**, 120 (*synth, ir*)
Iwashita, Y. *et al*, *Inorg. Chem.*, 1969, **8**, 1179 (*synth, ir, pmr*)
Iwashita, Y. *et al*, *Spectrochim. Acta, Part A*, 1971, **27**, 271 (*ir, nmr*)
Gambino, O. *et al*, *Org. Mass Spectrom.*, 1972, **6**, 723 (*ms*)
Aime, S. *et al*, *Inorg. Chim. Acta*, 1977, **22**, 135 (*nmr*)
Kettle, S.F.A. *et al*, *Inorg. Chem.*, 1977, **16**, 753 (*raman*)
Van Dam, H. *et al*, *J. Electron Spectrosc. Relat. Phenom.*, 1980, **21**, 47 (*pe*)

$C_8H_5CoN_3^{\ominus}$ Co-00030

Tris(cyano-*C*)(η^5-2,4-cyclopentadien-1-yl)cobaltate(1–), 9CI

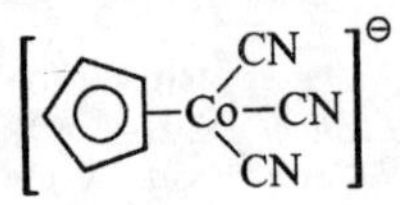

M 202.081 (ion)
K salt: [38600-67-8].
$C_8H_5CoKN_3$ M 241.179
Air-sensitive orange cryst. (MeOH). Mp 210° dec.
Benzyltriphenylphosphonium salt: [38600-68-9].
$C_{33}H_{27}CoN_3P$ M 555.503
Orange cryst. Mp 240° dec.

Dineen, J.A. *et al*, *J. Organomet. Chem.*, 1972, **43**, 209; 1974, **71**, 77 (*synth, ir, pmr*)

$C_8H_5CoN_5^{\ominus\ominus\ominus}$ Co-00031

Pentakis(cyano-*C*)-2-propenylcobaltate(3–), 9CI
Allylpentacyanocobaltate(3–), 8CI
[38887-47-7]

$[H_2C{=}CHCH_2Co(CN)_5]^{\ominus\ominus\ominus}$

M 230.094 (ion)
Olefin polymn. catalyst for aqueous media.
Tri-K salt: [32009-25-9].
$C_8H_5CoK_3N_5$ M 347.389
Yellow cryst. Mp 195° dec.

Lentzner, H.L. *et al*, *J. Polym. Sci., Part A-1*, 1971, **9**, 1401 (*synth, use*)
Funabiki, T. *et al*, *Bull. Chem. Soc. Jpn.*, 1972, **45**, 2945 (*ir*)
Dineen, J.A. *et al*, *J. Organomet. Chem.*, 1974, **71**, 87 (*synth, ir, pmr*)

$C_8H_{10}BrCo$ Co-00032

Bromo(η^5-2,4-cyclopentadien-1-yl)-η^3-2-propenylcobalt
η^3-Allylbromo(η^5-cyclopentadienyl)cobalt

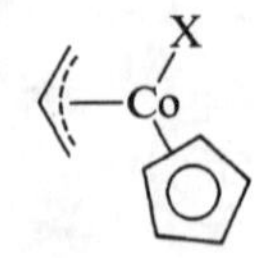

X = Br

M 245.004
Red cryst. (THF/pentane). Mp 96-8°.

Fischer, E.O. *et al*, *Z. Naturforsch., B*, 1961, **16**, 475 (*synth*)
Heck, R.F., *J. Org. Chem.*, 1963, **28**, 604 (*synth, pmr*)

$C_8H_{10}CoI$ Co-00033

(η^5-2,4-Cyclopentadien-1-yl)iodo-η^3-2-propenylcobalt
η^3-Allyl(η^5-cyclopentadienyl)iodocobalt
As Bromo(η^5-2,4-cyclopentadien-1-yl)-η^3-2-propenylcobalt, Co-00032 with

X = I

M 292.005
Dark-red cryst. (THF/pentane). Mp 95.5-97°.

Heck, R.F., *J. Org. Chem.*, 1963, **28**, 604 (*synth, ir, pmr*)

$C_8H_{12}Br_2CoN_4$ Co-00034

Dibromotetrakis(isocyanomethane)cobalt
Dibromotetrakis(methyl isonitrile)cobalt
[22512-42-1]

$(H_3CCN)_4CoBr_2$

M 382.951
Cryst. (CH_2Cl_2/pet. ether).

Maher, J.P., *J. Chem. Soc. (A)*, 1968, 2918 (*synth, esr*)

$C_8H_{12}CoI_2N_4$ Co-00035

Diiodotetrakis(isocyanomethane)cobalt
Diiodotetrakis(methyl isocyanide)cobalt
[22512-43-2]

$(MeNC)_4CoI_2$

M 476.952
Brown cryst. Mp 250° dec.

Maher, J.P., *J. Chem. Soc. (A)*, 1968, 2918 (*synth*)

$C_8H_{12}CoN_4^{\oplus\oplus}$ Co-00036

Tetrakis(isocyanomethane)cobalt(2+), 9CI
Tetrakis(methyl isocyanide)cobalt(2+)
[19223-28-0]

$[Co(CNMe)_4]^{\oplus\oplus}$

M 223.143 (ion)

Tetrachlorocobaltate(2−):
$C_8H_{12}Cl_4Co_2N_4$ M 423.888
Green cryst. Mp 220° dec.
Tetrabromocadmiate(2−):
$C_8H_{12}Br_4CdCoN_4$ M 655.169
Pink solid (MeOH). Mp 245° dec.
Tetraiodocadmiate(2−):
$C_8H_{12}CdCoI_4N_4$ M 843.171
Light-brown solid (MeOH). Mp 260°.
Tetrathiocyanatocobaltate(2−):
$C_{12}H_{12}Co_2N_8S_4$ M 514.387
Green solid. Mp 158° dec.

Sacco, A. *et al*, *Gazz. Chim. Ital.*, 1959, **89**, 1800 (*synth*)
Sacco, A. *et al*, *J. Am. Chem. Soc.*, 1960, **82**, 2983 (*synth, ir, uv*)
Malatesta, L. *et al*, *Isocyanide Complexes of Metals*, 1969, Wiley-Interscience, London, 140.

$C_8H_{24}CoNOP_2$ Co-00037

Dimethylnitrosylbis(trimethylphosphine)cobalt, 9CI
[59459-70-0]

$CoMe_2(NO)(PMe_3)_2$

M 271.164
Violet cryst. (pentane). Mp 25-8° (dec., under N_2). Dimerises on heating in soln. ir ν_{NO} 1712, 1657 cm^{-1} (nujol).

Klein, H.-F. *et al*, *Chem. Ber.*, 1976, **109**, 1453 (*synth, ir, pmr, ms*)

$C_9HCo_3O_9$ Co-00038

Tri-μ-carbonylhexacarbonyl-μ_3-hydrotricobalt, 10CI
Nonacarbonylhydridotricobalt
[71204-90-5]

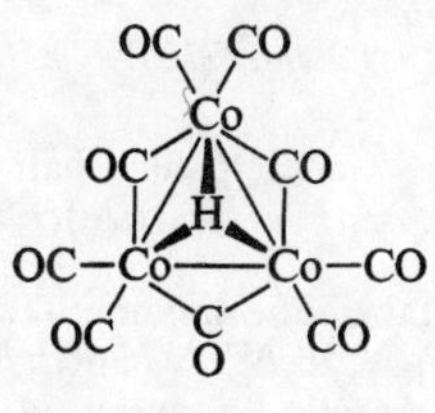

M 429.901
Blackish-green cryst. (hexane/toluene). Thermally unstable giving $Co_4(CO)_{12}$ above 30°.

Fachinetti, G. *et al*, *Angew. Chem., Int. Ed. Engl.*, 1979, **18**, 619; 1981, **20**, 204 (*synth, ir, ms, cryst struct*)

$C_9H_3Co_2O_6^{\oplus}$ Co-00039

Hexacarbonyl[μ-[(2,3-η:2,3-η)-2-propynylium]]dicobalt(1+)
(Propargyl)dicobalthexacarbonyl(1+)

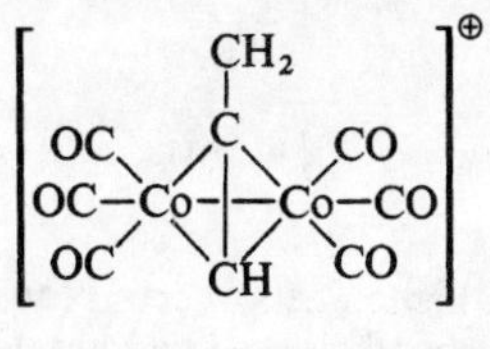

M 324.986 (ion)
Propargylating agent for aromatics, β-dicarbonyl compds., ketones and allyl silanes e.g. $R^1COCH_2R^2 \rightarrow R^1COCHR^2CH_2C{\equiv}CH$.
Tetrafluoroborate: [62867-00-9].
$C_9H_3BCoF_4O_6$ M 352.856
Red-black solid. Sol. liq. SO_2, CH_2Cl_2, Et_2O. ir ν_{CO} 2130, 2105, 2085 cm^{-1} (CH_2Cl_2). Solid hydrolysed in the atm. over several hours to give 1-propynol complex. Hydrolysis rapid in soln. (Et_2O/5% $HCO_3^{\ominus}$ aq.).

Connor, R.E. *et al*, *J. Organomet. Chem.*, 1977, **125**, C45 (*synth, ir, pmr*)
Nicholas, K.M. *et al*, *J. Am. Chem. Soc.*, 1980, **102**, 2508 (*use*)
O'Boyle, J.E. *et al*, *Tetrahedron Lett.*, 1980, **21**, 1595 (*use*)

$C_9H_4Co_2O_6$ Co-00040

Hexacarbonyl[(1,2-η:1,2-η)propyne]dicobalt
Propynehexacarbonyldicobalt
[41626-24-8]

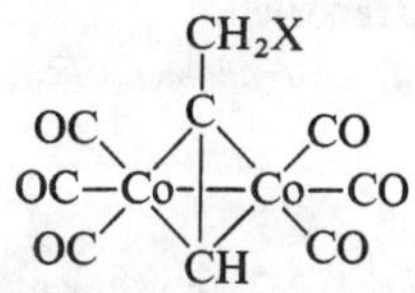

X = H

M 325.993
Synthetic reagent, e.g. for methylcyclopentenones. Red-brown oil. Sol. hydrocarbons. Reacts with $H^{\oplus}$ to give $Co_3(CO)_9(\mu_3{-}CC_2H_5)$.

Cetini, G. *et al*, *J. Organomet. Chem.*, 1967, **8**, 149 (*synth, ir, props*)
Bor, G. *et al*, *Inorg. Chim. Acta*, 1976, **18**, L18 (*synth, ir*)
Aime, S. *et al*, *Inorg. Chim. Acta*, 1977, **22**, 135 (*nmr*)
Pauson, P.L. *et al*, *Ann. N.Y. Acad. Sci.*, 1977, **295**, 2 (*props*)

$C_9H_4Co_2O_7$ Co-00041

Hexacarbonyl[μ-[(2,3-η:2,3-η)-2-propyn-1-ol]]dicobalt, 10CI
Propargyl alcohol dicobalt hexacarbonyl
[12264-12-9]
As Hexacarbonyl[(1,2-η:1,2-η)propyne]dicobalt, Co-00040 with

X = OH

M 341.993
Precursor of $[(HC{\equiv}CCH_2)Co_2(CO)_6]^{\oplus}$. Orange-red, air-sensitive needles (pet. ether). Sol. liq. SO_2, CCl_4, Et_2O, EtOH. Mp 52-5°. ir ν_{CO} 2090, 2050, 2025 cm^{-1} (CH_2Cl_2). Reacts with $HBF_4{\cdot}Et_2O$ to give Hexacarbonyl[μ-[(2,3-η:2,3-η)-2-propynylium]]dicobalt(1+), Co-00039 .

Greenfield, H. *et al*, *J. Am. Chem. Soc.*, 1956, **78**, 120 (*synth, props*)
Connor, R.E. *et al*, *J. Organomet. Chem.*, 1977, **125**, C45 (*ir, pmr, props*)

$C_9H_5CoN_2S_2$ Co-00042

(η^5-2,4-Cyclopentadien-1-yl)[2,3-dimercapto-2-butenedinitrilato(2−)-S,S′]cobalt
(π-Cyclopentadienyl)(1,2-dicyanoethene-1,2-dithiolato)cobalt
[12082-04-1]

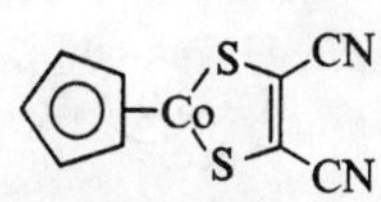

M 264.205
Purple cryst. (Me_2CO aq.). Sol. $CHCl_3$, Me_2CO, $MeNO_2$. Mp 282° dec. Reacts with tertiary phosphorus compds. to give stable 1:1 adducts.

Locke, J. *et al*, *Inorg. Chem.*, 1966, **5**, 1157 (*synth, ir, pmr*)

Churchill, M.R. *et al*, *Inorg. Chem.*, 1968, 7, 1123 (*cryst struct*)
Henderson, S.P. *et al*, *J. Organomet. Chem.*, 1979, **179**, 43.

$C_9H_8CoO_3^{\oplus}$ **Co-00043**

Tricarbonyl[(1,2,3,4-η)-1,3-cyclohexadiene]cobalt(1+), 9CI

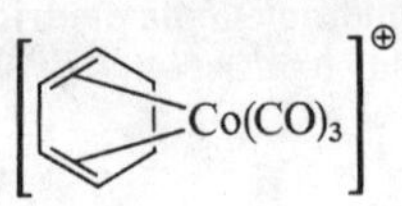

M 223.094 (ion)

Hexafluorophosphate: [53418-94-3].
$C_9H_8CoF_6O_3P$ M 368.058
No phys. props. reported.

Chaudary, F.M. *et al*, *J. Organomet. Chem.*, 1974, **69**, C31.

C_9H_9Co **Co-00044**

(η^4-1,3-Cyclobutadiene)(η^5-2,4-cyclopentadien-1-yl)cobalt, 10CI

[12306-95-5]

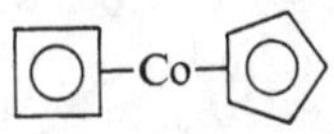

M 176.103

Yellow cryst. Mp 89° (83-5°).

Wells, D. *et al*, *J. Am. Chem. Soc.*, 1972, **94**, 1239 (*synth, pmr, uv*)
Riley, P.E. *et al*, *J. Organomet. Chem.*, 1976, **113**, 157 (*cryst struct*)
Clack, P.W. *et al*, *J. Organomet. Chem.*, 1978, **162**, 183 (*struct*)

$C_9H_9CoO_2$ **Co-00045**

Dicarbonyl[(1,2,3,4,5-η)-2,4-cycloheptadien-1-yl]cobalt, 8CI

[32758-06-8]

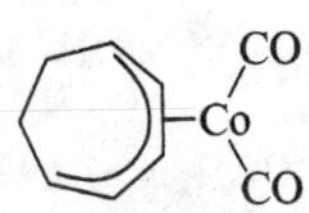

M 208.102

Dark-red needles (pentane). Mp 0-5°. ir ν_{co} 2050, 1988, 1950 cm^{-1} (nujol).

Otsuka, S. *et al*, *J. Chem. Soc. (A)*, 1971, 579 (*synth, ir, pmr*)

$C_9H_{10}CoO^{\oplus}$ **Co-00046**

Carbonyl(η^5-2,4-cyclopentadien-1-yl)(η^3-2-propenyl)cobalt(1+), 10CI

η^3-Allylcarbonyl(η^5-cyclopentadienyl)cobalt(1+)

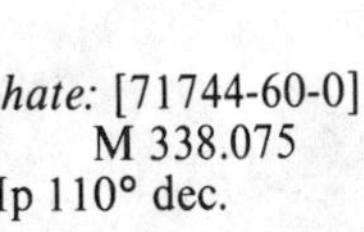

M 193.111 (ion)

Hexafluorophosphate: [71744-60-0].
$C_9H_{10}CoF_6OP$ M 338.075
Yellow solid. Mp 110° dec.

Fischer, E.O., *Z. Naturforsch., B*, 1961, **16**, 475 (*synth*)
Aviles, T. *et al*, *J. Chem. Soc., Dalton Trans.*, 1979, 1116 (*pmr*)

$C_9H_{15}BrCoO_3P$ **Co-00047**

Bromotricarbonyl(triethylphosphine)cobalt

[15977-01-2]

$Et_3PCoBr(CO)_3$

M 341.027

ir ν_{co} 1978 cm^{-1} (hexane).

Pankowski, M. *et al*, *C.R. Hebd. Seances Acad. Sci.*, 1967, **264**, 1382 (*synth, ir*)

$C_9H_{15}ClCoO_3P$ **Co-00048**

Tricarbonylchloro(triethylphosphine)cobalt

[15977-02-3]

$Et_3PCoCl(CO)_3$

M 296.576

ir ν_{co} 1976 cm^{-1} (hexane).

Pankowski, M. *et al*, *C.R. Hebd. Seances Acad. Sci.*, 1967, **264**, 1382 (*synth, ir*)

$C_9H_{15}Co$ **Co-00049**

Tris(η^3-2-propenyl)cobalt, 10CI, 9CI

Tri-π-allylcobalt, 8CI

[12082-45-0]

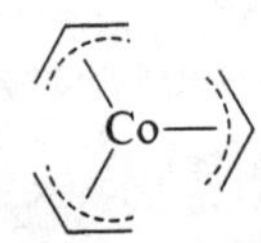

M 182.151

Catalyst for oligomerization of dienes. Orange-red prisms (butane at −100°). Dec. at ca. −50°.

Bönnemann, H., *Angew. Chem., Int. Ed. Engl.*, 1973, **12**, 964 (*rev, synth, pmr, use*)
Vasil'ev, V.A. *et al*, *J. Polym. Sci., Polym. Chem. Ed.*, 1973, **11**, 2489 (*use*)

$C_9H_{17}B_2CoF_4N_4O_5$ **Co-00050**

Aqua[bis[μ-[(2,3-butanedione dioximato)(2−)-*O*:*O*′]]][tetrafluorodiborato(2−)-*N*,*N*′,*N*″,*N*‴]methylcobalt, 10CI

[Tetrafluorobis[dimethylglyoximato(2−)]diborato]-aquomethylcobalt

[53729-64-9]

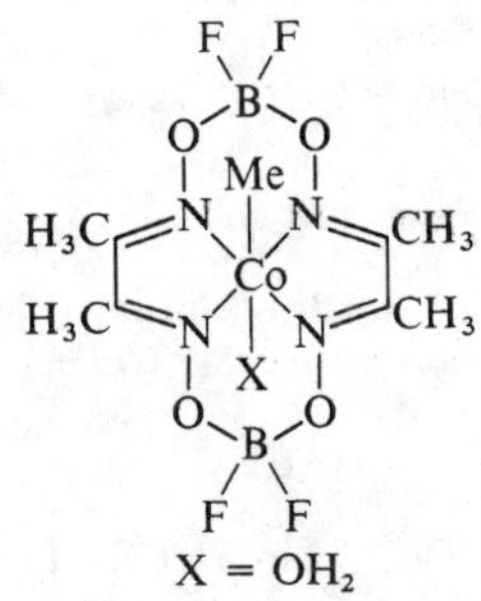

X = OH_2

M 417.804

Vitamin B_{12} analogue. Yellow cryst. (MeOH aq.). Sol. H_2O. Co−Me bond cleaved by light, OH^- and Hg^{2+}. H_2O ligand easily replaced by Lewis bases, e.g. Py.

Schrauzer, G.N. *et al*, *J. Am. Chem. Soc.*, 1966, **88**, 3738 (*synth*)
Adin, A. *et al*, *J. Chem. Soc., Chem. Commun.*, 1971, 653.
Cartano, A.V. *et al*, *Bioinorg. Chem.*, 1977, **7**, 351 (*synth*)
Brown, K.L., *J. Am. Chem. Soc.*, 1979, **101**, 6600 (*pmr, props*)

$C_9H_{19}CoN_4O_5$ Co-00051

Aquabis[(2,3-butanedione dioximato)(1−)-*N,N'*]methylcobalt, 10CI

Bis(dimethylglyoximato)aquamethylcobalt. Aquamethylbis(dimethylglyoximato)cobalt. Methylaquacobaloxime

[25360-55-8]

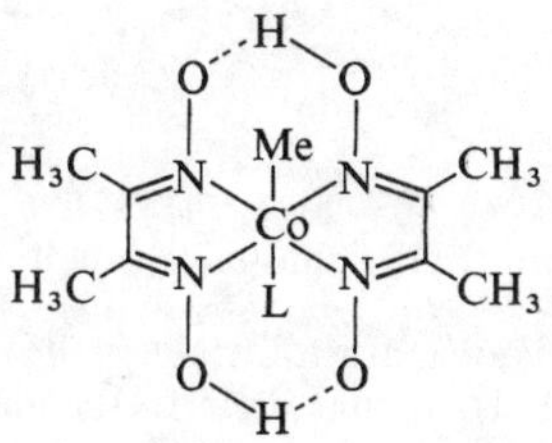

L = OH_2

M 322.206

Model compd. for vitamin B_{12}. Dark-orange cryst. (H_2O). Sol. H_2O. Mp 215-20° dec. Light-sensitive.

Monocation: [57105-04-1].

$C_9H_{19}CoN_4O_5^{\oplus}$ M 322.206 (ion)

Characterised spectroscopically.

Schrauzer, N. *et al*, *J. Am. Chem. Soc.*, 1966, **88**, 3738 (*synth*)
Inorg. Synth., 1969, **11**, 61 (*synth*)
McFadden, D.L. *et al*, *J. Chem. Soc., Dalton Trans.*, 1974, 363 (*struct*)
Reichert, B.E. *et al*, *J. Chem. Soc., Chem. Commun.*, 1974, 177 (*nmr*)
Halpern, J. *et al*, *J. Am. Chem. Soc.*, 1975, **97**, 1606 (*esr*)
Hoshino, M. *et al*, *Inorg. Chem.*, 1982, **21**, 89 (*uv, esr*)

$C_9H_{21}CoO_8P_2$ Co-00052

Dicarbonylmethylbis(trimethyl phosphite)cobalt, 9CI

[52556-64-6]

$[(MeO)_3P]_2CoMe(CO)_2$

M 378.141

Homogeneous hydrogenation catalyst for terminal olefins. Pale-yellow cryst. (THF/pentane). Mp 50°. ir ν_{CO} 1996, 1934 cm^{-1} (CCl_4).

Heck, R.F., *J. Am. Chem. Soc.*, 1963, **85**, 1220 (*synth, ir*)
Janik, T.S. *et al*, *J. Mol. Catal.*, 1981, **11**, 33 (*use*)

$C_{10}BrCo_3O_9$ Co-00053

[μ_3-(Bromomethylidyne)]nonacarbonyltricobalt, 10CI, 9CI

Nonacarbonyl[μ_3-(bromomethylidyne)tricobalt, 8CI

[19439-14-6]

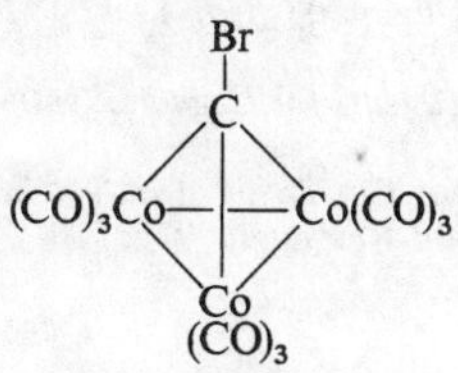

M 520.808

Initiator for radical polymerisation. Purple cryst. Mp 130-2° dec. Sublimes at 35-40°/0.5 mm.

Raffaele, E. *et al*, *Chim. Ind. (Milan)*, 1962, **44**, 1344 (*synth, props*)
Palyi, G. *et al*, *Inorg. Chim. Acta Rev.*, 1970, **4**, 109 (*synth, props*)
Dolby, R. *et al*, *J. Chem. Soc., Dalton Trans.*, 1973, 1794.
Aime, S. *et al*, *Inorg. Chim. Acta*, 1976, **18**, 9 (*cmr*)
Howard, M.W. *et al*, *Spectrochim. Acta, Part A*, 1981, **37**, 473 (*ir, raman*)

$C_{10}ClCo_3O_9$ Co-00054

Nonacarbonyl[μ_3-(chloromethylidyne)tricobalt], 10CI, 9CI

Chloromethylidenenonacarbonyltricobalt

[13682-02-5]

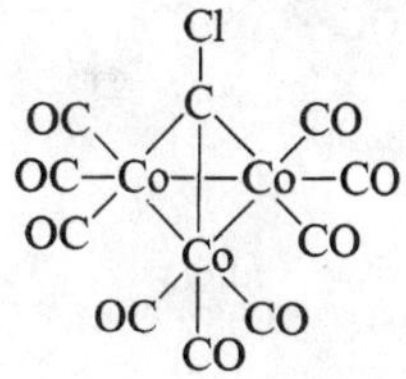

M 476.357

Catalyst for hydroformylation and dimerisation of norbornadiene. Useful precursor for derivs. of the $CCo_3(CO)_9$ cluster. Purple or black-violet solid by subl. *in vacuo*. Mp 131-3°. Subl. at 45°/0.1 mm.

Bor, G. *et al*, *Chem. Ber.*, 1962, **95**, 333 (*synth, ir*)
Robinson, B.H. *et al*, *J. Chem. Soc. (A)*, 1968, 1784 (*ms*)
Palyi, G. *et al*, *Inorg. Chim. Acta Rev.*, 1970, **4**, 109 (*rev*)
Aime, S. *et al*, *Inorg. Chim. Acta*, 1976, **18**, 9 (*cmr*)
Booth, B.L. *et al*, *J. Chem. Soc., Dalton Trans.*, 1980, 403 (*synth*)
Inorg. Synth., 1980, **20**, 234 (*synth, ir*)

$C_{10}CoF_5O_4$ Co-00055

Tetracarbonyl(pentafluorophenyl)cobalt, 10CI

[15671-73-5]

$(C_6F_5)Co(CO)_4$

M 338.033

Precursor for complexes of type $(C_6F_5)Co(CO)_3PR_3$. Pale-yellow cryst. by subl. ir ν_{CO} 2122, 2060, 2040 cm^{-1} (cyclohexane).

Wilford, J.B., *J. Chem. Soc.*, 1965, 6519 (*synth, ir*)
Royo, P. *et al*, *J. Organomet. Chem.*, 1981, **205**, 223.

$C_{10}Co_2F_4O_8$ Co-00056

Octacarbonyl[μ-(1,1,2,2-tetrafluoro-1,2-ethanediyl)]dicobalt, 9CI

[18128-84-2]

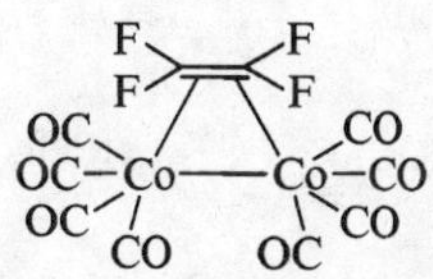

M 441.965

Orange solid, unstable in soln. Mp 97-8°. ir ν_{CO} 2116, 2110, 2087, 2055, 2050, 2016, 2008 cm^{-1} (hexane).

Hoehn, H.H. *et al*, *J. Chem. Soc.*, 1961, 2738 (*synth, ir, nmr*)
Booth, B.L. *et al*, *J. Chem. Soc. (A)*, 1969, 691 (*synth, ir*)

$C_{10}Co_2F_6O_7$ **Co-00057**

μ-Carbonylhexacarbonyl[μ-[2,2,2-trifluoro-1-(trifluoromethyl)ethyl]]dicobalt, 8CI

μ-Carbonyl-μ-bis(trifluoromethyl)methylenehexacarbonyldicobalt. Heptacarbonyl[bis(trifluoromethyl)-methylene]dicobalt

[18115-09-8]

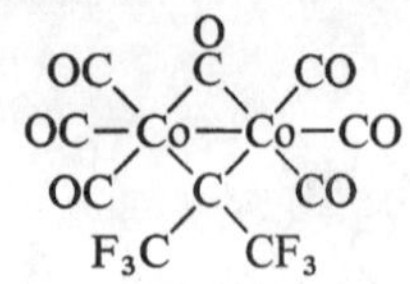

M 463.963

Air- and light-sensitive red cryst. by subl. $Bp_{10.3}$ 20° subl. ir ν_{co} 2124m, 2091s, 2068s, 2061s, 2052sh, 2025w, 1867m cm^{-1} (cyclohexane).

Cooke, J. *et al*, *J. Chem. Soc. (A)*, 1969, 1872 (*synth, ir, nmr*)
Booth, B.L. *et al*, *J. Chem. Soc., Dalton Trans.*, 1975, 1449 (*synth, ir, nmr*)

$C_{10}Co_3O_{10}^{\ominus}$ **Co-00058**

Tri-μ-carbonyl-μ_3-carbonylhexacarbonyltricobaltate(1−), 10CI

Decacarbonyltricobaltate(1−)

[12557-91-4]

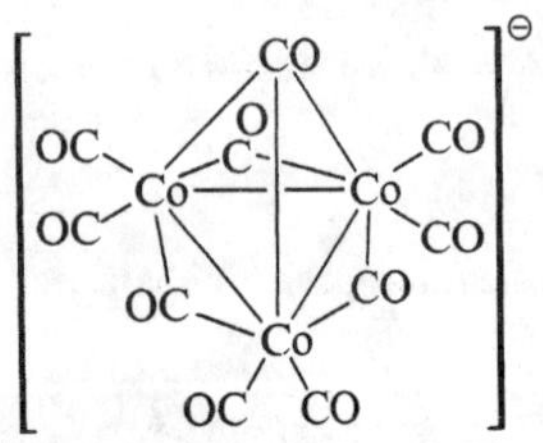

M 456.904 (ion)

Li salt:

$C_{10}Co_3LiO_{10}$ M 463.845

Black cryst. (Et_2O) forming red soln. ir ν_{CO} 2080-2000, 1850, 1600 cm^{-1} (THF).

Fieldhouse, S.A. *et al*, *J. Chem. Soc., Chem. Commun.*, 1970, 181 (*synth, ir*)

$C_{10}HCo_3O_9$ **Co-00059**

Nonacarbonyl-μ_3-methylidynetricobalt, 10CI, 9CI, 8CI

Methylidenenonacarbonyltricobalt

[15664-75-2]

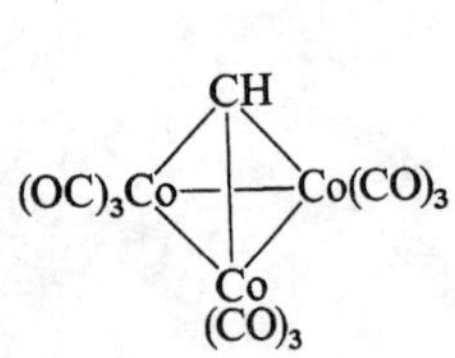

M 441.912

Catalyses olefin hydroformylation and isomerization, and the dimerization of norbornadiene. Black-violet cryst. (hexane). Mp 105-6°.

Raffaele, E. *et al*, *Chim. Ind (Milan)*, 1962, **44**, 1344 (*synth*)
Palyi, G. *et al*, *Inorg. Chim. Acta Rev.*, 1970, **4**, 109 (*props*)
Seyferth, D. *et al*, *J. Organomet. Chem.*, 1973, **50**, 265 (*use*)
Aime, S. *et al*, *Inorg. Chim. Acta*, 1976, **18**, 9.
Aime, S. *et al*, *Transition Met. Chem.*, 1976, **1**, 177 (*cmr*)
Howard, M. *et al*, *J. Chem. Soc., Faraday Trans. 2*, 1981, **77**, 397.
Leung, P. *et al*, *Acta Crystallogr., Sect. B*, 1981, **37**, 1347.

$C_{10}H_6Co_2O_6$ **Co-00060**

[μ-[(2,3-η:2,3-η)-2-Butyne]]hexacarbonyldicobalt, 10CI

[37726-81-1]

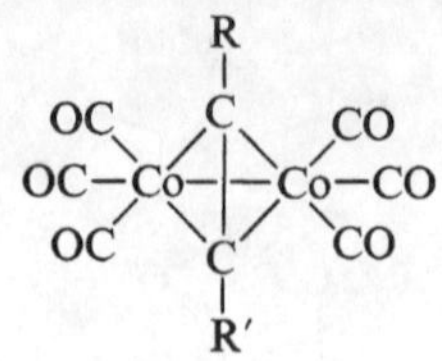

R = R′ = CH_3

M 340.020

CO ligands fluxional at r.t. Catalyses the reaction of alkynes with B_5H_9 to give the corresponding 2-(RCH=CR′)B_5H_8. Dark-red oil.

Greenfield, H. *et al*, *J. Am. Chem. Soc.*, 1956, **78**, 120 (*synth*)
Aime, S. *et al*, *Inorg. Chim. Acta*, 1977, **22**, 135 (*pmr, cmr*)
Wilczynski, R. *et al*, *Inorg. Chem.*, 1981, **20**, 3955 (*use*)

$C_{10}H_9Co_2O_{10}P$ **Co-00061**

Heptacarbonyl(trimethyl phosphite)dicobalt

$(OC)_4Co{-}Co(CO)_3P(OMe)_3$

M 438.015

ir ν_{CO} 2085.5, 2033, 2018, 1999.5, 1977 cm^{-1} (hexane).

Szabo, P. *et al*, *J. Organomet. Chem.*, 1968, **12**, 245 (*synth, ir*)

$C_{10}H_{10}Co$ **Co-00062**

Cobaltocene, 10CI, 9CI

Di-π-cyclopentadienylcobalt, 8CI. Bis(cyclopentadienyl)-cobalt

[1277-43-6]

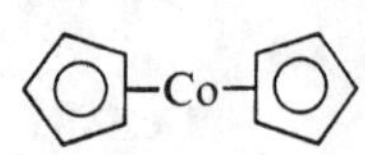

M 189.122

Catalyses synth. of pyridines from alkynes and nitriles. Black-purple cryst. Mp 173-4°. V. air-sensitive, paramagnetic. Readily oxid. → cobaltocenium cation. Subl. at 40°/0.1 mm.

▷Toxic. GG0350000.

Wilkinson, G. *et al*, *J. Am. Chem. Soc.*, 1954, **76**, 1970 (*synth, ir, uv*)
Friedman, L. *et al*, *J. Am. Chem. Soc.*, 1955, **92**, 3689 (*ms*)
Furlani, L. *et al*, *Chem. Ber.*, 1962, **95**, 2928 (*synth*)
Hunt, C.C. *et al*, *Inorg. Nucl. Chem. Lett.*, 1966, **2**, 283 (*synth*)
Koehler, F.H., *J. Organomet. Chem.*, 1976, **110**, 235 (*synth, cmr, pmr*)
Sheats, J.E., *J. Organomet. Chem. Libr.*, 1979, **7**, 461 (*props, rev*)
Fieser, M. *et al*, *Reagents for Organic Synthesis*, Wiley, 1967-83, **6**, 128 (*use*)
Sax, N.I., *Dangerous Properties of Industrial Materials*, 5th Ed., Van Nostrand-Reinhold, 1979, 510.

Having problems with locating a compound? Have you checked the indexes?

$C_{10}H_{10}Co^{\oplus}$ Co-00063

Cobaltocenium, 10CI, 9CI

Di-π-cyclopentadienylcobalt(1+), 8CI. *Bis(cyclopentadienyl)cobalt*

[12241-42-8]

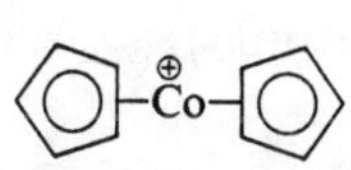

M 189.122 (ion)

Diamagnetic.

Chloride: [12170-26-2].
$C_{10}H_{10}ClCo$ M 224.575
Pale-yellow. Hygroscopic.

Tribromide: [12146-10-0].
$C_{10}H_{10}Br_3Co$ M 428.834
Orange-yellow.

Tetrafluoroborate: [52314-53-1].
$C_{10}H_{10}BCoF_4$ M 275.926
Pale-yellow.

Hartley, D. *et al*, *J. Chem. Soc. (A)*, 1969, 138 (*synth, ir, raman*)
van den Akker, M. *et al*, *Recl. Trav. Chim. Pays-Bas*, 1971, **90**, 1101 (*synth*)
Nesmeyanov, A.N. *et al*, *J. Organomet. Chem.*, 1972, **39**, 173 (*pmr*)
van den Akker, M. *et al*, *Recl. Trav. Chim. Pays-Bas*, 1972, **91**, 75 (*struct*)
Sheats, J.E. *et al*, *Synth. Inorg. Met.-Org. Chem.*, 1973, **3**, 59 (*synth*)
El Murr, N., *J. Organomet. Chem.*, 1976, **112**, 189 (*uv*)
Geiger, W.E. *et al*, *Inorg. Chem.*, 1979, **18**, 2358 (*synth*)
Sheats, J.E., *J. Organomet. Chem. Libr.*, 1979, **7**, 461 (*props, rev*)
Materikova, R.B. *et al*, *J. Organomet. Chem.*, 1981, **219**, 259 (*pmr, cmr*)

$C_{10}H_{10}CoI_2N$ Co-00064

(η^5-2,4-Cyclopentadien-1-yl)diiodo(pyridine)cobalt, 9CI

Diiodocyclopentadienyl(pyridine)cobalt

[12146-14-4]

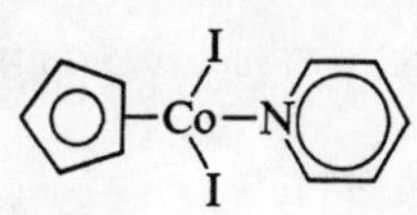

M 456.938

Dark-purple cryst. ($CHCl_3/Et_2O$). Mp 185° dec.

Heck, R.F. *et al*, *Inorg. Chem.*, 1965, **4**, 855 (*synth, ir*)
King, R.B., *Inorg. Chem.*, 1966, **5**, 82 (*synth, ir*)
Bailey, W.I. *et al*, *J. Organomet. Chem.*, 1979, **173**, 317 (*synth*)
Byers, L.P. *et al*, *Inorg. Chem.*, 1980, **19**, 277 (*synth, cryst struct, ir, ms*)

$C_{10}H_{10}Co_2N_2O_2$ Co-00065

Bis(η^5-2,4-cyclopentadien-1-yl)di-μ-nitrosyldicobalt, 10CI

Bis(η^5-2,4-cyclopentadien-1-yl)bis[μ-(nitrosyl-N:N)-dicobalt, 9CI. *Di-π-cyclopentadienyldi-μ-nitrosyldicobalt*, 8CI. *Di-μ-nitrosylbis(η^5-cyclopentadienyl)dicobalt. Di-μ-nitrosyldi-π-cyclopentadienyldicobalt*

[51862-20-5]

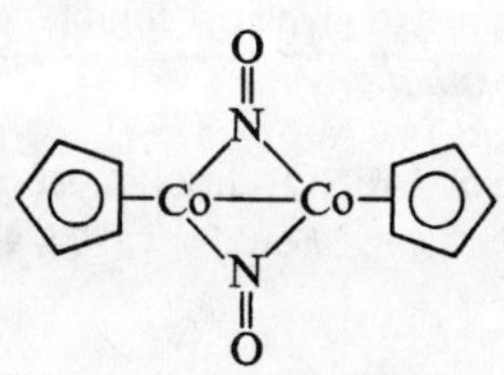

M 308.068

Reagent for vic-diamination of olefins. Black cryst. by subl. Bp 90-100° subl. *in vacuo*. ir ν_{NO} 1585, 1525 cm^{-1} (KBr).

Brunner, H., *J. Organomet. Chem.*, 1968, **12**, 517 (*synth, ir, ms, pmr, struct*)
Bernal, J. *et al*, *J. Organomet. Chem.*, 1977, **139**, 321 (*cryst struct*)
Herrmann, W.A. *et al*, *Angew. Chem., Int. Ed. Engl.*, 1977, **89**, 186 (*synth, ir*)
Hames, B.W. *et al*, *Inorg. Chem.*, 1980, **19**, 1565 (*synth*)
Connelly, N.G. *et al*, *J. Chem. Soc., Dalton Trans.*, 1983, 295 (*synth*)
Fieser, M. *et al*, *Reagents for Organic Synthesis*, Wiley, 1967-83, **10**, 116 (*use*)

$C_{10}H_{11}Co$ Co-00066

[(1,2,3,4-η)-1,3,-Cyclopentadiene](η^5-2,4-cyclopentadien-1-yl)cobalt, 10CI, 9CI

Cyclopentadienylcyclopentadienecobalt

[33032-03-0]

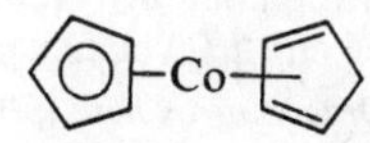

M 190.130

Wine-red cryst. Mp 98-9°. Sublimes *in vacuo*. Air-sensitive.

Green, M.L.H. *et al*, *J. Chem. Soc.*, 1959, 3753 (*ir, uv, pmr, synth*)
U.S.P., 3 088 960, (*1963*); *CA*, **59**, 10127 (*synth*)
Court, T.L. *et al*, *J. Organomet. Chem.*, 1974, **65**, 245 (*pmr*)
Geiger, W.E. *et al*, *Inorg. Chem.*, 1979, **18**, 2358 (*synth, props*)

$C_{10}H_{11}CoN$ Co-00067

Aminocobaltocenium, 9CI

(Amino-π-cyclopentdienyl)-π-cyclopentadienylcobalt(1+), 8CI. *Aminocobalticinium*

[40698-34-8]

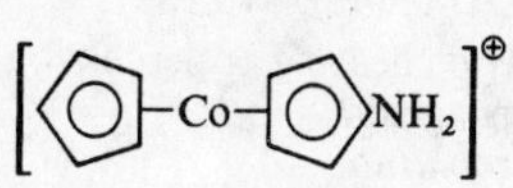

M 204.137

pK_a 15.6.

Hexafluorophosphate: [12427-43-9].
$C_{10}H_{11}CoF_6NP$ M 349.101
Cryst. ($Me_2CO/CHCl_3$). Mp 324-5° dec.

Tetraphenylborate: [12427-55-3].
$C_{34}H_{31}BCoN$ M 523.369
Cryst. ($Me_2CO/CHCl_3$). Mp 199-202° dec.

Sheats, J.E. *et al*, *J. Org. Chem.*, 1970, **35**, 3245 (*synth, ir, pmr, uv*)
El Murr, N., *J. Organomet. Chem.*, 1976, **112**, 189 (*ir, uv*)
Sheats, J.E. *et al*, *J. Organomet. Chem.*, 1976, **121**, 73 (*nmr*)

$C_{10}H_{12}CoN_2^{\oplus}$ Co-00068

1,1′-Diaminocobaltocenium, 9CI

Bis(amino-π-cyclopentadienyl)cobalt(1+), 8CI. *1,1′-Diaminocobalticinium*

[40759-59-9]

M 219.151 (ion)

Hexafluorophosphate: [43021-04-1].
$C_{10}H_{12}CoF_6N_2P$ M 364.116
Cryst. ($Me_2CO/CHCl_3$). Mp 275° dec.

Tetraphenylborate: [33635-67-5].

$C_{34}H_{32}BCoN_2$ M 538.383
Cryst. Mp 204-6°.

El Murr, N. *et al*, *C.R. Hebd. Seances Acad. Sci.*, 1971, **272**, 1989 (*synth, uv*)
Sheats, J.E. *et al*, *J. Organomet. Chem.*, 1975, **91**, 97 (*synth, ir, pmr*)
El Murr, N., *J. Organomet. Chem.*, 1976, **112**, 177 (*synth, ir, uv, nmr*)

$C_{10}H_{13}CoO_2Si$ Co-00069

Dicarbonyl[(1,2,3,4,5-η)-1-(trimethylsilyl)-2,4-cyclopentadien-1-yl]cobalt, 10CI
[32614-15-6]

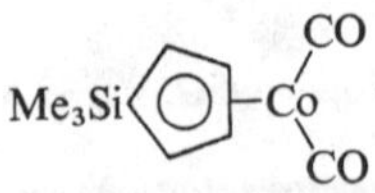

M 252.230
Precursor for generation of cobalt radical anions. Red liq. ir ν_{CO} 2025, 1966 cm^{-1} (cyclohexane).

Abel, E.W. *et al*, *J. Organomet. Chem.*, 1971, **28**, 211 (*synth, pmr, ir*)
Schore, N.E., *J. Organomet. Chem.*, 1979, **173**, 301.

$C_{10}H_{15}CoN_5^{\oplus}$ Co-00070

Pentakis(isocyanomethane)cobalt(1+), 10CI
Pentakis(methylisonitrile)cobalt(1+). Pentakis(methyl isocyanide)cobalt(1+)
[14024-66-9]

$(MeNC)_5Co^{\oplus}$

M 264.195 (ion)
Perchlorate:
$C_{10}H_{15}ClCoN_5O_4$ M 363.646
Yellow cryst. (MeOH). Mp 152°.
▷Dec. violently on heating or percussion or in contact with oxidising agents
Tetracarbonylcobaltate:
$C_{14}H_{15}Co_2N_5O_4$ M 435.170
Yellow cryst. Dec. violently on heating, percussion or contact with oxidising agents.
Hexafluorophosphate:
$C_{10}H_{15}CoF_6N_5P$ M 409.159
Yellow cryst.

Hieber, W. *et al*, *Chem. Ber.*, 1954, **87**, 789 (*synth*)
Sacco, A. *et al*, *Gazz. Chim. Ital.*, 1959, **89**, 1800 (*synth*)
Cotton, F.A. *et al*, *Inorg. Chem.*, 1965, **4**, 318 (*cryst struct*)
Boorman, P.M. *et al*, *Can. J. Chem.*, 1970, **48**, 838 (*ir, raman*)
Balch, A.L., *J. Am. Chem. Soc.*, 1976, **98**, 285.

$C_{10}H_{17}CoOP^{\oplus}$ Co-00071

Carbonyl(η^5-2,4-cyclopentadien-1-yl)methyl(trimethylphosphine)cobalt(1+), 10CI

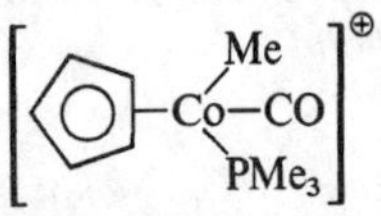

M 243.151 (ion)
Iodide: [70824-63-4].
$C_{10}H_{17}CoIOP$ M 370.055
Unstable yellow solid. Dec. to $CoEt(COCH_3)(PMe_3)I$.
Hexafluorophosphate: [70739-45-6].
$C_{10}H_{17}CoF_6OP_2$ M 388.115
Yellow solid ($MeOH/MeNO_2$). ir ν_{CO} 2040 cm^{-1}.

Spencer, A., *J. Organomet. Chem.*, 1979, **171**, 219 (*synth, pmr, ir*)

$C_{10}H_{21}CoO_9P_2$ Co-00072

Acetyldicarbonylbis(trimethyl phosphite-*P*)cobalt, 10CI
[38268-98-3]

$[(MeO)_3P]_2Co(CO)_2Ac$

M 406.151
Catalyst for homogeneous hydrogenation of terminal olefins. Pale-yellow cryst. (pentane). Mp 35°. ir ν_{CO} 2006, 1995, 1944, 1938, 1660 cm^{-1}; multiple CO frequencies due to presence of stereoisomers.

Attali, S. *et al*, *Inorg. Chim. Acta*, 1972, **6**, 478 (*synth, ir, pmr*)
Janik, T.S. *et al*, *J. Mol. Catal.*, 1981, **11**, 33 (*use*)

$C_{10}H_{27}ClCoOP_3$ Co-00073

Carbonylchlorotris(trimethylphosphine)cobalt, 9CI
[55449-08-6]

$(Me_3P)_3CoCl(CO)$

M 350.630
Air-sensitive wine-red cryst. (Et_2O). Spar. sol. pentane. Mp 94° dec. ir ν_{CO} 1877 cm^{-1} (nujol).

Klein, H.-F. *et al*, *Inorg. Chem.*, 1975, **14**, 473 (*synth, ir, pmr*)

$C_{10}H_{27}CoIOP_3$ Co-00074

Carbonyliodotris(trimethylphosphine)cobalt, 9CI
[53432-14-7]

$(Me_3P)_3CoI(CO)$

M 442.082
Air-sensitive red cryst. (pentane or Et_2O). Spar. sol. Et_2O. Mp 130-2°. ir ν_{CO} 1882 cm^{-1} (nujol). Dec. >170°.

Klein, H.-F. *et al*, *Inorg. Chem.*, 1975, **14**, 473 (*synth, ir, pmr*)

$C_{11}Co_3O_{10}^{\oplus}$ Co-00075

Nonacarbonyl[μ_3-(oxoethenylidene)]tricobalt(1+), 9CI
[61587-47-1]

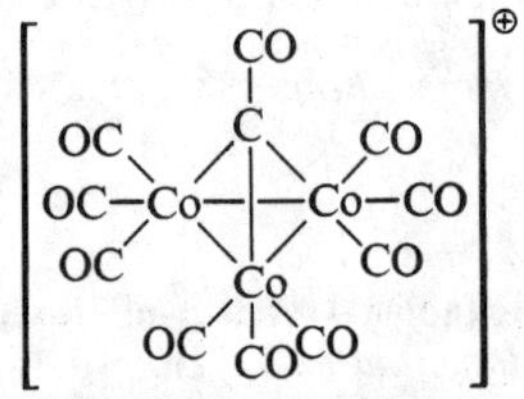

M 468.915 (ion)
Moisture-sensitive. Reduced by $SiEt_3H$ to $Co_3(CO)_9CCH_3$ and $Co_3(CO)_9CH$. Reacts with nucleophiles and arenes to give acyl complexes.
Hexafluorophosphate:
$C_{11}Co_3F_6O_{10}P$ M 613.879
Black solid. Sol. $MeNO_2$, insol. most other solvs.

Seyferth, D. *et al*, *J. Am. Chem. Soc.*, 1974, **96**, 1730 (*synth, struct*)

$C_{11}H_2Co_2O_9$ Co-00076

μ-Carbonylhexacarbonyl[μ-(5-oxo-2(5H)-furanylidene)]dicobalt, 10CI

Heptacarbonyl(3-carboxy-2-hydroxyallylidene)dicobalt γ-lactone. (μ_2-2-Buten-4-olide-4-ylidene)-μ_2-carbonylhexacarbonyldicobalt

[18177-59-8]

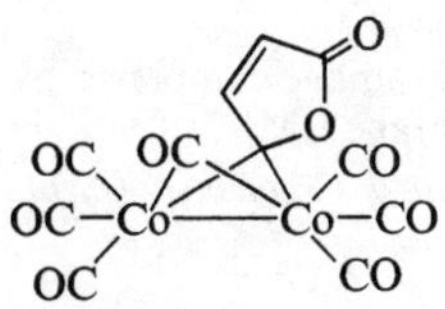

M 395.998

Catalyst for the production of bifurandiones by carbonylation of alkynes. Red cryst. (Et_2O). Mp 114-6°.

Sternberg, W.H. *et al*, *J. Am. Chem. Soc.*, 1959, **81**, 2339 (*synth, ir*)
Heck, R.F., *J. Am. Chem. Soc.*, 1963, **85**, 657.
Mills, O.S. *et al*, *Inorg. Chim. Acta*, 1967, **1**, 61 (*struct*)
Palyi, G. *et al*, *J. Organomet. Chem.*, 1975, **90**, 85 (*ir, pmr, synth*)
Guthrie, D.J.S. *et al*, *J. Organomet. Chem.*, 1975, **90**, 93 (*use*)
Varadi, G. *et al*, *J. Organomet. Chem.*, 1981, **206**, 119.

$C_{11}H_3Co_3O_9$ Co-00077

Nonacarbonyl-μ_3-ethylidynetricobalt, 9CI

[13682-04-7]

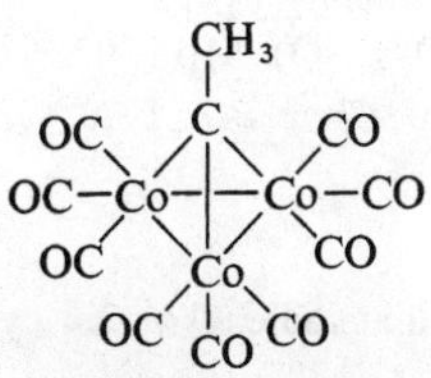

M 455.939

Catalyst for polymerisation of acrylonitrile, hydroformylation of alkenes, and hydrogenation of CO. Red cryst. Mp 183-4° dec. ir ν_{co} 2018, 2011, 2008 cm^{-1} (hexane).

Sutton, P.W. *et al*, *J. Am. Chem. Soc.*, 1967, **89**, 261 (*cryst struct*)
Palyi, G. *et al*, *Inorg. Chim. Acta Rev.*, 1970, **4**, 109 (*synth, ir, pmr*)
Aime, S. *et al*, *Transition. Met. Chem.*, 1976, **1**, 177 (*cmr*)
Ryan, R.C. *et al*, *J. Am. Chem. Soc.*, 1977, **99**, 1986 (*use*)
King, R.B. *et al*, *J. Mol. Catal.*, 1980, **10**, 75 (*use*)
Murata, K. *et al*, *Bull. Chem. Soc. Jpn.*, 1980, **53**, 214 (*use*)

$C_{11}H_5CoFeO_6$ Co-00078

Dicarbonyl(η^5-2,4-cyclopentadien-1-yl)(tetracarbonylcobalt)-iron, 9CI

Dicarbonyl-π-cyclopentadienyl(tetracarbonylcobaltio)iron. Cyclopentadienylirondicarbonyltetracarbonylcobalt. (η^5-Cyclopentadienyl)dicarbonylirontetracarbonylcobalt. Hexacarbonyl(η^5-cyclopentadienyl)-cobaltiron

[34778-25-1]

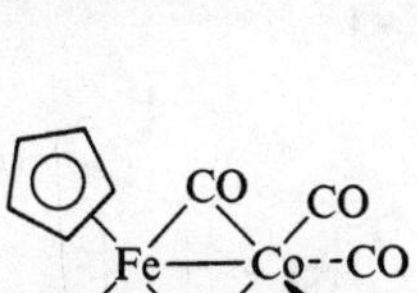

M 347.937

In soln. exists as a mixt. of CO bridged and non-bridged isomers. Red or brown cryst. (pentane). Mp 95-6°.

Joshi, K.K. *et al*, *Z. Naturforsch., B*, 1962, **17**, 565 (*synth*)
Manning, A.R., *J. Chem. Soc. (A)*, 1971, 2321 (*synth, ir, struct*)
Campbell, I.L.C. *et al*, *J. Chem. Soc., Dalton Trans.*, 1975, 22 (*cryst struct*)
Madach, T. *et al*, *Chem. Ber.*, 1980, **113**, 2675 (*pmr, ir*)

$C_{11}H_5CoN_5^{\ominus\ominus\ominus}$ Co-00079

Pentakis(cyano-*C*)phenylcobaltate(3−)

Pentacyano(phenyl)cobaltate(3−)

$[PhCo(CN)_5]^{\ominus\ominus\ominus}$

M 266.127 (ion)

Tri-K salt:

$C_{11}H_5CoK_3N_5$ M 383.422

Cryst. (H_2O). Eliminates PhCN on treatment with acids.

Kwiatek, J. *et al*, *J. Organomet. Chem.*, 1965, **3**, 421 (*synth, ir*)

$C_{11}H_5CoO_6Ru$ Co-00080

Dicarbonyl(η^5-2,4-cyclopentadien-1-yl)(tetracarbonylcobalt)-ruthenium

Cyclopentadienylrutheniumdicarbonyltetracarbonylcobalt. (η^5-Cyclopentadienyl)-dicarbonylrutheniumtetracarbonylcobalt

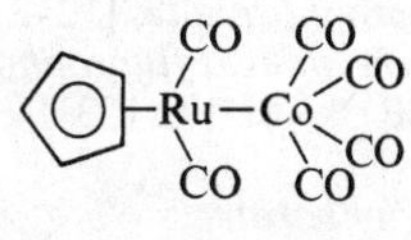

M 393.160

Light-orange cryst. (pentane). Mp 60-2°. Sublimes.

Blackmore, T. *et al*, *J. Chem. Soc. (A)*, 1968, 2931 (*synth, ms, ir*)
Bruce, M.I., *Int. J. Mass Spectrom. Ion Phys.*, 1968, **1**, 141 (*ms*)
Manning, A.R., *J. Chem. Soc. (A)*, 1971, 2321 (*ir*)

$C_{11}H_8Co_2O_6$ Co-00081

Di-μ-carbonyltetracarbonyl[(1,2,3,4-η)-2-methyl-1,3-butadiene]dicobalt, 9CI

Di-μ-carbonyltetracarbonyl(η^4-isoprene)dicobalt. Hexacarbonyl(2-methyl-1,3-butadiene)dicobalt

[50405-09-9]

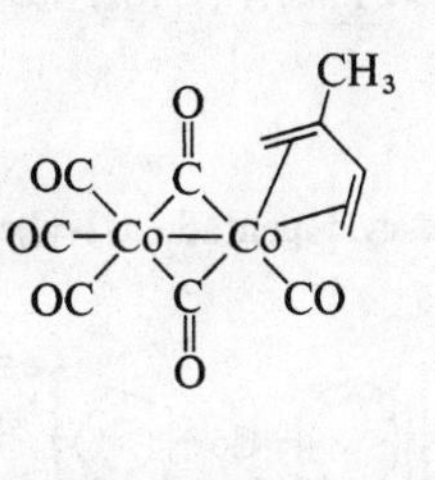

M 354.047

Not obt. pure. Orange-red needles (hexane). ir ν_{co} 2080, 2063, 2040, 1834 cm^{-1} (CCl_4).

Winkhouse, G. *et al*, *J. Chem. Soc.*, 1961, 602 (*synth, ir*)
Newman, J. *et al*, *J. Chem. Soc., Dalton Trans.*, 1973, 1593 (*ir*)

$C_{11}H_9Co_2O_6P$ Co-00082

Hexacarbonyl[μ-[η^2:η^2-(2,2-dimethylpropylidyne)phosphine]]dicobalt

[81523-74-2]

$(H_3C)_3CC{\equiv}PCo_2(CO)_6$

M 386.029

Light- and air-sensitive blood-red oil. ir ν_{CO} 2090, 2050, 2020 cm^{-1}.

Burckett-St. Laurent, J.C.T.R. *et al*, *J. Chem. Soc., Chem. Commun.*, 1981, 1141 (*synth, ir*)

$C_{11}H_{10}CoO_2^{\oplus}$ Co-00083

Carboxycobaltocenium, 9CI

(Carboxy-π-cyclopentadienyl)-π-cyclopentadienylcobalt(1+), 8CI. Carboxycobalticinium. Cobaltoceniumcarboxylic acid

[40759-57-7]

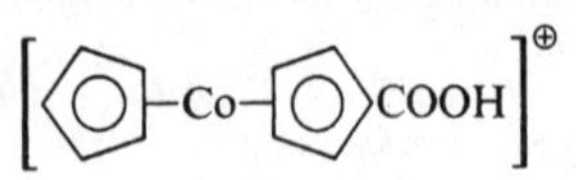

M 233.132 (ion)

Hexafluorophosphate: [41576-48-1].
$C_{11}H_{10}CoF_6O_2P$ M 378.096
Yellow flakes (Me_2CO).

Tetraphenylborate: [12427-47-3].
$C_{35}H_{30}BCoO_2$ M 552.364
Cryst. ($Me_2CO/CHCl_3$). Mp 234-5° dec.

Chloride, hexafluorophosphate: [12427-45-1]. *Chlorocarbonylcobaltocenium hexafluorophosphate.*
$C_{11}H_9ClCoF_6OP$ M 396.542
Moisture-sensitive yellow cryst.

Me ester, hexafluorophosphate: [12427-50-8]. *Carbomethoxycobaltocenium hexafluorophosphate.*
$C_{12}H_{12}CoF_6O_2P$ M 392.123
Cryst.

Me ester, tetraphenylborate:
$C_{36}H_{32}BCoO_2$ M 566.391
Cryst. ($Me_2CO/CHCl_3$). Mp 215° dec.

Amide, hexafluorophosphate: [60293-76-7]. *(Aminocarbonyl)cobaltocenium hexafluorophosphate. [(Aminocarbonyl)-π-cyclopentadienyl]-π-cyclopentadienylcobalt hexafluorophosphate.*
$C_{11}H_{11}CoF_6NOP$ M 377.111
Cryst.

Sheats, J.E. *et al*, *J. Org. Chem.*, 1970, **35**, 3245 (*synth, ir, pmr, uv*)
Sheats, J.E. *et al*, *Synth. React. Inorg. Metal-Org. Chem.*, 1973, **3**, 59 (*synth*)
El Murr, N., *J. Organomet. Chem.*, 1976, **112**, 189 (*ir, uv*)
Sheats, J.E. *et al*, *J. Organomet. Chem.*, 1976, **121**, 73 (*nmr*)
Riley, P.E. *et al*, *J. Organomet. Chem.*, 1978, **152**, 209 (*cryst struct*)

$C_{11}H_{11}Co^{\oplus\oplus}$ Co-00084

(η^6-Benzene)(η^5-2,4-cyclopentadien-1-yl)cobalt(2+), 10CI

[72339-93-6]

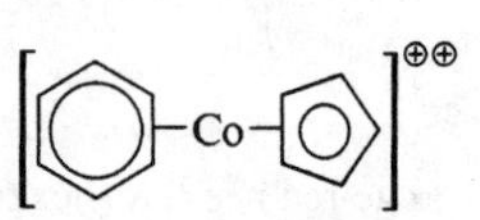

M 202.141 (ion)

Bis(tetrafluoroborate): [72339-94-7].
$C_{11}H_{11}B_2CoF_8$ M 375.748
Yellow cryst. ($MeNO_2/Et_2O$).

Fischer, E.O. *et al*, *Chem. Ber.*, 1965, **98**, 175.
Fairhurst, G. *et al*, *J. Chem. Soc., Dalton Trans.*, 1979, 1531 (*synth, ir, nmr*)
Lai, Y.-H. *et al*, *J. Organomet. Chem.*, 1981, **216**, 97 (*synth*)

$C_{11}H_{12}BrCo$ Co-00085

[(1,2,3,4-η)-5-Bromomethyl-1,3-cyclopentadiene](η^5-2,4-cyclopentadien-1-yl)cobalt, 10CI

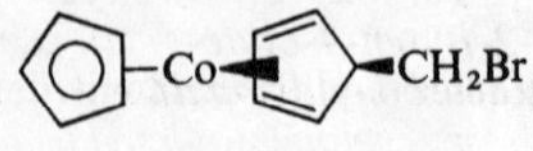

M 283.053

exo-form [31870-78-7]
Forms η^4-cyclohexadiene complexes by ring expansion. Red cryst. (pentane). Mp 50°.

Herberich, G.E. *et al*, *J. Organomet. Chem.*, 1969, **17**, 445 (*synth, pmr, ir*)
El Murr, N. *et al*, *Can. J. Chem.*, 1976, **54**, 3350.

$C_{11}H_{12}Co^{\oplus}$ Co-00086

[(1,2,3,4,5-η)-2,4-Cyclohexadien-1-yl](η^5-2,4-cyclopentadien-1-yl)cobalt(1+), 10CI

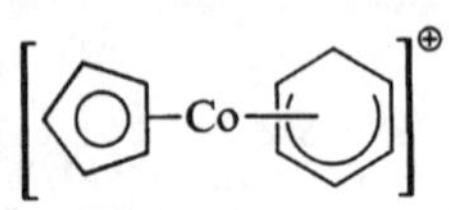

M 203.149 (ion)

Iodide: [32824-43-4].
$C_{11}H_{12}CoI$ M 330.054
Deep-red cryst. ($MeOH/Et_2O$). Mp 100° dec.

Hexafluorophosphate: [32825-27-7].
$C_{11}H_{12}CoF_6P$ M 348.113
Cryst. (Me_2CO/Et_2O). Mp 180° dec.

Tetraphenylborate:
$C_{35}H_{32}BCo$ M 522.381
Cryst. (Me_2CO/Et_2O). Mp 160°.

Herberich, G.E. *et al*, *Chem. Ber.*, 1970, **103**, 2016 (*synth, ir, pmr*)

$C_{11}H_{13}Co$ Co-00087

[(1,2,3,4-η)-1,3-Cyclohexadiene](η^5-2,4-cyclopentadien-1-yl)cobalt, 9CI

Cyclopentadienyl(1,3-cyclohexadiene)cobalt. (1,3-Cyclohexadiene)cyclopentadienylcobalt

[38959-22-7]

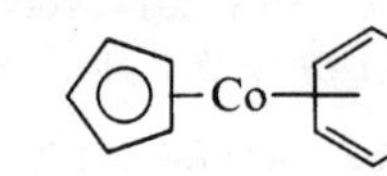

M 204.157

Orange cryst. (pentane). Mp 40-2°. Sublimes at 50°/0.1 mm. Air-sensitive.

Fischer, E.O. *et al*, *Z. Naturforsch., B*, 1961, **16**, 556.
King, R.B. *et al*, *J. Am. Chem. Soc.*, 1961, **83**, 3593 (*synth, ir, pmr*)
Herberich, G.E. *et al*, *J. Organomet. Chem.*, 1970, **22**, 723 (*synth*)
Herberich, G.E. *et al*, *Chem. Ber.*, 1970, **103**, 2016, 3615.
Yasufuku, K. *et al*, *Org. Mass Spectrom.*, 1970, **3**, 23 (*ms*)
Gmelin Suppl., **5**, 341.

$C_{11}H_{13}Co$ Co-00088

(η^5-2,4-Cyclopentadien-1-yl)[(1,2,3,4-η)-5-methyl-1,3-cyclopentadiene]cobalt, 9CI

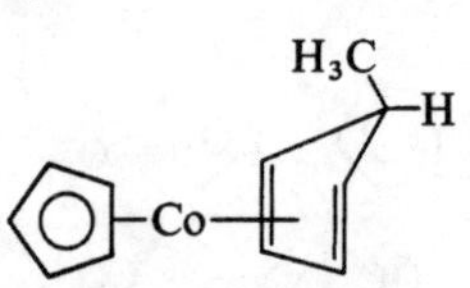

M 204.157

endo-form [52193-26-7]
Mp −5°.

Green, M.L.H. *et al*, *J. Chem. Soc.*, 1959, 3753 (*synth, ir, pmr*)

El Murr, N. *et al*, *J. Organomet. Chem.*, 1974, **67**, C77 (*synth*)

$C_{11}H_{14}CoN_3^{\oplus\oplus}$ Co-00089

(η^5-2,4-Cyclopentadien-1-yl)tris(acetonitrile)cobalt(2+), 10CI

Trisacetonitrile(η^5-cyclopentadienyl)cobalt(2+)

[73065-79-9]

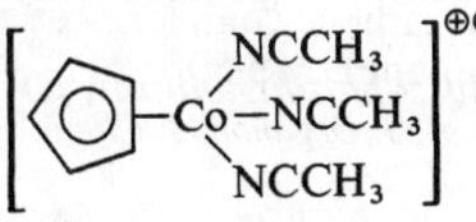

M 247.185 (ion)

Bis(perchlorate): [73065-80-2].
$C_{11}H_{14}Cl_2CoN_3O_8$ M 446.086
Dark-red cryst. (CH_2Cl_2).

Koelle, U., *J. Organomet. Chem.*, 1980, **184**, 379 (*synth, nmr, uv*)

$C_{11}H_{15}B_2Co$ Co-00090

(η^5-2,4-Cyclopentadien-1-yl)[(2,3,5,6-η)-1,4-dihydro-1,4-dimethyl-1,4-diborin]cobalt, 10CI

Cyclopentadienyl(1,4-dimethyl-1,4-dibora-2,5-cyclohexadiene)cobalt

[75039-09-7]

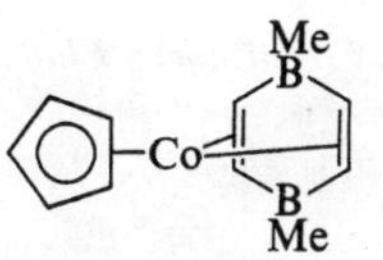

M 227.793

Red cryst. (pentane). Mp 122°. Air stable; thermally stable to 300°.

Herberich, G.E. *et al*, *J. Organomet. Chem.*, 1980, **192**, 421 (*synth, struct*)

$C_{11}H_{15}CoI_2O$ Co-00091

Carbonyldiiodo[(1,2,3,4,5-η)-1,2,3,4,5-pentamethyl-2,4-cyclopentadien-1-yl)cobalt, 9CI

[35886-64-7]

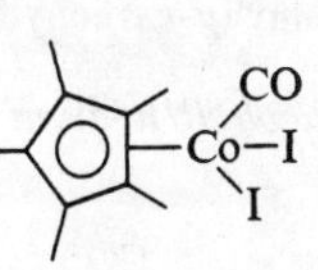

M 475.981

Starting material for pentamethylcyclopentadienylcobalt compounds. Black cryst. Mp 175-7°. ir ν_{CO} 2060 cm^{-1} (CH_2Cl_2), pmr δ 2.16 (C_5Me_5) ($CDCl_3$).

Roe, D.M. *et al*, *J. Chem. Soc. (A)*, 1971, 3173 (*synth, ir, pmr*)
King, R.B. *et al*, *J. Organomet. Chem.*, 1973, **56**, 345 (*synth, ir, pmr*)
Fairhurst, G. *et al*, *J. Chem. Soc., Dalton Trans.*, 1979, 1524 (*props*)

$C_{11}H_{18}CoN_3O_2$ Co-00092

Carbonylbis(2-isocyano-2-methylpropane)nitrosylcobalt, 10CI

*Bis(*tert*-butyl isocyanide)carbonylnitrosylcobalt*, 8CI.

*Nitrosylbis(*tert*-butyl isocyanide)carbonylcobalt*

[14518-25-3]

$[(H_3C)_3CNC]_2Co(CO)NO$

M 283.215

Air-sensitive orange-red cryst. Mp 70-2°. ir ν_{CN} 2141, 2108, ν_{CO} 1975, ν_{NO} 1733 cm^{-1} (cyclohexane).

Horrocks, W.D. *et al*, *Inorg. Chem.*, 1963, **2**, 723 (*synth, ir*)
Horrocks, W.D. *et al*, *Spectrochim. Acta*, 1965, **21**, 399 (*ir*)
Gutmann, V. *et al*, *Monatsh. Chem.*, 1977, **108**, 429 (*ir*)

$C_{11}H_{23}CoP_2$ Co-00093

(η^5-2,4-Cyclopentadien-1-yl)bis(trimethylphosphine)cobalt, 10CI, 9CI

Bis(trimethylphosphine)cyclopentadienylcobalt

[63413-01-4]

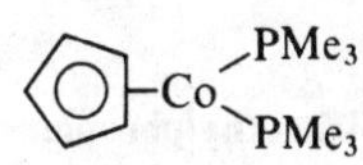

M 276.183

Red-brown needles. Mp 55-7°. Nucleophilic metal centre.

Werner, H. *et al*, *Angew. Chem.*, 1977, **89**, 835; 1978, **90**, 496 (*props*)
Werner, H. *et al*, *Chem. Ber.*, 1977, **110**, 3481 (*synth, ms, pmr*)
Werner, H. *et al*, *Chem. Ber.*, 1981, **114**, 2681 (*props*)

$C_{11}H_{24}CoP_2^{\oplus}$ Co-00094

(η^5-2,4-Cyclopentadien-1-yl)hydrobis(trimethylphosphine)-cobalt(1+), 10CI

(Cyclopentadienyl)hydridobis(trimethylphosphine)cobalt(1+)

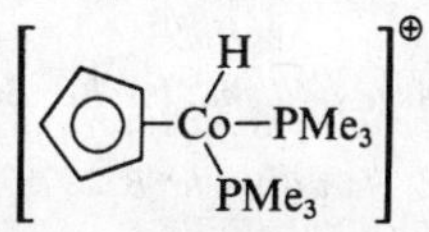

M 277.191 (ion)

Hexafluorophosphate: [64508-16-3].
$C_{11}H_{24}CoF_6P_3$ M 422.155
Yellow cryst.

Tetrafluoroborate: [65606-19-1].
$C_{11}H_{24}BCoF_4P_2$ M 363.995
Yellow cryst.

Werner, H. *et al*, *Chem. Ber.*, 1977, **110**, 3481 (*synth, ir, pmr*)
Dey, K. *et al*, *Chem. Ber.*, 1979, **112**, 823 (*synth*)

$C_{11}H_{26}CoN_3^{\oplus\oplus}$ Co-00095

Triammine[(1,2,3,4,5-η)-1-ethyl-2,3,4,5-tetramethyl-2,4-cyclopentadien-1-yl]cobalt(2+)

Tris(ammonia)(η^5-ethyltetramethylcyclopentadienyl)-cobalt(2+)

[71251-88-2]

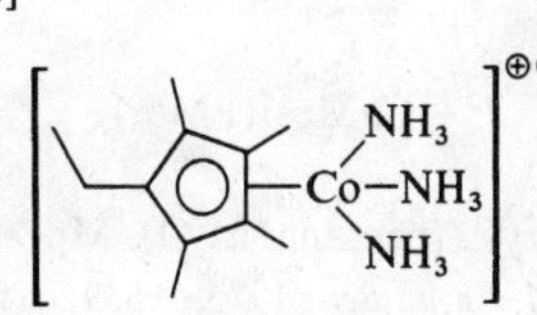

M 259.280 (ion)

Bis(hexafluorophosphate):
$C_{11}H_{26}CoF_{12}N_3P_2$ M 549.208
Red cryst. (Me_2CO/EtOH). V. sol., Me_2CO, MeCN. Reacts with Me_2CO on standing.

Green, M.L.H. *et al*, *J. Chem. Soc., Dalton Trans.*, 1979, 355 (*synth, pmr*)

$C_{11}H_{27}CoO_{11}P_3^{\oplus}$ **Co-00096**

Dicarbonyltris(trimethyl phosphite-*P*)cobalt(1+), 10CI

[45291-03-0]

$[(MeO)_3P]_3Co(CO)_2^{\oplus}$

M 487.182 (ion)

Thought to have fluxional trigonal bipyramidal struct. in soln.

Tetraphenylborate: [38670-40-5].

$C_{35}H_{47}BCoO_{11}P_3$ M 806.414

White cryst. (Me_2CO). Mp 160° dec. ir ν_{CO} 2037, 1982 cm^{-1}.

Attali, S. *et al*, *Inorg. Chim. Acta*, 1972, **6**, 475 (*synth*, *ir*, *pmr*)

$C_{11}H_{32}CoP_3$ **Co-00097**

(η^2-Ethene)hydrotris(trimethylphosphine)cobalt, 10CI

(η^2-Ethylene)hydridotris(trimethylphosphine)cobalt

[68830-26-2]

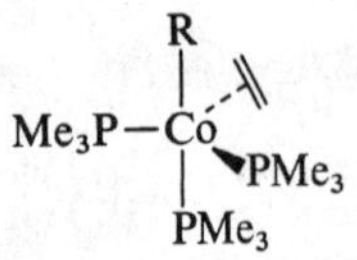

R = H

M 316.228

In equilib. with $CoEt(PMe_3)_3$. Model compd. for alkene isomerisation catalysts. Greenish-yellow, air-sensitive leaflets (pentane). $Bp_{0.008}$ 35-40° subl. dec. Stable up to 95° under Ar.

Klein, H.F. *et al*, *Angew. Chem., Int. Ed. Engl.*, 1980, **19**, 309 (*synth*, *props*)

Klein, H.F. *et al*, *Z. Naturforsch., B*, 1978, **33**, 1267 (*synth*)

$C_{11}H_{33}BrCoP_3$ **Co-00098**

Bromodimethyltris(trimethylphosphine)cobalt, 9CI

[55449-10-0]

$CoMe_2Br(PMe_3)_3$

M 397.140

Light-brown cryst. (pentane/Et_2O). Mp 116-8°.

Klein, H.-F. *et al*, *Chem. Ber.*, 1975, **108**, 956 (*synth*, *ir*, *nmr*, *ms*)

$C_{11}H_{33}CoIP_3$ **Co-00099**

Iododimethyltris(trimethylphosphine)cobalt, 9CI

[55449-11-1]

$CoMe_2I(PMe_3)_3$

M 444.141

Light-brown cryst. (pentane/Et_2O). Mp >118°.

Klein, H.-F. *et al*, *Chem. Ber.*, 1975, **108**, 956 (*synth*, *nmr*, *ir*, *ms*)

$C_{11}H_{33}CoP_3$ **Co-00100**

Dimethyltris(trimethylphosphine)cobalt, 9CI

[59400-73-6]

$CoMe_2(PMe_3)_3$

M 317.236

Orange-brown cryst. (pentane). Mp >87° dec. Undergoes CO and NO insertions. Reacts with hydroxylic solvs. and with halogens.

Klein, H.-F. *et al*, *Chem. Ber.*, 1976, **109**, 1453 (*synth*, *ir*, *pmr*)

$C_{12}Co_2O_{12}Rh_2$ **Co-00101**

Tri-μ-carbonyltetracarbonyl(pentacarbonyldicobalt)dirhodium, 9CI

Dodecacarbonyldicobaltdirhodium, 8CI

[50696-78-1]

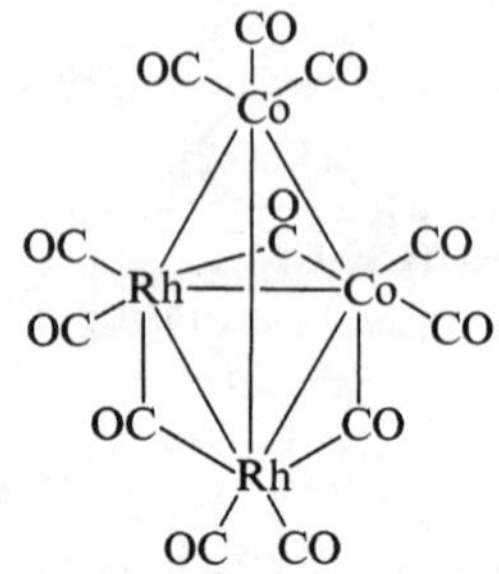

M 659.802

Proposed struct. shown. Hydrogenation and hydroformylation catalyst. Brown cryst. (pentane). Sol. hydrocarbons. Dec. at 120°.

Martinengo, S. *et al*, *J. Organomet. Chem.*, 1973, **59**, 379 (*synth*, *ir*)

Michelin-Lausarot, P. *et al*, *Inorg. Chim. Acta*, 1979, **35**, 227 (*ms*)

$C_{12}Co_3ErO_{12}$ **Co-00102**

Tris(tetracarbonylcobalto)erbium, 8CI

$Er[Co(CO)_4]_3$

M 680.184

Tetrakis-THF complex: [34216-63-2].

$C_{28}H_{32}Co_3ErO_{16}$ M 968.611

Struct. not yet established. Dark-red solid. Mp 138-40° dec. Air- and water-sensitive.

Marianelli, R.S. *et al*, *J. Organomet. Chem.*, 1971, **32**, C41 (*synth*, *ir*)

$C_{12}Co_3O_{12}Rh$ **Co-00103**

Di-μ-carbonyldicarbonyl(μ-carbonylheptacarbonyltricobalt)rhodium, 9CI

Dodecacarbonyltricobaltrhodium, 8CI

[50696-79-2]

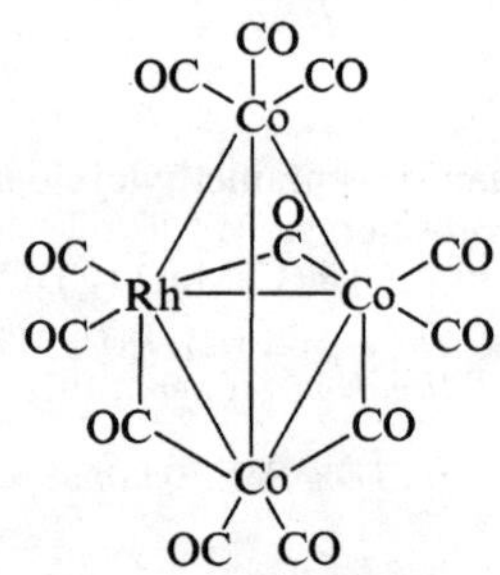

M 615.830

Fluxional at r.t. with CO exchange. Hydrogenation and hydroformylation catalyst. Brown cryst. (CH_2Cl_2/cyclohexane). Sol. hydrocarbons. Dec. at 130°.

Martinengo, S. *et al*, *J. Organomet. Chem.*, 1973, **59**, 379 (*synth*, *ir*)

Johnson, B.F.G. *et al*, *J. Chem. Soc., Chem. Commun.*, 1974, 441 (*cmr, struct*)
Labroue, D. *et al*, *J. Mol. Catal.*, 1977, **2**, 329 (*use*)
Evans, J. *et al*, *J. Chem. Soc., Dalton Trans.*, 1978, 626 (*cmr*)
Ichikawa, M., *J. Catal.*, 1979, **59**, 67 (*use*)
Michelin-Lausarot, P. *et al*, *Inorg. Chim. Acta*, 1979, **35**, 227 (*ms*)

$C_{12}Co_4O_{12}$ Co-00104

Tri-μ-carbonylnonacarbonyltetracobalt, 10CI, 9CI
Dodecacarbonyltetracobalt
[17786-31-1]

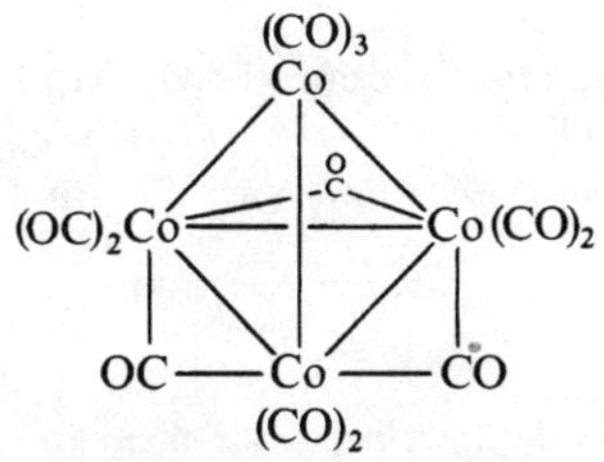

M 571.858
Catalyses methanation of CO, hydroformylation and hydroesterification of alkenes, and dimerization of norbornadiene. Purple-black orthorhombic cryst.

Friedel, R.A. *et al*, *J. Am. Chem. Soc.*, 1955, **77**, 3951 (*synth, ir*)
Natta, G. *et al*, *Chem. Ind. (Milan)*, 1955, **36**, 3 (*synth*)
Bor, G. *et al*, *Spectrochim. Acta*, 1963, **19**, 1209 (*synth, struct*)
Wei, C.H., *Inorg. Chem.*, 1969, **8**, 2384 (*struct*)
Evans, J. *et al*, *J. Chem. Soc., Dalton Trans.*, 1978, 626 (*cmr*)
Lennertz, A.M. *et al*, *J. Organomet. Chem.*, 1979, **171**, 203 (*use*)
Michelin-Lausarot, P. *et al*, *Inorg. Chim. Acta*, 1979, **35**, 227 (*ms*)

$C_{12}H_4Co_2O_2S_4$ Co-00105

Dicarbonyl[μ-[1,4,5,8-naphthalenetetrathiolato(4−)-S^1,S^8:S^4,S^5]]dicobalt, 10CI
Dicarbonyltetrathionaphthalenedicobalt

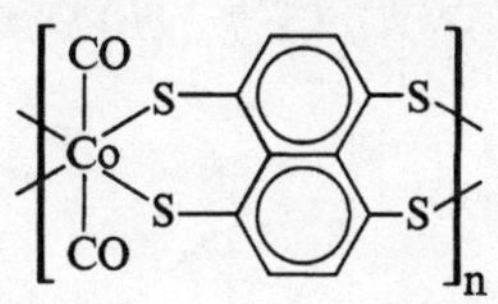

M 426.269
Polymeric. One dimensional conductor. Red solid. Insol. most org. solvs. ir ν_{CO} 2010s, br cm^{-1} (CsI).

Teo, B.-K. *et al*, *J. Am. Chem. Soc.*, 1977, **99**, 4862 (*synth, ir*)

$C_{12}H_5CoMoO_7$ Co-00106

Tricarbonyl(η^5-2,4-cyclopentadien-1-yl)(tetracarbonylcobalt)molybdenum, 10CI
[62015-65-0]

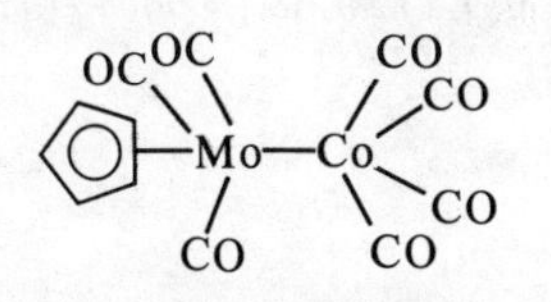

M 416.041
Cryst. (pentane). Mp 63-5° (sealed under Ar).

Schmid, E. *et al*, *Z. Naturforsch., B*, 1977, **32**, 1277 (*synth, props*)
Abrahamson, H.B. *et al*, *Inorg. Chem.*, 1978, **17**, 1003 (*synth, props*)

$C_{12}H_6Co^{\ominus\ominus\ominus\ominus}$ Co-00107

Hexaethynylcobaltate(4−), 9CI

$[Co(C{\equiv}CH)_6]^{\ominus\ominus\ominus\ominus}$

M 209.113 (ion)
See also Hexakis(phenylethynyl)cobaltate(4−), Co-00342 . Tetraphenylarsonium and phosphonium salts more stable than the pyrophoric alkali metal salts.

▷Alkynyl complexes potentially pyrophoric or explosive

Sodium tris(tetraphenylarsonium) salt: [59120-24-0].
$C_{84}H_{66}As_3CoNa$ M 1382.133
Orange cryst.
Sodium tris(tetraphenylphosphonium) salt: [59120-23-9].
$C_{84}H_{66}CoNaP_3$ M 1250.290
Orange cryst.

Nast, R. *et al*, *Chem. Ber.*, 1976, **109**, 455 (*synth, ir*)

$C_{12}H_7CoN_5^{\ominus\ominus\ominus}$ Co-00108

Benzylpentacyanocobaltate(3−)
Pentakis(cyano-C)(phenylmethyl)cobaltate(3−)

$(NC)_5CoCH_2Ph^{\ominus\ominus\ominus}$

M 280.154 (ion)
Stable in absence of oxygen.
Tri-Na salt:
$C_{12}H_7CoN_5Na_3$ M 349.123
Yellow cryst. (EtOH). Sol. H_2O, MeOH, EtOH, insol. Et_2O, Me_2CO, hydrocarbons.

Halpern, J. *et al*, *J. Am. Chem. Soc.*, 1964, **86**, 2311 (*synth, ir, pmr, uv*)
Kwiatek, J. *et al*, *J. Organomet. Chem.*, 1965, **3**, 421 (*synth, ir*)

$C_{12}H_{10}CoO_4^{\oplus}$ Co-00109

1,1′-Dicarboxycobaltocenium, 9CI
Bis(carboxy-π-cyclopentadienyl)cobalt(1+), 8CI. 1,1′-Dicarboxycobalticinium. Cobaltocenium-1,1′-dicarboxylic acid
[40698-35-9]

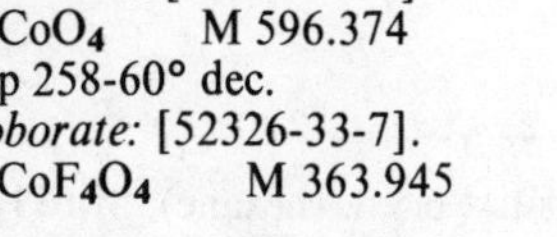

M 277.142 (ion)
Hexafluorophosphate: [33638-89-0].
$C_{12}H_{10}CoF_6O_4P$ M 422.106
Yellow cryst.
Tetraphenylborate: [52149-68-5].
$C_{36}H_{30}BCoO_4$ M 596.374
Cryst. Mp 258-60° dec.
Tetrafluoroborate: [52326-33-7].
$C_{12}H_{10}BCoF_4O_4$ M 363.945
Cryst.
Di-Me ester, hexafluorophosphate:
$C_{14}H_{14}CoF_6O_4P$ M 450.160
Cryst. Mp 176-7°.

Sheats, J.E. *et al*, *J. Org. Chem.*, 1970, **35**, 3245 (*synth, ir, pmr, uv*)
El Murr, N. *et al*, *C.R. Hebd. Seances Acad. Sci.*, 1971, **272**, 1989 (*synth, ir, uv*)
Sheats, J.E. *et al*, *Synth. React. Inorg. Metal-Org. Chem.*, 1973, **3**, 59 (*synth*)
El Murr, N., *J. Organomet. Chem.*, 1976, **112**, 117 (*synth*)

El Murr, N., *J. Organomet. Chem.*, 1976, **112**, 189 (*ir, uv, pmr*)

$C_{12}H_{10}Co_2O_2$ Co-00110

Di-μ-carbonylbis(η⁵-2,4-cyclopentadien-1-yl)dicobalt, 9CI

[58496-39-2]

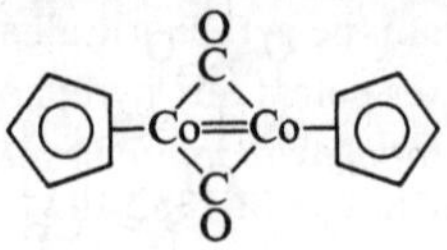

M 304.076

Green solid.

Lee, W.S. *et al*, *J. Organomet. Chem.*, 1977, **127**, 87 (*synth, ir, pmr*)

Pinhas, A. *et al*, *Inorg. Chem.*, 1979, **18**, 654 (*cryst struct*)

$C_{12}H_{12}CoN_4O_4^{\ominus\ominus\ominus}$ Co-00111

Tetrakis(cyano-*C*)[(2,3-η)-diethyl-2-butenedioate]cobaltate(3−)

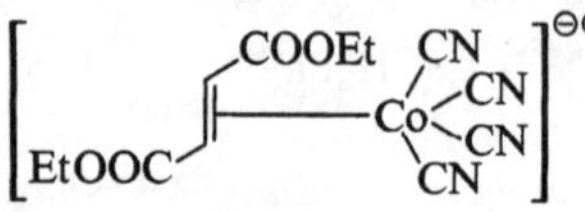

(*E*)-*form*

M 335.184 (ion)

(E)-form [81130-59-8]

*Tetrakis(cyano-*C*)(diethyl fumarate)cobaltate(3−)*

Tri-K salt: [73464-59-2].

$C_{12}H_{12}CoK_3N_4O_4$ M 452.479

Characterised spectroscopically. Isomerises to (*Z*)-form.

(Z)-form [81130-61-2]

*Tetrakis(cyano-*C*)(diethyl maleate)cobaltate(3−)*

Tri-K salt: [73382-86-2]. Characterised spectroscopically. Isomerises to (*E*)-form.

Funabiki, T. *et al*, *J. Chem. Soc., Dalton Trans.*, 1981, 2529 (*nmr, uv, synth*)

$C_{12}H_{12}Co_2O_4$ Co-00112

Bis(η⁴-1,3-butadiene)di-μ-carbonyldicarbonyldicobalt, 10CI

Bis(1,3-butadiene)tetracarbonyldicobalt

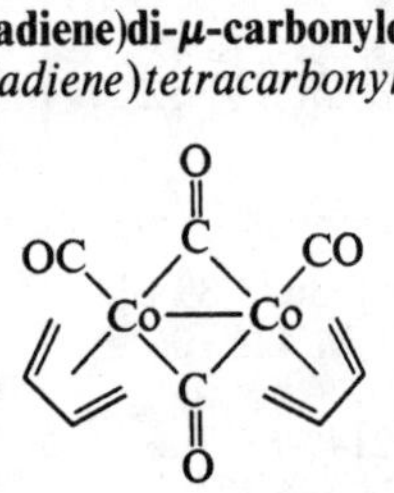

cis-form

M 338.091

cis-form [31904-76-4]

Brown air-sensitive cryst. (hexane). Mp 118°. Exists in soln. as equil. mixt. of *cis-trans* tautomers.

trans-form [31869-12-2]

Formed in solns. of *cis*-form.

Fischer, E.O. *et al*, *Z. Naturforsch., B*, 1961, **16**, 138 (*synth, ir*)

Jones, R.O. *et al*, *Z. Kristallogr.*, 1966, **123**, 330 (*cryst struct*)

McArdle, P. *et al*, *J. Chem. Soc. (A)*, 1970, 2123 (*struct, ir*)

$C_{12}H_{13}BCoF_3N_4$ Co-00113

(η⁵-2,4-Cyclopentadien-1-yl)[dihydrobis(1*H*-pyrazolato-*N*¹)borato(1−)-*N*²,*N*²′](trifluoromethyl)cobalt, 9CI

[52470-75-4]

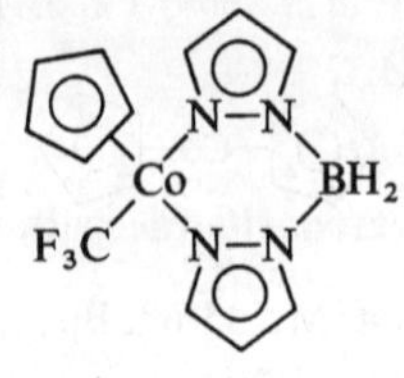

M 340.000

Air-stable red cryst. (hexane/Et_2O). Mp 147°. ^{19}F nmr δ 11.9s (Et_2O).

King, R.B. *et al*, *J. Am. Chem. Soc.*, 1974, **96**, 1334 (*synth, nmr*)

$C_{12}H_{13}Co$ Co-00114

[(2,3,5,6-η)-Bicyclo[2.2.1]hepta-2,5-diene](η⁵-2,4-cyclopentadien-1-yl)cobalt, 10CI

η⁵-Cyclopentadienyl-η⁴-norbornadienecobalt. Cyclopentadienylcobalt-norbornadiene

[12295-13-5]

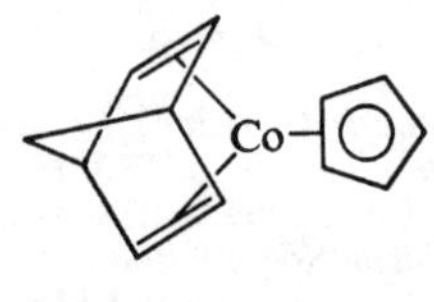

M 216.168

Orange cryst. Mp 60-1°.

King, R.B. *et al*, *J. Am. Chem. Soc.*, 1961, **83**, 3593 (*synth, ir*)

Cook, D.J. *et al*, *J. Chem. Soc. (A)*, 1967, 1547 (*synth, ir, pmr*)

$C_{12}H_{13}CoO_2$ Co-00115

(η⁵-2,4-Cyclopentadien-1-yl)[(2,3,4,5-η)methyl-2,4-cyclopentadiene-1-carboxylate]cobalt, 10CI

(η-Cyclopentadienyl)(4,5-η-2-methoxycarbonylcyclopent-4-en-1,3-ylene)cobalt

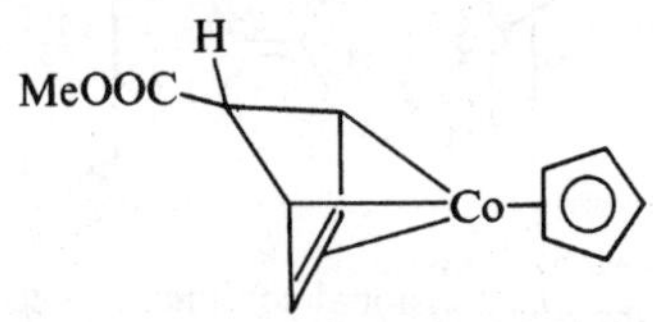

M 248.167

Cryst. struct. more consistent with a pent-4-en-1,3-ylene compd. rather than a cyclopentadiene compd.

exo-form [55565-82-7]

Red cryst. (Et_2O/pet. ether). Mp 90-2°.

El Murr, N. *et al*, *Can. J. Chem.*, 1976, **54**, 3350 (*synth, ir, pmr*)

El Murr, N. *et al*, *J. Chem. Soc., Dalton Trans.*, 1979, 901 (*synth, nmr, cryst struct*)

$C_{12}H_{14}Co$ Co-00116

1,1′-Dimethylcobaltocene, 10CI
Bis[(1,2,3,4,5-η)-1-methyl-2,4-cyclopentadien-1-yl-)]cobalt, 9CI
[12146-91-7]

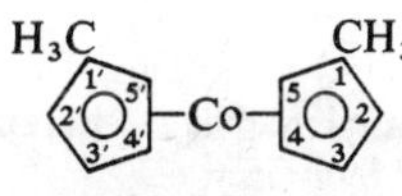

M 217.176
Lustrous black cryst. Mp 35-6°. $Bp_{0.37}$ 56-57.5°.

Riemschneider, R. *et al*, *Z. Naturforsch.*, *B*, 1959, **14**, 811 (*synth*, *ir*)
Fritz, H.P. *et al*, *Z. Naturforsch.*, *B*, 1968, **23**, 298 (*pmr*)
Rettig, M.F. *et al*, *J. Am. Chem. Soc.*, 1969, **91**, 1361 (*synth*, *nmr*)
Anderson, S.E. *et al*, *Chem. Phys. Lett.*, 1972, **13**, 150 (*cmr*)
Evans, S. *et al*, *J. Chem. Soc.*, *Faraday Trans. 2*, 1974, 356 (*synth*, *pe*)
Köhler, F.H. *et al*, *J. Organomet. Chem.*, 1976, **110**, 235; 1978, **160**, 299 (*synth*, *nmr*)

$C_{12}H_{15}CoO_2$ Co-00117

Dicarbonyl[(1,2,3,4,5-η)-1,2,3,4,5-pentamethyl-2,4-cyclopentadien-1-yl]cobalt, 9CI
(Pentamethylcyclopentadienyl)cobalt dicarbonyl
[12129-77-0]

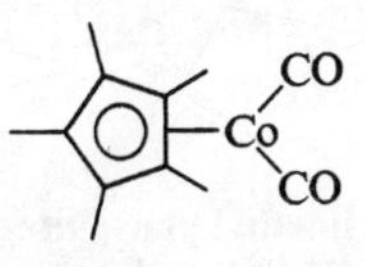

M 250.183
Starting material for pentamethylcyclopentadienyl cobalt compounds. Air-sensitive red cryst. by subl. Mp 56-8°. Bp_1 30-50° subl. ir ν_{CO} 2011, 1949 cm^{-1} (cyclohexane). Turns blue on exposure to air. Readily undergoes oxidative addition.

King, R.B. *et al*, *J. Organomet. Chem.*, 1967, **8**, 287 (*synth*, *ir*, *pmr*)
King, R.B. *et al*, *J. Organomet. Chem.*, 1973, **56**, 345 (*pmr*, *props*)
Kölle, V. *et al*, *Angew. Chem.*, *Int. Ed. Engl.*, 1982, **21**, 131 (*synth*)

$C_{12}H_{16}B_2Co$ Co-00118

Bis(1-methylborinato)cobalt
1-Methylboratabenzene cobalt complex, 10CI, 9CI.
Bis[(1,2,3,4,5,6-η)-1-methylboratabenzene]cobalt
[36534-27-7]

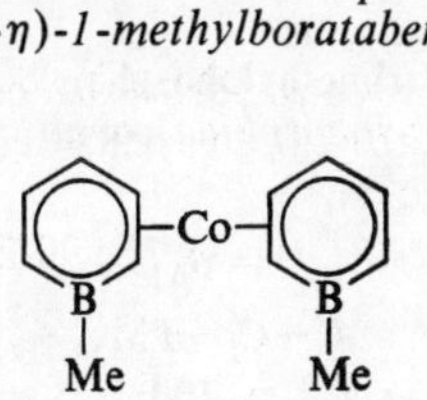

M 240.812
Solid. Mp 98°.

Herberich, G.E. *et al*, *J. Chem. Soc.*, *Chem. Commun.*, 1971, 1328 (*synth*, *props*)
Huttner, G. *et al*, *Chem. Ber.*, 1972, **105**, 3424 (*struct*)
Koelle, U., *J. Organomet. Chem.*, 1978, **152**, 225 (*props*)
Clack, D.W. *et al*, *Inorg. Chem.*, 1979, **18**, 513 (*props*)
Herberich, G.E. *et al*, *J. Organomet. Chem.*, 1980, **197**, 1.

$C_{12}H_{19}CoN_4O_3$ Co-00119

Carbonyl[[3,3′-(1,3-propanediyldinitrilo)bis[2-butanonato]](1−)]cobalt, 10CI
Carbonyl[[3,3′-(trimethylenedinitrilo)di-2-butanone dioximato](1−)]cobalt
[36514-48-4]

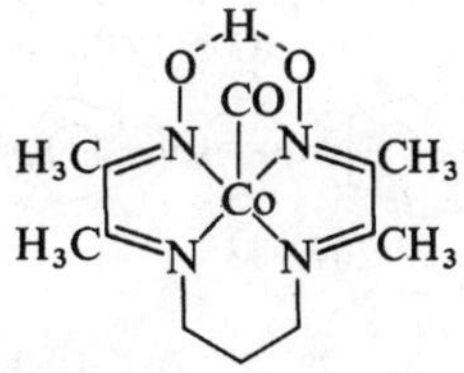

M 326.240
Important starting material for vitamin B_{12} analogues. Cryst. obt. from blue, air-sensitive soln. in aq. THF. Solns. stable several days under N_2. Sol. C_6H_6, Et_2O, THF, spar. sol. EtOH, insol. H_2O. ir ν_{CO} 1950 cm^{-1}. Readily oxidatively adds alkyl halides with concomitant displacement of CO.

Costa, G. *et al*, *J. Chem. Soc.*, *Dalton Trans.*, 1972, 450 (*synth*, *props*, *ir*)

$C_{12}H_{20}Br_2Co_2$ Co-00120

Di-μ-bromotetrakis(η^3-2-propenyl)dicobalt, 10CI
Bis(di-η^3-allylbromocobalt). Tetraallyldibromodicobalt

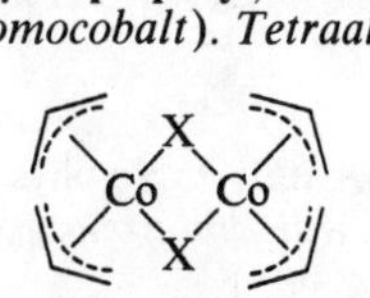

X = Br

M 441.964
Bridged dimer. Red cryst., thermally sensitive. Extremely insol.
Monomer: Bromobis(η^3-2-propenyl)cobalt. Diallylbromocobalt.
$C_6H_{10}BrCo$ M 220.982
Unknown.

Bönneman, H. *et al*, *Angew. Chem.*, *Int. Ed. Engl.*, 1973, **12**, 964.

$C_{12}H_{20}Cl_2Co_2$ Co-00121

Di-μ-chlorotetrakis(η^3-2-propenyl)dicobalt, 10CI
Bis(di-η^3-allylchlorocobalt). Tetraallyldichlorodicobalt
[65106-69-6]
As Di-μ-bromotetrakis(η^3-2-propenyl)dicobalt, Co-00120 with

X = Cl

M 353.062
Bridged dimer. Catalyst precursor in oligomerisation of alkenes and dienes. Red cryst., thermally sensitive. Extremely insol.
Monomer: Chlorobis(η^3-2-propenyl)cobalt. Diallylchlorocobalt.
$C_6H_{10}ClCo$ M 176.531
Unknown.

Belg. Pat., 651 596, (*1965*); *CA*, **64**, 9928
Bönneman, H. *et al*, *Angew. Chem.*, *Int. Ed. Engl.*, 1973, **12**, 964.

$C_{12}H_{20}Co_2I_2$ Co-00122

Di-μ-iodotetrakis(η^3-2-propenyl)dicobalt, 10CI
Bis(di-η^3-allyliodocobalt). Tetraallyldiiododicobalt

[65140-02-5]
As Di-μ-bromotetrakis(η^3-2-propenyl)dicobalt, Co-00120 with

X = I

M 535.965
Bridged dimer. Catalyst precursor for polymerisation of alkenes and 1,3-dienes. Red cryst., thermally sensitive. Extremely insol.
Monomer: Iodobis(η^3-2-propenyl)cobalt. Diallyliodocobalt.
$C_6H_{10}CoI$ M 267.983
Unknown.

Netherlands Pat., 6 409 179, (*1965*); *CA*, **63**, 11617
Bonneman, H. *et al*, *Angew. Chem., Int. Ed. Engl.*, 1973, **12**, 964.

$C_{12}H_{21}Co$ Co-00123

Tris[(1,2,3-η)-1-butenyl]cobalt
Tris(1-methyl-π-allyl)cobalt, 8CI
[12295-18-0]

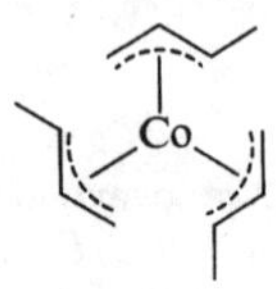

M 224.231
Butadiene polymn. catalyst. No phys. props. recorded.
Tinyakova, E.I. *et al*, *CA*, 1971, **75**, 65000t (*synth, use*)

$C_{12}H_{24}CoN_4O_3^{\oplus}$ Co-00124

Aquamethyl[[3,3′-(1,3-propanediyldinitrilo)bis[2-butanone]-dioximato](1−)-N,N′,N″,N‴]cobalt(1+), 10CI, 9CI
Aquamethyl[3,3′-(trimethylenedinitrilo)di-2-butanone dioximato(1−)]cobalt(1+), 8CI. *Aquamethyl[1,3-bis-(diacetylmonoximeimino)propane]cobalt(1+)*
[26334-78-1]

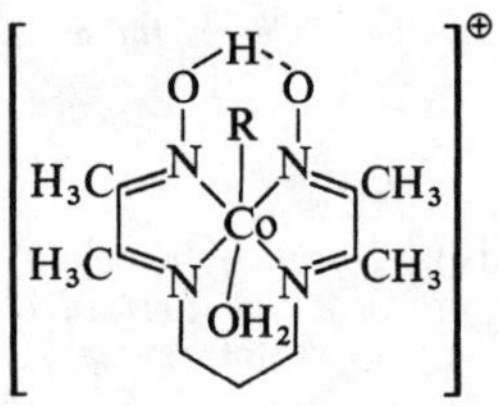

R = Me

M 331.280 (ion)
Analogue of methylcobalamin. Co−Me bond cleaved by light. Methylates Hg^{2+}.
Perchlorate: [23940-46-7].
$C_{12}H_{24}ClCoN_4O_7$ M 430.730
Red-orange, air-stable cryst. (Me_2CO aq. or H_2O). Sol. H_2O. Conductivity 86 ohm^{-1} cm^2.
Tetraphenylborate: [24111-03-3].
$C_{36}H_{44}BCoN_4O_3$ M 650.512
Orange-yellow solid (MeOH aq.). Sol. MeOH, Me_2CO, $CHCl_3$, insol. H_2O.
Hexafluorophosphate:
$C_{12}H_{24}CoF_6N_4O_3P$ M 476.244
Deep-red microcryst. (MeOH aq.). Sol. Me_2CO, MeOH, insol. H_2O.

Brückner, S. *et al*, *Inorg. Chim. Acta*, 1969, **3**, 278 (*cryst struct*)
Costa, G. *et al*, *Inorg. Chim. Acta*, 1969, **3**, 323 (*synth, uv*)
Dodd, D. *et al*, *J. Organomet. Chem.*, 1973, **52**, 1 (*props, rev*)
Magnuson, V.E. *et al*, *J. Organomet. Chem.*, 1975, **92**, 233 (*pmr, props*)
Allen, R.J. *et al*, *Bioinorg. Chem.*, 1976, **5**, 311 (*props*)
Finke, R.G. *et al*, *Inorg. Chem.*, 1981, **20**, 687 (*synth, pmr, props*)

$C_{12}H_{25}CoN_6O_4$ Co-00125

[Bis(methylamino)methylene]bis[2,3-butanedione dioximato(1−)-N,N′]methylcobalt, 9CI
Bis(dimethylglyoximato)[bis(methylamino)methylene]-methylcobalt. [Bis(methylamino)methylene]-methylcobaloxime
[59412-93-0]

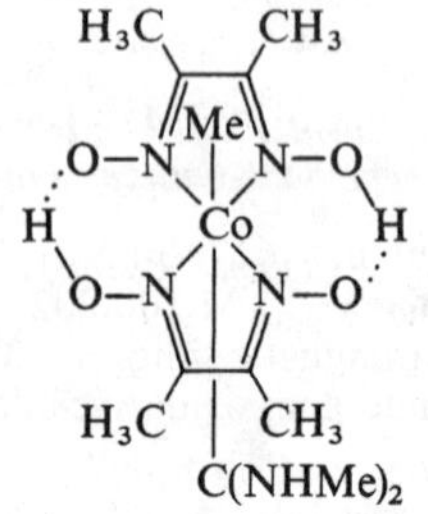

M 376.301
Yellow cryst. ($CHCl_3$/hexane).
Doonan, D.J. *et al*, *J. Am. Chem. Soc.*, 1976, **98**, 2129 (*synth, ir, pmr*)

$C_{12}H_{32}CoO_9P_3$ Co-00126

(η^3-2-Propenyl)tris(trimethyl phosphite-P)cobalt, 9CI
η^3-Allyltris(trimethyl phosphite)cobalt
[42603-27-0]

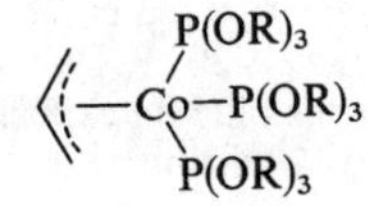

R = Me

M 472.234
Hydrogenation catalyst for arenes. Orange cryst. (pentane). Mp 154° dec. Absorbs CO.
Muetterties, E.L. *et al*, *J. Am. Chem. Soc.*, 1974, **96**, 7920 (*synth, nmr*)
Hirsekorn, F.J. *et al*, *J. Am. Chem. Soc.*, 1975, **97**, 237 (*use*)

$C_{12}H_{32}CoP_3$ Co-00127

(η^3-2-Propenyl)tris(trimethylphosphine)cobalt, 10CI
Allyltris(trimethylphosphine)cobalt
[59830-92-1]

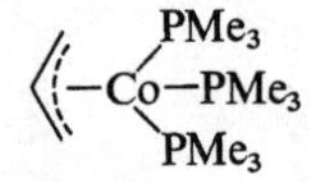

M 328.239
Homogeneous catalyst for hydrogenation of arenes. Red-orange cryst. by subl. Mp >140°. Discolours at 110°.
Rakowski, M.C. *et al*, *Inorg. Chem.*, 1976, **15**, 2379 (*synth, pmr, ir*)

$C_{12}H_{34}CoP_3$ Co-00128

(η^2-Ethene)methyltris(trimethylphosphine)cobalt, 10CI
(η^2-Ethylene)methyltris(trimethylphosphine)cobalt

[73557-04-7]

As (η^2-Ethene)hydrotris(trimethylphosphine)cobalt, Co-00097 with

R = Me

M 330.255

Fluxional in soln. Brownish-yellow air-sensitive cryst. (pentane). May be stored under Ar at 20°. Mp 115-7° dec. $Bp_{0.08}$ 60-5° subl.

Klein, H.F. *et al*, *Chem. Ber.*, 1983, **116**, 1441 (*synth*, *pmr*, *ir*)

$C_{12}H_{36}CoO_{12}P_4^{\oplus}$ Co-00129

Tetrakis(trimethyl phosphite-*P*)cobalt(1+), 10CI

[60790-64-9]

$Co[P(OMe)_3]_4^{\oplus}$

M 555.237 (ion)

Hexafluorophosphate: [60790-65-0]. $C_{12}H_{36}CoF_6O_{12}P_5$ M 700.202

Deep-blue paramagnetic solid. Sol. CH_2Cl_2. Mp 192° dec. Readily undergoes oxidative addition.

Muetterties, E.L. *et al*, *J. Am. Chem. Soc.*, 1978, **100**, 6978 (*synth*)

$C_{12}H_{36}CoP_3$ Co-00130

Trimethyltris(trimethylphosphine)cobalt, 9CI

$CoMe_3(PMe_3)_3$

M 332.271

Orange-yellow cryst. (Et_2O). Sol. polar org. solvs. Mp >115°.

Klein, H.-F. *et al*, *Chem. Ber.*, 1975, **108**, 956 (*synth*, *ir*, *nmr*, *ms*)

$C_{12}H_{37}CoO_{12}P_4$ Co-00131

Hydrotetrakis(trimethyl phosphite-*P*)cobalt, 10CI

[29189-85-3]

$HCo[P(OMe)_3]_4$

M 556.245

Yellow cryst. (hexane). Mp 212° dec. Protonated by strong acids and deprotonated by alkali metal hydrides.

Muetterties, E.L. *et al*, *J. Am. Chem. Soc.*, 1974, **96**, 7920; 1982, **104**, 2940 (*synth*, *nmr*)

$C_{13}H_8Co_2O_6$ Co-00132

[(2,3,5,6-η)-Bicyclo[2.2.1]hepta-2,5-diene]di-μ-carbonyltetracarbonyldicobalt, 9CI

Hexacarbonyl(2,5-norbornadiene)dicobalt. Di-μ-carbonyltetracarbonyl(π-norbornadiene)dicobalt

[60004-08-2]

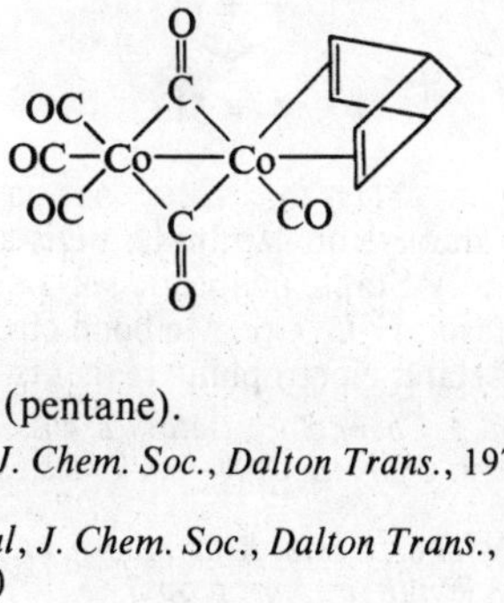

M 378.069

Orange cryst. (pentane).

Stephens, F.S., *J. Chem. Soc., Dalton Trans.*, 1972, 1754 (*cryst struct*)

Newman, J. *et al*, *J. Chem. Soc., Dalton Trans.*, 1973, 1593 (*synth*, *ir*, *uv*)

Chia, L.S. *et al*, *Inorg. Chem.*, 1975, **14**, 2975 (*nqr*, *struct*)

$C_{13}H_{10}BCoO_2$ Co-00133

Dicarbonyl[(1,2,3,4,5,6-η)-1-phenylboratabenzene]cobalt, 9CI

[53852-42-9]

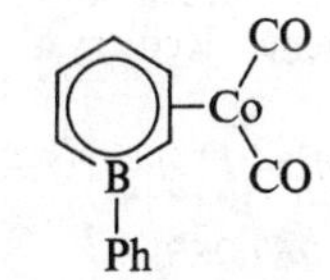

M 267.964

Cryst. (pentane). ir ν_{co} 2043, 1990 cm^{-1} (hexane). ^{11}B nmr −20.5 ppm ($BF_3 \cdot Et_2O$ ref.).

Herberich, G.E. *et al*, *Z. Naturforsch.*, *B*, 1974, **29**, 439 (*synth*, *ir*, *nmr*, *ms*)

$C_{13}H_{13}Co$ Co-00134

[(1,2,3,4-η)-1,3,5,7-Cyclooctatetraene](η^5-2,4-cyclopentadien-1-yl)cobalt, 10CI

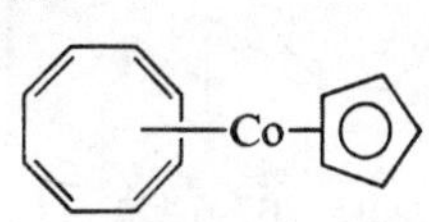

M 228.179

(1,2,3,4-η)-form [70810-90-1]

Minor component in solns. of the (1,2,5,6-η)-isomer. Not isol. pmr (hexane) δ 5.1 (C_5H_5), 4.7 (C_8H_8).

(1,2,5,6-η)-form [12110-49-5]

More thermodynamically stable isomer. Brown cryst. Mp 81-2°. pmr (hexane) δ 4.5 (C_5H_5), 3.5, 5.4 (C_8H_8). Subl. at 60°/0.1 mm.

Nakamura, A. *et al*, *Bull. Chem. Soc. Jpn.*, 1960, **33**, 425 (*synth*, *ir*, *pmr*)

Fritz, H.P. *et al*, *Chem. Ber.*, 1962, **95**, 158 (*synth*, *ir*, *uv*, *pmr*)

Otsuka, S. *et al*, *Inorg. Chem.*, 1966, **5**, 2059 (*pmr*)

Evans, J. *et al*, *J. Chem. Soc., Dalton Trans.*, 1974, 2375 (*props*)

Moraczewski, J. *et al*, *J. Am. Chem. Soc.*, 1979, **101**, 3407; 1981, **103**, 4779, 4787 (*synth*, *pmr*)

Reger, D.L. *et al*, *J. Organomet. Chem.*, 1980, **187**, 243 (*synth*, *props*)

$C_{13}H_{13}Co$ Co-00135

[(1,2,5,6-η)-1,3,5,7-Cyclooctatetraene](η^5-2,4-cyclopentadien-1-yl)cobalt, 10CI

[12110-49-5]

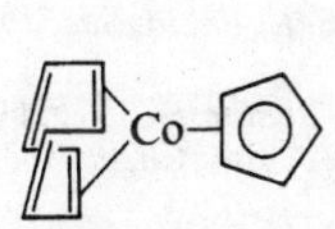

M 228.179

Brown cryst. by subl. Mp 79-81°. $Bp_{0.1}$ 60° subl. In soln. exists in equil. with its 1,2,3,4-η-isomer.

Nakamura, A. *et al*, *Bull. Chem. Soc. Jpn.*, 1960, **37**, 425 (*synth*, *ir*)

Fritz, H.P. *et al*, *Chem. Ber.*, 1962, **95**, 158 (*synth*, *ir*, *nmr*, *uv*)

Otsuka, S. *et al*, *Inorg. Chem.*, 1966, **5**, 2059 (*nmr*)

Moraczewski, J. *et al*, *J. Am. Chem. Soc.*, 1981, **103**, 4779 (*synth*, *pmr*, *struct*)

$C_{13}H_{16}Co^{\oplus}$ **Co-00136**

[(1,2,3,4,5-η)-2,4-Cyclooctadien-1-yl](η⁵-2,4-cyclopentadien-1-yl)cobalt(1+), 9CI

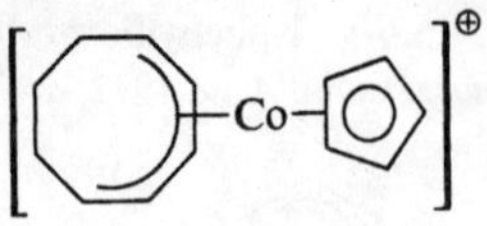

M 231.203 (ion)

Hexafluorophosphate: [41444-76-2].
$C_{13}H_{16}CoF_6P$ M 376.167
Characterised spectroscopically.

Evans, J. *et al*, *J. Chem. Soc.*, *Dalton Trans.*, 1972, 2668 (*synth*, *pmr*)

$C_{13}H_{16}Co^{\oplus}$ **Co-00137**

[(1,2,3,5,6-η)-2,5-Cyclooctadien-1-yl](η⁵-2,4-cyclopentadien-1-yl)cobalt(1+), 9CI

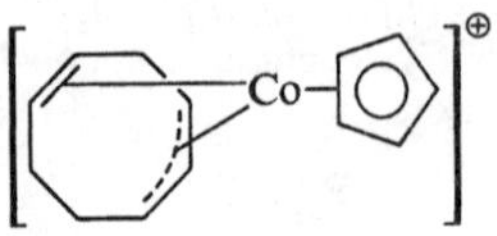

M 231.203 (ion)

Tetrafluoroborate: [57207-62-2].
$C_{13}H_{16}BCoF_4$ M 318.006
Crimson cryst. No sharp Mp.

Picrate:
$C_{19}H_{18}CoN_3O_7$ M 459.300
Red needles (Me_2CO). Darkens >145°, no sharp Mp.

Lewis, J. *et al*, *J. Chem. Soc.* (*A*), 1967, 1150 (*synth*, *nmr*)
Games, D.E. *et al*, *J. Organomet. Chem.*, 1975, **88**, 345 (*ms*)

$C_{13}H_{17}Co$ **Co-00138**

[(1,2,5,6-η)-1,5-Cyclooctadiene](η⁵-2,4-cyclopentadien-1-yl-)cobalt, 9CI

[12184-35-9]

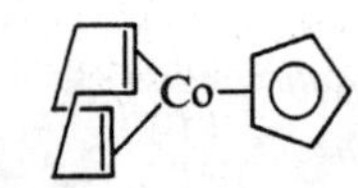

M 232.211

Catalyst for synth. of pyridines from 1 nitrile + 2 alkynes. Golden-yellow cryst. by subl. Mp 102-4°. $Bp_{0.1}$ 76° subl.

King. R.B. *et al*, *J. Am. Chem. Soc.*, 1961, **83**, 3593 (*synth*, *ir*, *pmr*)
Nakamura, A. *et al*, *Bull. Chem. Soc. Jpn.*, 1961, **34**, 452 (*synth*)
Bönnemann, H. *et al*, *Synthesis*, 1975, 600 (*use*)
Moraczewski, J. *et al*, *J. Am. Chem. Soc.*, 1981, **103**, 4779 (*synth*)

$C_{13}H_{17}Co$ **Co-00139**

[(1,2,3,4-η)-1,3-Cyclooctadiene](η⁵-2,4-cyclopentadien-1-yl-)cobalt, 10CI

[78063-06-6]

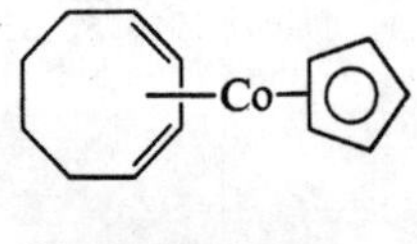

M 232.211

No phys. props. recorded.

Moraczewski, J. *et al*, *J. Am. Chem. Soc.*, 1981, **103**, 4779 (*synth*, *pmr*)

$C_{13}H_{17}CoO_2$ **Co-00140**

Dicarbonyl[(1,2,3,4,5-η)-1-ethyl-2,3,4,5-tetramethyl-2,4-cyclopentadien-1-yl]cobalt

[68494-68-8]

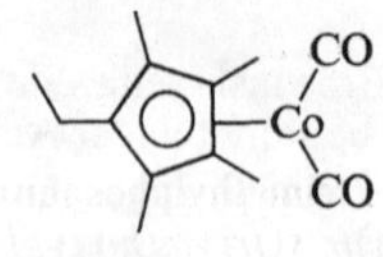

M 264.209

Precursor for formation of η⁶-arene Co complexes. Air-sensitive red liq. $Bp_{0.2}$ 92-5°. ir ν_{CO} 2011, 1953 cm^{-1} (cyclohexane).

Dooley, T. *et al*, *Transition. Met. Chem.*, 1978, **3**, 299 (*synth*, *ir*, *pmr*, *cmr*)
Fairhurst, G. *et al*, *J. Chem. Soc.*, *Dalton Trans.*, 1979, 1531.

$C_{13}H_{21}CoN_2O_2$ **Co-00141**

[[4,4′-(1,2-Ethanediyldinitrilo)bis[2-pentanonato]](2−)-*N*,*N*′-,*O*,*O*′]methylcobalt, 10CI

[4,4′-(Ethylenedinitrilo)di-2-pentanonato(2−)]methylcobalt, 8CI

[41744-14-3]

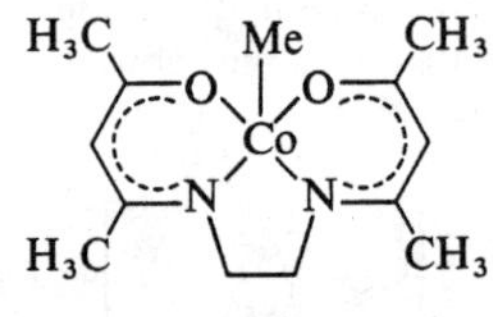

M 296.254

Analogue of coenzyme B_{12}. Green cryst. (C_6H_6). Mp 194° dec. Undergoes transmethylation with Hg and other Co complexes.

Costa, G. *et al*, *J. Organomet. Chem.*, 1966, **6**, 181 (*synth*, *uv*)
Brückner, S. *et al*, *Inorg. Chim. Acta*, 1969, **3**, 308 (*cryst struct*)
La Rossa, R.A. *et al*, *J. Am. Chem. Soc.*, 1974, **96**, 2072 (*nmr*)
Mestroni, G. *et al*, *Transition. Met. Chem.*, 1976, **1**, 32.

$C_{13}H_{25}CoN_4O_2$ **Co-00142**

Dimethyl[3,3′-(1,3-propanediyldinitrilo)bis[2-butanone]dioximato(1−)-*N*,*N*′,*N*″,*N*‴]cobalt, 10CI

Dimethyl[[3,3′-(trimethylenedinitrilo)di-2-butanone dioximato](1−)]cobalt, 8CI

[33569-60-7]

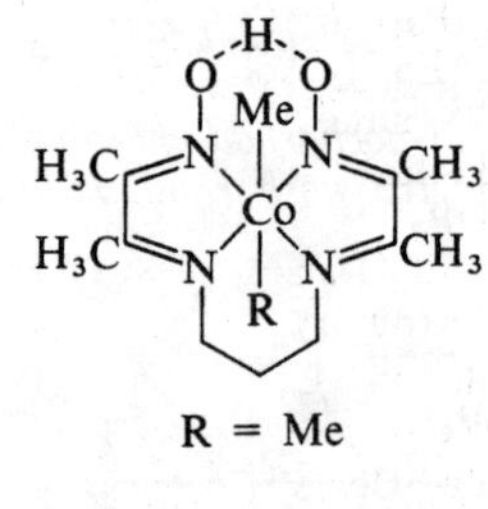

R = Me

M 328.299

Analogue of methylcobalamin. Orange, air-stable cryst. (Me_2CO aq.). Stable in aprotic solvs. Sol. Et_2O, C_6H_6, Me_2CO, insol. H_2O. Co−Me bond cleaved by light or $H^{\oplus}$. Methylates electrophilic reagents, e.g. $Hg^{\oplus\oplus}$.

Costa, G. *et al*, *J. Chem. Soc.*, *Dalton Trans.*, 1972, 450 (*synth*)
Calligaris, M., *J. Chem. Soc.*, *Dalton Trans.*, 1974, 1628 (*cryst struct*)
Witman, M.W. *et al*, *Synth. React. Inorg. Metal-Org. Chem.*, 1977, **7**, 143 (*synth*, *ir*, *pmr*, *props*)

Tamblyn, W.H. *et al*, *J. Am. Chem. Soc.*, 1981, **103**, 3160 (*synth, uv, pmr, props*)
Elliott, C.M. *et al*, *J. Am. Chem. Soc.*, 1981, **103**, 5558 (*synth, pmr, props*)
Dimmit, J.H. *et al*, *Inorg. Chem.*, 1982, **21**, 700 (*uv, pmr, props*)

$C_{13}H_{33}CoO_9P_3^{\oplus}$ Co-00143

(η^4-1,3-Butadiene)tris(trimethylphosphite-*P*)cobalt(1+), 10CI
*(η^4-1,3-Butadiene)tris(trimethoxyphosphine-*P*)-cobalt(1+)*

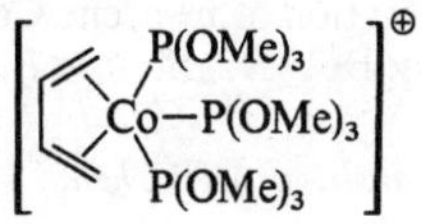

M 485.253 (ion)
Hexafluorophosphate:
$C_{13}H_{33}CoF_6O_9P_4$ M 630.217
Pale-green cryst. ($MeOH/Et_2O$). Mp 204° dec.

Muetterties, E.L. *et al*, *J. Am. Chem. Soc.*, 1978, **100**, 6978 (*synth, ir, nmr*)

$C_{13}H_{33}CoP_3^{\oplus}$ Co-00144

(η^4-1,3-Butadiene)tris(trimethylphosphine)cobalt(1+)
[65836-16-0]

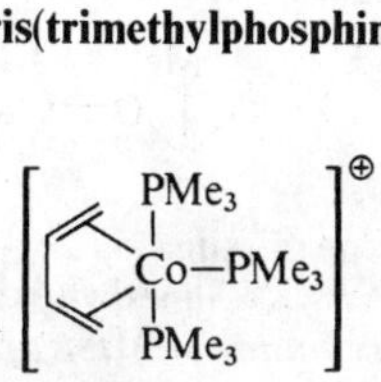

M 341.258 (ion)
Tetraphenylborate: [65836-17-1].
$C_{37}H_{53}BCoP_3$ M 660.490
Yellow-orange cryst. (THF/2-propanol). Mp 189°.

Bressan, M. *et al*, *J. Organomet. Chem.*, 1978, **144**, 215 (*synth, nmr*)

$C_{13}H_{36}CoOP_4^{\oplus}$ Co-00145

Carbonyltetrakis(trimethylphosphine)cobalt(1+), 9CI

$[(Me_3P)_4CoCO]^{\oplus}$

M 391.255 (ion)
Unstable in soln. above 10°.
Bromide: [53432-16-9].
$C_{13}H_{36}BrCoOP_4$ M 471.159
Orange cryst. (Et_2O). Sol. CH_2Cl_2. Mp >115° dec. ir ν_{CO} 1908 cm^{-1} (nujol).
Chloride: [53432-15-8].
$C_{13}H_{36}ClCoOP_4$ M 426.708
Yellow cryst. (Et_2O). Sol. CH_2Cl_2. Mp >110° dec. ir ν_{CO} 1901 cm^{-1} (nujol).
Iodide: [53432-17-0].
$C_{13}H_{36}CoIOP_4$ M 518.160
Orange cryst. (Et_2O). Sol. CH_2Cl_2. Mp >115° dec. ir ν_{CO} 1909 cm^{-1} (nujol).

Klein, H.F. *et al*, *Inorg. Chem.*, 1975, **14**, 473 (*synth, ir, pmr*)

$C_{13}H_{36}CoO_{13}P_4^{\oplus}$ Co-00146

Carbonyltetrakis(trimethyl phosphite-*P*)cobalt(1+), 9CI
[45303-88-6]

$[(MeO)_3P]_4Co(CO)^{\oplus}$

M 583.248 (ion)
Tetraphenylborate: [82323-13-3].
$C_{37}H_{56}BCoO_{13}P_4$ M 902.480
Yellow solid (Me_2CO). Mp 170° dec. ir ν_{CO} 1970 cm^{-1}.

Attali, S. *et al*, *Inorg. Chim. Acta*, 1972, **6**, 475 (*synth, ir, pmr*)

$C_{13}H_{39}CoP_4$ Co-00147

Methyltetrakis(trimethylphosphine)cobalt, 9CI
[53432-25-0]

$CoMe(PMe_3)_4$

M 378.279
Orange-red cryst. (pentane/Et_2O). Mp 85° dec.

Klein, H.-F. *et al*, *Chem. Ber.*, 1975, **108**, 944 (*synth, ir, nmr*)

$C_{14}Co_6O_{14}^{\ominus\ominus\ominus\ominus}$ Co-00148

Octa-μ_3-carbonylhexacarbonylcobaltate(4−), 9CI
Tetradecacarbonylhexacobaltate(4−)
[51635-26-8]

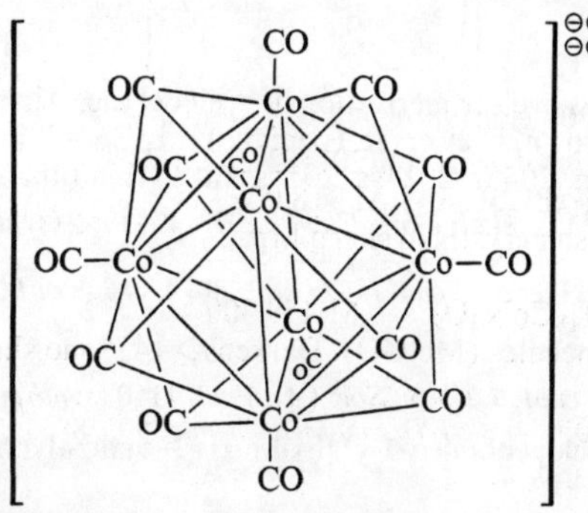

M 745.745 (ion)
Tetra-K salt:
$C_{14}Co_6K_4O_{14}$ M 902.138
Dark-red cryst. Sol. H_2O, insol. THF. ir ν_{CO} 1990w, 1940s, 1910s, 1660s cm^{-1} (MeOH).
Tetrakis(tetrabutylammonium) salt:
$C_{78}H_{144}Co_6N_4O_{14}$ M 1715.613
Red-brown solid (MeOH). Insol. H_2O, sl. sol. THF, MeOH. ir ν_{CO} 2040, 1890, 1680 cm^{-1} (nujol).
Tetra(cobaltocenium) salt:
$C_{54}H_{40}Co_{10}O_{14}$ M 1502.234
Dark-brown solid (THF). Insol. Me_2CO, MeOH, THF, H_2O. ir ν_{CO} 1905, 1865, 1708, 1677 cm^{-1} (nujol).

Albano, V. *et al*, *J. Organomet. Chem.*, 1969, **16**, 461 (*cryst struct*)
Chini, P. *et al*, *J. Organomet. Chem.*, 1969, **16**, 471 (*synth, ir, props*)
Benfield, R.E. *et al*, *J. Chem. Soc., Dalton Trans.*, 1980, 1743 (*struct*)

$C_{14}H_6Co_2O_6$ Co-00149

Hexacarbonyl(η^2-phenylacetylene)dicobalt
Hexacarbonyl[μ-[[(1,2-η:1,2-η)-ethynyl]benzene]]dicobalt, 9CI. Phenylacetylenedicobalt hexacarbonyl
[12154-91-5]

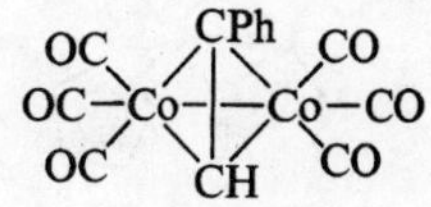

M 388.064

Red oil or dark-violet cryst. ir ν_{CO} 2095, 2058, 2032, 2028, 2016, 2011 cm^{-1} (hexane). Dec. at 165° to Co + CO + 1,2,4-triphenylbenzene.

Greenfield, H. *et al*, *J. Am. Chem. Soc.*, 1956, **78**, 120 (*synth*, *ir*)
Krüerke, U. *et al*, *Chem. Ber.*, 1961, **94**, 2829 (*synth*)
Dickson, R.S. *et al*, *Aust. J. Chem.*, 1970, **23**, 229; 1975, **28**, 1957 (*synth*, *props*)
Bor, G. *et al*, *Inorg. Chim. Acta*, 1976, **18**, L18 (*ir*)
Aime, S. *et al*, *Inorg. Chim. Acta*, 1977, **22**, 135 (*nmr*)
Pauson, P.L. *et al*, *Ann. N.Y. Acad. Sci.*, 1977, **295**, 2 (*props*)

$C_{14}H_{12}Co_2O_6$ Co-00150

Hexacarbonyl[μ-[(1,2-η:1,2-η)-cyclooctyne]dicobalt, 10CI
μ-Cyclooctynehexacarbonyldicobalt
[66213-35-2]

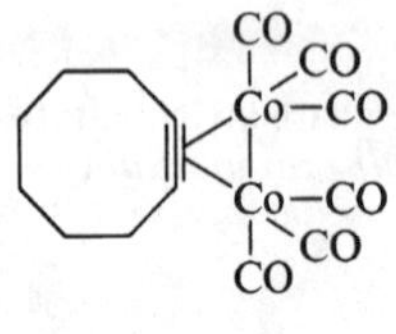

M 394.112
Catalyses cyclotrimerisation of cyclooctyne. Air-sensitive red-brown waxy cryst. (pentane). Mp 36-8°. ir ν_{CO} 2086m, 2045s, 2024vs, 2010s cm^{-1} (thin film).

Bennett, M.A. *et al*, *Inorg. Chem.*, 1978, **17**, 1995 (*synth*, *ir*, *ms*, *nmr*)
Petersen, H. *et al*, *Nouv. J. Chim.*, 1980, **4**, 687 (*synth*, *ir*, *pmr*)

$C_{14}H_{15}BCoN_6^{\oplus}$ Co-00151

(η^5-2,4-Cyclopentadien-1-yl)[hydrotris(1-pyrazolyl)borato]-cobalt(1+), 9CI

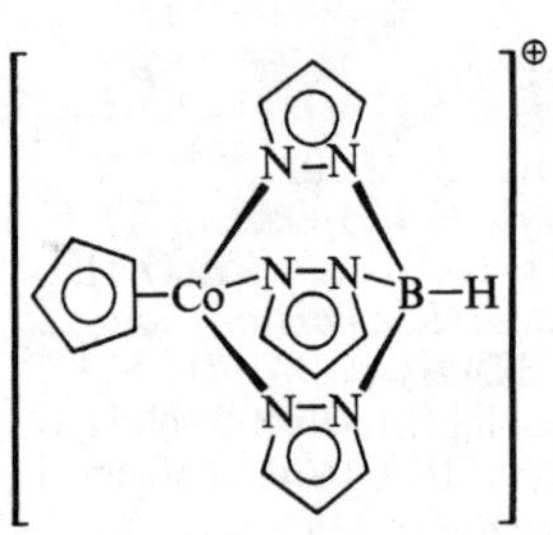

M 337.056 (ion)
Iodide: [51839-61-3].
$C_{14}H_{15}BCoIN_6$ M 463.960
No phys. props. reported.

O'Sullivan, D.J. *et al*, *J. Organomet. Chem.*, 1973, **57**, C58 (*synth*)

$C_{14}H_{16}Co_2O_2$ Co-00152

Di-μ-carbonylbis(2,4-cyclopentadien-1-yl)dimethyldicobalt, 10CI
Bis[μ-carbonyl-η^5-cyclopentadienylmethylcobalt]
[61276-02-6]

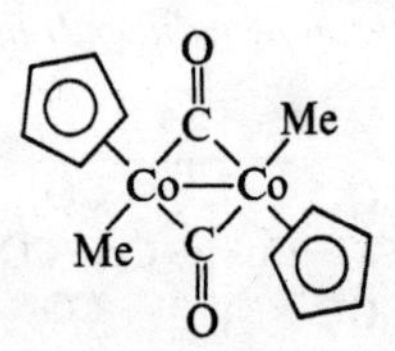

M 334.146
Unstable in soln. at r.t. ir ν_{CO} 1820 cm^{-1} (THF).

Schore, N.E. *et al*, *J. Am. Chem. Soc.*, 1976, **98**, 7436 (*synth*, *ir*, *nmr*)

$C_{14}H_{17}CoO$ Co-00153

(η^5-2,4-Cyclopentadien-1-yl)[(2,3,4,5-η)-2,3,4,5-tetramethyl-2,4-cyclopentadien-1-one]cobalt, 10CI
[1272-06-6]

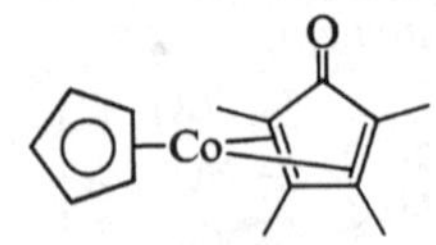

M 260.221
Orange cryst. (Et_2O). Mp 178-80°. pK_a 4.30 (H_2O). ir ν_{CO} 1559 cm^{-1} (CH_2Cl_2). May be protonated at oxygen to give $CpCo(C_5Me_4OH)^{\oplus}$.

Markby, R. *et al*, *Chem. Ind. (London)*, 1959, 1381 (*synth*, *ir*)
Dahl, L.F. *et al*, *J. Am. Chem. Soc.*, 1961, **83**, 752 (*struct*)
Schrauzer, G.N. *et al*, *J. Organomet. Chem.*, 1964, **2**, 336 (*uv*)
Sheats, J.E. *et al*, *J. Organomet. Chem.*, 1975, **96**, 115 (*uv*, *ir*, *pmr*)
Lee, W.S. *et al*, *J. Organomet. Chem.*, 1977, **127**, 93.

$C_{14}H_{20}B_2CoF_4N_5O_4$ Co-00154

[Bis[μ-[(2,3-butanedione dioximato)(2−)-*O*:*O′*]]tetrafluorodiborato(2−)-*N*,*N′*,*N″*,*N‴*]methyl(pyridine)cobalt, 9CI
[Bis[dimethylglyoximato(2−)]tetrafluorodiborato]-methyl(pyridine)cobalt
[28627-87-4]

As Aqua[bis[μ-[(2,3-butanedione dioximato)(2−)-*O*:*O′*-]]][tetrafluorodiborato(2−)-*N*,*N′*,*N″*,*N‴*]methylcobalt, Co-00050 with

X = Py

M 478.890
Vitamin B_{12} analogue. Yellow cryst. (Me_2CO). Light-sensitive in soln.
▷Explodes at 245°

Schrauzer, G.N. *et al*, *J. Am. Chem. Soc.*, 1966, **88**, 3738 (*synth*, *props*)
Giannotti, C. *et al*, *J. Organomet. Chem.*, 1975, **91**, 357 (*props*)

$C_{14}H_{22}CoN_5O_4$ Co-00155

Bis[(2,3-butanedione dioximato)(1−)-*N*,*N′*-methyl(pyridine)-cobalt, 10CI
Bis(dimethylglyoximato)methyl(pyridine)cobalt.
Methylpyridinatocobaloxime
[23642-14-0]

As Aquabis[(2,3-butanedione dioximato)(1−)-*N*,*N′*]-methylcobalt, Co-00051 with

L = Py

M 383.292
Model compd. for vitamin B_{12}. Orange cryst. (Me_2CO). ir ν_{CN} 1557 cm^{-1}, ν_{NO} 1235 cm^{-1}.
Monocation: [61754-28-7].
$C_{14}H_{22}CoN_5O_4^{\oplus}$ M 383.292 (ion)
Characterised spectroscopically.

Schrauzer, N. *et al*, *J. Am. Chem. Soc.*, 1966, **88**, 3738 (*synth*)

Inorg. Synth., 1969, **11**, 61 (*synth*)
Hill, H.A.O. *et al*, *J. Chem. Soc.* (*A*), 1969, 554 (*synth*, *ir*, *pmr*)
La Rossa, R.A. *et al*, *J. Am. Chem. Soc.*, 1974, **96**, 2072 (*nmr*)
Bigotto, A. *et al*, *J. Chem. Soc., Dalton Trans.*, 1976, 96 (*struct*)
Halpern, J. *et al*, *Inorg. Chim. Acta*, 1976, **20**, L21 (*esr*)
Huet, J. *et al*, *Org. Magn. Reson.*, 1981, **15**, 347 (*cmr*)

$C_{14}H_{23}Co$ Co-00156

Bis(η^2-ethene)[(1,2,3,4,5-η)-1,2,3,4,5-pentamethyl-2,4-cyclopentadien-1-yl]cobalt, 10CI
Bis(ethylene)(pentamethylcyclopentadienyl)cobalt
[80848-36-8]

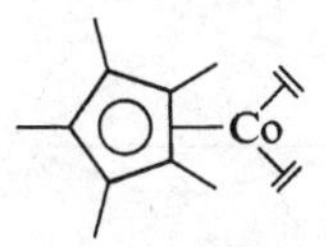

M 250.269
Starting material for $Co(C_5Me_5)$ complexes and heteronuclear metal clusters. Orange cryst. (pet. ether at −78°). Solid may be handled in air for short periods but is stored under ethylene at 5°. Ethylene easily displaced by alkynes, phosphines etc.

Beevor, R.G. *et al*, *J. Organomet. Chem.*, 1981, **221**, C25 (*synth*, *nmr*, *props*)
Brun, P. *et al*, *J. Chem. Soc., Chem. Commun.*, 1982, 926 (*props*)
Kölle, U. *et al*, *Angew. Chem., Int. Ed. Engl.*, 1982, **21**, 131; *Angew. Chem. Suppl.*, 1982, 230 (*synth*)

$C_{14}H_{23}CoN_4O_3$ Co-00157

Carbonyl[[2,2′-(1,3-propanediyldinitrilo)bis[3-pentanone]dioximato](1−)-*N,N′,N″,N‴*]cobalt
[75504-41-5]

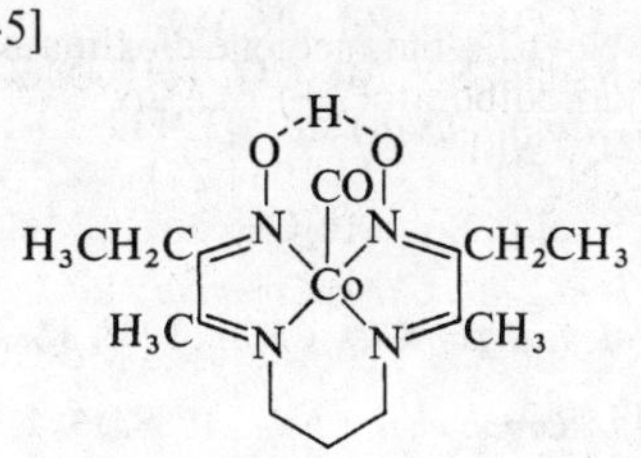

M 354.294
Readily adds alkyl halide oxidatively to displace CO and give vitamin B_{12} analogues. Red-purple, air-sensitive microcryst. (deep-blue in C_6H_6 soln.). ir ν_{CO} 1972 cm^{-1} (THF). Coordinated CO readily exchanges with ^{13}CO.

Finke, R.G. *et al*, *J. Chem. Soc., Chem. Commun.*, 1980, 460 (*props*)
Finke, R.G. *et al*, *Inorg. Chem.*, 1981, **20**, 687 (*synth*, *ir*, *pmr*)

$C_{14}H_{24}CoOP$ Co-00158

Carbonyl[(1,2,3,4,5-η)-1,2,3,4,5-pentamethyl-2,4-cyclopentadien-1-yl](trimethylphosphine)cobalt
[84879-73-2]

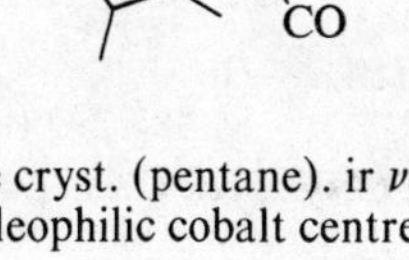

M 298.250
Red, air-sensitive cryst. (pentane). ir ν_{CO} 1915 cm^{-1} (hexane). Nucleophilic cobalt centre.

Werner, H. *et al*, *J. Organomet. Chem.*, 1982, **240**, 179 (*synth*, *pmr*, *ir*, *props*)

$C_{14}H_{26}CoIN_4O_2$ Co-00159

Iodomethyl[[2,2′-(1,3-propanediyldinitrilo)bis[3-pentanone]-dioximato](1−)-*N,N′,N″,N‴*]cobalt
[75962-05-9]

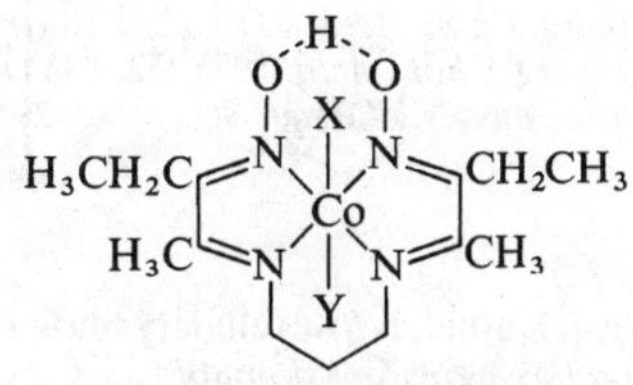

X = Me,Y = I

M 468.223
Easily synthesised vitamin B_{12} analogue. Orange solid (Et_2O/THF). Spar. sol. C_6H_6. Co−Me bond readily cleaved by light.

Finke, R.G. *et al*, *Inorg. Chem.*, 1981, **20**, 687 (*synth*, *pmr*)

$C_{14}H_{26}CoN_4O^{\oplus}$ Co-00160

[2-[1-[(2-Aminoethyl)imino]ethyl]phenolato-*N,N′,O*](1,2-ethanediamine-*N,N′*)ethylcobalt(1+), 10CI
[N-(*2-Aminoethyl*)*-7-methylsalicylideneaminato*](*ethylenediamine*)*ethylcobalt*(*1*+)

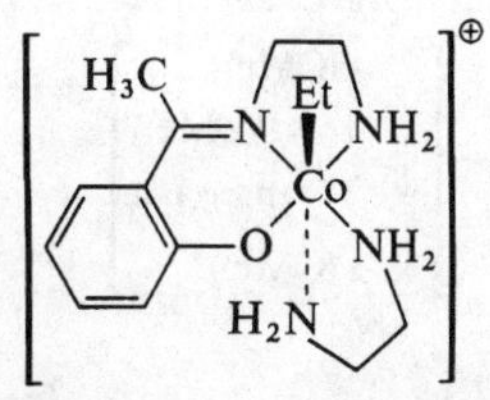

M 325.319 (ion)
Photolysis give C_2H_6, C_2H_4 and C_4H_{10}.
Bromide: [76375-20-7].
$C_{14}H_{26}BrCoN_4O$ M 405.223
Red prisms (MeOH aq.). Sol. MeOH, DMSO, spar. sol H_2O.
Perchlorate: [79199-99-8].
$C_{14}H_{26}ClCoN_4O_5$ M 424.769
No phys. props. reported.

Levitin, I. *et al*, *J. Chem. Soc., Chem. Commun.*, 1981, 440 (*synth*, *cryst struct*, *ir*, *props*)

$C_{14}H_{27}Co_2O_5$ Co-00161

Di-μ-carbonyltricarbonyltris(trimethylphosphine)dicobalt, 10CI
Pentacarbonyltris(trimethylphosphine)dicobalt
[22784-67-4]

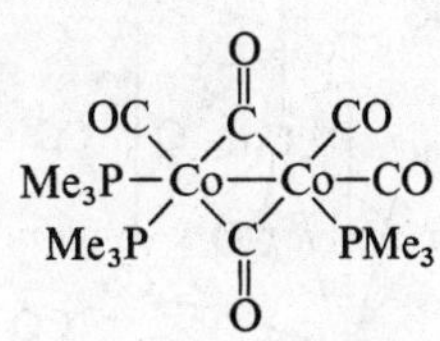

M 393.231
Cryst. (pentane). Mp 96°. ir ν_{CO} 2000, 1947, 1772, 1753 cm^{-1}.

Pegot, C. *et al*, *C.R. Hebd. Seances Acad. Sci.*, 1969, **268**, 955 (*synth*, *ir*)
Arabi, M.S. *et al*, *J. Organomet. Chem.*, 1974, **67**, 109 (*synth*, *ir*)

$C_{14}H_{28}CoN_4O_3^{\oplus}$ **Co-00162**

Aquamethyl[[2,2′-(1,3-propanediyldinitrilo)bis[3-pentanone] dioximato](1−)-*N*,*N*′,*N*″,*N*‴]cobalt(1+)

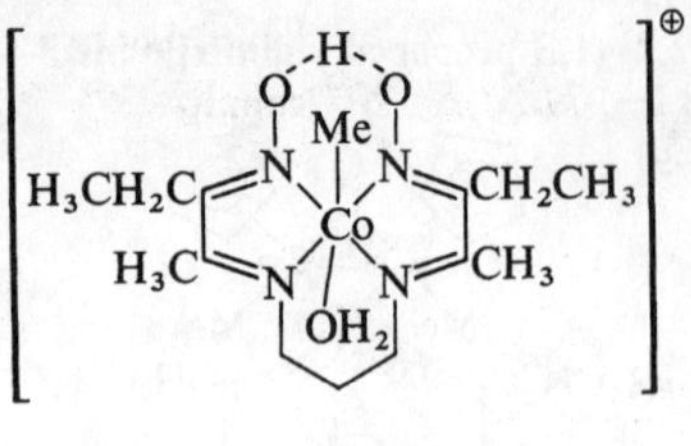

M 359.333 (ion)

Electrochemical mimic of vitamin B_{12} derivs. Co−Me bond cleaved by light. Coordinated H_2O readily replaced by other bases e.g. Py.

Hexafluorophosphate: [75962-12-8].

$C_{14}H_{28}CoF_6N_4O_3P$ M 504.298

Deep-red microcryst. (MeOH aq.).

Elliott, C.M. *et al*, *J. Am. Chem. Soc.*, 1981, **103**, 5558 (*props*)
Finke, R.G. *et al*, *Inorg. Chem.*, 1981, **20**, 687 (*synth, pmr*)

$C_{14}H_{40}CoO_{12}P_4^{\oplus}$ **Co-00163**

(η^2-Ethene)tetrakis(trimethyl phosphite-*P*)cobalt(1+), 9CI

η^2-Ethylenetetrakis(trimethyl phosphite)cobalt(1+)

[60823-99-6]

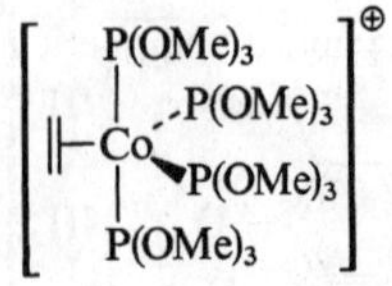

M 583.291 (ion)

Hexafluorophosphate: [68831-63-0].

$C_{14}H_{40}CoF_6O_{12}P_5$ M 728.255

Pale-yellow solid. Mp 188-94° dec.

Muetterties, E.L. *et al*, *J. Am. Chem. Soc.*, 1978, **100**, 6978 (*synth, ir, nmr*)

$C_{15}Co_6NO_{15}^{\ominus}$ **Co-00164**

Nona-μ-carbonylhexacarbonyl-μ_6-nitridohexacobaltate(1−), 10CI

Pentadecacarbonylnitridohexacobaltate(1−)

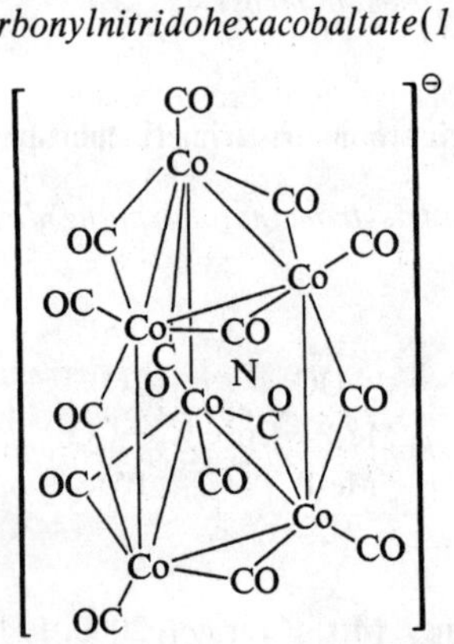

M 787.762 (ion)

K salt:

$C_{15}Co_6KNO_{15}$ M 826.860

Orange-red.

Bis(triphenylphosphine)iminium salt: [72782-99-1].

$C_{51}H_{30}Co_6N_2O_{15}P_2$ M 1326.349

Obt. by metathesis of the K salt in 2-propanol.

Martinengo, S. *et al*, *J. Am. Chem. Soc.*, 1979, **101**, 7095 (*synth, cryst struct, nmr, ir*)

$C_{15}Co_6O_{15}^{\ominus\ominus}$ **Co-00165**

Pentadecacarbonylhexacobaltate(2−)

Tri-μ-carbonyltri-μ_3-carbonylnonacarbonylhexacobaltate(2−), 8CI

[54340-33-9]

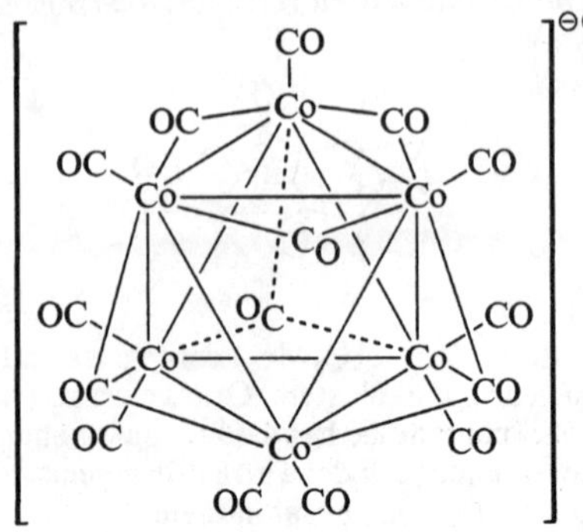

M 773.755 (ion)

All derivatives air-sensitive. ir ν_{CO} 2040, 1980, 1960, 1780, 1735, 1680 cm^{-1} (THF).

Di-K salt: [12111-52-3].

$C_{15}Co_6K_2O_{15}$ M 851.952

Dark-green solid. Sol. Me_2CO, THF, MeOH, H_2O.

Di-Cs salt: [12111-51-2].

$C_{15}Co_6Cs_2O_{15}$ M 1039.566

Dark-green solid. Sol. Me_2CO, THF, MeOH.

Bis(tetramethylammonium) salt: [12111-49-8].

$C_{23}H_{24}Co_6H_2O_{15}$ M 896.049

Dark-green solid. Sol. Me_2CO, THF, MeOH.

Bis(tetraethylammonium) salt: [12111-50-1].

$C_{31}H_{40}Co_6N_2O_{15}$ M 1034.261

Dark-green solid. Sol. Me_2CO, THF, MeOH.

Chini, P., *J. Chem. Soc., Chem. Commun.*, 1967, 29 (*synth, ir*)
Albano, V. *et al*, *J. Organomet. Chem.*, 1968, **15**, 423 (*cryst struct*)
Chini, P. *et al*, *J. Organomet. Chem.*, 1968, **15**, 433 (*synth, props*)

$C_{15}H_6Co_4O_9$ **Co-00166**

(η^6-Benzene)tri-μ-carbonylhexacarbonyltetracobalt, 9CI

(η^6-Benzene)nonacarbonyltetracobalt

[32699-48-2]

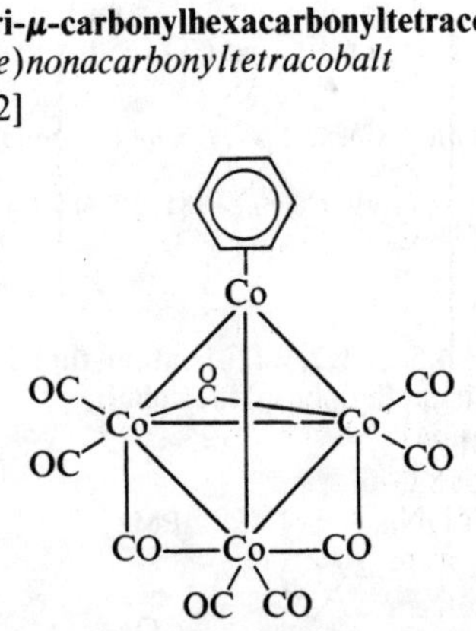

M 565.940

Black cryst.

Khaud, I.V. *et al*, *J. Chem. Soc., Chem. Commun.*, 1971, 36 (*synth*)

Bor, G. *et al*, *J. Organomet. Chem.*, 1972, **46**, 357 (*synth, ir*)
Khaud, I.V. *et al*, *J. Chem. Soc., Perkin Trans. 1*, 1973, 975 (*synth, ir, pmr*)
Bird, P.H. *et al*, *J. Organomet. Chem.*, 1974, **73**, 103 (*struct*)
Alper, H. *et al*, *J. Organomet. Chem.*, 1978, **144**, C18 (*use*)

$C_{15}H_7CoFeO_6$ Co-00167

Di-μ-carbonylcarbonyl[(1,2,3,3a,7a-η)-1H-inden-1-yl](tricarbonylcobalt)iron, 9CI

Di-μ-carbonylcarbonyl-π-indenyl(tricarbonylcobaltio)iron, 8CI. Hexacarbonyl-π-indenylcobaltiron

[34779-05-0]

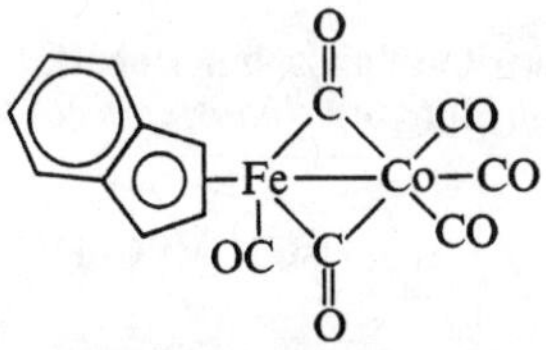

M 397.997

Brown cryst. (pentane). Mp 74-6°.

Manning, A.R., *J. Chem. Soc. (A)*, 1971, 2321 (*synth, ir*)
Stephens, F.S., *J. Chem. Soc., Dalton Trans.*, 1974, 13 (*struct*)

$C_{15}H_{10}Co_2O_4$ Co-00168

Tetracarbonyl[μ-[(1,2,3,4,5-η:1′,2′,3′,4′,5′-η)-methylenedi-2,4-cyclopentadien-1-ylidene]]dicobalt

[71674-53-8]

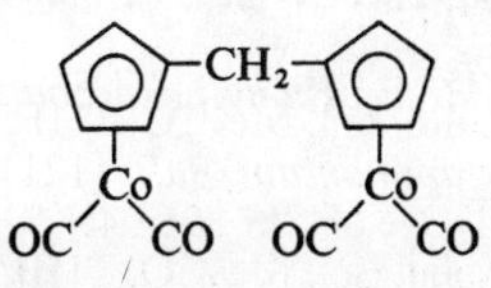

M 372.108

No phys. props. reported.

Bryndza, H.E. *et al*, *J. Am. Chem. Soc.*, 1979, **101**, 4766 (*synth*)

$C_{15}H_{15}CoN_6^{\oplus\oplus}$ Co-00169

(η^5-2,4-Cyclopentadien-1-yl)[1,1′,1″-methylidynetris[1H-pyrazole]-N^2,$N^{2'}$,$N^{2''}$]cobalt(2+), 9CI

η^5-Cyclopentadienyl[tris(1-pyrazolyl)methane]cobalt(2+)

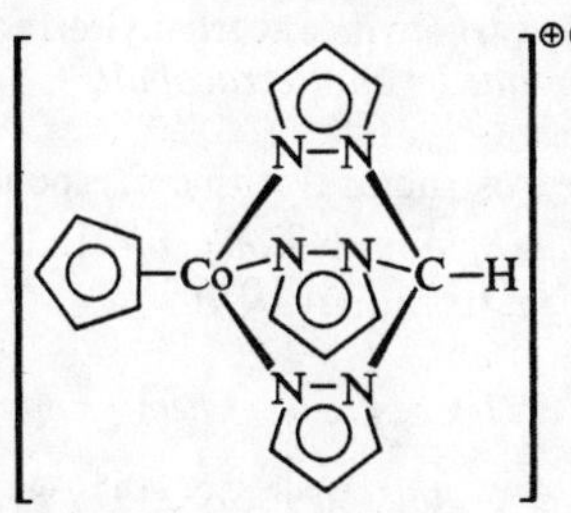

M 338.257 (ion)

Diiodide: [51839-63-5].

$C_{15}H_{15}CoI_2N_6$ M 592.066

No phys. props. reported.

O'Sullivan, D.J. *et al*, *J. Organomet. Chem.*, 1973, **57**, C58 (*synth*)

$C_{15}H_{16}Co_2O_2$ Co-00170

Di-μ-carbonyldimethyl[μ-[(1,2,3,4,5-η:1′,2′,3′,4′,5′-η)methylenedi-2,4-cyclopentadien-1-ylidene]]dicobalt, 10CI

[71729-49-2]

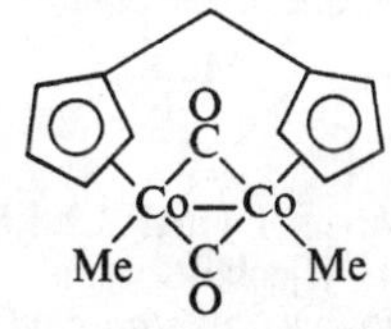

M 346.157

No phys. props. reported.

Bryndza, H.E. *et al*, *J. Am. Chem. Soc.*, 1979, **101**, 4766 (*synth*)

$C_{15}H_{17}CoO_2$ Co-00171

(η^5-2,4-Cyclopentadien-1-yl)[(2,3,5,6-η)-2,3,5,6-tetramethyl-2,5-cyclohexadiene-1,4-dione]cobalt, 9CI

Cyclopentadienyl(duroquinone)cobalt

[41611-29-4]

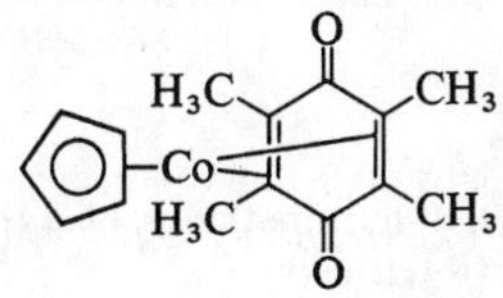

M 288.231

Red-brown hygroscopic diamagnetic cryst. Sl. sol. org. solvs., sol. H_2O. Duroquinone ligand readily protonated.

Schrauzer, G.N. *et al*, *Angew. Chem., Int. Ed. Engl.*, 1963, **2**, 478 (*synth, pmr, ir, uv*)
Uchtman, V.A. *et al*, *J. Organomet. Chem.*, 1972, **40**, 403 (*cryst struct*)
Bodner, G.M. *et al*, *J. Organomet. Chem.*, 1975, **88**, 391 (*cmr*)

$C_{15}H_{21}Co$ Co-00172

[(1,2,3,4,5-η)-2,4-Cycloheptadien-1-yl][(1,2,5,6-η)-1,5-cyclooctadiene]cobalt, 9CI

[32798-84-8]

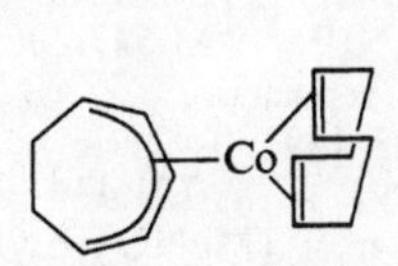

M 260.264

Fluxional molecule. Red-brown needles (pentane), air- and heat-stable. Sol. hydrocarbons. Mp 63-5°.

Otsuka, S. *et al*, *J. Chem. Soc. (A)*, 1971, 579, 583 (*synth, ir, pmr*)

$C_{15}H_{22}CoO^{\oplus}$ Co-00173

Carbonyl[(1,2,3,4,5-η)-1-ethyl-2,3,4,5-tetramethyl-2,4-cyclopentadien-1-yl](η^3-2-propenyl)cobalt(1+)

(η^3-Allyl)carbonyl(η^5-ethyltetramethylcyclopentadienyl)cobalt(1+)

[71280-85-8]

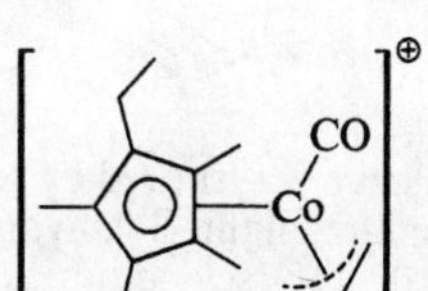

M 277.271 (ion)

Hexafluorophosphate: [71280-86-9].
$C_{15}H_{22}CoF_6OP$ M 422.236
Orange-yellow cryst. (EtOH). ir ν_{CO} 2055 cm^{-1} (nujol).

Green, M.L.H. *et al*, *J. Chem. Soc.*, *Dalton Trans.*, 1979, 355 (*synth*, *ir*, *pmr*)

$C_{15}H_{25}Co$ Co-00174

Bis(η^2-ethene)[(1,2,3,4,5-η)-1-ethyl-2,3,4,5-tetramethyl-2,4-cyclopentadien-1-yl]cobalt, 10CI
(η^5-Ethyltetramethylcyclopentadienyl)bis(ethylene)cobalt
[71251-97-3]

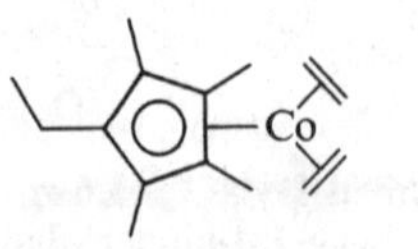

M 264.296
V. air-sensitive brown oil. $Bp_{0.05}$ 130°.

Green, M.L.H. *et al*, *J. Chem. Soc.*, *Dalton Trans.*, 1979, 355 (*synth*, *pmr*)

$C_{15}H_{27}BrCoN_4^{\oplus}$ Co-00175

Bromomethyl(2,3,9,10-tetramethyl-1,4,8,11-tetraazacyclotetradeca-1,3,8,10-tetraene-N^1,N^4,N^8,N^{11})cobalt(1+), 9CI

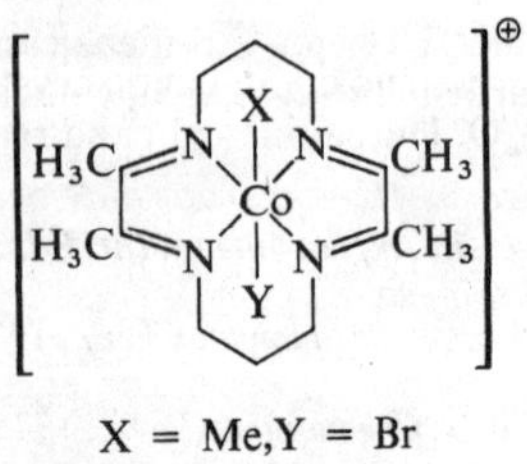

X = Me,Y = Br

M 402.242 (ion)
Vitamin B_{12} analogue. Stable in the solid state but light-sensitive in soln. in the presence of air. Bromide ligand rapidly displaced by MeCN.
Hexafluorophosphate: [39175-75-2].
$C_{15}H_{27}BrCoF_6N_4P$ M 547.206
No phys. props. reported.
Tetraphenylborate: [29190-42-9].
$C_{39}H_{47}BBrCoN_4$ M 721.474
Red, air-stable cryst. ($MeNO_2$/EtOH).

Farmery, K. *et al*, *Inorg. Chem.*, 1972, **11**, 2901 (*synth*, *pmr*, *uv*)

$C_{15}H_{27}CoN_4$ Co-00176

Methyl(2,3,9,10-tetramethyl-1,4,8,11-tetraazacyclotetradeca-1,3,8,10-tetraene)cobalt
[29190-41-8]

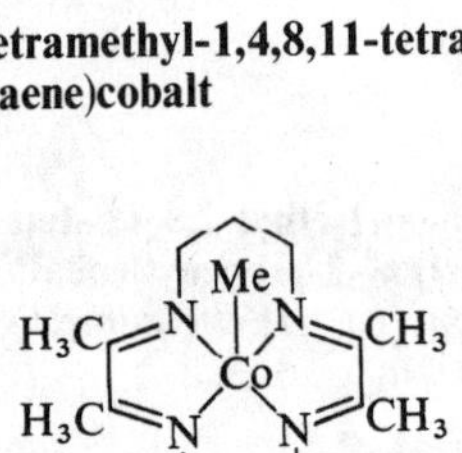

M 322.338
Vitamin B_{12} analogue. Cryst., red by reflected light, green by transmitted light. Sol. hydrocarbons, Et_2O, alcohols. Mp 109-11° (sealed tube).

Farmery, K. *et al*, *J. Chem. Soc.*, *Chem. Commun.*, 1970, 1091 (*synth*, *pmr*)

$C_{15}H_{27}CoN_4O$ Co-00177

Tris(2-isocyano-2-methylpropane)nitrosylcobalt, 10CI
*Tri-*tert*-butylisonitrilenitrosylcobalt*
[73382-46-4]

$[(H_3C)_3CNC]_3CoNO$

M 338.338
Air-stable red cryst. (hexane). Mp 107° dec. ir ν_{NC} 2137, 2101, 2054, ν_{NO} 1704 cm^{-1} (hexane).

Carroll, W.E. *et al*, *J. Chem. Soc.*, *Dalton Trans.*, 1980, 80 (*synth*, *ir*, *pmr*)

$C_{15}H_{28}CoO_3P$ Co-00178

Tricarbonylhydro(tributylphosphine)cobalt, 10CI
Tricarbonylhydridotributylphosphinecobalt
[20161-43-7]

$(H_3CCH_2CH_2CH_2)_3PCoH(CO)_3$

M 346.291
Hydroformylation and alkene hydrogenation catalyst. Air-sensitive white cryst. ir ν_{CO} 2050m, 1970s, 1933w cm^{-1} (heptane). Dec. at 20° in air liberating H_2.

Piacenti, F. *et al*, *Chim. Ind.* (*Milan*), 1967, **49**, 245 (*synth*, *ir*, *use*)
Whyman, R., *J. Organomet. Chem.*, 1975, **94**, 303 (*ir*)
Mirbach, M.J. *et al*, *J. Am. Chem. Soc.*, 1981, **103**, 7590, 7594 (*use*)

$C_{16}Co_6O_{15}^{\ominus\ominus}$ Co-00179

[μ_6-(Carbon)]nona-μ-carbonylhexacarbonylhexacobaltate(2−), 9CI
μ_6-Carbonpentadecacarbonylhexacobaltate(2−)
[53323-13-0]

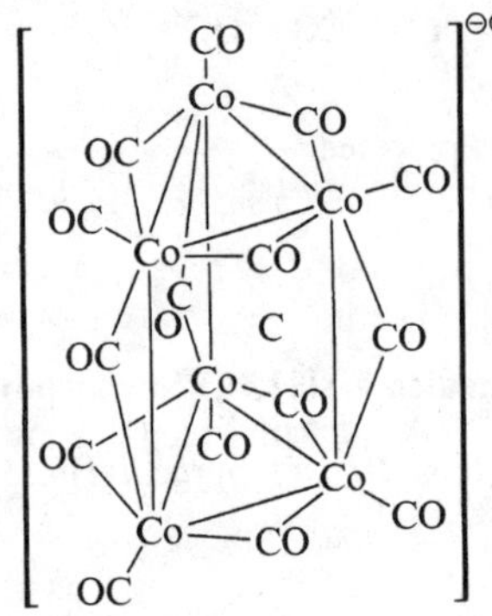

M 785.766 (ion)
Proposed to be isostructural with corresponding Rh deriv.
Bis(benzyltrimethylammonium) salt: [53509-52-7].
$C_{36}H_{32}Co_6N_2O_{15}$ M 1086.252
Black cryst.

Albano, V.G. *et al*, *J. Chem. Soc.*, *Chem. Commun.*, 1974, 299 (*synth*, *ir*)
Albano, V.G. *et al*, *J. Am. Chem. Soc.*, 1976, **98**, 5027 (*cmr*, *props*)

$C_{16}Co_6O_{16}$ Co-00180

Hexadecacarbonylhexacobalt, 8CI

Hexacobalt hexadecacarbonyl

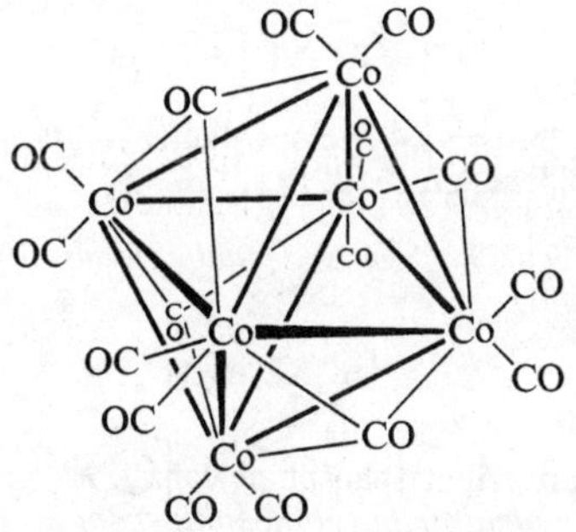

M 801.766

Struct. assumed to be analogous to Tetra-μ_3-carbonyldodecacarbonylhexarhodium, Rh-00184. Black solid. Sl. sol. hydrocarbons. Slowly dec. in air, dec. at 70-80°/0.001 mm.

Chini, P. *et al*, *Inorg. Chem.*, 1969, **8**, 1206 (*synth*, *ir*)

$C_{16}H_5Co_3O_9^{\ominus}$ Co-00181

Nonacarbonyl[μ_3-(phenylmethylidyne)]tricobaltate(1−), 10CI

[61024-80-4]

As Nonacarbonyl[μ_3-(phenylmethylidyne)]tricobalt, Co-00183 with

n = 1

M 518.010 (ion)

Radical anion, formed in electrochemical reduction of Nonacarbonyl[μ_3-(phenylmethylidyne)]tricobalt, Co-00183 . Not isol.

Bond, A.M. *et al*, *Inorg. Chem.*, 1977, **16**, 410 (*esr*)

$C_{16}H_5Co_3O_9^{\ominus\ominus}$ Co-00182

Nonacarbonyl[μ_3-(phenylmethylidyne)]tricobaltate(2−), 10CI

[61024-74-6]

As Nonacarbonyl[μ_3-(phenylmethylidyne)]tricobalt, Co-00183 with

n = 2

M 518.010 (ion)

Formed by electrochemical reduction of Nonacarbonyl[μ_3-(phenylmethylidyne)]tricobalt, Co-00183 . Deep-red soln. Not isol.

Bond, A.M. *et al*, *Inorg. Chem.*, 1977, **16**, 410 (*esr*)

$C_{16}H_5Co_3O_9$ Co-00183

Nonacarbonyl[μ_3-(phenylmethylidyne)]tricobalt, 10CI

[13682-03-6]

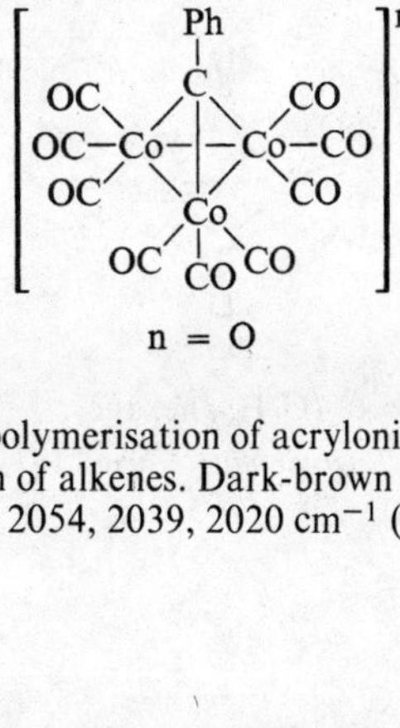

n = O

M 518.010

Catalyst for polymerisation of acrylonitrile and hydroformylation of alkenes. Dark-brown cryst. Mp 105-7°. ir ν_{co} 2101, 2054, 2039, 2020 cm^{-1} (hexane).

Palyi, G. *et al*, *Inorg. Chem.-Acta Rev.*, 1970, **4**, 109 (*synth*, *ir*)
Czajlik, I., *J. Organomet. Chem.*, 1973, **49**, C85 (*use*)
Aime, S. *et al*, *Transition. Met. Chem.*, 1976, **1**, 177 (*cmr*)
Ryan, R.C. *et al*, *J. Am. Chem. Soc.*, 1977, **99**, 1986 (*use*)

$C_{16}H_6Co_2FeO_9$ Co-00184

Tricarbonyl(hexacarbonyldicobalt)hydro[μ_3-(phenylmethylidyne)]iron, 10CI

Nonacarbonylhydro(phenylmethylidyne)dicobaltiron

[69493-77-2]

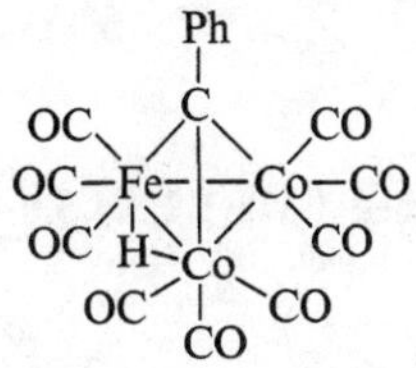

M 515.931

Position of hydride not definitely determined but thought to be bridging. Unstable under N_2, stored at 0° under CO. Pmr −9.00 ppm (bridging hydride).

Epstein, R.A. *et al*, *Inorg. Chem.*, 1979, **18**, 942 (*synth*, *ir*, *uv*, *pmr*, *ms*)

$C_{16}H_{13}CoFeO_4$ Co-00185

[[(2,3,5,6-η)-Bicyclo[2.2.1]hepta-2,5-diene]carbonylcobalt-]di-μ-carbonylcarbonyl(η^5-2,4-cyclopentadien-1-yl)iron, 9CI

Di-μ-carbonyl(carbonyl-π-norbornadienecobaltio)carbonyl-π-cyclopentadienyliron. π-Norbornadiene-π-cyclopentadienyltetracarbonylcobaltiron

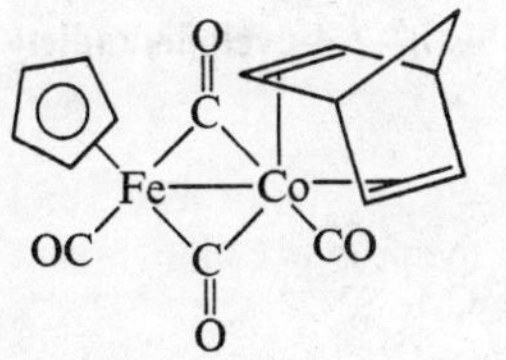

cis-form

M 384.057

Exists as a mixt. of *cis-* and *trans-*isomers in soln.

cis-form [36344-67-9]

Purple cryst. Mp 138-40°.

trans-form [36180-29-7]

Formed from *cis*-isomer in soln.

Manning, A.R., *J. Chem. Soc.*, *Dalton Trans.*, 1972, 821 (*synth*, *ir*)
Campbell, I.L.C. *et al*, *J. Chem. Soc.*, *Dalton Trans.*, 1975, 226 (*cryst struct*)

$C_{16}H_{14}CoN_2O_2$ Co-00186

[[2,2′-[1,2-Ethanediylbis[nitrilomethylidyne]]bis[phenolato]](2−)-*N*,*N′*,*O*,*O′*]cobalt, 10CI, 9CI

[Bis(salicylaldehyde)ethylenediimine]cobalt. N,N′-*Ethylenebis(salicylideneiminato)cobalt. Salcomine*

[14167-18-1]

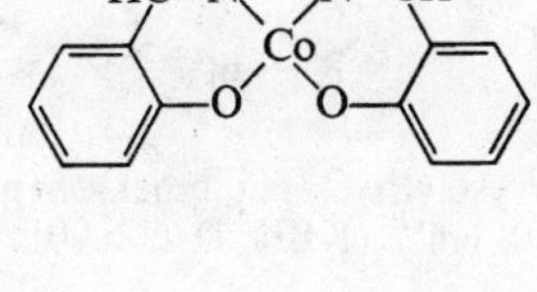

M 325.232

Autoxidn. catalyst. Red cryst. (DMF); maroon cryst. + C_6H_6 (C_6H_6). Ability to bind reversibly with dioxygen v. dependent upon crystalline form.

▷GG0590000.

Inorg. Synth., 1950, **3**, 196 (*synth*)
Busetto, C. *et al*, *J. Chem. Soc., Dalton Trans.*, 1973, 1712 (*esr, ir, uv*)
West, B.O. *et al*, *J. Organomet. Chem.*, 1974, **64**, 125 (*synth*)
McAllister, R.M. *et al*, *J. Organomet. Chem.*, 1974, **77**, 91 (*synth*)
Aymes, D. *et al*, *Bull. Soc. Chim. Fr.*, 1976, 1717 (*synth*)
Fieser, M. *et al*, *Reagents for Organic Synthesis*, Wiley, 1967-83, **7**, 316 (*use*)

$C_{16}H_{15}BCo^{\oplus}$ Co-00187

(η^5-2,4-Cyclopentadien-1-yl)[(1,2,3,4,5,6-η)-1-phenylboratabenzene]cobalt(1+), 10CI

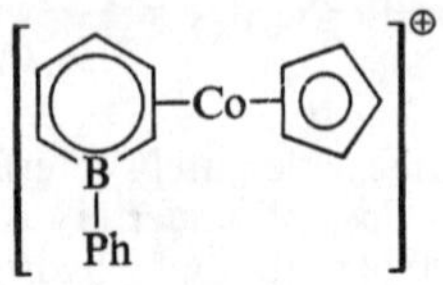

M 277.038 (ion)

Hexafluorophosphate: [51151-11-2].
$C_{16}H_{15}BCoF_6P$ M 422.002
Orange-red cryst. (Me_2CO aq.). Mp 139° dec. ^{11}B nmr δ −23.3 ($(CD_3)_2CO$, reference $BF_3{\cdot}Et_2O$). Reacts with nucleophiles to give stable adducts. Undergoes ring-contraction with amines, OH^- and CN^-.

Herberich, G.E. *et al*, *Chem. Ber.*, 1975, **108**, 3413 (*synth, ms*)
Herberich, G.E. *et al*, *Chem. Ber.*, 1979, **112**, 607 (*synth, nmr*)

$C_{16}H_{16}Co_2O_4$ Co-00188

Diacetyldi-μ-carbonylbis(η^5-2,4-cyclopentadien-1-yl)dicobalt, 10CI

[73465-73-3]

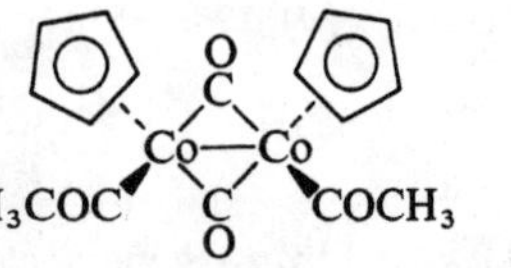

cis-f

M 390.166

Isol. as a mixt. of *cis* and *trans* isomers. Thermally unstable brown solid. Dec. in the presence of CO to give $Co(C_5H_5)(CO)_2$, Me_2CO and Et_2CO.

White, M.A. *et al*, *J. Chem. Soc., Chem. Commun.*, 1979, 1056 (*synth*)

$C_{16}H_{16}Co_2O_4$ Co-00189

Dicarbonylbis(η^5-2,4-cyclopentadien-1-yl)[μ-(2-ethoxy-2-oxoethylidene)]dicobalt, 10CI

μ-(Ethoxycarbonyl)methylenedicarbonylbis(η^5-cyclopentadienyl)dicobalt

[66484-10-4]

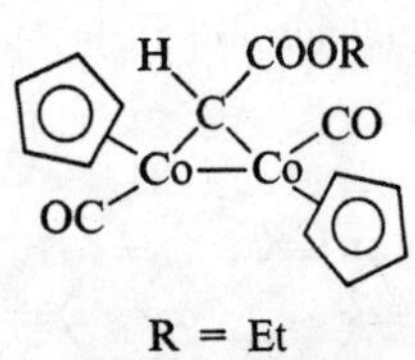

R = Et

M 390.166

Black-brown cryst. (Et_2O/pet. ether). Mp >250°. ir ν_{co} 1955vs, 1670s cm^{-1} (KBr). Dec. >90°.

Herrmann, W.A., *Chem. Ber.*, 1978, **111**, 1077 (*synth, ir, nmr*)
Herrmann, W.A. *et al*, *J. Organomet. Chem.*, 1979, **165**, C17 (*synth, struct*)
Mayer, K.K. *et al*, *J. Organomet. Chem.*, 1979, **182**, 361 (*ms*)

$C_{16}H_{16}Co_2O_4$ Co-00190

Di-μ-carbonylbis(η^5-2,4-cyclopentadien-1-yl)[μ-(2-ethoxy-2-oxyethylidene)]dicobalt, 10CI

[68586-88-9]

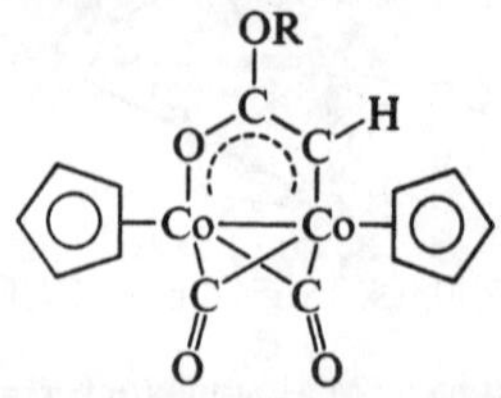

R = Et

M 390.166

Undergoes irreversible intramolecular rearrangement to Dicarbonylbis(η^5-2,4-cyclopentadien-1-yl)[μ-(2-ethoxy-2-oxoethylidene)]dicobalt, Co-00189 in soln. Green cryst. (CH_2Cl_2/Et_2O). Dec. >150° without melting. ir ν_{CO} 1843vs, 1582s, 1552vs, 1528s, cm^{-1} (KBr).

Herrmann, W.A., *Chem. Ber.*, 1978, **111**, 177 (*synth, ir, nmr*)
Herrmann, W.A. *et al*, *J. Organomet. Chem.*, 1979, **165**, C17 (*synth, struct*)
Mayer, K.K. *et al*, *J. Organomet. Chem.*, 1979, **182**, 361 (*ms*)

$C_{16}H_{16}Co_2O_4$ Co-00191

Di-μ-carbonyldicarbonylbis[(1,2,3,4-η)-1,3-cyclohexadiene]dicobalt, 9CI

Tetracarbonylbis[(1,2,3,4-η)-1,3-cyclohexadiene]dicobalt

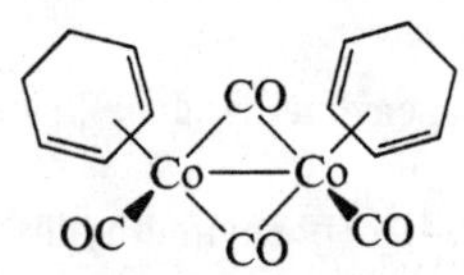

M 390.166

Exists in soln. predominantly as the *cis*-isomer. ir ν_{CO} 2012, 1984, 1822, 1777 cm^{-1} (KBr).

***cis*-form** [31798-56-8]
Golden-yellow plates.

McArdle, P. *et al*, *J. Chem. Soc. (A)*, 1970, 2123 (*synth, ir*)
Stephens, F.S., *J. Chem. Soc., Dalton Trans.*, 1972, 1752 (*cryst, struct*)

$C_{16}H_{16}Co_2O_4$ Co-00192

Tetracarbonylbis(1,4-cyclohexadiene)dicobalt, 9CI

[51913-83-8]

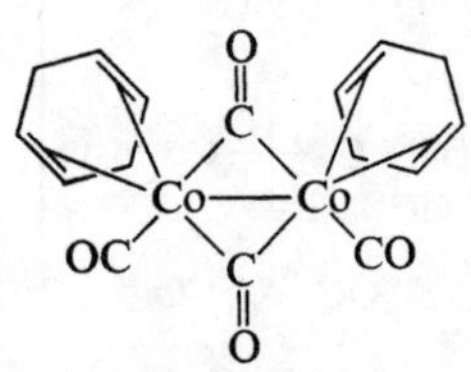

M 390.166

Orange-yellow cryst. (C_6H_6/hexane).

Kitamura, T. *et al*, *J. Organomet. Chem.*, 1974, **65**, 235 (*synth*)

$C_{16}H_{20}Co_2O_4$ **Co-00193**

Di-μ-carbonyldicarbonylbis[(1,2,3,4-η)-2,3-dimethyl-1,3-butadiene]dicobalt, 9CI

[33009-59-5]

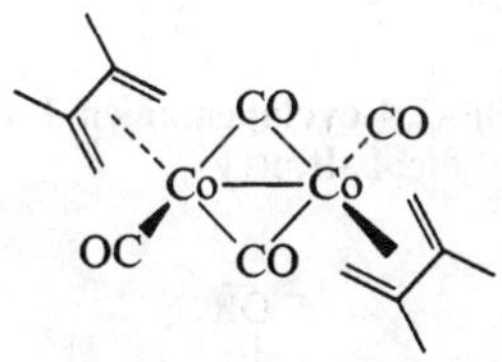

M 394.198

trans-form [31884-82-9]

Brown plates (C_6H_6/hexane) or red needles (Et_2O). ir ν_{CO} 2025sh, 2005vs, 1805s cm^{-1} (CCl_4).

Winkhaus, G. *et al*, *J. Chem. Soc.*, 1961, 602 (*synth, ir*)
Stephens, F.S., *J. Chem. Soc.* (*A*), 1970, 2745 (*cryst struct*)
Newman, J. *et al*, *J. Chem. Soc., Dalton Trans.*, 1973, 1593 (*synth*)

$C_{16}H_{21}Co^{\oplus\oplus}$ **Co-00194**

(η^6-Benzene)[(1,2,3,4,5-η)-1,2,3,4,5-pentamethyl-2,4-cyclopentadien-1-yl]cobalt(2+), 10CI

[72339-83-4]

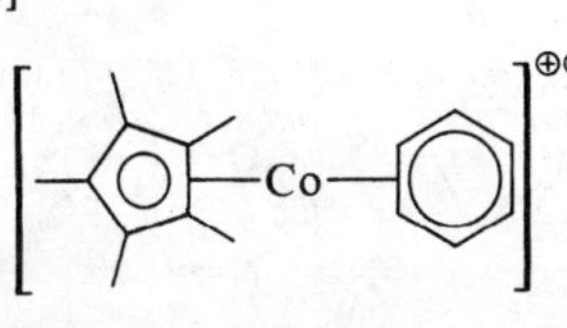

M 272.275 (ion)

Bis(tetrafluoroborate): [72339-84-5].
$C_{16}H_{21}B_2CoF_8$ M 445.882
Yellow cryst. ($MeNO_2/Et_2O$).

Fairhurst, G. *et al*, *J. Chem. Soc., Dalton Trans.*, 1979, 1531 (*synth, ir, nmr*)

$C_{16}H_{24}CoN_3^{\oplus\oplus}$ **Co-00195**

Tris(acetonitrile)[(1,2,3,4,5-η)-1,2,3,4,5-pentamethyl-2,4-cyclopentadien-1-yl]cobalt(2+)

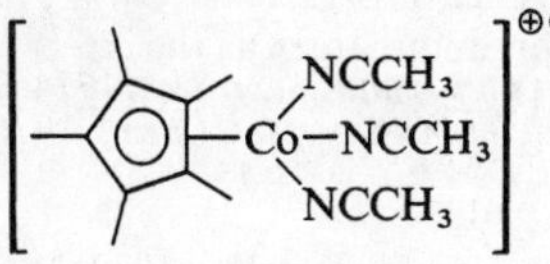

M 317.319 (ion)

Bis(hexafluorophosphate): [72339-57-2].
$C_{16}H_{22}CoF_{12}N_3P_2$ M 605.231
Dark-red cryst. (MeCN/Et_2O). ir ν_{CN} 2327, 2302 cm^{-1} (nujol). Acetonitrile readily displaced by other donor ligands.

Fairhurst, G. *et al*, *J. Chem. Soc., Dalton Trans.*, 1979, 1524 (*synth, ir, nmr*)

$C_{16}H_{25}BrCoN_4^{\oplus}$ **Co-00196**

Bromo(2,12-dimethyl-3,7,11,17-tetraazabicyclo[11.3.1]heptadeca-1(17),2,11,13,15-pentaene)methylcobalt(1+), 8CI

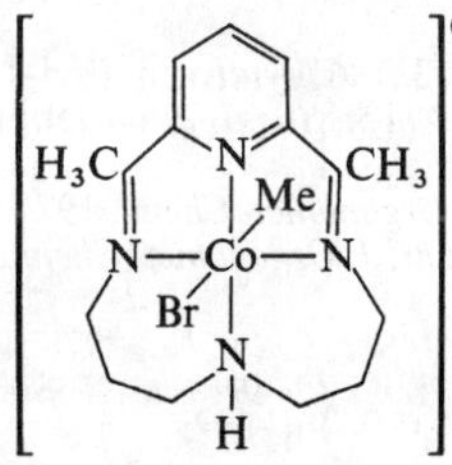

M 412.238 (ion)

Vitamin B_{12} analogue. Co–Me bond cleaved by light. Coordinated Br easily replaced by H_2O.

Hexafluorophosphate: [21707-19-7]. Red, light-sensitive cryst. (MeOH aq.).

Ochiai, E. *et al*, *J. Am. Chem. Soc.*, 1969, **91**, 3201 (*synth, pmr, uv*)

$C_{16}H_{25}Co$ **Co-00197**

[(1,2,5,6-η)-1,5-Cyclooctadiene][(1,2,3-η)-2-cycloocten-1-yl]cobalt

π-Cyclooctenyl-π-1,5-cyclooctadienecobalt

[34829-55-5]

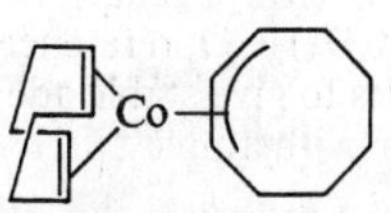

M 276.307

Olefin polymn. catalyst. Catalyses pyridine formation from 1 nitrile + 2 alkynes. Valuable reagent for the prepn. of low-oxidn.-state cobalt compds. Air-sensitive black cryst. (pentane). Sol. pentane, THF, aromatic solvs. Cryst. collapse at ca. 60-5° but do not melt <100°. Forms air-sensitive solns.

Otsuka, S. *et al*, *J. Chem. Soc.* (*A*), 1968, 2630 (*synth, ms, ir, pmr*)
Koda, S. *et al*, *Bull. Chem. Soc. Jpn.*, 1971, **44**, 653 (*cryst struct*)
Lehmkuhl, H. *et al*, *J. Organomet. Chem.*, 1971, **30**, 407 (*synth*)
Otsuka, S. *et al*, *J. Chem. Soc., Dalton Trans.*, 1972, 1879 (*props*)
Bönnemann, H. *et al*, *Synthesis*, 1974, 575 (*use*)
Inorg. Synth., 1977, **17**, 112 (*synth*)

$C_{16}H_{25}Co$ **Co-00198**

[(1,2,5,6-η)-1,5-Cyclooctadiene][(1,4,5-η)-4-cycloocten-1-yl]cobalt, 9CI

[53530-06-6]

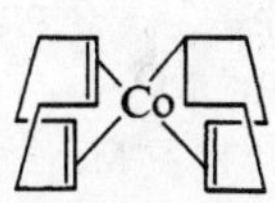

M 276.307

Exists in solution in equilibrium with $Co(COD)_2H$. Catalyses pyridine formation from 1 nitrile + 2 alkynes. Black lustrous cryst.

▷Ignites spontaneously in air

Grard, C., *Dissertation, Universität Bochum*, 1967 (*synth*)
Wilke, G., *Chem. Ind.* (*Kagaku Kogyo*), 1967, **20**, 1308
Bönnemann, H., *Angew. Chem., Int. Ed. Engl.*, 1973, **12**, 964 (*props*)
Bönnemann, H. *et al*, *Synthesis*, 1974, 575 (*use*)

$C_{16}H_{30}CoN_4^{\oplus}$ Co-00199

Dimethyl(2,3,9,10-tetramethyl-1,4,8,11-tetraazacyclotetradeca-1,3,8,10-tetraene-N^1,N^4,N^8,N^{11})cobalt(1+), 9CI

[47100-11-8]

As Bromomethyl(2,3,9,10-tetramethyl-1,4,8,11-tetraazacyclotetradeca-1,3,8,10-tetraene-N^1,N^4,N^8,N^{11})cobalt(1+), Co-00175 with

X = Y = Me

M 337.373 (ion)

Vitamin B_{12} analogue. Transfers a methyl group to electrophiles e.g. $H_3O^{\oplus}$, $PhHg^{\oplus}$.

Tetraphenylborate: [29190-43-0].
$C_{40}H_{50}BCoN_4$ M 656.605
Bright-red, air-stable, light-sensitive cryst. Sol. DMSO.

Perchlorate: [77310-45-3].
$C_{16}H_{30}ClCoN_4O_4$ M 436.824
Bright-red, air-stable, light-sensitive cryst. + $1H_2O$ (MeOH aq.). Sol. MeCN.

Farmery, K. *et al*, *J. Chem. Soc., Chem. Commun.*, 1970, 1091 (*synth, pmr*)
Espenson, J.H. *et al*, *Inorg. Chem.*, 1976, **15**, 906 (*props*)
Dimmit, J.H. *et al*, *Inorg. Chem.*, 1982, **21**, 700 (*synth, uv, pmr, props*)

$C_{16}H_{33}CoP_2$ Co-00200

[(1,2,3,4,5-η)-1,2,3,4,5-Pentamethyl-2,4-cyclopentadien-1-yl]bis(trimethylphosphine)cobalt

[84879-68-5]

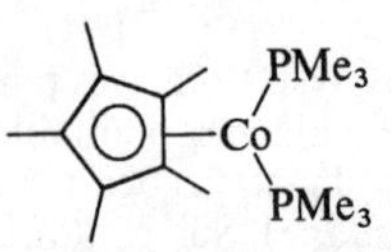

M 346.317

Dark-brown, air-sensitive cryst. (pentane).

Werner, H. *et al*, *J. Organomet. Chem.*, 1982, **240**, 179 (*synth, pmr, props*)

$C_{16}H_{36}Co_2O_4P_4$ Co-00201

Di-μ-carbonyldicarbonyltetrakis(trimethylphosphine)dicobalt, 10CI

Tetracarbonyltetrakis(trimethylphosphine)dicobalt

[25513-11-5]

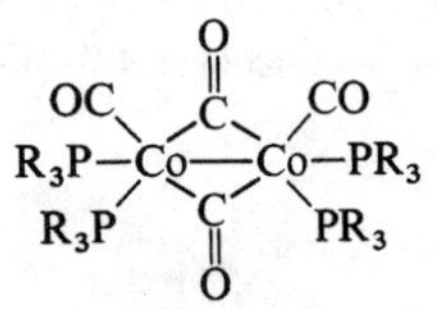

R = Me

M 534.219

Red cryst (pentane). Mp 152° dec. ir ν_{CO} 1938, 1917, 1735, 1719 cm^{-1} (nujol).

Klein, H-F. *et al*, *Chem. Ber.*, 1975, **108**, 944 (*synth, ir, nmr*)

$C_{17}H_{16}Co_2O_6$ Co-00202

Dicarbonylbis(η^5-2,4-cyclopentadien-1-yl)[μ-[2-methoxy-1-(methoxycarbonyl)-2-oxoethylidene]]dicobalt, 10CI

μ-Bis(methoxycarbonyl)methylenedicarbonylbis(η^5-cyclopentadienyl)dicobalt

[66559-34-0]

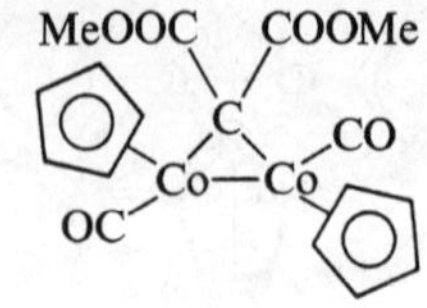

M 434.176

Dark-brown cryst. (Et_2O). Mp >250°. ir ν_{co} 1987 vs, 1678s cm^{-1} (KBr). Dec. >170°.

Herrmann, W.A., *Chem. Ber.*, 1978, **111**, 1077 (*ir, synth, pmr*)
Herrmann, W.A., *Chem. Ber.*, 1979, **112**, 1731.
Mayer, K.K. *et al*, *J. Organomet. Chem.*, 1979, **182**, 361 (*ms*)

$C_{17}H_{17}CoN_2O_2$ Co-00203

[[2,2′-[1,2-Ethanediylbis(nitrilomethylidyne)]bis[phenolato]](2−)-N,N′,O,O′]methylcobalt, 10CI

[[α,α′-(Ethylenedinitrilo)di-o-cresolato](2−)]methylcobalt, 8CI

[15053-54-0]

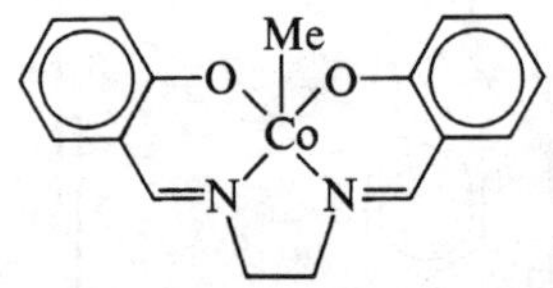

M 340.267

Obt. by dehydration of Aqua[[2,2′-[1,2-ethanediylbis(nitrilomethylidyne)]bis(phenolato)]-(2−)-*N,N′,O,O′*]methylcobalt, Co-00205 with P_2O_5. Model complex in studies of coenzyme B_{12}. Green solid. ^{59}Co nmr −7125 ppm ($CO(CN)_6^{3-}$ reference). Undergoes transmethylation with Hg and other Co complexes.

Floriani, C. *et al*, *J. Organomet. Chem.*, 1968, **12**, 209 (*synth, ir, nmr*)
McAllister, R.M. *et al*, *J. Organomet. Chem.*, 1973, **55**, C85 (*synth*)
La Rossa, R.A. *et al*, *J. Am. Chem. Soc.*, 1974, **96**, 2072 (*nmr*)
Mestroni, G. *et al*, *Transition. Met. Chem.*, 1976, **1**, 32.

$C_{17}H_{19}Co$ Co-00204

[(1,2,5,6-η)-1,5-Cyclooctadiene][(1,2,3,3a,7a-η)-1H-inden-1-yl]cobalt, 10CI

(π-Indenyl)(π-1,5-cyclooctadiene)cobalt

[84180-58-5]

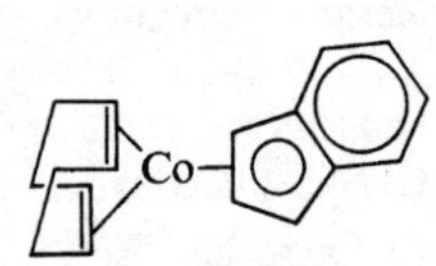

M 282.270

Catalyst for pyridine synth. from 2 alkynes + 1 nitrile.

Can. Pat., 1 130 289, (*1982*); *CA*, **98**, 72884a (*synth, use*)

$C_{17}H_{19}CoN_2O_3$ **Co-00205**

Aqua[[2,2′-[1,2-ethanediylbis(nitrilomethylidyne)]bis(phenolato)](2−)-*N,N′,O,O′*]methylcobalt

Aqua[[α,α′-(*ethylenedinitrilo*)*di*-o-*cresolato*](2−)]-*methylcobalt*

[43043-84-1]

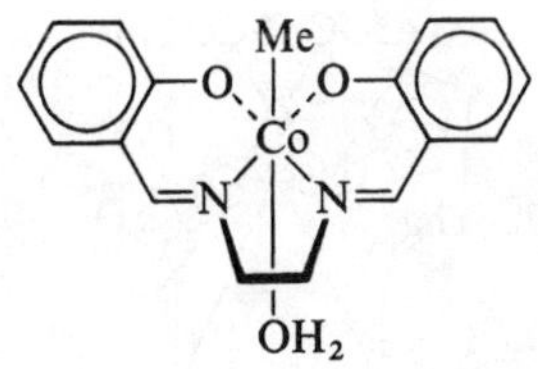

M 358.282

Model complex in studies of coenzyme B_{12}. Red-orange, air-stable, light-sensitive solid. Sol. Me_2CO, MeOH, sl. sol. THF.

Floriani, C. *et al*, *J. Organomet. Chem.*, 1968, **12**, 209 (*synth, ir, nmr*)
Reichert, B.E. *et al*, *J. Chem. Soc., Chem. Commun.*, 1974, 177 (*nmr*)

$C_{17}H_{23}Co^{\oplus\oplus}$ **Co-00206**

(η[5]-2,4-Cyclopentadien-1-yl)[(1,2,3,4,5,6-η)-hexamethylbenzene]cobalt(2+), 10CI

[72339-88-9]

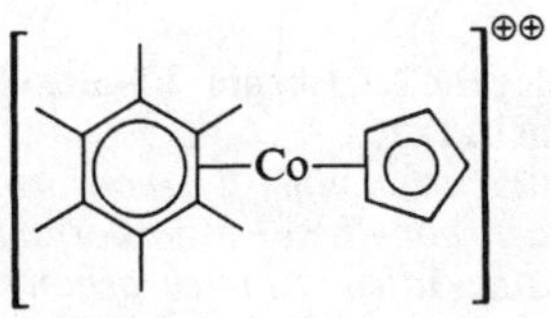

M 286.302 (ion)

Bis(tetrafluoroborate):

$C_{17}H_{23}B_2CoF_8$ M 459.909

Yellow cryst. ($MeNO_2/Et_2O$).

Fairhurst, G. *et al*, *J. Chem. Soc., Dalton Trans.*, 1979, 1531 (*synth, ir, nmr*)

$C_{17}H_{24}Co^{\oplus}$ **Co-00207**

[(1,2,3,4,5-η)-2,4-Cyclohexadien-1-yl-][(1,2,3,4,5-η)-1-ethyl-2,3,4,5-tetramethyl-2,4-cyclopentadien-1-yl]cobalt(1+), 10CI

[71251-96-2]

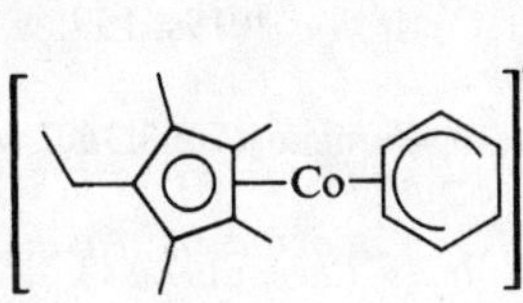

M 287.310 (ion)

Hexafluorophosphate:

$C_{17}H_{24}CoF_6P$ M 432.274

Orange cryst. (CH_2Cl_2/EtOH).

Green, M.L.H. *et al*, *J. Chem. Soc., Dalton Trans.*, 1979, 355 (*synth, nmr*)

$C_{17}H_{26}CoN_4O_3^{\oplus}$ **Co-00208**

Aquaphenyl[[3,3′-(1,3-propanediyldinitrilo)bis[2-butanone]-dioximato](1−)-*N,N′,N″,N‴*]cobalt(1+), 10CI

Aquaphenyl[(3,3′-*trimethylenedinitrilo*)*di-2-butanone dioximato*(1−)]*cobalt*(1+), 8CI. *Aquaphenyl*[1,3-*bis*(*diacetylmonoximeimino*)*propane*]*cobalt*(1+)

[26334-77-0]

As Aquamethyl[[3,3′-(1,3-propanediyldinitrilo)bis[2-butanone]dioximato](1−)-*N,N′,N″,N‴*]cobalt(1+), Co-00124 with

R = Ph

M 393.351 (ion)

Vitamin B_{12} analogue. Phenylates Hg^{2+}.

Perchlorate: [23868-71-5].

$C_{17}H_{26}ClCoN_4O_7$ M 492.801

Orange-yellow solid (Me_2CO aq.).

Hexafluorophosphate: [59692-13-6].

$C_{17}H_{26}CoF_6N_4O_3P$ M 538.315

Orange cryst. (MeOH aq.).

Costa, G. *et al*, *Inorg. Chim. Acta*, 1969, **3**, 323 (*synth, uv*)
Dodd, D. *et al*, *J. Organomet. Chem.*, 1973, **52**, 1 (*props, rev*)
Tauzher, G. *et al*, *J. Organomet. Chem.*, 1974, **81**, 107 (*props*)
Mestroni, G. *et al*, *Transition Met. Chem.*, 1976, **1**, 32 (*synth*)
Tauzher, G. *et al*, *Inorg. Chem.*, 1980, **19**, 3790 (*props*)

$C_{17}H_{28}CoN_4^{\oplus}$ **Co-00209**

(2,12-Dimethyl-3,7,11,17-tetraazabicyclo[11.3.1]heptadeca-1(17),2,11,13,15-pentaene-*N*[3],*N*[7],*N*[11],*N*[17])dimethylcobalt(1+), 9CI

[47176-21-6]

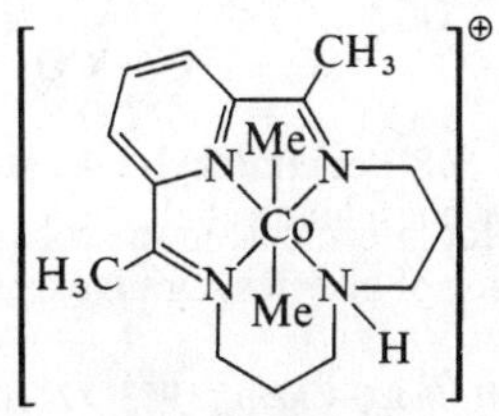

M 347.368 (ion)

Vitamin B_{12} analogue. Transfers a methyl group to electrophiles e.g. $H_3O^{\oplus}$ and $PhHg^{\oplus\oplus}$.

Tetraphenylborate: [29190-45-2].

$C_{41}H_{48}BCoN_4$ M 666.600

Orange-brown, light-sensitive, air-stable cryst. Sol. DMSO.

Farmery, K. *et al*, *J. Chem. Soc., Chem. Commun.*, 1970, 1091 (*synth, pmr*)
Espenson, J.H. *et al*, *Inorg. Chem.*, 1976, **15**, 906 (*props*)

$C_{17}H_{36}CoP_3$ **Co-00210**

(η[2]-Ethene)phenyltris(trimethylphosphine)cobalt, 10CI

(η[2]-*Ethylene*)*phenyltris*(*trimethylphosphine*)*cobalt*

[74965-07-4]

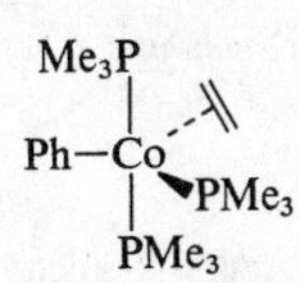

M 392.326

Fluxional in soln. Orange-yellow, air-sensitive cryst. (pentane). Stored under Ar at 20°. Mp >99° dec.

Klein, H.F. *et al*, *Angew. Chem., Int. Ed. Engl.*, 1980, **19**, 809 (*synth, cryst struct*)
Klein, H.F. *et al*, *Chem. Ber.*, 1983, **116**, 1441 (*synth, nmr, ir*)

$C_{17}H_{39}CoO_9P_3^{\oplus}$ **Co-00211**

[(1,2,3,4-η)-1,3-Cyclooctadiene]tris(trimethyl phosphite-*P*)-cobalt(1+), 10CI

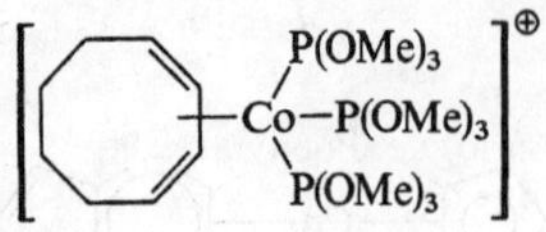

M 539.344 (ion)

Hexafluorophosphate: [68644-07-5].
$C_{17}H_{39}CoF_6O_9P_4$ M 684.308
Yellow solid. Mp 198-200° dec.

Muetterties, E.L. *et al*, *J. Am. Chem. Soc.*, 1978, **100**, 6978 (*synth, pmr*)

$C_{18}H_7BrCo_2O_8$ **Co-00212**

[μ-(Bromophenylethenylidene)]hexacarbonyl[μ-(5-oxo-2(5*H*)-furanylidene)]dicobalt, 10CI

(μ_2-2-Buten-4-olide-4-ylidene)(μ_2-1′,1′-bromophenylethene-2-ylidene)hexacarbonyldicobalt

[73363-31-2]

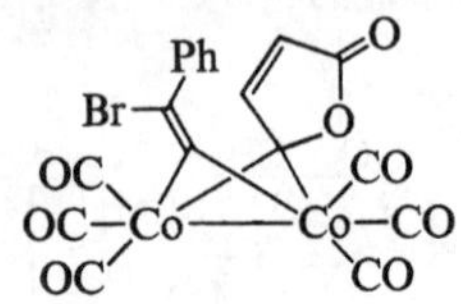

M 549.019

Model compd. for intermeds. in the cobalt-catalysed oligomerisation of alkynes. Orange cryst. (hexane/Et_2O).

Horvath, I.T. *et al*, *Inorg. Chem.*, 1983, **22**, 1049 (*synth, ir, ms, pmr*)

$C_{18}H_{10}Co_2FeO_6$ **Co-00213**

Hexacarbonyl[μ-[[(1,2-η:1,2-η)-ethynyl]ferrocene]]dicobalt, 9CI

[56665-21-5]

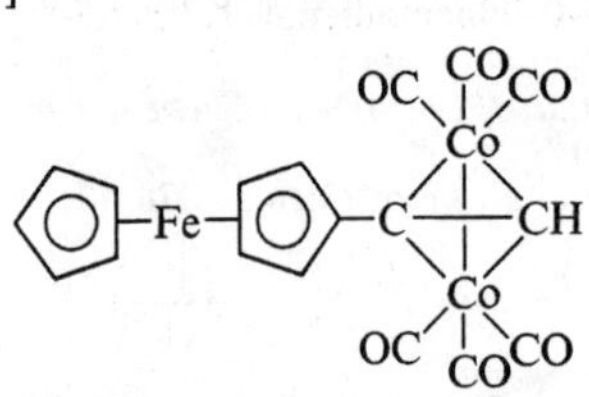

M 495.987

Dark-green air-stable cryst. (MeOH under N_2). Mp 64-6°.

Pittman, C.U. *et al*, *J. Organomet. Chem.*, 1975, **90**, 203 (*synth, ms, ir, pmr*)

$C_{18}H_{14}Co$ **Co-00214**

Bis[(1,2,3,3*a*,7*a*-η)-1*H*-inden-1-yl]cobalt, 10CI

Di-π-indenylcobalt, 8CI

[33056-04-1]

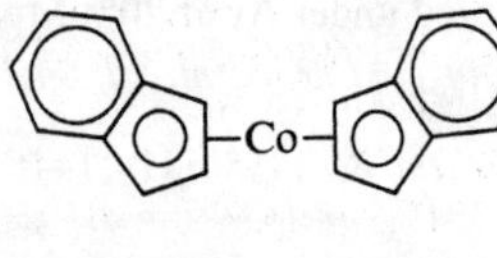

M 289.242

Air-sensitive black-brown cryst.

Fritz, H.P. *et al*, *J. Organomet. Chem.*, 1969, **16**, P14 (*synth, pmr*)

$C_{18}H_{15}Co_3O_3$ **Co-00215**

Tri-μ-carbonyltris(η^2-2,4-cyclopentadien-1-yl)tricobalt, 9CI

Tricarbonyltri-π-cyclopentadienyltricobalt, 8CI. Tris(η^5-cyclopentadienyl)tricarbonyltricobalt. Tris(carbonylcyclopentadienylcobalt)

[53450-15-0]

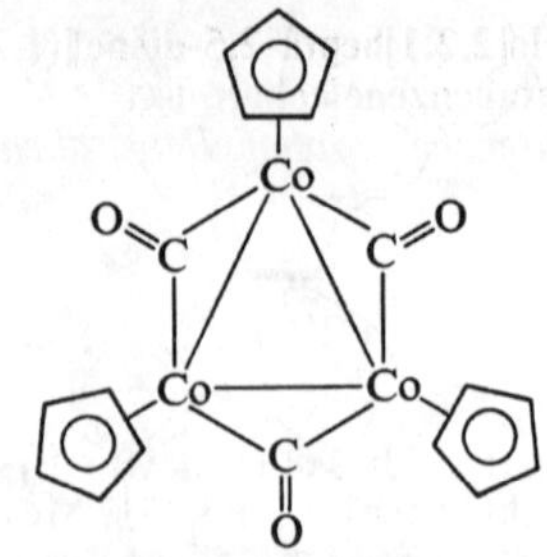

M 456.114

Catalyses the trimerisation of acetylenes. Black cryst. (Et_2O). ir ν_{CO} 1675 cm^{-1} (KBr).

King, R.B., *Inorg. Chem.*, 1966, **5**, 2227 (*synth, ir, ms, pmr*)
Vollhardt, K.P.C. *et al*, *J. Organomet. Chem.*, 1975, **97**, 283 (*synth, ir, ms, pmr*)
Cotton, F.A. *et al*, *J. Am. Chem. Soc.*, 1976, **98**, 1273 (*synth, ir, cryst struct*)
Lee, W.-S. *et al*, *J. Organomet. Chem.*, 1977, **127**, 87 (*synth*)

$C_{18}H_{16}Co_2O_4$ **Co-00216**

Bis[(2,3,5,6-η)-bicyclo[2.2.1]hepta-2,5-diene]di-μ-carbonyldicarbonyldicobalt, 10CI

Di-μ-carbonyldicarbonylbis(2,5-norbornadiene)dicobalt, 8CI. Tetracarbonylbis(2,5-norbornadiene)dicobalt. Bis(norbornadiene)dicobalt tetracarbonyl. Bis(norbornadiene)di-μ-carbonyldicarbonyldicobalt

[50404-93-8]

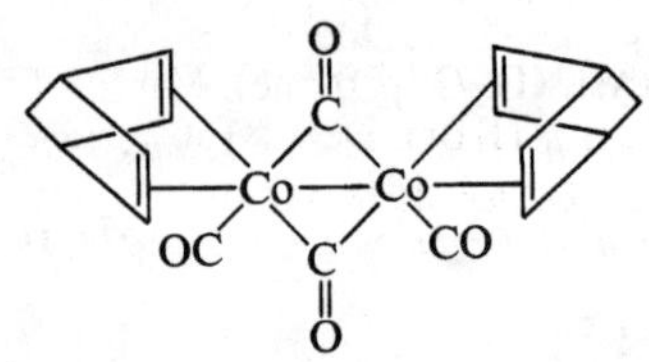

cis-form

M 414.188

Exists as *cis*-form in soln. and in solid.

cis-form [31666-51-0]

Red needles (Et_2O). ir ν_{CO} 2019s, 1995w, 1821w, 1804m cm^{-1} (heptane).

Winkhaus, G. *et al*, *J. Chem. Soc.*, 1961, 602 (*synth, ir*)
McArdle, P. *et al*, *J. Chem. Soc. (A)*, 1970, 2123 (*synth, ir, struct*)
Cullen, W.R. *et al*, *Inorg. Chem.*, 1975, **14**, 2975 (*nqr*)
Swaminathan, S. *et al*, *Cryst. Struct. Commun.*, 1978, **7**, 621 (*cryst struct*)

$C_{18}H_{16}Co_2O_4$ **Co-00217**

[μ-[(1,2,3,4,5-η:1′,2′,3′,4′,5′-η)-1,4-Butanediyldi-2,4-cyclopentadien-1-ylidene]]tetracarbonyldicobalt, 10CI

[73409-81-1]

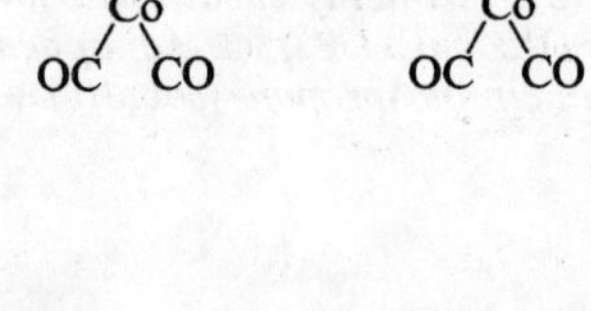

M 414.188

Dark reddish-brown oil. ir ν_{CO} 2010, 1945 cm^{-1} (neat).

Eilbracht, P. *et al*, *J. Organomet. Chem.*, 1980, **185**, C25 (*synth, ir, pmr*)

$C_{18}H_{18}BCo$ Co-00218

[(2,3,5,6-η)Bicyclo[2.2.1]hepta-2,5-diene][(1,2,3,4,5,6-η-)-1-phenylboratabenzene]cobalt, 10CI

η⁴-Norbornadiene(η⁶-1-phenylboratabenzene)cobalt

[69021-84-7]

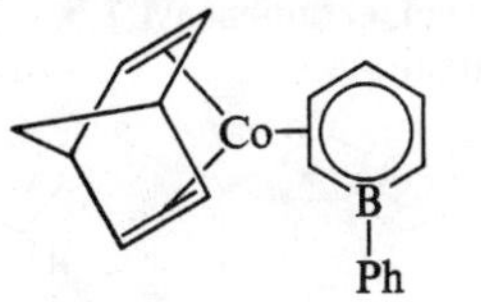

M 304.083

Red-brown cryst. Mp 189-90°. ^{11}B nmr δ +21.2 (C_6D_6, reference BF_3Et_2O).

Herberich, G.E. *et al*, *J. Organomet. Chem.*, 1978, **160**, 17 (*synth, ms, nmr*)

$C_{18}H_{20}Co_2O_4$ Co-00219

Dicarbonylbis(η^5-2,4-cyclopentadien-1-yl)[μ-[2-(1,1-dimethylethoxy)-2-oxoethylidene]]dicobalt, 10CI

μ-(tert-Butoxycarbonyl)methylenedicarbonylbis(η^5-cyclopentadienyl)dicobalt

[66484-12-6]

As Dicarbonylbis(η^5-2,4-cyclopentadien-1-yl)[μ-(2-ethoxy-2-oxoethylidene)]dicobalt, Co-00189 with

R = $(H_3C)_3C$–

M 418.220

V. dark-brown cryst. (Et_2O/pentane). Mp >250°. ir ν_{co} 1987 vs, 1672 cm^{-1} (KBr). Dec. >150°.

Herrmann, W.A., *Chem. Ber.*, 1978, **111**, 1077 (*synth, ir, pmr*)
Herrmann, W.A. *et al*, *J. Organomet. Chem.*, 1979, **165**, C17 (*synth*)
Mayer, K.K. *et al*, *J. Organomet. Chem.*, 1979, **182**, 361 (*ms*)

$C_{18}H_{20}Co_2O_4$ Co-00220

Di-μ-carbonylbis(η^5-2,4-cyclopentadien-1-yl)[μ-[2-(1,1-dimethylethoxy)-2-oxoethylidene]]dicobalt, 10CI

[68586-89-0]

As Di-μ-carbonylbis(η^5-2,4-cyclopentadien-1-yl-)[μ-(2-ethoxy-2-oxyethylidene)]dicobalt, Co-00190 with

R = $(H_3C)_3C$–

M 418.220

Undergoes irreversible intramolecular rearrangement to Dicarbonylbis(η^5-2,4-cyclopentadien-1-yl)[μ-[2-(1,1-dimethylethoxy)-2-oxoethylidene]]dicobalt, Co-00219 in soln. Green cryst. (CH_2Cl_2/Et_2O). Dec. >150° without melting. ir ν_{CO} 1831vs, 1588vs, 1561s, 1525s cm^{-1} (KBr).

Herrmann, W.A., *Chem. Ber.*, 1978, **111**, 1077 (*synth, ir, pmr*)
Herrmann, W.A. *et al*, *J. Organomet. Chem.*, 1979, **165**, C17 (*synth*)
Mayer, K.K. *et al*, *J. Organomet. Chem.*, 1979, **182**, 361 (*ms*)

$C_{18}H_{22}CoN^{\oplus\oplus}$ Co-00221

[(3-a,4,5,6,7,7a-η)-1H-indole][(1,2,3,4,5-η)-1,2,3,4,5-pentamethyl-2,4-cyclopentadien-1-yl]cobalt(2+), 10CI

[72339-81-2]

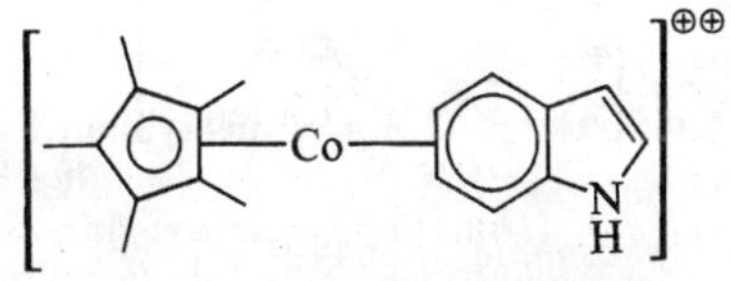

M 311.312 (ion)

Bis(hexafluorophosphate): [72339-82-3].
$C_{18}H_{22}CoF_{12}NP_2$ M 601.240
Orange cryst. (Me_2CO/Et_2O).

Fairhurst, G. *et al*, *J. Chem. Soc., Dalton Trans.*, 1979, 1531 (*synth, ir, nmr*)

$C_{18}H_{22}Co_2$ Co-00222

Bis(η^5-2,4-cyclopentadien-1-yl)[μ-[(1,2,3,4-η:1,4-η)-1,2,3,4-tetramethyl-1,3-butadiene-1,4-diyl]]dicobalt, 10CI

[63150-97-0]

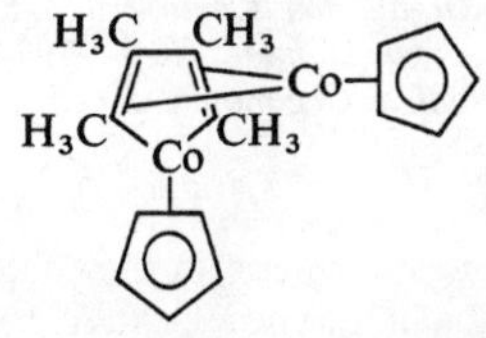

M 356.238

Catalyst for cyclotrimerisation of alkynes. Dark-green solid by subl.

Lee, W.-S. *et al*, *J. Organomet. Chem.*, 1977, **127**, 93 (*synth, ir, pmr, use*)

$C_{18}H_{27}Co$ Co-00223

[(1,2,5,6-η)-1,5-Cyclooctadiene][(1,2,3,4,5-η)-1,2,3,4,5-pentamethyl-2,4-cyclopentadien-1-yl]cobalt

[82595-75-3]

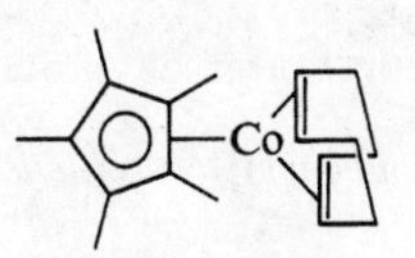

M 302.345

Solid. Sol. C_6H_6. Mp 132°. Undergoes reversible one electron oxidation at $E°$ = +0.07V (CH_2Cl_2).

Kölle, U. *et al*, *Angew. Chem., Int. Ed. Engl.*, 1982, **21**, 131; *Angew. Chem. Suppl.*, 1982, 230 (*synth, pmr*)
Moraczewski, J. *et al*, *Organometallics*, 1982, **1**, 1385.

$C_{18}H_{27}CoN_4O_2$ Co-00224

Methylphenyl[[3,3′-(1,3-propanediyldinitrilo)bis[2-butanone]-dioximato](1−)-N,N′,N″,N‴]cobalt, 9CI

Methylphenyl[[3,3′-(trimethylenedinitrilo)di-2-butanone dioximato](1−)]cobalt, 8CI

[36542-16-2]

As Dimethyl[3,3′-(1,3-propanediyldinitrilo)bis[2-butanone]dioximato(1−)-N,N′,N″,N‴]cobalt, Co-00142 with

R = Ph

M 390.370

Orange, air-stable cryst. (Me_2CO aq.).

Costa, G. *et al*, *J. Chem. Soc.*, *Dalton Trans.*, 1972, 450 (*synth*)
Dreos, R. *et al*, *J. Organomet. Chem.*, 1975, **92**, 227 (*props*)

$C_{18}H_{28}CoN_4O_3^{\oplus}$ Co-00225

Aqua(phenylmethyl)[[3,3′-(1,3-propanediyldinitrilo)bis[2-butanone]dioximato](1−)-*N,N′,N″,N‴*]cobalt(1+), 10CI, 9CI

Aquabenzyl[3,3′-(trimethylenedinitrilo)di-2-butanone dioximato(1−)]cobalt(1+), 8CI. *Aquabenzyl[1,3-bis(diacetylmonoximeimino)propane]cobalt(1+)*

[26317-24-8]

As Aquamethyl[[3,3′-(1,3-propanediyldinitrilo)bis[2-butanone]dioximato](1−)-*N,N′,N″,N‴*]cobalt(1+), Co-00124 with

R = CH_2Ph

M 407.377 (ion)

Vitamin B_{12} analogue.

Perchlorate: [23940-48-9].
$C_{18}H_{28}ClCoN_4O_7$ M 506.828
Orange-red cryst. (Me_2CO aq. or H_2O).

Costa, G. *et al*, *Inorg. Chim. Acta*, 1969, **3**, 323 (*synth, uv*)
Dodd, D. *et al*, *J. Organomet. Chem.*, 1973, **52**, 1 (*props, rev*)
Tauzher, G. *et al*, *J. Organomet. Chem.*, 1974, **81**, 107 (*props*)
Magnuson, V.E. *et al*, *J. Organomet. Chem.*, 1975, **92**, 233 (*pmr, uv, props*)

$C_{18}H_{41}CoO_{12}P_4$ Co-00226

Phenyltetrakis(trimethyl phosphite-*P*)cobalt, 9CI

[54204-76-1]

$PhCo[P(OMe)_3]_4$

M 632.343

Yellow cryst. (pentane).

Muetterties, E.L. *et al*, *J. Am. Chem. Soc.*, 1974, **96**, 7920 (*synth, nmr*)

$C_{19}H_8CoF_{10}$ Co-00227

[(1,2,3,4,5,6-η)-Methylbenzene]bis(pentafluorophenyl)cobalt, 10CI

[Bis(pentafluorophenyl)](η⁶-toluene)cobalt

[60528-58-7]

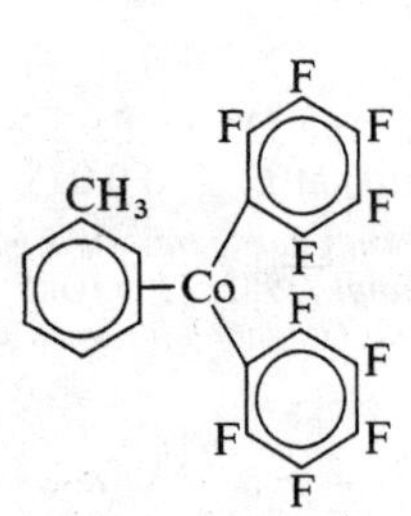

M 485.189

Dark orange-red, air-sensitive cryst. Sol. toluene. Mp 133-8°. Darkens at 92°. Paramagnetic.

Anderson, B.B. *et al*, *J. Am. Chem. Soc.*, 1976, **98**, 5390 (*synth, cryst struct, ir*)
Martin, W.J. *et al*, *CA*, 1980, **92**, 42095q (*props*)
Radonovich, L.J. *et al*, *Inorg. Chem.*, 1980, **19**, 1221 (*cryst struct*)

$C_{19}H_{15}CoF_3N_2O_5P$ Co-00228

Dinitrosyl(trifluoromethyl)(triphenyl phosphite)cobalt

$(PhO)_3PCo(NO)_2CF_3$

M 498.240

Red-brown cryst. (C_6H_6/pet. ether). Mp 58°. ir ν_{NO} 1847, 1789 cm^{-1} (CCl_4).

Beck, W. *et al*, *Chem. Ber.*, 1965, **98**, 2657 (*ir*)
Hieber, W. *et al*, *Chem. Ber.*, 1965, **98**, 3924 (*synth*)

$C_{19}H_{22}BCo$ Co-00229

[(1,2,5,6-η)-1,5-Cyclooctadiene][(1,2,3,4,5,6-η)-1-phenylboratabenzene]cobalt, 10CI

[69021-83-6]

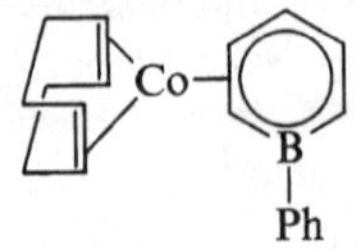

M 320.126

Yellow-brown cryst. Mp 168-9°. ^{11}B nmr δ +20.9 (C_6D_6, reference $BF_3{\cdot}Et_2O$).

Herberich, G.E. *et al*, *J. Organomet. Chem.*, 1978, **160**, 17 (*synth, ms, nmr*)

$C_{19}H_{24}CoN_5O_4$ Co-00230

Bis[(2,3-butanedione dioximato)(1−)-*N,N′*]phenyl(pyridine)cobalt, 10CI, 9CI

Phenyl(pyridine)bis(dimethylglyoximato)cobalt. Phenyl(pyridine)cobaloxime

[29130-85-6]

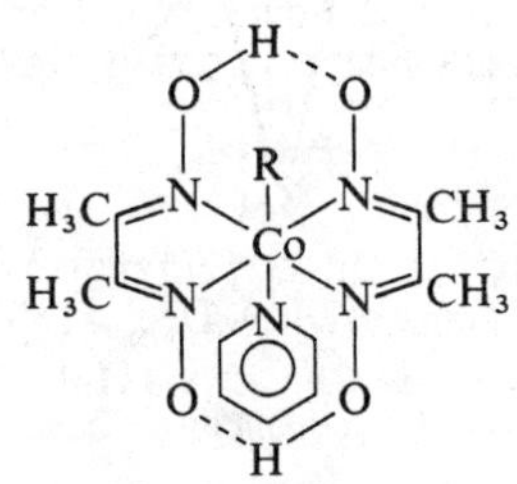

R = Ph

M 445.363

Analogue of vitamin B_{12}. Orange, air-stable solid (MeOH or 1,1,1-trifluoroethanol). Spar. sol. most org. solvs. Mp 276° dec.

Schrauzer, G.N. *et al*, *Chem. Ber.*, 1964, **97**, 3056 (*synth*)
Inorg. Synth., 1968, **11**, 61 (*synth*)
Schrauzer, G.N. *et al*, *J. Am. Chem. Soc.*, 1970, **92**, 2997 (*uv*)
Palmer, J.M. *et al*, *Inorg. Chem.*, 1975, **14**, 17 (*pmr*)

$C_{19}H_{29}Co$ Co-00231

[(1,2,5,6-η)-1,5-Cyclooctadiene][(1,2,3,4,5-η)-1-ethyl-2,3,4,5-tetramethyl-2,4-cyclopentadien-1-yl]cobalt, 10CI

[71251-98-4]

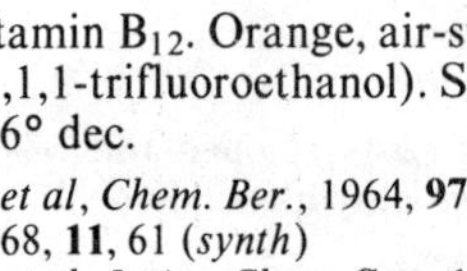

M 316.371

Waxy orange solid (CH_2Cl_2/EtOH). Mp 50°. $Bp_{0.05}$ 125°.

Green, M.L.H. *et al*, *J. Chem. Soc.*, *Dalton Trans.*, 1979, 355 (*synth, nmr*)

$C_{19}H_{43}CoO_2P_2$ **Co-00232**

Dimethyl(2,4-pentanedionato-*O-O'*)bis(triethylphosphine)-cobalt, 10CI

(Acetylacetonato)dimethylbis(triethylphosphine)cobalt

[58658-90-5]

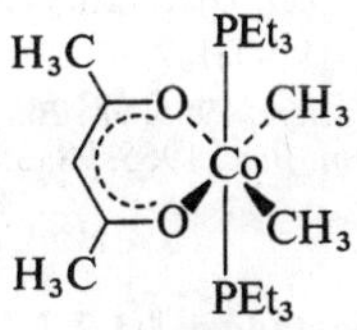

M 424.428

Orange-yellow cryst. (Et_2O). Sol. most org. solvs. Mp 78-81°. Stable to air and moisture at r.t.

Ikariya, T. *et al*, *J. Organomet. Chem.*, 1976, **118**, 101 (*synth, nmr*)

$C_{20}Co_4H_{24}$ **Co-00233**

Tetrakis[(η^5-2,4-cyclopentadien-1-yl)tetra-μ_3-hydrotetracobalt, 10CI

Tetrakis[(μ_3-hydrido)η^5-cyclopentadienylcobalt]

[51831-76-6]

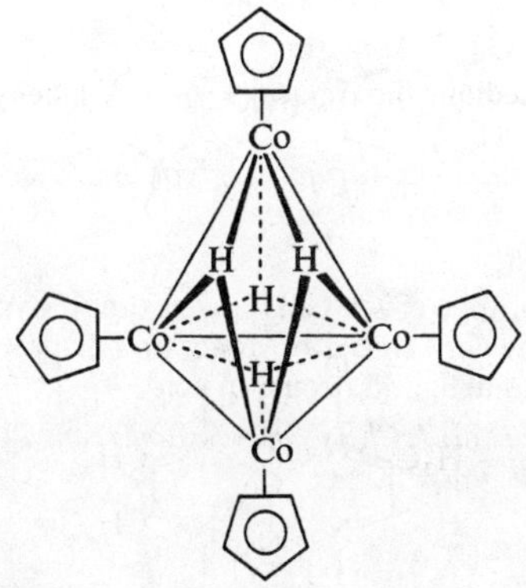

M 500.142

Air-stable dark-violet shiny cryst. (pentane/THF). Sol. pentane, C_6H_6, Et_2O. Mp >300°. Air-sensitive in soln.

Müller, J. *et al*, *Angew. Chem., Int. Ed. Engl.*, 1973, **12**, 843 (*synth, ir, pmr, struct*)

Andrews, J.A. *et al*, *Inorg. Chem.*, 1980, **19**, 3033 (*ir*)

$C_{20}H_9Co_3FeO_9$ **Co-00234**

Nonacarbonyl[μ_3-(ferrocenylmethylidyne)]tricobalt

[52215-96-0]

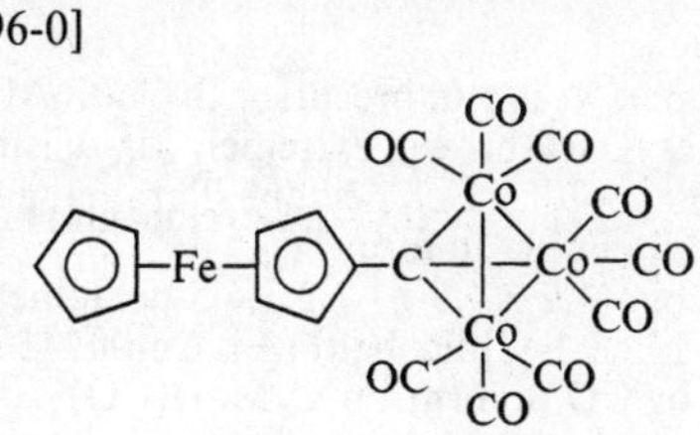

M 625.932

Black cryst. (hexane). Mp 75-7°.

Seyferth, D. *et al*, *J. Organomet. Chem.*, 1974, **65**, 99 (*synth, ir, pmr*)

Seyferth, D. *et al*, *J. Am. Chem. Soc.*, 1976, **98**, 6724; 1981, **103**, 3320.

$C_{20}H_{10}ClCo_3HfO_{10}$ **Co-00235**

Chlorobis(η^5-2,4-cyclopentadien-1-yl)[μ_4-[methanolato(4−)-*C:C:C:O*]](nonacarbonyltricobalt)dihafnium, 10CI

Chlorobis(η^5-cyclopentadienyl)[μ_3-(oxymethylidyne)-cyclotris(tricarbonylcobalt)hafnium

[67422-56-4]

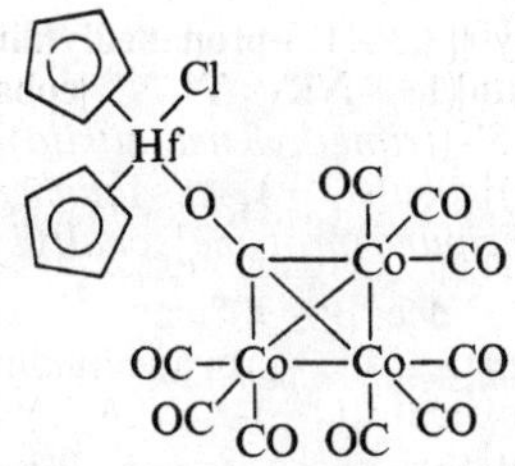

M 801.036

Dark-red cryst. (toluene). Spar. sol. aromatic, v. spar. sol. aliphatic solvs. Mp 139-42°.

Stutte, B. *et al*, *Chem. Ber.*, 1978, **111**, 1603 (*synth, ir, nmr, cryst struct*)

Ishii, M. *et al*, *Ber. Bunsenges Phys. Chem.*, 1979, **83**, 1026 (*synth, ir, nmr*)

$C_{20}H_{10}ClCo_3O_{10}Zr$ **Co-00236**

Chlorobis(η^5-2,4-cyclopentadien-1-yl)[μ_4-[methanolato(4−)-*C:C:C:O*]](nonacarbonyltricobalt)zirconium, 10CI

Chlorobis(η^5-cyclopentadienyl)[μ_3-(oxymethylidyne)-cyclotris(tricarbonylcobalt)]zirconium

[67422-57-5]

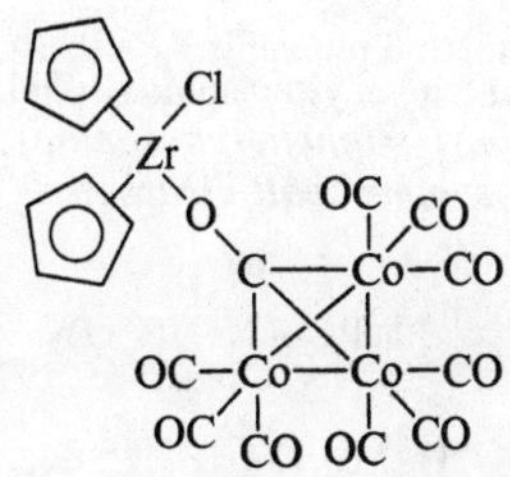

M 713.766

Dark-red cryst. (toluene). Spar. sol. aromatic, v. spar. sol. aliphatic solvs. Mp 138-41°.

Stutte, B. *et al*, *Chem. Ber.*, 1978, **111**, 1603 (*synth,ir, nmr, cryst struct*)

Ishii, M. *et al*, *Ber. Bunsenges. Phys. Chem.*, 1979, **83**, 1026 (*synth, ir, nmr*)

$C_{20}H_{10}Co_2MoO_8$ **Co-00237**

Dicarbonyl(η^5-2,4-cyclopentadien-1-yl)(hexacarbonyldicobalt)[μ_3-(phenylmethylidyne)]molybdenum, 10CI

[68185-45-5]

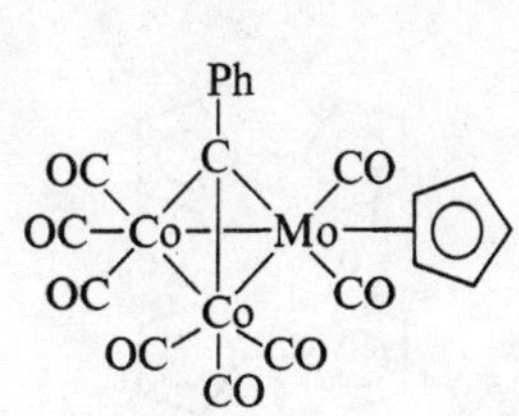

M 592.101

Black cryst. Mp 165° dec. pmr δ 5.22 (s, C_5H_5) 7.21 (m, C_6H_5-)ppm (CCl_4).

Beurich, H., *Angew. Chem., Int. Ed. Engl.*, 1978, **17**, 863 (*synth, ir, pmr*)

$C_{20}H_{10}Co_2O_6$ **Co-00238**

Hexacarbonyl(diphenylacetylene)dicobalt

Hexacarbonyl[μ-[1,1′-(η²:η²-1,2-ethynediyl)bis[benzene]]]dicobalt, 10CI

[14515-69-6]

As [μ-[(2,3-η:2,3-η)-2-Butyne]]hexacarbonyldicobalt, Co-00060 with

R = Ph

M 464.162

CO ligands fluxional at r.t. Homogeneous catalyst for hydrogenation of CO to alcohols. Dark-purple cryst. Mp 109.5-110° dec. Bp_1 90° subl.

Greenfield, H. *et al*, *J. Am. Chem. Soc.*, 1956, **78**, 120 (*synth*, *ir*)
Gambino, O. *et al*, *Org. Mass Spectrom.*, 1972, **6**, 723 (*ms*)
Aime, S. *et al*, *Inorg. Chim. Acta*, 1977, **22**, 135 (*pmr*, *cmr*, *struct*)
King, R.B. *et al*, *J. Mol. Catal.*, 1980, **10**, 75 (*use*)

$C_{20}H_{10}Co_2O_8Pb$ **Co-00239**

Bis(tetracarbonylcobalt)diphenyllead

Octacarbonyl(diphenylplumbylene)dicobalt

$Ph_2Pb[Co(CO)_4]_2$

M 703.361

Red cryst. Dec. at 103-6°.

Hein, F. *et al*, *Justus Liebigs Ann. Chem.*, 1965, **684**, 4.

$C_{20}H_{15}CoNO_3P$ **Co-00240**

Dicarbonylnitrosyl(triphenylphosphine)cobalt, 9CI

Nitrosyldicarbonyl(triphenylphosphine)cobalt. Nitrosyltriphenylphosphinecobalt dicarbonyl

[13681-96-4]

$Ph_3PCo(NO)(CO)_2$

M 407.250

Red cryst. (MeOH). Mp 134-135.5° dec. ir ν_{CO} 2035, 1981, ν_{NO} 1756 cm^{-1} (CCl_4).

Heck, R.F. *et al*, *J. Am. Chem. Soc.*, 1963, **85**, 657 (*synth*, *ir*)
Hieber, W. *et al*, *Chem. Ber.*, 1963, **96**, 1643 (*synth*, *ir*)
Foffani, A. *et al*, *Spectrochim. Acta, Part A*, 1968, **24**, 1437 (*ir*)
Albano, V.G. *et al*, *J. Organomet. Chem.*, 1972, **38**, 155 (*cryst struct*)
Ward, D.L. *et al*, *Acta Crystallogr., Sect. B*, 1972, **29**, 1949 (*cryst struct*)

$C_{20}H_{16}Co_2$ **Co-00241**

1,1″:1′,1‴-Bicobaltocene, 10CI

Bis(fulvalene)dicobalt

[69365-57-7]

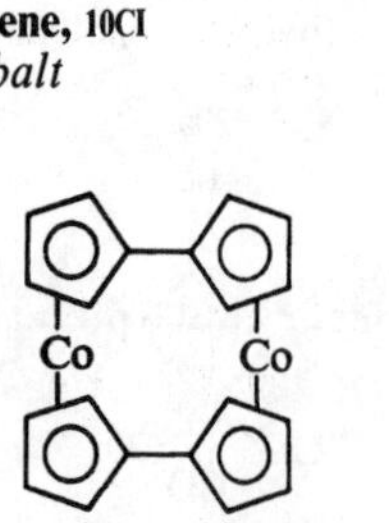

M 374.213

Dark-red cryst. by subl. Air-sensitive.

Davison, A. *et al*, *J. Organomet. Chem.*, 1973, **49**, C43 (*synth*)
Bakke, A.A. *et al*, *Inorg. Chem.*, 1979, **18**, 1343.

$C_{20}H_{26}CoN_3^{\oplus\oplus}$ **Co-00242**

[(1,2,3,4,5-η)-1-Ethyl-2,3,4,5-tetramethyl-2,4-cyclopentadien-1-yl]tris(2-propenenitrile)cobalt(2+)

Tris(acrylonitrile)(η-ethyltetramethylcyclopentadienyl)cobalt(2+)

[71280-84-7]

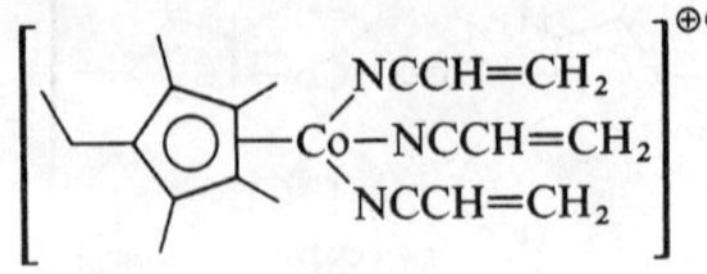

M 367.379 (ion)

Red cryst. (acrylonitrile/EtOH).

Green, M.L.H. *et al*, *J. Chem. Soc., Dalton Trans.*, 1979, 355 (*synth*, *pmr*)

$C_{20}H_{27}CoO_2$ **Co-00243**

[(1,2,3,4,5-η)-1,2,3,4,5-Pentamethyl-2,4-cyclopentadien-1-yl][(2,3,5,6-η)-2,3,5,6-tetramethyl-2,5-cyclohexadiene-1,4-dione]cobalt, 10CI

(η⁴-Duroquinone)(η⁵-pentamethylcyclopentadienyl)cobalt

[68706-57-0]

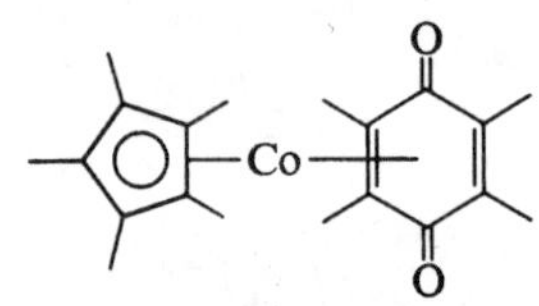

M 358.365

Hydrated orange cryst. ($CHCl_3$/hexane), hygroscopic brown powder (anhyd.). Mono- and di-protonation of duroquinone ligand occurs in acid.

Fairhurst, G. *et al*, *J. Chem. Soc., Dalton Trans.*, 1979, 1531 (*synth*, *nmr*, *ir*)

$C_{20}H_{30}Br_2Co_2$ **Co-00244**

Di-μ-bromobis[(1,2,3,4,5-η)-1,2,3,4,5-pentamethyl-2,4-cyclopentadien-1-yl]dicobalt

Bis[bromo(pentamethylcyclopentadienyl)cobalt]

[82595-79-7]

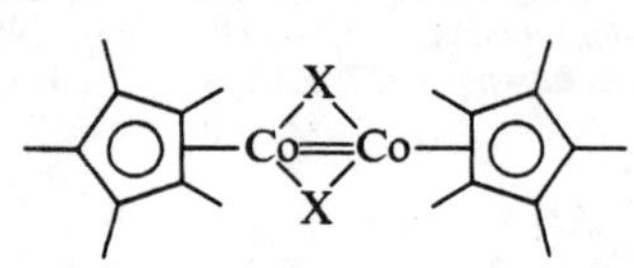

X = Br

M 548.131

Bridged dimer. V. useful precursor to $Co(C_5Me_5)$ complexes. Dark-brown, extremely air-sensitive cryst. (pentane at −70°). Sol. pentane. Paramagnetic. Oxidised by $FeBr_3$ to form Tri-μ-bromobis[(1,2,3,4,5-η)-1,2,3,4,5-pentamethyl-2,4-cyclopentadien-1-yl]dicobalt(1+), Co-00245 and reduced by CO to form $Co(C_5Me_5)(CO)_2$.

Monomer: Bromo(pentamethylcyclopentadienyl)cobalt.
$C_{10}H_{15}BrCo$ M 274.066
Unknown.

Kölle, U. *et al*, *Angew. Chem., Int. Ed. Engl.*, 1982, **21**, 131; *Angew. Chem. Suppl.*, 1982, 230 (*synth*, *ms*, *esr*, *props*)

$C_{20}H_{30}Br_3Co_2^{\oplus}$ Co-00245

Tri-μ-bromobis[(1,2,3,4,5-η)-1,2,3,4,5-pentamethyl-2,4-cyclopentadien-1-yl]dicobalt(1+)

[82595-91-3]

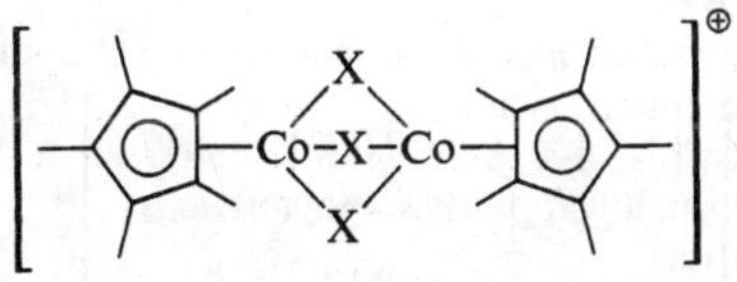

X = Br

M 628.035 (ion)

Useful precursor to $Co(C_5Me_5)$ complexes. pmr 1.37 ppm (CD_2Cl_2).

Hexafluorophosphate:

$C_{20}H_{30}Br_3Co_2F_6P$ M 773.000

No phys. props. reported.

Köelle, U. *et al*, *Angew. Chem., Int. Ed. Engl.*, 1982, **21**, 131; *Angew. Chem. Suppl.*, 1982, 230 (*synth, pmr*)

$C_{20}H_{30}Br_4Co_2$ Co-00246

Di-μ-bromodibromobis[(1,2,3,4,5-η)-1,2,3,4,5-pentamethyl-2,4-cyclopentadien-1-yl]dicobalt, 10CI

Bis[dibromo(pentamethylcyclopentadienyl)cobalt]. Tetrabromobis(pentamethylcyclopentadienyl)dicobalt

[82595-78-6]

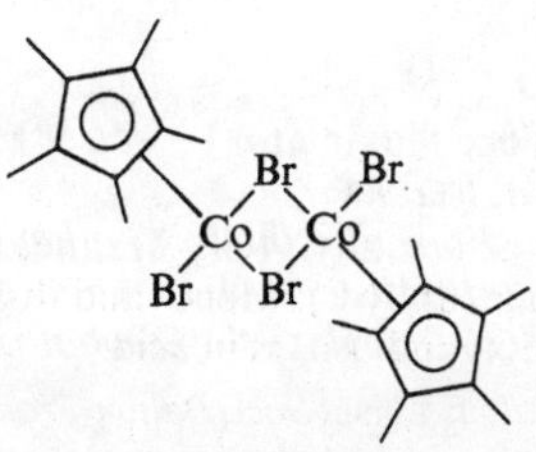

M 707.939

Bridged dimer. Useful precursor to $Co(C_5Me_5)$ complexes. pmr 1.25 ppm (CD_2Cl_2).

Monomer: Dibromo(pentamethylcyclopentadienyl)-cobalt.

$C_{10}H_{15}Br_2Co$ M 353.970

Unknown.

Köelle, U. *et al*, *Angew. Chem., Int. Ed. Engl.*, 1982, **21**, 131; *Angew. Chem. Suppl.*, 1982, 230 (*synth, pmr*)

$C_{20}H_{30}ClCoN_4O_2$ Co-00247

Benzylchloro[2,2′-(1,3-propanediyldinitrilo)bis[3-pentanone]-dioximato](1−)-*N,N′,N″,N‴*]cobalt

*Chloro(phenylmethyl)[[2,2′-(1,3-propanediyldinitrilo)-bis[3-pentanone]dioximato](1−)-*N,N′,N″,N‴*]cobalt, 10CI*

[75962-07-1]

As Iodomethyl[[2,2′-(1,3-propanediyldinitrilo)bis[3-pentanone]dioximato](1−)-*N,N′,N″,N‴*]cobalt, Co-00159 with

X = $-CH_2Ph$, Y = Cl

M 452.869

Easily synthesised vitamin B_{12} analogue. Red-orange solid (THF/Et_2O). Co−benzyl bond readily cleaved by light; chloride ligand partially dissociates in polar solvs. e.g. MeOH.

Finke, R.G. *et al*, *Inorg. Chem.*, 1981, **20**, 687 (*synth, pmr*)

$C_{20}H_{30}Cl_2Co_2$ Co-00248

Di-μ-chlorobis[(1,2,3,4,5-η)-1,2,3,4,5-pentamethyl-2,4-cyclopentadien-1-yl]dicobalt

Bis[chloro(pentamethylcyclopentadienyl)cobalt]

[74353-89-2]

As Di-μ-bromobis[(1,2,3,4,5-η)-1,2,3,4,5-pentamethyl-2,4-cyclopentadien-1-yl]dicobalt, Co-00244 with

X = Cl

M 459.229

Bridged dimer. V. useful precursor to $Co(C_5Me_5)$ complexes. Dark-brown, extremely air-sensitive cryst. (pentane at −70°). Sol. pentane. Paramagnetic. Oxidised by $FeCl_3$ to form Tri-μ-chlorobis[(1,2,3,4,5-η)-1,2,3,4,5-pentamethyl-2,4-cyclopentadien-1-yl]dicobalt(1+), Co-00249 reduced by CO to $Co(C_5Me_5)(CO)_2$.

Monomer: Chloro(pentamethylcyclopentadienyl)cobalt.

$C_{10}H_{15}ClCo$ M 229.615

Unknown.

Halbert, T.R. *et al*, *J. Am. Chem. Soc.*, 1980, **102**, 5101 (*synth, ms*)

Kölle, U. *et al*, *Angew. Chem., Int. Ed. Engl.*, 1982, **21**, 131; *Angew. Chem. Suppl.*, 1982, 230 (*synth, ms, esr, props*)

$C_{20}H_{30}Cl_3Co_2^{\oplus}$ Co-00249

Tri-μ-chlorobis[(1,2,3,4,5-η)-1,2,3,4,5-pentamethyl-2,4-cyclopentadien-1-yl]dicobalt(1+)

[82595-90-2]

As Tri-μ-bromobis[(1,2,3,4,5-η)-1,2,3,4,5-pentamethyl-2,4-cyclopentadien-1-yl]dicobalt(1+), Co-00245 with

X = Cl

M 494.682 (ion)

Useful precursor to $Co(C_5Me_5)$ complexes. pmr 1.17 ppm (CD_2Cl_2).

Hexafluorophosphate:

$C_{20}H_{30}Cl_3Co_2F_6P$ M 639.647

No phys. props. reported.

Köelle, U. *et al*, *Angew. Chem., Int. Ed. Engl.*, 1982, **21**, 131; *Angew. Chem. Suppl.*, 1982, 230 (*synth, pmr*)

$C_{20}H_{30}Co$ Co-00250

Decamethylcobaltocene, 10CI

Bis(η^5-pentamethylcyclopentadienyl)cobalt

[74507-62-3]

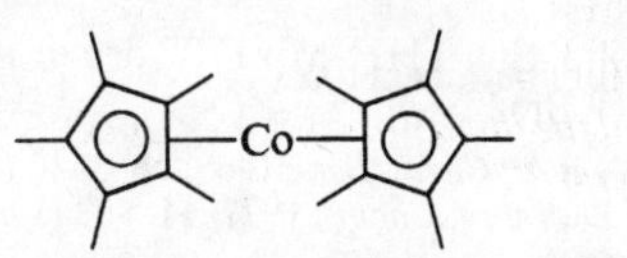

M 329.390

Black cryst. by subl. Mp 300°. Bp_{13} 130-40° subl.

Kölle, U. *et al*, *Chem. Ber.*, 1981, **114**, 2929 (*synth, ms, pmr*)

Cauletti, C. *et al*, *J. Electron Spectrosc. Relat. Phenom.*, 1980, **19**, 327 (*pe*)

$C_{20}H_{30}Co_2I_4$ **Co-00251**

Di-μ-iododiiodobis[(1,2,3,4,5-η)-1,2,3,4,5-pentamethyl-2,4-cyclopentadien-1-yl]dicobalt

Bis[diiodo(pentamethylcyclopentadienyl)cobalt]. Tetraiodobis(pentamethylcyclopentadienyl)dicobalt

[72339-52-7]

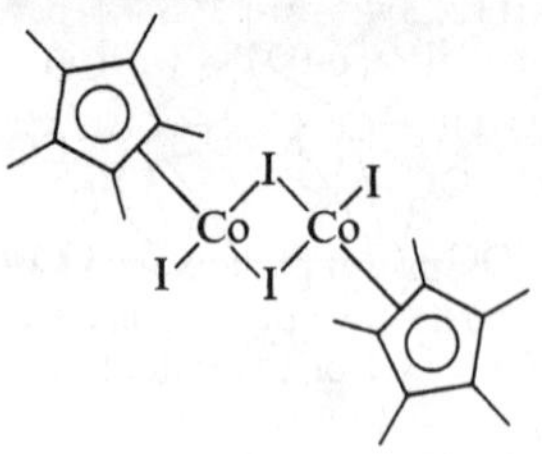

M 895.941

Bridged dimer. Starting material for $Co(C_5Me_5)$ compds. Black cryst. (CH_2Cl_2/hexane). Mp 207-10° dec. pmr 1.83 ppm ($CDCl_3$).

Monomer: Diiodo(pentamethylcyclopentadienyl)cobalt. $C_{10}H_{15}CoI_2$ M 447.971 Unknown.

Roe, D.M. *et al*, *J. Chem. Soc. (A)*, 1971, 3173 (*synth, pmr*)
Fairhurst, G. *et al*, *J. Chem. Soc., Dalton Trans.*, 1979, 1524 (*props*)

$C_{20}H_{36}Co_4O_{23}P_4$ **Co-00252**

Octacarbonyltetrakis(trimethyl phosphite-*P*)tetracobalt, 9CI

[37308-85-3]

$Co_4(CO)_8[P(OMe)_3]_4$

M 1004.118

Black cryst. ir ν_{CO} 2018, 1987, 1980, 1950 cm^{-1}. Fluxional in soln.

Labroue, D. *et al*, *Inorg. Chim. Acta*, 1972, **6**, 387 (*synth, ir, pmr*)

$C_{21}H_{15}CoIO_3P$ **Co-00253**

Tricarbonyliodo(triphenylphosphine)cobalt, 9CI

[36158-67-5]

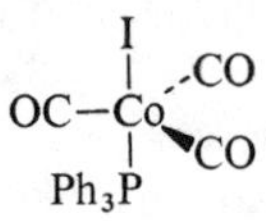

M 532.159

Brown solid. Sol. THF. Mp 170° dec. ir ν_{CO} 2020, 1950 cm^{-1} (KBr).

Hieber, W. *et al*, *Chem. Ber.*, 1965, **98**, 2530 (*synth, ir*)
Curtis, M.D., *Inorg. Chem.*, 1972, **11**, 802 (*synth, ir*)

$C_{21}H_{16}CoO_3$ **Co-00254**

Tricarbonylhydro(triphenylphosphine)cobalt, 9CI

Tricarbonylhydridotriphenylphosphinecobalt. Cobalt hydrotricarbonyl triphenylphosphine

[19537-79-2]

$HCo(CO)_3PPh_3$

M 375.289

Hydroformylation catalyst. Air-sensitive bright-yellow cryst. Dec. rapidly above 20°.

Hieber, W. *et al*, *Chem. Ber.*, 1961, **94**, 1417 (*synth*)
Heck, R.F. *et al*, *J. Am. Chem. Soc.*, 1963, **85**, 657 (*synth*)
Hieber, W. *et al*, *Chem. Ber.*, 1965, **98**, 2933 (*pmr*)

$C_{21}H_{24}Co_2O_5$ **Co-00255**

[μ-[(1,1′,2,2′-η:2,2′-η)Bi-1-cycloocten-1-yl-2,2′-diyl]]pentacarbonyldicobalt, 10CI

Dicarbonyl[2-5η-2,3:4,5-bis(hexamethylene)(tricarbonylcobalta)cyclopentadiene]cobalt

[66213-40-9]

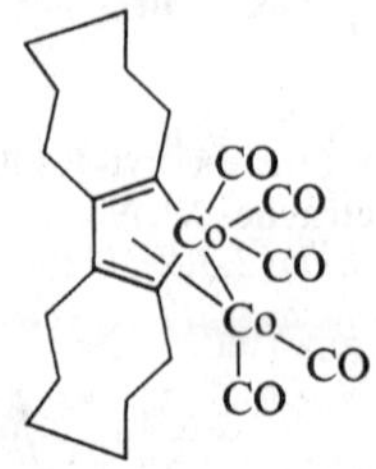

M 474.284

Intermed. in $Co_2(CO)_8$-catalysed cyclotrimerisation of cyclooctyne. Air-stable orange cryst. (pentane). Mp 101-3°.

Bennett, M.A. *et al*, *Inorg. Chem.*, 1978, **17**, 1995 (*synth, ir, nmr, cryst struct*)

$C_{21}H_{26}CoN_5O_4$ **Co-00256**

Bis[(2,3-butanedione dioximato)(1−)-*N,N′*](1-phenylethenyl)-(pyridine)cobalt, 10CI, 9CI

Bis(2,3-butanedione dioximato)(1-phenylvinyl)(pyridine)cobalt, 8CI. Bis(dimethylglyoximato)(pyridine)(α-styryl)cobalt. (α-Styryl)pyridinecobaloxime

[56498-38-5]

As Bis[(2,3-butanedione dioximato)(1−)-*N,N′*]-phenyl(pyridine)cobalt, Co-00230 with

R = H₂C=CPh—

M 471.401

Analogue of vitamin B_{12}. Red-orange solid (MeOH). Co—C bond cleaved by light, especially in air.

Nguyen-Van Duong, K. *et al*, *J. Organomet. Chem.*, 1970, **22**, 473 (*synth, pmr*)
Naumberg, M. *et al*, *J. Organomet. Chem.*, 1970, **25**, 231 (*synth, pmr*)
Johnson, M.D. *et al*, *J. Chem. Soc. (B)*, 1971, 185 (*synth, pmr*)
Bied-Charreton, C. *et al*, *Org. Magn. Reson.*, 1975, **7**, 116 (*cmr*)
Dodd, D. *et al*, *J. Chem. Soc., Perkin Trans. 2*, 1976, 1261 (*props*)

$C_{21}H_{26}CoN_5O_4$ **Co-00257**

Bis[(2,3-butanedione dioximato)(1−)-*N,N′*](2-phenylethenyl)-(pyridine)cobalt

β-Styrylbis(dimethylglyoximato)pyridinecobalt. (β-Styryl)pyridinecobaloxime. Bis(dimethylglyoximato)-pyridine-β-styrylcobalt

As Bis[(2,3-butanedione dioximato)(1−)-*N,N′*]-phenyl(pyridine)cobalt, Co-00230 with

R = PhCH=CH— (*E*- or *Z*-)

M 471.401

Analogue of vitamin B_{12}.

(*E*)-***form*** [29628-57-7]
Yellow solid (CH_2Cl_2/Et_2O); orange solid (MeOH).
(*Z*)-***form*** [56710-85-1]
Yellow solid (CH_2Cl_2/Et_2O); reddish-orange solid (MeOH).

Nguyen-Van Duong, K. *et al*, *J. Organomet. Chem.*, 1970, **22**, 473 (*synth, pmr*)
Johnson, M.D. *et al*, *J. Chem. Soc.* (*B*), 1971, 185 (*synth, pmr*)
Bied-Charreton, C. *et al*, *Org. Magn. Reson.*, 1975, **7**, 116 (*cmr*)
Dodd, D. *et al*, *J. Chem. Soc., Perkin Trans. 2*, 1976, 1261 (*props*)

$C_{21}H_{31}ClCoN_4^{\oplus}$ Co-00258

Benzylchloro(2,3,9,10-tetramethyl-1,4,8,11-tetraazacyclotetradeca-1,3,8,10-tetraene-*N*[1],*N*[4],*N*[8],*N*[11])cobalt(1+)
*Chloro(phenylmethyl)(2,3,9,10-tetramethyl-1,4,8,11-tetraazocyclotetradeca-1,3,8,10-tetraene-*N[1],N[4],N[8],N[11]*)cobalt(1+)*, 9CI

As Bromomethyl(2,3,9,10-tetramethyl-1,4,8,11-tetraazacyclotetradeca-1,3,8,10-tetraene-*N*[1],*N*[4],*N*[8],*N*[11])cobalt(1+), Co-00175 with

X = $PhCH_2$—, Y = Cl

M 433.889 (ion)
Vitamin B_{12} analogue. Stable in the solid state but light-sensitive in soln. in the presence of air. Chloride ligand rapidly displaced by MeCN.
Hexafluorophosphate: [38887-55-7].
$C_{21}H_{31}ClCoF_6N_4P$ M 578.853
Orange-brown, air-stable cryst. ($MeNO_2$/EtOH).
Tetraphenylborate: [38887-54-6].
$C_{45}H_{51}BClCoN_4$ M 753.121
Orange-brown, air-stable cryst. ($MeNO_2$/EtOH).

Farmery, K. *et al*, *Inorg. Chem.*, 1972, **11**, 2901 (*synth, uv*)

$C_{21}H_{33}Co$ Co-00259

[(1,2,3,4-η)-1,2,3,4,5,5-Hexamethyl-1,3-cyclopentadiene][(1,2,3,4,5-η)-1,2,3,4,5-pentamethyl-2,4-cyclopentadien-1-yl]cobalt
[74552-21-9]

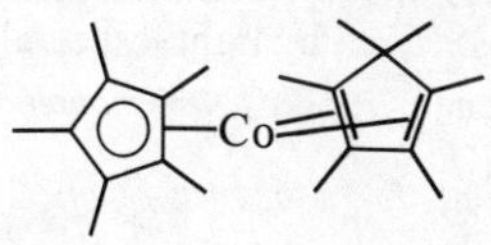

M 344.425
Red cryst. (MeOH/pentane). Mp 198-200°.

Koelle, U. *et al*, *Chem. Ber.*, 1981, **114**, 2929 (*synth, pmr, ms*)

$C_{22}H_{10}Co_4O_{10}P_2$ Co-00260

Di-μ-carbonyloctacarbonylbis[μ-(phenylphosphinidene)]tetracobalt, 10CI
Decacarbonylbis[μ-(phenylphosphinidene)]tetracobalt
[58092-22-1]

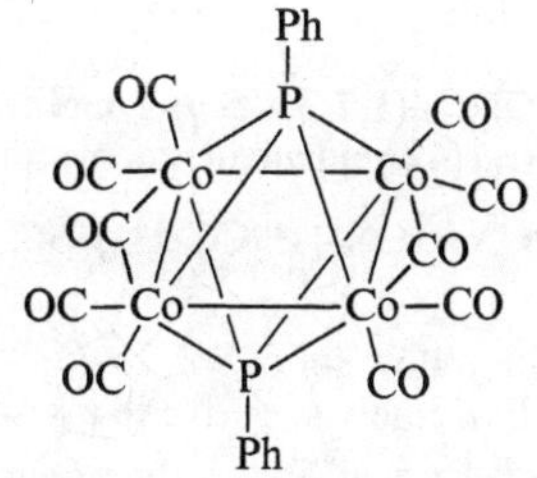

M 731.995
First example of a five-coordinate phosphorus atom bonded to four metal atoms. Homogeneous catalyst for hydroformylation of alkenes and for hydrogenation of CO to alcohols. Dark-red solid. ir ν_{CO} 2040vs, 2032s, 2016s, 1866w cm^{-1} (CH_2Cl_2).

King, R.B. *et al*, *J. Mol. Catal.*, 1980, **10**, 75 (*use*)
Pittman, C.U. *et al*, *Angew. Chem., Int. Ed. Engl.*, 1980, **19**, 478.
Ryan, R.C. *et al*, *J. Organomet. Chem.*, 1980, **193**, 247 (*synth, ir, pmr, cryst struct*)

$C_{22}H_{15}CoO_4P^{\oplus}$ Co-00261

Tetracarbonyl(triphenylphosphine)cobalt(1+), 9CI

$Ph_3PCo(CO)_4^{\oplus}$

M 433.265 (ion)
Stable below −15°, air- and moisture-sensitive.
Hydrogen dichloride: [40892-10-2].
$C_{22}H_{16}CoCl_2O_4P$ M 505.179
Pale-yellow cryst. Sol. polar solvs. ir ν_{CO} 2145, 2099, 2053 cm^{-1} ($CHCl_3$). Reacts with R_2NH to give $Co(CO)_3(PPh_3)(CONR_2)$.

Hieber, W. *et al*, *Chem. Ber.*, 1965, **98**, 1744 (*synth, ir*)
Krohberger, H. *et al*, *J. Organomet. Chem.*, 1972, **46**, 139.

$C_{22}H_{15}CoO_4Pb$ Co-00262

Tetracarbonyl(triphenylplumbyl)cobalt, 10CI
(*Triphenylplumbyl*)*tetracarbonylcobalt*
[16561-04-9]

$Ph_3PbCo(CO)_4$

M 609.491
Orange-yellow cryst.

Hein, F. *et al*, *Naturwissenschaften*, 1957, **44**, 34.
Hein, F. *et al*, *Justus Liebigs Ann. Chem.*, 1965, **684**, 4.

$C_{22}H_{20}B_2Co$ Co-00263

Bis[(1,2,3,4,5,6-η)-1-phenylboratabenzene]cobalt, 9CI
Bis(1-phenylborabenzene)cobalt
[36534-31-3]

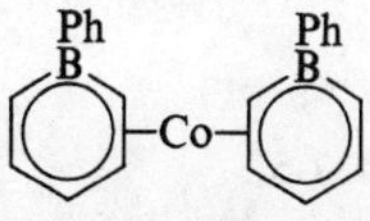

M 364.953
Dark-red plates (hexane) or dark-brown cryst. Mp 70-1°, 191-2°. Paramagnetic.

Herberich, G.E. *et al*, *Chem. Ber.*, 1972, **105**, 3413 (*synth, ms*)
Herberich, G.E. *et al*, *J. Chem. Soc., Dalton Trans.*, 1975, 985 (*esr*)
Herberich, G.E. *et al*, *Chem. Ber.*, 1979, **112**, 607.

$C_{22}H_{21}CoO_4Sm$ Co-00264

[μ-(Carbonyl-*C,O*)]tris[(1,2,3,4,5-η)-1-methyl-2,4-cyclopentadien-1-yl](tetracarbonylcobalt)samarium

$Co(CO)_4 \cdot [H_3CC_5H_4]_3Sm$

M 558.699

Dimeric. Bright-yellow solid. Sol. Et_2O, THF, insol. non-polar solvs. Extremely sensitive to air and moisture.

Crease, A.E. *et al*, *J. Chem. Soc., Dalton Trans.*, 1973, 1501 (*synth, ir*)

$C_{22}H_{22}CoN_3O_2$ Co-00265

[[2,2'-[1,2-ethanediylbis[nitrilomethylidyne]]bis[phenolato]](2−)-*N,N',O,O'*]methyl(pyridine)cobalt, 10CI

[[α,α'-(Ethylenedinitrilo)di-o-cresolato](2−)methyl-(pyridine)cobalt, 8CI. *[Bis(salicylaldehyde)-ethylenediimine]methyl(pyridine)cobalt*

[52759-66-7]

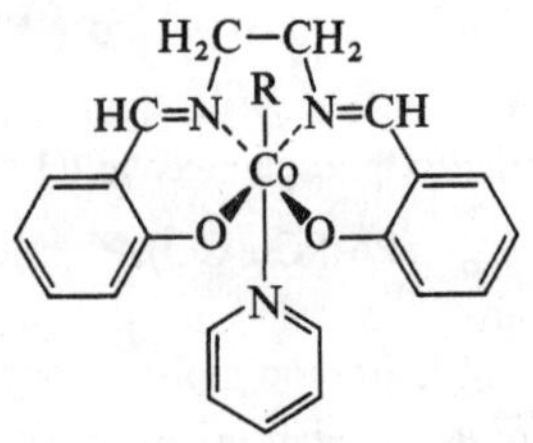

R = Me

M 419.368

Model compd. for vitamin B_{12}. Red-orange cryst. (Py/pet. ether/Et_2O). Mp 150°. Homolytic cleavage of cobalt-carbon bond catalysed by light.

Costa, G. *et al*, *J. Organomet. Chem.*, 1967, **7**, 493 (*synth, uv*)
Floriani, C. *et al*, *J. Organomet. Chem.*, 1968, **12**, 209 (*synth, ir, nmr*)

$C_{22}H_{22}CoN_4^{\oplus}$ Co-00266

Bis(2,2'-bipyridine-*N,N'*)dimethylcobalt(1+), 9CI

Bis(2,2'-bipyridyl)dimethylcobalt(1+), 8CI

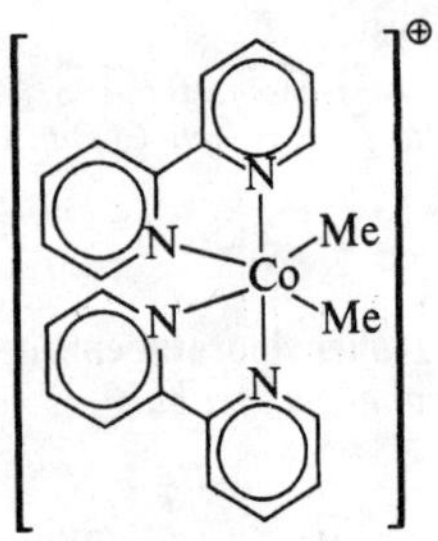

M 401.376 (ion)

Chloride: [71697-37-5].
$C_{22}H_{22}ClCoN_4$ M 436.829
Needles (CH_2Cl_2).

Iodide: [63851-21-8].
$C_{22}H_{22}CoIN_4$ M 528.280
No phys. props. reported.

Perchlorate: [28964-65-0].
$C_{22}H_{22}ClCoN_4O_4$ M 500.826
Air-sensitive red cryst.

Tetraphenylborate: [63851-17-2].
$C_{46}H_{42}BCoN_4$ M 720.608
Red cryst. Mp 191-5°.

Mestroni, G. *et al*, *J. Organomet. Chem.*, 1970, **24**, 775 (*synth, pmr*)
Yamamoto, T. *et al*, *Chem. Lett.*, 1977, 833 (*synth, ir, pmr*)
Komiya, S. *et al*, *J. Organomet. Chem.*, 1979, **174**, 343 (*synth, ir, pmr*)

$C_{22}H_{29}BrCoN_4^{\oplus}$ Co-00267

Benzylbromo[2,12-dimethyl-3,7,11,17-tetraazabicyclo[11.3.1]heptadeca-1(17),2,11,13,15-pentaene]cobalt(1+)

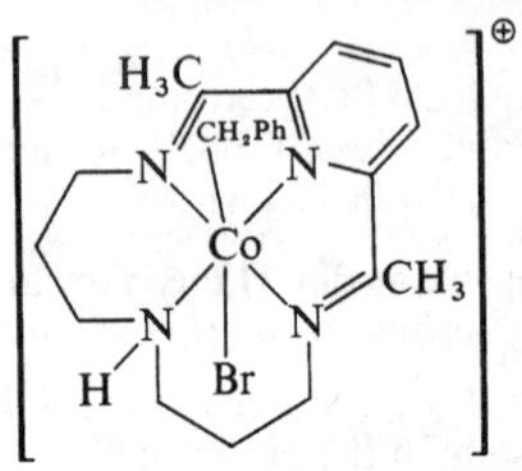

M 488.335 (ion)

Vitamin B_{12} analogue. Co−benzyl bond cleaved by light and coordinated Br easily replaced by H_2O.

Hexafluorophosphate: [22784-71-0].
$C_{22}H_{29}BrCoF_6N_4P$ M 633.299
Brownish-red, light-sensitive cryst. + $1H_2O$ (MeOH aq.). Sol. MeOH.

Ochiai, E. *et al*, *J. Am. Chem. Soc.*, 1969, **91**, 3201 (*synth, ir, uv*)

$C_{22}H_{30}Co_2O_2$ Co-00268

Di-μ-carbonylbis[(1,2,3,4,5-η)-1,2,3,4,5-pentamethyl-2,4-cyclopentadien-1-yl]dicobalt, 9CI

Bis(pentamethylcyclopentadienyl)dicarbonyldicobalt. *Bis[(μ-carbonyl)-η⁵-pentamethylcyclopentadienylcobalt]*

[69657-52-9]

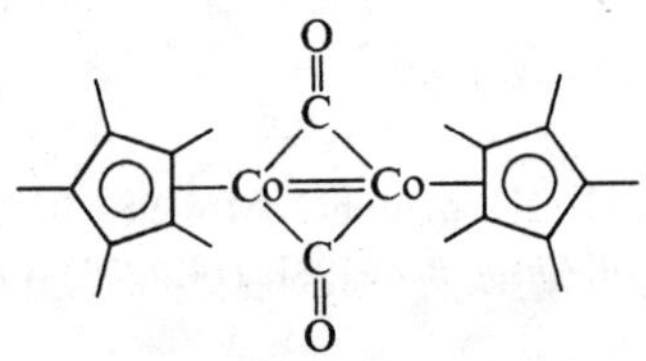

M 444.344

Air-sensitive purple-black cryst. (pentane). Mp 272-4° dec. (sealed tube). ir ν_{CO} 1756 cm^{-1} (nujol), pmr δ 1.46(s) ppm (C_6D_6). Reacts with diazoalkanes to give bridged alkylidene complexes.

Bailey, W.I. *et al*, *J. Organomet. Chem.*, 1979, **165**, 373 (*synth, cryst struct, ir, ms, pmr*)
Ginsbury, R.E. *et al*, *J. Chem. Soc., Chem. Commun.*, 1979, 468 (*synth, cryst struct, ir, ms, pmr*)
Schore, N.E., *J. Organomet. Chem.*, 1979, **173**, 301 (*ir*)
Beevor, G. *et al*, *J. Organomet. Chem.*, 1981, **221**, C25 (*synth*)
Herrmann, W.A. *et al*, *J. Organomet. Chem.*, 1982, **226**, C59 (*synth, props*)

$C_{22}H_{32}Co_2O$ Co-00269

μ-Carbonyl-μ-methylenebis[(1,2,3,4,5-η)-1,2,3,4,5-pentamethyl-2,4-cyclopentadien-1-yl]dicobalt

[74331-28-5]

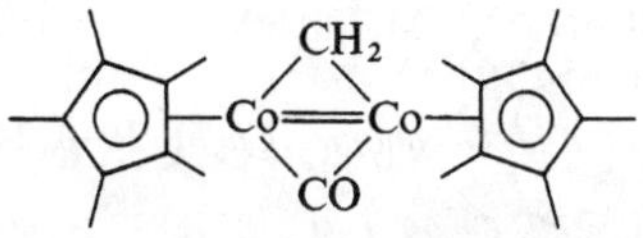

M 430.361

Blue-black air-sensitive cryst. (hexane at −20°). Sol. THF, hexane to give deep royal-blue extremely air-sensitive solns. ir ν_{CO} 1758 cm^{-1} (KBr).

Halbert, T.R. *et al*, *J. Am. Chem. Soc.*, 1980, **102**, 5102 (*synth, cryst struct, nmr, ms, ir*)
Herrmann, W.A. *et al*, *J. Organomet. Chem.*, 1982, **226**, C59 (*synth, ir, props*)

$C_{22}H_{34}Cl_4Co_2$ Co-00270

Di-μ-chlorodichlorobis[(1,2,3,4,5-η)-1-ethyl-2,3,4,5-tetramethyl-2,4-cyclopentadien-1-yl]dicobalt, 10CI

Bis[dichloro(η^5-ethyltetramethylcyclopentadienyl)cobalt]. Tetrachlorobis(η^5-ethyltetramethylcyclopentadienyl)dicobalt

[71238-34-1]

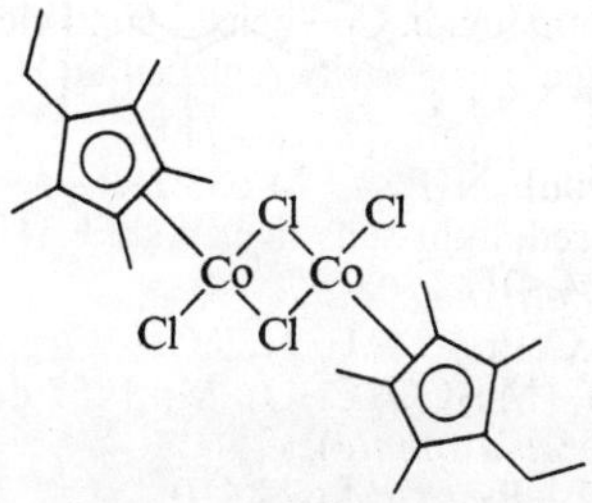

M 558.189

Bridged dimer. Green cryst. (chlorobenzene/pet. ether).

Monomer: Dichloro(ethyltetramethylcyclopentadienyl)cobalt.
$C_{11}H_{17}Cl_2Co$ M 279.095
Unknown.

Green, M.L.H. *et al*, *J. Chem. Soc., Dalton Trans.*, 1979, 355 (*synth, pmr, props*)

$C_{23}H_{20}CoI_2P$ Co-00271

(η^5-2,4-Cyclopentadien-1-yl)diiodo(triphenylphosphine)cobalt, 9CI

Diiodocyclopentadienyl(triphenylphosphine)cobalt. Triphenylphosphine-π-cyclopentadienylcobalt diiodide. Diiodo(triphenylphosphine)cyclopentadienylcobalt

[12194-27-3]

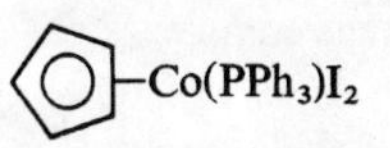

M 640.127

Useful starting material for cyclopentadienylcobalt compds. Dark-green cryst. (CH_2Cl_2/toluene) or blue-black cryst. Mp 176-7°.

King, R.B., *Inorg. Chem.*, 1966, **5**, 82 (*synth, ir*)
Yamazaki, H. *et al*, *J. Organomet. Chem.*, 1970, **21**, 431 (*synth, pmr*)
Roe, D.M. *et al*, *J. Chem. Soc. (A)*, 1971, 3173 (*synth, pmr*)

$C_{23}H_{20}CoO_2P$ Co-00272

Dicarbonyl(η^3-2-propenyl)(triphenylphosphine)cobalt, 10CI

π-Allyldicarbonyl(triphenylphosphine)cobalt, 8CI. Dicarbonyltriphenylphosphine-π-allylcobalt. Allylcobalt dicarbonyl triphenylphosphine

[12149-50-7]

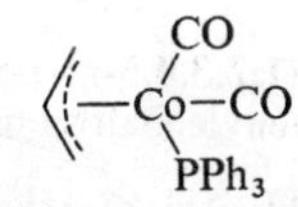

M 418.317

Yellow-orange cryst. (CH_2Cl_2/pentane). Mp 150-2° (132° dec.). ir ν_{CO} 2000, 1948 cm^{-1} (cyclohexane).

Heck, R.F. *et al*, *J. Am. Chem. Soc.*, 1961, **83**, 1097 (*synth, ir*)
Heck, R.F., *J. Am. Chem. Soc.*, 1968, **90**, 317 (*synth*)
Seyferth, D. *et al*, *J. Am. Chem. Soc.*, 1969, **91**, 3037 (*synth, ir*)
Clarke, H.L. *et al*, *J. Organomet. Chem.*, 1974, **66**, 119 (*ir, nmr*)
Rinze, P.V. *et al*, *Chem. Ber.*, 1979, **112**, 1973 (*cryst struct, nmr*)

$C_{23}H_{22}BrCoNO_2P$ Co-00273

Bromodicarbonyl[(dimethylamino)methylene]triphenylphosphinecobalt, 10CI

[67407-29-8]

$Co(CHNMe_2)(CO)_2Br(PPh_3)$

M 514.243

Stereochemistry unknown. Yellow cryst. (THF/Et_2O). Mp 185°. ir ν_{co} 1991m, 1925s cm^{-1} (CH_2Cl_2).

Hartshorn, A.J. *et al*, *J. Chem. Soc., Dalton Trans.*, 1978, 348 (*synth, pmr, ir*)

$C_{23}H_{22}ClCoNO_2P$ Co-00274

Dicarbonylchloro[(dimethylamino)methylene]triphenylphosphinecobalt, 10CI

[58938-34-4]

$Co(CHNMe_2)(CO)_2Cl(PPh_3)$

M 469.792

Stereochemistry unknown. Yellow cryst. (THF/Et_2O). Mp 178-80°. ir ν_{co} 1991m, 1925s cm^{-1} (CH_2Cl_2).

Hartshorn, A.J. *et al*, *J. Chem. Soc., Dalton Trans.*, 1978, 348 (*synth, ir, pmr*)

$C_{24}H_{15}CoO_3$ Co-00275

Tricarbonyl[(1,2,3-η)-1,2,3-triphenylcyclopropen-1-yl]cobalt, 10CI

[69745-64-8]

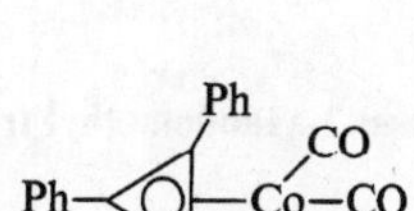

M 410.314

Yellow cryst. (MeOH). Mp 134-5°. ir ν_{CO} 2090, 2056, 2028 cm^{-1} (hexane).

Chiang, T. *et al*, *Inorg. Chem.*, 1979, **18**, 1687 (*synth, nmr, ir, ms, cryst struct*)

$C_{24}H_{16}Co_2N_4O_4$ **Co-00276**

Bis(2,2′-bipyridine)di-μ-carbonyldicarbonyldicobalt, 8CI

Bis(2,2′-bipyridine)tetracarbonyldicobalt

[28855-47-2]

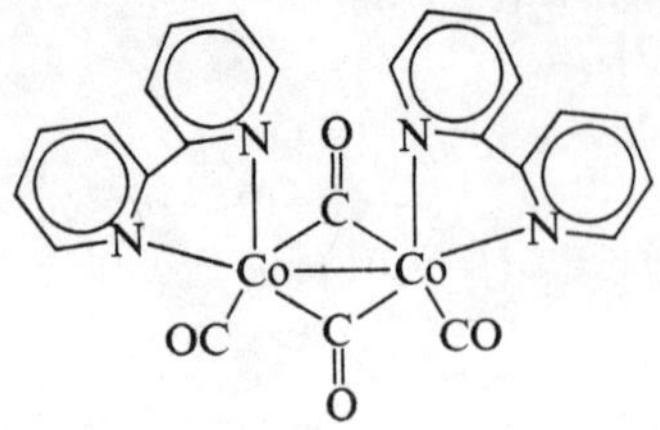

M 542.281

cis-form

Black cryst. (C_6H_6). ir ν_{CO} 1937.5, 1897, 1717 cm^{-1} (KBr).

Behrens, U. *et al*, *Z. Anorg. Allg. Chem.*, 1967, **356**, 8 (*synth, ir*)

$C_{24}H_{18}CoO_4$ **Co-00277**

Acetyltricarbonyltriphenylphosphinecobalt, 10CI

[14054-68-3]

Co(CH_3CO)(CO)$_3$(PPh$_3$)

M 429.337

Alkene oligomerisation catalyst. Mod. air-stable cryst. Mp 123° dec.

Heck, R.F. *et al*, *J. Am. Chem. Soc.*, 1962, **84**, 2499 (*synth*)

$C_{24}H_{20}CoOP$ **Co-00278**

Carbonyl(η^5-2,4-cyclopentadien-1-yl)(triphenylphosphine)-cobalt, 9CI

π-Cyclopentadienylcarbonyl(triphenylphosphine)cobalt

[12203-85-9]

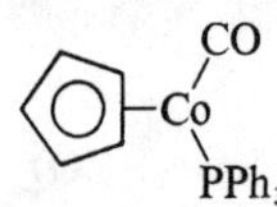

M 414.328

Red-black cryst. Mp 140-3°. ir ν_{CO} 1905 cm^{-1} (KBr); turns black after 1 week in a closed vial.

King, R.B. *et al*, *Inorg. Chem.*, 1966, **5**, 82 (*synth, ir*)

Schuster-Woldan, H.G. *et al*, *J. Am. Chem. Soc.*, 1966, **88**, 1657 (*synth, ir*)

Yamazaki, H. *et al*, *Bull. Chem. Soc. Jpn.*, 1971, **44**, 2260 (*synth, ir, pmr*)

McDonnell Bushnell, L.P. *et al*, *J. Organomet. Chem.*, 1978, **157**, 445 (*synth, ir, pmr*)

$C_{24}H_{23}CoIP$ **Co-00279**

(η^5-2,4-Cyclopentadien-1-yl)iodomethyl(triphenylphosphine)-cobalt

[12203-93-9]

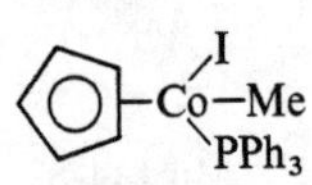

M 528.257

Brown solid (C_6H_6/hexane). Mp 112° dec.

Yamazaki, H. *et al*, *J. Organomet. Chem.*, 1970, **21**, 431 (*synth, pmr*)

$C_{24}H_{30}CoO_4Yb$ **Co-00280**

[μ-(Carbonyl-*C,O*)]bis[(1,2,3,4,5-η)-1,2,3,4,5-pentamethyl-2,4-cyclopentadien-1-yl](tricarbonylcobalt)ytterbium

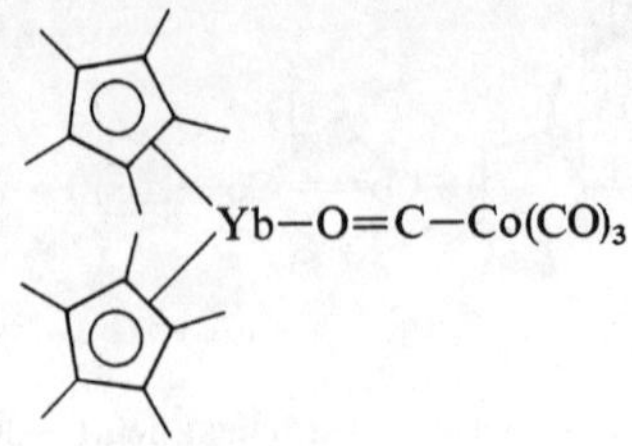

M 614.472

Blue ppt.

THF adduct: [80758-23-2].

$C_{28}H_{38}CoO_5Yb$ M 686.578

Blue prisms. μ_{eff} = 4.1μ_B.

Tilley, T.D. *et al*, *J. Chem. Soc., Chem. Commun.*, 1981, 985 (*synth, struct, ir, nmr*)

$C_{24}H_{36}Co^{⊕⊕}$ **Co-00281**

Bis(η^6-hexamethylbenzene)cobalt(2+), 9CI

[53382-65-3]

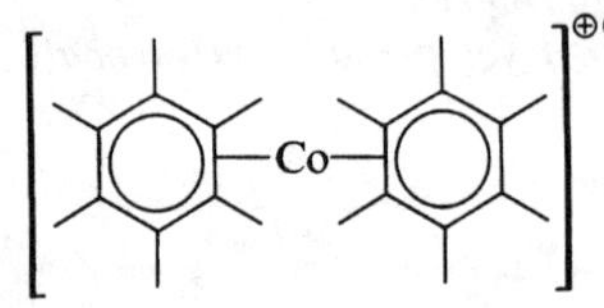

M 383.482 (ion)

Air-stable.

Hexachloroplatinate:

$C_{24}H_{36}Cl_6CoPt$ M 791.280

Olive cryst. (Me_2CO/Et_2O). Mp 140° dec.

Bis(hexafluorophosphate):

$C_{24}H_{36}CoF_{12}P_2$ M 673.410

Yellow cryst.

Fischer, E.O. *et al*, *J. Organomet. Chem.*, 1964, **1**, 307 (*synth, ir*)

$C_{24}H_{72}Co_2O_{24}P_8$ **Co-00282**

Octakis(trimethyl phosphite)dicobalt

[54204-71-6]

$Co_2[P(OMe)_3]_8$

M 1110.475

Brown cryst. (pentane), unstable in air. Mp 131.4° dec. Readily converted to CoH[P(OMe)$_3$] by water and alumina.

Muetterties, E.L. *et al*, *J. Am. Chem. Soc.*, 1974, **96**, 7920 (*synth, ir, pmr*)

The symbol ▷ in Entries highlights hazard or toxicity information

$C_{25}H_{15}CoO_4$ Co-00283

Tricarbonyl[(1,2,3-η)-4-oxo-1,2,3-triphenyl-2-cyclobuten-1-yl]cobalt, 9CI

Tricarbonyl(η³-triphenylcyclobutenoyl)cobalt

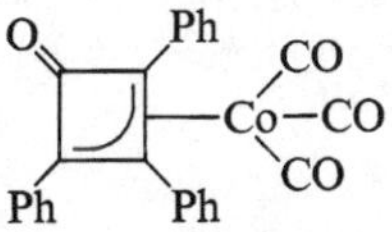

M 438.324

Yellow cryst. (MeCN). ir ν_{CO} 2080, 2015, 1710 cm^{-1} (KBr).

Coffey, C.E., *J. Am. Chem. Soc.*, 1962, **84**, 118 (*synth, ir*)
Potenza, J. *et al*, *J. Organomet. Chem.*, 1974, **64**, C13 (*cryst struct*)

$C_{25}H_{15}Co_2O_7P$ Co-00284

Heptacarbonyl(triphenylphosphine)dicobalt

[15906-55-5]

$Co_2(CO)_7(PPh_3)$

M 576.229

Air-sensitive dark-brown cryst. Spar. sol. org. solvs. ir ν_{CO} 2079, 2026, 2010, 1996, 1964 cm^{-1} (hexane).

Bor, G. *et al*, *Chem. Ind.* (*London*), 1963, 912 (*synth, ir*)

$C_{25}H_{20}Co_2O$ Co-00285

μ-Carbonylbis(η⁵-cyclopentadienyl)(diphenylacetylene)dicobalt

μ-Carbonylbis(η⁵-2,4-cyclopentadien-1-yl-)[μ-[1,1′-(η²:η²-ethynediyl)bis[benzene]]]dicobalt, 10CI.
μ-Carbonyl-μ-η-(diphenylacetylene)bis(η-cyclopentadienyl)dicobalt

[63150-96-9]

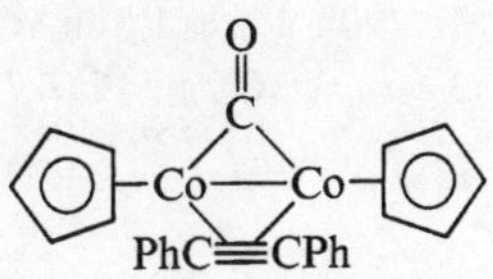

M 454.299

Dark-green solid by subl. $Bp_{0.001}$ 80° subl. ir ν_{CO} 1800 cm^{-1} (nujol).

Lee, W.-S. *et al*, *J. Organomet. Chem.*, 1977, **127**, 93 (*synth, ir, pmr, ms*)

$C_{25}H_{24}CoP$ Co-00286

(η⁵-2,4-Cyclopentadien-1-yl)(η²-ethene)(triphenylphosphine)-cobalt, 10CI

Cyclopentadienylethylene(triphenylphosphine)cobalt

[71163-83-2]

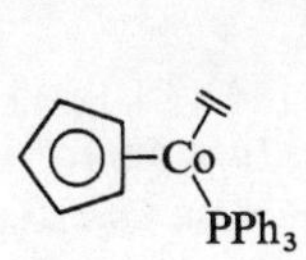

M 414.372

Important intermed. in model studies of Ziegler-Natta catalysis of polymerisation of ethylene. Impure purple-red cryst. (C_6H_6/hexane). V. labile.

Evitt, E.R. *et al*, *J. Am. Chem. Soc.*, 1980, **102**, 7003 (*synth, nmr*)

$C_{25}H_{25}CoN_5^{\oplus}$ Co-00287

(16,17-Dihydro-7*H*-dibenzo[*f,m*][1,4,8,12]tetraazacyclopentadecinato-*N*⁵,*N*⁹,*N*¹⁵,*N*¹⁸)methyl(pyridine)cobalt(1+), 10CI

(Dibenzocorrominato)methyl(pyridine)cobalt(1+)

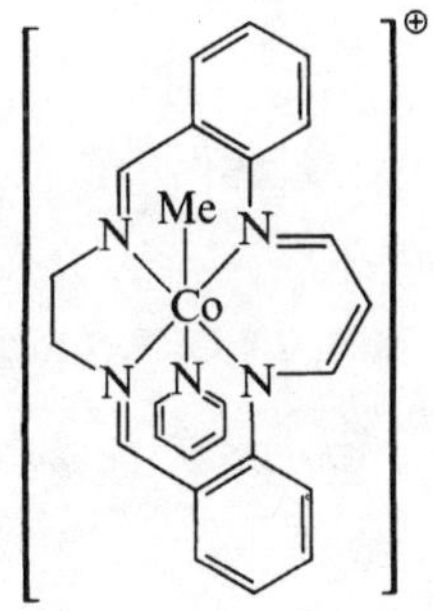

M 454.439 (ion)

Solns. extremely light-sensitive.

Hexafluorophosphate: [65554-63-4].
$C_{25}H_{25}CoF_6N_5P$ M 599.403
Dark-green cryst. (EtOH aq./Py). Sol. THF, DMSO.

Black, D.S.C. *et al*, *Aust. J. Chem.*, 1977, **30**, 2553 (*synth, ir, pmr, uv, props*)

$C_{25}H_{26}CoP$ Co-00288

(2,4-Cyclopentadien-1-yl)dimethyl(triphenylphosphine)cobalt

[12115-18-3]

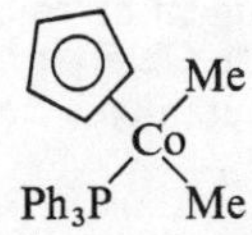

M 416.387

Erroneously listed in 10CI as Bis(2,4-cyclopentadienyl)-dimethyl(triphenylphosphine) cobalt. Alkylates alkynes. Important in model studies of homogenous catalyst systems. Air-stable, red cryst. (C_6H_6/hexane). Mp 142-3°.

Yamazaki, H. *et al*, *J. Organomet. Chem.*, 1970, **21**, 431 (*synth, pmr*)
Evitt, E.R. *et al*, *J. Am. Chem. Soc.*, 1981, **102**, 7003.

$C_{25}H_{45}CoN_5^{\oplus}$ Co-00289

Pentakis(2-isocyano-2-methylpropane)cobalt(1+), 10CI

Pentakis(tert-*butylisonitrile*)*cobalt*(*1*+)

[41517-72-0]

$[(H_3C)_3CNC]_5Co^{\oplus}$

M 474.597 (ion)

Catalyst for trimerisation of malononitrile.

Hexafluorophosphate:
$C_{25}H_{45}CoF_6N_5P$ M 619.561
Golden-yellow cryst. (EtOH). Mp 113°. Undergoes substitution reactions with tertiary phosphines and arsines.

Perchlorate:
$C_{25}H_{45}ClCoN_5O_4$ M 574.048
Yellow cryst.

King, R.B. *et al*, *Inorg. Chem.*, 1972, **11**, 2112 (*synth, ir*)
Cronin, D.L. *et al*, *J. Magn. Reson.*, 1975, **17**, 353 (*cmr*)
Kawakami, K. *et al*, *J. Inorg. Nucl. Chem.*, 1979, **4**, 1501 (*synth, ir, pmr*)
Yamamoto, Y. *et al*, *Bull. Chem. Soc. Jpn.*, 1981, **54**, 787.

$C_{26}H_{23}CoNP$ Co-00290

(η^5-2,4-Cyclopentadien-1-yl)[(2,3-η)-2-propenenitrile](triphenylphosphine)cobalt

Acrylonitrile(cyclopentadienyl)(triphenyphosphine)cobalt

[77023-80-4]

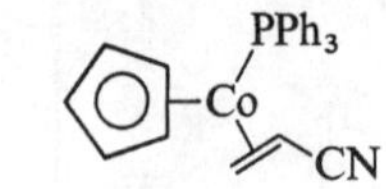

M 439.381

Red oil. Reacts with $H_2C{=}CHCN$ to give cobaltacyclopentanes.

Wakatsuki, Y. *et al*, *J. Chem. Soc., Dalton Trans.*, 1982, 1923 (*synth*)

$C_{26}H_{28}CoP$ Co-00291

(η^4-1,3-Butadiene)[(1,2,3-η)-2-butenyl](triphenylphosphine)cobalt, 9CI

(1,3-Butadiene)(1-methyl-π-allyl)(triphenylphosphine)cobalt, 8CI. 1,3-Butadiene(crotyl)-triphenylphosphinecobalt

[33136-97-9]

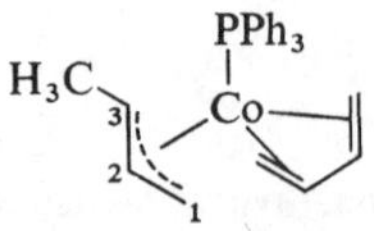

syn-form

M 430.414

Stereochemical assignment based on the position of CH_3 relative to H on C(2). Catalyst for the acyclic oligomerisation of butadiene. Dec. in chlorinated solvs. to give $CoCl_2$.

syn-form

Orange-red cryst. (hexane). Mp 94° (under N_2). Dec. >100°.

anti-form

Red-orange cryst. (pentane). Mp 110° dec. Rapidly transformed in soln. into the syn-isomer by PPh_3, Py, Me_2SO.

Porri, L. *et al*, *J. Chem. Soc., Chem. Commun.*, 1969, 276 (*cryst struct*)
Rinze, P.V. *et al*, *J. Organomet. Chem.*, 1971, **30**, 115 (*synth, ir, nmr*)
Vitulli, G. *et al*, *J. Chem. Soc. (A)*, 1971, 3246 (*synth, pmr, props*)
Cairns, M.A. *et al*, *J. Organomet. Chem.*, 1975, **87**, 109 (*props*)
Kerrinnes, H.J. *et al*, *J. Prakt. Chem.*, 1975, **317**, 503 (*use*)

$C_{26}H_{55}CoO_2P_2$ Co-00292

Dicarbonylhydrobis(tributylphosphine)cobalt, 9CI

Hydridodicarbonylbis(tributylphosphine)cobalt

[22392-50-3]

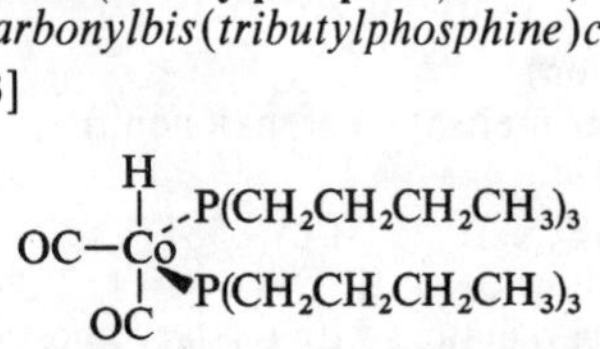

M 520.600

Catalyst for the hydrogenation of alkynes and aldehydes and for the isomerisation and hydroformylation of alkenes. Air-sensitive yellow-orange oil. Exists in soln. as a mixt. of stereoisomers.

Pregaglia, G. *et al*, *J. Organomet. Chem.*, 1971, **30**, 387 (*synth, ir, pmr, use*)

$C_{27}H_{24}CoNO_2P_2$ Co-00293

Carbonyl[1,2-ethanediylbis[diphenylphosphine]-*P,P'*]nitrosylcobalt

Carbonyl[ethylenebis(diphenylphosphine)]nitrosylcobalt, 9CI. Nitrosylcarbonyl[ethylenebis(diphenylphosphine)]cobalt. Carbonyl[1,2-bis(diphenylphosphino)-ethane]nitrosylcobalt

[21476-09-5]

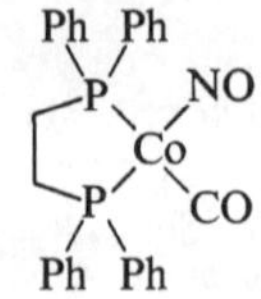

M 515.373

Air-stable orange cryst. (CH_2Cl_2/hexane). Mp 166°. ir ν_{CO} 1965, ν_{NO} 1712 cm^{-1} (C_6H_6).

Mawby, R.J. *et al*, *Inorg. Chem.*, 1966, **5**, 27 (*synth, ir*)
Van Hecke, G.R. *et al*, *Inorg. Chem.*, 1966, **5**, 1960 (*synth, ir*)
Foffani, A. *et al*, *Spectrochim. Acta, Part A*, 1968, **24**, 1437 (*ir*)

$C_{27}H_{24}CoN_3O_3$ Co-00294

[[2,2'-[1,2-ethanediylbis[nitrilomethylidyne]]bis[phenolato]](2−)-*N,N',O,O'*]phenyl(pyridine)cobalt, 10CI

[α,α-(Ethylenedinitrilo)di-o-cresolato](2−)]phenyl-(pyridine)cobalt, 8CI. [Bis(salicylaldehyde)-ethylenediimine]phenyl(pyridine)cobalt

[15488-74-1]

As [[2,2'-[1,2-ethanediylbis[nitrilomethylidyne]]-bis[phenolato]](2−)-*N,N',O,O'*]methyl(pyridine)cobalt, Co-00265 with

R = Ph

M 497.438

Model compd. for vitamin B_{12}. Orange-yellow needles (Py/EtOH). Mp 290°. Loses Py on heating.

Costa, G. *et al*, *J. Organomet. Chem.*, 1967, **7**, 493 (*synth, uv*)
Floriani, C. *et al*, *J. Organomet. Chem.*, 1968, **12**, 209 (*synth*)
Reisenhofer, E. *et al*, *Inorg. Chim. Acta*, 1981, **49**, 121 (*props*)

$C_{27}H_{28}CoP$ Co-00295

1,4-Butanediyl(η^5-2,4-cyclopentadien-1-yl)(triphenylphosphine)cobalt, 10CI

η^5-Cyclopentadienyl(triphenylphosphine)cobaltacyclopentane

[66416-06-6]

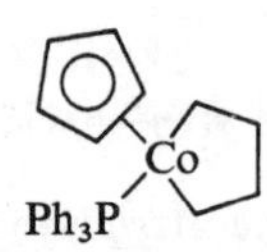

M 442.425

Reacts with ethylene to give 1-hexene and 1-butene. Orange-red needles (pentane).

Diversi, P. *et al*, *J. Chem. Soc., Chem. Commun.*, 1977, 811 (*synth, cryst struct*)
Wakatsuki, Y. *et al*, *J. Chem. Soc., Perkin Trans. 2*, 1980, 1344.

$C_{28}H_{18}Co_3O_8P$ Co-00296

Octacarbonyl-μ_3-ethylidyne(triphenylphosphine)tricobalt, 9CI

[15633-24-6]

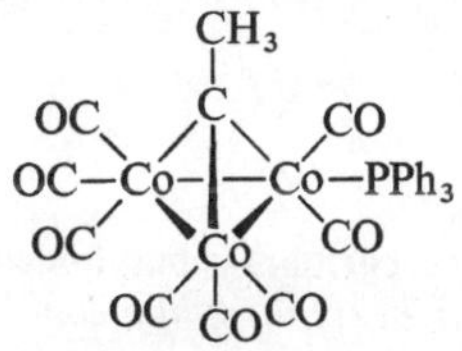

M 690.219

PPh_3 ligand equatorial in cryst. Fluxional in soln. Mod. air-stable dark-maroon cryst. (pet. ether). Sol. C_6H_6, $CHCl_3$. Mp >80° dec.

Brice, M.D. *et al*, *Inorg. Chem.*, 1970, **9**, 362 (*cryst struct*)
Matheson, T.W. *et al*, *J. Chem. Soc. (A)*, 1971, 1457 (*synth, ms, pmr, ir*)
Matheson, T.W. *et al*, *J. Organomet. Chem.*, 1975, **88**, 367 (*cmr, nmr*)

$C_{28}H_{20}Br_2CoN_4$ Co-00297

Dibromotetrakis(isocyanobenzene)cobalt

Dibromotetrakis(phenylisonitrile)cobalt. Dibromotetrakis(phenyl isocyanide)cobalt

$CoBr_2(CNPh)_4$

M 631.234

Green cryst. ($CHCl_3$).

Malatesta, L. *et al*, *Z. Anorg. Allg. Chem.*, 1953, **273**, 247.

$C_{28}H_{20}CoI_2N_4$ Co-00298

Diiodotetrakis(isocyanobenzene)cobalt, 10CI

Diiodotetrakis(phenyl isocyanide)cobalt

[61825-59-0]

$(PhNC)_4CoI_2$

M 725.235

α-form

Green cryst. (EtOH/$CHCl_3$).

β-form

Black cryst. (Me_2CO/diisopropyl ether).

Malatesta, L. *et al*, *Z. Anorg. Allg. Chem.*, 1953, **273**, 247.

$C_{29}H_5Co_7O_{24}Ti$ Co-00299

(η^5-2,4-Cyclopentadien-1-yl)bis[μ_4-[methanolato(4−)-*C*:*C*:*C*:*O*]]bis(nonacarbonyltricobalt)(tetracarbonylcobalt)-titanium, 10CI

(η-Cyclopentadienyl)bis[μ_3-(oxymethylidyne)cyclotris-(tricarbonylcobalt)](tetracarbonylcobaltio)titanium

[67421-95-8]

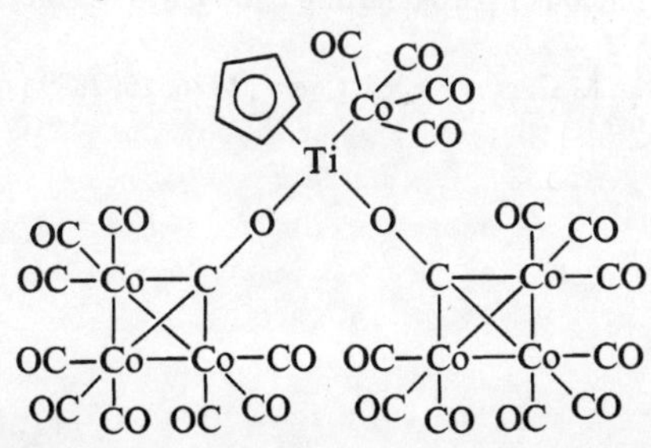

M 1197.757

Air-sensitive cryst. Black cryst. (toluene). Spar. sol. aromatic solvs., almost insol. aliphatic hydrocarbons. Mp 125° dec.

Schmid, G. *et al*, *Chem. Ber.*, 1978, **111**, 1239 (*synth, ir, nmr, cryst struct*)

$C_{29}H_{15}Co_4O_{11}P$ Co-00300

Undecacarbonyl(triphenylphosphine)tetracobalt, 9CI

[12336-57-1]

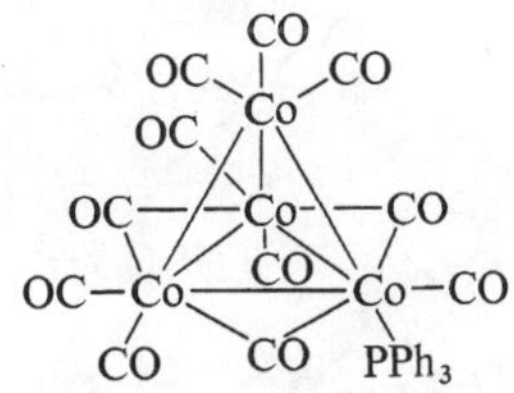

M 806.137

Black cryst. (heptane). ir ν_{CO} 2082, 2041, 2036, 2027, 2003, 1990 cm^{-1}.

Labroue, D. *et al*, *Inorg. Chim. Acta*, 1972, **6**, 387 (*synth, ir*)
Darensbourg, D.T. *et al*, *Inorg. Chem.*, 1981, **20**, 1911 (*cryst struct*)

$C_{29}H_{26}CoO_4P$ Co-00301

(η^5-2,4-Cyclopentadien-1-yl)[(2,3-η)-dimethyl 2-butynedioate](triphenylphosphine)cobalt, 10CI

η^5-Cyclopentadienyl(η^2-dimethyl acetylenedicarboxylate)triphenylphosphinecobalt

[50277-79-7]

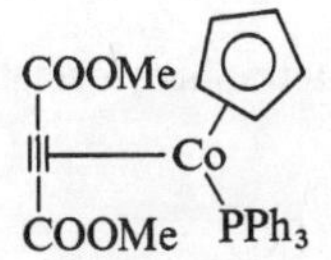

M 528.429

Chocolate-brown cryst. Mp 147-8°.

Wakatsuki, Y. *et al*, *J. Am. Chem. Soc.*, 1973, **95**, 5781 (*synth*)
Yasufuku, K. *et al*, *J. Organomet. Chem.*, 1976, **121**, 405 (*synth, ir, pmr*)

$C_{30}H_{10}Co_6HfO_{20}$ Co-00302

Bis(η^5-2,4-cyclopentadien-1-yl)bis[μ_4-[methanolato(4−)-*C*:*C*:*C*:*O*]]bis(nonacarbonyltricobalt)dihafnium, 10CI

Bis(η^5-cyclopentadienyl)di[μ_3-(oxymethylidyne)cyclotris(tricarbonylcobalt)]hafnium

[67422-55-3]

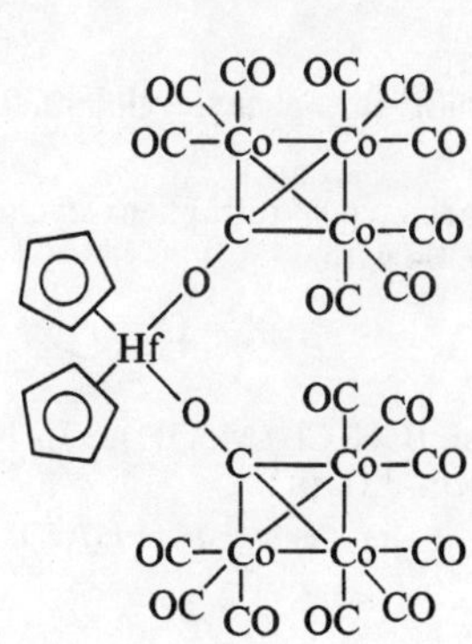

M 1222.486

Red cryst. (toluene).

Ishii, M. *et al*, *Ber. Bunsenges. Phys. Chem.*, 1978, **83**, 1026 (*cryst struct*)
Stutte, B. *et al*, *Chem. Ber.*, 1978, **111**, 1603 (*synth, ir, nmr, cryst struct*)

$C_{30}H_{10}Co_6O_{20}Zr$ Co-00303

Bis(η^5-2,4-cyclopentadien-1-yl)bis[μ_4-[methanolato(4–)-*C*:*C*:*C*:*O*]]bis(nonacarbonyltricobalt)dizirconium, 10CI

Bis(η-cyclopentadienyl)di[μ_3-(oxymethylidyne)cyclo-tris(tricarbonylcobalt)]zirconium

[67422-53-1]

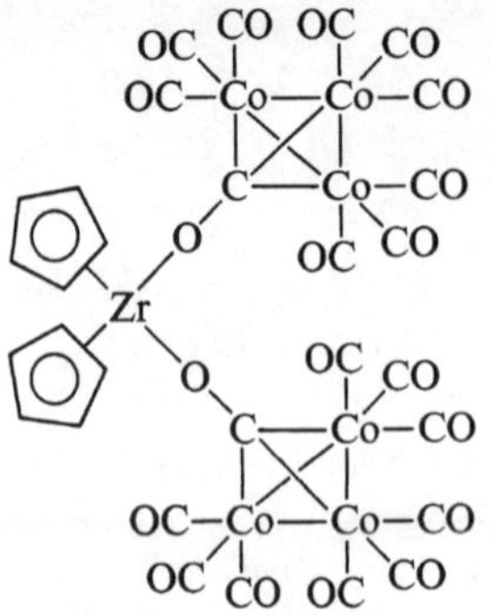

M 1135.216

Red cryst. (toluene). Mp 145° dec.

Ishii, M. *et al*, *Ber. Bunsenges. Phys. Chem.*, 1978, **83**, 1026 (*cryst struct*)
Stutte, B. *et al*, *Chem. Ber.*, 1978, **111**, 1603 (*synth, ir, nmr, cryst struct*)

$C_{30}H_{20}BrCoO_2$ Co-00304

Bromodicarbonyl(η^4-tetraphenylcyclobutadiene)cobalt, 8CI

[12243-59-3]

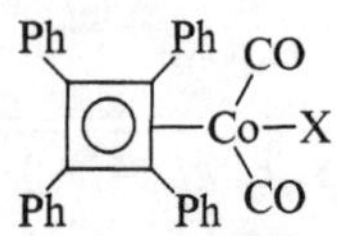

X = Br

M 551.324

Dark-red cryst. (CH_2Cl_2/MeOH). Mp 258-60° dec. ir ν_{CO} 2020, 2055cm^{-1} (C_6H_6 soln.).

Efraty, A. *et al*, *J. Am. Chem. Soc.*, 1967, **89**, 3744 (*synth, pmr, ir*)

$C_{30}H_{20}ClCoO_2$ Co-00305

Dicarbonylchloro(η^4-tetraphenylcyclobutadiene)cobalt, 8CI

[12243-49-1]

As Bromodicarbonyl(η^4-tetraphenylcyclobutadiene)cobalt, Co-00304 with

X = Cl

M 506.873

Dark-red cryst. (CH_2Cl_2/MeOH). Mp 200° dec. ir ν_{CO} 2028, 2062cm^{-1} (KBr).

Efraty, A. *et al*, *J. Am. Chem. Soc.*, 1967, **89**, 3744 (*synth, pmr, ir*)

$C_{30}H_{20}CoIO_2$ Co-00306

Dicarbonyliodo(η^4-tetraphenylcyclobutadiene)cobalt, 8CI

As Bromodicarbonyl(η^4-tetraphenylcyclobutadiene)cobalt, Co-00304 with

X = I

M 598.325

Dark-red cryst. (CH_2Cl_2/pet. ether). Mp 268-71° dec. ir ν_{CO} 2028, 2062cm^{-1} (KBr).

Efraty, A. *et al*, *J. Am. Chem. Soc.*, 1967, **89**, 3744 (*synth, pmr, ir*)

$C_{30}H_{54}Co_2HgO_6P_2$ Co-00307

Hexacarbonyl-μ-mercuriobis(tributylphosphine)dicobalt, 8CI

Bis(tricarbonyltributylphosphinocobalt)mercury

[18115-13-4]

$(H_3CCH_2CH_2CH_2)_3PCo(CO)_3$–Hg–$Co(CO)_3P(CH_2CH_2CH_2CH_3)_3$

M 891.157

Yellow solid (Me_2CO/MeOH). Mp 108-9°. ir ν_{CO} 2017, 1987.5, 1944.5, 1931 cm^{-1} (hexane).

Bonati, F. *et al*, *J. Chem. Soc. (A)*, 1967, 932 (*synth, ir*)
Van Reutergem, M. *et al*, *J. Mol. Struct.*, 1982, **82**, 43 (*ir*)

$C_{30}H_{54}Co_2O_6P_2$ Co-00308

Hexacarbonylbis(tributylphosphine)dicobalt, 10CI

Bis[tricarbonyltributylphosphinecobalt]

[14911-28-5]

$(H_3CCH_2CH_2CH_2)_3P$–$Co(CO)_3$–$Co(CO)_3$–$P(CH_2CH_2CH_2CH_3)_3$

M 690.567

Catalyst for hydroformylation of alkenes, alkene isomerisation, dehydration of formic acid. Catalyst percursor for isomerisation of pentene. Air-sensitive red cryst. (2-methylbutane). Mp 134-5°. ir ν_{CO} 1968.5, 1948.0 cm^{-1} (CCl_4).

Bower, L.M. *et al*, *J. Organomet. Chem.*, 1968, **13**, 235 (*ir, raman*)
Slaugh, L.H. *et al*, *J. Organomet. Chem.*, 1968, **13**, 469 (*synth, use*)
Ibers, J.A., *J. Organomet. Chem.*, 1968, **14**, 423 (*cryst struct*)
Kneise, W. *et al*, *J. Organomet. Chem.*, 1969, **17**, 133 (*use*)

$C_{30}H_{68}CoO_9P_3$ Co-00309

(η^3-2-Propenyl)tris(triisopropyl phosphite)cobalt

*(η^3-2-Propenyl)tris[tris(1-methylethyl phosphite-*P*)]-cobalt*, 9CI. *η^3-Allyltris(triisopropyl phosphite)cobalt*

[59830-93-2]

As (η^3-2-Propenyl)tris(trimethyl phosphite-*P*)cobalt, Co-00126 with

R = $(H_3C)_2CH$–

M 724.716

Homogeneous hydrogenation catalyst for arenes. Cryst. (Et_2O).

Rakowski, M.C. *et al*, *Inorg. Chem.*, 1976, **15**, 2379 (*synth, pmr, use*)

$C_{31}H_{22}As_2Co_2O_6$ **Co-00310**

Di-μ-carbonyltetracarbonyl[μ-[methylenebis[diphenylarsine]-*As,As'*]]dicobalt, 10CI

Hexacarbonyl[μ-bis(diphenylarsino)methane]dicobalt

[52615-17-5]

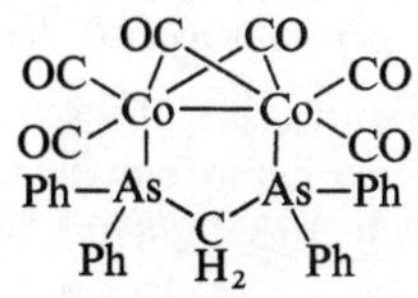

M 758.221

Orange-form

Orange cryst. (C_6H_6/hexane). Mp 143-5°. ir ν_{CO} 2050, 2016, 2004, 1969, 1832, 1770 cm^{-1} (nujol).

Red-form

Red cryst. (CH_2Cl_2/pet. ether). Mp 138-40°. ir ν_{CO} 2047, 2043, 2005, 2000, 1985, 1975, 1810, 1795 cm^{-1} (nujol).

Fukumoto, T. *et al*, *J. Organomet. Chem.*, 1974, **69**, 437 (*synth, ir, pmr, props*)

$C_{31}H_{22}Co_2O_6P_2$ **Co-00311**

Di-μ-carbonyltetracarbonyl[μ-[methylenebis[diphenylphosphine]-*P,P'*]]dicobalt, 9CI

Hexacarbonyl[μ-bis(diphenylphosphino)methane]dicobalt

[52615-19-7]

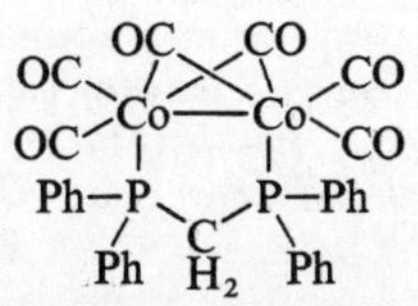

M 670.325

Orange cryst. (CH_2Cl_2/pet. ether). Mp 158-60°. ir ν_{CO} 2049, 2017, 1991, 1842, 1793 cm^{-1} (cyclohexane).

Fukumoto, T. *et al*, *J. Organomet. Chem.*, 1974, **69**, 437 (*synth, pmr, ir, props*)
Chia, L.S. *et al*, *Inorg. Chem.*, 1975, **14**, 482 (*synth, pmr, ir*)
Hanson, B.E. *et al*, *Organometallics*, 1983, **2**, 126 (*cmr*)
Laws, W.J. *et al*, *J. Chem. Soc., Chem. Commun.*, 1983, 1020 (*props*)

$C_{31}H_{32}CoP$ **Co-00312**

(η^5-2,4-Cyclopentadien-1-yl)(1,2,3,4-tetramethyl-1,3-butadiene-1,4-diyl)triphenylphosphinecobalt

[62745-35-1]

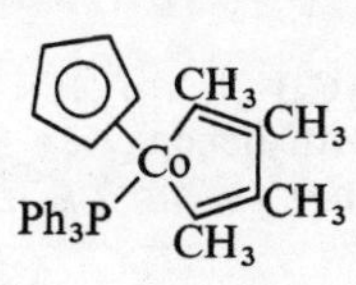

M 494.501

Intermed. in Co-catalysed cyclomerisation of alkynes. Yellow-brown solid (C_6H_6).

McAlister, D.R. *et al*, *J. Am. Chem. Soc.*, 1977, **99**, 1666 (*synth, pmr*)

$C_{33}H_{20}Co_3O_8P$ **Co-00313**

Octacarbonyl[μ_3-(phenylmethylidyne)](triphenylphosphine)tricobalt, 10CI

[13681-99-7]

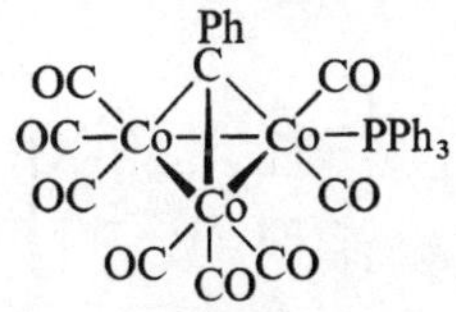

M 752.290

Mod. air-stable dark-brown cryst. (pentane at −10°). Sol. C_6H_6, $CHCl_3$. Mp >80° dec.

Matheson, T.W. *et al*, *J. Chem. Soc.* (*A*), 1971, 1457 (*synth, ir, pmr*)

$C_{33}H_{25}Co$ **Co-00314**

η^5-Cyclopentadienyl(η^4-tetraphenylcyclobutadiene)cobalt

[1,1',1'',1'''-(η^4-Cyclobutadiene-1,2,3,4-tetrayl)tetrakis[benzene]](η^5-2,4-cyclopentadien-1-yl)cobalt, 10CI

[1278-02-0]

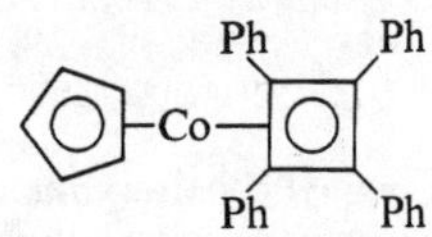

M 480.494

Previously thought to be $Co(PhC{\equiv}CPh)_2(C_5H_5)$. Intermed. in synth. of cyclobutadienes and alkyne cyclotrimerisation using cyclopentadienylcobalt(I) catalysts. Yellow cryst. (C_6H_6/heptane). Mp 262-4°. pmr δ 7.17-7.68 (Ph), δ 4.68 (C_5H_5).

Nakamura, A. *et al*, *Bull. Chem. Soc. Jpn.*, 1961, **34**, 452 (*synth*)
Rausch, M.D. *et al*, *J. Org. Chem.*, 1970, **35**, 3888 (*synth, ir, nmr*)
Nicholas, K.M. *et al*, *Transition Metal Organometallics in Organic Synth.*, Alper, H. (*Ed.*), Academic Press, New York, 1978, II, 1

$C_{34}H_{26}Co^{\oplus}$ **Co-00315**

η^6-Benzene(η^4-tetraphenylcyclobutadiene)cobalt(1+)

(η^6-Benzene)[1,1',1'',1''-(η^4-1,3-cyclobutadiene-1,2,3,4-tetrayl)tetrakis[benzene]]cobalt(1+), 10CI

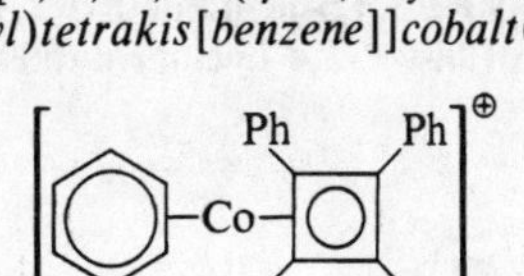

M 493.513 (ion)

Bromide:

$C_{34}H_{26}BrCo$ M 573.417

Yellow cryst. Mp 180-90° dec. pmr δ 7.5 (Ph-), 6.49 (π-C_6H_6).

Hexafluorophosphate: [12243-55-9].

$C_{34}H_{26}F_6CoP$ M 638.477

Yellow cryst. (MeOH aq.). Mp 370-8° dec.

Efraty, A. *et al*, *J. Am. Chem. Soc.*, 1967, **89**, 3744 (*synth, pmr*)
Seyferth, D. *et al*, *J. Organomet. Chem.*, 1970, **160**, 275 (*synth*)

$C_{34}H_{26}CoO^{\oplus}$ Co-00316

1-Hydroxy-2,3,4,5-tetraphenylcobaltocenium(1+), 10CI
π-(Hydroxytetraphenylcyclopentadienyl)(π-cyclopentadienyl)cobalt(1+)
[51155-56-7]

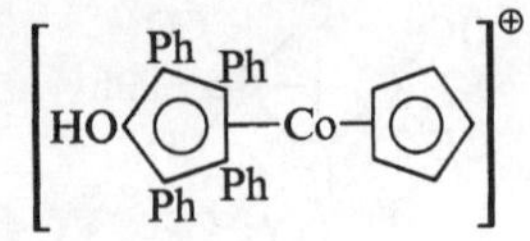

M 509.512 (ion)
pK_a 2.62 (50% EtOH aq.).
Bromide: [12427-59-7].
$C_{34}H_{26}BrCoO$ M 589.416
Yellow solid. Mp 289-90°.
Perchlorate: [12427-58-6].
$C_{34}H_{26}ClCoO_5$ M 608.963
Yellow solid. Mp 253-5°.
Tetrafluoroborate:
$C_{34}H_{26}BCoF_4O$ M 596.316
Yellow solid. Mp 245-50°.

Sheats, J.E. *et al*, *J. Org. Chem.*, 1970, **35**, 3245 (*synth, nmr, uv, ir*)
Sheats, J.E. *et al*, *J. Organomet. Chem.*, 1979, **173**, 107.

$C_{34}H_{28}CoP_2^{\oplus}$ Co-00317

1,1′-Bis(diphenylphosphinyl)cobaltocenium(1+), 10CI
Bis(η^5-diphenylphosphinocyclopentadienyl)cobalt(1+). Bis[η^5-(cyclopentadienyldiphenylphosphine)]-cobalt(1+)
[67292-48-2]

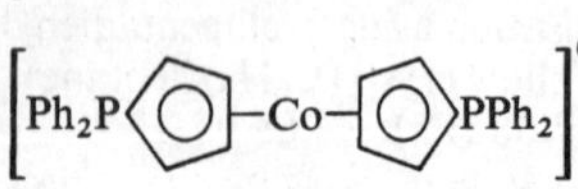

M 557.476 (ion)
Unusual cationic chelating ligand.
Hexafluorophosphate:
$C_{34}H_{28}CoF_6P_3$ M 702.440
Bright-yellow cryst. (CH_2Cl_2/Et_2O). Mp 189-91° dec.

Rudie, A.W. *et al*, *Inorg. Chem.*, 1978, **17**, 2859 (*synth, pmr, ir*)

$C_{34}H_{40}Co_2Mo_2O_4$ Co-00318

Di-μ-carbonyl[μ_4-(η^2-carbonyl-*C:C*)][μ-carbonyl-bis[(1,2,3,4,5-η)-1,2,3,4,5-pentamethyl-2,4-cyclopentadien-1-yl]dicobalt]bis(η^5-2,4-cyclopentadien-1-yl)dimolybdenum
[84474-41-9]

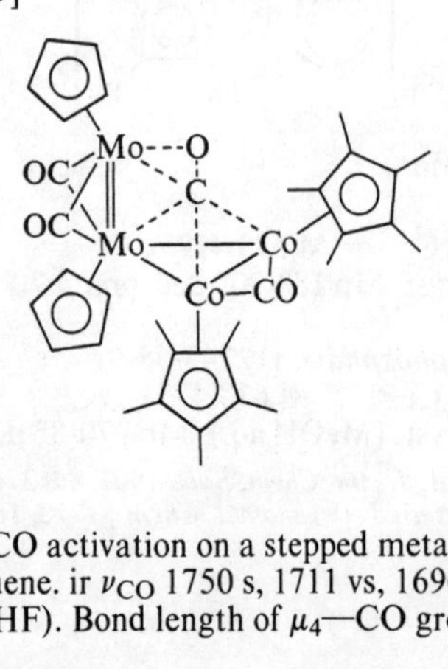

M 822.434
Model for CO activation on a stepped metal surface. Isol. from toluene. ir ν_{CO} 1750 s, 1711 vs, 1696 vs, 1633 vw cm^{-1} (THF). Bond length of μ_4—CO group 1.283(3) Å.

Brun, P. *et al*, *J. Chem. Soc., Chem. Commun.*, 1982, 926 (*synth, cryst struct, nmr, ir*)

$C_{35}H_{25}CoN_5^{\oplus}$ Co-00319

Pentakis(isocyanobenzene)cobalt(1+), 9CI
Pentakis(phenyl isocyanide)cobalt(1+)

$Co(NCPh)_5^{\oplus}$

M 574.549 (ion)
Perchlorate: [26544-45-6].
$C_{35}H_{25}ClCoN_5O_4$ M 674.000
Golden-yellow cryst. + $CHCl_3$ ($CHCl_3$). Mp 144-7°.
Tetrafluoroborate: [56063-73-1].
$C_{35}H_{25}BCoF_4N_5$ M 661.353
Cryst. Mp 109-12°. Trimorphic.

Becker, C.A.L., *J. Inorg. Nucl. Chem.*, 1973, **35**, 1875; 1975, **37**, 707 (*synth, pmr, ir*)

$C_{35}H_{25}CoN_5^{\oplus\oplus}$ Co-00320

Pentakis(isocyanobenzene)cobalt(2+), 10CI
Pentakis(phenyl isocyanide)cobalt(2+)

$[Co(CNPh)_5]^{\oplus\oplus}$

M 574.549 (ion)
Diperchlorate: [15271-38-2].
$C_{35}H_{25}Cl_2CoN_5O_8$ M 773.450
Yellow powder.
Diperchlorate, 1,2-dichloroethane adduct: [62206-99-9].
Green cryst. (1,2-dichloroethane). Loses $C_2H_4Cl_2$ on heating.

Sacco, A. *et al*, *Gazz Chim. Ital.*, 1959, **89**, 1800 (*synth*)
Jurnak, F.A. *et al*, *Inorg. Chem.*, 1975, **14**, 2585 (*cryst struct*)
Endicott, J.F. *et al*, *J. Am. Chem. Soc.*, 1979, **99**, 429 (*struct*)

$C_{35}H_{32}CoO_8P$ Co-00321

(η^5-2,4-Cyclopentadien-1-yl)[1,2,3,4-tetrakis(methoxycarbonyl)-1,3-butadiene-1,4-diyl](triphenylphosphine)cobalt, 9CI
Cyclopentadienyl(triphenylphosphine)-2,3,4,5-tetramethoxycarbonylcobaltacyclopentadiene
[55410-88-3]

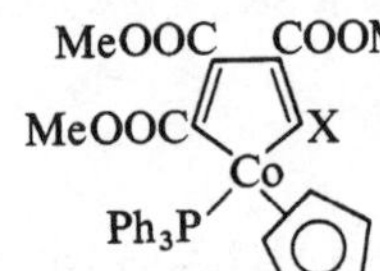

X = COOMe

M 670.540
Orange-red cryst. (C_6H_6). Mp 216-7°. Undergoes Diels-Alder reaction with dienophiles.

Yamazaki, H. *et al*, *J. Organomet. Chem.*, 1977, **139**, 157, 169 (*synth, nmr*)

$C_{36}H_{30}CoN_2O_2P_2^{\oplus}$ Co-00322

Dinitrosylbis(triphenylphosphine)cobalt(1+), 9CI
Bis(triphenylphosphine)dinitrosylcobalt(1+)

$(Ph_3P)_2Co(NO)_2^{\oplus}$

M 643.526 (ion)
Structure intermed. between tetrahedral and square planar. Catalyst for the reaction $2NO + CO \rightarrow CO_2 + N_2O$.

Chloride: [69890-22-8].
$C_{36}H_{30}ClCoN_2O_2P_2$ M 678.979
No phys. props. reported.
Hexafluorophosphate: [37739-55-2].
$C_{36}H_{30}CoF_6N_2O_2P_3$ M 788.490
Black cryst. (CH_2Cl_2/MeOH). ir ν_{NO} 1840, 1785 cm^{-1} (nujol).
Perchlorate: [24504 01 6].
$C_{36}H_{30}ClCoN_2O_6P_2$ M 742.977
Black cryst. Sol. Me_2CO, $CHCl_3$, insol. C_6H_6. Dec. at 224°.
Tetraphenylborate: [24507-62-8].
$C_{60}H_{50}BCoN_2O_2P_2$ M 962.758
Dark-brown cryst. (CH_2Cl_2/MeOH). Dec. at 145°.

Jackson, T.B. *et al*, *Inorg. Chem.*, 1966, **5**, 2046 (*synth*)
Bianco, T. *et al*, *Inorg. Chim. Acta*, 1969, **3**, 443 (*synth, ir*)
Kudak, J.A. *et al*, *Inorg. Chem.*, 1975, **14**, 3070 (*cryst struct*)
Reichert, B.E., *Acta Crystallogr., Sect. B*, 1976, **32**, 1934 (*cryst struct*)
Inorg. Synth., 1976, **16**, 16 (*synth*)
Ungermann, C.B. *et al*, *J. Am. Chem. Soc.*, 1976, **98**, 3862 (*synth, ir*)
Ballivet, D. *et al*, *Inorg. Chem.*, 1977, **16**, 945 (*synth*)
Bhaduri, S. *et al*, *J. Chem. Soc., Dalton Trans.*, 1977, 2085 (*synth*)

$C_{36}H_{34}CoO_3P$ Co-00323

(η^5-2,4-Cyclopentadien-1-yl)(1,2,3,4-tetraphenyl-1,3-butadiene-1,4-diyl)(trimethyl phosphite-*P*)cobalt, 10CI
Cyclopentadienyl(trimethyl phosphite)cobaltatetraphenylcyclopentadiene
[74658-32-5]

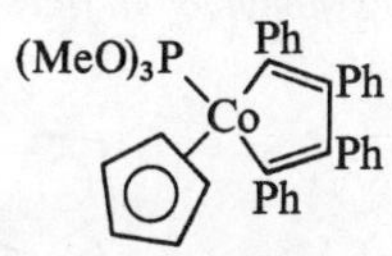

M 604.570
Red cryst. (C_6H_6/CH_2Cl_2/hexane). Mp 189-90°.

Yasufuku, K. *et al*, *J. Am. Chem. Soc.*, 1980, **102**, 4363 (*synth, pmr, ir*)

$C_{37}H_{30}CoNO_2P_2$ Co-00324

Carbonylnitrosylbis(triphenylphosphine)cobalt, 9CI
Nitrosylcarbonylbis(triphenylphosphine)cobalt. Bis(triphenylphosphine)nitrosylcarbonylcobalt
[14515-77-6]

$(Ph_3P)_2Co(NO)(CO)$

M 641.530
Red-brown leaflets. Mp 177° dec. ir ν_{CO} 1970, ν_{NO} 1670 cm^{-1} (nujol).

Hieber, W. *et al*, *Chem. Ber.*, 1963, **96**, 1643 (*synth, ir*)
Horrocks, W.D. *et al*, *Inorg. Chem.*, 1963, **2**, 723 (*synth, ir*)
Beck, W. *et al*, *Chem. Ber.*, 1965, **98**, 2657 (*ir*)
Albano, V.G. *et al*, *J. Organomet. Chem.*, 1972, **38**, 155 (*cryst struct*)
Dolcetti, G. *et al*, *Inorg. Chim. Acta*, 1972, **6**, 531 (*synth, ir*)
Finn, P. *et al*, *Inorg. Chem.*, 1972, **11**, 893 (*pe*)

$C_{37}H_{30}CoP$ Co-00325

Cyclopentadienyl(diphenylacetylene)triphenylphosphinecobalt, 8CI
(η^5-2,4-Cyclopentadien-1-yl)[1,1′-(η^2-1,2-ethynediyl)bis[benzene]](triphenylphosphine)cobalt, 10CI
[12172-41-7]

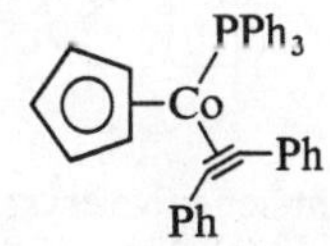

M 564.551
Catalyst for condensation of acetylenes with nitriles.
Black cryst. (C_6H_6/hexane). Mp 140° dec. ir $\nu_{C\equiv C}$ 1818 cm^{-1} (KBr).

Yamazaki, H. *et al*, *J. Organomet. Chem.*, 1967, **7**, P22; 1970, **21**, 431 (*synth, ir, pmr*)
Wakatsuki, Y. *et al*, *Tetrahedron Lett.*, 1973, 3383 (*use*)
Yamazaki, H. *et al*, *J. Organomet. Chem.*, 1977, **139**, 157 (*synth*)
Wakatsuki, Y. *et al*, *J. Am. Chem. Soc.*, 1979, **101**, 1123 (*use*)

$C_{37}H_{34}CoP$ Co-00326

Dibenzyl-π-cyclopentadienyl(triphenylphosphine)cobalt, 8CI
(η^5-2,4-Cyclopentadien-1-yl)bis(phenylmethyl)(triphenylphosphine)cobalt, 9CI
[12216-71-6]

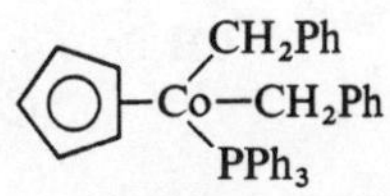

M 568.583
Dark-red cryst. (C_6H_6/hexane). Mp 102° dec.

Yamazaki, H. *et al*, *Bull. Chem. Soc. Jpn.*, 1965, **38**, 2212 (*synth, pmr*)
Yamazaki, H. *et al*, *J. Organomet. Chem.*, 1970, **21**, 431 (*synth, pmr*)

$C_{37}H_{82}CoOP_3$ Co-00327

Carbonylhydrotris(tributylphosphine)cobalt, 8CI
Hydridocarbonyltris(tributylphosphine)cobalt
[22829-69-2]

$[(H_3CCH_2CH_2CH_2)_3P]_3CoH(CO)$

M 694.909
Homogeneous catalyst for alkene and aldehyde hydrogenation and for alkene isomerisation. Green-yellow oil. Sol. most org. solvs. ir ν_{CO} 1883, ν_{CoH} 1908 cm^{-1} (heptane).

Pregaglia, G.F. *et al*, *J. Organomet. Chem.*, 1971, **30**, 387 (*synth, ir, pmr, use*)

$C_{38}H_{30}ClCoO_2P_2$ Co-00328

Dicarbonylchlorobis(triphenylphosphine)cobalt, 9CI
Chlorodicarbonylbis(triphenylphosphine)cobalt
[14564-33-1]

$(Ph_3P)_2CoCl(CO)_2$

M 674.988

Speier, G. *et al*, *Acta Chim. Acad. Sci. Hung.*, 1974, **82**, 25 (*synth*)

$C_{38}H_{30}CoIO_2P_2$ **Co-00329**

Dicarbonyliodobis(triphenylphosphine)cobalt, 9CI

[36210-11-4]

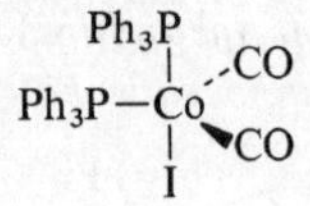

M 766.439

Chestnut-brown, light-sensitive cryst. (CH_2Cl_2/EtOH). Sol. $CHCl_3$, insol. Et_2O. Mp >170° dec. ir ν_{CO} 1985, 1920 cm^{-1} (KBr).

Heiber, W. *et al*, *Chem. Ber.*, 1965, **98**, 2530 (*synth, ir*)
Curtis, M.D., *Inorg. Chem.*, 1972, **11**, 802 (*synth, ir*)

$C_{38}H_{30}CoN_4^{\oplus}$ **Co-00330**

Dibenzylbis(1,10-phenanthroline)cobalt(1+), 8CI

Bis(1,10-phenanthroline)bis(phenylmethyl)cobalt(1+)

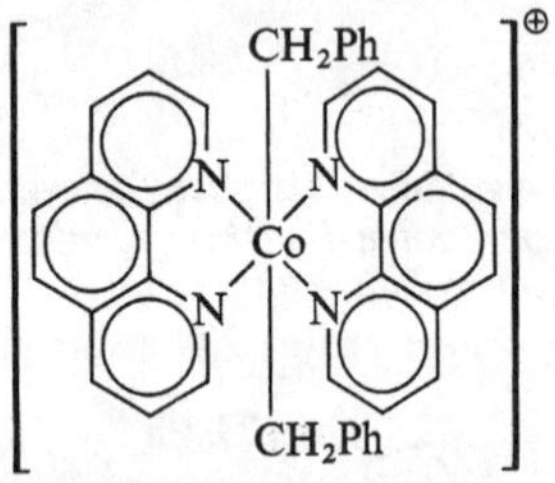

M 601.615 (ion)

Iodide: [28855-46-1].
$C_{38}H_{30}CoIN_4$ M 728.520
No phys. props. reported.

Mestroni, G. *et al*, *J. Organomet. Chem.*, 1970, **24**, 775 (*synth*)

$C_{39}H_{30}CoO_3P_2^{\oplus}$ **Co-00331**

Tricarbonylbis(triphenylphosphine)cobalt(1+), 10CI

[14243-08-4]

$[Co(CO)_3(PPh_3)_2]^{\oplus}$

M 667.545 (ion)

Tetrafluoroborate:
$C_{39}H_{30}BCoF_4O_3P_2$ M 754.349
Orange-yellow solid. Mp 221° dec. ir ν_{CO} 2010 cm^{-1}.

Attali, S. *et al*, *Inorg. Chim. Acta*, 1972, **6**, 475 (*synth, ir*)

$C_{39}H_{36}CoO_6P$ **Co-00332**

(η^5-2,4-Cyclopentadien-1-yl)(triphenylphosphine)[2,3,4-tris-(methoxycarbonyl)-1-phenyl-1-butene-1,4-diyl]cobalt

Cyclopentadienyl(triphenylphosphine)-2-phenyl-3,4,5-trimethoxycarbonylcobaltacyclopentadiene

[53559-30-1]

As (η^5-2,4-Cyclopentadien-1-yl)[1,2,3,4-tetrakis(methoxycarbonyl)-1,3-butadiene-1,4-diyl](triphenylphosphine)cobalt, Co-00321 with

X = Ph

M 690.617

Intermed. in Co-catalysed oligomerisation of alkynes. Red-brown cryst. (C_6H_6/hexane). Mp 177-8° dec.

Wakatsuki, Y. *et al*, *J. Am. Chem. Soc.*, 1979, **101**, 1123 (*synth, ir, cryst struct*)

$C_{40}H_{20}Co_6Hf_2O_{21}$ **Co-00333**

Tetrakis(η^5-2,4-cyclopentadien-1-yl)bis[μ_4-[methanolato(4−)-*C*:*C*:*C*:*O*]bis(nonacarbonyltricobalt)-μ-oxodihafnium, 10CI

μ-Oxobis[bis(η^5-cyclopentadienyl)[μ_3-(oxymethylidyne)cyclotris(tricarbonylcobalt)]hafnium]

[67422-50-8]

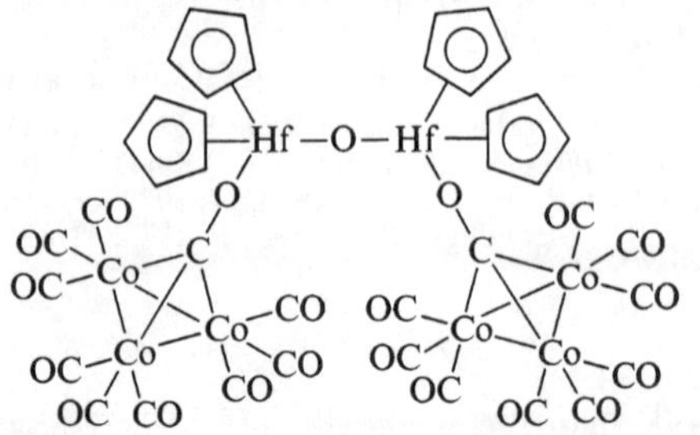

M 1547.165

Red cryst. (toluene). Mp 156-60°.

Stutte, B. *et al*, *Chem. Ber.*, 1978, **111**, 1603 (*synth, ir, nmr*)

$C_{40}H_{20}Co_6O_{21}Zr_2$ **Co-00334**

Tetrakis(η^5-2,4-cyclopentadien-1-yl)bis[μ_4-[methanolato(4−)-*C*:*C*:*C*:*O*]bis(nonacarbonyltricobalt)-μ-oxodizirconium, 10CI

μ-Oxobis[bis(η^5-cyclopentadienyl)[μ_3-(oxymethylidyne)cyclotris(tricarbonylcobalt)]zirconium]

[67422-52-0]

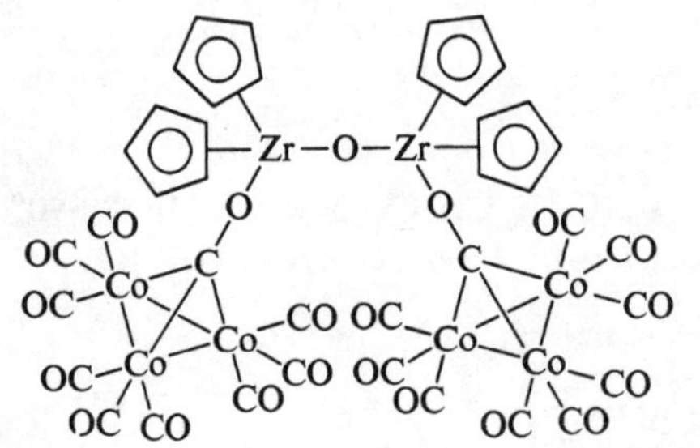

M 1372.625

Mp 160-4° dec.

Stutte, B. *et al*, *Chem. Ber.*, 1978, **111**, 1603 (*synth, ir, nmr*)

$C_{40}H_{72}Co_2N_8$ **Co-00335**

Octakis(2-isocyano-2-methylpropane)dicobalt, 10CI

*Octa(*tert*-butylisonitrile)dicobalt*

[64315-51-1]

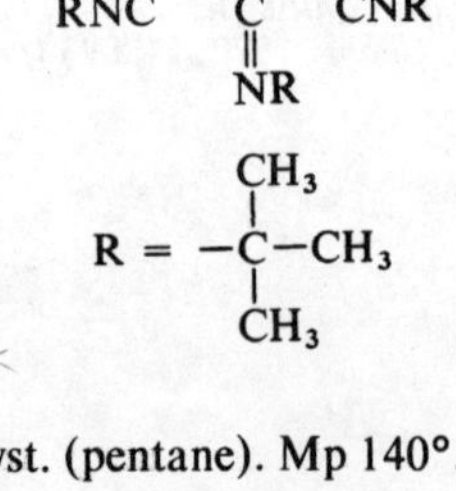

M 782.929

Orange-red cryst. (pentane). Mp 140°.

Carrol, W.E. *et al*, *J. Chem. Soc., Dalton Trans.*, 1980, 80 (*synth, ir, nmr, cryst struct*)

$C_{41}H_{35}CoP_2$ Co-00336

η^5-2,4-Cyclopentadien-1-ylbis(triphenylphosphine)cobalt, 9CI

[32993-07-0]

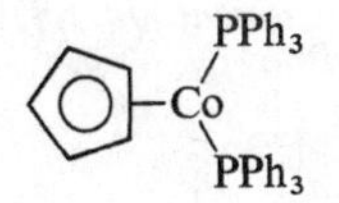

M 648.608

Air-sensitive cryst. (toluene).

Rinze, P.V. *et al*, *J. Organomet. Chem.*, 1969, **19**, 399 (*synth, ir*)
Yamazaki, H. *et al*, *Bull. Chem. Soc. Jpn.*, 1971, **44**, 2260 (*synth, ir, pmr*)

$C_{42}H_{30}Co_2O_6P_2$ Co-00337

Hexacarbonylbis(triphenylphosphine)dicobalt, 10CI, 9CI, 8CI

Bis(tricarbonyltriphenylphosphinecobalt)

[24212-54-2]

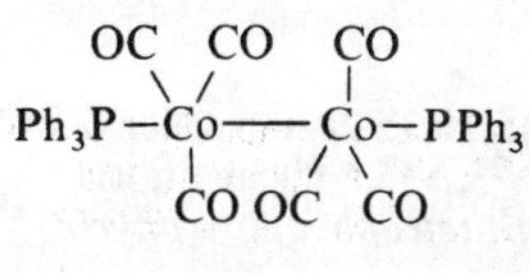

M 810.509

Catalyses hydroformylation of olefins, hydrogenation of olefins, aldehydes and ketones, cross-linking of epoxyresins, and the reaction of methanol with synthesis gas to give ethanol. Orange cryst. Mp 130° dec.

Hieber, W. *et al*, *Chem. Ber.*, 1958, **91**, 1230 (*synth*)
Vohler, O., *Chem. Ber.*, 1958, **91**, 1235 (*ir*)
Bower, L.M. *et al*, *J. Organomet. Chem.*, 1968, **13**, 235 (*ir*)
Manning, A.R., *J. Chem. Soc. (A)*, 1968, 1135 (*synth, ir*)
Speier, G. *et al*, *Inorg. Chim. Acta*, 1969, **3**, 126 (*formn*)
Kogami, K. *et al*, *CA*, 1973, **79**, 92400n (*use*)
Behrens, H. *et al*, *Z. Naturforsch., B*, 1977, **32**, 1217 (*synth*)
Poe, A.J. *et al*, *Inorg. Chem.*, 1978, **17**, 2330 (*uv*)

$C_{42}H_{30}Co_2O_{12}P_2$ Co-00338

Hexacarbonylbis(triphenyl phosphite)dicobalt, 10CI

[21118-36-5]

$Co_2(CO)_6[P(OPh)_3]_2$

M 906.506

Precursor for homogeneous hydrogenation catalyst. Air-stable red solid. Slow dec. in soln. ir ν_{CO} 1998, 1979 cm^{-1} ($CHCl_3$).

Manning, A.R., *J. Chem. Soc. (A)*, 1968, 1135 (*synth, ir*)
Reichel, C.L. *et al*, *J. Am. Chem. Soc.*, 1979, **101**, 6771 (*use*)

$C_{43}H_{32}Co_2O_4P_2$ Co-00339

Tetracarbonyl[μ-[1,1′-(η^2:η^2-1,2-ethynediyl)bis[benzene]][μ--[methylenebis[diphenylphosphine]-*P,P′*]]dicobalt, 10CI, 9CI

[μ-Diphenylethyne][μ-bis(diphenylphosphino)methane]-tetracarbonyldicobalt. [μ-Bis(diphenylphosphino)methane][μ-tolan]tetracarbonyldicobalt

[52659-27-5]

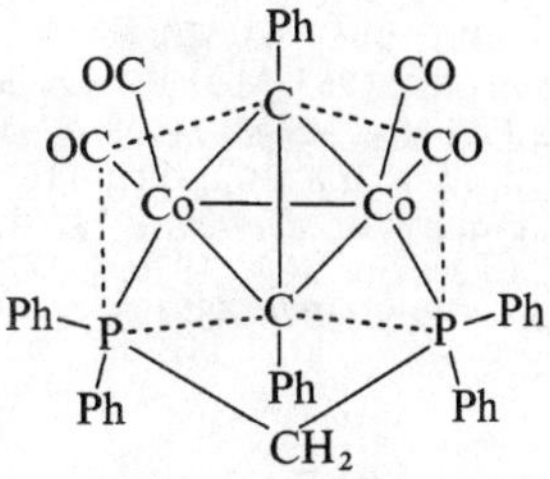

M 792.537

Brown or red solid (CH_2Cl_2/hexane). Mp 225° (205-7°). ir ν_{CO} 2025, 1999, 1972 cm^{-1} ($CHCl_3$).

Fukumoto, T. *et al*, *J. Organomet. Chem.*, 1974, **69**, 437 (*synth, ir, nmr*)
Chia, L.S. *et al*, *Inorg. Chem.*, 1975, **14**, 2521 (*synth, ir, pmr*)
Bird, P.H. *et al*, *Inorg. Chem.*, 1977, **16**, 1923 (*synth, ir, cryst struct*)

$C_{43}H_{41}CoO_2P_2$ Co-00340

(η^2-Ethene)(2,4-pentanedionato-*O,O′*)bis(triphenylphosphine)cobalt, 9CI

(Acetylacetonato)ethylenebis(triphenylphosphine)cobalt

[55471-71-1]

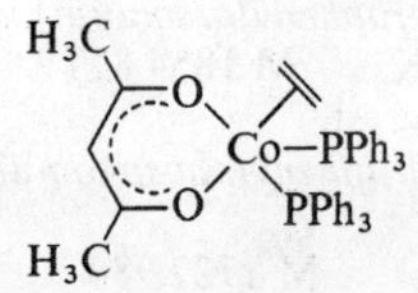

M 710.676

Air-sensitive yellow-brown cryst. Mp 85-7°. Dec. in aromatic solvs. or THF liberating ethylene.

Kubo, Y. *et al*, *J. Organomet. Chem.*, 1975, **84**, 369 (*synth, ir*)

$C_{45}H_{60}CoN_4O_{14}^{\oplus\oplus}$ Co-00341

Cobyrinic acid

[33593-50-9]

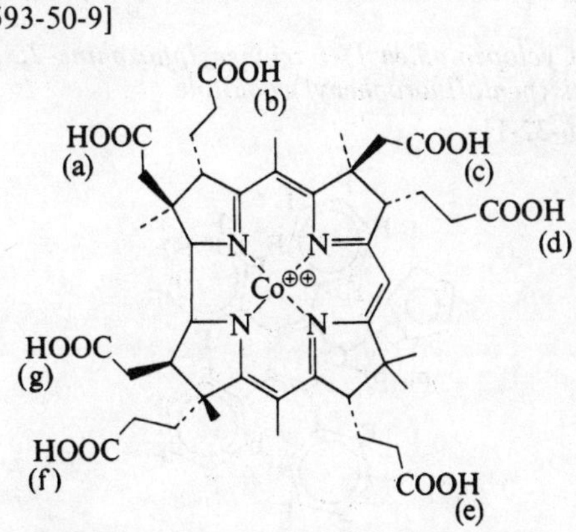

M 939.921 (ion)

Struct. approved as the basis of the nomenclature of Vitamin B_{12} derivs. Can be obtained by acid hydrol. of Vitamin B_{12}. Dark-red cryst.

abcdeg-*Hexaamide:* [14708-92-0]. *Cobyric acid. Factor V_{1a}.*

$C_{45}H_{66}CoN_{10}O_8$ M 934.012
Can be obt. by acid hydrol. of Vitamin B_{12}, also isol. from sewage sludge. Dark-red cryst. [M] −41 000 (405), +5 300 (436), 2 160 (578). Dec. >200°. λ_{max} 275, 319, 353, 403, 495, 527 nm (pH 5) (H_2O). Isol. in monocyano, dicyano or monocyano, monoaquo form.

IUPAC, *J. Am. Chem. Soc.*, 1960, **82**, 5582.
Bernhauer, K. *et al*, *Helv. Chim. Acta*, 1960, **43**, 696.
Bonnett, R., *Chem. Rev.*, 1963, **63**, 573.
Smith, E.L., *Vitamin B_{12}*, 1965, Methuen, London, 3rd. Ed., 62.
Woodward, R.B., *Pure Appl. Chem.*, 1968, **17**, 519 (*synth*)
Bonnett, R. *et al*, *J. Chem. Soc. (C)*, 1969, 1163.
Eschenmoser, A., *Naturwissenchaften*, 1974, **61**, 513.
Eschenmoser, A., *Chem. Soc. Rev.*, 1976, **5**, 377 (*synth*)
IUPAC, *Pure Appl. Chem.*, 1976, **48**, 495.

$C_{48}H_{30}Co^{\ominus\ominus\ominus\ominus}$ Co-00342

Hexakis(phenylethynyl)cobaltate(4−), 9CI

$[Co(C{\equiv}CPh)_6]^{\ominus\ominus\ominus\ominus}$

M 665.698 (ion)
See also Hexaethynylcobaltate(4−), Co-00107.
▷Alkynyl complexes potentially pyrophoric or explosive

Sodium tris(tetraphenylarsonium) salt: [59120-27-3].
$C_{120}H_{90}As_3CoNa$ M 1838.719
Light-brown cryst.

Sodium tris(tetraphenylphosphonium) salt: [59154-36-8].
$C_{120}H_{90}CoNaP_3$ M 1706.875
Orange-brown cryst.

Potassium tris(tetraphenylarsonium) salt: [59120-28-4].
$C_{120}H_{90}As_3CoK$ M 1854.827
Orange-brown cryst.

Potassium tris(tetraphenylphosphonium) salt: [59120-26-2].
$C_{120}H_{90}CoKP_3$ M 1722.984
Brown cryst.

Nast, R. *et al*, *Chem. Ber.*, 1976, **109**, 455 (*synth, ir*)

$C_{51}H_{20}CoF_{20}P$ Co-00343

(η^5-2,4-Cyclopentadien-1-yl)[1,2,3,4-tetrakis(pentafluorophenyl)-1,3-butadiene-1,4-diyl](triphenylphosphine)cobalt, 9CI

1-(π-Cyclopentadienyl)-1-triphenylphosphine-2,3,4,5-tetrakis(pentafluorophenyl)cobaltole

[61136-32-1]

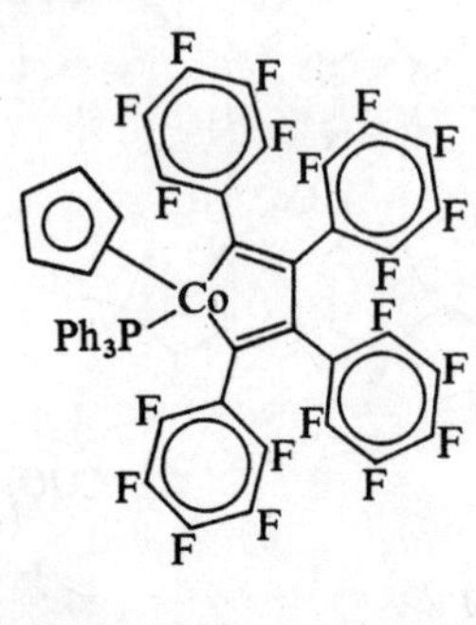

M 1102.594
Dark-red cryst. (pet. ether); stable in refluxing C_6H_6. Mp 259-61° (under N_2).

Gastinger, R.G. *et al*, *J. Am. Chem. Soc.*, 1976, **98**, 719 (*synth, nmr, ir, ms, cryst struct*)

$C_{51}H_{40}CoP$ Co-00344

(η^5-2,4-Cyclopentadien-1-yl)(1,2,3,4-tetraphenyl-1,3-butadien-1,4-diyl)(triphenylphosphine)cobalt, 10CI, 9CI

Cyclopentadienyl(triphenylphosphine)cobaltatetraphenylcyclopentadiene

[12124-09-3]

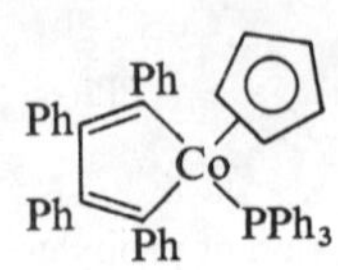

M 742.784
Catalyst for cocyclisation of 2 acetylenes and 1 nitrile to give substituted pyridines. Air-stable red-brown cryst. Mp 193-4° dec.

Yamazaki, H. *et al*, *J. Organomet. Chem.*, 1967, **7**, P22 (*synth, pmr*)
Yamazaki, H. *et al*, *J. Organomet. Chem.*, 1977, **139**, 157 (*synth*)
Yamazaki, H. *et al*, *J. Chem. Soc., Dalton Trans.*, 1978, 1278 (*use*)
Fieser, M. *et al*, *Reagents for Organic Synthesis*, Wiley, 1967-83, **5**, 173 (*use*)

$C_{52}H_{35}CoN_4$ Co-00345

(1-Phenylethenyl)[5,10,15,20-tetraphenyl-21*H*,23*H*-porphinato(2−)-N^{21},N^{22},N^{23},N^{24}]cobalt, 10CI

(α-Styryl)cobalt tetraphenylporphyrin

[75262-17-8]

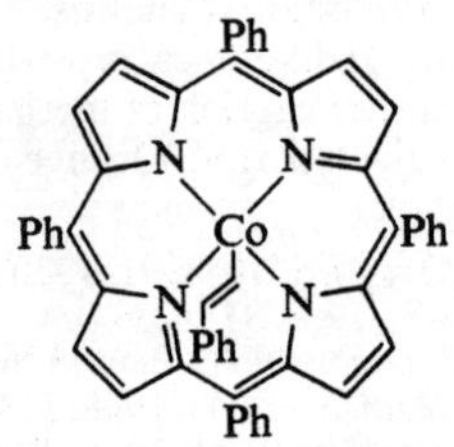

M 774.809
Model for the *in vivo* transformation of cytochrome P-450. Solid (CH_2Cl_2/MeOH). Rearranges to *N*-α-styrylporphyrin on acid treatment.

Callot, H.J. *et al*, *J. Organomet. Chem.*, 1980, **193**, 111 (*synth, pmr, uv*)
Callot, H.J. *et al*, *Tetrahedron Lett.*, 1980, **21**, 1335 (*props*)

$C_{54}H_{45}ClCoP_3$ Co-00346

Chlorotris(triphenylphosphine)cobalt

[26305-75-9]

$CoCl(PPh_3)_3$

M 881.257
Catalyst for cyclodimerisation of butadiene and for dimerisation of alkenes. Green solid (EtOH). Mp 177° dec. Stable in solid state, dec. in soln.

Aresta, M. *et al*, *Inorg. Chim. Acta*, 1969, **3**, 227 (*synth, ir*)
Cairns, M.A. *et al*, *J. Organomet. Chem.*, 1974, **64**, C19.
Kawakami, K. *et al*, *Bull. Chem. Soc. Jpn.*, 1978, **51**, 21.
Kanai, H. *et al*, *Bull. Chem. Soc. Jpn.*, 1981, **54**, 1015.

$C_{54}H_{46}CoN_2P_3$ Co-00347

(Dinitrogen)hydrotris(triphenylphosphine)cobalt, 10CI, 9CI
Hydrido(dinitrogen)tris(triphenylphosphine)cobalt
[16920-54-0]

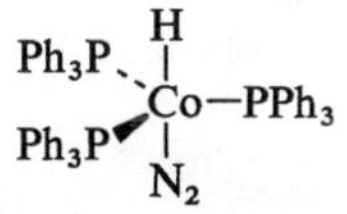

M 874.825

Catalyst for dimerisation of propylene and isom. of olefins. Reagent for formation of ketones from RCOCl and olefins. Air-sensitive orange cryst. (THF). Dec. at 80°→N_2 + H_2. ir ν_{N_2} 2088 cm^{-1} (C_6H_6).

Davis, B.R. *et al*, *Inorg. Chem.*, 1969, **8**, 2719 (*cryst struct*)
Misono, A. *et al*, *Inorg. Chem.*, 1969, **8**, 168 (*synth, ir, pmr*)
Inorg. Synth., 1970, **12**, 12, 18 (*synth*)
Yamamoto, A. *et al*, *J. Am. Chem. Soc.*, 1971, **93**, 371 (*synth*)
Kachapina, L.M. *et al*, *Zh. Strukt. Khim.*, 1974, **15**, 37 (*ir, raman*)
Petit, F. *et al*, *J. Organomet. Chem.*, 1980, **202**, 319 (*use*)
Fieser, M. *et al*, *Reagents for Organic Synthesis*, Wiley, 1967-83, **5**, 331 (*use*)

$C_{54}H_{48}CoP_3$ Co-00348

Trihydrotris(triphenylphosphine)cobalt, 9CI
Trihydridotris(triphenylphosphine)cobalt. Tris(triphenylphosphine)cobalt trihydride
[21329-68-0]

$H_3Co(PPh_3)_3$

M 848.828

Olefin hydrogenation and isomerisation catalyst. Air-sensitive pale-yellow needles (C_6H_6/EtOH or pet. ether). Mp 80° dec. ir $\nu_{Co\text{-}H}$ 1933m, 1745s cm^{-1} (nujol). Releases H_2 on dec.

Sacco, A. *et al*, *Inorg. Chim. Acta*, 1968, **2**, 127 (*synth, ir*)
Inorg. Synth., 1970, **12**, 18 (*synth, ir*)
Yamamoto, A. *et al*, *J. Am. Chem. Soc.*, 1971, **93**, 371 (*synth*)

$C_{55}H_{48}CoP_3$ Co-00349

Methyltris(triphenylphosphine)cobalt, 10CI
[30677-57-7]

$CoMe(PPh_3)_3$

M 860.839

Alkylating agent for alkynes. Homogeneous catalyst for aldol condensation. Yields biphenyls on reaction with PhCl in THF. Orange cryst. (THF/Et_2O). Stable in THF soln. Mp 78-81° dec.

Yamamoto, A. *et al*, *J. Am. Chem. Soc.*, 1971, **93**, 371 (*synth, ir*)
Michman, M. *et al*, *J. Organomet. Chem.*, 1976, **122**, 77 (*synth*)
Nussbaum, S. *et al*, *J. Organomet. Chem.*, 1979, **182**, 555.
Michman, M. *et al*, *J. Organomet. Chem.*, 1981, **205**, 111 (*use*)

$C_{56}H_{48}Co_2O_4P_4$ Co-00350

Di-μ-carbonyldicarbonylbis[1,2-ethanediylbis[diphenylphosphine]-*P,P'*]dicobalt, 9CI
Bis[1,2-bis(diphenylphosphino)ethane]tetracarbonyldicobalt
[18757-99-8]

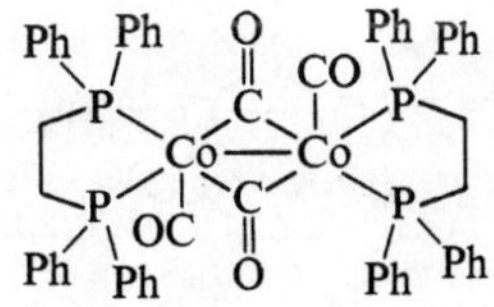

M 1026.754

Dark-red cryst. + $2CH_2Cl_2$ (MeCN/CH_2Cl_2). Mp 223-8°. ir ν_{CO} 1970, 1920, 1753, 1728 cm^{-1} (CH_2Cl_2).

Pedersen, S.E. *et al*, *Inorg. Chem.*, 1975, **14**, 2360 (*synth, ir*)
Behrens, H. *et al*, *Z. Anorg. Allg. Chem.*, 1967, **356**, 8 (*synth, ir*)

$C_{56}H_{49}CoNO_9P_3$ Co-00351

Acetonitrilehydrotris(triphenyl phosphite-*P*)cobalt, 10CI
[58527-60-9]

$(H_3CCN)CoH[P(OPh)_3]_3$

M 1031.859

Alkene hydrogenation catalyst giving good selectivity with little isomerisation. Yellow solid (C_6H_6/MeCN), unstable in air. Mp 145° dec.

Gosser, L.W., *Inorg. Chem.*, 1976, **15**, 1348 (*synth, nmr*)

$C_{56}H_{49}CoP_3$ Co-00352

(η^2-Ethene)tris(triphenylphosphine)cobalt, 9CI
(Ethylene)tris(triphenylphosphine)cobalt, 8CI. Bis[ethylenetris(triphenylphosphine)cobalt]
[12304-55-1]

$Co(H_2C{=}CH_2)(PPh_3)_3$

M 873.858

Struct. unknown. Monomeric or dimeric. Catalyst for ethylene dimerisation. Orange solid. V. air-sensitive. Mp 81-3° dec. Dec. in aromatic solvs. or THF at r.t. Paramagnetic.

Yamamoto, A. *et al*, *J. Organomet. Chem.*, 1975, **84**, 369 (*synth, ir, props*)
Speier, G., *Hung. J. Ind. Chem.*, 1975, **3**, 449 (*use*)

$C_{58}H_{41}CoO_2$ Co-00353

(1-Hydroxy-2,3,4,5-tetraphenylcyclopentadienyl)(tetraphenylcyclopentadienone)cobalt
Hydrobis(tetraphenylcyclopentadienone)cobalt. (1-Hydroxy-2,3,4,5-tetraphenylcyclopentadienyl)(tetracyclone)cobalt

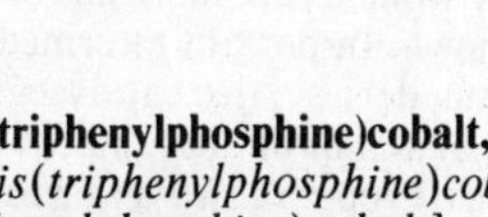

M 828.894

Red cryst. ($CHCl_3$/pet. ether). Mp 295-300° dec. Forms salts on treatment with RONa.

Weiss, E. *et al*, *J. Inorg. Nucl. Chem.*, 1959, **11**, 42 (*synth, ir*)
Sneider, A., *Z. Phys. Chem.*, 1962, **31**, 249.

$C_{60}H_{45}Co_3O_6P_3$ Co-00354

Tri-μ-carbonyltricarbonyltris(triphenylphosphine)tricobalt, 9CI

Tris(dicarbonyltriphenylphosphinecobalt). Hexacarbonyltris(triphenylphosphine)tricobalt

[23777-31-3]

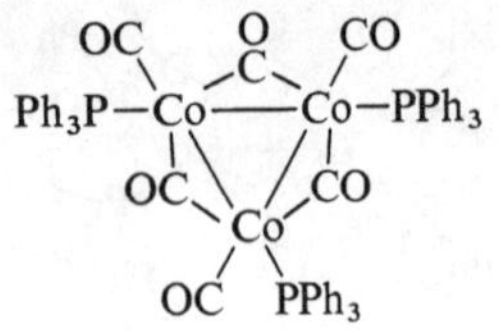

M 1131.733

Dark-green air-sensitive paramagnetic cryst. (toluene). Sol. C_6H_6, toluene. Mp 141° dec.

Pregaglia, G.F. *et al*, *J. Organomet. Chem.*, 1971, **33**, 73 (*synth, ir*)

$C_{62}H_{89}CoN_{13}O_{14}P$ Co-00355

Vitamin B_{12s}

Cobinamide-Co(I) *dihydrogen phosphate(ester) 3′-ester with 5,6-dimethyl-1-α-D-ribofuranosyl-1*H*-benzimidazole*, 10CI. *Hydridocobalamin. Cob(I)alamin*

[18534-66-2]

As Vitamin B_{12}, Co-00359 with

R = H, R′ = $CONH_2$

M 1330.371

Under neutral conditions is not protonated and exists as $[Co(I)]^{\ominus}$ which is one of the most powerful nucleophiles known. Important intermed. in the synth. of alkylcobalamin derivs. Also catalyses the asymmetric reduction of olefins and α,β-unsaturated carbonyl compds. Prepd. *in situ* as a grey-green air-sensitive soln. in H_2O or alcohols. pK_a 1. Diamagnetic. Dec. into vitamin B_{12r} and hydrogen ($t_{1/2}$ 355m at pH 10.0, 87m at pH 8.0).

Das, P.K. *et al*, *Biochim. Biophys. Acta*, 1967, **161**, 646 (*props*)
Schrauzer, G.N. *et al*, *J. Am. Chem. Soc.*, 1969, **91**, 3341 (*props, use*)
Dolphin, D. *et al*, *Methods Enzymol.*, 1971, **18C**, 34; *CA*, **75**, 76999b (*synth, props*)
Schrauzer, G.N. *et al*, *J. Am. Chem. Soc.*, 1971, **93**, 4060 (*synth, uv, props*)
Moskophidis, M., *Z. Naturforsch., C*, 1979, **34**, 689; *CA*, **92**, 71440d (*synth, uv, cd*)
Espenson, J.H. *et al*, *Inorg. Chem.*, 1980, **19**, 3549 (*synth, props*)
Fischli, A. *et al*, *Helv. Chim. Acta*, 1980, **63**, 529 (*use*)

$C_{62}H_{90}CoN_{15}O_{16}P$ Co-00356

Vitamin B_{12c}

Nitrosocobalamin. Nitritocobalamin. Nitrocobalamin

[20623-13-6]

As Vitamin B_{12}, Co-00359 with

R = ONO, R′ = $CONH_2$

M 1391.391

Antipernicious anaemia factor isol. from *Streptomyces griseus*. Red cryst. (Me_2CO aq.).

Bonnett, R., *Chem. Rev.*, 1963, **63**, 573.

$C_{62}H_{92}CoN_{13}O_{14}P$ Co-00357

Vitamin B_{12b}

Hydroxocobalamin. Aquocobalamin

As Vitamin B_{12}, Co-00359 with

R = OH, R′ = $CONH_2$

M 1333.394

Red cryst. (Me_2CO aq.).

Bonnett, R., *Chem. Rev.*, 1963, **63**, 573.

$C_{63}H_{87}CoN_{13}O_{15}P$ Co-00358

Vitamin B_{12} monocarboxylic acid

Cobinic acid abcdg-*pentamide*, 9CI

[26264-28-8]

As Vitamin B_{12}, Co-00359 with

R = CN, R′ = OH

M 1356.365

Prod. by mild acid hydrol. of Vitamin B_{12}. Also formed as a metab. in *Propionibacterium shermanii*. Red cryst. (H_2O). λ_{max} 277, 360 and 545 nm.

Armitage, J.B. *et al*, *J. Chem. Soc.*, 1953, 3849.
Bernhauer, K. *et al*, *Biochem. Z.*, 1966, **344**, 289.
Moore, F.M. *et al*, *Nature (London)*, 1967, **214**, 130 (*cryst struct*)

$C_{63}H_{88}CoN_{14}O_{14}P$ Co-00359

Vitamin B_{12}

Cyanocobalamin, BAN. *α-(5,6-Dimethylbenzimidazoyl-)cobamide cyanide. Numerous proprietary synonyms*

[68-19-9]

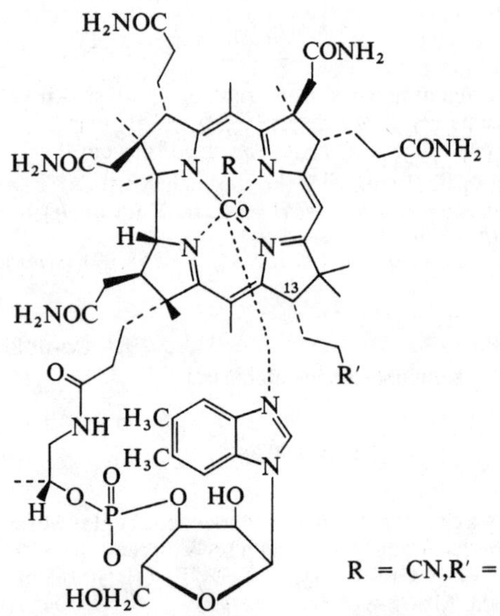

R = CN, R′ = $CONH_2$

M 1355.381

Antipernicious anaemia factor isol. from liver extracts; now obt. comly. from fermentation liquors of *Streptomyces griseus* and other microorganisms, e.g. *Propionibacterium shermanii*. Red needles. Sol. H_2O, alcohols, AcOH. Mp >300°, darkens at 210-20°. $[\alpha]^{23}_{656}$−59°. λ_{max} 279, 306, 322, 361, 520, 550 nm. Isotopic variants contg. ^{57}Co ($t_{1/2}$ 270d), ^{58}Co ($t_{1/2}$ 71.3d) and ^{60}Co ($t_{1/2}$ 5.26y) are prepd. by fermentation in the presence of radiocobalt and are used for diagnostic purposes.

▷GG3750000.

13-Epimer: [32627-38-8]. *Cyanoneocobalamin. Neovitamin* B_{12}. Formed by acid-catalysed epimerisation. Red cryst. (Me_2CO aq.). CD; λ_{max} 330 nm, $\Delta\epsilon$ + 8.4 (0.1M KCN). Dec. at ca. 200° without melting.

Bonnett, R., *Chem. Rev.*, 1963, **63**, 573.
Nockolds, C.K. *et al*, *Nature (London)*, 1967, **214**, 129 (*cryst struct*)
Pratt, J.M., *Inorg. Chem. of Vit. B_{12}*, 1972, Academic Press (*rev*)
Stöckli-Evans, H. *et al*, *J. Chem. Soc., Perkin Trans. 2*, 1972, 605 (*cryst struct*)
Bonnett, R., *Philos. Trans. R. Soc. London*, 1976, **273**, 295 (*epimer*)
Rajoria, D.S. *et al*, *J. Inorg. Nucl. Chem.*, 1977, **39**, 1291 (*ir*)
Schrauzer, G.N., *Angew. Chem.*, 1977, **89**, 239.
Florent, J., *Microbial Technology*, Peppler, H.J. *et al*, Eds., Academic Press, 1979, 2nd Ed., **1**, 497 (*rev, manuf*)
Toraya, T., *Vitamins*, 1977, **51**, 87.
Anton, D.L. *et al*, *J. Am. Chem. Soc.*, 1980, **102**, 2215 (*cmr*)
Battersby, A.R. *et al*, *Nature (London)*, 1980, **285**, 17 (*biosynth*)

$C_{63}H_{91}CoN_{13}O_{14}P$ Co-00360

Methylcobalamin

Methylcobinamide hydroxide dihydrogen phosphate(ester) inner salt 3′-ester with 5,6-dimethyl-1-α-D-ribofuranosyl-1H-benzimidazole, IOCI

[13422-55-4]

As Vitamin B_{12}, Co-00359 with

$$R = Me, R' = CONH_2$$

M 1344.398

Microorganisms effect Me transfer to heavy metals, e.g. Hg. Red needles (Me_2CO aq.), stable at r.t. in the absence of strong light. Sol. H_2O, alcohols. pK_a 3.3-3.5 (2.72). Co-methyl bond cleaved by light. In aq. soln. in equilibrium with methylcobinamide, i.e. axial coordination site occupied by H_2O not 5,6-dimethylbenzimidazole.

13-Epimer: [53701-54-5]. *Methyl-13-epicobalamin.* $C_{63}H_{91}CoN_{13}O_{14}P$ M 1344.398
As methylcobalamin with the stereochemistry at C13 reversed. Red cryst. (MeOH aq.). pK_a 2.2.

Hogenkamp, H.P.C. *et al*, *J. Biol. Chem.*, 1965, **240**, 3641 (*synth, props*)
Dolphin, D. *et al*, *Methods Enzymol.*, 1971, **18C**, 34; *CA*, **75**, 76999b (*synth, uv, props*)
Tkachuck, R.D. *et al*, *Biochemistry*, 1974, **13**, 2645 (*epimer*)
Hogenkamp, H.P.C. *et al*, *J. Chem. Soc., Dalton Trans.*, 1975, 2628 (*cmr*)
Hensens, O.D. *et al*, *Philos. Trans. R. Soc. London, Ser. B*, 1976, **273**, 353; *CA*, **84**, 160891d (*pmr*)
Rajoria, D.S. *et al*, *J. Inorg. Nucl. Chem.*, 1977, **39**, 1291 (*ir*)
Chemaly, S.M. *et al*, *J. Chem. Soc., Dalton Trans.*, 1980, 2259, 2267 (*props, uv*)
Wood, J.M., *CA*, 1982, **96**, 138055k (*props*)

$C_{63}H_{93}CoN_{13}O_{15}P$ Co-00361

Methylcobinamide a

α-Methylcobinamide

[61849-79-4]

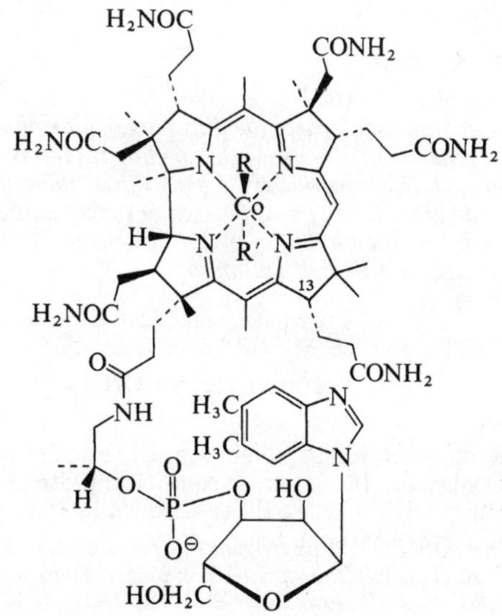

$$R = OH_2, R' = Me$$

M 1362.413

Yellow cryst. Sol. H_2O. Converted to methylcobalamin on heating at 95° for several hours in O_2-free aq. medium; light-sensitive.

Wilhelm, F. *et al*, *Z. Naturforsch., B*, 1968, **23**, 1119 (*synth, ir*)
Friedrich, W. *et al*, *Z. Naturforsch., B*, 1969, **24**, 588 (*synth, uv*)
Moskophidis, M., *Z. Naturforsch., C*, 1981, **36**, 497 (*cd, ord*)

$C_{63}H_{93}CoN_{13}O_{15}P$ Co-00362

Methylcobinamide b

Methylcobinamide hydroxide monohydrate dihydrogen phosphate(ester) inner salt 3′-ester with 5,6-dimethyl-1-α-D-ribofuranosyl-1H-benzimidazole, IOCI.
β-Methylcobinamide

[18662-90-3]

As Methylcobinamide a, Co-00361 with

$$R = Me, R' = OH_2$$

M 1362.413

In soln. in equilibrium with methylcobalamin. Sol. H_2O, alcohols. Heating at 95° for several hours in O_2-free aq. medium gives several percent of Methylcobinamide a, Co-00361 .

Firth, R.A. *et al*, *Biochemistry*, 1967, **6**, 2178 (*cd, uv*)
Matsumoto, J. *et al*, *CA*, 1976, **84**, 8884w (*synth*)
Brown, K.L. *et al*, *Vitam. B_{12} Proc. Eur. Symp., 3rd*, 1979, 199; *CA*, **92**, 176128m (*props*)

$C_{64}H_{89}CoN_{13}O_{14}P$ Co-00363

Ethynylcobalamin

Ethynylcobinamide hydroxide dihydrogen phosphate(ester) inner salt 3′-ester with 5,6-dimethyl-1-α-D-ribofuranosylbenzimidazole, IOCI

[15041-10-8]

As Vitamin B_{12}, Co-00359 with

$$R = -C{\equiv}CH, R' = CONH_2$$

M 1354.393

Deep-red needles (Me_2CO aq.), light-sensitive esp. in soln. Sol. H_2O, alcohols. In aq. soln. in equilibrium with ethynylcobinamide, i.e. axial site occupied by H_2O not 5,6-dimethylbenzimidazole.

Firth, R.A. *et al*, *Biochemistry*, 1967, **6**, 2178 (*cd, uv*)
Firth, R.A. *et al*, *J. Chem. Soc.* (*A*), 1968, 2419 (*uv, props*)
Dolphin, D. *et al*, *Methods Enzymol.*, 1971, **18C**, 34; *CA*, **75**, 76999b (*synth, uv*)

$C_{64}H_{91}CoN_{13}O_{14}P$ Co-00364

Vinylcobalamin

Ethenylcobinamide hydroxide dihydrogen phosphate(ester) inner salt 3′-ester with 5,6-dimethyl-1-α-D-ribofuranosyl-1H-benzimidazole, 10CI. *Vinylcobinamide hydroxide dihydrogen phosphate(ester) inner salt 3′-ester with 5,6-dimethyl-1-α-D-ribofuranosyl-1H-benzimidazole*, 8CI. *Ethenylcobalamin*

[15053-74-4]

As Vitamin B_{12}, Co-00359 with

R = –CH=CH$_2$, R′ = CONH$_2$

M 1356.409

Red needles (Me_2CO aq.), light-sensitive esp. in soln. Sol. H_2O, alcohols. In aq. soln. in equilibrium with vinylcobinamide, i.e. axial site occupied by H_2O not 5,6-dimethylbenzimidazole.

Firth, R.A. *et al*, *Biochemistry*, 1967, **6**, 2178 (*cd, uv*)
Hill, H.A.O. *et al*, *J. Chem. Soc.* (*A*), 1968, 564 (*pmr, uv*)
Firth, R.A. *et al*, *J. Chem. Soc.* (*A*), 1968, 2419 (*uv, props*)
Dolphin, D. *et al*, *Methods Enzymol.*, 1971, **18C**, 34; *CA*, **75**, 76999b (*synth, uv, props*)

$C_{72}H_{61}CoO_{12}P_4$ Co-00365

Hydrotetrakis(triphenyl phosphite-*P*)cobalt, 9CI

Hydrotetrakis(phosphorous acid)cobalt dodecaphenyl ester, 8CI

[24651-64-7]

HCo[P(OPh)$_3$]$_4$

M 1301.095

Olefin isomerisation and hydrogenation catalyst. Air-sensitive pale-yellow platelets (C_6H_6/MeOH). Mp 165-8°.

Levison, J.J. *et al*, *J. Chem. Soc.* (*A*), 1970, 96 (*synth, ir, nmr*)
Inorg. Synth., 1971, **13**, 105 (*synth*)
Gosser, L.W., *Inorg. Chem.*, 1977, **16**, 430 (*synth*)

$C_{72}H_{72}Co_2N_8$ Co-00366

Bis[μ-[[(2,6-dimethylphenyl)imino]methylene]hexakis(2-isocyano-1,3-dimethylbenzene)]dicobalt, 10CI

Octa(2,6-dimethylphenyl isocyanide)dicobalt

[64707-28-4]

As Octakis(2-isocyano-2-methylpropane)dicobalt, Co-00335 with

R = 2,6-dimethylphenyl

M 1167.281

Red-brown cryst. (C_6H_6/hexane). Mp 194-6°.

Yamamoto, Y. *et al*, *Inorg. Chem.*, 1978, **17**, 3111 (*synth, ir, pmr, uv*)

$C_{72}H_{100}CoN_{18}O_{17}P$ Co-00367

Coenzyme B_{12}

(5′-Deoxyadenosine-5′)cobinamide hydroxide dihydrogen phosphate(ester) inner salt 3′-ester with 5,6-dimethyl-1-α-D-ribofuranosyl-1H-benzimidazole, 10CI. *5′-Deoxyadenosylcobalamin*

[13870-90-1]

As Vitamin B_{12}, Co-00359 with

R =

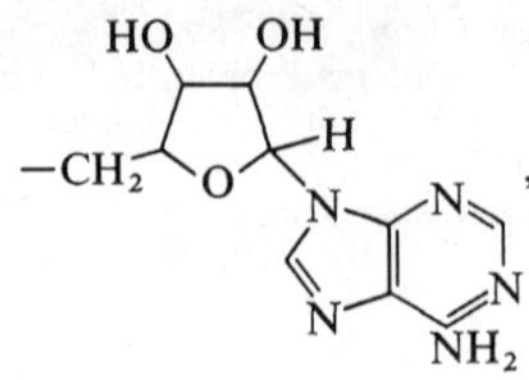

,R′ = CONH$_2$

M 1579.599

See also Vitamin B_{12}, Co-00359 for refs. Red, flattened diamond (needles) cryst. (Me_2CO). Stable to light only in the absence of oxygen (10^{-6} mm). Sol. H_2O, DMSO. pK_a 3.52. Co—C bond cleaved by CN^-, dil. H^+ or light. λ_{max} 522 (ε8400), 374 (11,000), 340 (12,000), 305 (13,000) nm.

13-Epimer: [54594-49-9]. *5′-Deoxyadenosyl-13-epicobalamin.*
$C_{72}H_{100}CoN_{18}O_{17}P$ M 1579.599
Has the structure of coenzyme B_{12} with the stereochemistry at C13 reversed. Red cryst. (Me_2CO aq.). pK_a 2.8.

Lenhert, P.G. *et al*, *Nature* (*London*), 1961, **192**, 937 (*cryst struct*)
Tkachuck, R.D. *et al*, *Biochemistry*, 1974, **13**, 2645 (*epimer, synth, uv, ord, cmr*)
Rajoria, D.S. *et al*, *J. Inorg. Nucl. Chem.*, 1977, **39**, 1291 (*ir*)
Florent, J. *et al*, *Microbiol. Technology*, Peppler, H.J. and Perlmann, D. (*Eds.*) 1979, Academic Press, London, 2nd. Ed., Vol. 1, 497 (*manuf, rev*)
Chemaly, S.M. *et al*, *J. Chem. Soc., Dalton Trans.*, 1980, 2259 (*uv*)
Cheung, A. *et al*, *J. Am. Chem. Soc.*, 1980, **102**, 384 (*pmr*)
Dolphin, D., *Adv. Chem. Ser.*, 1980, **191**, 65 (*rev*)

$C_{76}H_{60}Co_2O_{16}P_4$ Co-00368

Tetracarbonyltetrakis(triphenyl phosphite)dicobalt

As Di-μ-carbonyldicarbonyltetrakis(trimethylphosphine)dicobalt, Co-00201 with

R = OPh

M 1471.062

Unstable yellow-brown oil. ir ν_{CO} 2000(br), 1780(br) cm^{-1}.

Booth, B.L. *et al*, *J. Chem. Soc., Chem. Commun.*, 1969, 1388 (*synth, ir*)

$C_{110}H_{90}Co_2O_2P_6$ Co-00369

Dicarbonylhexakis(triphenylphosphine)dicobalt, 8CI

$Co_2(CO)_2(PPh_3)_6$

M 1747.629

Brown solid. Mp 158-60° dec. ir ν_{CO} 1877 cm^{-1}.

Yamamoto, A. *et al*, *J. Am. Chem. Soc.*, 1971, **93**, 371 (*synth, ir*)

$CoF_{12}HP_4$ Co-00370

Hydrotetrakis(phosphorous trifluoride)cobalt, 10CI

Hydridotetrakis(trifluorophosphine)cobalt. Hydrogen tetrakis(trifluorophosphine)cobaltate(1−)

[19454-38-7]

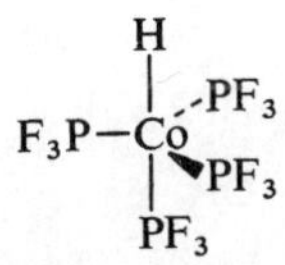

M 411.817

Highly volatile pale-yellow liq. Mp −51°. Bp_{730} 80°.

Kruck, T. *et al*, *Angew. Chem., Int. Ed. Engl.*, 1965, **4**, 148; 1967, **6**, 53 (*synth, ir, pmr, bibl*)
Kruck, T. *et al*, *Chem. Ber.*, 1968, **101**, 138 (*synth*)
Saalfield, F.E. *et al*, *J. Am. Chem. Soc.*, 1968, **90**, 3684 (*ms*)
Aylett, B.J. *et al*, *J. Chem. Soc. (A)*, 1969, 1920 (*synth, ir*)
Ibers, J.A. *et al*, *Inorg. Chem.*, 1970, **9**, 2403 (*cryst struct*)
Meakin, P. *et al*, *J. Am. Chem. Soc.*, 1972, **94**, 5271 (*nmr, struct*)
Head, R.A. *et al*, *J. Chem. Soc., Dalton Trans.*, 1975, 2054 (*pe*)

Ir Iridium

C. White

Iridium (Fr., Ger.), Iridio (Sp., Ital.), Иридий (Iridii) (Russ.), イリジウム (Japan.)

Atomic Number. 77

Atomic Weight. 192.22 ± 0.03

Electronic Configuration. [Xe] $4f^{14}\ 5d^7\ 6s^2$

Oxidation States. The oxidation state of iridium in organometallic complexes ranges from −1 to +5 although the oxidation states +1 and +3 predominate. Compounds of Ir(−I) and Ir(0) are largely restricted to the air-sensitive carbonylate anions, polynuclear carbonyls and the phosphine-substituted carbonyls. The most studied Ir(I) compounds are *trans*-$IrCl(CO)(PPh_3)_2$ and its analogues which are of interest because they readily undergo a range of oxidative addition and simple addition reactions. Only a small number of Ir(II) compounds has been reported; these are either monomeric paramagnetic compounds or diamagnetic polynuclear compounds with iridium—iridium bonds. In contrast, there is a large number of Ir(III) compounds, a significant proportion of which contain hydrido- or alkyl-ligands together with phosphine or arsine ligands. The Ir(III) compounds undergo reductive elimination reactions to form Ir(I) compounds; normally, however, such reactions do not occur as readily as with the corresponding rhodium(III) analogues and this accounts for the limited use of iridium compounds in catalysis relative to rhodium compounds. Iridium(IV) compounds are not uncommon but very few of these could be classed as organometallic. The organometallic chemistry of Ir(V) is also limited and is largely restricted to polyhydrides.

Coordination Number. Tetrahedral $[Ir(CO)_3PPh_3]^-$ is a typical 18-electron complex ion of iridium(−I). In the oxidation state 0 iridium has a d^9 configuration and therefore in order to satisfy the 18-electron rule Ir(0) compounds such as $Ir_4(CO)_{12}$ normally exist as polynuclear species containing iridium—iridium bonds. Organometallic compounds of Ir(I) (d^8) have a strong tendency to be square planar such as *trans*-$IrCl(CO)(PPh_3)_2$ or to be five-coordinate such as the trigonal bipyramidal $IrH(CO)(PPh_3)_3$. Iridium(II) compounds, for example $[IrH(CO)PPh_3]_2(\mu\text{-}SBu^t)_2$ again tend to form compounds which contain iridium—iridium bonds and which satisfy the 18-electron rule. The vast majority of Ir(III) compounds are, like $IrH_2Cl(CO)(PPh_3)_2$, octahedral but with bulky phosphine ligands five-coordinate complexes may be stabilised. Thus $IrH_3(PPh_3)_2$ has a trigonal bipyramidal structure. 7-coordination is commonly found in compounds of iridium(V).

Colour. Organoiridium compounds are commonly yellow, orange or red. There are, however, several examples of colourless compounds, particularly iridium hydride complexes such as $IrH_2Cl(PPh_3)_3$.

Availability. The commonest starting material for organoiridium chemistry is iridium trichloride in the form of a black or very dark green hydrated solid having the composition $IrCl_3 \cdot nH_2O$ with n ~3. The anhydrous form is also available but is not recommended since it is usually insoluble and extremely unreactive.

Hydrogen hexachloroiridate(2−) hydrate, $H_2[IrCl_6]\cdot nH_2O$ (n = 4), often sold under the name of chloroiridic acid, is another useful starting material (black-brown hygroscopic crystals). The corresponding iridium(IV) and iridium(III) salts $M_2[IrCl_6]\cdot xH_2O$ and $M_3[IrCl_6]\cdot xH_2O$ where M = NH_4, K or Na are also commercially available.

For the preparation of carbonyl derivatives $H_2[Ir_4Cl_8(CO)_8]\cdot nH_2O$ (n = 8) and $Ir_4(CO)_{12}$ are useful precursors; they are commercially available but expensive. Iridium is a precious metal whose price fluctuates on the world market.

Handling. Organoiridium compounds are normally synthesised or handled in solution under an inert atmosphere. However, most iridium(III) compounds and many iridium(I) compounds are sufficiently air-stable in the solid state to be handled in air, at least for short periods.

Toxicity. Organometallic compounds of iridium except for carbonyl complexes are not normally regarded as being extremely toxic although the toxicity of the vast majority of such compounds has not been investigated in detail.

Isotopic Abundance. ^{191}Ir, 37.3%; ^{193}Ir, 62.7%.

Spectroscopy. Both isotopes of Ir have a nuclear spin of $I = \frac{3}{2}$ but it is virtually impossible to observe iridium nmr because of their large quadrupole moments, (^{191}Ir, $Q = 1.5 \times 10^{-28}$ cm^2; ^{193}Ir, $Q = 1.4 \times 10^{-28}$ cm^2), their low receptivity and their low observation frequencies. A compilation of ^{13}C nmr data of organoiridium compounds has been published and developments in this field are reviewed annually. See the introduction for references.

Analysis. Atomic absorption spectroscopy and emission spectroscopy are probably the methods of choice for the precise determination of iridium. The use of these and other spectroscopic methods for iridium has been reviewed. Gravimetric and titrimetric methods have also been surveyed. Iridium in organometallic compounds has been determined by combusting the compound in oxygen at 800°C followed by reduction of the oxides in a stream of hydrogen and then weighing the metallic residue. Alternatively, iridium may be determined by conversion to $IrCl_6^{2-}$ and measuring the intense absorption at 488 nm.

References

General

Griffith, W. P., *The Chemistry of the Rarer Platinum Metals*, Wiley-Interscience, London, 1967.

Collman, J. P. *et al.*, *Adv. Organomet. Chem.*, 1968, **7**, 53.

Livingstone, S. E., *The Chemistry of Ruthenium, Rhodium, Palladium, Osmium, Iridium and Platinum*, Pergamon, Oxford, 1973.

Geoffroy, G. L. *et al.*, *Adv. Inorg. Chem. Radiochem.*, 1977, **20,** 189.

Crabtree, R., *Acc. Chem. Res.*, 1979, **12,** 331.

Dickson, R. S., *Organometallic Chemistry of Rhodium and Iridium*, Academic Press, New York, 1983.

Analysis

Beamish, F. E. *et al.*, *Miner. Sci. Eng.*, 1972, **4,** 3.

Sykora, V. *et al.*, *Sb. Vys. Sk. Chem. Technol. Praze. Anal. Chem.*, 1973, **H9,** 183 (Engl.)

Beamish, F. E., *Analytical Chemistry of the Noble Metals*, Pergamon, Oxford, 1966.

Gel'man, N. E. *et al.*, *Usp. Anal. Khim.*, 1974, 297.

$Ir(CO)Br_5^{\ominus\ominus}$

Ir-00001

$Ir(CO)Cl_5^{\ominus\ominus}$

Ir-00002

$(Cl_3Sn)_2IrCl(CO)^{\ominus\ominus}$

Ir-00003

$Ir(CO)I_5^{\ominus\ominus}$

Ir-00004

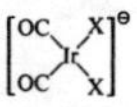

X = Br

Ir-00005

$[Ir(CO)_2Br_2]_n$

Ir-00006

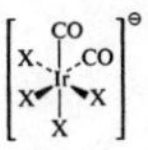

X = Br

Ir-00007

As Ir-00005 with X = Cl

Ir-00008

$[Ir(CO)_2Cl_2]_n$

Ir-00009

As Ir-00007 with X = Cl

Ir-00010

As Ir-00005 with X = I

Ir-00011

$[Ir(CO)_2I_2]_n$

Ir-00012

As Ir-00007 with X = I

Ir-00013

$[IrBr(CO)_3]_n$

Ir-00014

$[IrCl(CO)_3]_n$

Ir-00015

$IrMe(CO)_2I_3^{\ominus}$

Ir-00016

$IrI(CO)_3$

Ir-00017

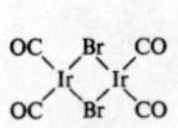

Ir-00018

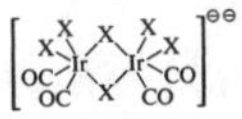

X = Br

Ir-00019

$Ir_2(CO)_4Cl_4^{\ominus}$

Ir-00020

$Ir_2(CO)_4Cl_5^{\ominus\ominus}$

Ir-00021

As Ir-00019 with X = Cl

Ir-00022

$HIr(CO)_4$

Ir-00023

$[IrHCl_2(CO)_2]_2$

Ir-00024

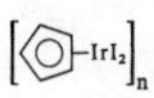

Ir-00025

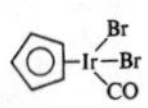

Ir-00026

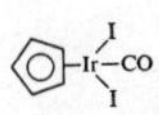

Ir-00027

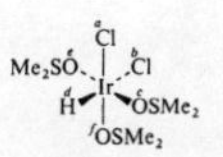

ab-dichloro-*d*-hydro-*cef*-trisulfoxide-*form*

Ir-00028

$IrH_5(PMe_3)_2$

Ir-00029

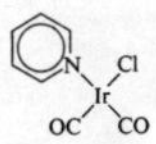

Ir-00030

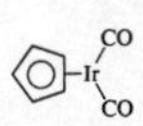

Ir-00031

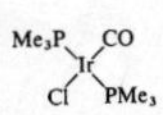

Ir-00032

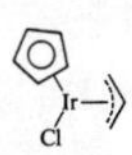

Ir-00033

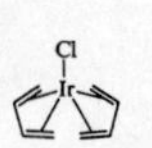

Ir-00034

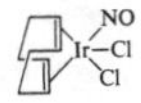

Ir-00035

$Ir(CNMe)_4^{\oplus}$

Ir-00036

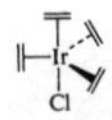

Ir-00037

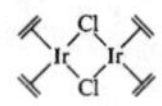

Ir-00038

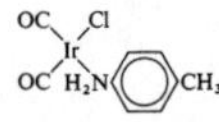

Ir-00039

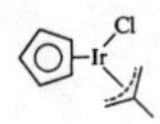

Ir-00040

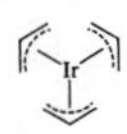

Ir-00041

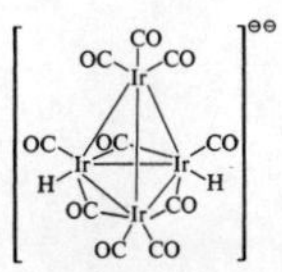

Ir-00042

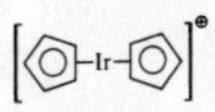

Ir-00043

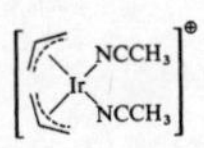

Ir-00044

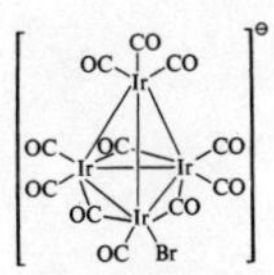

Ir-00045

$[Ir_4(CO)_{11}H]^{\ominus}$

Ir-00046

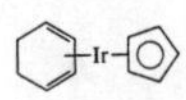

Ir-00047

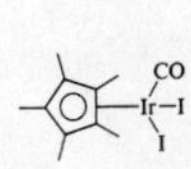

Ir-00048

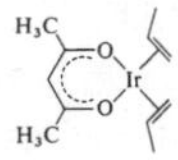

Ir-00049

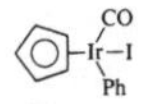

Ir-00050

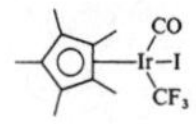

Ir-00051

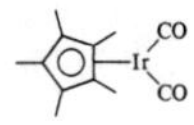

Ir-00052

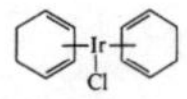

Ir-00053

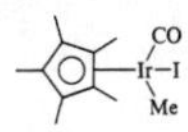

Ir-00054

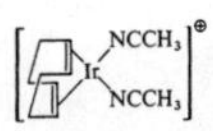

Ir-00055

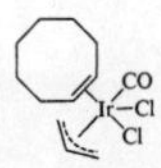

Ir-00056

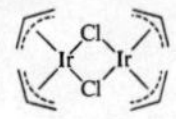

Ir-00057

$IrH_5(PEt_3)_2$

Ir-00058

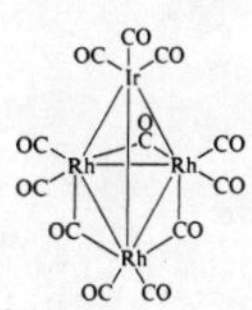

Ir-00059

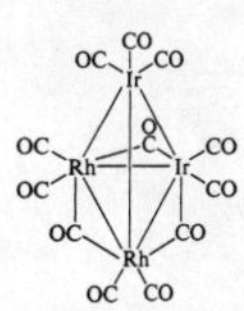

Ir-00060

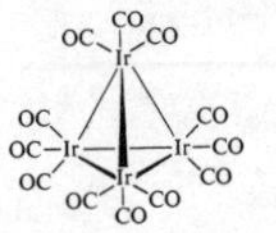

Ir-00061

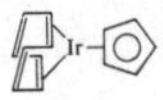

Ir-00062

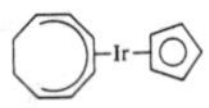

Ir-00063

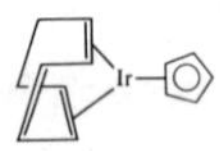

Ir-00064

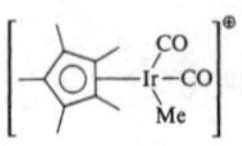

Ir-00065

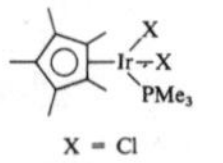

Ir-00066

As Ir-00066 with
X = H

Ir-00067

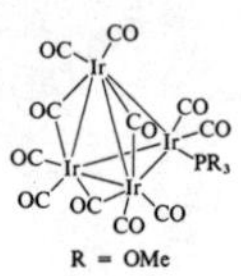

Ir-00068

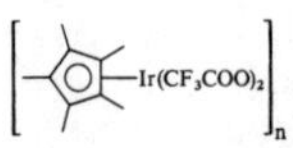

Ir-00069

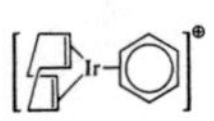

Ir-00070

$[C_5(CH_3)_5Ir(OAc)_2H_2O]_n$

Ir-00071

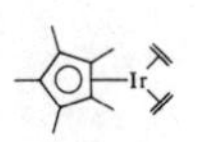

Ir-00072

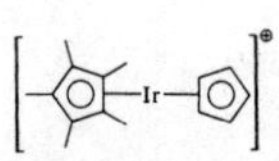

Ir-00073

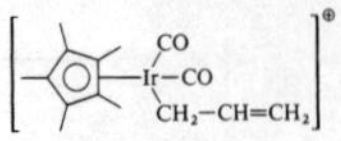

Ir-00074

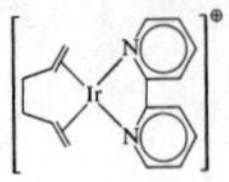

Ir-00075

$Ir_4(CO)_{10}[P(OMe)_3]_2$

Ir-00076

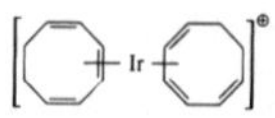

Ir-00077

Ir-00078

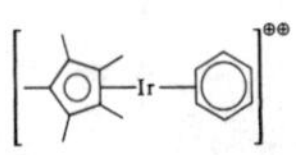

Ir-00079

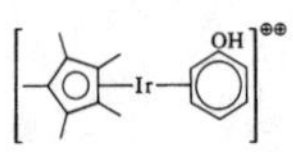

Ir-00080

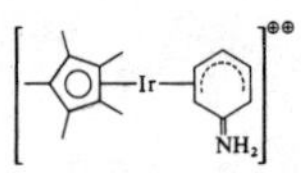

Ir-00081

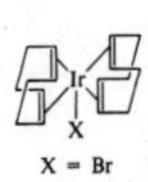

Ir-00082

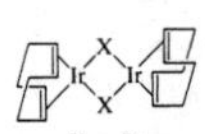

Ir-00083

As Ir-00082 with
X = Cl

Ir-00084

As Ir-00083 with
X = Cl

Ir-00085

As Ir-00083 with
X = I

Ir-00086

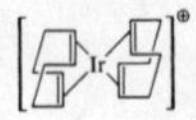

Ir-00087

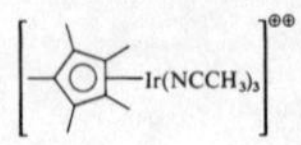

Ir-00088

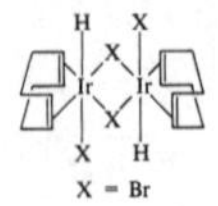

Ir-00089

As Ir-00089 with
X = Cl

Ir-00090

As Ir-00089 with
X = I

Ir-00091

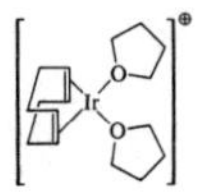

Ir-00092

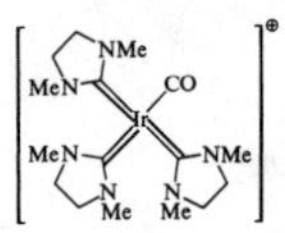

Ir-00093

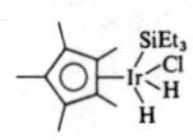

Ir-00094

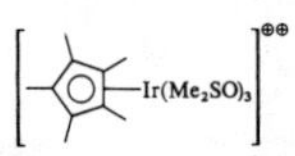

Ir-00095

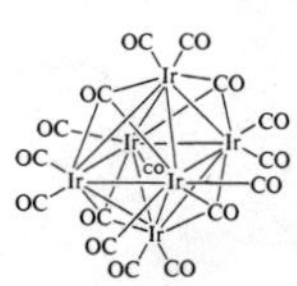

Ir-00096

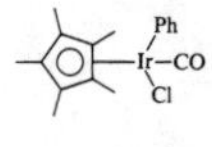

Ir-00097

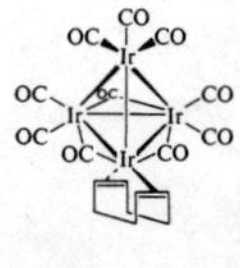

Ir-00098

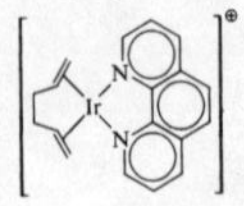

Ir-00099

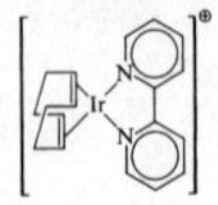

Ir-00100

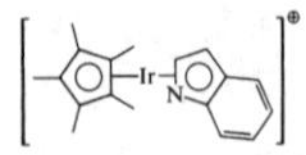

Ir-00101

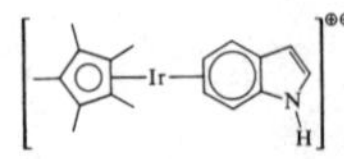

Ir-00102

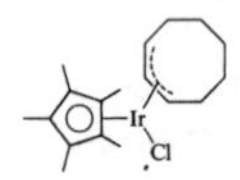

Ir-00103

As Ir-00083 with
X = OMe

Ir-00104

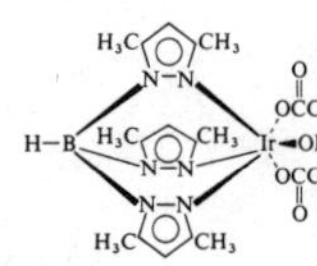

Ir-00105

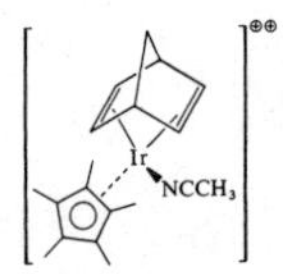

Ir-00106

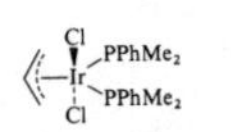

Ir-00107

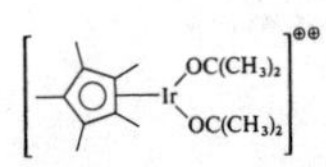

Ir-00108

$Ph_3AsIrCl(CO)_2$

Ir-00109

$Ph_3PIrCl(CO)_2$

Ir-00110

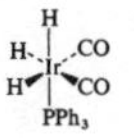

Ir-00111

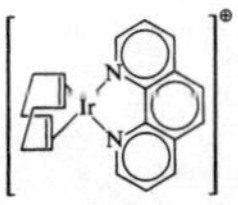

Ir-00112

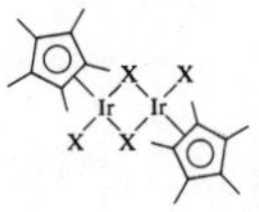

X = Br

Ir-00113

As Ir-00113 with
X = Cl

Ir-00114

As Ir-00113 with
X = I

Ir-00115

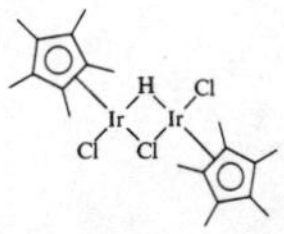

Ir-00116

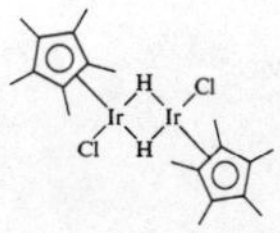

Ir-00117

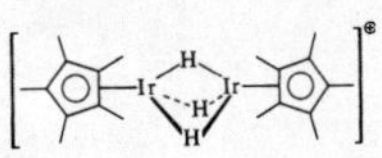

Ir-00118

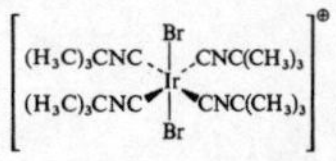

Ir-00119

$Ir[CNC(CH_3)_3]_4^{\oplus}$

Ir-00120

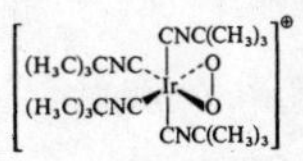

Ir-00121

$(Ph_3P)Ir(CO)_3^{\ominus}$

Ir-00122

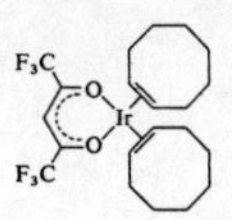

Ir-00123

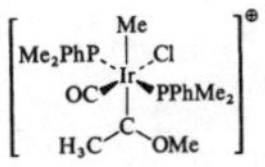

Ir-00124

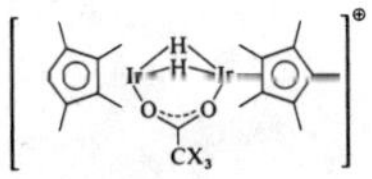

X = F

Ir-00125

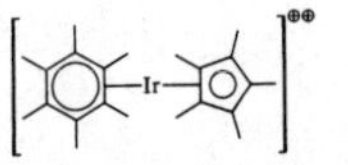

Ir-00126

As Ir-00125 with
X = H

Ir-00127

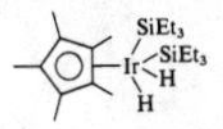

Ir-00128

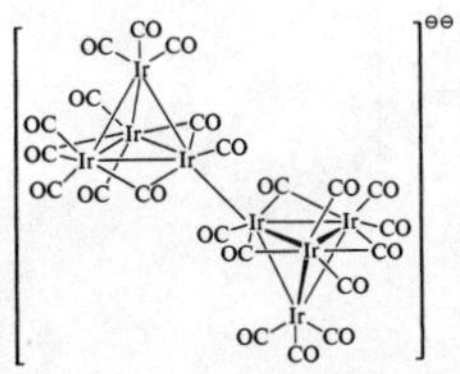

Ir-00129

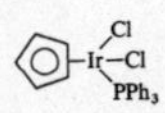

Ir-00130

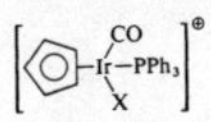

X = Br

Ir-00131

As Ir-00131 with
X = Cl

Ir-00132

As Ir-00131 with
X = I

Ir-00133

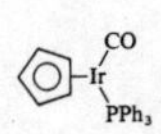

Ir-00134

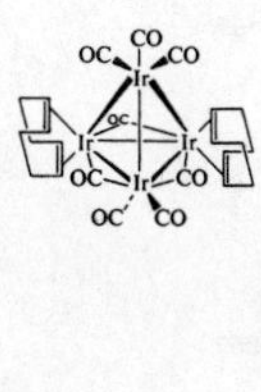

Ir-00135

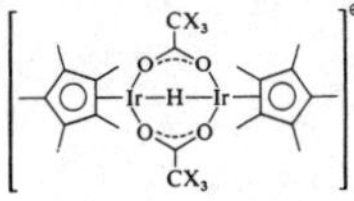

X = F

Ir-00136

As Ir-00136 with
X = H

Ir-00137

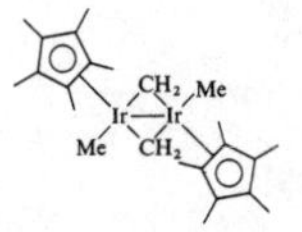

trans-form

Ir-00138

As Ir-00131 with
X = Me

Ir-00139

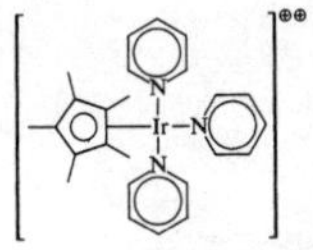

Ir-00140

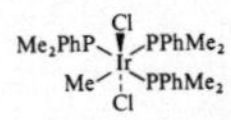

Ir-00141

$Ir(BH_4)(CO)[P(CH_2SiMe_3)_3]_2$

Ir-00142

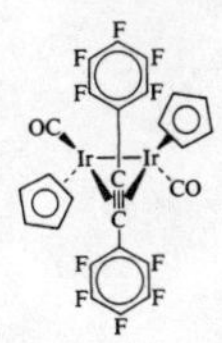

Ir-00143

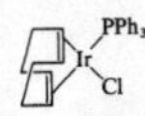

Ir-00144

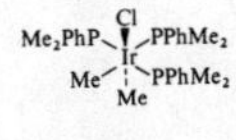

Ir-00145

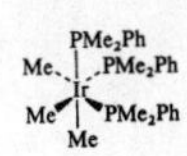

Ir-00146

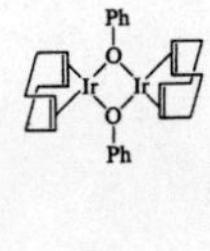

Ir-00147

As Ir-00068 with
R = Ph

Ir-00148

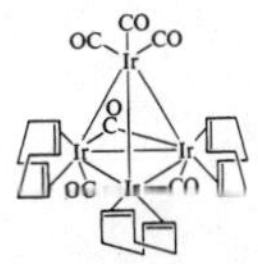

Ir-00149

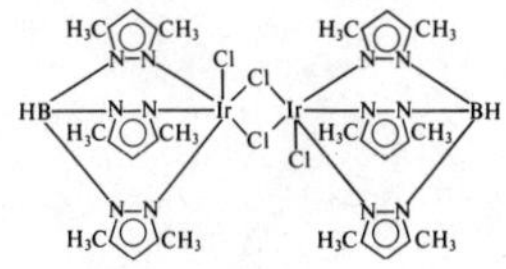

Ir-00150

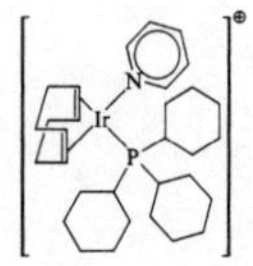

Ir-00151

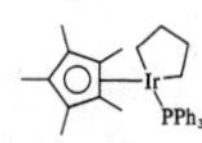

Ir-00152

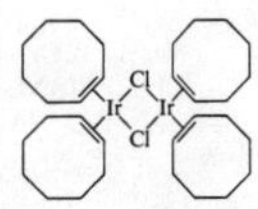

Ir-00153

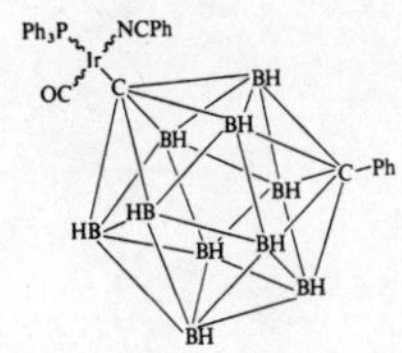

Ir-00154

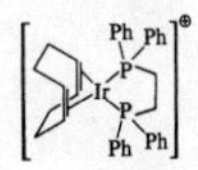

Ir-00155

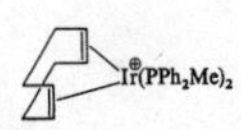

Ir-00156

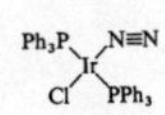

Ir-00157

$IrH_5(PPh_3)_2$

Ir-00158

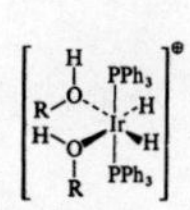

R = H

Ir-00159

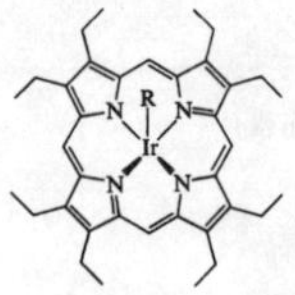

R = H

Ir-00160

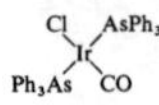

Ir-00161

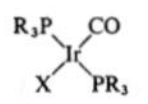

X = Br,R = Ph

Ir-00162

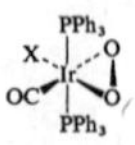

X = Br

Ir-00163

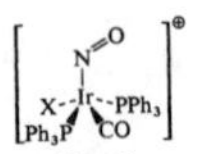

X = Cl

Ir-00164

$Ir(PPh_3)_2(CO)Cl$

Ir-00165

As Ir-00163 with
X = Cl

Ir-00166

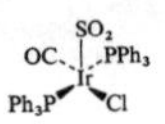

Ir-00167

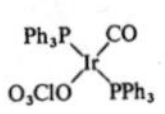

Ir-00168

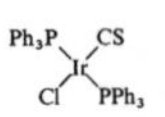

Ir-00169

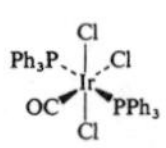

trans-form

Ir-00170

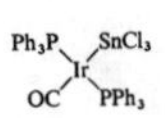

Ir-00171

As Ir-00162 with
X = F, R = Ph

Ir-00172

As Ir-00164 with
X = I

Ir-00173

As Ir-00162 with
X = I, R = Ph

Ir-00174

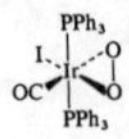

Ir-00175

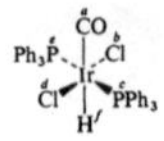

a-carbonyl-*bd*-dichloro-*f*-hydro-*form*

Ir-00176

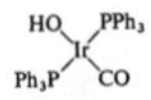

Ir-00177

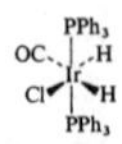

Ir-00178

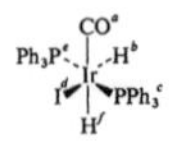

a-carbonyl-*bc*-dihydro-*d*-iodo-*ef*-diphosphine-*form*

Ir-00179

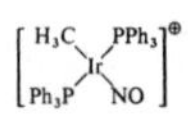

Ir-00180

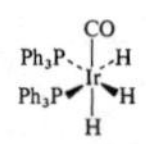

fac-form

Ir-00181

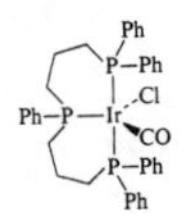

Ir-00182

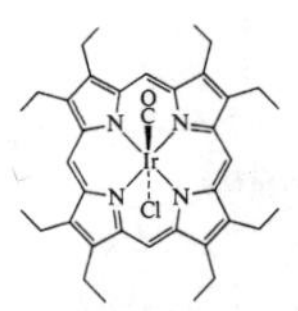

Ir-00183

As Ir-00160 with
R = Me

Ir-00184

As Ir-00162 with
X = Cl, R = cyclohexyl

Ir-00185

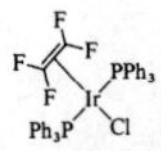

Ir-00186

$IrCl(CO)_2(NO)(PPh_3)_2^{\oplus}$

Ir-00187

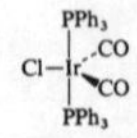

Ir-00188

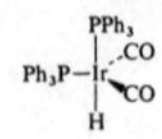

Ir-00189

As Ir-00164 with
X = Me

Ir-00190

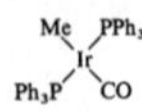

Ir-00191

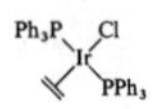

Ir-00192

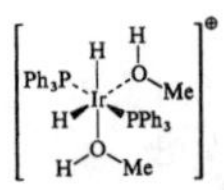

Ir-00193

As Ir-00162 with
X = F$_3$CCOO–, R = Ph

Ir-00194

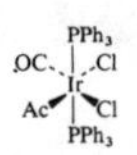

Ir-00195

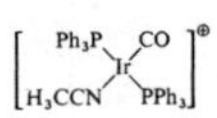

Ir-00196

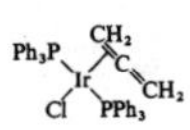

Ir-00197

Ir-00198

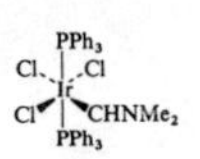

Ir-00199

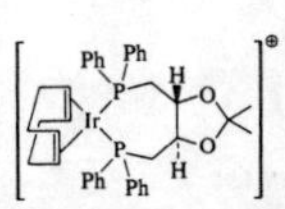

Ir-00200

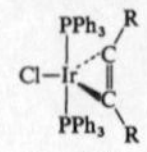

R = CF$_3$

Ir-00201

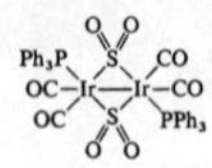

Ir-00202

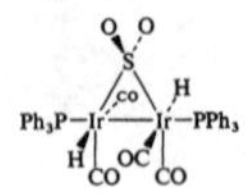

Ir-00203

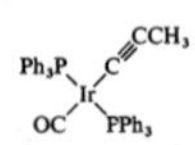

Ir-00204

$[(H_3C)_2CO]Ir(PPh_3)_2CO^{\oplus}$

Ir-00205

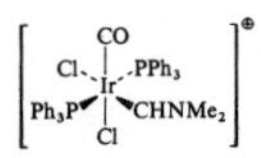

Ir-00206

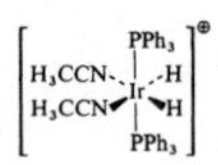

Ir-00207

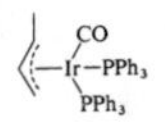

Ir-00208

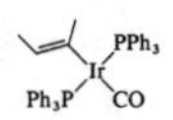

(E)-trans-form

Ir-00209

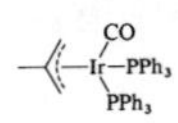

Ir-00210

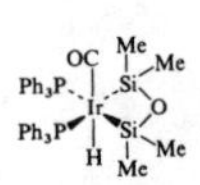

Ir-00211

$Ph_3PIr(CO)_3—SnCl_2—Ir(CO)_3PPh_3$

Ir-00212

$(Ph_3P)(OC)_3IrIr(CO)_3(PPh_3)$

Ir-00213

As Ir-00201 with
R = COOMe

Ir-00214

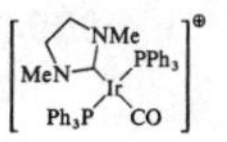

Ir-00215

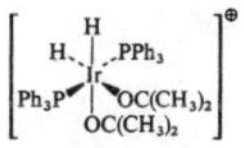

Ir-00216

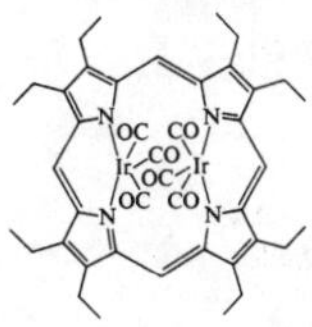

Ir-00217

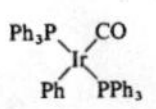

Ir-00218

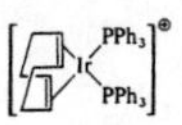

Ir-00219

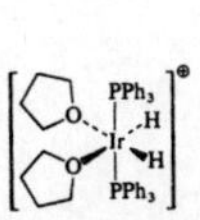

Ir-00220

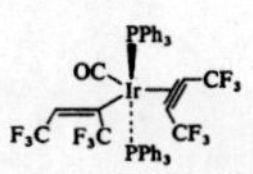

Ir-00221

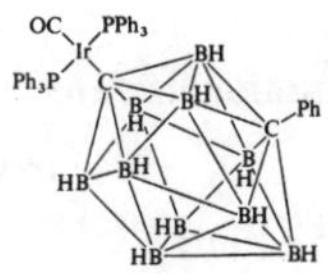

Ir-00222

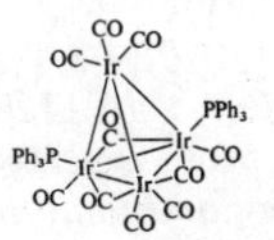

Ir-00223

As Ir-00201 with
R = Ph

Ir-00224

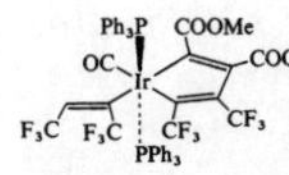

Ir-00225

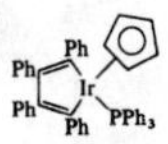

Ir-00226

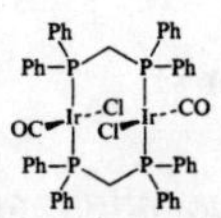

Ir-00227

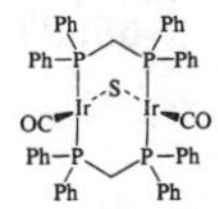

Ir-00228

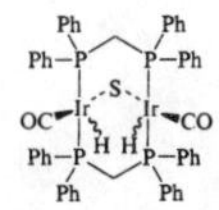

Ir-00229

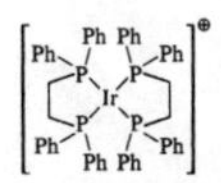

Ir-00230

$ClIr(PPh_3)_3$

Ir-00231

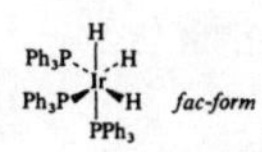

Ir-00232

$IrCl(CO)(SbPh_3)_3$

Ir-00233

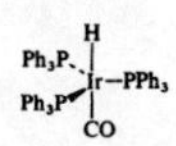

Ir-00234

$(Ph_3P)_3IrMe$ (s.p.)

Ir-00235

$Ir(CO)(C{\equiv}CH)(PPh_3)_3$

Ir-00236

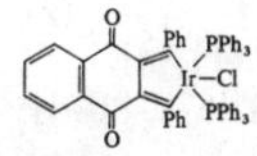

Ir-00237

$(CO)_8Ir_4(Ph_2PCH_2CH_2PPh_2)_2$

Ir-00238

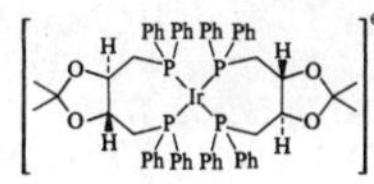

Ir-00239

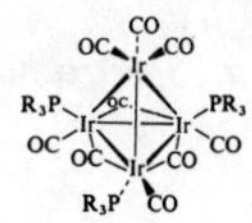

R = Ph

Ir-00240

$Ir_4(CO)_9[P(OPh)_3]_3$

Ir-00241

$Ir_4(CO)_8[P(OPh)_3]_4$

Ir-00242

$CBr_5IrO^{\ominus\ominus}$ **Ir-00001**

Pentabromocarbonyliridate(2−), 9CI

$Ir(CO)Br_5^{\ominus\ominus}$

M 619.750 (ion)

Di-K salt: [15928-66-2].
CBr_5IrK_2O M 697.947
Pink cryst. (H_2O).

Di-Cs salt: [18532-80-4].
CBr_5CS_2IrO M 695.881
Orange-brown solid. Sl. sol. H_2O. ir ν_{CO} 2038 cm^{-1} (nujol).

Bis(tetraethylammonium)salt: [38478-21-6].
$C_{17}H_{40}Br_5IrN_2O$ M 880.256
ir ν_{CO} 2033 cm^{-1} ($MeNO_2$).

Malatesta, L. *et al, J. Chem. Soc.*, 1964, 961 (*synth*)
Bonamico, M. *et al, Ric. Sci., Redn., Sez. A*, 1964, **7**, 613; *CA*, **63**, 12437g (*struct*)
Cleare, M.J. *et al, J. Chem. Soc. (A)*, 1969, 372 (*synth, ir, raman*)
Cleare, M.J. *et al, J. Chem. Soc. (A)*, 1970, 2788 (*raman*)
Forster, D. *et al, Synth. React. Inorg. Metal-Org. Chem.*, 1971, **1**, 221 (*synth, ir*)
Cleare, M.J. *et al, Spectrochim. Acta, Part A*, 1972, **28**, 2019 (*ir, raman*)

$CCl_5IrO^{\ominus\ominus}$ **Ir-00002**

Carbonylpentachloroiridate(2−), 9CI

$Ir(CO)Cl_5^{\ominus\ominus}$

M 397.495 (ion)

Di-K salt: [53779-01-4].
CCl_5IrK_2O M 475.692
Pink cryst. ($EtOH/Et_2O$). V. sol. H_2O.

Di-Cs salt: [15380-15-1].
CCl_5Cs_2IrO M 663.306
Pale-orange solid. Sl. sol. H_2O. ir ν_{CO} 2070 cm^{-1} (nujol).

Bis(tetraethylammonium)salt: [38478-19-2].
$C_{17}H_{40}Cl_5IrN_2O$ M 658.001
ir ν_{CO} 2039 cm^{-1} ($MeNO_2$).

Malatesta, L. *et al, J. Chem. Soc.*, 1964, 961 (*synth*)
Cleare, M.J. *et al, J. Chem. Soc. (A)*, 1969, 372 (*synth, ir, raman*)
Cleare, M.J. *et al, J. Chem. Soc. (A)*, 1970, 2788 (*raman*)
Forster, D. *et al, Synth. React. Inorg. Metal-Org. Chem.*, 1971, **1**, 221 (*synth, ir*)
Cleare, M.J. *et al, Spectrochim. Acta, Part A*, 1972, **28**, 2019 (*ir, raman*)

$CCl_7IrOSn_2^{\ominus\ominus}$ **Ir-00003**

Carbonylchlorobis(trichlorostannyl)iridate(2−), 9CI

[61347-92-0]

$(Cl_3Sn)_2IrCl(CO)^{\ominus\ominus}$

M 705.781 (ion)

Reacts with O_2, H_2 and CO.

Bis(tetraethylammonium) salt: [58146-94-4].
$C_{17}H_{40}Cl_7IrN_2OSn_2$ M 966.287
ir ν_{CO} 2020 cm^{-1}.

Yurchenko, E.N. *et al, Koord. Chim.*, 1975, **1**, 1406 (*möss-bauer, pe, ir*)
Elizarova, G.L. *et al, React. Kinet. Catal. Lett.*, 1975, **2**, 349 (*synth, props*)

$CI_5IrO^{\ominus\ominus}$ **Ir-00004**

Carbonylpentaiodoiridate(2−), 9CI

$Ir(CO)I_5^{\ominus\ominus}$

M 854.753 (ion)

Di-K salt: [53728-64-6].
CI_5IrK_2O M 932.950
Dark-red cryst.

Di-Cs salt: [30192-03-1].
CCs_2I_5IrO M 1120.564
No phys. props reported.

Bis(tetraethylammonium)salt: [28491-15-8].
$C_{17}H_{40}I_5IrN_2O$ M 1115.258
ir ν_{CO} 2019 cm^{-1} ($MeNO_2$).

Malatesta, L. *et al, J. Chem. Soc.*, 1964, 961 (*synth*)
Cleare, M.J. *et al, J. Chem. Soc. (A)*, 1970, 2788 (*synth, raman*)
Forster, D. *et al, Synth. React. Metal-Org. Chem.*, 1971, **1**, 221 (*synth, ir*)
Cleare, M.J. *et al, Spectrochim. Acta, Part A*, 1972, **28**, 2019 (*ir, raman*)

$C_2Br_2IrO_2^{\ominus}$ **Ir-00005**

Dibromodicarbonyliridate(1−), 10CI

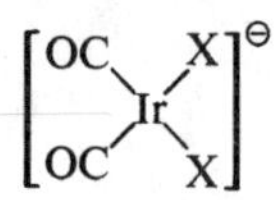

X = Br

M 408.049 (ion)

cis-form

Tetrabutylammonium salt: [73190-78-0].
$C_{18}H_{36}Br_2IrNO_2$ M 650.516
Yellow needles (EtOH). Mp 92°.

Tetraphenylarsonium salt: [30191-85-6].
$C_{26}H_{20}AsBr_2IrO_2$ M 791.392
Yellow cryst. (EtOH). Mp 228°. ir ν_{CO} 2055, 1975 cm^{-1} ($MeNO_2$).

Malatesta, L. *et al, J. Inorg. Nucl. Chem.*, 1961, **19**, 81 (*synth, ir*)
Forster, D., *Inorg. Nucl. Chem. Lett.*, 1969, **5**, 433 (*synth, ir*)
Cleare, M.J. *et al, J. Chem. Soc. (A)*, 1970, 2788 (*synth, ir*)
Piraino, P. *et al, Inorg. Nucl. Chem. Lett.*, 1973, **9**, 1237 (*synth, ir*)

$C_2Br_2IrO_2$ **Ir-00006**

Dibromodicarbonyliridium

$[Ir(CO)_2Br_2]_n$

M 408.049

Polymeric, n unknown. Pale-yellow cryst. Subl. at 150° in a stream of CO.

Hieber, W. *et al, Z. Anorg. Allg. Chem.*, 1941, **246**, 138 (*synth*)

$C_2Br_4IrO_2^{\ominus}$ Ir-00007

Tetrabromodicarbonyliridate(1–), 10CI

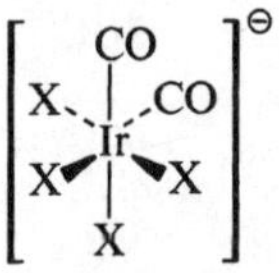

X = Br

M 567.857 (ion)

cis-form

K salt:
$C_2Br_4IrKO_2$ M 606.955
Yellow cryst.
Cs salt: [30191-97-0].
$C_2Br_4CsIrO_2$ M 700.762
Cryst.
Tetraethylammonium salt: [54032-04-1].
$C_{10}H_{20}Br_4IrNO_2$ M 698.110
Shiny orange cryst.
Tetraphenylarsonium salt: [30191-82-3].
$C_{26}H_{20}AsBr_4IrO_2$ M 951.200
Bright-yellow cryst. (Me_2CO). ir ν_{CO} 2133, 2087 cm^{-1} ($CHCl_3$).

Malatesta, L. *et al*, *J. Inorg. Nucl. Chem.*, 1961, **19**, 81 (*synth, ir*)
Cleare, M.J. *et al*, *J. Chem. Soc. (A)*, 1970, 2788 (*synth, ir*)
Forster, D., *Synth. React. Inorg. Metal-Org. Chem.*, 1971, **1**, 221 (*synth, ir*)
Forster, D., *Inorg. Chem.*, 1972, **11**, 473 (*synth*)
Malatesta, L., *J. Organomet. Chem.*, 1974, **73**, 265 (*synth, ir*)

$C_2Cl_2IrO_2^{\ominus}$ Ir-00008

Dicarbonyldichloroiridate(1–), 10CI

As Dibromodicarbonyliridate(1–), Ir-00005 with

X = Cl

M 319.147 (ion)

cis-form

Tetraphenylarsonium salt: [15152-88-2].
$C_{26}H_{20}AsCl_2IrO_2$ M 702.490
Yellow cryst. ir ν_{CO} 2056, 1975 cm^{-1} ($MeNO_2$).

Bonati, F. *et al*, *J. Organomet. Chem.*, 1967, **7**, 167 (*synth*)
Forster, D. *et al*, *Inorg. Nucl. Chem. Lett.*, 1969, **5**, 433 (*synth, ir*)
Forster, D., *Synth. React. Inorg. Metal-Org. Chem.*, 1971, **1**, 221 (*synth, ir*)
Piraino, P. *et al*, *Inorg. Nucl. Chem. Lett.*, 1973, **9**, 1237 (*synth, ir*)

$C_2Cl_2IrO_2$ Ir-00009

Dicarbonyldichloroiridium

$[Ir(CO)_2Cl_2]_n$

M 319.147

Polymeric, n unknown. Lustrous white needles. Mp 140° dec. Rapidly dec. by water but stable in dry air. Subl. at 150° in a stream of CO.

Manchot, W. *et al*, *Ber.*, 1925, **58**, 232 (*synth*)
Hieber, W. *et al*, *Z. Anorg. Allg. Chem.*, 1941, **246**, 138 (*synth*)

$C_2Cl_4IrO_2^{\ominus}$ Ir-00010

Dicarbonyltetrachloroiridate(1–)

As Tetrabromodicarbonyliridate(1–), Ir-00007 with

X = Cl

M 390.053 (ion)

cis-form

K salt: [53728-62-4].
$C_2Cl_4IrKO_2$ M 429.151
No phys. props. reported.
Tetraphenylarsonium salt: [36487-39-5].
$C_{26}H_{20}AsCl_4IrO_2$ M 773.396
Pale-yellow cryst. ir ν_{CO} 2143, 2094 cm ^{1} ($MeNO_2$).

Forster, D., *Synth. React. Inorg. Metal-Org. Chem.*, 1971, **1**, 221 (*synth, ir*)
Forster, D., *Inorg. Chem.*, 1972, **11**, 433 (*synth*)

$C_2I_2IrO_2^{\ominus}$ Ir-00011

Dicarbonyldiiodoiridate(1–), 10CI

As Dibromodicarbonyliridate(1–), Ir-00005 with

X = I

M 502.050 (ion)

cis-form

Tetraphenylarsonium salt: [29629-79-6].
$C_{26}H_{20}AsI_2IrO_2$ M 885.393
Yellow cryst. Mp 158°. ir ν_{CO} 2049, 1973 cm^{-1} ($MeNO_2$).

Malatesta, L. *et al*, *J. Chem. Soc.*, 1964, 961 (*synth*)
Forster, D. *et al*, *Inorg. Nucl. Chem. Lett.*, 1969, **5**, 433 (*synth, ir*)
Forster, D., *Synth. React. Inorg. Metal-Org. Chem.*, 1971, **1**, 221 (*synth, ir*)
Piraino, P. *et al*, *Inorg. Nucl. Chem. Lett.*, 1973, **9**, 1237 (*synth, ir*)
Forster, D., *J. Chem. Soc., Dalton Trans.*, 1979, 1639 (*ir*)

$C_2I_2IrO_2$ Ir-00012

Dicarbonyldiiodoiridium

$[Ir(CO)_2I_2]_n$

M 502.050

Polymeric, n unknown. Bright-yellow cryst. Subl. at 150° in a stream of CO.

Hieber, W. *et al*, *Z. Anorg. Allg. Chem.*, 1941, **246**, 138 (*synth*)

$C_2I_4IrO_2^{\ominus}$ Ir-00013

Dicarbonyltetraiodoiridate(1–)

Tetraiododicarbonyliridate(1–)

As Tetrabromodicarbonyliridate(1–), Ir-00007 with

X = I

M 755.859 (ion)

Formed during the iridium-catalysed carbonylation of methanol.

cis-form

K salt:
$C_2I_4IrKO_2$ M 794.957
Red cryst.
Cs salt: [30191-84-5].
$C_2CsI_4IrO_2$ M 888.764
Cryst. (Me_2CO/pet. ether).
Benzyltrimethylammonium salt: [38268-47-2].
$C_{12}H_{16}I_4IrNO_2$ M 906.102
Red cryst.
Tetraphenylarsonium salt: [30191-83-4].
$C_{26}H_{20}AsI_4IrO_2$ M 1139.202

Red cryst. Mp 210°. ir ν_{CO} 2109, 2074 cm^{-1} (nujol).

Malatesta, L. *et al*, *J. Chem. Soc.*, 1964, 961 (*synth, ir*)
Cleare, M.J. *et al*, *J. Chem. Soc.* (*A*), 1970, 2788 (*synth, ir*)
Forster, D., *Synth. React. Inorg. Metal-Org. Chem.*, 1971, **1**, 221 (*synth*)
Malatesta, L. *et al*, *J. Organomet. Chem.*, 1974, **73**, 265 (*synth*)
Forster, D., *J. Chem. Soc., Dalton Trans.*, 1979, 1639 (*ir, use*)

C_3BrIrO_3 Ir-00014

Bromotricarbonyliridium

[17836-58-7]

$[IrBr(CO)_3]_n$

M 356.155

Polymeric. Useful starting material for iridium bromo-carbonyl complexes. Cryst. One-dimensional conductor.

Malatesta, L. *et al*, *J. Inorg. Nucl. Chem.*, 1961, **19**, 81 (*synth*)
Canziani, F. *et al*, *Chim. Ind.* (*Milan*), 1967, **49**, 469 (*synth, use*)

C_3ClIrO_3 Ir-00015

Tricarbonylchloroiridium

$[IrCl(CO)_3]_n$

M 311.704

Polymeric. Crystals consist of staggered nonequiv. planar $Ir(CO)_3Cl$ molecules stacked forming parallel chains of Ir atoms. Useful starting material for iridium chloro-carbonyl complexes; catalyst for water gas shift reaction. Highly conducting brown-violet needles with metallic lustre. Subl. in CO.

Fischer, E.O. *et al*, *Z. Naturforsch., B*, 1962, **17**, 774 (*synth*)
Farone, F. *et al*, *J. Chem. Soc., Dalton Trans.*, 1973, 1625 (*use*)
Aderjan, R. *et al*, *Z. Naturforsch., A*, 1974, **29**, 1367 (*uv*)
Dahl, L.F. *et al*, *Prog. Inorg. Chem.*, 1976, **20**, 1 (*ms*)
Reis, A.H. *et al*, *J. Am. Chem. Soc.*, 1977, **99**, 4184 (*synth, cryst struct, bibl*)
Ford, P.C. *et al*, *Adv. Chem. Ser.*, 1979, **173**, 81 (*use*)

$C_3H_3I_3IrO_2^{\ominus}$ Ir-00016

Dicarbonyltriiodomethyliridate(1–), 9CI

[44769-63-3]

$IrMe(CO)_2I_3^{\ominus}$

M 643.989 (ion)

Catalytic species in carbonylation of methanol and in water-gas shift reaction.

Tetraphenylarsonium salt: [36542-09-3].
$C_{27}H_{23}AsI_3IrO_2$ M 1027.333
Yellow solid. Mp 123°. ir ν 2102, 2049 cm^{-1} ($MeNO_2$).

Forster, D. *et al*, *Inorg. Chem.*, 1972, **11**, 473 (*synth, ir*)
Forster, D. *et al*, *J. Chem. Soc., Dalton Trans.*, 1979, 1639 (*use, ir*)

C_3IIrO_3 Ir-00017

Tricarbonyliodoiridium

$IrI(CO)_3$

M 403.156

Possibly polymeric. Useful starting material for iridium iodo-carbonyl complexes. Brown solid. ir ν_{CO} 2073, 2046 cm^{-1} (nonanoic acid). One-dimensional conductor.

Hieber, W. *et al*, *Z. Anorg. Chem.*, 1941, **246**, 138 (*synth*)
Forster, D., *J. Chem. Soc., Dalton Trans.*, 1979, 1639 (*synth, ir*)

$C_4Br_2Ir_2O_4$ Ir-00018

Di-μ-bromotetracarbonyldiiridium, 8CI

[28529-98-8]

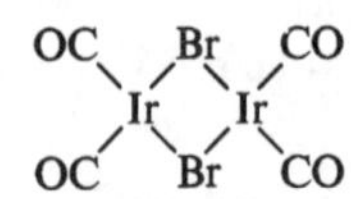

M 656.290

Bridged dimer. Catalyst for the carboxylation of olefins. No phys. props. reported.

Monomer: Bromodicarbonyliridium.
C_2BrIrO_2 M 328.145
Unknown.

Ger. Pat., 1 941 501, (*1970*); *CA*, **72**, 110811j (*use*)

$C_4Br_6Ir_2O_4^{\ominus\ominus}$ Ir-00019

Di-μ-bromotetrabromotetracarbonyldiiridate(2–), 8CI

Hexabromotetracarbonyldiiridate(2–)

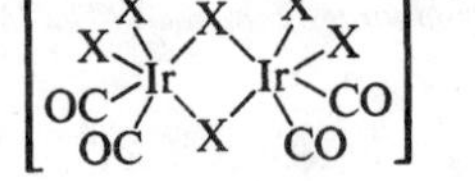

X = Br

M 975.906 (ion)

Bis(tetraphenylarsonium) salt: [31390-54-2].
$C_{52}H_{40}As_2Br_6Ir_2O_4$ M 1742.593
Yellow cryst. (Me_2CO/pet. ether). ir ν_{CO} 2090, 2057 cm^{-1} (Me_2CO).

Cleare, M.J. *et al*, *J. Chem. Soc.* (*A*), 1970, 2788 (*synth, ir, struct*)

$C_4Cl_4Ir_2O_4^{\ominus}$ Ir-00020

Tetracarbonyltetrachlorodiiridate(1–), 10CI

Tetrachlorotetracarbonyldiiridate(1–)

$Ir_2(CO)_4Cl_4^{\ominus}$

M 638.294 (ion)

Solid contains conducting linear chains of *cis*-$Ir(CO)$-$Cl_2^{0.6\ominus}$ units.

K salt: [12703-90-1].
$C_4Cl_4Ir_2KO_4$ M 677.392
Golden hair-like needles.

Benzyltrimethylammonium salt:
$C_{14}H_{16}Cl_4Ir_2NO_4$ M 788.537
Brown cryst. (MeOH).

Tetraphenylarsonium salt:
$C_{28}H_{20}AsIr_2O_4$ M 879.825
Brown needles. Insol. all solvs. Mp 225° dec.

Malatesta, L. *et al*, *J. Inorg. Nucl. Chem.*, 1961, **19**, 81 (*synth*)
Cleare, M.J. *et al*, *J. Chem. Soc.* (*A*), 1970, 2788 (*synth, ir*)
Ginsberg, A.P. *et al*, *Inorg. Chem.*, 1976, **15**, 514 (*synth, ir, mössbauer, struct*)
Kaplunov, M.G. *et al*, *Koord. Khim.*, 1976, **2**, 1268 (*pe, ir*)

$C_4Cl_5Ir_2O_4^{\ominus\ominus}$ Ir-00021

Tetracarbonylpentachlorodiiridate(2−), 9CI

$Ir_2(CO)_4Cl_5^{\ominus\ominus}$

M 673.747 (ion)

Di-K salt: [12703-91-2].
$C_4Cl_5Ir_2K_2O_4$ M 751.943
Hair-like cryst. with golden lustre (Me_2CO). Cryst. contain conducting linear chains of *cis*-$Ir(CO)_2Cl_2^{0.6\ominus}$ units with interstitial KCl.

Tetraphenylarsonium salt:
$C_{28}H_{20}AsCl_5Ir_2O_4$ M 1057.090
Golden-brown powder.

Malatesta, L. *et al*, *J. Inorg. Nucl. Chem.*, 1961, **19**, 81 (*synth*)
Ginsberg, A.P. *et al*, *Inorg. Chem.*, 1976, **15**, 514 (*synth, ir, mössbauer, struct*)
Kaplunov, M.G. *et al*, *Koord. Khim.*, 1976, **2**, 1268 (*ir, pe*)

$C_4Cl_6Ir_2O_4^{\ominus\ominus}$ Ir-00022

Tetracarbonyldi-μ-chlorotetrachlorodiiridate(2−), 9CI
Tetracarbonylhexachlorodiiridate(2−)

As Di-μ-bromotetrabromotetracarbonyldiiridate(2−), Ir-00019 with

X = Cl

M 709.200 (ion)

Bis(tetraphenylarsonium) salt: [31471-23-5].
$C_{52}H_{40}As_2Cl_6Ir_2O_4$ M 1475.887
Yellow cryst. (Me_2CO/pet. ether). ir ν_{CO} 2092, 2051 cm^{-1} (Me_2CO).

Cleare, M.J. *et al*, *J. Chem. Soc. (A)*, 1970, 2788 (*synth, ir, struct*)
Ginsberg, A.P. *et al*, *Phys. Lett.*, 1976, **38**, 310 (*mössbauer*)

C_4HIrO_4 Ir-00023

Tetracarbonylhydroiridium, 10CI
Hydridotetracarbonyliridium. Tetracarbonyliridium hydride
[25282-12-6]

$HIr(CO)_4$

M 305.270

Formed under high pressure from $Ir_4(CO)_{12}$, CO and H_2. Dec. to $Ir_4(CO)_{12}$ on removal of pressure. ir ν_{CO} 2054m, 2031s, 1999w cm^{-1} (pet. ether).

Whyman, R., *J. Chem. Soc., Dalton Trans.*, 1972, 2294 (*synth, ir*)
Vidal, J.L. *et al*, *Inorg. Chem.*, 1981, **20**, 249 (*synth, ir*)

$C_4H_2Cl_4Ir_2O_4$ Ir-00024

Tetracarbonyldi-μ-chlorodichlorodihydrodiiridium
Di-μ-chlorodichlorodihydrotetracarbonyldiiridium. Tetracarbonyltetrachlorodihydrodiiridium

$[IrHCl_2(CO)_2]_2$

M 640.309

Bronze needles (EtOH), probably dihydrate. Mp >192° dec. ir ν_{IrH} 2105, ν_{CO} 2077, 2044 cm^{-1} (KCl).

Monomer: Dicarbonyldichlorohydridoiridium.
$C_2HCl_2IrO_2$ M 320.155
Unknown.

Shaw, B.L. *et al*, *J. Chem. Soc. (A)*, 1967, 1683 (*synth, ir*)

$C_5H_5I_2Ir$ Ir-00025

(η[5]-2,4-Cyclopentadien-1-yl)diiodoiridium, 9CI
Diiodo(η[5]-cyclopentadienyl)iridium. Cyclopentadienyliridium diiodide
[32660-87-0]

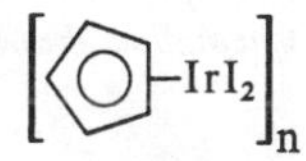

M 511.124

Polymeric. Brown cryst. Insol.

Yamazaki, H., *Bull. Chem. Soc. Jpn.*, 1971, **44**, 582 (*synth*)

$C_6H_5Br_2IrO$ Ir-00026

Dibromocarbonyl(η[5]-2,4-cyclopentadien-1-yl)iridium, 9CI
[64867-78-3]

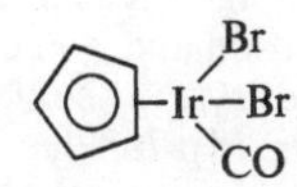

M 445.133

Yellow air-stable solid. ir ν_{CO} 2065 cm^{-1} (KBr).

Rausch, M.D. *et al*, *J. Am. Chem. Soc.*, 1977, **99**, 7870 (*synth, ir, pmr, ms*)

$C_6H_5I_2IrO$ Ir-00027

Carbonyl(η[5]-2,4-cyclopentadien-1-yl)diiodoiridium, 9CI
Diiodocarbonyl-π-cyclopentadienyliridium. π-Cyclopentadienyliridium carbonyl diiodide
[32660-90-5]

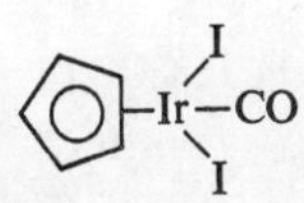

M 539.134

Red cryst. Mp >180° dec. ir ν_{CO} 2040cm^{-1} (KBr), 2070cm^{-1} (CH_2Cl_2).

Yamazaki, H., *Bull. Chem. Soc. Jpn.*, 1971, **44**, 582 (*synth, ir, pmr*)
Gardner, S.A. *et al*, *J. Organomet. Chem.*, 1973, **60**, 179 (*synth, ir, pmr*)

$C_6H_{19}Cl_2IrO_3S_3$ Ir-00028

Dichlorotris(dimethyl sulfoxide)hydroiridium
Dichlorohydrotris[sulfinylbis[methane]-O]iridium, 10CI.
Dichlorohydrotris(methyl sulfoxide)iridium, 8CI
[21861-65-4]

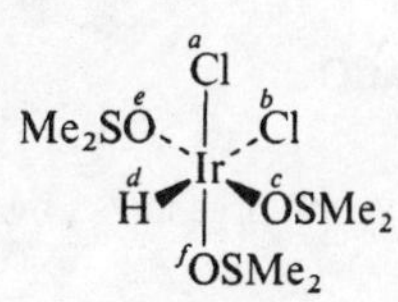

ab-dichloro-*d*-hydro-*cef*-trisulfoxide-*form*

M 498.520

Catalyst for the redn. of αβ-unsatd. ketones and cyclohexanones, the *cis*-hydrogenation of acetylenes, and the air oxidn. of sulfoxides to sulfones.

ab*-dichloro-*d*-hydro-*cef*-trisulfoxide-*form [52611-14-0]
mer,cis-*form*
Yellow cryst. Mp 160-5°. Unstable in solid state and in soln.

af-dichloro-b-hydro-cde-trisulfoxide-form [52611-13-9]
mer,trans-*form*
Yellow prisms (DMSO/2-propanol/1M HCl). Mp 184.5-185°.

Haddad, Y.M.Y. *et al*, *J. Chem. Soc.*, *Perkin Trans. 1*, 1974, 592 (*synth*, *ir*, *pmr*, *struct*)
Henbest, H.B. *et al*, *J. Chem. Soc.*, *Perkin Trans. 1*, 1974, 596, 601, 607 (*use*)

$C_6H_{23}IrP_2$ Ir-00029

Pentahydrobis(trimethylphosphine)iridium, 9CI
Pentahydridobis(trimethylphosphine)iridium
[29497-68-5]

$IrH_5(PMe_3)_2$

M 349.415
Initially incorrectly formulated as $IrH_3(PMe_3)_2$. Catalyses the exchange between D_2 and aromatics, and also the hydrogenation and isomerisation of alkenes. White air-sensitive cryst. by subl. at r.t. under vacuum. V. sol. all org. solvs. Mp 70° dec. ir ν_{IrH} 1920 cm^{-1}. Slowly dec. in C_6H_6 to liberate H_2.

Barefield, E.K. *et al*, *J. Am. Chem. Soc.*, 1970, **92**, 5234 (*synth*, *nmr*, *use*)
Klabunde, U. *et al*, *J. Am. Chem. Soc.*, 1972, **94**, 9081 (*use*)
Inorg. Synth., 1974, **15**, 34 (*synth*, *nmr*, *props*)
Clerici, M.G. *et al*, *J. Organomet. Chem.*, 1975, **84**, 379 (*use*)

$C_7H_5ClIrNO_2$ Ir-00030

Dicarbonylchloropyridineiridium, 9CI
Chlorodicarbonylpyridineiridium. Chloropyridinedicarbonyliridium

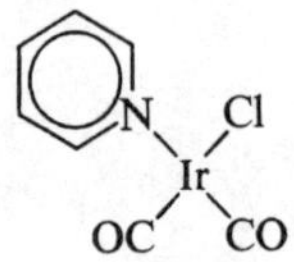

M 362.795
cis-form [55780-99-9]
Iridescent green cryst. (Et_2O/pet. ether). Mp 73° dec. ir ν_{CO} 2072, 1987 cm^{-1} (THF).

Hieber, W. *et al*, *Chem. Ber.*, 1966, **99**, 2607 (*synth*, *ir*)
Jeter, D.Y. *et al*, *J. Coord. Chem.*, 1974, **4**, 107 (*cryst struct*)

$C_7H_5IrO_2$ Ir-00031

Dicarbonyl(η^5-2,4-cyclopentadien-1-yl)iridium, 9CI
[12192-96-0]

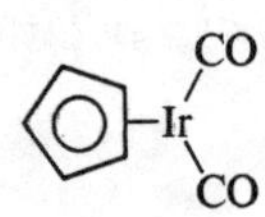

M 313.335
Stable in air for a short time. V. sol. org. solvs., insol. H_2O. μ = 3.45D.

Fischer, E.O. *et al*, *Z. Naturforsch.*, *B*, 1962, **17**, 774 (*synth*, *ir*, *pmr*, *uv*)
Fritz, H.P. *et al*, *Z. Naturforsch.*, *B*, 1963, **18**, 435 (*ir*)
Gardner, S.A. *et al*, *Inorg. Chem.*, 1973, **12**, 2396 (*props*)

$C_7H_{18}ClIrOP_2$ Ir-00032

Carbonylchlorobis(trimethylphosphine)iridium, 9CI
Carbonylbis(trimethylphosphine)iridium chloride

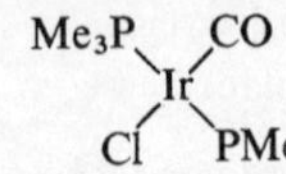

M 407.839
trans-form [21209-86-9]
Air-sensitive yellow prisms (MeOH). Mp >140° dec. $Bp_{0.1}$ 130° subl. ir ν_{CO} 1938 cm^{-1} (nujol).

Deeming, A.J. *et al*, *J. Chem. Soc.* (*A*), 1968, 1887 (*synth*, *ir*, *pmr*)
Inorg. Synth., 1978, **18**, 62 (*synth*, *ir*, *pmr*)

$C_8H_{10}ClIr$ Ir-00033

Chloro(η^5-2,4-cyclopentadien-1-yl)(η^3-2-propenyl)iridium, 9CI
Allylchloro(π-cyclopentadienyl)iridium
[54566-49-3]

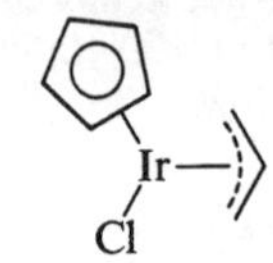

M 333.840

Rubezhov, A.Z. *et al*, *Izv. Akad. Nauk SSSR, Ser. Khim.*, 1974, 2651; *CA*, **82**, 57910n (*synth*)

$C_8H_{12}ClIr$ Ir-00034

Bis(η^4-1,3-butadiene)chloroiridium, 9CI
[39732-20-2]

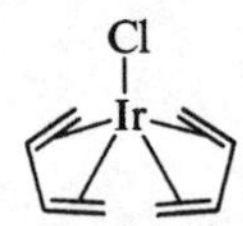

M 335.856
Effective catalyst for oxygenation of butadiene. White prisms (CH_2Cl_2). Mp 178-81° dec.

Onderdelinden, A.L. *et al*, *Inorg. Chim. Acta*, 1972, **6**, 420 (*synth*, *ir*, *pmr*)
Van Soest, T.C. *et al*, *Cryst. Struct. Commun.*, 1973, **2**, 527 (*struct*)
Mares, F. *et al*, *J. Org. Chem.*, 1978, **43**, 4631 (*use*)

$C_8H_{12}Cl_2IrNO$ Ir-00035

Dichloro[(1,2,5,6-η)-1,5-cyclooctadiene]nitrosyliridium
Dichloronitrosyl-1,5-cyclooctadieneiridium
[31741-46-5]

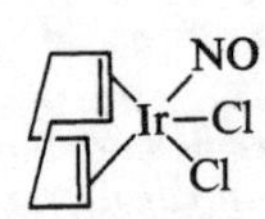

M 401.315
Green solid. ir ν_{NO} 1685s, 1710sh cm^{-1} (nujol).

Crooks, G.R. *et al*, *J. Chem. Soc.* (*A*), 1970, 1662 (*synth*, *ir*, *pmr*)

$C_8H_{12}IrN_4^{\oplus}$ Ir-00036

Tetrakis(isocyanomethane)iridium(1+), 9CI
Tetrakis(methylisonitrile)iridium(1+). Tetrakis(methyl isocyanide)iridium(1+)

$Ir(CNMe)_4^{\oplus}$

M 356.430 (ion)

Chloride: [40226-52-6].
$C_8H_{12}ClIrN_4$ M 391.883
Air-sensitive blue-black cryst.

Iodide: [40226-55-9].
$C_8H_{12}IIrN_4$ M 483.334
Air-sensitive cryst.

Hexafluorophosphate: [40226-53-7].
$C_8H_{12}F_6IrN_4P$ M 501.394
Air-sensitive cryst.

Perchlorate: [40226-64-0].
$C_8H_{12}ClIrN_4O_4$ M 455.880
Air-sensitive cryst.

Tetraphenylborate: [40226-54-8].
$C_{32}H_{32}BIrN_4$ M 675.662
Air-sensitive cryst.

Bedford, W.M. *et al*, *J. Chem. Soc., Chem. Commun.*, 1972, 1224 (*synth, ir, pmr*)
Geoffrey, G.L. *et al*, *Inorg. Chem.*, 1978, **17**, 777 (*synth*)

$C_8H_{16}ClIr$ Ir-00037

Chlorotetrakis(η^2-ethene)iridium, 9CI
Chlorotetrakis(ethylene)iridium, 8CI
[39722-80-0]

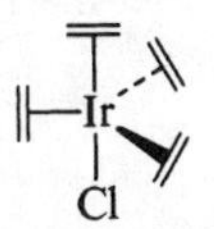

M 339.887
White needles, stable under ethylene <0° or *in vacuo* <−50°. Above 50° loses two moles of ethylene and dimerises.

Onderdelinden, A.L. *et al*, *Inorg. Chem.*, 1972, **6**, 420 (*synth, ir, pmr*)
van Putte, K. *et al*, *Inorg. Chim. Acta*, 1973, **7**, 497 (*nmr, struct*)
Howard, J. *et al*, *J. Chem. Soc., Faraday Trans. 2*, 1978, 1275 (*raman*)

$C_8H_{16}Cl_2Ir_2$ Ir-00038

Di-μ-chlorotetrakis(η^2-ethene)diiridium, 9CI
Di-μ-chlorotetrakis(η^2-ethylene)diiridium
[39722-81-1]

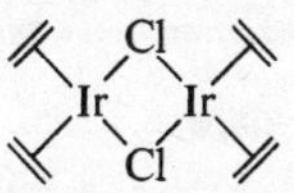

M 567.560
Bridged dimer. Red cryst. which dec. slowly in air at r.t. Mp >90° dec.

Monomer: Chlorobis(ethylene)iridium.
C_4H_8ClIr M 283.780
Unknown.

Onderdelinden, A.L. *et al*, *Inorg. Chim. Acta*, 1972, **6**, 420 (*synth, ir, pmr*)
Howard, J. *et al*, *J. Chem. Soc., Faraday Trans. 2*, 1978, 1275 (*raman*)

$C_9H_9ClIrNO_2$ Ir-00039

Dicarbonylchloro(4-methylaniline)iridium
Dicarbonylchloro(4-methylbenzenamine)iridium, 9CI.
*Dicarbonylchloro(*p*-toluidine)iridium. Chlorodicarbonyl(*p*-toluidine)iridium*
[14243-22-2]

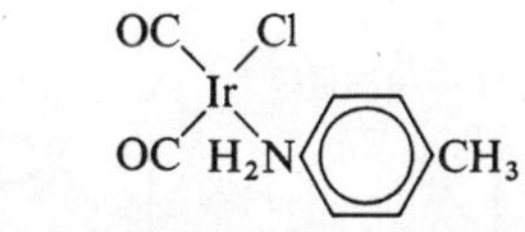

M 390.849

cis-form [59952-74-8]
Violet cryst. (Et_2O/pet. ether). Mp 165° dec. ir ν_{CO} 2088, 2027 cm^{-1} (nujol). Forms sl. air-sensitive solns. in polar org. solvs.

Angoletta, M., *Gazz. Chim. Ital.*, 1959, **89**, 2359 (*synth, ir*)
Hieber, W. *et al*, *Chem. Ber.*, 1966, **99**, 2607 (*synth, ir*)
Inorg. Synth., 1974, **15**, 82 (*synth, ir, pmr*)
Aderjan, R. *et al*, *J. Chem. Phys.*, 1976, **64**, 3748 (*mössbauer*)

$C_9H_{12}ClIr$ Ir-00040

Chloro(η^5-2,4-cyclopentadien-1-yl)[(1,2,3-η)-2-methyl-2-propenyl]iridium, 9CI
Chloro(cyclopentadienyl)(2-methylallyl)iridium
[54566-48-2]

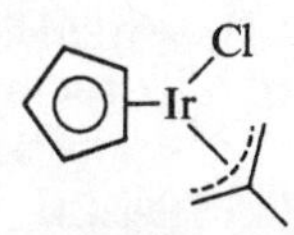

M 347.867
Mp 152-4°.

Nesmeyanov, A.N. *et al*, *J. Organomet. Chem.*, 1978, **164**, 259 (*synth, ir, pmr, struct*)

$C_9H_{15}Ir$ Ir-00041

Tris(η^3-2-propenyl)iridium, 9CI
Tri-π-allyliridium, 8CI. *Tris(η^3-allyl)iridium*
[12108-64-4]

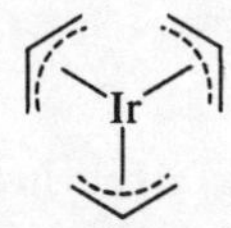

M 315.438
White cryst. with camphoraceous odour by subl. Mp 65° dec. Bp_{15} 50° subl.

Chini, P. *et al*, *Inorg. Chem.*, 1967, **6**, 837 (*synth, ir, ms, pmr*)
Andrews, D.C. *et al*, *J. Organomet. Chem.*, 1973, **55**, 382 (*ir*)

$C_{10}H_2Ir_4O_{10}^{\ominus\ominus}$ Ir-00042

Tri-μ-carbonylheptacarbonyldihydrotetrairidate(2−)
Decacarbonyldihydrotetrairidate(2−)

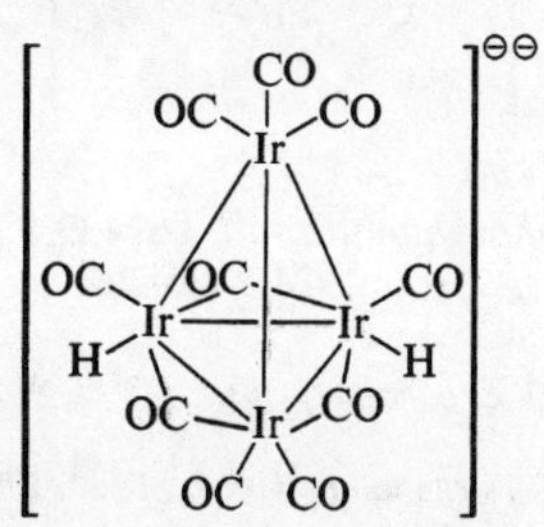

M 1051.000 (ion)
Bis[bis(triphenylphosphine)iminium]salt: [67388-82-3].

$C_{82}H_{62}Ir_4N_2O_{10}P_4$ M 2128.174
Orange-brown prisms.

Ciani, G. *et al*, *J. Organomet. Chem.*, 1978, **150**, C17 (*ir, pmr, struct*)

$C_{10}H_{10}Ir^{\oplus}$ Ir-00043

Iridocenium(1+)
Biscyclopentadienyliridium(1+)

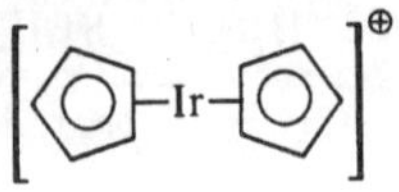

M 322.409 (ion)
Tribromide:
$C_{10}H_{10}Br_3Ir$ M 562.121
Golden-yellow powder. Mp >185° dec.
Hexafluorophosphate:
$C_{10}H_{10}F_6IrP$ M 467.373
Colourless cryst.

Cotton, F.A. *et al*, *J. Am. Chem. Soc.*, 1953, **75**, 3586 (*synth, uv*)
Fischer, E.O., *J. Organomet. Chem.*, 1966, **5**, 559 (*synth, ir, uv, pmr*)

$C_{10}H_{16}IrN_2^{\oplus}$ Ir-00044

Bis(acetonitrile)bis(η^3-2-propenyl)iridium(1+), 9CI
Bis(acetonitrile)bis(π-allyl)iridium(1+)

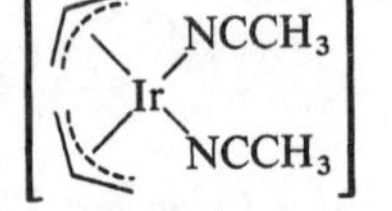

M 356.470 (ion)
Stereochemically non-rigid in soln.
Tetrafluoroborate: [52445-18-8].
$C_{10}H_{16}BF_4IrN_2$ M 443.273
White cryst. Mp 157° dec.

Green, M. *et al*, *J. Chem. Soc., Dalton Trans.*, 1974, 333 (*synth, ir, pmr*)

$C_{11}BrIr_4O_{11}^{\ominus}$ Ir-00045

Bromotri-μ-carbonyloctacarbonyltetrairidate(1–), 9CI
Bromoundecacarbonyltetrairidate(1–)
[67234-38-2]

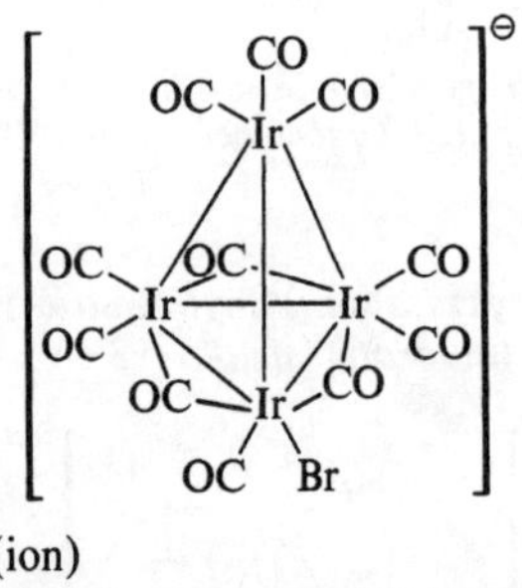

M 1156.898 (ion)
Tetraphenylphosphonium salt: [67429-93-0].
$C_{35}H_{20}BrIr_4O_{11}P$ M 1496.294
Orange cryst.

Chini, P. *et al*, *J. Organomet. Chem.*, 1978, **152**, C35 (*synth, ir, struct*)
Ciani, G. *et al*, *J. Organomet. Chem.*, 1980, **199**, 271 (*struct*)

$C_{11}HIr_4O_{11}^{\ominus}$ Ir-00046

Undecacarbonylhydrotetrairidate(1–), 9CI
Tri-μ-carbonyloctacarbonylhydrotetrairidate(1–), 10CI
[56938-69-3]

$$[Ir_4(CO)_{11}H]^{\ominus}$$

M 1078.002 (ion)
Struct. is said to contain two bridging carbonyls; 10CI name prob. inappropriate.
Benzyltrimethylammonium salt: [57108-18-6].
$C_{21}H_{17}Ir_4NO_{11}$ M 1228.245
Yellow powder. Diamagnetic. Λ 22.1 Ω^{-1} cm^2 mol.$^{-1}$.

Malatesta, L. *et al*, *J. Chem. Soc., Chem. Commun.*, 1967, 420 (*synth, pmr, ir*)
Angoletta, M. *et al*, *J. Organomet. Chem.*, 1975, **94**, 99 (*synth, ir*)
Pruett, R. *et al*, *J. Organomet. Chem.*, 1979, **182**, C57.
Garlaschelli, L. *et al*, *J. Organomet. Chem.*, 1981, **213**, 379.

$C_{11}H_{13}Ir$ Ir-00047

[(1,2,3,4-η)-1,3-Cyclohexadiene](η^5-2,4-cyclopentadien-1-yl)iridium, 9CI
[39015-37-7]

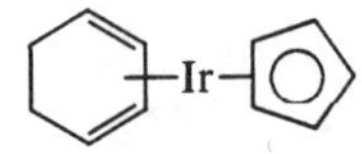

M 337.444
White cryst.

Johnson, B.F.G. *et al*, *J. Chem. Soc., Dalton Trans.*, 1972, 2084 (*synth, pmr*)

$C_{11}H_{15}I_2IrO$ Ir-00048

Carbonyldiiodo(η^5-pentamethylcyclopentadienyl)iridium
[32677-79-5]

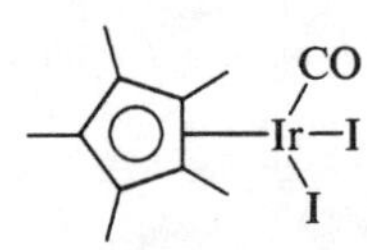

M 609.268
Red cryst. (CH_2Cl_2/hexane). Mod. air stable (dec. slowly to 97238-1). Mp 305-6°. ir υ_{CO} 2040cm^{-1} (KBr).

Maitlis, P.M. *et al*, *J. Organomet. Chem.*, 1971, **26**, 393 (*ir, pmr*)
King, R.B. *et al*, *J. Organomet. Chem.*, 1971, **27**, 409 (*synth, pmr, ir, ms*)

$C_{11}H_{19}IrO_2$ Ir-00049

(2,4-Pentanedionato-*O,O'*)bis[(1,2-η)-1-propene]iridium, 9CI
(Acetylacetonato)bis(propylene)iridium
[66467-05-8]

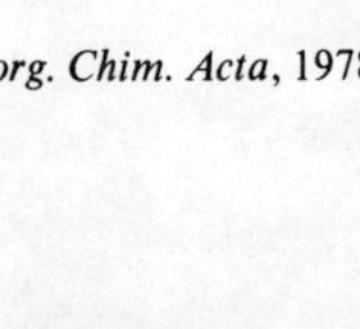

M 375.490
Yellow cryst.

Jesse, A.C. *et al*, *Inorg. Chim. Acta*, 1978, **26**, 129 (*synth, ir, cmr*)

Van Dam, H. *et al*, *Inorg. Chem.*, 1980, **19**, 3448 (*pe*)

$C_{12}H_{10}IIrO$ Ir-00050

Carbonyl(η^5-2,4-cyclopentadien-1-yl)iodophenyliridium, 9CI
[64867-79-4]

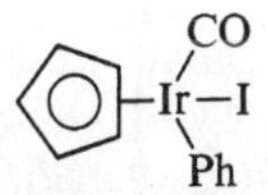

M 489.335
Brown air-stable solid. ir ν_{CO} 2025 cm^{-1} (KBr disc).

Rausch, M.D. *et al*, *J. Am. Chem. Soc.*, 1977, **99**, 7870 (*synth, pmr, ir, ms*)

$C_{12}H_{15}F_3IIrO$ Ir-00051

Carbonyliodo(η^5-pentamethylcyclopentadienyl)(trifluoromethyl)iridium
[32661-01-1]

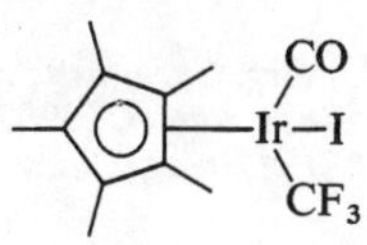

M 551.370
Air-stable, yellow-orange cryst. (CH_2Cl_2/hexane). Mp 228-30°. ir ν_{CO} 2050cm^{-1} (CH_2Cl_2).

King, R.B. *et al*, *J. Organomet. Chem.*, 1971, **27**, 409 (*synth, ir, nmr, ms*)

$C_{12}H_{15}IrO_2$ Ir-00052

Dicarbonyl[(1,2,3,4,5-η)-1,2,3,4,5-pentamethyl-2,4-cyclopentadien-1-yl]iridium, 9CI
(Pentamethylcyclopentadienyl)iridium dicarbonyl
[32660-96-1]

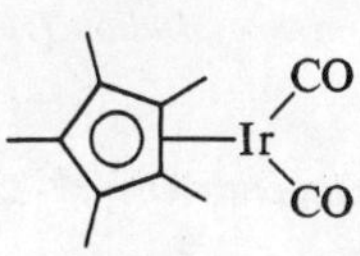

M 383.469
Yellow cryst. (C_6H_6), air-stable for several days. V. sol. org. solvs. Mp 145-8° dec. ir ν_{CO} 1925(s) 2000(s) cm^{-1}.

Kang, J.W. *et al*, *J. Am. Chem. Soc.*, 1969, **91**, 5970 (*synth, purifn, ir, pmr*)
Kang, J.W. *et al*, *J. Organomet. Chem.*, 1971, **26**, 393 (*props*)
King, R.B. *et al*, *J. Organomet. Chem.*, 1971, **27**, 409 (*ms*)

$C_{12}H_{16}ClIr$ Ir-00053

Chlorobis[(1,2,3,4-η)-1,3-cyclohexadiene]iridium
[12109-91-0]

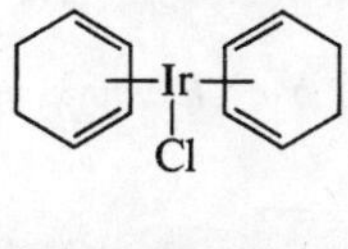

M 387.931
Needles (CH_2Cl_2/pet. ether). Mp 135-40° dec.

Winkhaus, G. *et al*, *Chem. Ber.*, 1966, **99**, 3610 (*synth, ir, pmr*)

$C_{12}H_{18}IIrO$ Ir-00054

Carbonyliodomethyl(η^5-pentamethylcyclopentadienyl)iridium
[32661-04-4]

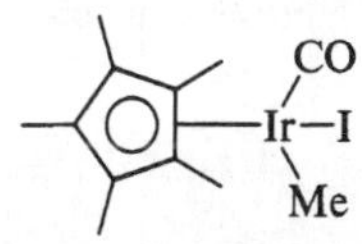

M 497.398
Yellow-orange cryst. (CH_2Cl_2/hexane). Mp 190-1° dec. ir υ_{CO} 2002cm^{-1} (CH_2Cl_2).

King, R.B. *et al*, *J. Organomet. Chem.*, 1971, **27**, 409 (*synth, ir, pmr, ms*)

$C_{12}H_{18}IrN_2^{\oplus}$ Ir-00055

Bis(acetonitrile)[(1,2,5,6-η)-1,5-cyclooctadiene]iridium(1+), 9CI

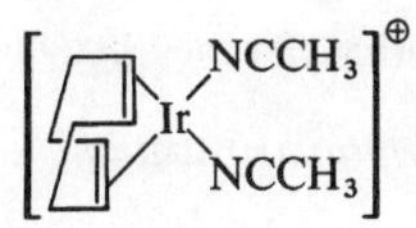

M 382.508 (ion)
Tetrafluoroborate: [32679-03-1].
$C_{12}H_{18}BF_4IrN_2$ M 469.311
Hexene isomerisation catalyst. Bright-yellow cryst. Mp 107-8°, ca. 150°.

Green, M. *et al*, *J. Chem. Soc. (A)*, 1971, 2334 (*synth, ir, pmr, use*)
Sievert, A.C. *et al*, *Inorg. Chem.*, 1981, **20**, 489 (*synth, ir, pmr*)

$C_{12}H_{19}Cl_2IrO$ Ir-00056

Carbonyldichloro(η^2-cyclooctene)(η^3-2-propenyl)iridium
π-Allylcarbonyldichloro(cyclooctene)iridium, 8CI. *Dichloro(allyl)carbonyl(cyclooctene)iridium*

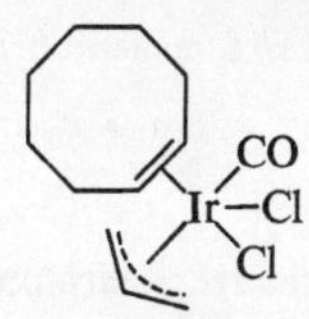

M 442.408
White prisms. Mp >114° dec. ir ν_{CO} 2056 cm^{-1} (KBr).

Shaw, B.L. *et al*, *J. Chem. Soc. (A)*, 1967, 1683 (*synth, ir, pmr*)

$C_{12}H_{20}Cl_2Ir_2$ Ir-00057

Di-μ-chlorotetrakis(η^3-2-propenyl)diiridium, 10CI
Tetra-π-allyldi-μ-chlorodiiridium. Di-[μ-chlorobis(η^3-allyl)]diiridium
[52488-76-3]

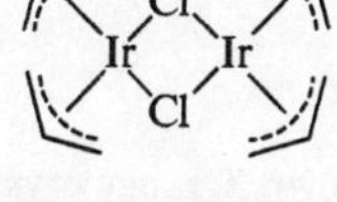

M 619.636
Bridged dimer. Yellow cryst. Mp 157°.
Monomer: Chlorobis(η^3-2-propenyl)iridium. Diallylchloroiridium. Bis(allyl)chloroiridium.
$C_6H_{10}ClIr$ M 309.818
Unknown.

Green, M. *et al*, *J. Chem. Soc., Dalton Trans.*, 1974, 333 (*synth, ir, ms, pmr*)

$C_{12}H_{35}IrP_2$ Ir-00058

Pentahydrobis(triethylphosphine)iridium, 10CI

[32490-60-1]

$IrH_5(PEt_3)_2$

M 433.576

Originally incorrectly formulated as $IrH_3(PEt_3)_2$. Catalyst for the exchange reaction of H_2 with D_2 in C_6H_6 soln. White, air-sensitive prisms (MeOH). V. sol. all org. solvs. Mp 33-4°. ir ν_{IrH} 1931 cm^{-1} (nujol).

Chatt, J. *et al*, *J. Chem. Soc.*, 1965, 7391 (*synth*)
Barefield, E.K. *et al*, *J. Am. Chem. Soc.*, 1970, **92**, 5234 (*use*)
Mann, B.E. *et al*, *J. Inorg. Nucl. Chem.*, 1971, **33**, 2195 (*nmr, props*)
Lehner, H. *et al*, *J. Organomet. Chem.*, 1981, **213**, C46 (*props*)

$C_{12}IrO_{12}Rh_3$ Ir-00059

Tricarbonyl(tri-μ-carbonylhexacarbonyltrirhodium)iridium, 9CI

Dodecacarbonyliridiumtrirhodium

[50864-90-9]

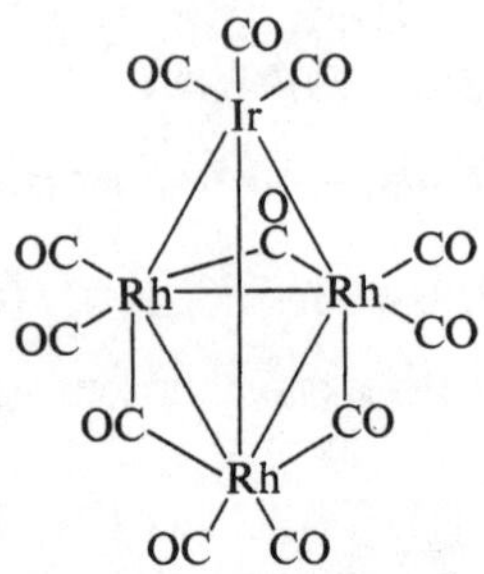

M 837.061

Proposed struct. shown. Orange cryst. Sol. hydrocarbons. Dec. at 130°.

Martinengo, S. *et al*, *J. Organomet. Chem.*, 1973, **59**, 379 (*synth, ir, ms*)

$C_{12}Ir_2O_{12}Rh_2$ Ir-00060

Di-μ-carbonylpentacarbonyl(μ-carbonyltetracarbonyldirhodium)diiridium, 9CI

Dodecacarbonyldiiridiumdirhodium

[50696-81-6]

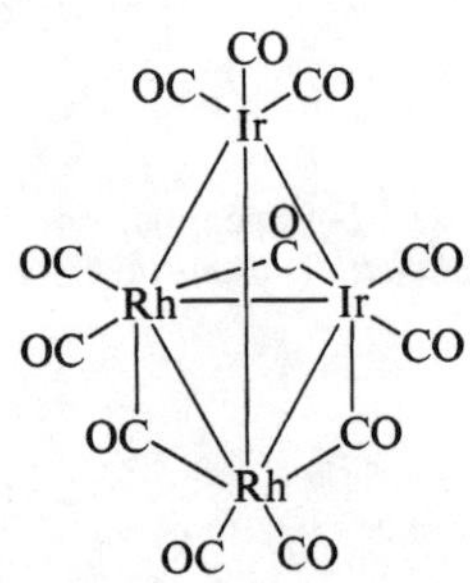

M 926.376

Proposed struct. shown. Orange cryst. Sol. hydrocarbons. Dec. ca. 150°.

Martinengo, S. *et al*, *J. Organomet. Chem.*, 1973, **59**, 379 (*synth, ir, uv, ms*)

$C_{12}Ir_4O_{12}$ Ir-00061

Dodecacarbonyltetrairidium, 9CI

Tetrairidium dodecacarbonyl

[18827-81-1]

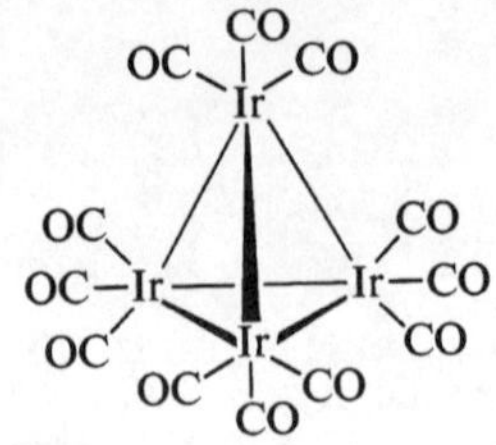

M 1105.005

Widely useful catalyst (water-gas shift reaction, hydrogenation, hydroformylation etc.). Canary-yellow cryst., sl. air-sensitive, inert to water. V. sl. sol. most org. solvs. Mp 170° dec. (in air). Purified by subl. at 120° under CO. ir ν_{CO} 2084, 2053, 2020 cm^{-1} (solid).

Inorg. Synth., 1972, **13**, 95 (*synth, ir*)
Whyman, R., *J. Chem. Soc., Dalton Trans.*, 1972, **20**, 2294 (*synth, ir, props*)
Churchill, M.R. *et al*, *Inorg. Chem.*, 1978, **17**, 3528 (*struct*)
Plummer, E.W. *et al*, *Phys. Rev. B*, 1978, **18**, 1673 (*pe*)
Michelin-Lausarot, P. *et al*, *Inorg. Chim. Acta*, 1979, **35**, 227 (*ms*)

$C_{13}H_{13}Ir$ Ir-00062

[(1,2,5,6-η)-1,3,5,7-Cyclooctatetraene](η⁵-2,4-cyclopentadien-1-yl)iridium, 9CI

[33360-54-2]

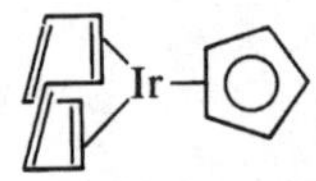

M 361.466

Mp 114-6°.

Yamazaki, H. *et al*, *Bull. Chem. Soc. Jpn.*, 1971, **44**, 582 (*synth, pmr, ir*)
Evans, J. *et al*, *J. Chem. Soc., Dalton Trans.*, 1974, 2375 (*pmr*)

$C_{13}H_{15}Ir$ Ir-00063

[(1,2,5,6-η)-1,3,5-Cyclooctatriene](η⁵-2,4-cyclopentadien-1-yl)iridium, 9CI

[41371-58-8]

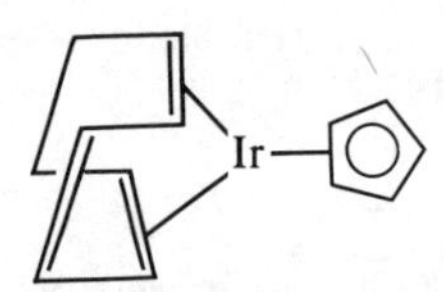

M 363.482

V. pale-green cryst. $Bp_{0.02}$ 80° subl.

Evans, J. *et al*, *J. Chem. Soc., Dalton Trans.*, 1972, 2668 (*synth, ms, ir, uv, pmr*)

$C_{13}H_{17}Ir$ Ir-00064

[(1,2,5,6-η)-1,5-Cyclooctadiene](η⁵-2,4-cyclopentadien-1-yl)iridium, 9CI

Cyclopentadienyl(1,5-cyclooctadiene)iridium

[12154-82-4]

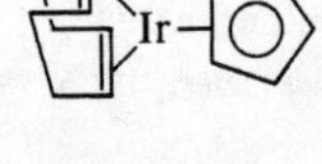

M 365.497

Stable white prisms (EtOH). Sol. org. solvs. Mp 122-7°. Volatile.

Robinson, S.D. *et al*, *J. Chem. Soc.*, 1965, 4997 (*synth*, *pmr*, *ir*)
King, R.B. *et al*, *Appl. Spectrosc.*, 1969, **23**, 148 (*ms*)
Pannetier, G. *et al*, *J. Less-Common Metals*, 1971, **24**, 470 (*struct*)

$C_{13}H_{18}IrO_2^{\oplus}$ Ir-00065

Dicarbonylmethyl(η^5-pentamethylcyclopentadienyl)iridium(1+)

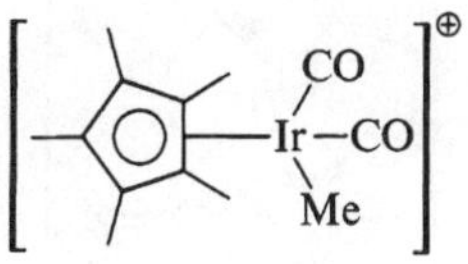

M 398.504 (ion)

Iodide: [32626-70-3].
$C_{13}H_{18}IIrO_2$ M 525.409
Pale-yellow cryst. (Et_2O). Dec. to Dicarbonyl[(1,2,3,4,5-η)-1,2,3,4,5-pentamethyl-2,4-cyclopentadien-1-yl]iridium, Ir-00052 ir υ_{CO} 2100, 2050cm^{-1} (Nujol).

Tetraphenylborate: [32825-77-7].
$C_{37}H_{38}BIrO_2$ M 717.736
White cryst. (MeOH). Mp 155-8° dec. ir υ_{CO} 2095, 2050cm^{-1} (KBr).

Maitlis, P.M. *et al*, *J. Organomet. Chem.*, 1971, **26**, 393 (*pmr*, *ir*, *synth*)

$C_{13}H_{24}Cl_2IrP$ Ir-00066

Dichloro[(1,2,3,4,5-η)-1,2,3,4,5-pentamethyl-2,4-cyclopentadien-1-yl](trimethylphosphine)iridium, 10CI

[80298-81-3]

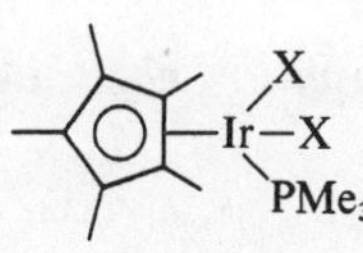

X = Cl

M 474.432
Air-stable pale orange-yellow solid. Sol. CH_2Cl_2, $CHCl_3$, C_6H_6, toluene, spar. sol. H_2O, Et_2O, insol. pet. ether.

Isobe, K. *et al*, *J. Chem. Soc., Dalton Trans.*, 1981, 2003 (*synth*, *ir*, *nmr*)

$C_{13}H_{26}IrP$ Ir-00067

Dihydro[(1,2,3,4,5-η)-1,2,3,4,5-pentamethyl-2,4-cyclopentadien-1-yl](trimethylphosphine)iridium, 10CI

[80146-01-6]

As Dichloro[(1,2,3,4,5-η)-1,2,3,4,5-pentamethyl-2,4-cyclopentadien-1-yl](trimethylphosphine)iridium, Ir-00066 with

X = H

M 405.542
Air-sensitive.

Isobe, K. *et al*, *J. Chem. Soc., Dalton Trans.*, 1981, 2003 (*synth*, *ir*, *pmr*, *ms*)
Janowicz, A.H. *et al*, *J. Am. Chem. Soc.*, 1982, **104**, 352 (*synth*)

$C_{14}H_9Ir_4O_{14}P$ Ir-00068

Tri-μ-carbonyloctacarbonyl(trimethyl phosphite-*P*)tetrairidium, 9CI

Undecacarbonyl(trimethyl phosphite)tetrairidium

[59532-81-9]

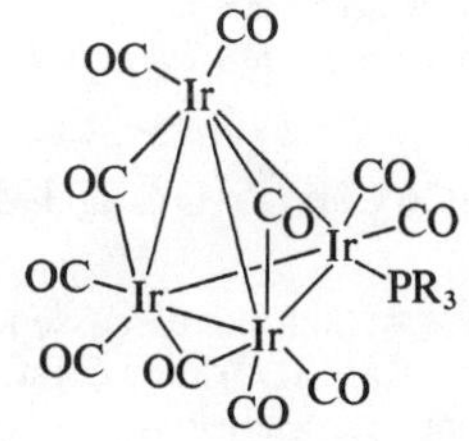

R = OMe

M 1201.070
Cryst. Mp 147-8°.

Stuntz, G.F. *et al*, *Inorg. Chem.*, 1976, **15**, 1994 (*synth*, *ir*, *struct*)

$C_{14}H_{15}F_6IrO_4$ Ir-00069

[(1,2,3,4,5-η)-1,2,3,4,5-Pentamethyl-2,4-cyclopentadien-1-yl]bis(trifluoroacetato-*O*)iridium, 9CI

Bis(trifluoroacetato)(pentamethylcyclopentadienyl)iridium

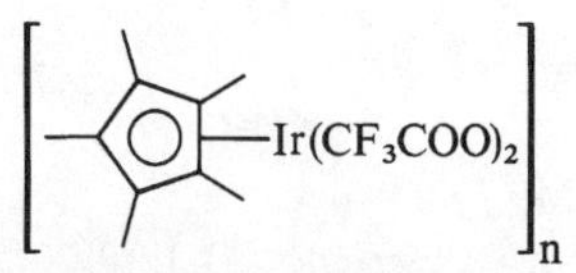

M 553.481
Exists as monohydrate.

Hydrate: [47378-85-8]. *Aqua(η^5-pentamethylcyclopentadienyl)bis(trifluoroacetato)iridium.*
$C_{14}H_{17}F_6IrO_5$ M 571.496
Homogenous hydrogenation cat. Mod. air-stable. Polymeric, struct. unknown.

Kang, J.W. *et al*, *J. Organomet. Chem.*, 1971, **30**, 127 (*synth*, *pmr*)
White, C. *et al*, *J. Chem. Soc., Dalton Trans.*, 1973, 1901 (*use*)

$C_{14}H_{18}Ir^{\oplus}$ Ir-00070

(η^6-Benzene)[(1,2,5,6-η)-1,5-cyclooctadiene]iridium(1+), 10CI

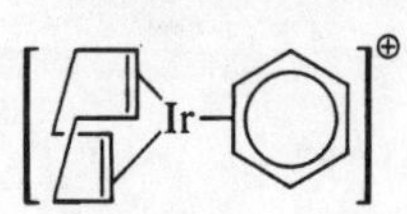

M 378.516 (ion)

Hexafluorophosphate: [76742-03-5].
$C_{14}H_{18}F_6IrP$ M 523.480
Cryst.

Tetrafluoroborate: [74238-34-9].
$C_{14}H_{18}BF_4Ir$ M 465.320
Cryst.

4-Methylbenzenesulfonate: [75751-42-7].
$C_{21}H_{25}IrO_3S$ M 549.707
Yellow powder (CH_2Cl_2/C_6H_6/heptane).

Trifluoromethanesulfonate: [75751-52-9].
$C_{15}H_{18}F_3IrO_3S$ M 527.581
White cryst.

Trifluoroacetate: [76706-15-5].
$C_{16}H_{18}F_3IrO_2$ M 491.532
White needles. Mp 178° dec.

Bianchi, F. *et al*, *J. Organomet. Chem.*, 1980, **202**, 99 (*synth*, *ir*, *pmr*)

Sievert, A.C. *et al*, *Inorg. Chem.*, 1981, **20**, 489 (*synth, ir, pmr*)

$C_{14}H_{21}IrO_4$ Ir-00071

Bis(acetato-*O*)[(1,2,3,4,5-η)-1,2,3,4,5-pentamethyl-2,4-cyclopentadien-1-yl]iridium, 9CI
Diacetato(η^5-pentamethylcyclopentadienyl)iridium
[34738-19-7]

$[C_5(CH_3)_5Ir(OAc)_2H_2O]_n$

M 445.538
Polymeric hydrate, struct. not discussed. Olefin hydrogenation catalyst. Orange cryst. + $1H_2O$. Sol. org. solvs. V. moisture sensitive.

Kang, J.W. *et al*, *J. Organomet. Chem.*, 1971, **30**, 127 (*synth, pmr*)

$C_{14}H_{23}Ir$ Ir-00072

Bis(ethylene)(pentamethyl-π-cyclopentadienyl)iridium, 8CI
η^5-Pentamethylcyclopentadienylbis(ethylene)iridium
[32613-71-1]

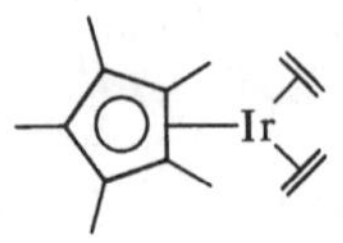

M 383.556
White cryst. Mp 79-84° dec.

Moseley, K. *et al*, *J. Chem. Soc. (A)*, 1970, 2875 (*synth, pmr*)

$C_{15}H_{20}Ir^{\oplus}$ Ir-00073

η^5-Cyclopentadienyl(η^5-pentamethylcyclopentadienyl)iridium(1+)

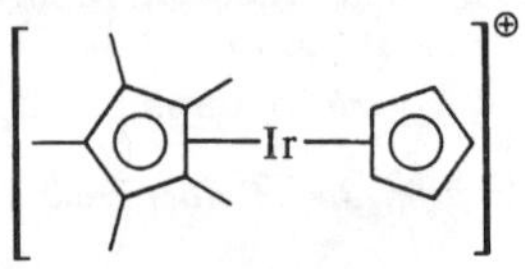

M 392.543 (ion)
Hexafluorophosphate: [32679-06-4].
$C_{15}H_{20}F_6IrP$ M 537.507
Mp >260°.

Moseley, K. *et al*, *J. Chem. Soc. (A)*, 1970, 2875 (*synth, pmr*)

$C_{15}H_{20}IrO_2^{\oplus}$ Ir-00074

Dicarbonyl(η^5-pentamethyl-2,4-cyclopentadien-1-yl)(η^1-1-propenyl)iridium(1+)
(η^1-Allyl)dicarbonyl(η^5-pentamethylcyclopentadienyl)-iridium(1+)

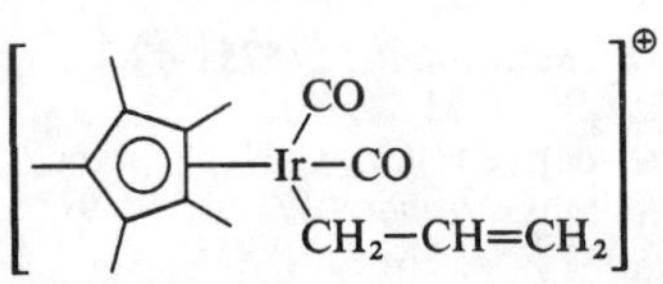

M 424.542 (ion)
Iodide: [32661-05-5].
$C_{15}H_{20}IIrO_2$ M 551.446
Yellow solid, dec. slowly on standing. Mp 125-7°. ir υ_{CO} 2103, 2063cm^{-1} (CH_2Cl_2).

King, R.B. *et al*, *J. Organomet. Chem.*, 1971, **27**, 409 (*synth, ir*)

$C_{16}H_{18}IrN_2^{\oplus}$ Ir-00075

(2,2′-Bipyridine-*N,N′*)[(1,2,5,6-η)-1,5-hexadiene]iridium(1+), 9CI
1,5-Hexadiene(2,2′-bipyridyl)iridium(1+)

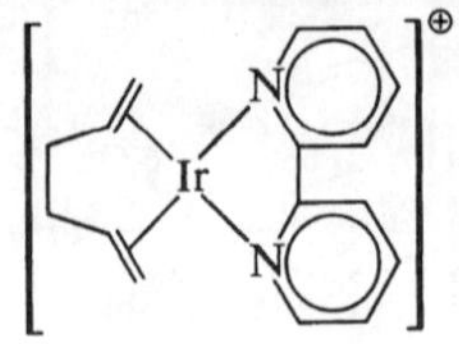

M 430.552 (ion)
Ionic complexes (except chloride) readily take up molecular hydrogen and catalyse hydrogenation of 1-hexene.
Chloride: [65060-37-9].
$C_{16}H_{18}ClIrN_2$ M 466.005
Violet solid. Sol. MeOH.
Hexafluorophosphate: [62630-34-6].
$C_{16}H_{18}F_6IrN_2P$ M 575.516
Green microcryst. solid. Insol. MeOH.
Tetraphenylborate: [62630-41-5].
$C_{40}H_{38}BIrN_2$ M 749.784
Green microcryst. solid. Insol. MeOH.

Zassinovich, G. *et al*, *J. Mol. Catal.*, 1977, **2**, 63 (*use*)
Mestroni, G. *et al*, *J. Organomet. Chem.*, 1977, **140**, 63 (*synth, use*)

$C_{16}H_{18}Ir_4O_{16}P_2$ Ir-00076

Tri-μ-carbonylheptacarbonylbis(trimethyl phosphite)tetrairidium, 9CI
Decacarbonylbis(trimethyl phosphite)tetrairidium
[59532-54-6]

$Ir_4(CO)_{10}[P(OMe)_3]_2$

M 1297.136
Air- and moisture-stable yellow-orange solid (pentane). Mp 140-3°. Stereochemically nonrigid in soln.

Stuntz, G.F. *et al*, *Inorg. Chem.*, 1976, **15**, 1994 (*synth, ir*)

$C_{16}H_{20}Ir^{\oplus}$ Ir-00077

[(1,2,3,4-η)-1,3,5-Cyclooctatriene][(1,2,3,4,5,6-η)-1,3,5-cyclooctatriene]iridium(1+), 9CI
Bis(1,3,5-cyclooctatriene)iridium(1+)

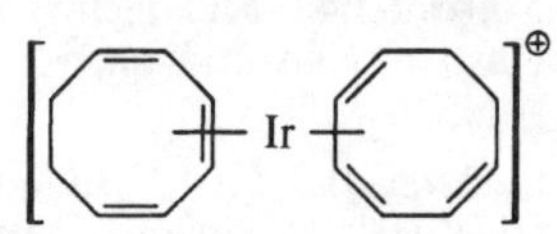

M 404.554 (ion)
Tetrafluoroborate: [64331-08-4].
$C_{16}H_{20}BF_4Ir$ M 491.358
No phys. props. reported.

Draggett, P.T. *et al*, *J. Organomet. Chem.*, 1977, **135**, C60 (*synth, pmr*)

$C_{16}H_{20}IrN_2^{\oplus}$ Ir-00078

Bis(η^3-2-propenyl)bis(pyridine)iridium(1+), 9CI
Bis[(η^3-allyl)pyridine]iridium(1+)

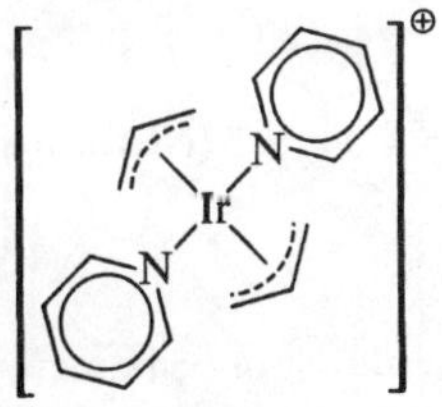

M 432.567 (ion)
Tetrafluoroborate: [52471-94-0].
$C_{16}H_{20}BF_4IrN_2$ M 519.371
White cryst. Mp 115-6°.

Green, M. *et al*, *J. Chem. Soc., Dalton Trans.*, 1974, 333 (*synth, ir, pmr*)

$C_{16}H_{21}Ir^{\oplus\oplus}$ Ir-00079

(η^6-Benzene)[(1,2,3,4,5-η)-1,2,3,4,5-pentamethyl-2,4-cyclopentadien-1-yl]iridium(2+), 9CI

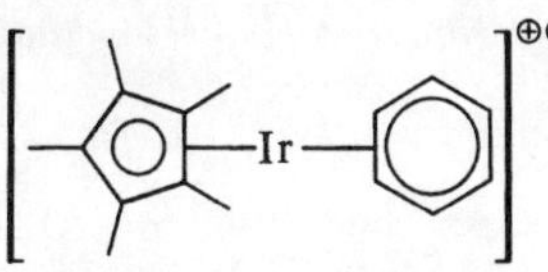

M 405.562 (ion)
Bis(hexafluorophosphate): [12715-71-8].
$C_{16}H_{21}F_{12}IrP_2$ M 695.490
Sol. Me_2SO. pmr (DMSO-d_6) δ 2.35, 7.41s.

White, C. *et al*, *J. Chem. Soc. (A)*, 1971, 3322 (*synth, pmr*)

$C_{16}H_{21}IrO^{\oplus\oplus}$ Ir-00080

[(1,2,3,4,5-η)-1,2,3,4,5-Pentamethyl-2,4-cyclopentadien-1-yl](η^6-phenol)iridium(2+), 9CI

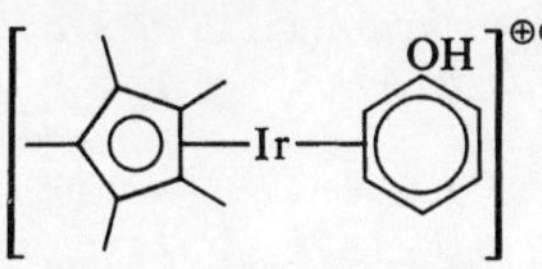

M 421.561 (ion)
Deprotonates readily to give hydrogen-bonded dimer.
Bis(hexafluorophosphate): [62837-16-5].
$C_{16}H_{21}F_{12}IrOP_2$ M 711.490
Unstable white solid (Me_2CO/Et_2O).
Deprotonated dimer, tris(hexafluorophosphate): [62837-10-9].
$C_{22}H_{41}F_{18}Ir_2O_4P_3$ M 1188.896
Stable white cryst. (Me_2CO/Et_2O).

White, C. *et al*, *J. Organomet. Chem.*, 1..77, **127**, 415 (*pmr, ir, synth*)

$C_{16}H_{22}IrN^{\oplus\oplus}$ Ir-00081

(η^6-Benzenamine)[(1,2,3,4,5-η)-1,2,3,4,5-pentamethyl-2,4-cyclopentadien-1-yl]iridium(2+), 9CI
Aniline(pentamethylcyclopentadienyl)iridium(2+)

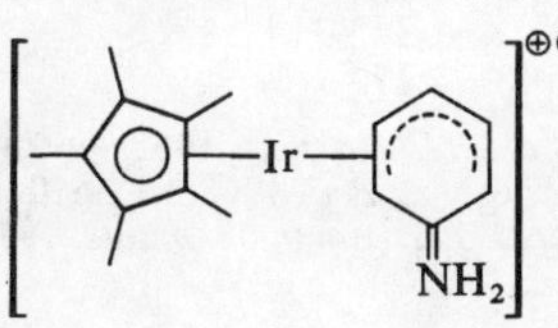

M 420.577 (ion)
C_6 ring is η^5 bonded to iridium. Air-stable.
Bis(hexafluorophosphate): [75583-66-3].
$C_{16}H_{22}F_{12}IrNP_2$ M 710.505
Air-stable.

Espinet, P. *et al*, *J. Chem. Soc., Dalton Trans.*, 1980, 1048 (*synth, struct, pmr, cmr, ir*)

$C_{16}H_{24}BrIr$ Ir-00082

Bromobis[(1,2,5,6-η)-1,5-cyclooctadiene]iridium, 9CI
Bis(cyclooctadiene)bromoiridium
[39732-28-0]

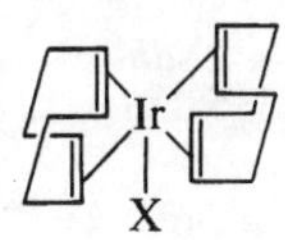

X = Br

M 488.490
Brown cryst. Loses a cyclooctadiene ligand at 97° or in soln. to form Di-μ-bromobis[(1,2,5,6-η)-1,5-cyclooctadiene]diiridium, Ir-00083 .

Onderdelinden, A.L. *et al*, *Inorg. Chim. Acta*, 1972, **6**, 420 (*synth, ir*)

$C_{16}H_{24}Br_2Ir_2$ Ir-00083

Di-μ-bromobis[(1,2,5,6-η)-1,5-cyclooctadiene]diiridium, 9CI
Bis(cyclooctadiene)dibromodiiridium. Bis[μ-bromo(η^4-1,5-cyclooctadiene)iridium]
[12245-73-7]

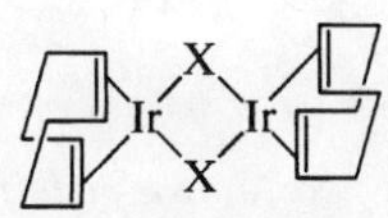

X = Br

M 760.614
Bridged dimer. Dark-red cryst. Mp 200-10°.
Monomer: Bromo(1,5-cyclooctadiene)iridium.
$C_8H_{12}BrIr$ M 380.307
Unknown.

Ainscough, E.W. *et al*, *J. Chem. Soc. (A)*, 1971, 3413 (*synth*)
Onderdelinden, A.L. *et al*, *Inorg. Chim. Acta*, 1972, **6**, 420 (*synth, ir*)

$C_{16}H_{24}ClIr$ Ir-00084

Chlorobis[(1,2,5,6-η)-1,5-cyclooctadiene]iridium, 9CI
Bis(cyclooctadiene)chloroiridium
[39732-27-9]
As Bromobis[(1,2,5,6-η)-1,5-cyclooctadiene]iridium, Ir-00082 with

X = Cl

M 444.039
Brown-yellow cryst. Loses a cyclooctadiene ligand at 70° or in soln. to form Di-μ-chloro[(1,2,5,6-η)-1,5-cyclooctadiene]diiridium, Ir-00085 .

Onderdelinden, A.L. *et al*, *Inorg. Chim. Acta*, 1972, **6**, 420 (*synth, ir*)

$C_{16}H_{24}Cl_2Ir_2$ **Ir-00085**

Di-μ-chloro[(1,2,5,6-η)-1,5-cyclooctadiene]diiridium, 10CI, 9CI

Bis(1,5-cyclooctadiene-μ-chloroiridium)

[12112-67-3]

As Di-μ-bromobis[(1,2,5,6-η)-1,5-cyclooctadiene]-diiridium, Ir-00083 with

X = Cl

M 671.712

Bridged dimer. Catalyst for the selective hydrogenation of α,β-unsaturated acids to saturated acids and for the polymerisation of norbornadiene. Orange cryst. Mp >200° dec.

Monomer: (1,5-Cyclooctadiene)chloroiridium.

$C_8H_{12}ClIr$ M 335.856

Unknown.

Winkhaus, G. *et al*, *Chem. Ber.*, 1966, **99**, 3610 (*synth, pmr, ir*)
Inorg. Synth., 1974, **15**, 18 (*synth*)
Bonnaire, R. *et al*, *J. Organomet. Chem.*, 1976, **104**, 107 (*cmr*)
Bezman, S.A. *et al*, *Inorg. Chem.*, 1980, **19**, 3755 (*synth, props*)
Sievert, A.C. *et al*, *Inorg. Chem.*, 1981, **20**, 489 (*synth, props*)
Fieser, M. *et al*, *Reagents for Organic Synthesis*, Wiley, 1967-83, **5**, 113 (*use*)

$C_{16}H_{24}I_2Ir_2$ **Ir-00086**

Bis[(1,2,5,6-η)-1,5-cyclooctadiene]di-μ-iododiiridium, 9CI

Di-μ-iodobis(η⁴-1,5-cyclooctadiene)diiridium. Bis[μ-iodo(η⁴-1,5-cyclooctadiene)iridium]

[60255-25-6]

As Di-μ-bromobis[(1,2,5,6-η)-1,5-cyclooctadiene]diiridium, Ir-00083 with

X = I

M 854.615

Bridged dimer.

Monomer: (1,5-Cyclooctadiene)iodoiridium. (1,5-Cyclooctadiene)iridium iodide.

$C_8H_{12}IIr$ M 427.307

Unknown.

Forster D., *J. Chem. Soc., Dalton Trans.*, 1979, 1639 (*synth*)

$C_{16}H_{24}Ir^{\oplus}$ **Ir-00087**

Bis[(1,2,5,6-η)-1,5-cyclooctadiene]iridium(1+), 9CI

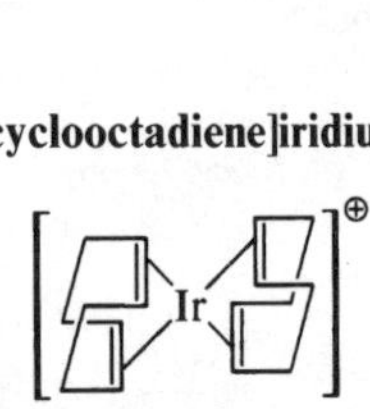

M 408.586 (ion)

Useful starting material.

Tetrafluoroborate: [35138-23-9].

$C_{16}H_{24}BF_4Ir$ M 495.389

Deep purple-brown cryst. Mp 140-5°.

Hexafluorophosphate: [70471-95-3].

$C_{16}H_{24}F_6IrP$ M 553.550

No phys. props. reported.

Green, M. *et al*, *J. Chem. Soc. (A)*, 1971, 2334 (*synth, ir, pmr*)

$C_{16}H_{24}IrN_3^{\oplus\oplus}$ **Ir-00088**

Tris(acetonitrile)[(1,2,3,4,5-η)-1,2,3,4,5-pentamethyl-2,4-cyclopentadien-1-yl]iridium(2+), 9CI

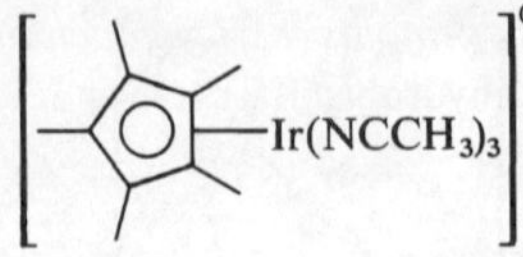

M 450.606 (ion)

Useful starting material for pentamethylcyclopentadienyl iridium complexes.

Bis(hexafluorophosphate): [59738-32-8].

$C_{16}H_{24}F_{12}IrN_3P_2$ M 740.534

Yellow cryst. (MeCN/Et_2O).

Thompson, S.J. *et al*, *Angew. Chem., Int. Ed. Engl.*, 1976, **15**, 490 (*synth*)
White, C. *et al*, *J. Chem. Soc., Dalton Trans.*, 1977, 1654 (*synth, ir, pmr*)

$C_{16}H_{26}Br_4Ir_2$ **Ir-00089**

Tetrabromodihydrobis(1,5-cyclooctadiene)diiridium

Di-μ-bromobis[bromo(η⁴-1,5-cyclooctadiene)hydridoiridium. Di-μ-bromodibromodihydridobis(1,5-cyclooctadiene)diiridium

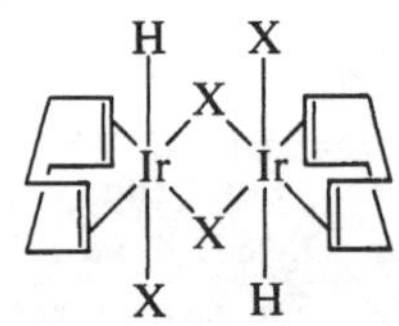

X = Br

M 922.437

Bridged dimer. Cream cryst. Mp >200° dec. ir ν_{IrH} 2239 cm^{-1} (nujol).

Monomer: Dibromohydro(1,5-cyclooctadiene)iridium.

$C_8H_{13}Br_2Ir$ M 461.219

Unknown.

Robinson, S.D. *et al*, *J. Chem. Soc.*, 1965, 4997 (*synth, ir*)

$C_{16}H_{26}Cl_4Ir_2$ **Ir-00090**

Di-μ-chlorodichlorobis[(1,2,5,6-η)-1,5-cyclooctadiene]dihydrodiiridium, 9CI

Di-μ-chlorobis[chloro(η⁴-1,5-cyclooctadiene)hydridoiridium]. Tetrachlorobis(1,5-cyclooctadiene)-dihydrodiiridium

[12148-12-8]

As Tetrabromodihydrobis(1,5-cyclooctadiene)diiridium, Ir-00089 with

X = Cl

M 744.633

Bridged dimer. Model for olefin hydrogenation catalysts. Yellow-cream stable microcryst. Insol. Mp 198°, >200°. ir ν_{IrH} 2261 cm^{-1} (nujol).

Monomer: Dichloro(1,5-cyclooctadiene)hydridoiridium.

$C_8H_{13}Cl_2Ir$ M 372.317

Unknown.

Robinson, S.D. *et al*, *J. Chem. Soc.*, 1965, 4997 (*synth, ir*)
Winkhaus, G. *et al*, *Chem. Ber.*, 1966, **99**, 3610 (*synth, pmr*)
Pannetier, G. *et al*, *J. Less-Common Metals*, 1970, **21**, 103 (*synth, ir*)

$C_{16}H_{26}I_4Ir_2$ **Ir-00091**

Dihydrotetraiodobis(1,5-cyclooctadiene)diiridium

Di-μ-iodobis(η^4-1,5-cyclooctadiene)hydridoiodoiridium. Di-μ-iododiiododihydridobis(1,5-cyclooctadiene)diiridium

As Tetrabromodihydrobis(1,5-cyclooctadiene)diiridium, Ir-00089 with

X = I

M 1110.439

Bridged dimer. Light-brown cryst. ν_{IrH} 2188 cm^{-1} (nujol).

Monomer: Hydrodiiodo-1,5-cyclooctadieneiridium.
$C_8H_{13}IIr$ M 428.315
Unknown.

Robinson, S.D. *et al*, *J. Chem. Soc.*, 1965, 4997 (*synth*, *ir*)

$C_{16}H_{28}IrO_2^{\oplus}$ **Ir-00092**

[(1,2,5,6-η)-1,5-Cyclooctadiene]bis(tetrahydrofuran)iridium(1+), 9CI

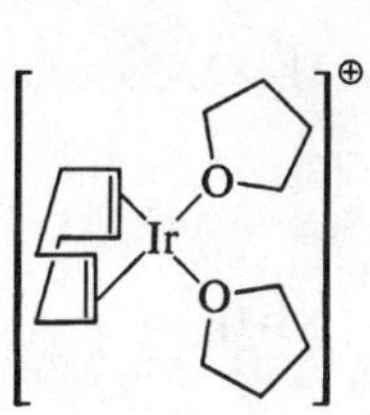

M 444.616 (ion)

Synthetic starting material. THF readily replaced by phosphines or other Lewis bases.

Hexafluoroantimonate: [38816-42-1].
$C_{16}H_{28}F_6IrO_2Sb$ M 680.356
No phys. props. reported.

Haines, R.J. *et al*, *J. Chem. Soc., Chem. Commun.*, 1972, 990 (*use*)

$C_{16}H_{30}IrN_6O^{\oplus}$ **Ir-00093**

Carbonyltris(1,3-dimethyl-2-imidazolidinylidene)iridium(1+), 9CI

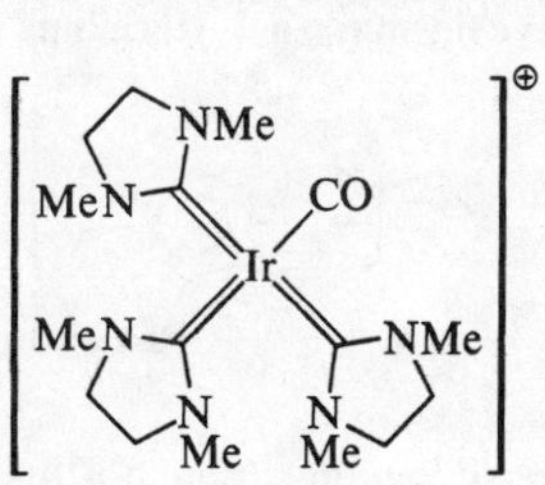

M 514.673 (ion)

Tetrafluoroborate: [41654-75-5].
$C_{16}H_{30}BF_4IrN_6O$ M 601.476
Yellow solid (CH_2Cl_2/Et_2O). Mp 204-6°. ir ν_{CO} 1960 cm^{-1} (nujol).

Cetinkaya, B. *et al*, *J. Chem. Soc., Dalton Trans.*, 1974, 1827 (*synth*, *pmr*, *ir*)

$C_{16}H_{32}ClIrSi$ **Ir-00094**

Chlorodihydro[(1,2,3,4,5-η)-1,2,3,4,5-pentamethyl-2,4-cyclopentadien-1-yl](triethylsilyl)iridium

Chlorodihydrido(triethylsilyl)(pentamethylcyclopentadienyl)iridium

[84029-38-9]

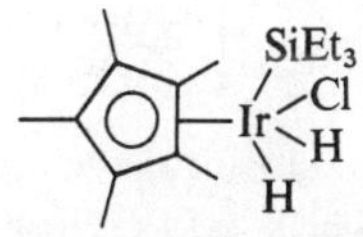

M 480.187

Colourless solid. ir ν_{IrH} 2135 cm^{-1}.

Fernandez, M.J. *et al*, *Organometallics*, 1983, **2**, 164 (*synth*, *ir*, *pmr*)

$C_{16}H_{33}IrO_3S_3^{\oplus\oplus}$ **Ir-00095**

[(1,2,3,4,5-η)-1,2,3,4,5-Pentamethyl-2,4-cyclopentadien-1-yl]tris[sulfinylbis[methane]]iridium(2+), 9CI

Tris(dimethyl sulfoxide)(η^5-pentamethylcyclopentadienyl)iridium(2+)

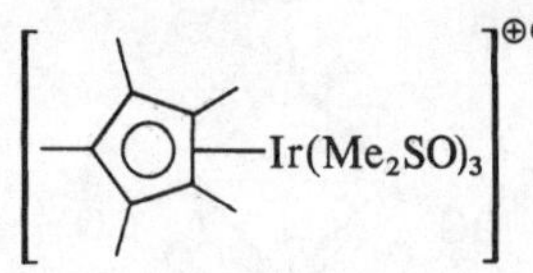

M 561.835 (ion)

Contains both O- and S-bonded Me_2SO.

Bis(hexafluorophosphate): [59738-06-6].
$C_{16}H_{33}F_{12}IrO_3P_2S_3$ M 851.763

Thompson, S.J. *et al*, *Angew. Chem., Int. Ed. Engl.*, 1976, **15**, 490 (*synth*, *ir*)

White, C. *et al*, *J. Chem. Soc., Dalton Trans.*, 1977, 1654 (*pmr*, *ir*)

$C_{16}Ir_6O_{16}$ **Ir-00096**

Hexadecacarbonylhexairidium

Iridium carbonyl ($Ir_6(CO)_{16}$), 10CI. *Tetra-μ_3-carbonyldodecacarbonylhexairidium*, 8CI. *Hexairidium hexadecacarbonyl*

[56801-74-2]

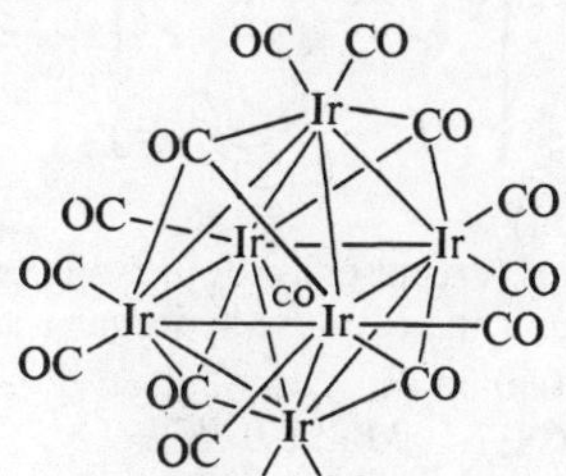

M 1601.486

Red cryst. V. sl. sol. org. solvs.

Malatesta, L. *et al*, *J. Chem. Soc., Chem. Commun.*, 1970, 532 (*synth*, *ir*)

Angoletta, M. *et al*, *J. Organomet. Chem.*, 1975, **94**, 99 (*synth*, *ir*)

Demartin, F. *et al*, *J. Organomet. Chem.*, 1982, **234**, C39; 1983, **243**, C10.

$C_{17}H_{20}ClIrO$ **Ir-00097**

Carbonylchloro(η^5-pentamethylcyclopentadienyl)(phenyl)iridium
[32609-75-9]

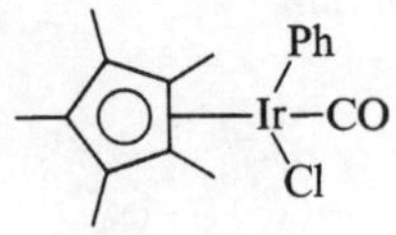

M 468.017
Yellow solid (Et_2O/hexane). Mp 150-3°. ir ν_{CO} 2025cm^{-1} (KBr).

Maitlis, P.M. *et al*, *J. Organomet. Chem.*, 1971, **26**, 393 (*synth, ir, pmr*)

$C_{18}H_{12}Ir_4O_{10}$ **Ir-00098**

Tri-μ-carbonylheptacarbonyl[(1,2,5,6-η)-1,5-cyclooctadiene]tetrairidium, 9CI
Decacarbonyl(η^4-1,5-cyclooctadiene)tetrairidium
[66809-80-1]

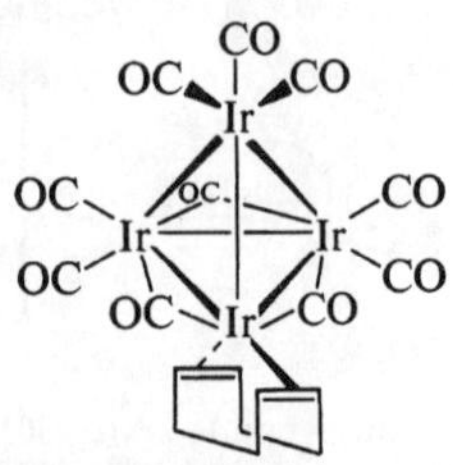

M 1157.167
Yellow oil. Sol. CH_2Cl_2, pentane.

Stantz, G.F. *et al*, *Inorg. Chem.*, 1978, **17**, 2596 (*synth, ir, nmr, ms*)

$C_{18}H_{18}IrN_2^{\oplus}$ **Ir-00099**

[(1,2,5,6-η)-1,5-Hexadiene](1,10-phenanthroline-N^1,N^{10})iridium(1+), 9CI

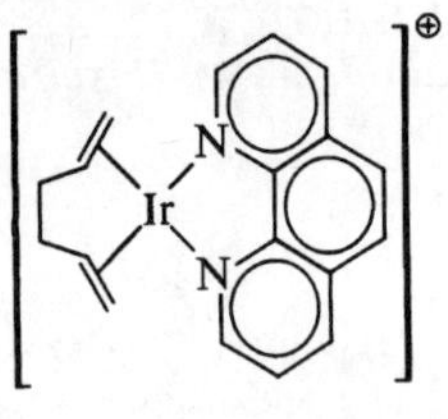

M 454.574 (ion)
Ionic complexes (except chloride) readily take up molecular hydrogen, and catalyse hydrogenation of 1-hexene.
Chloride: [65060-38-0].
$C_{18}H_{18}ClIrN_2$ M 490.027
Violet microcryst. Sol. MeOH.
Hexafluorophosphate: [62630-36-8].
$C_{18}H_{18}F_6IrN_2P$ M 599.538
Green microcryst. Insol. MeOH.
Tetraphenylborate: [62630-42-6].
$C_{42}H_{38}BIrN_2$ M 773.806
Green microcryst. Insol. MeOH.

Zassinovich, G. *et al*, *J. Mol. Catal.*, 1977, **2**, 63 (*use*)
Mestroni, G. *et al*, *J. Organomet. Chem.*, 1977, **140**, 63 (*synth, use*)

$C_{18}H_{20}IrN_2^{\oplus}$ **Ir-00100**

(2,2′-Bipyridine-N,N')[(1,2,5,6-η)-1,5-cyclooctadiene]iridium(1+), 10CI
Bipyridyl(1,5-cyclooctadiene)iridium(1+)
[48185-98-4]

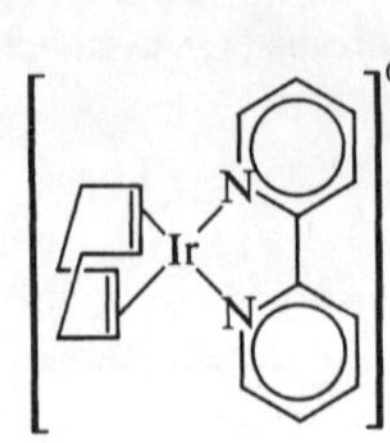

M 456.589 (ion)
Bromide: [53505-21-8].
$C_{18}H_{20}BrIrN_2$ M 536.493
Red-violet cryst.
Chloride: [53505-20-7].
$C_{18}H_{20}ClIrN_2$ M 492.042
Violet-blue cryst.
Iodide:
$C_{18}H_{20}IIrN_2$ M 583.494
Violet cryst.
Tetrafluoroborate: [34692-05-2].
$C_{18}H_{20}BF_4IrN_2$ M 543.393
Dark-green cryst. Mp 260°.
Perchlorate: [41444-85-3].
$C_{18}H_{20}ClIrN_2O_4$ M 556.040
Red-violet cryst.
Hexafluorophosphate: [41444-84-2].
$C_{18}H_{20}F_6IrN_2P$ M 601.554
Red-violet cryst.
Thiocyanate: [41396-73-0].
$C_{19}H_{20}IrN_3S$ M 514.667
Violet cryst.

Green, M. *et al*, *J. Chem. Soc. (A)*, 1971, 2334 (*synth, ir, pmr*)
Mestroni, G. *et al*, *J. Organomet. Chem.*, 1974, **73**, 119 (*synth*)
Louw, W.J. *et al*, *Inorg. Chem.*, 1980, **19**, 7 (*uv*)

$C_{18}H_{21}IrN^{\oplus}$ **Ir-00101**

[(1,2,3,3a,7a-η)-1*H*-indol-1-yl][(1,2,3,4,5-η)-1,2,3,4,5-pentamethyl-2,4-cyclopentadien-1-yl]iridium(1+)

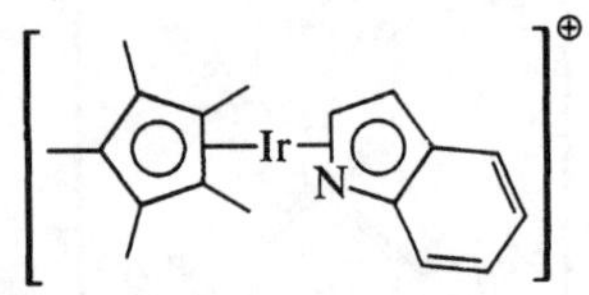

M 443.591 (ion)
Undergoes reversible protonation to η^6-indole complex [(3*a*,4,5,6,7,7*a*-η)-1*H*-Indole][(1,2,3,4,5-η-)-1,2,3,4,5-pentamethyl-2,4-cyclopentadien-1-yl]iridium(2+), Ir-00102.
Hexafluorophosphate: [60937-06-6].
$C_{18}H_{21}F_6IrNP$ M 588.555
Yellow solid. Insol. EtOH, Et_2O.

White, C. *et al*, *J. Chem. Soc., Dalton Trans.*, 1977, 1654 (*synth, nmr*)

$C_{18}H_{22}IrN^{\oplus\oplus}$ **Ir-00102**

[(3*a*,4,5,6,7,7*a*-η)-1*H*-Indole][(1,2,3,4,5-η)-1,2,3,4,5-pentamethyl-2,4-cyclopentadien-1-yl]iridium(2+), 9CI

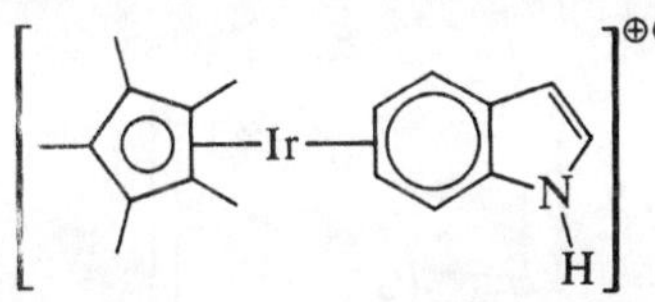

M 444.599 (ion)

Undergoes reversible deprotonation to η^5-indolyl complex.

Bis(hexafluorophosphate): [60936-98-3].
$C_{18}H_{22}F_{12}IrNP_2$ M 734.527
Yellow.

White, C. *et al*, *J. Chem. Soc.*, *Dalton Trans.*, 1977, 1654 (*synth*, *pmr*, *cmr*, *ir*)

$C_{18}H_{28}ClIr$ **Ir-00103**

Chloro[(1,2,3-η)-2-cycloocten-1-yl][(1,2,3,4,5-η)-1,2,3,4,5-pentamethyl-2,4-cyclopentadien-1-yl]iridium, 9CI

[32757-78-1]

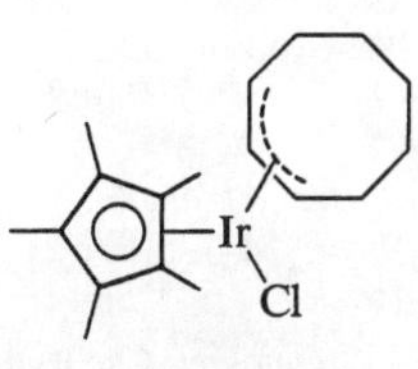

M 472.092

Yellow cryst. Mp 95° dec.

Lee, H.B. *et al*, *J. Chem. Soc.*, *Dalton Trans.*, 1975, 2322 (*synth*, *pmr*)

$C_{18}H_{30}Ir_2O_2$ **Ir-00104**

Bis[(1,2,5,6-η)-1,5-cyclooctadiene]di-μ-methoxydiiridium, 10CI

Bis[μ-methoxyl(η^4-1,5-cyclooctadiene)iridium]

[12148-71-9]

As Di-μ-bromobis[(1,2,5,6-η)-1,5-cyclooctadiene]diiridium, Ir-00083 with

X = OMe

M 662.874

Bridged dimer. Yellow cryst. (hexane or CH_2Cl_2/MeOH). Mp 180° dec.

Monomer: *(1,5-Cyclooctadiene)methoxyiridium.*
$C_9H_{15}IrO$ M 331.437
Unknown.

Robinson, S.D. *et al*, *J. Chem. Soc.*, 1965, 4997 (*synth*, *ir*)
Green, M. *et al*, *J. Chem. Soc. (A)*, 1971, 2334 (*synth*, *pmr*)
Tabrizi, D. *et al*, *J. Less-Common Metals*, 1971, **23**, 443 (*cryst struct*)
Tabrizi, D. *et al*, *J. Less-Common Metals*, 1971, **24**, 83 (*synth*, *ir*, *nmr*)

$C_{19}H_{24}BF_6IrN_6O_5$ **Ir-00105**

Aquabis(trifluoroacetato-*O*)[tris(3,5-dimethyl-1*H*-pyrazolato-N^1)hydroborato(1−)-N^2,$N^{2'}$,$N^{2''}$]iridium, 10CI

[Hydrotris(3,5-dimethylpyrazolyl)borato]bis(trifluoroacetato)aquoiridium

[73117-36-9]

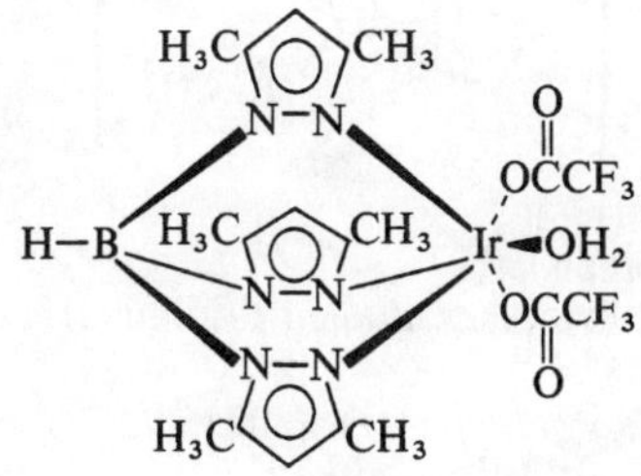

M 733.456

Pale-yellow needles (toluene). Mp 166-7°.

May, S. *et al*, *Inorg. Chem.*, 1980, **19**, 1582 (*synth*, *pmr*)

$C_{19}H_{26}IrN^{\oplus\oplus}$ **Ir-00106**

(Acetonitrile)[(2,3,5,6-η)-bicyclo[2.2.1]hepta-2,5-diene][(1,2,3,4,5-η)-1,2,3,4,5-pentamethyl-2,4-cyclopentadien-1-yl]iridium(2+), 9CI

Acetonitrile(η^4-norbornadiene)(η^5-pentamethylcyclopentadienyl)iridium(2+)

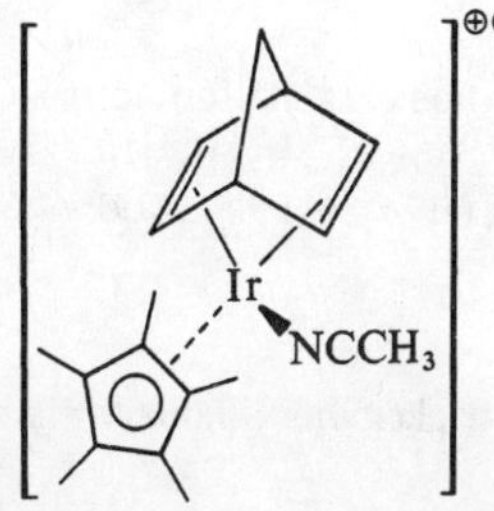

M 460.641 (ion)

Bis(hexafluorophosphate): [69210-09-9].
$C_{19}H_{26}F_{12}IrNP_2$ M 750.569
Yellow cryst. (Me_2CO/Et_2O), stable to air.

White, C. *et al*, *J. Chem. Soc.*, *Dalton Trans.*, 1978, 1305 (*synth*, *pmr*)

$C_{19}H_{27}Cl_2IrP_2$ **Ir-00107**

Dichlorobis(dimethylphenylphosphine)(η^3-propenyl)iridium

π-Allyldichlorobis(dimethylphenylphosphine)iridium, 8CI

[12308-85-9]

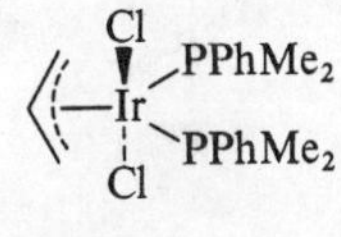

M 580.496

Yellow cryst. (MeOH). Mp 140-2° (138-41°).

Powell, J. *et al*, *J. Chem. Soc. (A)*, 1968, 780 (*synth*, *ir*, *pmr*)

$C_{19}H_{33}IrO_3^{\oplus\oplus}$ **Ir-00108**

[(1,2,3,4,5-η)-1,2,3,4,5-Pentamethyl-2,4-cyclopentadien-1-yl]tris(2-propanone)iridium(2+), 9CI

Tris(acetone)pentamethylcyclopentadienyliridium(2+)

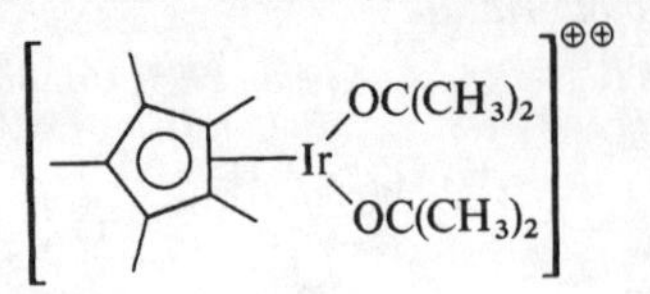

M 501.688 (ion)

Useful in prepn. of pentamethylcyclopentadienyliridium(III) arene complexes.

Bis(hexafluorophosphate): [60936-92-7].

$C_{19}H_{33}F_{12}IrO_3P_2$ M 791.616

Not isol. as solid. Rearranges to η^5-2-hydroxy-4-methylpentadienyl complex on healing.

White, C. *et al*, *J. Chem. Soc., Dalton Trans.*, 1977, 1654 (*synth, pmr*)
White, C. *et al*, *J. Organomet. Chem.*, 1977, **134**, 319.

$C_{20}H_{15}AsClIrO_2$ **Ir-00109**

Dicarbonylchloro(triphenylarsine)iridium, 9CI

[59809-93-7]

$Ph_3AsIrCl(CO)_2$

M 589.932

Hydrocarboxylation catalyst for olefins. Grey solid. ir ν_{CO} 2077s, 1996s cm^{-1} ($CHCl_3$).

U.S.P., 3 948 962, (*1976*); *CA*, **85**, 33191w (*synth, use*)

$C_{20}H_{15}ClIrO_2P$ **Ir-00110**

Dicarbonylchloro(triphenylphosphine)iridium, 9CI

[59810-02-5]

$Ph_3PIrCl(CO)_2$

M 545.984

Hydrocarboxylation catalyst for olefins. Grey-green solid. ir ν_{CO} 2085s, 2002s cm^{-1} ($CHCl_3$).

U.S.P., 3 948 962, (*1976*); *CA*, **85**, 33191w (*synth, use*)

$C_{20}H_{18}IrO_2P$ **Ir-00111**

Dicarbonyltrihydro(triphenylphosphine)iridium, 9CI

Dicarbonyltrihydrido(triphenylphosphine)iridium

[54082-20-1]

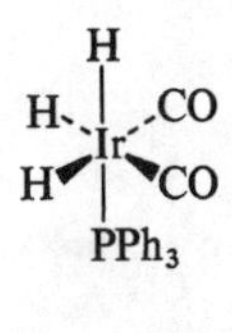

M 513.555

Cream-white solid (not purified). ir ν_{IrH} 2100, 2080, ν_{CO} 1980s, 1970s cm^{-1}.

Malatesta, L. *et al*, *J. Organomet. Chem.*, 1971, **33**, C43; 1974, **73**, 265 (*synth, ir, pmr*)

$C_{20}H_{20}IrN_2^{\oplus}$ **Ir-00112**

[(1,2,5,6-η)-1,5-Cyclooctadiene](1,10-phenanthroline-N^1,N^{10})iridium(1+), 10CI

[41895-12-6]

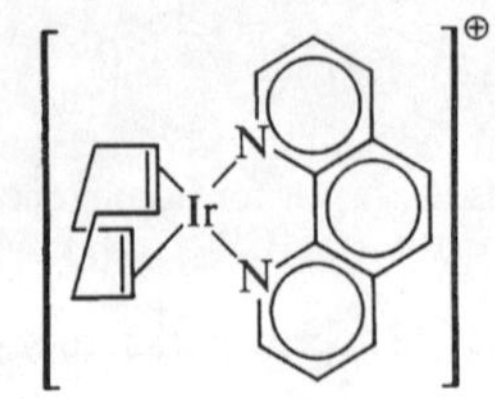

M 480.611 (ion)

Bromide: [53522-12-6].

$C_{20}H_{20}BrIrN_2$ M 560.515

Green cryst.

Chloride: [53522-11-5].

$C_{20}H_{20}ClIrN_2$ M 516.064

Green cryst.

Iodide:

$C_{20}H_{20}IIrN_2$ M 607.516

Violet cryst.

Thiocyanate:

$C_{21}H_{20}IrN_3S$ M 538.689

Red-violet cryst.

Perchlorate: [41392-67-0].

$C_{20}H_{20}ClIrN_2O_4$ M 580.062

Green cryst.

Hexafluorophosphate: [41392-66-9].

$C_{20}H_{20}F_6IrN_2P$ M 625.576

Green cryst.

Mestroni, G. *et al*, *J. Organomet. Chem.*, 1974, **73**, 119 (*synth*)

$C_{20}H_{30}Br_4Ir_2$ **Ir-00113**

Di-μ-bromodibromobis[(1,2,3,4,5-η)-1,2,3,4,5-pentamethyl-2,4-cyclopentadien-1-yl]diiridium, 9CI

Di-μ-bromobis[bromo(pentamethylcyclopentadienyl)-iridium]. Tetrabromobis(η^5-pentamethylcyclopentadienyl)diiridium

[55971-89-6]

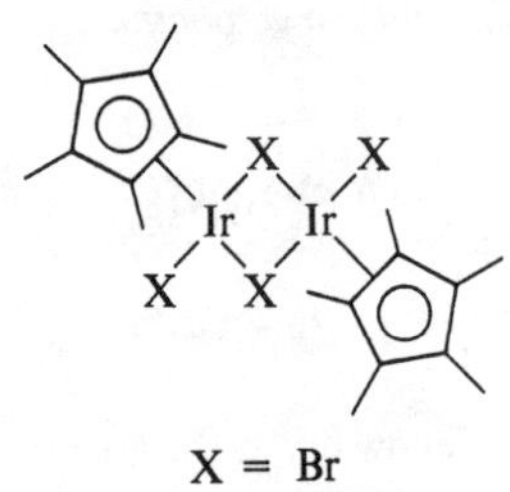

X = Br

M 974.513

Homogeneous hydrogenation catalyst.

Monomer: Dibromo(pentamethylcyclopentadienyl)iridium. (Pentamethylcyclopentadienyl)iridium dibromide.

$C_{10}H_{15}Br_2Ir$ M 487.257

Unknown.

Gill, D.S. *et al*, *J. Organomet. Chem.*, 1975, **87**, 359 (*synth, pmr*)
Gill, D.S. *et al*, *J. Chem. Soc., Dalton Trans.*, 1978, 617 (*use*)
Julis, S.A., *Inorg. Chem.*, 1979, **18**, 1215 (*struct*)

$C_{20}H_{30}Cl_4Ir$ **Ir-00114**

Di-μ-chlorodichlorobis[(1,2,3,4,5-η)-1,2,3,4,5-pentamethyl-2,4-cyclopentadien-1-yl]diiridium, 9CI

Tetrachlorobis(η^5-pentamethylcyclopentadienyl)diiridium

[12354-84-6]

As Di-μ-bromodibromobis[(1,2,3,4,5-η)-1,2,3,4,5-pentamethyl-2,4-cyclopentadien-1-yl]diiridium, Ir-00113 with

X = Cl

M 604.489

Starting material for pentamethylcyclopentadienyl-iridium compounds. Catalyst for homogeneous hydrogenations. Orange cryst. ($CHCl_3/C_6H_6$). Mp >230° dec., >300°.

Monomer: Dichloropentamethylcyclopentadienyliridium. (Pentamethylcyclopentadienyl)iridium dichloride.

$C_{10}H_{15}Cl_2Ir$ M 398.355

Unknown.

Booth, B.L. *et al*, *J. Organomet. Chem.*, 1969, **16**, 491 (*pmr, ir*)
Kang, J.W. *et al*, *J. Am. Chem. Soc.*, 1969, **91**, 5970 (*synth, pmr*)
Churchill, M.R. *et al*, *Inorg. Chem.*, 1976, **16**, 1488 (*struct*)
Gill, D.S. *et al*, *J. Chem. Soc., Dalton Trans.*, 1978, 617 (*use*)

$C_{20}H_{30}I_4Ir_2$ Ir-00115

Di-μ-iododiiodobis[(1,2,3,4,5-η)-1,2,3,4,5-pentamethyl-2,4-cyclopentadien-1-yl]diiridium, 9CI

Tetraiodobis(η⁵-pentamethylcyclopentadienyl)diiridium

[33040-12-9]

As Di-μ-bromodibromobis[(1,2,3,4,5-η)-1,2,3,4,5-pentamethyl-2,4-cyclopentadien-1-yl]diiridium, Ir-00113 with

X = I

M 1162.515

Homogeneous hydrogenation catalyst. Red-brown cryst. ($CHCl_3$). Mp >250°.

Monomer: Diiodo(pentamethylcyclopentadienyl)iridium. (Pentamethylcyclopentadienyl)iridium diiodide.

$C_{10}H_{15}I_2Ir$ M 581.258

Unknown.

Booth, B.L. *et al*, *J. Organomet. Chem.*, 1969, **16**, 491 (*synth, pmr, ir*)
Gill, D.S. *et al*, *J. Organomet. Chem.*, 1975, **87**, 359 (*synth, pmr*)
Gill, D.S. *et al*, *J. Chem. Soc., Dalton Trans.*, 1978, 617 (*use*)
Julis, S.A., *Inorg. Chem.*, 1979, **18**, 1215 (*struct*)

$C_{20}H_{31}Cl_3Ir_2$ Ir-00116

μ-Chlorodichloro-μ-hydrobis[(1,2,3,4,5-η)-1,2,3,4,5-pentamethyl-2,4-cyclopentadien-1-yl]diiridium, 9CI

μ-Chloro-μ-hydridodichlorobis(η⁵-pentamethylcyclopentadienyl)diiridium. μ-Chloro-μ-hydridobis[chloro(pentamethylcyclopentadienyl)iridium]

[41380-03-4]

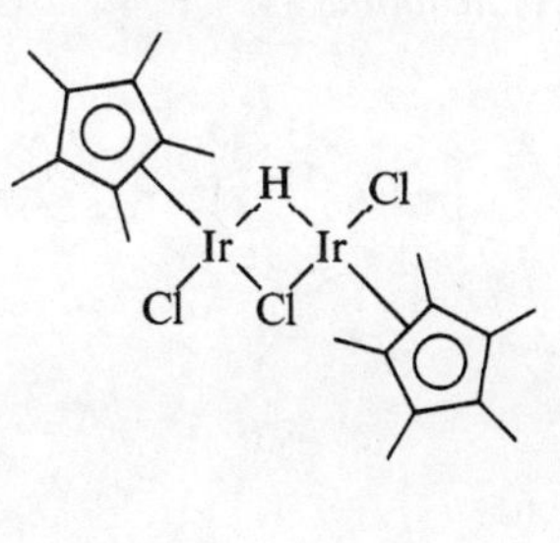

M 762.264

Homogeneous hydrogenation catalyst. Dark-red solid, stable in air, mod. stable in soln. V. sol.

White, C. *et al*, *J. Chem. Soc., Chem. Commun.*, 1971, 734 (*synth, pmr, ir*)
White, C. *et al*, *J. Chem. Soc., Dalton Trans.*, 1973, **18**, 1901 (*synth, pmr*)
Churchill, M.R. *et al*, *Inorg. Chem.*, 1977, **16**, 1488 (*struct*)
Gill, D.S. *et al*, *J. Chem. Soc., Dalton Trans.*, 1978, 617 (*use*)

$C_{20}H_{32}Cl_2Ir_2$ Ir-00117

Dichlorodi-μ-hydrobis[(1,2,3,4,5-η)-1,2,3,4,5-pentamethyl-2,4-cyclopentadien-1-yl]diiridium, 9CI

Di-μ-hydridodichlorobis(η⁵-pentamethylcyclopentadienyl)diiridium. Di-μ-hydridobis[chloro(pentamethylcyclopentadienyl)iridium]

[56086-50-1]

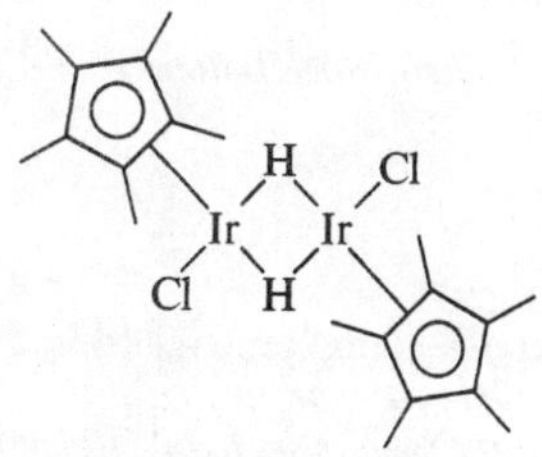

M 727.819

V. effective catalyst for hydrogenation and isomerisation of olefins. Dark-blue cryst. Mp 210° dec.

Monomer: Chlorohydro(pentamethylcyclopentadienyl)iridium.

$C_{10}H_{16}ClIr$ M 363.909

Unknown.

Gill, D.S. *et al*, *J. Organomet. Chem.*, 1975, **87**, 359 (*synth, pmr*)
Gill, D.S. *et al*, *J. Chem. Soc., Dalton Trans.*, 1978, 617 (*use*)
Maitlis, P.M., *Acc. Chem. Res.*, 1978, **11**, 301 (*use*)

$C_{20}H_{33}Ir_2^{\oplus}$ Ir-00118

Tri-μ-hydrobis[(1,2,3,4,5-η)-1,2,3,4,5-pentamethyl-2,4-cyclopentadien-1-yl]diiridium(1+), 9CI

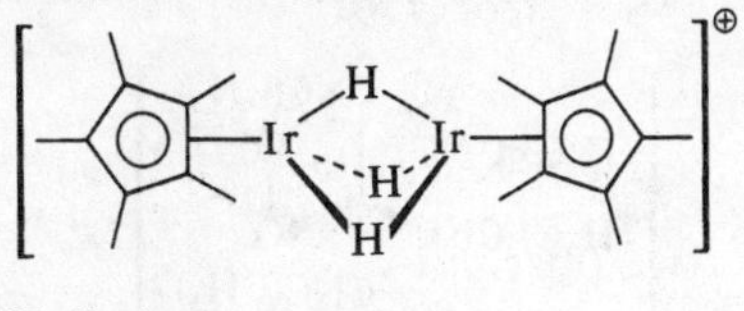

M 657.921 (ion)

Hexafluorophosphate: [39385-15-4].

$C_{20}H_{33}F_6Ir_2P$ M 802.885

Yellow solid. Mp >228° dec.

White, C. *et al*, *J. Chem. Soc., Dalton Trans.*, 1973, 1901 (*synth, pmr*)
Bau, R. *et al*, *J. Am. Chem. Soc.*, 1977, **99**, 3872 (*struct*)
Maitlis, P.M., *Chem. Soc. Rev.*, 1981, **10**, 1 (*struct, props*)

Having problems with locating a compound? Have you checked the indexes?

$C_{20}H_{36}Br_2IrN_4^{\oplus}$ Ir-00119

Dibromotetrakis(2-isocyano-2-methylpropane)iridium(1+), 9CI

Dibromotetrakis(tert-*butyl isocyanide*)*iridium*(*1*+)

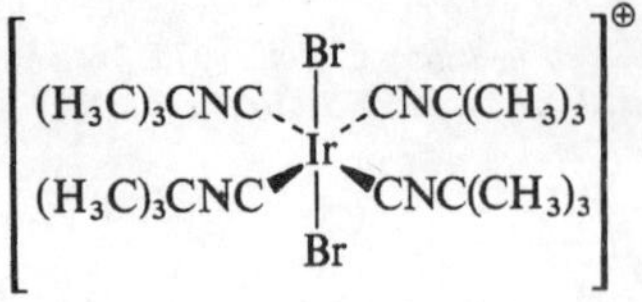

M 684.559 (ion)

trans-form

Probable config.

Hexafluorophosphate: [51139-16-3].

$C_{20}H_{36}Br_2F_6IrN_4P$ M 829.523

White cryst. ir ν_{CN} 2220 cm^{-1} (KBr). Λ 83 Ω^{-1} cm^2 mol.$^{-1}$ (10^{-4}M in $MeNO_2$).

Dart, J.W. *et al*, *J. Chem. Soc., Dalton Trans.*, 1973, 2039 (*synth, ir, pmr*)

$C_{20}H_{36}IrN_4^{\oplus}$ Ir-00120

Tetrakis(2-isocyano-2-methylpropane)iridium(1+), 9CI

Tetrakis(tert-*butyl isocyanide*)*iridium*(*1*+), *8CI.*

Tetrakis(tert-*butylisonitrile*)*iridium*(*1*+)

$$Ir[CNC(CH_3)_3]_4^{\oplus}$$

M 524.751 (ion)

Chloride: [34389-90-7].

$C_{20}H_{36}ClIrN_4$ M 560.204

No phys. props. reported.

Hexafluorophosphate: [51196-95-3].

$C_{20}H_{36}F_6IrN_4P$ M 669.715

Air-sensitive yellow cryst.

Dart, J.W. *et al*, *J. Chem. Soc., Dalton Trans.*, 1973, 2039 (*synth, ir, pmr*)

$C_{20}H_{36}IrN_4O_2^{\oplus}$ Ir-00121

Tetrakis(2-isocyano-2-methylpropane)peroxyiridium(1+)

Tetrakis(tert-*butylisonitrile*)*peroxoiridium*(*1*+). *Dioxygentetrakis*(*2-isocyano-2-methylpropane*)*iridium*

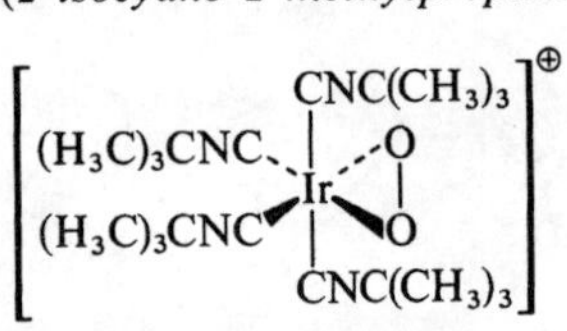

M 556.750 (ion)

Air oxidn. prod. of Tetrakis(2-isocyano-2-methylpropane)iridium(1+), Ir-00120 .

Hexafluorophosphate:

$C_{20}H_{36}F_6IrN_4O_2P$ M 701.714

No phys. props. reported.

Dart, J.W. *et al*, *J. Chem. Soc., Dalton Trans.*, 1973, 2039 (*synth*)

$C_{21}H_{15}IrO_3P^{\ominus}$ Ir-00122

Tricarbonyl(triphenylphosphine)iridate(1−), 10CI

[52352-68-8]

$$(Ph_3P)Ir(CO)_3^{\ominus}$$

M 538.541 (ion)

Strong base. Stable under CO.

Na salt: [19399-76-9].

$C_{21}H_{15}IrNaO_3P$ M 561.531

Pale green. ir ν_{CO} 1925, 1895, 1860 cm^{-1} (THF).

Tris(*phenanthroline*)*nickel salt:*

$C_{78}H_{54}Ir_2N_6NiO_6P_2$ M 1676.399

Red cryst.

Collman, J.P. *et al*, *J. Am. Chem. Soc.*, 1968, **90**, 2282 (*synth, ir*)

Fraser, P.J. *et al*, *J. Chem. Soc., Dalton Trans.*, 1974, 760 (*synth*)

Malatesta, L. *et al*, *J. Organomet. Chem.*, 1974, **73**, 265 (*synth, ir*)

$C_{21}H_{29}F_6IrO_2$ Ir-00123

Bis[(1,2-η)-cyclooctene](1,1,1,5,5,5-hexafluoro-2,4-pentanedionato-*O,O'*)iridium, 9CI

Bis(*cyclooctene*)(*hexafluoroacetylacetone*)*iridium*

[58616-58-3]

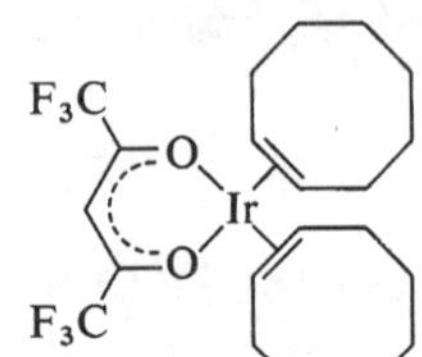

M 619.669

Red cryst. (pentane). Mp 96-7°.

Diversi, P. *et al*, *J. Organomet. Chem.*, 1977, **125**, 253 (*synth, ir, pmr*)

$C_{21}H_{31}ClIrO_2P_2^{\oplus}$ Ir-00124

Carbonylchlorobis(dimethylphenylphosphine)(1-methoxyethylidene)methyliridium(1+), 8CI

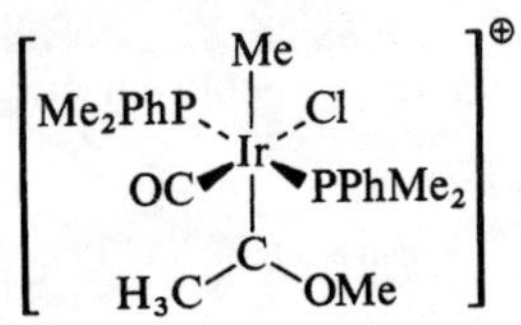

M 605.095 (ion)

Hexafluorophosphate: [41121-62-4].

$C_{21}H_{31}ClF_6IrO_2P_3$ M 750.059

Characterised spectroscopically.

Clark, H.C. *et al*, *J. Organomet. Chem.*, 1973, **47**, C17 (*synth, pmr*)

$C_{22}H_{32}F_3Ir_2O_2^{\oplus}$ Ir-00125

Di-μ-hydrobis[(1,2,3,4,5-η)-1,2,3,4,5-pentamethyl-2,4-cyclopentadien-1-yl][μ-(trifluoroacetato-*O:O'*)]diiridium(1+), 9CI

Di-μ-hydrido-μ-trifluoroacetatobis(*η-pentamethylcyclopentadienyl*)*diiridium*(*1*+)

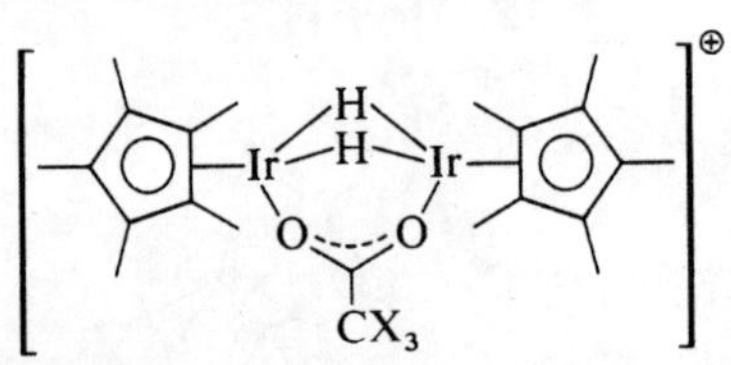

X = F

M 769.929 (ion)

Hexafluorophosphate: [39385-18-7].

$C_{22}H_{32}F_9Ir_2O_2P$ M 914.893
Mod. air-stable orange-brown solid (CH_2Cl_2/Et_2O). Mp 205°.
Hydrogenbis(trifluoroacetate): [51529-59-0].
$C_{26}H_{33}F_9Ir_2O_6$ M 996.969
Mod. air-stable red cryst. (CH_2Cl_2/Et_2O). Mp 145-146.5°.

White, C. *et al*, *J. Chem. Soc., Dalton Trans.*, 1973, 1901 (*synth, nmr*)
Gill, D.S. *et al*, *J. Organomet. Chem.*, 1975, **87**, 359 (*synth*)

$C_{22}H_{33}Ir^{\oplus\oplus}$ Ir-00126

[(1,2,3,4,5,6-η)-Hexamethylbenzene][(1,2,3,4,5-η-)-1,2,3,4,5-pentamethyl-2,4-cyclopentadien-1-yl]iridium(2+), 9CI

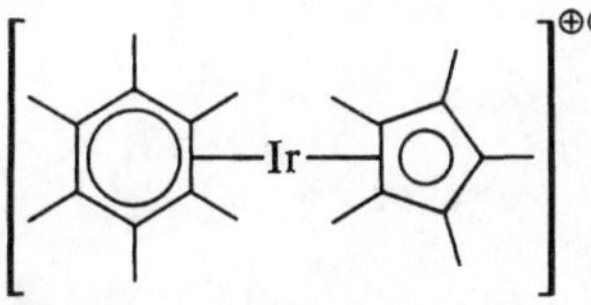

M 489.723 (ion)
Bis(hexafluorophosphate): [12715-83-2].
$C_{22}H_{33}F_{12}IrP_2$ M 779.651
Colourless microcryst. (Me_2CO). Mp >250°.

White, C. *et al*, *J. Chem. Soc. (A)*, 1971, 3322 (*synth, pmr*)

$C_{22}H_{35}Ir_2O_2^{\oplus}$ Ir-00127

[μ-(Acetato-*O*:*O'*)]di-μ-hydrobis[(1,2,3,4,5-η)-1,2,3,4,5-pentamethyl-2,4-cyclopentadien-1-yl]diiridium(1+), 9CI

As Di-μ-hydrobis[(1,2,3,4,5-η)-1,2,3,4,5-pentamethyl-2,4-cyclopentadien-1-yl][μ-(trifluoroacetato-*O*:*O'*)]diiridium(1+), Ir-00125 with

X = H

M 715.957 (ion)
Bis(hexafluorophosphate): [55971-92-1].
$C_{22}H_{35}F_{12}Ir_2O_2P_2$ M 1005.886
Mod. air-stable cryst. (CH_2Cl_2/pet. ether).

Gill, D.S. *et al*, *J. Organomet. Chem.*, 1975, **87**, 359 (*synth, pmr*)

$C_{22}H_{47}IrSi_2$ Ir-00128

Dihydro[(1,2,3,4,5-η)-1,2,3,4,5-pentamethyl-2,4-cyclopentadien-1-yl]bis(triethylsilyl)iridium
Dihydridobis(triethylsilyl)pentamethylcyclopentadienyliridium
[84029-39-0]

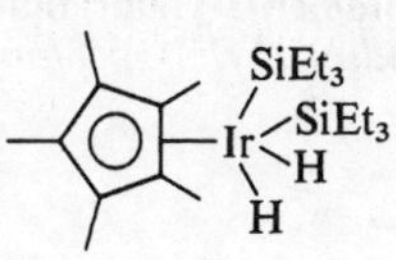

M 560.004
ir ν_{IrH} 2160 cm^{-1}.

Fernandez, M.J. *et al*, *Organometallics*, 1983, **2**, 164 (*synth, ir, ms, pmr*)

$C_{22}Ir_8O_{22}^{\ominus\ominus}$ Ir-00129

Hexa-μ-carbonylhexadecacarbonyloctairidate(2−), 10CI
Docosacarbonyloctairidate(2−), 9CI

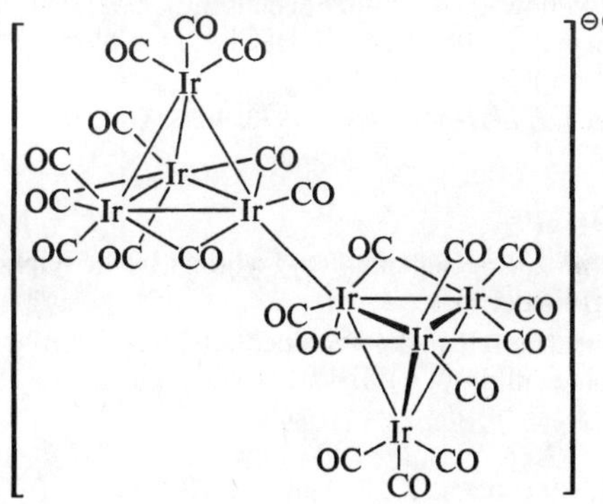

M 2153.989 (ion)
Other isomers may also exist.
Bis(tetramethylammonium) salt: [57108-07-3].
$C_{30}H_{24}Ir_8N_2O_{22}$ M 2302.280
Red powder. Λ 30.4 Ω^{-1} cm^2 $mol.^{-1}$.
Bis(tetraphenylphosphonium) salt: [79233-34-4].
$C_{70}H_{40}Ir_8O_{22}P_2$ M 2832.780
Dark-red air- and light-sensitive cryst.

Angoletta, M. *et al*, *J. Organomet. Chem.*, 1975, **94**, 99 (*synth, ir*)
Demartin, F. *et al*, *J. Chem. Soc., Chem. Commun.*, 1981, 528 (*synth, ir, struct*)

$C_{23}H_{20}Cl_2IrP$ Ir-00130

Dichloro(η⁵-2,4-cyclopentadien-1-yl)(triphenylphosphine)iridium
[32649-44-8]

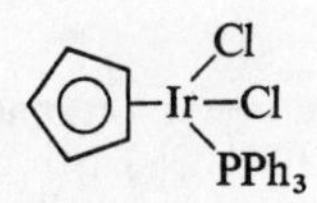

M 590.511
Yellow cryst. (toluene). Mp 290°.

Graham, W.A.G. *et al*, *Inorg. Chem.*, 1970, **9**, 2653 (*synth, pmr*)

$C_{24}H_{20}BrIrOP^{\oplus}$ Ir-00131

Bromocarbonyl(η⁵-2,4-cyclopentadien-1-yl)triphenylphosphineiridium(1+)

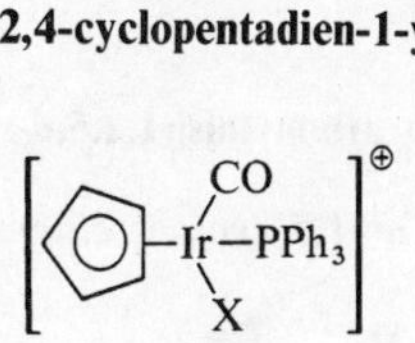

X = Br

M 627.519 (ion)
Bromide: [32824-55-8].
$C_{24}H_{20}Br_2IOP$ M 642.108
Stable yellow cryst. (CH_2Cl_2/hexane). Spar. sol. most org. solvs. Mp 140°. ir ν_{CO} 2085 cm^{-1} (CH_2Cl_2).

Oliver, A.J. *et al*, *Inorg. Chem.*, 1970, **9**, 2653 (*synth, ir*)

$C_{24}H_{20}ClIrOP^{\oplus}$ Ir-00132

Carbonylchloro(η⁵-2,4-cyclopentadien-1-yl)triphenylphosphineiridium(1+)

As Bromocarbonyl(η⁵-2,4-cyclopentadien-1-yl)triphenylphosphineiridium(1+), Ir-00131 with

X = Cl

M 583.068 (ion)
Chloride: [32660-49-4].
$C_{24}H_{20}Cl_2IrOP$ M 618.521
Pale-yellow cryst. (toluene/pentane). Spar. sol. org. solvs. ir ν_{CO} 2088 cm^{-1} (CH_2Cl_2). Dec. slowly in solid state, rapidly on htg.

Oliver, A.J. *et al*, *Inorg. Chem.*, 1970, **9**, 2653 (*synth, ir*)

$C_{24}H_{20}IIrOP^{\oplus}$ Ir-00133

Carbonyl(η^5-2,4-cyclopentadien-1-yl)iodo(triphenylphosphine)iridium(1+), 8CI

As Bromocarbonyl(η^5-2,4-cyclopentadien-1-yl)triphenylphosphineiridium(1+), Ir-00131 with

X = I

M 674.520 (ion)
Iodide: [32662-76-3].
$C_{24}H_{20}I_2IrOP$ M 801.424
Stable orange cryst. (CH_2Cl_2/hexane). Spar. sol. most org. solvs. Mp 110°. ir ν_{CO} 2079 cm^{-1} (CH_2Cl_2).

Oliver, A.J. *et al*, *Inorg. Chem.*, 1970, **9**, 2653 (*synth, ir*)

$C_{24}H_{20}IrOP$ Ir-00134

Carbonyl(η^5-2,4-cyclopentadien-1-yl)(triphenylphosphine)iridium, 9CI

(Triphenylphosphine)carbonyl-π-cyclopentadienyliridium

[32612-68-3]

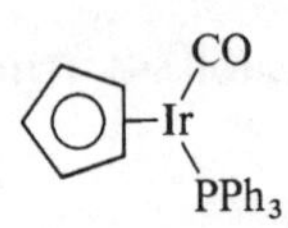

M 547.615
Bright orange (yellow) cryst. (hexane). Mp 160°, 180-3°. ir ν_{CO} 1926 cm^{-1} (CH_2Cl_2). Strongly nucleophilic compd., reacts with halogens, alkyl halides, Lewis acids, etc.

Oliver, A.J. *et al*, *Inorg. Chem.*, 1970, **9**, 2653 (*synth, ir, pmr*)
Yamazaki, H. *et al*, *Bull. Chem. Soc. Jpn.*, 1971, **44**, 582 (*synth, ir, pmr*)
Bennett, M.J. *et al*, *Inorg. Chem.*, 1974, **13**, 2408 (*struct*)

$C_{24}H_{24}Ir_4O_8$ Ir-00135

Tri-μ-carbonylpentacarbonylbis[(1,2,5,6-η)-1,5-cyclooctadiene]tetrairidium, 9CI

Octacarbonylbis(η^4-1,5-cyclooctadiene)tetrairidium

[66809-81-2]

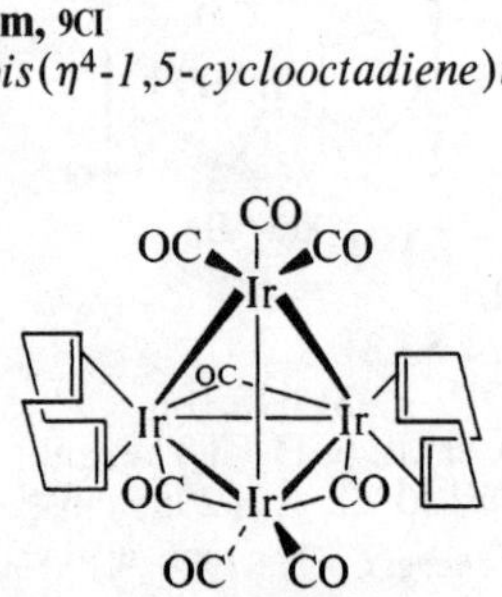

M 1209.329
Orange oil. Sol. CH_2Cl_2, pentane.

Stantz, G.F. *et al*, *Inorg. Chem.*, 1978, **17**, 2596 (*synth, ir, nmr, ms*)

$C_{24}H_{31}F_6Ir_2O_4^{\oplus}$ Ir-00136

μ-Hydrobis[(1,2,3,4,5-η)-1,2,3,4,5-pentamethyl-2,4-cyclopentadien-1-yl]bis[μ-(trifluoroacetato-*O*:*O'*)]diiridium(1+), 9CI

μ-Hydridobis(μ-trifluoroacetato)bis(η^5-pentamethylcyclopentadienyl)diiridium(1+)

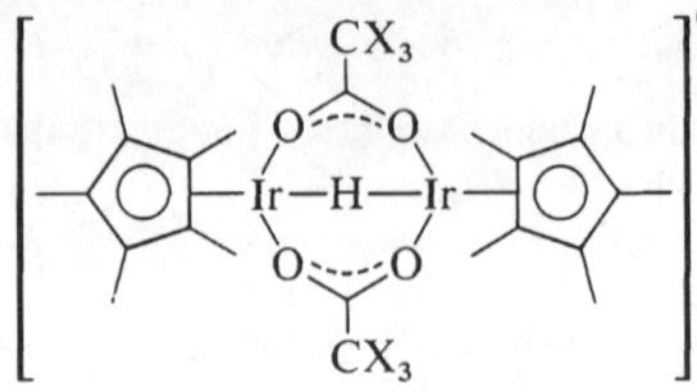

X = F

M 881.937 (ion)
Hexafluorophosphate: [39385-27-8].
$C_{24}H_{31}F_{12}Ir_2O_4P$ M 1026.901
Mod. air-stable yellow cryst. (CH_2Cl_2/Et_2O). Mp 227-8°.
Hydrogenbis(trifluoroacetate): [51552-49-9].
$C_{28}H_{32}F_{12}Ir_2O_8$ M 1108.977
Mod. air-stable yellow cryst. (CH_2Cl_2/hexane). Mp 169-71°.

White, C. *et al*, *J. Chem. Soc., Dalton Trans.*, 1973, 1901 (*synth, nmr*)

$C_{24}H_{37}Ir_2O_4^{\oplus}$ Ir-00137

Bis[μ-(acetato-*O*:*O'*)]-μ-hydrobis[(1,2,3,4,5-η)-1,2,3,4,5-pentamethyl-2,4-cyclopentadien-1-yl]diiridium(1+), 9CI

As μ-Hydrobis[(1,2,3,4,5-η)-1,2,3,4,5-pentamethyl-2,4-cyclopentadien-1-yl]bis[μ-(trifluoroacetato-*O*:*O'*)]diiridium(1+), Ir-00136 with

X = H

M 773.994 (ion)
Hexafluorophosphate: [51471-90-0].
$C_{24}H_{37}F_6Ir_2O_4P$ M 918.958
Mod. air-stable yellow solid (CH_2Cl_2/Et_2O).
Hydrogen diacetate: [51614-73-4].
$C_{28}H_{44}Ir_2O_6$ M 861.092
Mod. air-stable yellow solid (Et_2O/pet. ether).

White, C. *et al*, *J. Chem. Soc., Dalton Trans.*, 1973, 1901 (*synth, pmr*)
Maitlis, P.M., *Acc. Chem. Res.*, 1978, **11**, 301 (*struct*)

$C_{24}H_{40}Ir_2$ Ir-00138

Dimethyldi-μ-methylenebis[(1,2,3,4,5-η)-1,2,3,4,5-pentamethyl-2,4-cyclopentadien-1-yl]diiridium, 10CI

Bis(μ-methylene)dimethylbis(pentamethylcyclopentadienyl)diiridium

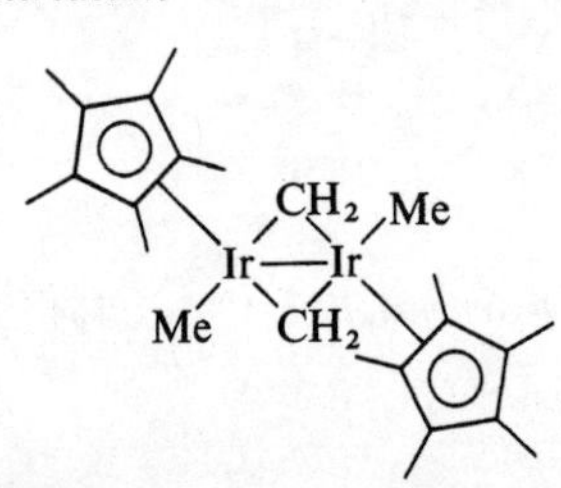

trans-form

M 713.020

***cis*-form** [83582-31-4]
Rapidly isomerises to *trans*-form in soln.
***trans*-form** [83603-54-7]

Vázquez de Miguel, A. *et al*, *J. Chem. Soc., Chem. Commun.*, 1982, 758 (*synth*)

$C_{25}H_{23}IrOP^{\oplus}$ Ir-00139

Carbonyl(η^5-2,4-cyclopentadien-1-yl)methyl(triphenylphosphine)iridium(1+)

As Bromocarbonyl(η^5-2,4-cyclopentadien-1-yl)triphenylphosphineiridium(1+), Ir-00131 with

X = Me

M 562.650 (ion)
Chloride: [32629-28-0].
$C_{25}H_{23}ClIOP$ M 532.787
White cryst. + $\frac{1}{2}CH_2Cl_2$ (CH_2Cl_2/Et_2O). Mp 170°. ir ν_{CO} 2049 cm^{-1} (CH_2Cl_2).
Bromide: [32629-29-1].
$C_{25}H_{23}BrIOP$ M 577.238
White cryst. + $\frac{1}{2}CH_2Cl_2$ (CH_2Cl_2/hexane). Mp 180°. ir ν_{CO} 2049 cm^{-1} (CH_2Cl_2).
Iodide: [32629-30-4].
$C_{25}H_{23}IIrOP$ M 689.554
White cryst. + $\frac{1}{2}CH_2Cl_2$ (CH_2Cl_2/hexane). Mp 180°. ir ν_{CO} 2050 cm^{-1} (CH_2Cl_2).

Oliver, A.J. *et al*, *Inorg. Chem.*, 1970, **9**, 2653 (*synth, ir, pmr*)

$C_{25}H_{30}IrN_3^{\oplus\oplus}$ Ir-00140

[(1,2,3,4,5-η)-1,2,3,4,5-Pentamethyl-2,4-cyclopentadien-1-yl]tris(pyridine)iridium(2+), 9CI

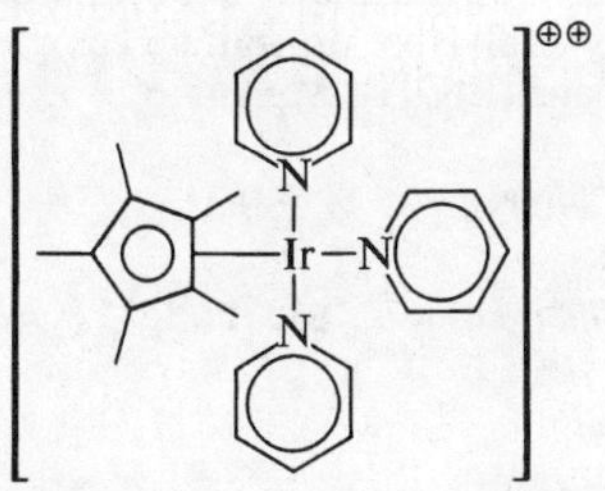

M 564.752 (ion)
Bis(hexafluorophosphate): [59738-34-0].
$C_{25}H_{30}F_{12}IrN_3P_2$ M 854.680
Relatively stable.

White, C. *et al*, *J. Chem. Soc., Dalton Trans.*, 1977, 1654 (*synth, pmr, cmr*)

$C_{25}H_{36}Cl_2IrP_3$ Ir-00141

Dichlorotris(dimethylphenylphosphine)methyliridium, 9CI

[36522-73-3]

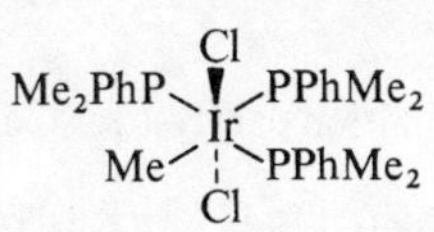

M 692.607
Yellow cryst. (CH_2Cl_2/MeOH or Et_2O). Mp 200-10° dec. (190-5°). Dipole moment 3.6D.

Shaw, B.L. *et al*, *J. Chem. Soc. (A)*, 1967, 1047 (*synth, pmr, ir*)
Mann, B.E. *et al*, *J. Chem. Soc., Dalton Trans.*, 1972, 704 (*nmr*)
Smith, L.R. *et al*, *J. Am. Chem. Soc.*, 1977, **99**, 3302 (*synth, pmr, ir*)

$C_{25}H_{70}BIrOP_2Si_6$ Ir-00142

Carbonyl[tetrahydroborato(1−)-*H,H'*]bis[tris[(trimethylsilyl)methyl]phosphine]iridium, 9CI

Tetrahydroborato(carbonyl)bis[tris(trimethylsilylmethyl)phosphine]iridium

[38437-59-1]

$Ir(BH_4)(CO)[P(CH_2SiMe_3)_3]_2$

M 820.318
Pale-yellow plates (C_6H_6/MeOH). Mp 105-25° dec. ir ν_{CO} 1922 cm^{-1} (nujol).

Hsieh, A.T.T. *et al*, *J. Chem. Soc., Dalton Trans.*, 1972, 1966 (*synth, ir, pmr*)

$C_{26}H_{10}F_{10}Ir_2O_2$ Ir-00143

Dicarbonylbis(η^5-2,4-cyclopentadien-1-yl)[1,1′-[(1,2-η:1,2-η)-1,2-ethynediyl]bis[2,3,4,5,6-pentafluorobenzene]]diiridium, 9CI

μ-[Bis(pentafluorophenyl)acetylene]bis(cyclopentadienylcarbonyliridium)

[39361-90-5]

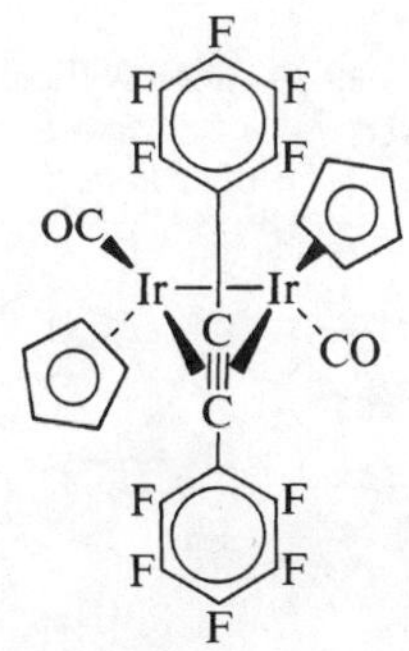

M 928.788
Yellow solid (Et_2O/hexane). Mp 238-40°.

Gardner, S.A. *et al*, *Inorg. Chem.*, 1973, **12**, 2396 (*synth, ir*)

$C_{26}H_{27}ClIrP$ Ir-00144

Chloro[(1,2,5,6-η)-1,5-cyclooctadiene](triphenylphosphine)iridium, 10CI

[12115-50-3]

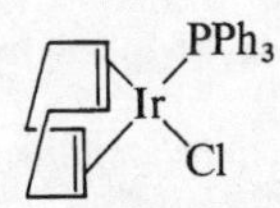

M 598.146
Olefin hydrogenation catalyst. Orange-red cryst. (CH_2Cl_2/pet. ether). Mp 175-8°.

Winkhaus, G. *et al*, *Chem. Ber.*, 1966, **99**, 3610 (*synth*)
Vrieze, K. *et al*, *J. Organomet. Chem.*, 1968, **14**, 185 (*pmr*)
Haszeldine, R.N. *et al*, *J. Chem. Soc. (A)*, 1971, 3711 (*ir*)
Denise, B. *et al*, *J. Organomet. Chem.*, 1978, **148**, 155 (*synth, pmr*)
De Aquino, W. *et al*, *J. Organomet. Chem.*, 1978, **154**, 159 (*use*)

$C_{26}H_{39}ClIrP_3$ Ir-00145

Chlorotris(dimethylphenylphosphine)dimethyliridium, 8CI

[15927-50-1]

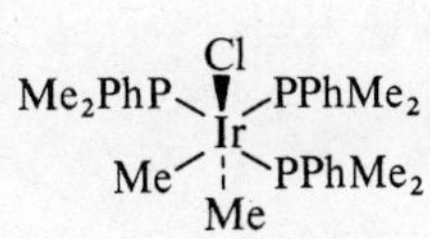

M 672.188

Colourless cryst. (MeOH). Mp 137-141.5° dec.

Shaw, B.L. *et al*, *J. Chem. Soc.* (*A*), 1967, 1047 (*synth*, *pmr*, *ir*)

$C_{27}H_{42}IrP_3$ Ir-00146

Tris(dimethylphenylphosphine)trimethyliridium, 10CI

Trimethyltris(dimethylphenylphosphine)iridium

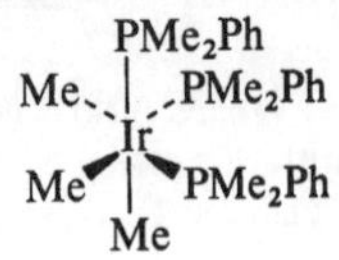

M 651.770

fac-form [15927-48-7]

Colourless prisms (CH_2Cl_2/MeOH). Mp 191-5°.

Shaw, B.L. *et al*, *J. Chem. Soc.* (*A*), 1967, 1047 (*synth*, *ir*, *pmr*, *struct*)
Mann, B.E. *et al*, *J. Chem. Soc., Dalton Trans.*, 1972, 704 (*nmr*)

$C_{28}H_{34}Ir_2O_2$ Ir-00147

Bis(1,5-cyclooctadiene)di-μ-phenoxydiiridium, 8CI

Di-μ-phenoxy-bis(π-1,5-cyclooctadiene)diiridium.
Bis[μ-phenoxy(1,5-cyclooctadiene)iridium]

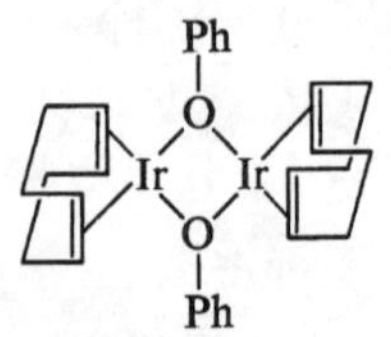

M 787.015

Bridged dimer. Greenish-yellow air-sensitive solid. Sol. CH_2Cl_2, less sol. other solvs.

Monomer: (*1,5-Cyclooctadiene*)*phenoxyiridium.*
$C_{14}H_{17}IrO$ M 393.508
Unknown.

Pannetier, G. *et al*, *J. Less-Common Met.*, 1971, **24**, 83 (*synth*, *ir*)

$C_{29}H_{15}Ir_4O_{11}P$ Ir-00148

Tri-μ-carbonyloctacarbonyl(triphenylphosphine)tetrairidium, 10CI

Undecacarbonyltriphenylphosphinetetrairidium

[53565-22-3]

As Tri-μ-carbonyloctacarbonyl(trimethyl phosphite-*P*)-tetrairidium, Ir-00068 with

R = Ph

M 1339.285

Mp 239-45°.

Karel, K.J. *et al*, *J. Am. Chem. Soc.*, 1974, **96**, 6812 (*synth*, *ir*)
Stuntz, G.F. *et al*, *Inorg. Chem.*, 1976, **15**, 1994 (*synth*, *ir*)

$C_{30}H_{36}Ir_4O_6$ Ir-00149

Tri-μ-carbonyltricarbonyltris[(1,2,5,6-η)-1,5-cyclooctadiene]tetrairidium, 10CI

Hexacarbonyltris(η⁴-1,5-cyclooctadiene)tetrairidium

[66809-79-8]

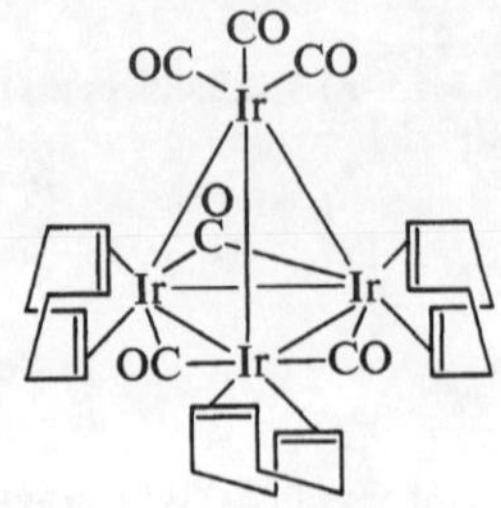

M 1261.491

Orange cryst.

Stuntz, G.F. *et al*, *Inorg. Chem.*, 1978, **17**, 2596 (*synth*, *ir*, *ms*, *pmr*, *struct*)

$C_{30}H_{44}B_2Cl_4Ir_2N_{12}$ Ir-00150

Di-μ-chlorodichlorobis[tris(3,5-dimethyl-1*H*-pyrazolato-*N*¹)hydroborato(1−)-*N*²,*N*²′,*N*²″]diiridium, 10CI

μ-Dichlorobis[[hydrotris(3,5-dimethylpyrazolyl)borato]chloriridium]

[73117-58-5]

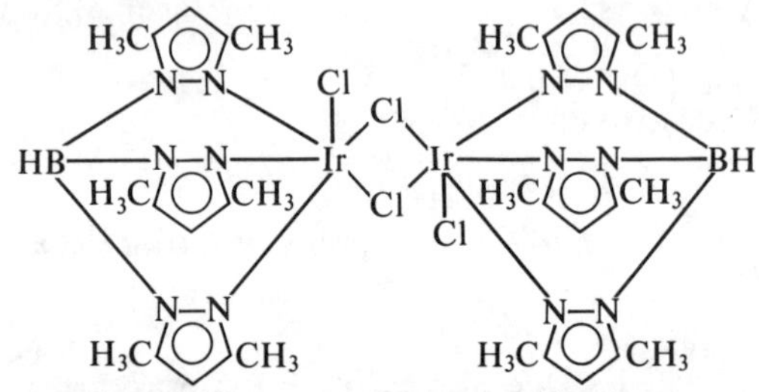

M 1120.630

Bridged dimer. Starting material for hydrotris(3,5-dimethylpyrazolyl)borato-iridium complexes. Pale-orange prisms (EtOH). Mp 245-5°.

Monomer:
$C_{15}H_{22}BCl_2IrN_6$ M 560.315
Unknown.

May, S. *et al*, *Inorg. Chem.*, 1980, **19**, 1582 (*synth*, *pmr*)

$C_{31}H_{50}IrNP^{\oplus}$ Ir-00151

[(1,2,5,6-η)-1,5-Cyclooctadiene](pyridine)(tricyclohexylphosphine)iridium(1+), 10CI

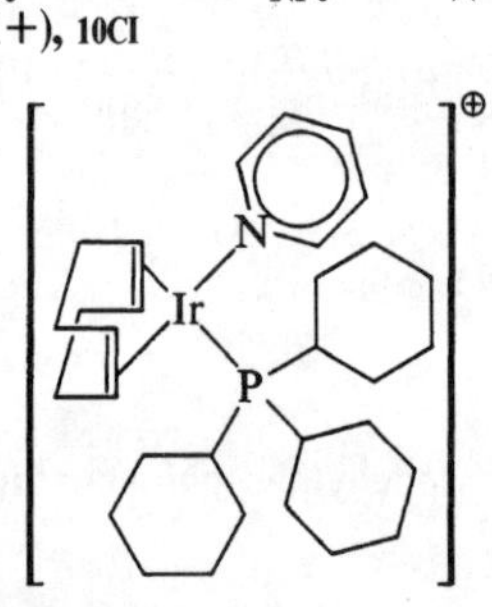

M 659.936 (ion)

V. active hydrogenation catalyst for olefins, esp. for hindered olefins.

Hexafluorophosphate: [64536-78-3].
$C_{31}H_{50}F_6IrNP_2$ M 804.901
Orange cryst. (CH_2Cl_2/Et_2O).

Crabtree, R.H. *et al*, *J. Organomet. Chem.*, 1977, **135**, 395 (*synth*, *pmr*)
Crabtree, R.H. *et al*, *J. Organomet. Chem.*, 1977, **141**, 205; 1979, **168**, 183 (*use*)
Suggs, J.W. *et al*, *Tetrahedron* Lett., 1981, **22**, 303 (*use*)
Chodosh, D.F. *et al*, *Inorg. Chem.*, 1982, **21**, 1307 (*use*)

$C_{32}H_{38}IrP$ Ir-00152

1,4-Butanediyl[(1,2,3,4,5-η)-1,2,3,4,5-pentamethyl-2,4-cyclopentadien-1-yl](triphenylphosphine)iridium, 9CI

(η⁵-Pentamethylcyclopentadienyl)(triphenylphosphine)-(1,4-butanediyl)iridium

[66416-07-7]

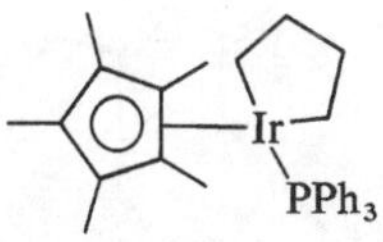

M 645.846

Yellow-green cryst. (pentane). Mp 209° dec.

Diversi, P. *et al*, *J. Chem. Soc., Chem. Commun.*, 1977, **22**, 811.
Diversi, P. *et al*, *Inorg. Chem.*, 1980, **19**, 3590 (*synth, struct, pmr, ms*)
Cuccuru, A. *et al*, *J. Organomet. Chem.*, 1981, **204**, 123.

$C_{32}H_{56}Cl_2Ir_2$ Ir-00153

Di-μ-chlorotetrakis[(1,2-η)-cyclooctene]diiridium, 9CI

[12246-51-4]

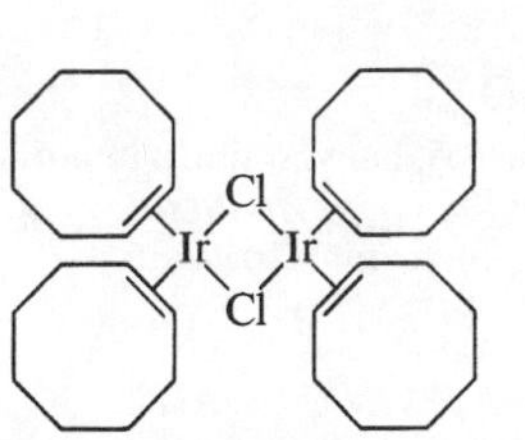

M 896.140

Bridged dimer. Olefin hydrogenation, isomerisation and polymerisation catalyst. Air-sensitive yellow cryst. Mp 168° dec.

Monomer: Chlorobis(cyclooctene)iridium.
$C_{16}H_{28}ClIr$ M 448.070
Unknown.

Rinehart, R.E., *J. Polym. Sci., Part C*, 1966, 7 (*synth, use*)
Herde, J.L. *et al*, *Inorg. Nucl. Chem. Lett.*, 1971, **7**, 1029 (*synth, ir*)
Onderdelinden, A.L. *et al*, *Inorg. Chim. Acta*, 1972, **6**, 420 (*synth, ir, nmr*)
Inorg. Synth., 1974, **15**, 18 (*synth*)

$C_{34}H_{35}B_{10}IrNOP$ Ir-00154

Benzonitrilecarbonyl(triphenylphosphine)[7-phenyl-1,7-dicarba-*closo*-dodecaborane(12)]iridium, 9CI

[70700-96-8]

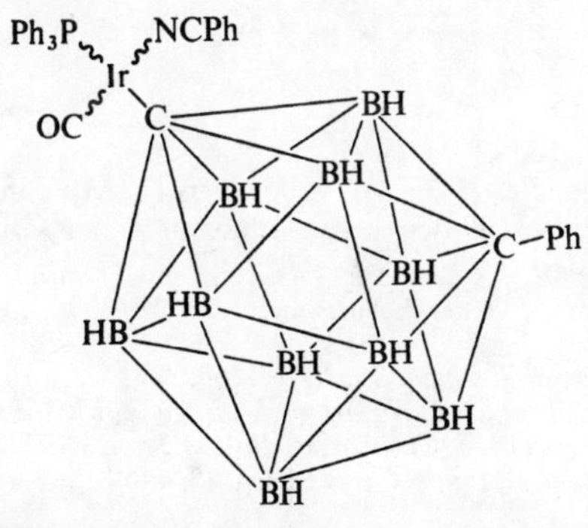

M 804.950

Stereochem. not assigned. Effective catalyst for homogeneous hydrogenation of terminal olefins and acetylenes. Yellow cryst. (C_6H_6/hexane). Mp 191°. ir ν_{CN} 2258w, ν_{CO} 1980vs cm^{-1} (nujol).

Longato, B. *et al*, *Inorg. Chim. Acta*, 1979, **33**, 189 (*synth, ir*)
Morandini, F. *et al*, *J. Organomet. Chem.*, 1982, **239**, 377 (*use*)

$C_{34}H_{36}IrP_2^{\oplus}$ Ir-00155

[(1,2,5,6-η)-1,5-Cyclooctadiene][1,2-ethanediylbis[diphenylphosphine]-*P,P'*]iridium(1+), 9CI

[1,2-Bis(diphenylphosphino)ethane](1,5-cycloctadiene)-iridium(1+)

[47778-82-5]

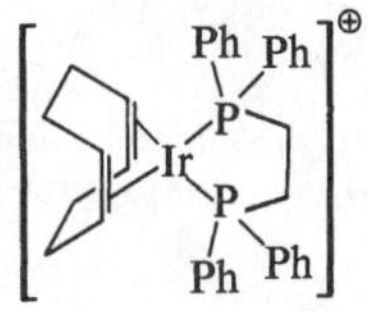

M 698.826 (ion)

Olefin hydrogenation catalyst.

Hexafluorophosphate: [61817-48-9].
$C_{34}H_{36}F_6IrP_3$ M 843.790
Dark-red cryst. (CH_2Cl_2/Et_2O). Forms complex with H_2.

Tetrafluoroborate: [34692-04-1].
$C_{34}H_{36}BF_4IrP_2$ M 785.630
Red-violet cryst. Mp 184-6°.

Green, M. *et al*, *J. Chem. Soc. (A)*, 1971, 2334 (*synth, ir, pmr*)
Crabtree, R.H. *et al*, *J. Organomet. Chem.*, 1979, **168**, 183 (*use*)
Fordyce, W.A. *et al*, *Inorg. Chem.*, 1982, **21**, 1455 (*uv, synth*)

$C_{34}H_{38}IrP_2^{\oplus}$ Ir-00156

(1,2,5,6-η)-1,5-Cyclooctadienebis(methyldiphenylphosphine)iridium(1+), 10CI, 9CI

[38465-86-0]

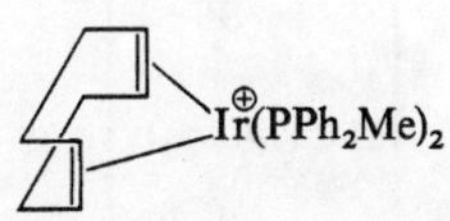

M 700.842 (ion)

Reagent for isomerisation of allylic alcohols and allyl ethers and conversion of allylic alcohols to carbonyl compds. Homogeneous cyclooctadiene hydrogenation catalyst.

Hexafluorophosphate:
$C_{34}H_{38}F_6IrP_3$ M 845.806
Red solid. Air-stable.

Crabtree, R.H. *et al*, *J. Organomet. Chem.*, 1977, **135**, 395 (*synth*)
Baudry, D. *et al*, *J. Chem. Soc., Chem. Commun.*, 1978, 694 (*use*)
Baudry, P. *et al*, *Nouv. J. Chim.*, 1978, **2**, 355 (*use*)
Crabtree, R.H. *et al*, *J. Organomet. Chem.*, 1979, **168**, 183 (*use*)
Fieser, M. *et al*, *Reagents for Organic Synthesis*, Wiley, 1967-83, **10**, 115 (*use*)

$C_{36}H_{30}ClIrN_2$ Ir-00157

Chloro(dinitrogen)bis(triphenylphosphine)iridium, 10CI

Chloronitrogenbis(triphenylphosphine)iridium

[15695-36-0]

M 718.319

trans-form [21414-18-6]

Yellow cryst. Sol. $CHCl_3$, C_6H_6, insol. Et_2O, MeOH. Mp 150-2° dec. ir ν_{NN} 2105 cm^{-1} ($CHCl_3$). Absorbs O_2 in soln. Unstable in $CHCl_3$ above 25°.

Collman, J.P. *et al*, *J. Am. Chem. Soc.*, 1968, **90**, 5430 (*synth, ir, struct*)
Inorg. Synth., 1970, **12**, 8 (*synth, ir*)
Darensburg, D.J. *et al*, *Inorg. Chem.*, 1971, **10**, 431 (*ir*)
Nefedov, V.I. *et al*, *Inorg. Chim. Acta*, 1973, **7**, 499 (*pe, ir*)
Kubota, M. *et al*, *Inorg. Chem.*, 1979, **18**, 3615 (*synth, ir*)

$C_{36}H_{35}IrP_2$ Ir-00158

Pentahydrobis(triphenylphosphine)iridium, 10CI
Pentahydridobis(triphenylphosphine)iridium
[53470-69-2]

$IrH_5(PPh_3)_2$

M 721.840
Originally incorrectly formulated as $IrH_3(PPh_3)_2$. Catalyses hydrogen transfer between monoolefins and also the hydrogenation of ketones and olefins. White air-sensitive powder. Insol. org. solvs. Mp 127°, 184-5° dec., *in vacuo*. ir ν_{IrH} 1948 cm^{-1} (nujol).

Malatesta, L. *et al*, *J. Chem. Soc.*, 1965, 6974 (*synth, ir*)
Chatt, J. *et al*, *J. Chem. Soc.*, 1965, 7391 (*synth*)
Clerici, M.G. *et al*, *J. Organomet. Chem.*, 1975, **84**, 379 (*use*)
Kalinkin, M.I. *et al*, *Izv. Akad. Nauk SSSR, Ser. Khim.*, 1981, 675; *CA*, **95**, 41917j (*use*)

$C_{36}H_{36}IrO_2P_2^{\oplus}$ Ir-00159

Diaquadihydrobis(triphenylphosphine)iridium(1+)
Diaquadihydridobis(triphenylphosphine)iridium(1+)

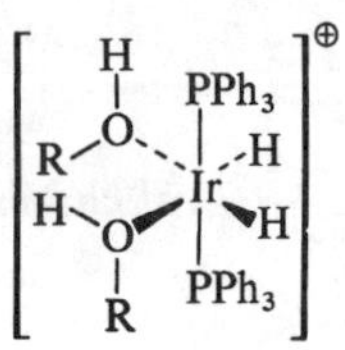

R = H

M 754.847 (ion)
Catalyst for olefin hydrogenation and alkane dehydrogenation. Tends to lose H_2 thermally, esp. in soln. H_2O readily displaced by ROH, Me_2CO or MeCN.
Tetrafluoroborate: [79792-57-7].
$C_{36}H_{36}BF_4IrO_2P_2$ M 841.650
Beige or light-yellow, air-stable cryst. + $1CH_2Cl_2$ (CH_2Cl_2/Et_2O).

Crabtree, R.H. *et al*, *J. Chem. Soc., Chem. Commun.*, 1981, 1217 (*use*)
Crabtree, R.H. *et al*, *J. Am. Chem. Soc.*, 1982, **104**, 107, 6994 (*synth, pmr, ir, use, props*)

$C_{36}H_{45}IrN_4$ Ir-00160

Hydro[2,3,7,8,12,13,17,18-octaethyl-21*H*,23*H*-porphinato(2−)-N^{21},N^{22},N^{23},N^{24}]iridium, 9CI
[68332-06-9]

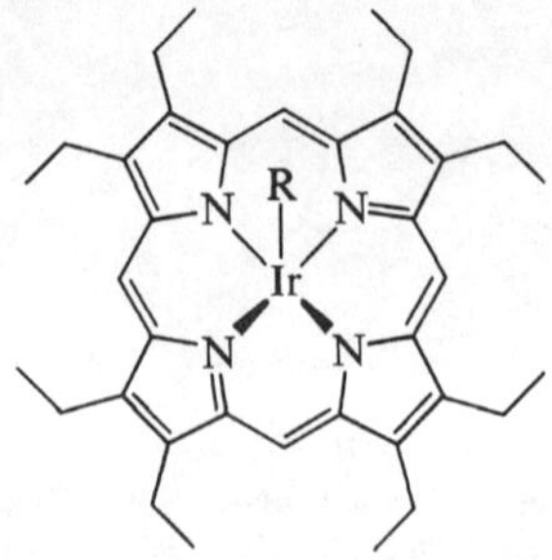

R = H

M 725.998
Amorph. red solid. pmr −40.27 ppm (hydride)(C_6D_6). ir ν_{IrH} 2330 cm^{-1} (KBr).

Ogoshi, H. *et al*, *J. Organomet. Chem.*, 1978, **159**, 317 (*synth, pmr, uv, ir*)

$C_{37}H_{30}As_2ClIrO$ Ir-00161

Carbonylchlorobis(triphenylarsine)iridium, 10CI
Chlorocarbonylbis(triphenylarsine)iridium. Chlorobis(triphenylarsine)carbonyliridium
[15682-62-9]

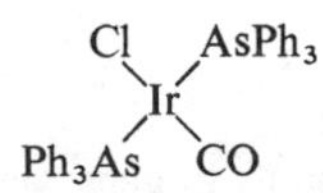

M 868.160
trans-form [23954-42-9]
Hydrogen transfer catalyst for dienes. Canary-yellow cryst. Mp 251° dec. ir ν_{CO} 1958 cm^{-1} (nujol).

Hieber, W. *et al*, *Chem. Ber.*, 1966, **99**, 2607 (*synth, ir*)
Kuyper, J. *et al*, *J. Organomet. Chem.*, 1975, **96**, 289 (*synth*)

$C_{37}H_{30}BrIrOP_2$ Ir-00162

Bromocarbonylbis(triphenylphosphine)iridium, 10CI
Bis(triphenylphosphine)bromocarbonyliridium
[14970-06-0]

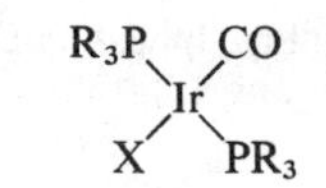

X = Br, R = Ph

M 824.715
trans-form [15842-08-7]
Yellow needles ($CHCl_3$), yellow prisms. Mp 258-80° dec., 290-319° dec. ir ν_{CO} 1966 cm^{-1} ($CHCl_3$).

Chatt, J. *et al*, *J. Chem. Soc. (A)*, 1967, 604 (*synth, ir*)
Strohmeier, W. *et al*, *Z. Naturforsch., B*, 1968, **23**, 1377 (*synth, ir*)
Chatt, J. *et al*, *J. Chem. Soc. (A)*, 1969, 2841 (*ir, struct*)
Schlodder, R. *et al*, *Z. Naturforsch., B*, 1972, **27**, 463 (*ir*)
Holsboer, J. *et al*, *J. Chem. Soc., Dalton Trans.*, 1973, 1838 (*pe*)
Brown, M. *et al*, *Inorg. Chem.*, 1976, **15**, 646 (*nmr*)

$C_{37}H_{30}BrIrO_3P_2$ Ir-00163

Bromocarbonyl(dioxygen)bis(triphenylphosphine)iridium

Bromocarbonylperoxybis(triphenylphosphine)iridium, 9CI

[29933-66-2]

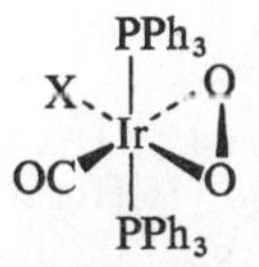

X = Br

M 856.714

ir ν_{CO} 1996, $\nu_{O\text{-}O}$ 860 cm^{-1} (nujol). Loses O_2 less readily than does corresponding chloro complex.

Chock, P.B. *et al*, *J. Am. Chem. Soc.*, 1966, **88**, 3511 (*synth*, *ir*)
Horn, R.W. *et al*, *Inorg. Chem.*, 1970, **9**, 2367 (*synth*, *ir*)

$C_{37}H_{30}ClIrNO_2P_2^{\oplus}$ Ir-00164

Carbonylchloronitrosylbis(triphenylphosphine)iridium(1+)

Chlorocarbonylnitrosylbis(triphenylphosphine)-iridium(1+)

[26846-35-5]

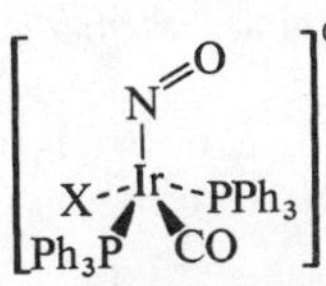

X = Cl

M 810.270 (ion)

Distorted tetragonal bipyramidal struct. with bent Ir–N=O group.

Perchlorate: [53093-67-7].
$C_{37}H_{30}Cl_2IrNO_6P_2$ M 909.721
No phys. props. reported.

Tetrafluoroborate: [22729-41-5].
$C_{37}H_{30}BClF_4IrNO_2P_2$ M 897.074
Violet cryst. ir ν_{CO} 2050, ν_{NO} 1680 cm^{-1} (nujol).

Hodgson, D.J. *et al*, *J. Am. Chem. Soc.*, 1968, **90**, 4486 (*synth*, *ir*, *cryst struct*)
Hodgson, D.J. *et al*, *Inorg. Chem.*, 1968, **7**, 2345 (*cryst struct*)
Reed, C.A. *et al*, *J. Chem. Soc., Chem. Commun.*, 1969, 1459 (*synth*, *ir*)
Holsboer, F. *et al*, *J. Chem. Soc., Dalton Trans.*, 1973, 1828 (*pe*)
Haymore, B.L. *et al*, *J. Am. Chem. Soc.*, 1974, **96**, 3325 (*synth*)
Faller, J.W. *et al*, *J. Organomet. Chem.*, 1975, **84**, 53 (*pe*)
Reed, J., *J. Inorg. Nucl. Chem.*, 1976, **38**, 2239 (*synth*)

$C_{37}H_{30}ClIrOP_2$ Ir-00165

Carbonylchlorobis(triphenylphosphine)iridium

Bis(triphenylphosphine)iridium carbonyl chloride. Carbonylbis(triphenylphosphine)chloroiridium. Vaska's compound

[14871-41-1]

$Ir(PPh_3)_2(CO)Cl$

M 780.264

Catalyses transfer-hydrogenation of α,β-enones, dehydrogenation of alkanes, and the selective isomerization of vinylcycloalkenes and alkanes to exocyclic double bond isomers.

***trans*-form**

Lemon-yellow cryst. Mp 327-8°. Stable in air but takes up O_2 in soln. ν_{CO} 1953 cm^{-1} (nujol).

Inorg. Synth., 1968, **11**, 101 (*synth*)
Chatt, J. *et al*, *J. Chem. Soc. (A)*, 1969, 2841.
Robinson, S. *et al*, *J. Chem. Soc. (A)*, 1970, 2947 (*synth*)
Williams, A.F. *et al*, *J. Chem. Soc., Dalton Trans.*, 1975, 1958 (*ir*, *pmr*)
Fieser, M. *et al*, *Reagents for Organic Synthesis*, Wiley, 1967-83, **10**, 79 (*use*)

$C_{37}H_{30}ClIrO_3P_2$ Ir-00166

Carbonylchloro(dioxygen)bis(triphenylphosphine)iridium, 9CI

Carbonylchloroperoxybis(triphenylphosphine)iridium

[29933-65-1]

As Bromocarbonyl(dioxygen)bis(triphenylphosphine)-iridium, Ir-00163 with

X = Cl

M 812.263

Catalyst for a wide variety of homogeneous oxidations and isomerisations etc. Air-stable, light-sensitive, orange cryst. (C_6H_6). Insol. H_2O, alcohols, sol. C_6H_6. ir ν_{CO} 2000, $\nu_{O\text{-}O}$ 858 cm^{-1} (nujol). Reversibly loses O_2 on heating. Diamagnetic nonelectrolyte.

Vaska, L., *Science*, 1963, **140**, 809 (*synth*, *ir*)
La Placa, S.J. *et al*, *J. Am. Chem. Soc.*, 1965, **87**, 2581 (*struct*)
Chock, P.B. *et al*, *J. Am. Chem. Soc.*, 1966, **88**, 3511 (*synth*, *ir*)
Mason, R. *et al*, *J. Chem. Soc., Dalton Trans.*, 1972, 1729 (*pe*)
Reed, C.A. *et al*, *J. Chem. Soc., Dalton Trans.*, 1973, 1370 (*synth*, *ir*)

$C_{37}H_{30}ClIrO_3P_2S$ Ir-00167

Carbonylchloro(sulfur dioxide-*S*)bis(triphenylphosphine)iridium, 10CI

[15692-64-5]

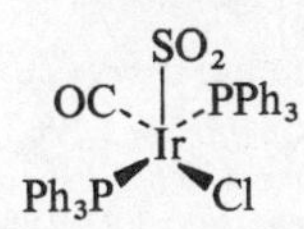

M 844.323

Can be used for extraction of SO_2 from flue gases. Green cryst. (C_6H_6). ir ν_{CO} 2021 cm^{-1} ($CHCl_3$).

La Placa, S.J. *et al*, *Inorg. Chem.*, 1966, **5**, 405 (*struct*)
Vaska, L. *et al*, *J. Am. Chem. Soc.*, 1966, **88**, 1333 (*synth*, *ir*)
Vaska, L. *Acc. Chem. Res.*, 1968, **1**, 335.
Moroni, E.C. *et al*, *J. Organomet. Chem.*, 1970, **21**, P23 (*use*)

$C_{37}H_{30}ClIrO_5P_2$ Ir-00168

Carbonyl(perchlorato-*O*)bis(triphenylphosphine)iridium, 9CI

[55821-24-4]

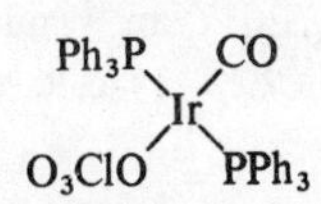

M 844.262

***trans*-form** [32356-67-5]

Air-sensitive yellow cryst. ir ν_{CO} 1982 cm^{-1} (C_6H_6).

Inorg. Synth., 1974, **15**, 68 (*synth, ir*)
Flynn, B.R. *et al*, *Inorg. Chem.*, 1976, **15**, 1485 (*uv*)

$C_{37}H_{30}ClIrP_2S$ Ir-00169

Chloro(thiocarbonyl)bis(triphenylphosphine)iridium, 8CI
Carbonothioylchlorobis(triphenylphosphine)iridium, 9CI. *Chlorobis(triphenylphosphine)(thiocarbonyl)iridium*
[61950-66-1]

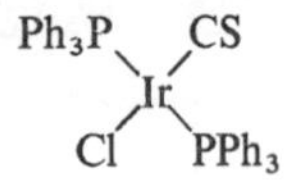

M 796.325
trans-form [30106-92-4]
Orange cryst. ($CHCl_3$/MeOH). Mp 260-70° dec.

Kubota, M. *et al*, *J. Organomet. Chem.*, 1970, **24**, 491 (*synth, ir*)
Nefedov, V.I. *et al*, *Inorg. Chim. Acta*, 1973, **7**, 499 (*pe*)
Collins, T.J. *et al*, *J. Organomet. Chem.*, 1976, **121**, C41 (*synth*)
Inorg. Synth., 1979, **19**, 206 (*synth, ir*)
Schramm, K.D. *et al*, *Inorg. Chem.*, 1980, **19**, 3183 (*ir, nmr*)

$C_{37}H_{30}Cl_3IrOP_2$ Ir-00170

Carbonyltrichlorobis(triphenylphosphine)iridium, 9CI
[14427-98-6]

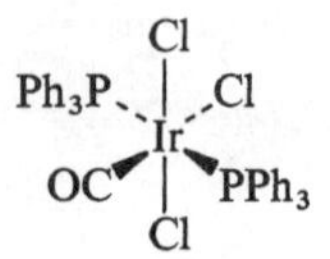
trans-form

M 851.170
White solid.
cis-form [24810-96-6]
Pale-yellow cryst. (C_6H_6/pet. ether). ir ν_{CO} 2100 cm^{-1} (nujol).
trans-form [22788-78-9]
White solid. Diamagnetic. ir ν_{CO} 2075 cm^{-1} (nujol).

Vaska, L. *et al*, *J. Am. Chem. Soc.*, 1962, **84**, 679 (*synth*)
Bennett, M.A. *et al*, *Inorg. Chem.*, 1967, **6**, 1647 (*ir*)
Dolcetti, G. *et al*, *J. Chem. Soc. (A)*, 1969, 1387 (*ir*)
Bennett, M.A. *et al*, *J. Am. Chem. Soc.*, 1969, **91**, 6983 (*synth, ir*)
Tripathi, S.C. *et al*, *Inorg. Chim. Acta*, 1976, **17**, 257.

$C_{37}H_{30}Cl_3IrOP_2Sn$ Ir-00171

Carbonyl(trichlorostannyl)bis(triphenylphosphine)iridium

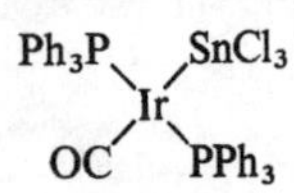

M 969.860
trans-form [28368-48-1]
Green needles. ir ν_{CO} 2017 cm^{-1} (nujol).

Camia, M. *et al*, *Inorg. Chem.*, 1970, **9**, 251 (*synth, ir, struct*)

$C_{37}H_{30}FIrOP_2$ Ir-00172

Carbonylfluorobis(triphenylphosphine)iridium, 10CI
Bis(triphenylphosphine)carbonylfluoroiridium. Fluorocarbonylbis(triphenylphosphine)iridium
[34247-65-9]
As Bromocarbonylbis(triphenylphosphine)iridium, Ir-00162 with

X = F, R = Ph

M 763.809
trans-form [32356-62-0]
Air-sensitive yellow cryst. (CH_2Cl_2/EtOH). Mp 208-11°. ir ν_{CO} 1957 cm^{-1} ($CHCl_3$).

Vaska, L. *et al*, *J. Chem. Soc., Chem. Commun.*, 1971, 418 (*synth, ir, struct*)
Clark, G.R. *et al*, *J. Chem. Soc., Chem. Commun.*, 1971, 758 (*synth, ir*)
Reed, C.A. *et al*, *J. Chem. Soc., Dalton Trans.*, 1973, 1370 (*synth, ir*)
Holsboer, J. *et al*, *J. Chem. Soc., Dalton Trans.*, 1973, 1838 (*pe*)
Inorg. Synth., 1974, **15**, 64 (*synth, ir*)
Reed, C.A. *et al*, *J. Chem. Soc., Dalton Trans.*, 1975, 1159 (*synth*)
Williams, A.F. *et al*, *J. Chem. Soc., Dalton Trans.*, 1975, 1958 (*mössbauer, nmr*)

$C_{37}H_{30}IIrNO_2P_2^{\oplus}$ Ir-00173

Carbonyliodonitrosylbis(triphenylphosphine)iridium(1+), 9CI
Iodocarbonylnitrosylbis(triphenylphosphine)-iridium(1+)
As Carbonylchloronitrosylbis(triphenylphosphine)-iridium(1+), Ir-00164 with

X = I

M 901.722 (ion)
Distorted tetragonal bipyramidal structure with bent Ir—N—O group.
Tetrafluoroborate: [38604-03-4].
$C_{37}H_{30}BF_4IIrNO_2P_2$ M 988.525
Violet cryst. +1C_6H_6. ir ν_{CO} 2070, ν_{NO} 1720 cm^{-1} (nujol).

Hodgson, D.J. *et al*, *Inorg. Chem.*, 1969, **8**, 1282 (*synth,ir, cryst struct*)
Mason, R. *et al*, *J. Chem. Soc., Dalton Trans.*, 1972, 1729 (*uv*)
Holsboer, F. *et al*, *J. Chem. Soc., Dalton Trans.*, 1973, 1828 (*pe*)

$C_{37}H_{30}IIrOP_2$ Ir-00174

Carbonyliodobis(triphenylphosphine)iridium, 10CI
Iodocarbonylbis(triphenylphosphine)iridium
[14970-07-1]
As Bromocarbonylbis(triphenylphosphine)iridium, Ir-00162 with

X = I, R = Ph

M 871.715
trans-form [19472-16-3]
Hydrogenation catalyst for unsatd. compds. Yellow cryst. Mp 270-296.5°, >300°. ir ν_{CO} 1967 cm^{-1} ($CHCl_3$).

Chock, P.B. *et al*, *J. Am. Chem. Soc.*, 1966, **88**, 3511 (*synth, ir*)
Chatt, J. *et al*, *J. Chem. Soc. (A)*, 1969, 2841 (*synth, ir*)
Malatesta, L. *et al*, *J. Chem. Soc. (A)*, 1970, 1836 (*synth, ir*)
Holsboer, J. *et al*, *J. Chem. Soc., Dalton Trans.*, 1973, 1838 (*pe*)
Williams, A.F. *et al*, *J. Chem. Soc., Dalton Trans.*, 1975, 1958 (*mössbauer, nmr*)
Kuyper, J. *et al*, *J. Organomet. Chem.*, 1975, **96**, 289 (*synth*)

$C_{37}H_{30}IIrO_3P_2$ Ir-00175

Carbonyliodoperoxybis(triphenylphosphine)iridium, 9CI

Carbonyldioxygeniodobis(triphenylphosphine)iridium, 8CI. *Carbonyliodoperoxobis(triphenylphosphine)iridium*

[15187-12-9]

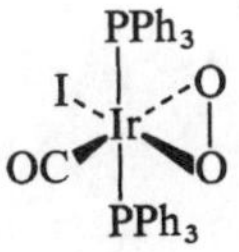

M 903.714

trans-form [18904-98-8]

Irreversibly formed from O_2 and $IrI(CO)(PPh_3)_2$. Brown powder (C_6H_6).

CH_2Cl_2 solvate: [15444-94-7]. Dark-brown needles (CH_2Cl_2). ir ν_{CO} 2005, $\nu_{Ir\text{-}O_2}$ 850 cm^{-1} (nujol).

Halpern, J. *et al*, *J. Am. Chem. Soc.*, 1966, **88**, 3511 (*synth*, *ir*, *struct*)
Reed, C.A. *et al*, *Inorg. Chem.*, 1967, **6**, 2243 (*synth*, *cryst struct*)
Mason, R. *et al*, *J. Chem. Soc., Dalton Trans.*, 1972, 1729 (*pe*)
Reed, C.A. *et al*, *J. Chem. Soc., Dalton Trans.*, 1973, 1370 (*ir*)

$C_{37}H_{31}Cl_2IrOP_2$ Ir-00176

Carbonyldichlorohydrobis(triphenylphosphine)iridium, 10CI

Carbonyldichlorohydridobis(triphenylphosphine)iridium. Dichlorohydridocarbonylbis(triphenylphosphine)iridium

[17000-10-1]

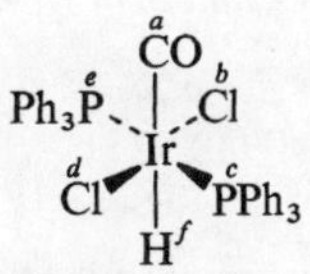

a-carbonyl-*bd*-dichloro-*f*-hydro-*form*

M 816.725

a-carbonyl-bd-dichloro-f-hydro-form [17000-10-1]

White cryst. Mp 315-20°. ir ν_{IrH} 2240, υ_{CO} 2025 cm^{-1} (nujol).

a-carbonyl-bf-dichloro-d-hydro-form [50409-74-0]

White cryst. ir ν_{IrH} 2245, υ_{CO} 2030 cm^{-1} (nujol).

Vaska, L. *et al*, *J. Am. Chem. Soc.*, 1961, **83**, 2784 (*synth*, *ir*)
Blake, D.M. *et al*, *Inorg. Chem.*, 1970, **9**, 989 (*synth*, *ir*, *cryst struct*)
Ball, M.C. *et al*, *J. Chem. Soc., Dalton Trans.*, 1973, 1802 (*bibl*, *synth*, *ir*)
Green, M.L.H. *et al*, *J. Organomet. Chem.*, 1974, **73**, 265 (*synth*, *ir*)

$C_{37}H_{31}IrO_2P_2$ Ir-00177

Carbonylhydroxybis(triphenylphosphine)iridium, 10CI

Hydroxocarbonylbis(triphenylphosphine)iridium

[34247-66-0]

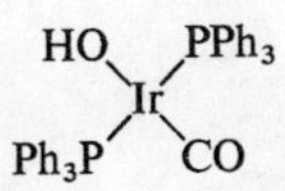

M 761.818

trans-form [32356-70-0]

Yellow cryst. (C_6H_6/EtOH). Mp 177-9°. ir ν_{CO} 1949 cm^{-1} ($CHCl_3$).

Vaska, L. *et al*, *J. Chem. Soc., Chem. Commun.*, 1971, 418 (*synth*, *ir*)
Reed, C.A. *et al*, *J. Chem. Soc., Dalton Trans.*, 1973, 1370 (*synth*, *ir*, *pmr*)
Williams, A.F. *et al*, *J. Chem. Soc., Dalton Trans.*, 1975, 1958 (*mössbauer*)

$C_{37}H_{32}ClIrOP_2$ Ir-00178

Carbonylchlorodihydridobis(triphenylphosphine)iridium, 10CI

Carbonylchlorodihydridobis(triphenylphosphine)iridium

[17000-11-2]

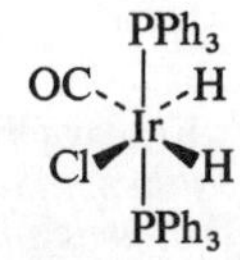

M 782.280

Disproportionation catalyst for olefins. Yellow-white cryst. Light sensitive in air. Mp 315-21° dec. ir ν_{IrH} 2196, 2100, ν_{CO} 1982 cm^{-1} (nujol).

Vaska, L. *et al*, *J. Am. Chem. Soc.*, 1962, **84**, 679 (*synth*, *ir*)
Halpern, J. *et al*, *J. Am. Chem. Soc.*, 1966, **88**, 3511 (*synth*, *ir*)
Vaska, L., *J. Am. Chem. Soc.*, 1966, **88**, 4100 (*ir*, *struct*)
Bennett, M.A. *et al*, *Inorg. Chem.*, 1967, 1647 (*ir*)

$C_{37}H_{32}IIrOP_2$ Ir-00179

Carbonyldihydroiodobis(triphenylphosphine)iridium, 9CI

[17125-90-5]

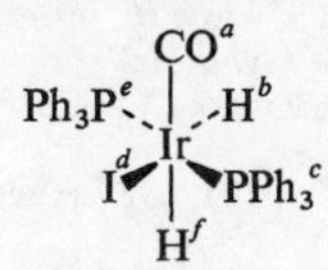

a-carbonyl-*bf*-dihydro-*d*-iodo-*ce*-diphosphine-*form*

M 873.731

Intermed. in $IrI(CO)(PPh_3)_2$-catalysed hydrogenations of olefins. Loses H_2 reversibly.

a-carbonyl-bf-dihydro-d-iodo-ce-diphosphine-form

cis,trans-*form*

Colourless cryst., v. air-sensitive. ir ν_{CO} 1995 cm^{-1} (nujol).

a-carbonyl-bc-dihydro-d-iodo-ef-diphosphine-form [27476-68-2]

cis,cis-*form*

Stereochemistry uncertain. May be identical to *cis-trans* form. Colourless cryst. Insol. Mp 242°. ir ν_{CO} 1987 cm^{-1} (nujol).

Vaska, L. *et al*, *J. Am. Chem. Soc.*, 1965, **87**, 4970 (*use*)
Chock, P.B. *et al*, *J. Am. Chem. Soc.*, 1966, **88**, 3511 (*synth*, *ir*)
Strohmeier, W. *et al*, *Z. Naturforsch*, *B*, 1969, **24**, 931 (*synth*)
Glockling, F. *et al*, *J. Chem. Soc. (A)*, 1970, 1675 (*synth*, *ir*)
Geoffroy, G.L. *et al*, *J. Am. Chem. Soc.*, 1975, **97**, 3933 (*synth*, *uv*)

$C_{37}H_{33}IrNOP_2^{\oplus}$ Ir-00180

Methylnitrosylbis(triphenylphosphine)iridium(1+), 10CI

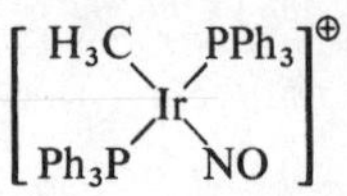

M 761.841 (ion)

***trans*-form**

Perchlorate:

$C_{37}H_{33}ClIrNO_5P_2$ M 861.292

Crimson cryst. + $\frac{1}{2}CH_2Cl_2$. Mp 146-8°. ir ν_{NO} 1850 cm^{-1} (nujol).

Reed, C.A. *et al*, *J. Chem. Soc., Dalton Trans.*, 1973, 1014 (*synth, ir, pmr*)

$C_{37}H_{33}IrOP_2$ Ir-00181

Carbonyltrihydrobis(triphenylphosphine)iridium, 10CI

Trihydrocarbonylbis(triphenylphosphine)iridium. Carbonyltris(hydrido)bis(triphenylphosphine)iridium

[17176-29-3]

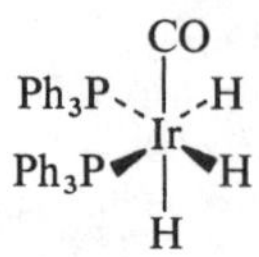

fac-form

M 747.835

***trans*-diphosphine-form** [16971-53-2]

White needles. Mp 146°. Exists in equilib. with *fac*-isomer in soln.

***fac*-form** [16971-55-4]

White cryst. Mp 138°. Exists in equilib. with *trans*-isomer in soln.

Malatesta, L. *et al*, *J. Chem. Soc.*, 1965, 6974 (*synth, ir*)
Malatesta, L. *et al*, *CA*, 1967, **66**, 111143f (*synth, ir*)
Vaska, L. *et al*, *J. Chem. Soc., Chem. Commun.*, 1971, 1615 (*synth*)
Harrod, J.F. *et al*, *J. Am. Chem. Soc.*, 1979, **101**, 3987 (*synth, ir, pmr, struct*)
Harrod, J.F. *et al*, *Inorg. Chem.*, 1981, **20**, 1156 (*synth, ir, nmr, struct*)

$C_{37}H_{37}ClIrOP_3$ Ir-00182

[Bis[3-(diphenylphosphino)propyl]phenylphosphine-*P,P',P''*-]carbonylchloroiridium, 10CI

[78505-57-4]

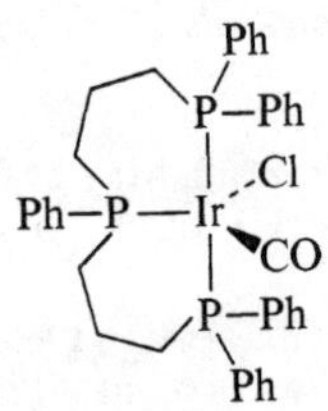

M 818.293

Pale-yellow solid (THF/hexane).

Arpac, E. *et al*, *Z. Naturforsch., B*, 1981, **36**, 672 (*synth, ir, nmr*)

$C_{37}H_{44}ClIrN_4O$ Ir-00183

Carbonylchloro[2,3,7,8,12,13,17,18-octaethyl-21*H*,23*H*-porphinato(2−)-*N*[21],*N*[22],*N*[23],*N*[24]]iridium, 9CI

[68324-58-3]

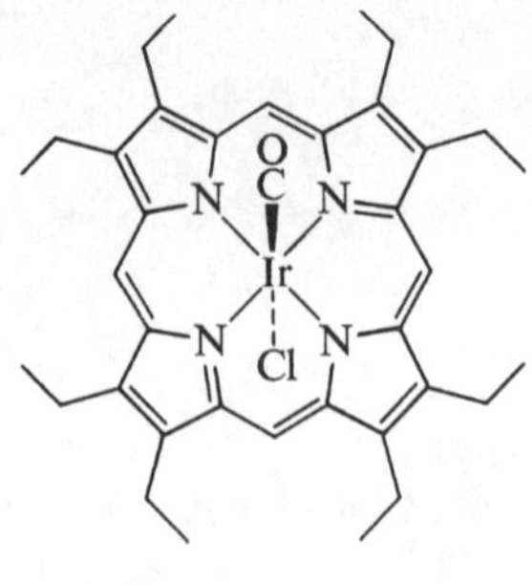

M 788.454

Violet cryst. (CH_2Cl_2/MeOH). ir ν_{CO} 2050 cm^{-1} (KBr).

Ogoshi, H. *et al*, *J. Organomet. Chem.*, 1978, **159**, 317 (*synth, ir, pmr, uv*)

$C_{37}H_{47}IrN_4$ Ir-00184

Methyl[2,3,7,8,12,13,17,18-octaethyl-21*H*,23*H*-porphinato(2−)-*N*[21],*N*[22],*N*[23],*N*[24]]iridium, 9CI

[68357-02-8]

As Hydro[2,3,7,8,12,13,17,18-octaethyl-21*H*,23*H*-porphinato(2−)-*N*[21],*N*[22],*N*[23],*N*[24]]iridium, Ir-00160 with

R = Me

M 740.025

Red solid.

Ogoshi, H. *et al*, *J. Organomet. Chem.*, 1978, **159**, 317 (*synth, pmr*)

$C_{37}H_{66}ClIrOP_2$ Ir-00185

Carbonylchlorobis(tricyclohexylphosphine)iridium, 10CI

Chlorocarbonylbis(tricyclohexylphosphine)iridium, 8CI. Bis(tricyclohexylphosphine)chlorocarbonyliridium

[15682-61-8]

As Bromocarbonylbis(triphenylphosphine)iridium, Ir-00162 with

X = Cl, R = cyclohexyl

M 816.548

***trans*-form** [34054-07-4]

Hydrogenation catalyst for unsatd. compds. Light-yellow cryst. (C_6H_6/pet. ether). Mp 237° dec. ir ν_{CO} 1931 cm^{-1} (THF).

Hieber, W. *et al*, *Chem. Ber.*, 1966, **99**, 2607 (*synth, ir*)
Strohmeier, W. *et al*, *Z. Naturforsch., B*, 1968, **23**, 1377 (*synth, ir*)
Strohmeier, W. *et al*, *Z. Naturforsch., B*, 1969, **24**, 770 (*uv*)
Moers, F.G. *et al*, *J. Inorg. Nucl. Chem.*, 1973, **35**, 1915 (*synth, ir*)

$C_{38}H_{30}ClF_4IrP_2$ **Ir-00186**
Chloro(η^2-tetrafluoroethene)bis(triphenylphosphine)iridium, 9CI
Chloro(η^2-tetrafluoroethylene)bis(triphenylphosphine)-iridium, 8CI

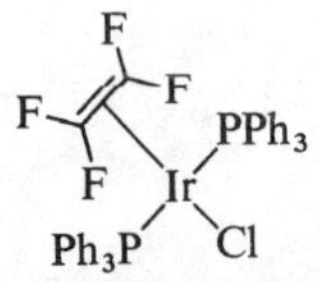

M 852.269
***trans*-form** [27709-91-7]
Yellow cryst. (C_6H_6/hexane). Mp 228° dec.

Clarke, B. *et al*, *J. Chem. Soc.* (*A*), 1970, 951 (*synth, ir, nmr*)
van Gaal, H.L.M. *et al*, *Inorg. Chim. Acta*, 1973, **7**, 653 (*synth, nmr, ir*)

$C_{38}H_{30}ClIrNO_3P_2{}^{\oplus}$ **Ir-00187**
Dicarbonylchloronitrosylbis(triphenylphosphine)iridium(1+), 10CI

$IrCl(CO)_2(NO)(PPh_3)_2{}^{\oplus}$

M 838.280 (ion)
Oxidised by air to the corresponding nitrato complex.
Tetrafluoroborate: [57108-50-6].
$C_{38}H_{30}BClF_4IrNO_3P_2$ M 925.084
Cryst. ir ν_{CO} 2070, 2023, 1980, ν_{NO} 1680 cm^{-1}.

Kubota, M. *et al*, *J. Am. Chem. Soc.*, 1975, **97**, 5637 (*synth, ir*)

$C_{38}H_{30}ClIrO_2P_2$ **Ir-00188**
Dicarbonylchlorobis(triphenylphosphine)iridium, 10CI
Chlorodicarbonylbis(triphenylphosphine)iridium
[20037-61-0]

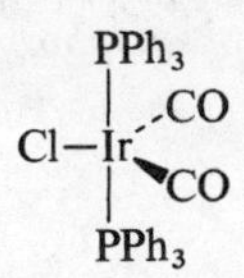

M 808.274
Pale-yellow cryst.

Vaska, L., *Science*, 1966, **152**, 769 (*synth, ir, struct*)
Benzoni, L. *et al*, *Chim. Ind.* (*Milan*), 1968, **50**, 347 (*synth, ir*)
Payne, N.C. *et al*, *Inorg. Chem.*, 1969, **8**, 2714 (*cryst struct*)
Mason, R. *et al*, *J. Chem. Soc., Dalton Trans.*, 1972, 1729 (*pe*)
Haymore, B.L. *et al*, *J. Am. Chem. Soc.*, 1974, **96**, 3325 (*synth*)

$C_{38}H_{31}IrO_2P_2$ **Ir-00189**
Dicarbonylhydrobis(triphenylphosphine)iridium, 10CI
Dicarbonylhydridobis(triphenylphosphine)iridium. Hydridodicarbonylbis(triphenylphosphine)iridium. Hydridobis(triphenylphosphine)dicarbonyliridium
[17250-59-8]

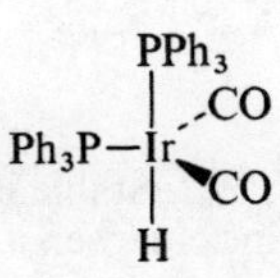

M 773.829
Stereochemically nonrigid in soln. Pale-yellow needles. Mp 135-6°. ir ν_{IrH} 2029, ν_{CO} 1970, 1915 cm^{-1} (nujol).

Collman, J.P. *et al*, *J. Am. Chem. Soc.*, 1968, **90**, 2282 (*synth, ir*)
Yagupsky, G. *et al*, *J. Chem. Soc.* (*A*), 1969, 725 (*synth, ir, nmr*)
Meakin, P. *et al*, *J. Am. Chem. Soc.*, 1972, **94**, 5271 (*nmr, struct*)
Ciechanowicz, M. *et al*, *Acta Crystallogr., Sect. B*, 1976, **32**, 1673 (*cryst struct*)
Tam, W. *et al*, *J. Am. Chem. Soc.*, 1979, **101**, 1589 (*synth*)

$C_{38}H_{33}IrNO_2P_2{}^{\oplus}$ **Ir-00190**
Carbonylmethylnitrosylbis(triphenylphosphine)iridium(1+), 10CI
Methylnitrosylcarbonylbis(triphenylphosphine)-iridium(1+). Methylcarbonylnitrosylbis(triphenylphosphine)-iridium(1+)
As Carbonylchloronitrosylbis(triphenylphosphine)-iridium(1+), Ir-00164 with

X = Me

M 789.852 (ion)
Complex has a bent Ir–N–O group.
Perchlorate: [42548-62-9].
$C_{38}H_{33}ClIrNO_6P_2$ M 889.302
Yellow-brown or red-brown cryst. +0.5 CH_2Cl_2. Mp 142-5° (solvate). ir ν_{CO} 2050, ν_{NO} 1680 cm^{-1} (nujol).

Reed, C.A. *et al*, *J. Chem. Soc., Dalton Trans.*, 1973, 1014 (*synth, ir, nmr, struct*)

$C_{38}H_{33}IrOP_2$ **Ir-00191**
Carbonylmethylbis(triphenylphosphine)iridium, 10CI
[53248-08-1]

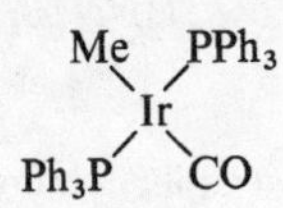

M 759.846
***trans*-form** [82190-06-5]
Canary-yellow cryst. (Et_2O/EtOH). ir ν_{CO} 1935 cm^{-1} (KBr). $E_{1/2}$ −2.66. V. vs. Ag/Ag$^{\oplus}$, 0.1M MeCN.

Dahlenburg, L. *et al*, *J. Organomet. Chem.*, 1974, **71**, C49 (*synth, ir*)
Zecchin, S. *et al*, *J. Organomet. Chem.*, 1976, **110**, C45 (*synth, ir*)
Schiavon, G. *et al*, *J. Organomet. Chem.*, 1976, **121**, 261 (*ir*)
Dahlenburg, L. *et al*, *Z. Naturforsch., B*, 1982, **37**, 310 (*synth, nmr, ir*)

$C_{38}H_{34}ClIrP_2$ **Ir-00192**
Chloro(η^2-ethene)bis(triphenylphosphine)iridium, 9CI
Chloro(ethylene)bis(triphenylphosphine)iridium, 8CI. *Chlorobis(triphenylphosphine)(ethylene)iridium*

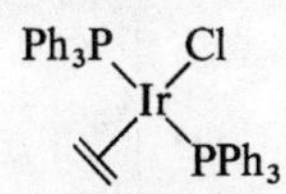

M 780.307

***trans*-form** [32761-43-6]
Orange cryst. or yellow-orange cryst. + $\frac{1}{2}C_6H_6$. Mp 172-4° (solvate 144-8°). Solns. and solvate sensitive to oxygen.

van der Ent, A. *et al*, *Inorg. Chim. Acta*, 1973, **7**, 203 (*synth, ir, pmr*)
Restivo, R.J. *et al*, *J. Organomet. Chem.*, 1975, **90**, 101 (*cryst struct*)
Clark, P.W. *et al*, *Inorg. Chem.*, 1979, **18**, 2067 (*cmr*)
Schramm, K. *et al*, *Inorg. Chem.*, 1980, **19**, 3183 (*ir, nmr*)

$C_{38}H_{40}IrO_2P_2^{\oplus}$ Ir-00193

Dihydrobis(methanol)bis(triphenylphosphine)iridium(1+)
Dihydridobis(methanol)bis(triphenylphosphine)iridium(1+)

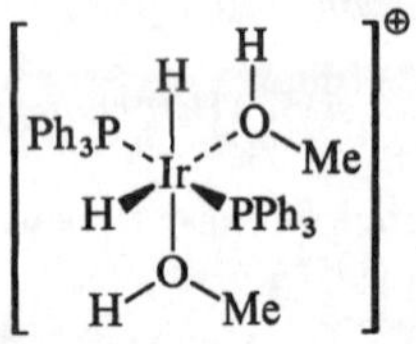

M 782.900 (ion)
Catalyst for olefin hydrogenation and alkane dehydrogenation. Tends to lose H_2 thermally, esp. in soln. MeOH displaced in MeCN.
Tetrafluoroborate: [83705-82-2].
$C_{38}H_{40}BF_4IrO_2P_2$ M 869.704
Beige solid (CH_2Cl_2/Et_2O).

Crabtree, R.H. *et al*, *J. Am. Chem. Soc.*, 1982, **104**, 6994 (*synth, pmr, ir, use*)

$C_{39}H_{30}F_3IrO_3P_2$ Ir-00194

Carbonyl(trifluoroacetato-*O*)bis(triphenylphosphine)iridium, 10CI
Trifluoroacetato(carbonyl)bis(triphenylphosphine)iridium
[41509-47-1]
As Bromocarbonylbis(triphenylphosphine)iridium, Ir-00162 with

X = F_3CCOO-, R = Ph

M 857.827
***trans*-form** [50442-04-1]
Yellow cryst. Mp 182-4°, 200-3° dec. ir ν_{CO} 1970vs, 1944w, 1695s cm^{-1} (nujol).

Brown, C.K. *et al*, *J. Chem. Soc., Dalton Trans.*, 1973, 929 (*synth, ir, pmr*)
Reed, C.A. *et al*, *J. Chem. Soc., Dalton Trans.*, 1973, 1370 (*synth, ir*)
Kuyper, J. *et al*, *J. Organomet. Chem.*, 1975, **96**, 289 (*synth, ir*)

$C_{39}H_{33}Cl_2IrO_2P_2$ Ir-00195

Acetylcarbonyldichlorobis(triphenylphosphine)iridium
[36447-22-0]

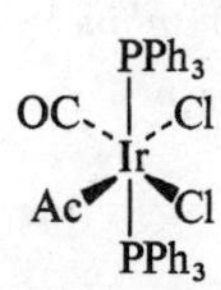

M 858.762

***a*-Acetyl-*b*-carbonyl-*d,f*-dichloro-*c,e*-diphosphine-form** [33394-10-4]
Colourless cryst. Dec. at 238-41°. ir ν_{CO} 2040 cm^{-1} (nujol).

Blum, J. *et al*, *J. Organomet. Chem.*, 1971, **33**, 227 (*synth, ir*)
Kubota, M. *et al*, *J. Am. Chem. Soc.*, 1971, **93**, 1368 (*synth, ir*)
Blake, D.M. *et al*, *J. Organomet. Chem.*, 1981, **204**, 257 (*synth, ir*)

$C_{39}H_{33}IrNOP_2^{\oplus}$ Ir-00196

Acetonitrilecarbonylbis(triphenylphosphine)iridium(1+), 10CI
Carbonyl(acetonitrile)bis(triphenylphosphine)iridium(1+)

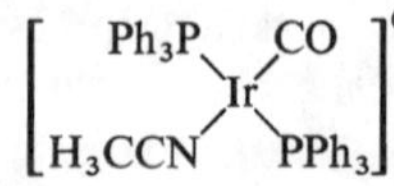

M 785.863 (ion)
***trans*-form**
ir ν_{CO} 1975 cm^{-1} (nujol).
Hexafluorophosphate: [51540-63-7].
$C_{39}H_{33}F_6IrNOP_3$ M 930.827
Cryst.
Perchlorate: [49795-87-1].
$C_{39}H_{33}ClIrNO_5P_2$ M 885.314
Yellow prisms (CH_2Cl_2/EtOH/cyclohexane). Mp 173-8°.
Tetrafluoroborate: [49745-75-7].
$C_{39}H_{33}BF_4IrNOP_2$ M 872.667
Cryst. Mp 153-6°.

Reed, C.A. *et al*, *J. Chem. Soc., Dalton Trans.*, 1973, 1365 (*synth, ir, pmr, struct*)
Williams, A.F. *et al*, *J. Chem. Soc., Dalton Trans.*, 1975, 1958 (*mössbauer, nmr*)

$C_{39}H_{34}ClIrP_2$ Ir-00197

Chloro[(1,2-η)-1,2-propadiene]bis(triphenylphosphine)iridium
(Allene)chlorobis(triphenylphosphine)iridium
[42590-19-2]

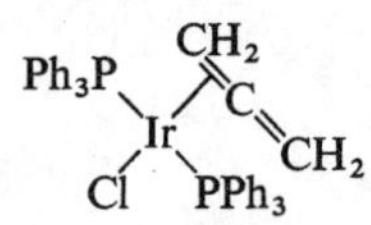

M 792.318
Air-sensitive yellow cryst. Mp 172-5° dec. Forms diamagnetic complexes with O_2 and SO_2.

Van der Ent, A. *et al*, *Inorg. Chim. Acta*, 1973, **7**, 203 (*synth, ir, pmr*)

$C_{39}H_{35}IrP_2$ Ir-00198

(η^3-2-Propenyl)bis(triphenylphosphine)iridium, 9CI
Allylbis(triphenylphosphine)iridium
[53337-56-7]

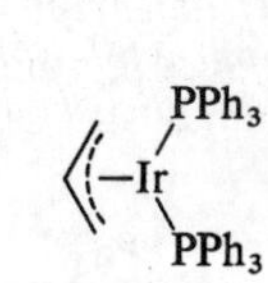

M 757.873
Pale-yellow cryst. (C_6H_6). Stable in an inert atm. in solid state but unstable in soln. even under N_2.

Clerici, M.G. *et al*, *J. Organomet. Chem.*, 1975, **84**, 379 (*ir, pmr, synth*)

$C_{39}H_{37}Cl_3IrNP_2$ **Ir-00199**

Trichloro[(dimethylamino)methylene]bis(triphenylphosphine)-iridium, 10CI

[38890-37-8]

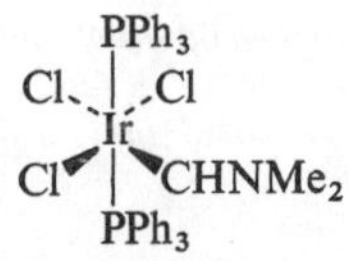

M 880.255

White powder. Mp 285-90° (274-6°). ir ν_{CN} 1584 cm^{-1} (nujol).

Cetinkaya, B. *et al*, *J. Chem. Soc., Chem. Commun.*, 1972, 851 (*synth*)

Hartshorn, A.J. *et al*, *J. Chem. Soc., Dalton Trans.*, 1978, 348 (*synth, ir, pmr*)

$C_{39}H_{44}IrO_2P_2^{\oplus}$ **Ir-00200**

[(1,2,5,6-η)-1,5-Cyclooctadiene][[(2,2-dimethyl-1,3-dioxolane-4,5-diyl)bis[methylene]]bis(diphenylphosphine)-*P*,*P'*]-iridium(1+), 10CI

(Cyclooctadiene)DIOP iridium(1+)

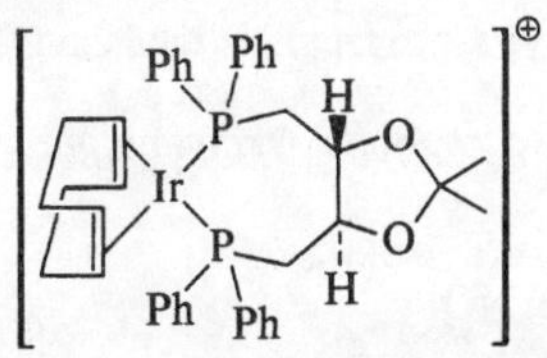

M 798.943 (ion)

Precursor of olefin hydrogenation catalyst.

(R,R)-form

Hexafluorophosphate: [64536-64-7].
$C_{39}H_{44}F_6IrO_2P_3$ M 943.907
Red.

Crabtree, R.H. *et al*, *J. Organomet. Chem.*, 1977, **135**, 395 (*synth, pmr*)

Crabtree, R.H. *et al*, *J. Organomet. Chem.*, 1979, **168**, 183.

$C_{40}H_{30}ClF_6IrP_2$ **Ir-00201**

Chloro[(2,3-η)-1,1,1,4,4,4-hexafluoro-2-butyne]bis(triphenylphosphine)iridium, 9CI

[27709-92-8]

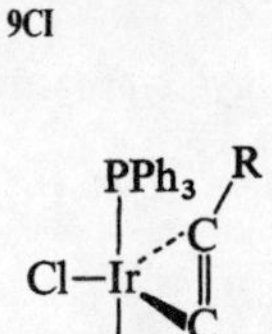

R = CF_3

M 914.288

Orange cryst. (C_6H_6/hexane). Mp 158-60° dec. Unstable towards water.

Clarke, B. *et al*, *J. Chem. Soc. (A)*, 1970, 951 (*synth, ir, nmr*)

Van Gaal, H.L. *et al*, *J. Organomet. Chem.*, 1977, **131**, 453 (*ir*)

$C_{40}H_{30}Ir_2O_8P_2S_2$ **Ir-00202**

Tetracarbonyldi-μ-sulfonylbis(triphenylphosphine)diiridium, 10CI

Tetracarbonylbis(sulfur dioxide)bis(triphenylphosphine)diiridium, 9CI. Di-μ-(sulfur dioxide)tetracarbonylbis(triphenylphosphine)diiridium. Di-μ-(sulfur dioxide)bis[dicarbonyl(triphenylphosphine)iridium]

[66908-68-7]

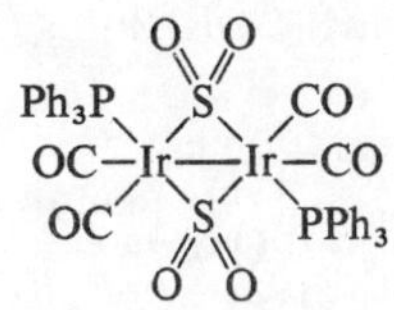

M 1149.180

Yellow cryst. ($CHCl_3$/pet. ether). Mp 205°. ir ν_{CO} 2118, 2080, 2043 cm^{-1} (nujol).

Monomer: Dicarbonyl(sulfur dioxide)(triphenylphosphine)iridium.
$C_{20}H_{15}IrO_4PS$ M 574.590
Has no independent existence.

Angoletta, M. *et al*, *J. Organomet. Chem.*, 1974, **81**, C40 (*ir*)

Angoletta, M. *et al*, *J. Chem. Soc., Dalton Trans.*, 1977, 2131 (*synth, ir*)

$C_{40}H_{32}Ir_2O_6P_2S$ **Ir-00203**

Tetracarbonyldihydro-μ-sulfonylbis(triphenylphosphine)diiridium, 10CI

Tetracarbonyldihydro[μ-(sulfur dioxide-S:S)]bis(triphenylphosphine)diiridium, 9CI. μ-Sulfur dioxidebis(dicarbonylhydridotriphenylphosphine)diiridium. μ-(Sulfur dioxide)-bis(hydridodicarbonyltriphenylphosphineiridium)

[54374-32-2]

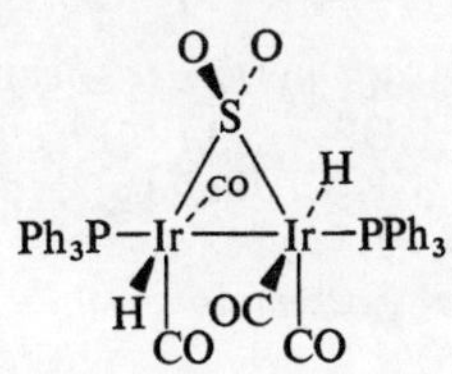

M 1087.137

Other structural isomers may be present in soln. V. stable white cryst. ($CHCl_3$). Mp 205°.

Angoletta, M. *et al*, *J. Organomet. Chem.*, 1974, **81**, C40 (*struct, ir*)

Angoletta, M. *et al*, *J. Chem. Soc., Dalton Trans.*, 1977, 2131 (*synth, ir, pmr*)

$C_{40}H_{33}IrOP_2$ **Ir-00204**

Carbonyl-1-propynylbis(triphenylphosphine)iridium, 9CI

[34720-36-0]

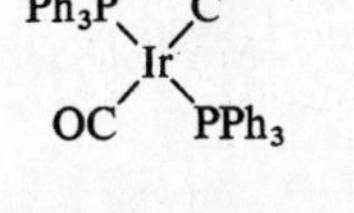

M 783.868

trans-form

Catalyst for polymerisation of alkynes. Orange-yellow solid (C_6H_6/EtOH). ir ν_{CO} 1968 cm^{-1} (nujol).

▷Undergoes a wide range of oxidative addn. reactions

Brown, C.K. *et al, J. Chem. Soc. (A)*, 1971, 3120 (*synth, ir, pmr*)

$C_{40}H_{36}IrO_2P_2^{\oplus}$ Ir-00205

Carbonyl(2-propanone)bis(triphenylphosphine)iridium(1+), 10CI

Acetonecarbonylbis(triphenylphosphine)iridium(1+)

$$[(H_3C)_2CO]Ir(PPh_3)_2CO^{\oplus}$$

M 802.891 (ion)
ir ν_{CO} 1970, 1630 cm^{-1} (nujol).
Hexafluorophosphate: [55493-09-9].
$C_{40}H_{36}F_6IrO_2P_3$ M 947.855
Yellow cryst. Mp >150°.
Tetrafluoroborate: [66540-34-9].
$C_{40}H_{36}BF_4IrO_2P_2$ M 889.694
Yellow cryst.

Clark, H.C. *et al, Inorg. Chem.*, 1975, **14**, 2133 (*synth, ir, pmr*)
Connelly, N.G. *et al, J. Chem. Soc., Dalton Trans.*, 1978, 50 (*synth*)

$C_{40}H_{37}Cl_2IrNOP_2^{\oplus}$ Ir-00206

Carbonyldichloro[(dimethylamino)methylene]bis(triphenylphosphine)iridium(1+), 9CI

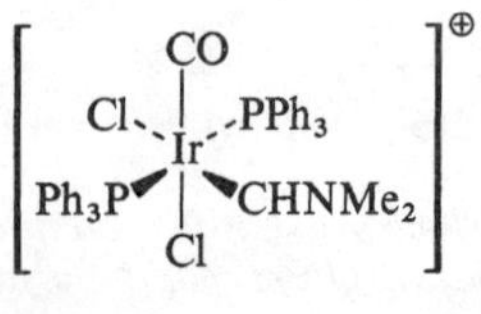

M 872.812 (ion)
Chloride: [67407-39-0].
$C_{40}H_{37}Cl_3IrNOP_2$ M 908.265
White powder + $CHCl_3$. Mp 176-85°. ir ν_{CO} 2073 cm^{-1} (nujol).
Hexafluorophosphate: [67407-41-4].
$C_{40}H_{37}Cl_2F_6IrNOP_3$ M 1017.776
White needles (CH_2Cl_2/hexane). Mp 245-8°. ir ν_{CO} 2080 cm^{-1} (nujol).

Hartshorn, A.J. *et al, J. Chem. Soc., Dalton Trans.*, 1978, 348 (*synth, ir, nmr*)

$C_{40}H_{38}IrN_2P_2^{\oplus}$ Ir-00207

Bis(acetonitrile)dihydrobis(triphenylphosphine)iridium(1+), 9CI

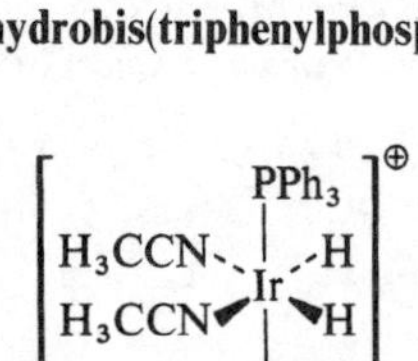

M 800.921 (ion)
Catalyst for hydrogenation of alkenes, alkynes and ketones, and for activation of alkane C—H bonds.
Hexafluorophosphate: [78990-54-2].
$C_{40}H_{38}F_6IrN_2P_3$ M 945.885
No phys. props. reported.
Tetrafluoroborate:
$C_{40}H_{38}BF_4IrN_2P_2$ M 887.725
Air-stable, pale buff-coloured cryst. ir ν_{IrH} 2207, ν_{CN} 2315 cm^{-1} (CH_2Cl_2). Loses H_2 on heating.

Howarth, O.W. *et al, J. Chem. Soc., Dalton Trans.*, 1981, 1481 (*synth, pmr*)

Crabtree, R.H. *et al, J. Am. Chem. Soc.*, 1982, **104**, 6994 (*synth, ir, pmr*)

$C_{41}H_{37}IrOP_2$ Ir-00208

[(1,2,3-η)-2-Butenyl]carbonylbis(triphenylphosphine)iridium, 9CI

Carbonyl(1-methyl-π-allyl)bis(triphenylphosphine)iridium, 8CI

[32802-84-9]

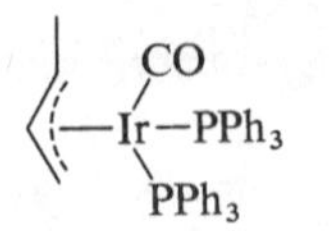

M 799.910
Air-stable pale-yellow cryst. (C_6H_6/EtOH); dec. slowly in $CHCl_3$ or CS_2. ir ν_{CO} 1930 cm^{-1} (nujol).

Brown, C.K. *et al, J. Chem. Soc. (A)*, 1971, 850 (*synth, pmr*)
Burnett, M.G. *et al, J. Chem. Soc., Dalton Trans.*, 1974, 1663 (*synth, pmr, ir*)

$C_{41}H_{37}IrOP_2$ Ir-00209

Carbonyl(1-methyl-1-propenyl)bis(triphenylphosphine)iridium, 9CI

(2-Buten-2-yl)carbonylbis(triphenylphosphine)iridium

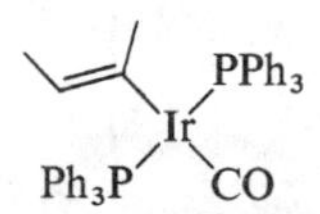

(*E*)-*trans-form*

M 799.910

(*E*)-*trans-form* [53770-15-3]
Golden yellow cryst. (Et_2O/EtOH). Sol. C_6H_6. ir ν_{CO} 1935 cm^{-1} (nujol). Isom. to [(1,2,3-η)-2-Butenyl]carbonylbis(triphenylphosphine)iridium, Ir-00208 on heating.

(*Z*)-*trans-form* [53797-90-3]
Sol. C_6H_6. Forms 2-butyne and $HIr(CO)(PPh_3)_3$ at r.t.

Schwartz, J. *et al, J. Am. Chem. Soc.*, 1974, **96**, 5613 (*synth, pmr, props*)

$C_{41}H_{37}IrOP_2$ Ir-00210

Carbonyl[(1,2,3-η)-2-methyl-2-propenyl]bis(triphenylphosphine)iridium, 9CI

Carbonyl(2-methyl-π-allyl)bis(triphenylphosphine)iridium, 8CI

[32662-74-1]

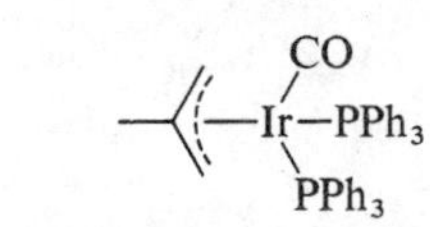

M 799.910
Air-stable yellow cryst. (toluene/EtOH); dec. slowly in $CHCl_3$ or CS_2. ir ν_{CO} 1946 cm^{-1} (cyclohexane).

Brown, C.K. *et al, J. Chem. Soc. (A)*, 1971, 850 (*synth, pmr*)

$C_{41}H_{43}IrO_2P_2Si_2$ Ir-00211

Carbonylhydro(1,1,3,3-tetramethyl-1,3-disiloxanediyl)bis-(triphenylphosphine)iridium, 9CI

Carbonylhydrido(tetramethyldisiloxanediyl)bis(triphenylphosphine)iridium

[68225-64-9]

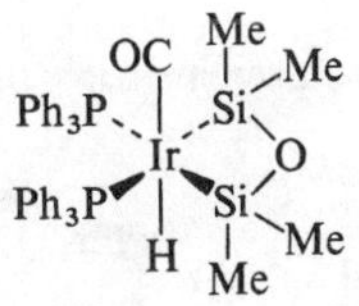

M 878.128

Catalyst for disproportionation of 1,1,3,3-tetramethyldisiloxane to dimethylsilane and higher siloxanes.

EtOH adduct: [69134-24-3].
$C_{43}H_{49}IrO_3P_2Si_2$ M 924.197
Colourless cryst. (CH_2Cl_2/EtOH). Sol. C_6H_6, CH_2Cl_2. Mp 150-4° dec. ir ν_{CO} 1950, ν_{IrH} 2060 (2030) cm^{-1}.

Greene, J. *et al*, *J. Am. Chem. Soc.*, 1977, **99**, 5176 (*synth, ir, pmr, struct*)
Curtis, M.D. *et al*, *J. Am. Chem. Soc.*, 1978, **100**, 6362 (*synth, ir, pmr, use*)
Curtis, M.D. *et al*, *J. Organomet. Chem.*, 1978, **164**, 371 (*struct*)

$C_{42}H_{30}Cl_2Ir_2O_6P_2Sn$ Ir-00212

Hexacarbonyl[μ-(dichlorostannylene)]bis(triphenylphosphine)diiridium, 9CI

Dichlorobis[tricarbonyl(triphenylphosphine)iridio]tin

[54032-03-0]

$Ph_3PIr(CO)_3—SnCl_2—Ir(CO)_3PPh_3$

M 1266.679

Yellow cryst. Mp 283-6°. ir ν_{CO} 2000w, 1970vs cm^{-1}.

Malatesta, L. *et al*, *J. Organomet. Chem.*, 1974, **73**, 265 (*synth, ir*)

$C_{42}H_{30}Ir_2O_6P_2$ Ir-00213

Hexacarbonylbis(triphenylphosphine)diiridium, 10CI

Bis[tricarbonyl(triphenylphosphine)iridium]. Bis(triphenylphosphine)hexacarbonyldiiridium

[29519-30-0]

$(Ph_3P)(OC)_3IrIr(CO)_3(PPh_3)$

M 1077.083

Yellow cryst. Almost insol. org. solvs. Mp 167-70°. ir ν_{CO} 1975, 1945, 1900 cm^{-1} (nujol).

Drakesmith, A.J. *et al*, *J. Chem. Soc., Dalton Trans.*, 1973, 362 (*ir, synth*)
Haymore, B.L. *et al*, *J. Am. Chem. Soc.*, 1974, **96**, 3325 (*synth*)
Malatesta, L. *et al*, *J. Organomet. Chem.*, 1974, **73**, 265 (*synth*)

$C_{42}H_{36}ClIrO_4P_2$ Ir-00214

Chlorobis(triphenylphosphine)(dimethyl acetylenedicarboxylate)iridium

Chloro(dicarboxyvinylene)bis(triphenylphosphine)iridium dimethyl ester, 8CI. Chloro(dimethyl 2-butynedioate)-bis(triphenylphosphine)iridium

[15412-68-7]

As Chloro[(2,3-η)-1,1,1,4,4,4-hexafluoro-2-butyne]bis(triphenylphosphine)iridium, Ir-00201 with

R = COOMe

M 894.365

Yellow cryst. ($CHCl_3$/MeOH). Mp 182-3° dec.

Collmann, J.P. *et al*, *J. Am. Chem. Soc.*, 1967, **89**, 844 (*synth, ir*)

$C_{42}H_{40}IrN_2OP_2^{\oplus}$ Ir-00215

Carbonyl(1,3-dimethyl-2-imidazolidinylidene)bis(triphenylphosphine)iridium(1+), 9CI

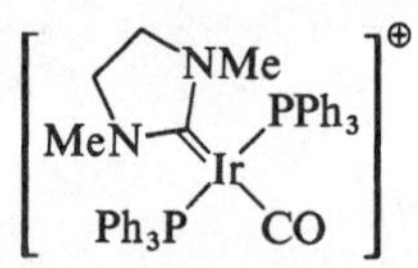

M 842.958 (ion)

trans-form

Tetrafluoroborate: [54352-44-2].
$C_{42}H_{40}BF_4IrN_2OP_2$ M 929.762
Yellow-orange cryst. (CH_2Cl_2/Et_2O). Mp 225-8° dec. ir ν_{CO} 1980 cm^{-1} (nujol).

Cetinkaya, B. *et al*, *J. Chem. Soc., Dalton Trans.*, 1974, 1827 (*synth, pmr, ir*)

$C_{42}H_{44}IrO_2P_2^{\oplus}$ Ir-00216

Dihydrobis(2-propanone)bis(triphenylphosphine)iridium(1+), 10CI, 9CI

Bis(acetone)dihydridobis(triphenylphosphine)-iridium(1+)

[24419-58-7]

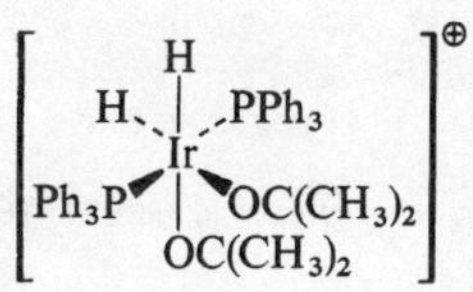

M 834.976 (ion)

Catalyst for olefin hydrogenation and alkane dehydrogenation. Tends to lose H_2 thermally, esp. in soln. Me_2CO displaced by EtOH<MeOH<MeCN.

Tetrafluoroborate: [72414-17-6].
$C_{42}H_{44}BF_4IrO_2P_2$ M 921.780
Pale buff-coloured cryst.

Hexafluorophosphate: [65804-43-5].
$C_{42}H_{44}F_6IrO_2P_3$ M 979.940
Colourless cryst. (Me_2CO/pet. ether).

Shapley, J.R. *et al*, *J. Am. Chem. Soc.*, 1969, **91**, 2816 (*synth, pmr, ir, use*)
Crabtree, R.H. *et al*, *J. Am. Chem. Soc.*, 1979, **101**, 7738 (*use*)
Howarth, O.W. *et al*, *J. Chem. Soc., Dalton Trans.*, 1981, 1481 (*synth, pmr, props*)
Crabtree, R.H. *et al*, *J. Am. Chem. Soc.*, 1982, **104**, 6994 (*use, ir, pmr*)

$C_{42}H_{44}Ir_2N_4O_6$ Ir-00217

Hexacarbonyl[μ-[2,3,7,8,12,13,17,18-octaethyl-21*H*,23*H*-porphinato(2−)-N^{21},N^{22},N^{23},N^{24}]]diiridium, 9CI

[68324-57-2]

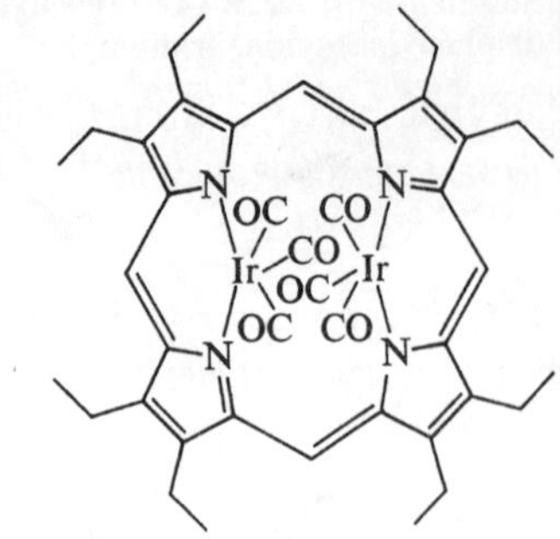

M 1085.273

Brownish-red cryst.

Ogoshi, H. *et al*, *J. Organomet. Chem.*, 1978, **159**, 317 (*synth, pmr, ir, uv*)

$C_{43}H_{35}IrOP_2$ Ir-00218

Carbonylphenylbis(triphenylphosphine)iridium, 10CI

Phenylcarbonylbis(triphenylphosphine)iridium

[25037-18-7]

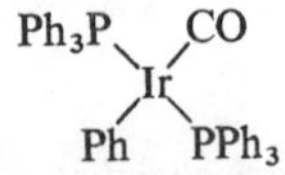

M 821.916

trans-form [42402-07-3]

Orange cryst. (C_6H_6/heptane). Mp 159° (sealed tube under N_2). ir ν_{CO} 1947 cm^{-1} ($CHCl_3$).

Yagupsky, G. *et al*, *J. Chem. Soc. (A)*, 1970, 1392 (*synth, ir*)
Rausch, M.D. *et al*, *Inorg. Chem.*, 1974, **13**, 11 (*synth*)
Dahlenburg, L. *et al*, *J. Organomet. Chem.*, 1976, **110**, 395 (*synth, ir, pmr*)
Zecchin, S. *et al*, *J. Organomet. Chem.*, 1976, **110**, C45 (*synth*)

$C_{44}H_{42}IrP_2^{\oplus}$ Ir-00219

[(1,2,5,6-η)-1,5-Cyclooctadiene]bis(triphenylphosphine)iridium(1+), 9CI

Bis(triphenylphosphine)(π-1,5-cyclooctadiene)-iridium(1+)

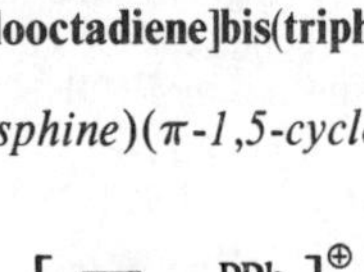

M 824.983 (ion)

Tetrafluoroborate: [38834-40-1].
$C_{44}H_{42}BF_4IrP_2$ M 911.787
Olefin hydrogenation catalyst. Cryst.

Perchlorate: [52657-94-0].
$C_{44}H_{42}ClIrO_4P_2$ M 924.434
Red cryst. Mp 130-2°.

Hexafluorophosphate: [61817-47-8].
$C_{44}H_{42}F_6IrP_3$ M 969.947
Cryst.

Tetraphenylborate: [32876-06-5].
$C_{68}H_{62}BIrP_2$ M 1144.215
Cryst. Mp 150° dec.

Haines, L.M., *Inorg. Nucl. Chem. Lett.*, 1969, **5**, 399 (*synth*)
Green, M. *et al*, *J. Chem. Soc., Chem. Commun.*, 1970, 1553 (*synth*)
Solodar, J., *J. Org. Chem.*, 1972, **37**, 1840 (*synth, use*)
Kawakami, K. *et al*, *J. Organomet. Chem.*, 1973, **60**, 363 (*synth, ir*)
Uson, R. *et al*, *J. Organomet. Chem.*, 1980, **193**, 127 (*synth*)

$C_{44}H_{48}IrO_2P_2^{\oplus}$ Ir-00220

Dihydrobis(tetrahydrofuran)bis(triphenylphosphine)iridium(1+)

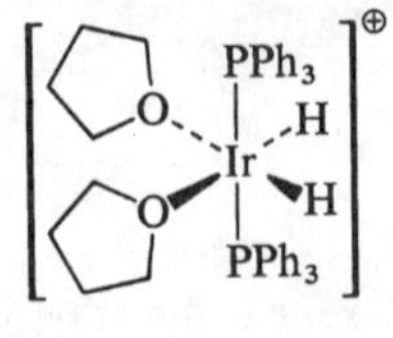

M 863.030 (ion)

Catalyst for hydrogenation of alkenes, alkynes and ketones, and for activation of alkane C—H bonds.

Tetrafluoroborate:
$C_{44}H_{48}BF_4IrO_2P_2$ M 949.833
Air-stable, pale buff-coloured cryst.

Crabtree, R.H. *et al*, *J. Am. Chem. Soc.*, 1982, **104**, 6994 (*synth, ir, pmr*)

$C_{45}H_{31}F_{12}IrOP_2$ Ir-00221

Carbonyl[(2,3-η)-1,1,1,4,4,4-hexafluoro-2-butyne][3,3,3-trifluoro-1-(trifluoromethyl)-1-propenyl]bis(triphenylphosphine)iridium, 9CI

[52611-63-9]

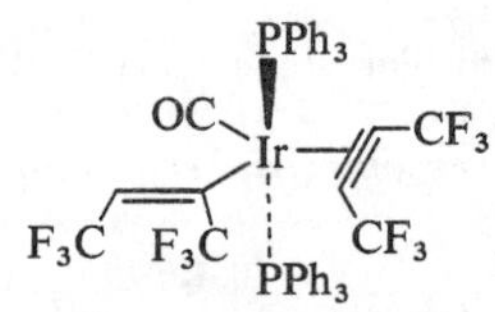

M 1069.888

Model for catalytic species in linear polymerisation of acetylenes. Off-white microcryst. (CH_2Cl_2/hexane). ir ν_{CO} 1991 cm^{-1} (nujol).

Baddles, W.H. *et al*, *J. Am. Chem. Soc.*, 1969, **91**, 3661 (*synth, ir*)
Baddles, W.H. *et al*, *J. Organomet. Chem.*, 1974, **67**, C16 (*use*)

$C_{45}H_{45}B_{10}IrOP_2$ Ir-00222

Carbonyl(7-phenyl-1,7-dicarbadodecaboran(12)-1-yl)bis(triphenylphosphine)iridium, 9CI

*Carbonylbis(triphenylphosphine)[7-phenyl-1,7-dicarba-*closo*-dodecaborane(12)]iridium*

[57527-11-4]

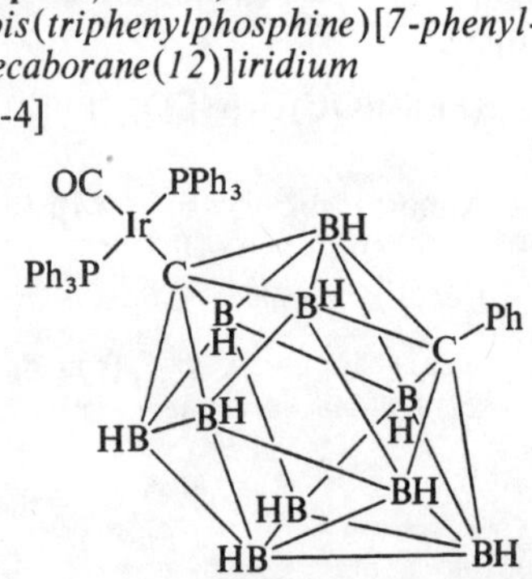

M 964.117

Orange cryst. (C_6H_6/hexane), solid and soln. stable under inert atm. Sol. C_6H_6, CH_2Cl_2. Mp 132-3° dec. ir ν_{CO} 1970 cm^{-1} (nujol). Readily takes up molecular hydrogen but is not an active hydrogenation catalyst.

Longato, B. *et al*, *Inorg. Chem.*, 1976, **15**, 650 (*synth*, *ir*)

$C_{46}H_{30}Ir_4O_{10}P_2$ **Ir-00223**

Tri-μ-carbonylheptacarbonylbis(triphenylphosphine)tetrairidium, 10CI

Decacarbonylbis(triphenylphosphine)tetrairidium, 8CI

[12279-27-5]

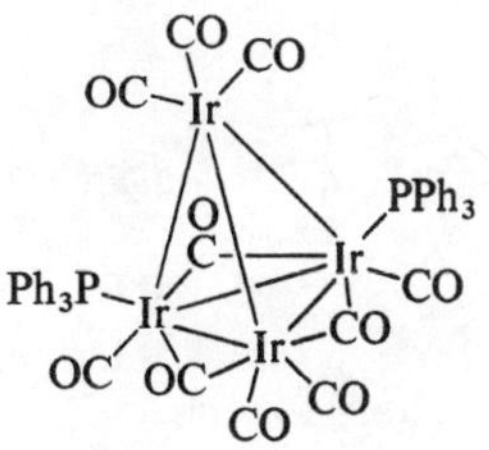

M 1573.565

Yellow cryst. Mp 251-3°.

Albano, V. *et al*, *J. Chem. Soc., Chem. Commun.*, 1967, 730 (*synth*, *cryst struct*)
Drakesmith, A.J. *et al*, *J. Chem. Soc., Dalton Trans.*, 1973, 362 (*synth*, *ir*)
Karel, K. *et al*, *J. Am. Chem. Soc.*, 1974, **96**, 6812 (*synth*)
Malatesta, L. *et al*, *J. Organomet. Chem.*, 1974, **73**, 265 (*synth*, *ir*)
Angoletta, M. *et al*, *J. Organomet. Chem.*, 1975, **94**, 99 (*synth*)
Stuntz, G.F. *et al*, *Inorg. Chem.*, 1976, **15**, 1994 (*synth*, *ir*)
Stuntz, G.F. *et al*, *J. Am. Chem. Soc.*, 1977, **99**, 607 (*nmr*, *struct*)

$C_{50}H_{40}ClIrP_2$ **Ir-00224**

Chlorobis(triphenylphosphine)(diphenylacetylene)iridium

Chloro[1,1′-[(1,2-η)-1,2-ethynediyl]bis[benzene]]bis(triphenylphosphine)iridium, 9CI. Chloro(diphenylvinylene)bis(triphenylphosphine)iridium, 8CI

[15380-66-2]

As Chloro[(2,3-η)-1,1,1,4,4,4-hexafluoro-2-butyne]bis(triphenylphosphine)iridium, Ir-00201 with

R = Ph

M 930.487

Yellow cryst. (C_6H_6/MeOH). Mp 193-5°.

Collmann, J.P. *et al*, *J. Am. Chem. Soc.*, 1967, **89**, 844 (*synth*, *ir*)
Nefedov, V.I. *et al*, *Inorg. Chim. Acta*, 1973, **7**, 499 (*pe*)
Bennett, M.A. *et al*, *Aust. J. Chem.*, 1977, **30**, 1213 (*ir*)
Van Gaal, H.L. *et al*, *J. Organomet. Chem.*, 1977, **131**, 453 (*ir*)

$C_{51}H_{37}F_{12}IrO_5P_2$ **Ir-00225**

[1,2-Bis(methoxycarbonyl)-3,4-bis(trifluoromethyl)-1,3-butadiene-1,4-diyl]carbonyl[3,3,3-trifluoro-1-(trifluoromethyl)-1-propenyl]bis(triphenylphosphine)iridium, 9CI

[52556-57-7]

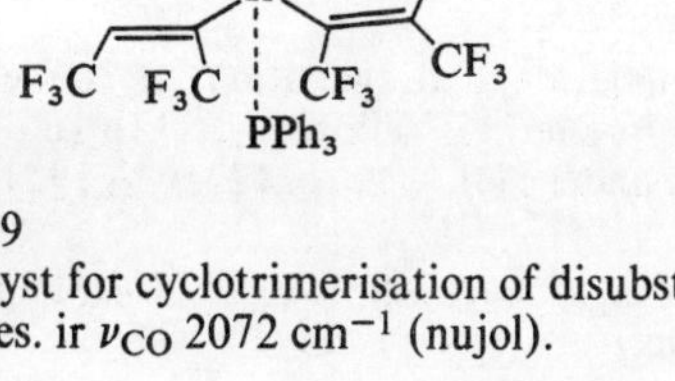

M 1211.999

Good catalyst for cyclotrimerisation of disubstituted acetylenes. ir ν_{CO} 2072 cm^{-1} (nujol).

Baddles, W.H. *et al*, *J. Organomet. Chem.*, 1974, **67**, C16 (*synth*, *ir*, *pmr*, *use*)

$C_{51}H_{40}IrP$ **Ir-00226**

(η^5-2,4-Cyclopentadien-1-yl)(1,2,3,4-tetraphenyl-1,3-butadiene-1,4-diyl)(triphenylphosphine)iridium, 9CI

1-(π-Cyclopentadienyl)-1-(triphenylphosphine)-2,3,4,5-tetraphenyliridole

[54873-30-2]

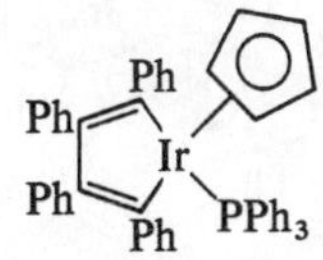

M 876.071

Yellow cryst. (C_6H_6/hexane).

Gardner, S.A. *et al*, *J. Organomet. Chem.*, 1974, **78**, 415 (*synth*, *ir*, *pmr*, *ms*)

$C_{52}H_{44}Cl_2Ir_2O_2P_4$ **Ir-00227**

Dicarbonyldichlorobis[μ-[methylenebis[diphenylphosphine]-*P*:*P*′]]diiridium, 9CI

Bis[μ-bis(diphenylphosphino)methane]dicarbonyldichlorodiiridium. Bis[[μ-bis(diphenylphosphino)methane]carbonylchloroiridium]

[66125-35-7]

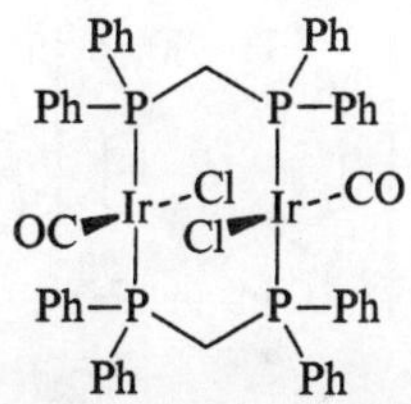

M 1280.159

Orange-red solid. ir ν_{CO} 2000s, 1960vs cm^{-1} (KBr).

Sanger, A.R. *et al*, *J. Chem. Soc., Dalton Trans.*, 1977, 1971 (*synth*, *ir*)

$C_{52}H_{44}Ir_2O_2P_4S$ **Ir-00228**

Dicarbonylbis[μ-[methylenebis[diphenylphosphine]-*P*:*P*′]]-μ-thioxodiiridium, 9CI

Dicarbonylbis[μ-bis(diphenylphosphino)methane][μ-sulfido]diiridium

[73972-13-1]

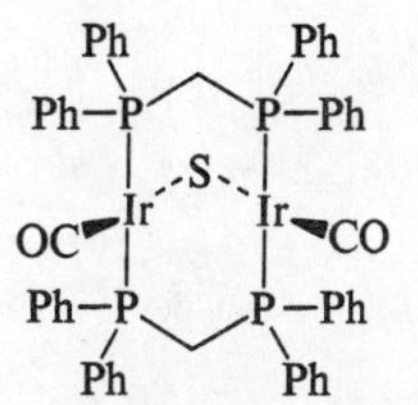

M 1241.313

Hydrogenation catalyst for simple unsaturated compds. Deep-purple cryst. (CH_2Cl_2/MeOH). ir ν_{CO} 1918s, 1902vs cm^{-1} (nujol). Adds H_2 reversibly forming Dicarbonyldihydrobis[μ-[methylenebis[diphenylphosphine]-*P*:*P*′]]-μ-thioxodiiridium, Ir-00229 .

Kubiak, C.P. *et al*, *Inorg. Chem.*, 1980, **19**, 2733 (*synth*, *ir*)

$C_{52}H_{46}Ir_2O_2P_4S$ Ir-00229

Dicarbonyldihydrobis[μ-[methylenebis[diphenylphosphine]-*P*:*P'*]]-μ-thioxodiiridium, 9CI
Bis[μ-bis(diphenylphosphino)methane]dicarbonyldihydrido(μ-sulfido)diiridium
[74033-51-5]

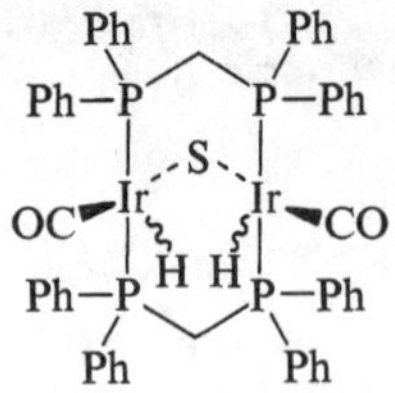

M 1243.329
At least two isomeric forms exist. Hydrogenation catalyst for simple unsaturated compds. Characterised spectroscopically. Loses H_2 on heating. ir ν_{CO} 2006, 1976, 1960, 1921 cm^{-1} (toluene).

Kubiak, C.P. *et al*, *Inorg. Chem.*, 1980, **19**, 2733 (*synth*, *pmr*, *ir*)

$C_{52}H_{48}IrP_4^{\oplus}$ Ir-00230

Bis[1,2-ethanediylbis[diphenylphosphine]-*P*,*P'*]iridium(1+), 10CI
Bis[ethylenebis[diphenylphosphine]]iridium(1+), 8CI. *Bis[(diphenylphosphino)ethane]iridium(1+)*
[29871-99-6]

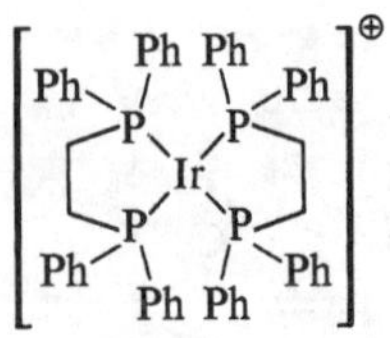

M 989.066 (ion)
Chloride: [15390-38-2].
$C_{52}H_{48}ClIrP_4$ M 1024.519
Bright-orange cryst. Mp 225° dec. Hygroscopic.
Perchlorate: [15685-14-0].
$C_{52}H_{48}ClIrO_4P_4$ M 1088.517
Orange cryst. (CH_2Cl_2/Et_2O).
Tetrafluoroborate: [15130-28-6].
$C_{52}H_{48}BF_4IrP_4$ M 1075.870
Orange cryst. Mp 270° dec.
Hexafluorophosphate: [41047-09-0].
$C_{52}H_{48}F_6IrP_5$ M 1134.030
Orange cryst. (Me_2CO/Et_2O).

Sacco, A. *et al*, *J. Chem. Soc., Chem. Commun.*, 1966, 589 (*synth*)
Vaska, L. *et al*, *J. Am. Chem. Soc.*, 1966, **88**, 5324 (*synth*, *ir*)
Brady, R. *et al*, *J. Chem. Soc., Chem. Commun.*, 1974, 393 (*uv*)
Williams, A.F. *et al*, *J. Chem. Soc., Dalton Trans.*, 1975, 1958 (*synth*, *mössbauer*)
Geoffroy, G.L. *et al*, *Inorg. Chem.*, 1977, **16**, 1950 (*cd*, *uv*)
Fordyce, W.A. *et al*, *Inorg. Chem.*, 1982, **21**, 1455 (*synth*, *uv*)

$C_{54}H_{45}ClIr$ Ir-00231

Chlorotris(triphenylphosphine)iridium, 10CI
Tris(triphenylphosphine)iridium chloride
[16070-58-9]

$ClIr(PPh_3)_3$

M 921.623
Air-sensitive orange solid. Forms an octahedral hydridoaryl complex on heating.

Collman, J.P. *et al*, *J. Am. Chem. Soc.*, 1968, **90**, 5430 (*synth*, *ir*)
Bennett, M.A. *et al*, *J. Am. Chem. Soc.*, 1969, **91**, 6983 (*synth*)
Roundhill, D.M. *et al*, *Inorg. Chem.*, 1980, **19**, 284 (*ir*, *nmr*)
Schramm, K.D. *et al*, *Inorg. Chem.*, 1980, **19**, 3183 (*ir*, *nmr*)

$C_{54}H_{48}IrP_3$ Ir-00232

Trihydrotris(triphenylphosphine)iridium, 10CI
Tris(hydridotriphenylphosphine)iridium
[16924-01-9]

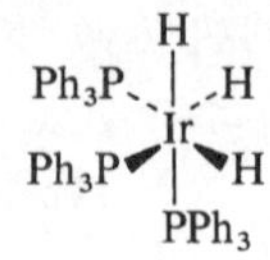

M 982.114
Hydrogenation catalyst for unsatd. compds. and carboxylation catalyst for olefins.

fac-form [18660-46-3]
cis-*form*
Cubes or white microcryst. (C_6H_6/MeOH). Mp 233-5°, 208-18° (sealed under N_2), 168-71° (in air). Isomerises in soln. to give a mixt. of *fac-* and *mer-*isomers.

mer-form [18660-47-4]
trans-*form*
White needles (THF), cryst. (CH_2Cl_2/hexane). Mp 228-30° (sealed under N_2), 192-3° (in air). Isomerises in soln. to give a mixt. of *fac-* and *mer-*isomers. Forms cryst. 1:1 C_6H_6 solvate.

Chatt, J. *et al*, *J. Chem. Soc.*, 1965, 7391 (*synth*, *ir*, *pmr*, *struct*, *bibl*)
Levison, J.J. *et al*, *J. Chem. Soc. (A)*, 1970, 2947 (*synth*, *ir*)
Ahmad, N. *et al*, *J. Chem. Soc., Dalton Trans.*, 1972, 843 (*synth*)
Clark, G.R. *et al*, *Inorg. Chim. Acta*, 1975, **12**, 235 (*cryst struct*)
Longato, B. *et al*, *J. Organomet. Chem.*, 1977, **132**, 291 (*synth*, *ir*)

$C_{55}H_{45}ClIrOSb_3$ Ir-00233

Chlorocarbonyltris(triphenylstibine)iridium
Carbonylchlorotris(triphenylstibine)iridium

$IrCl(CO)(SbPh_3)_3$

M 1314.883
Red cryst. (C_6H_6/pet. ether). Sol. C_6H_6, dioxan, THF. Mp 136° dec. ir ν_{CO} 1962 cm^{-1} (THF).

Hieber, W. *et al*, *Chem. Ber.*, 1966, **99**, 2607 (*synth*, *ir*)

$C_{55}H_{46}IrOP_3$ Ir-00234

Carbonylhydrotris(triphenylphosphine)iridium, 10CI
Carbonylhydridotris(triphenylphosphine)iridium. Hydrocarbonyltris(triphenylphosphine)iridium. Hydridocarbonyltris(triphenylphosphine)iridium
[17250-25-8]

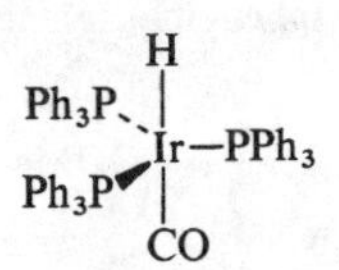

M 1008.109
Hydroformylation, hydrogenation and isomerisation catalyst for olefins. Yellow cryst. Mp 166-7° (in air), 216-8° (under N_2). ir ν_{IrH} 2123, ν_{CO} 1921 cm^{-1} (nujol).

Bath, S.S. *et al*, *J. Am. Chem. Soc.*, 1963, **85**, 3500 (*synth*, *ir*, *pmr*, *struct*)

Vaska, L., *J. Am. Chem. Soc.*, 1966, **88**, 4100 (*ir, struct*)
Ahmad, N. *et al*, *J. Chem. Soc., Dalton Trans.*, 1972, 843 (*synth*)
Inorg. Synth., 1972, **13**, 126 (*synth, ir, pmr*)

$C_{55}H_{48}IrP_3$ Ir-00235

Methyltris(triphenylphosphine)iridium, 9CI
Tris(triphenylphosphine)methyliridium
[38316-79-9]

$(Ph_3P)_3IrMe$ (sq. pl.)

M 994.125
Bright-red cryst. (Et_2O/pentane). Solid v. unstable, Et_2O soln. more stable. V. susceptible to oxidative addn.

Schwartz, J. *et al*, *J. Am. Chem. Soc.*, 1972, **94**, 6226 (*synth, ir*)

$C_{57}H_{46}IrOP_3$ Ir-00236

Carbonylethynyltris(triphenylphosphine)iridium, 9CI
[36154-02-6]

$Ir(CO)(C{\equiv}CH)(PPh_3)_3$

M 1032.131
Pale-yellow cryst. (C_6H_6/EtOH). Readily dissociates in soln., to $[Ir(CO)(C{\equiv}CH)(PPh_3)_2] + PPh_3$. Diamagnetic. ir ν_{CO} 1956 cm^{-1} (nujol). Stable to air and heat.

Nast, R. *et al*, *Chem. Ber.*, 1972, **105**, 1456 (*synth, ir, pmr*)

$C_{60}H_{44}ClIrO_2P_2$ Ir-00237

Chloro[(1,4-dihydro-1,4-dioxo-2,3-naphthalenediylidene)bis(phenylmethylidyne)]bis(triphenylphosphine)iridium, 9CI
[36153-90-9]

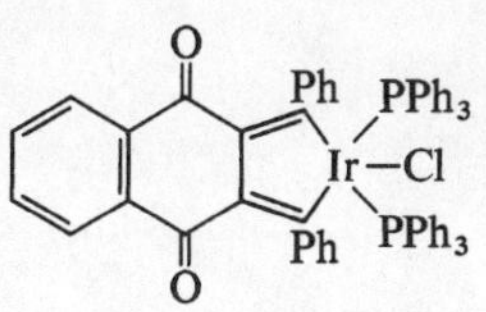

M 1086.627
Reacts with acetylenes to give organic cycloaddition products. Dark-red cryst. (C_6H_6). Mp 150-3°.

Mueller, E. *et al*, *Justus Liebigs Ann. Chem.*, 1971, **754**, 64 (*synth, ir*)
Mueller, E. *et al*, *Chem.-Ztg.*, 1972, **96**, 170 (*use*)

$C_{60}H_{48}Ir_4O_8P_4$ Ir-00238

Octacarbonylbis[μ-[1,2-ethanediylbis[diphenylphosphine]-*P*:*P'*]]tetrairidium, 9CI
Octacarbonylbis[1,2-bis(diphenylphosphino)ethane]tetrairidium
[51900-00-6]

$(CO)_8Ir_4(Ph_2PCH_2CH_2PPh_2)_2$

M 1789.809
Stereochemically nonrigid in soln. Struct. unknown. Orange-yellow air-stable solid. Mp 170-85° dec.

Cattermole, P.E. *et al*, *J. Chem. Soc., Dalton Trans.*, 1974, 328 (*synth, ir*)

$C_{62}H_{64}IrO_4P_4^{\oplus}$ Ir-00239

Bis[[(2,2-dimethyl-1,3-dioxolane-4,5-diyl)bis(methylene)]bis[diphenylphosphine]-*P,P'*]iridium(1+), 10CI
Bis[isopropylidene-2,3-dihydroxy-1,4-bis(diphenylphosphino)butane]iridium(1+). Bis[4,5-bis(diphenylphosphinomethyl)-2,2-dimethyldioxolan]iridium(1+). Bis(DIOP)iridium(1+)

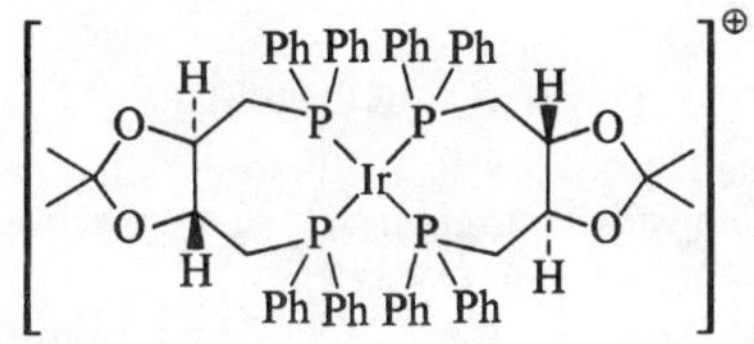

M 1189.300 (ion)
Precursor of optically active hydrido complexes.

(all-R)-form

Tetraphenylborate: [69108-22-1].
$C_{86}H_{84}BIrO_4P_4$ M 1508.532
Deep-red cryst. (Me_2CO/pet. ether). Sol. Me_2CO, CH_2Cl_2. Conductivity 104 ohm^{-1} cm^{-1} (10^{-3} M in Me_2CO).

Sanger, A.R. *et al*, *Inorg. Chim. Acta*, 1978, **31**, L439 (*synth, props*)

$C_{63}H_{45}Ir_4O_9P_3$ Ir-00240

Tri-μ-carbonylhexacarbonyltris(triphenylphosphine)tetrairidium, 10CI
Nonacarbonyltris(triphenylphosphine)tetrairidium, 9CI
[19631-23-3]

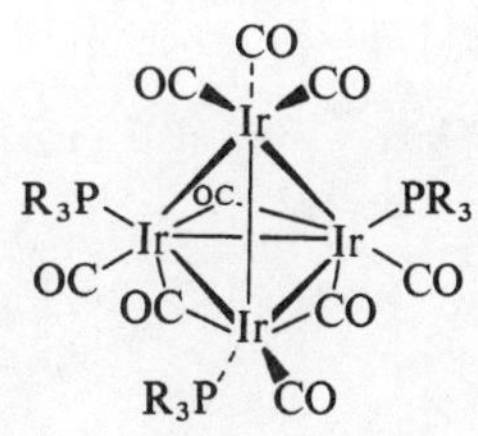

R = Ph

M 1807.844
Polymer supported analogue a catalyst for ethylene and cyclohexene hydrogenation. Yellow-orange cryst. Sol. C_6H_6, Me_2CO, sl. sol. other org. solvs. Diamagnetic.

Malatesta, L. *et al*, *J. Chem. Soc., Chem. Commun.*, 1967, 420 (*synth, ir*)
Albano, V. *et al*, *J. Chem. Soc., Chem. Commun.*, 1967, 730 (*struct*)
Drakesmith, A. *et al*, *J. Chem. Soc., Dalton Trans.*, 1973, 362 (*synth, ir*)
Stuntz, G.F. *et al*, *J. Am. Chem. Soc.*, 1977, **99**, 607 (*struct*)
Lieto, J. *et al*, *J. Catal.*, 1980, **62**, 149 (*use*)
Otero-Schipper, Z. *et al*, *CA*, 1980, **93**, 238307g (*use*)
Sonnenberger, D. *et al*, *Inorg. Chem.*, 1981, **20**, 3243.

$C_{63}H_{45}Ir_4O_{18}P_3$ Ir-00241

Tri-μ-carbonylhexacarbonyltris(triphenyl phosphite-*P*)tetrairidium, 9CI
Nonacarbonyltris(triphenyl phosphite)tetrairidium
[78128-49-1]

$Ir_4(CO)_9[P(OPh)_3]_3$

M 1951.839

No phys. props. recorded.

Sonnenberger, D. *et al*, *Inorg. Chem.*, 1981, **20**, 3243 (*synth*, *ir*)

$C_{80}H_{60}Ir_4O_{20}P_4$ **Ir-00242**

Octacarbonyltetrakis(triphenyl phosphite-*P*)tetrairidium, 10CI

[51900-03-9]

$Ir_4(CO)_8[P(OPh)_3]_4$

M 2234.117

Stereochemically nonrigid in soln. Contains bridging carbonyl groups. Cryst. Mp 178-9°.

Cattermole, P.E. *et al*, *J. Chem. Soc., Dalton Trans.*, 1974, 328 (*synth*, *ir*, *nmr*)

Sonnenberger, D. *et al*, *Inorg. Chem.*, 1981, **20**, 3243 (*synth*, *ir*)

Rh Rhodium

C. White

Rhodium (Fr., Ger.), Rodio (Sp., Ital.), Родий (Rodii) (Russ.), ロジウム (Japan.)

Atomic Number. 45

Atomic Weight. 102.9055

Electronic Configuration. [Kr] $4d^8\ 5s^1$

Oxidation States. Rhodium is normally encountered in organometallic compounds in the oxidation states +1 or +3 although compounds of Rh(0) such as polynuclear rhodium carbonyls and their Group V substituted derivatives are now quite numerous. Oxidation states down to −3 eg. $[Rh(CO)_3]^{3-}$ have been reported; such compounds, however, are usually extremely reactive anions only formed under strongly reducing conditions. Organometallic complexes of Rh(II) such as paramagnetic $Rh(\eta^5\text{-}C_5Me_5)_2$ are rare and compounds of +4 are even rarer. Organometallic complexes of Rh(V) are also rare although some interesting examples such as $[Rh(\eta^5\text{-}C_5Me_5)(CH_2)]_2(\mu\text{-}CH_2)_2$ have recently been reported.

An important feature of the chemistry of rhodium is the easy oxidative addition to four-coordinate rhodium(I) and the reverse reaction, reductive elimination from octahedral rhodium(III). The reversibility of such reactions connecting the Rh(I) and Rh(III) oxidation states is largely responsible for the ability of organorhodium compounds to catalyse a wide range of organic reactions.

Coordination Number. The d^{10} electronic configuration of Rh(−I) normally gives rise to tetrahedral anions such as $[Rh(CO)_4]^-$. Rh(0) (d^9) may form tetrahedral, monomeric, paramagnetic compounds; usually, however, such organometallic species eg. $Rh(CO)_4$ are unstable and oligomerization occurs via metal—metal bond formation and/or bridging carbonyl ligands to give diamagnetic polynuclear species eg. $Rh_4(CO)_{12}$. Rh(I) (d^8) normally gives rise to four-coordinate square planar compounds eg. $RhCl(PPh_3)_3$ or to five-coordinate trigonal bipyramidal structures eg. $RhH(CO)(PPh_3)_3$. Often interconversion is possible between such four and five coordinate complexes. Ligand dissociation from 18-electron five-coordinate Rh(I) frequently proposed as a means of generating an active site in catalytic reactions.

Organometallic compounds containing Rh(III) are usually octahedral.

Bulky ligands may cause deviations from normal coordination numbers and geometries. Thus, $RhH(PBu_3^t)_2$ is a rare example of a three-coordinate rhodium(I) complex.

Colour. Although there are striking exceptions such as the royal blue compound $[Rh(\eta^5\text{-}C_5Me_5)(\mu\text{-}CO)]_2$ and the dark green anion $[Rh_7(CO)_{16}]^{3-}$, most organorhodium compounds are yellow, orange, red or brown.

Availability. The most important starting material is rhodium trichloride. The anhydrous form tends to be insoluble and unreactive and the readily available dark red deliquescent material $RhCl_3{\cdot}nH_2O$ (n ~ 3) is preferred. $RhBr_3{\cdot}2H_2O$ and RhI_3 are also commercially available. A useful starting material where halide cannot be tolerated is tetrakis(μ-acetato)-dirhodium(II) $[Rh(OAc)_2]_2$. Tetracarbonyldi-μ-chlorodirhodium, $[RhCl(CO)_2]_2$, dodecacarbonyltetrarhodium $Rh_4(CO)_{12}$ and hexacarbonylhexarhodium $Rh_6(CO)_{16}$ are useful precursors to rhodium carbonyl derivatives; they are available commercially but are very expensive. Rhodium compounds used to catalyse organic reactions can now often be purchased; although expensive only small quantities are usually required. Those available include $RhCl(PPh_3)_3$, $HRh(CO)(PPh_3)_3$, $[RhCl(C_8H_{12})]_2$, $[RhCl(C_7H_8)]_2$ and dicarbonyl(3-trifluoroacetyldicamphorato)rhodium $Rh(C_{12}H_{14}F_3O_2)(CO)_2$.

Rh is an expensive precious metal whose price fluctuates on the world market. Recovery of rhodium residues and reconversion to $RhCl_3$ is often a worthwhile procedure: *Inorg. Synth.*, **7**, 214.

Handling. Most Rh(III) compounds and many Rh(I) compounds are air-stable. However, there are many instances of catalytic studies involving "air-stable" rhodium complexes in which irreproducible behaviour is reported. Hence it is strongly recommended that all rhodium compounds to be used for quantitative catalytic studies should be stored in an inert atmosphere and freshly recrystallised before use.

Handling in an inert atmosphere becomes essential for the lower states of rhodium.

Toxicity. The toxicity of the majority of rhodium compounds has not been investigated in detail. Rhodium compounds in general do not appear to be extremely toxic although a few compounds have been shown to exhibit anti-tumour activity; given the mode of action of such anti-tumour agents rhodium compounds should be regarded as potentially carcinogenic.

Isotopic Abundance. ^{103}Rh, 100%.

Spectroscopy. ^{103}Rh has nuclear spin $I = \frac{1}{2}$. In addition, most organorhodium compounds are diamagnetic and thus ^{103}Rh nmr is potentially a powerful spectroscopic tool. Unfortunately, the low value of the gyromagnetic ratio means that the resonance frequency is below that of most multinuclear spectrometers (ie. 3.185 MHz: ^1H TMS 100 MHz); also the receptivity of ^{103}Rh is only 0.177 relative to ^{13}C. To date therefore ^{103}Rh nmr has been rather limited although the situation is improving rapidly. Compilations of ^{103}Rh and ^{13}C nmr data of organorhodium compounds have been published and developments in these fields are reviewed annually.

Analysis. A variety of gravimetric and titrimetric methods have been used to analyse for rhodium but these have largely been replaced by spectroscopic methods, particularly atomic absorption and emission spectroscopy. A simple technique consists of combusting the organometallic compound at 800°C and weighing the Rh_2O_3 residue.

References

General

Griffith, W. P., *The Chemistry of the Rarer Platinum Metals*, Wiley-Interscience, London, 1967.

Collman, J. P. *et al.*, *Adv. Organomet. Chem.*, 1968, **7,** 53.

Livingstone, S. E., *The Chemistry of Ruthenium, Rhodium, Palladium, Osmium, Iridium and Platinum*, Pergamon, Oxford, 1973.

Geoffroy, G. L. *et al.*, *Adv. Inorg. Chem. Radiochem.*, 1977, **20,** 190.

Parshall, G. W., *Catal. Rev.*, 1981, **23,** 107.

Dickson, R. S., *Organometallic Chemistry of Rhodium and Iridium*, Academic Press, New York, 1983.

Analysis

Beamish, F. E., *Analytical Chemistry of the Noble Metals*, Pergamon, Oxford, 1966.

Van Loon, J. C., *Pure Appl. Chem.*, 1977, **49,** 1495.

Gel'man, N. E. *et al.*, *Usp. Anal. Khim.*, 1974, 297.

$RhBr_3CO$

Rh-00001

$RhBr_5CO^{\ominus\ominus}$ (O_h)

Rh-00002

$RhCl_3CO$

Rh-00003

$RhCl_5CO^{\ominus\ominus}$ (O_h)

Rh-00004

RhI_3CO

Rh-00005

$RhI_5CO^{\ominus\ominus}$ (O_h)

Rh-00006

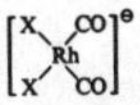

X = Br

Rh-00007

$RhBr_4(CO)_2^{\ominus}$ (O_h)

Rh-00008

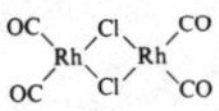

Rh-00009

As Rh-00007 with
X = Cl

Rh-00010

$RhCl_4(CO)_2^{\ominus}$ (O_h)

Rh-00011

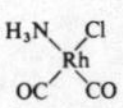

Rh-00012

As Rh-00007 with
X = I

Rh-00013

$RhI_4(CO)_2^{\ominus}$ (O_h)

Rh-00014

$Rh(CO)_3Br$

Rh-00015

$Rh(CO)_3Cl$

Rh-00016

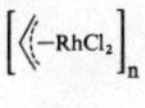

Rh-00017

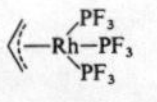

Rh-00018

$Rh(CO)_3I$

Rh-00019

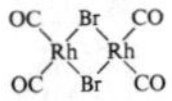

Rh-00020

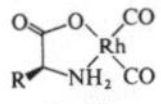

R = H

Rh-00021

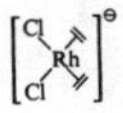

Rh-00022

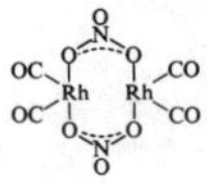

Rh-00023

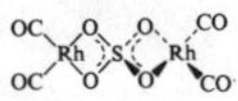

Rh-00024

$RhH(CN)_5^{\ominus\ominus\ominus}$

Rh-00025

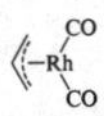

Rh-00026

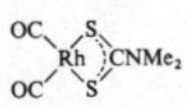

Rh-00027

As Rh-00021 with
R = CH_3

Rh-00028

$RhCl(CO)(SMe_2)_2$

Rh-00029

$(Me_2SO)_2RhCl(CO)$

Rh-00030

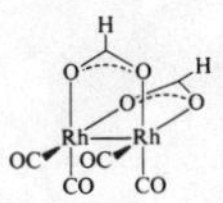

Rh-00031

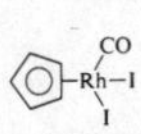

Rh-00032

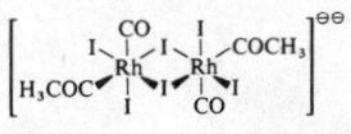

Rh-00033

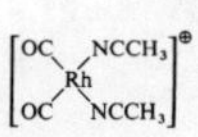

Rh-00034

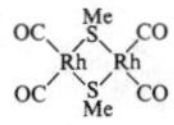

Rh-00035

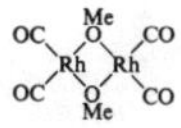

Rh-00036

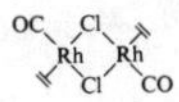

Rh-00037

$RhI_3(CNMe)_3$

Rh-00038

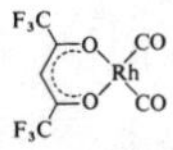

Rh-00039

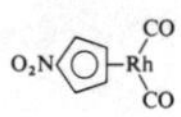

Rh-00040

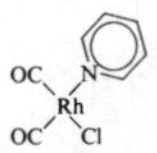

Rh-00041

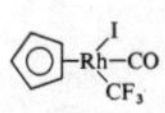

Rh-00042

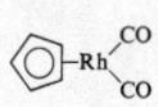

Rh-00043

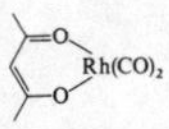

Rh-00044

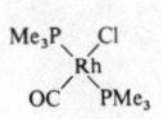

Rh-00045

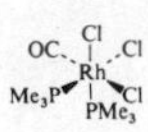

fac-form

Rh-00046

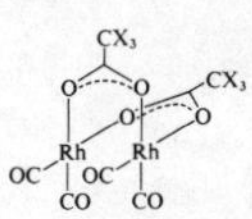

X = F

Rh-00047

As Rh-00047 with
X = H

Rh-00048

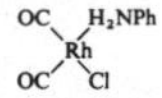

Rh-00049

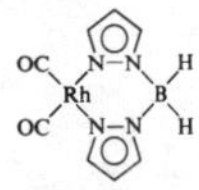

Rh-00050

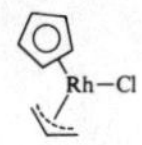

Rh-00051

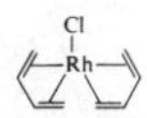

Rh-00052

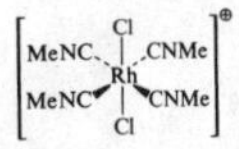

Rh-00053

$[Rh(CNMe)_4]^{\oplus}$ (D_{4h})

Rh-00054

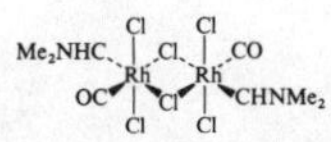

Rh-00055

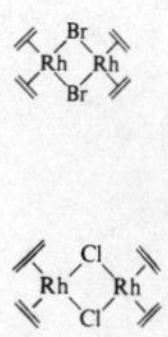

Rh-00056

Rh-00057

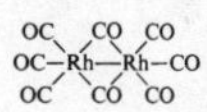

Rh-00058

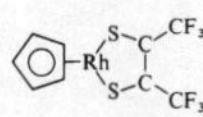

Rh-00059

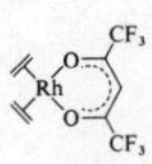

Rh-00060

Rh-00061

Rh-00062

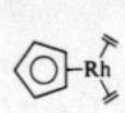

Rh-00063

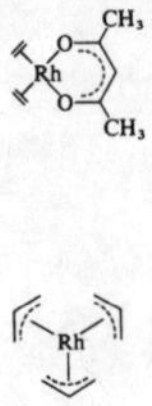

Rh-00064

Rh-00065

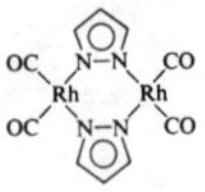

Rh-00066

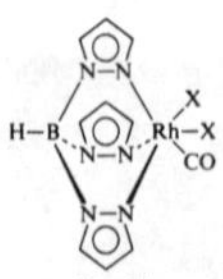

X = Cl

Rh-00067

As Rh-00067 with
X = I

Rh-00068

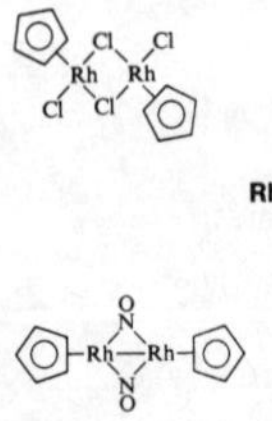

Rh-00069

Rh-00070

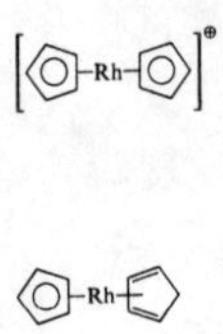

Rh-00071

Rh-00072

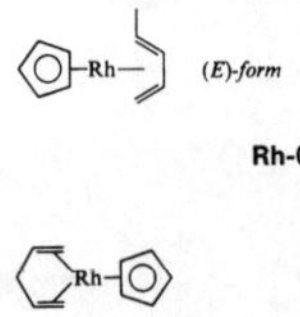

Rh-00073

Rh-00074

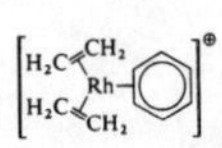

Rh-00075

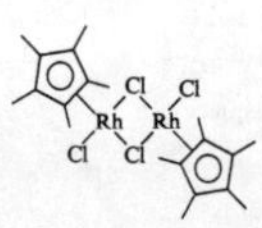

Rh-00076

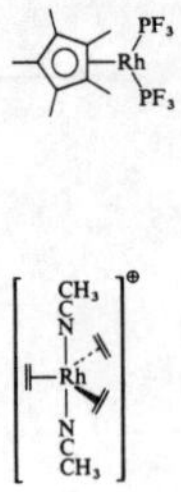

Rh-00077

Rh-00078

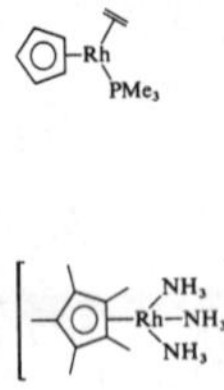

Rh-00079

Rh-00080

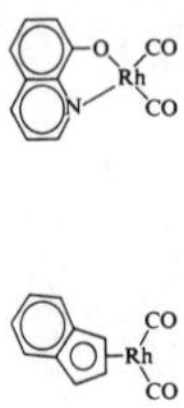

Rh-00081

Rh-00082

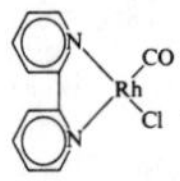

Rh-00083

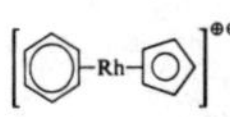

Rh-00084

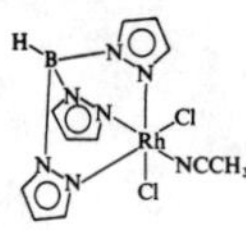

Rh-00085

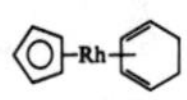

Rh-00086

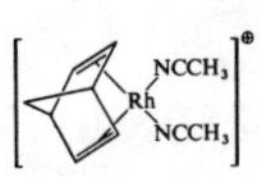

Rh-00087

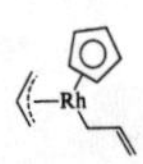

Rh-00088

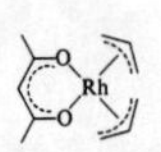

Rh-00089

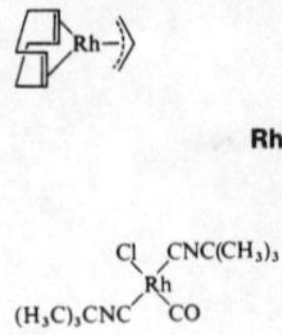

Rh-00090

Rh-00091

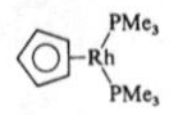

Rh-00092

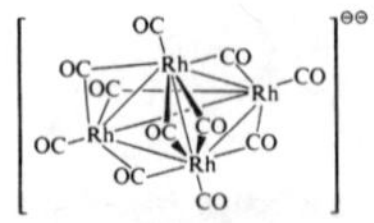

Rh-00093

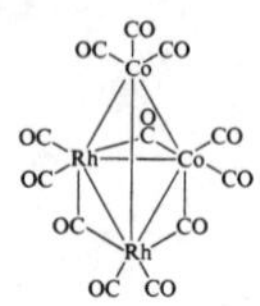

Rh-00094

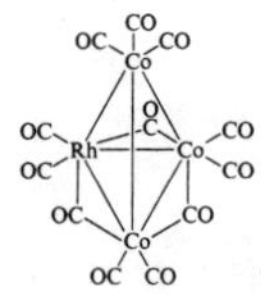

Rh-00095

$RhOs_2H(CO)_{12}$

Rh-00096

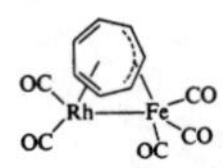

Rh-00097

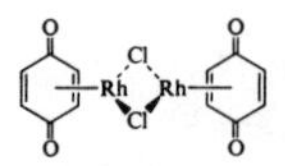

Rh-00098

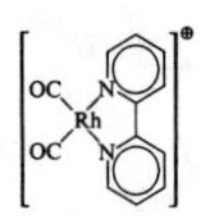

Rh-00099

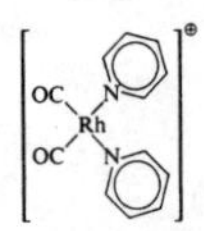

Rh-00100

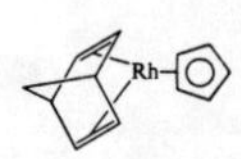

Rh-00101

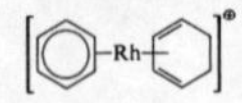

Rh-00102

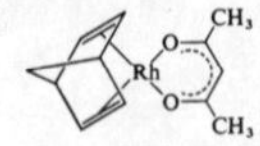

Rh-00103

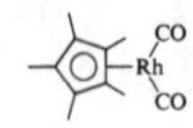

Rh-00104

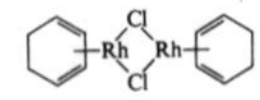

Rh-00105

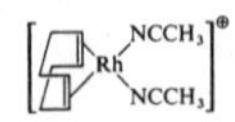

Rh-00106

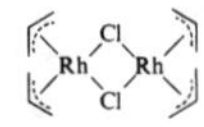

Rh-00107

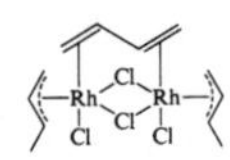

Rh-00108

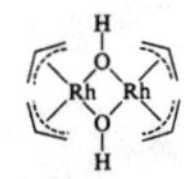

Rh-00109

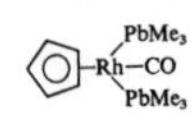

Rh-00110

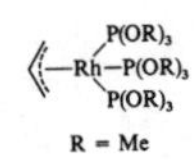

Rh-00111

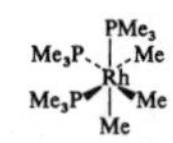

Rh-00112

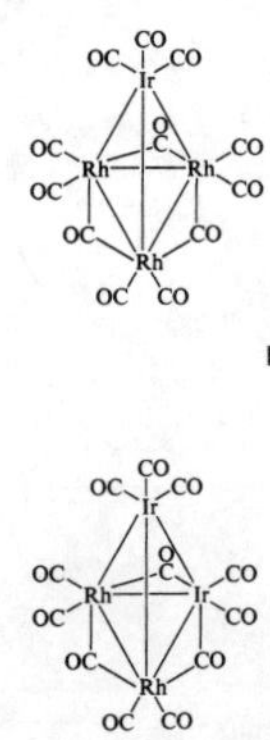

Rh-00113

Rh-00114

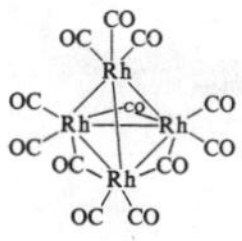

Rh-00115

$[(OC)_{11}Rh_4(COOMe)]^{\ominus}$

Rh-00116

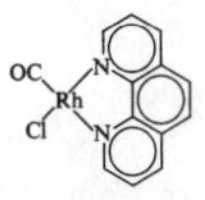

Rh-00117

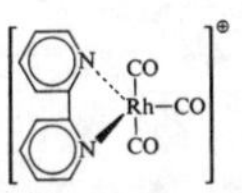

Rh-00118

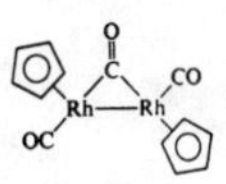

Rh-00119

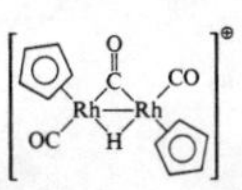

Rh-00120

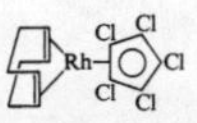

Rh-00121

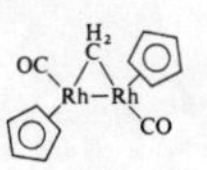

Rh-00122

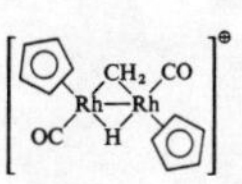

Rh-00123

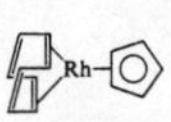

Rh-00124

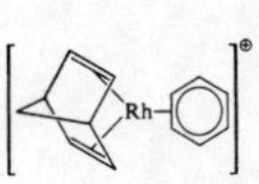

Rh-00125

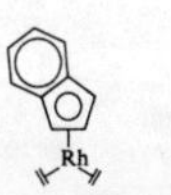

Rh-00126

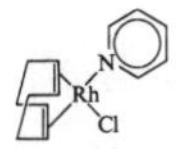

Rh-00127

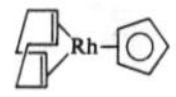

Rh-00128

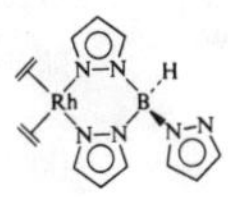

Rh-00129

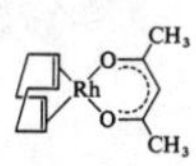

Rh-00130

$Rh(HCONMe_2)Py_2Cl_2(BH_4)$

Rh-00131

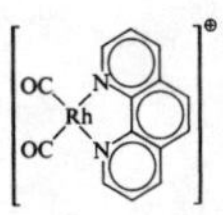

Rh-00132

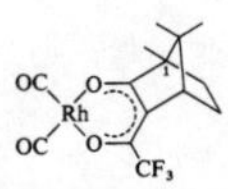

(1R)-form

Rh-00133

$[Rh(\eta^5\text{-}C_5(CH_3)_5)(OOCCF_3)_2H_2O]_n$

Rh-00134

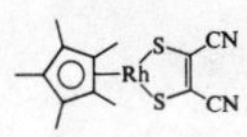

Rh-00135

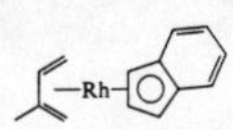

Rh-00136

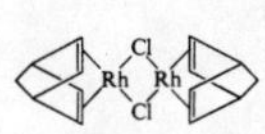

Rh-00137

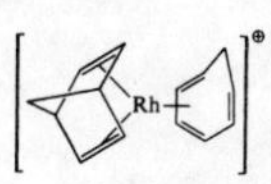

Rh-00138

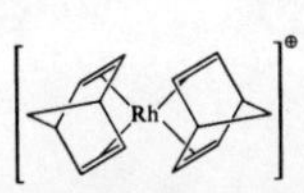

Rh-00139

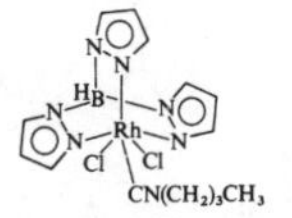

Rh-00140

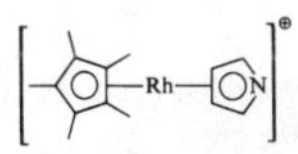

Rh-00141

$[Rh(\eta^5\text{-}C_5(CH_3)_5)(OAc)_2H_2O]_n$

Rh-00142

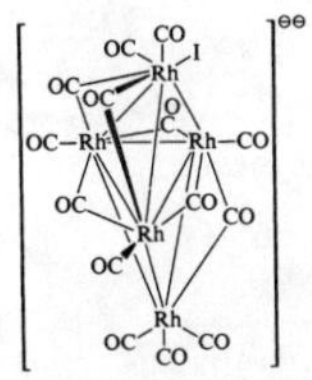

Rh-00143

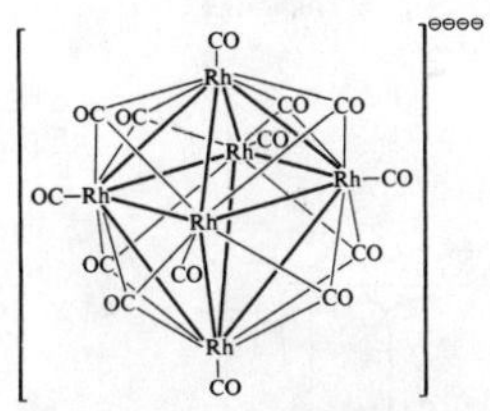

Rh-00144

As Rh-00146 with
X = Cl

Rh-00145

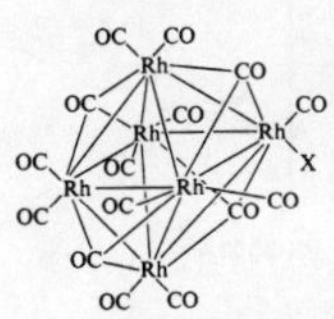

X = H

Rh-00146

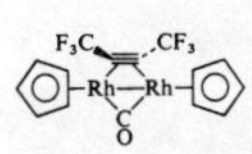

Rh-00147

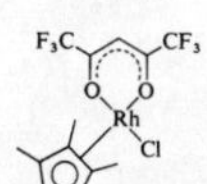

Rh-00148

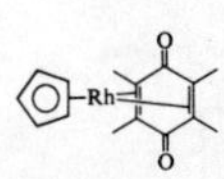

Rh-00149

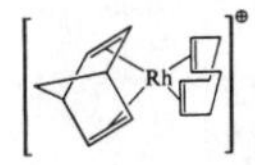

Rh-00150

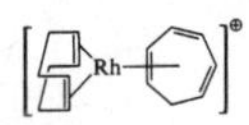

Rh-00151

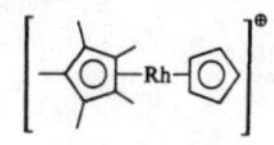

Rh-00152

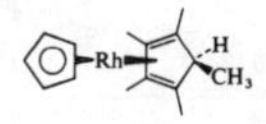

endo-form

Rh-00153

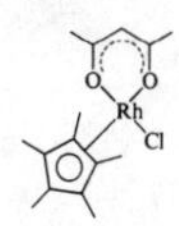

Rh-00154

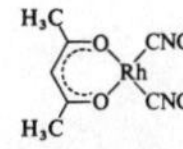

Rh-00155

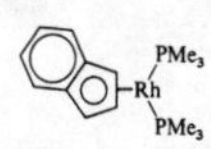

Rh-00156

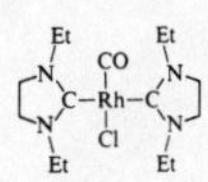

Rh-00157

$PhRh(PMe_3)_3$

Rh-00158

$Rh[P(OMe)_3]_5^{\oplus}$

Rh-00159

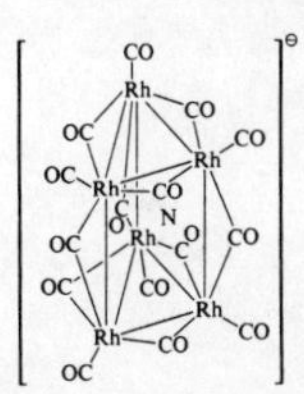

Rh-00160

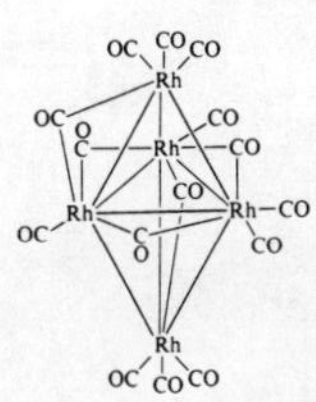

Rh-00161

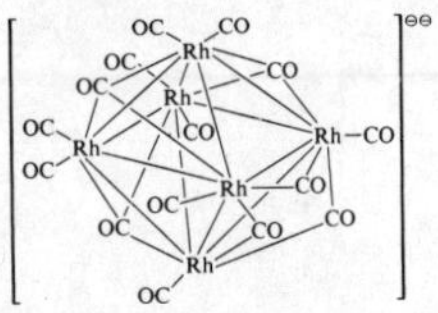

Rh-00162

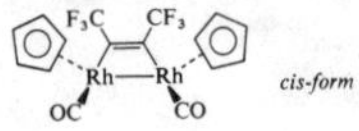

Rh-00163

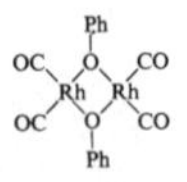

Rh-00164

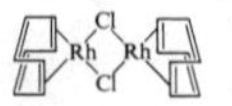

Rh-00165

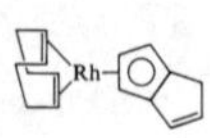

Rh-00166

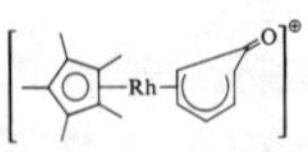

Rh-00168

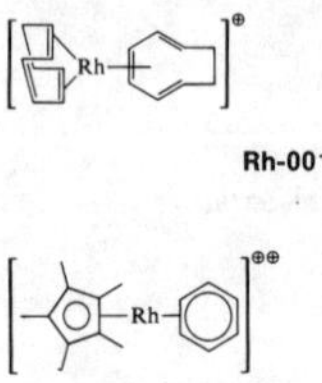

Rh-00169

Rh-00170

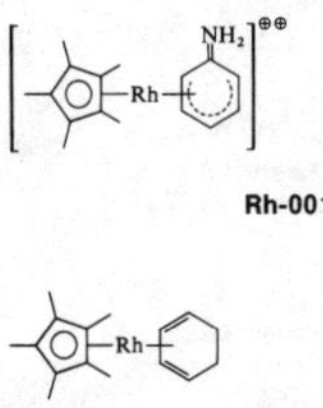

Rh-00171

Rh-00172

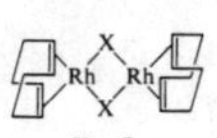

Rh-00173

As Rh-00173 with
X = Cl

Rh-00174

As Rh-00173 with
X = I

Rh-00175

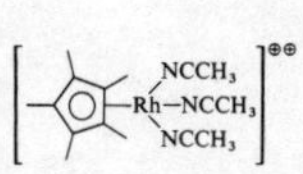

Rh-00176

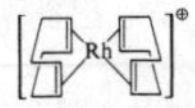

Rh-00177

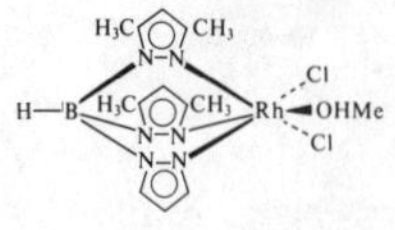

Rh-00178

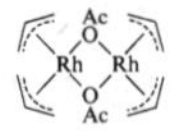

Rh-00179

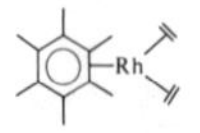

Rh-00180

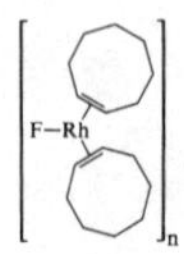

Rh-00181

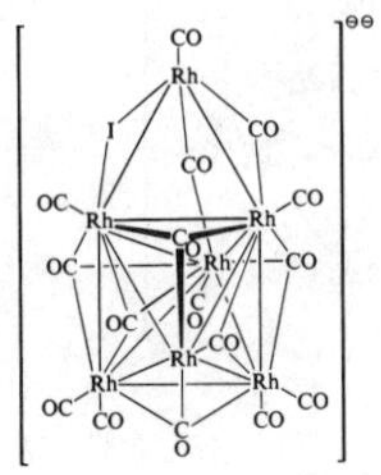

Rh-00182

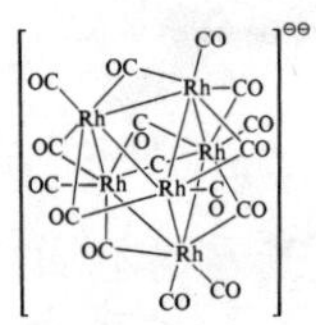

Rh-00183

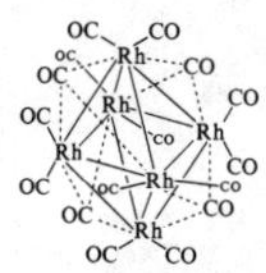

Rh-00184

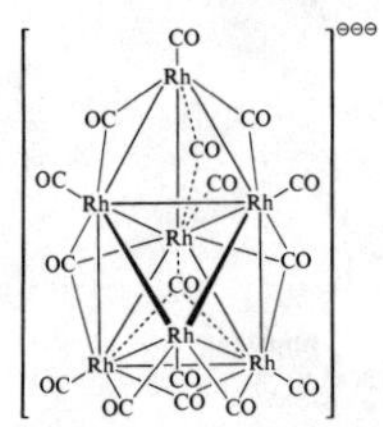

Rh-00185

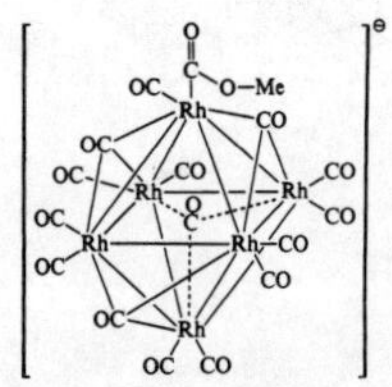

Rh-00186

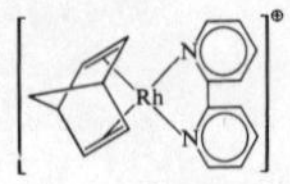

Rh-00187

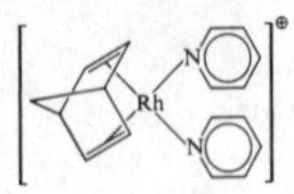

Rh-00188

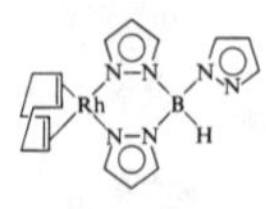

Rh-00189

$RhCl(PMe_2Ph)_2(CO_2)$

Rh-00190

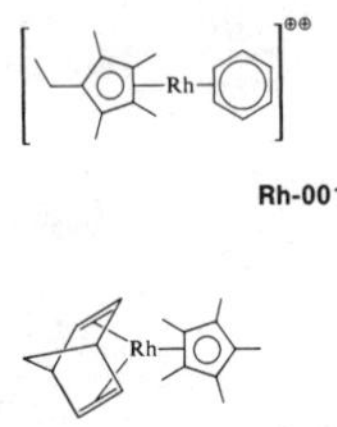

Rh-00191

Rh-00192

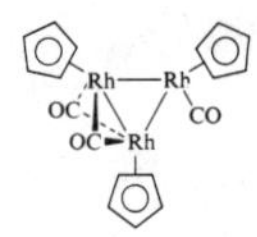

Rh-00193

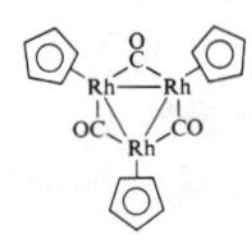

Rh-00194

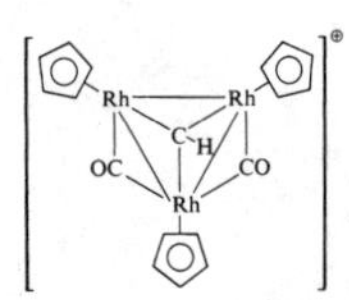

Rh-00195

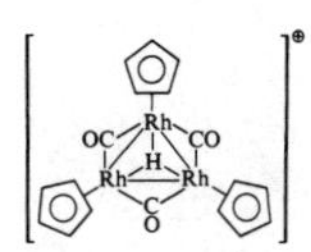

Rh-00196

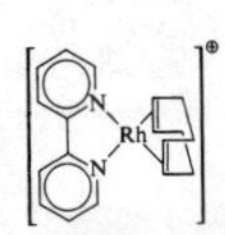

Rh-00197

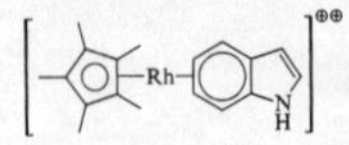

Rh-00198

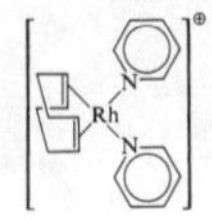

Rh-00199

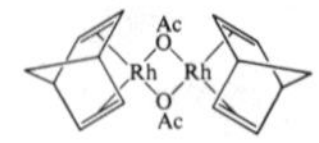

Rh-00200

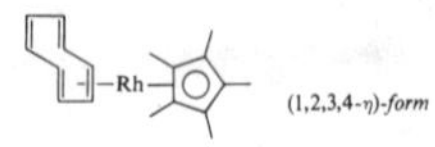

Rh-00201

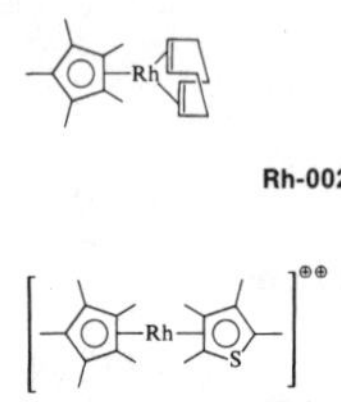

Rh-00202

Rh-00203

As Rh-00173 with
X = OMe

Rh-00204

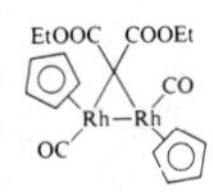

Rh-00205

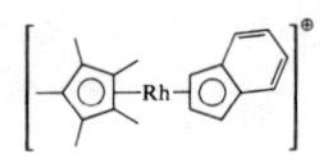

Rh-00206

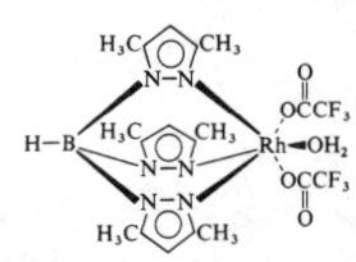

Rh-00207

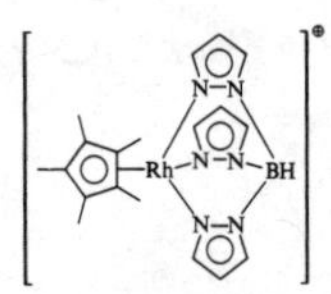

Rh-00208

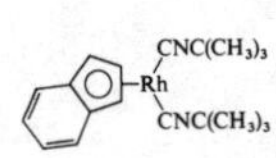

Rh-00209

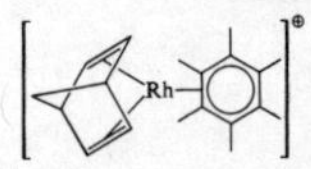

Rh-00210

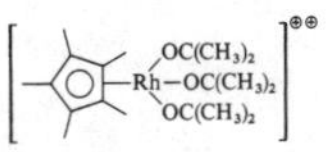

Rh-00211

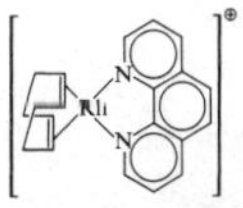

Rh-00212

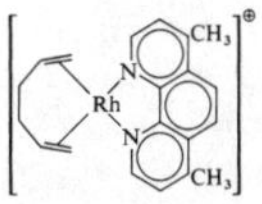

Rh-00213

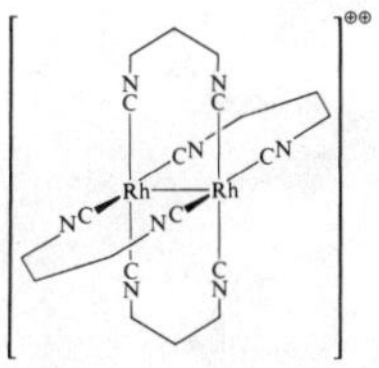

Rh-00214

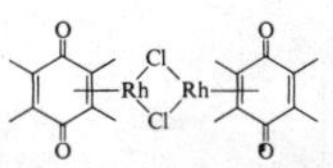

Rh-00215

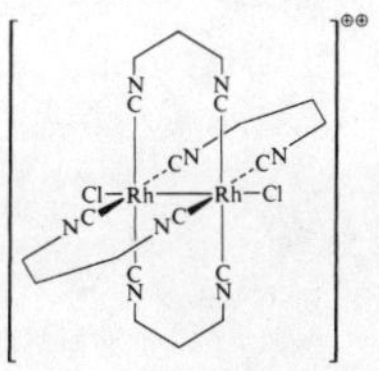

Rh-00216

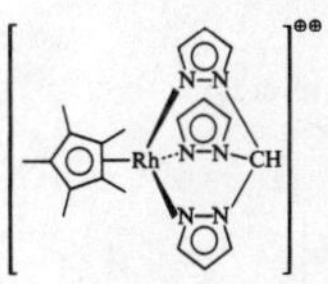

Rh-00217

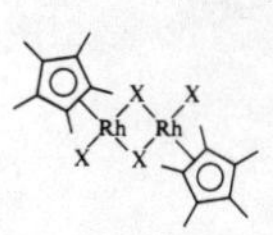

Rh-00218

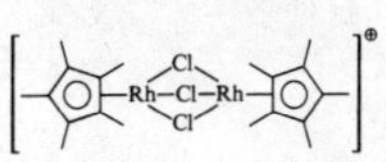

Rh-00219

As Rh-00218 with
X = I

Rh-00220

As Rh-00173 with
X = OAc

Rh-00221

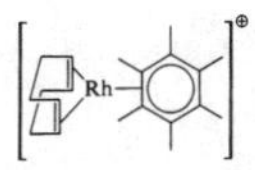

Rh-00222

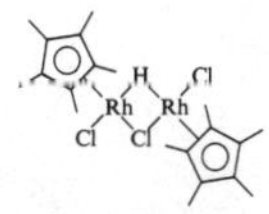

Rh-00223

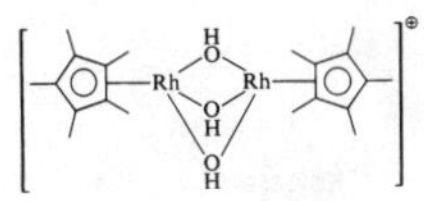

Rh-00224

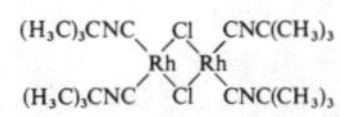

Rh-00225

$Rh[CNC(CH_3)_3]_4^{\oplus}$ (D_{4h})

Rh-00226

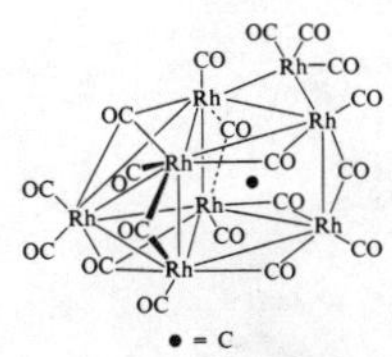

Rh-00227

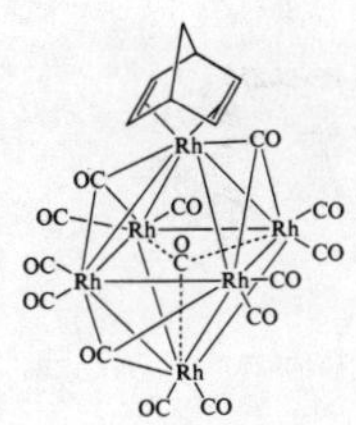

Rh-00228

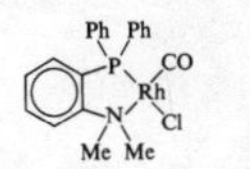

Rh-00229

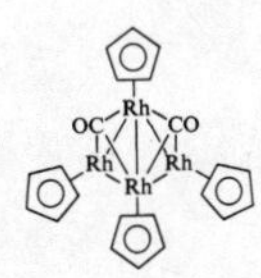

Rh-00230

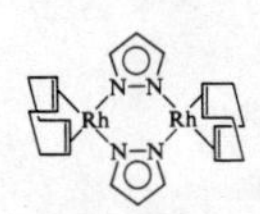

Rh-00231

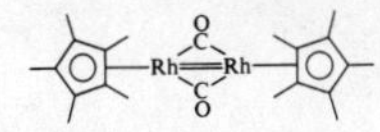

Rh-00232

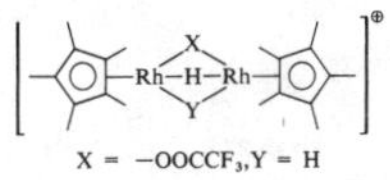

Rh-00233

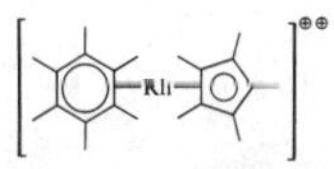

Rh-00234

As Rh-00233 with
X = OAc, Y = H

Rh-00235

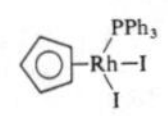

Rh-00236

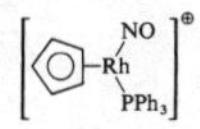

Rh-00237

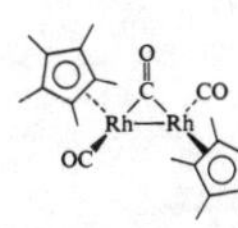

Rh-00238

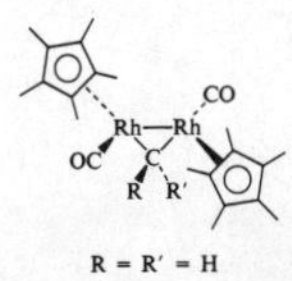

Rh-00239

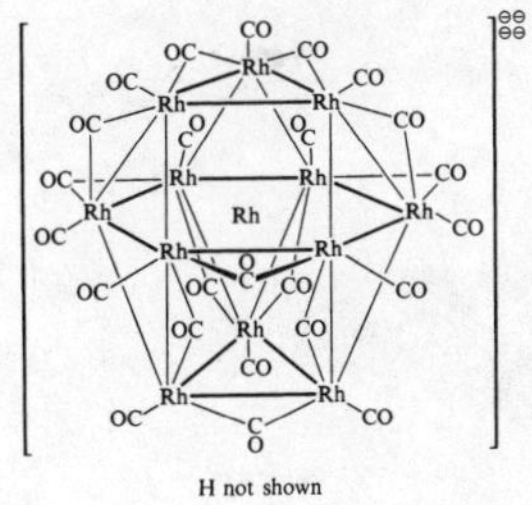

Rh-00240

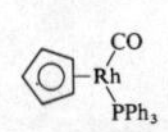

Rh-00243

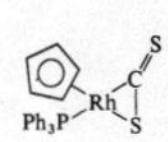

Rh-00244

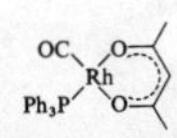

Rh-00245

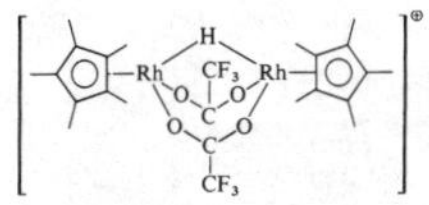

Rh-00246

As Rh-00239 with
R = CH_3, R′ = H

Rh-00247

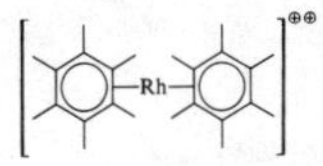

Rh-00248

As Rh-00233 with
X = Y = OAc

Rh-00249

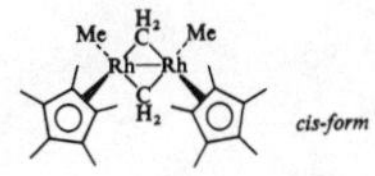

Rh-00250

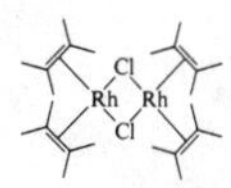

Rh-00251

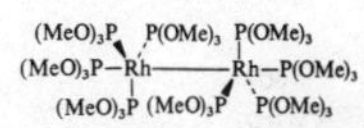

Rh-00252

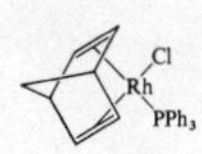

Rh-00253

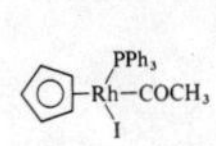

Rh-00254

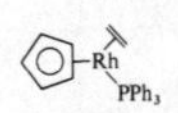

Rh-00255

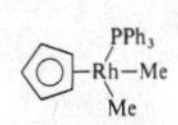

Rh-00256

As Rh-00239 with
R = R′ = CF_3

Rh-00257

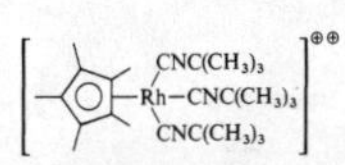

Rh-00258

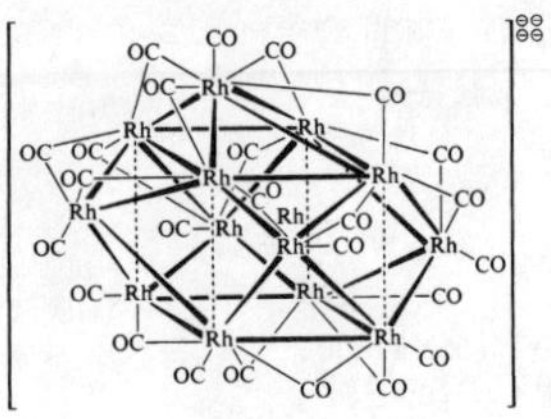

Rh-00259

Rh-00260

Rh-00261

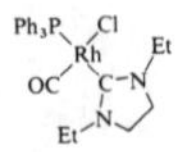

Rh-00262

As Rh-00261 with
R = cyclohexyl

Rh-00263

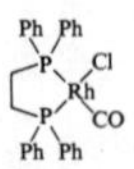

Rh-00264

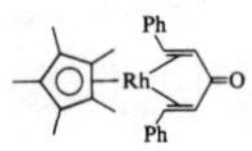

Rh-00265

HRh[P[CH(CH$_3$)$_2$]$_3$]$_3$

Rh-00266

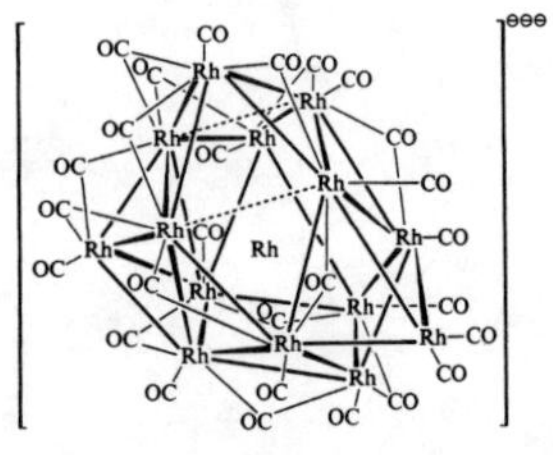

Rh-00267

[(PhNC)$_4$Rh—Rh(CNPh)$_4$]$^{\oplus\oplus}$ (D_{4d})

Rh-00268

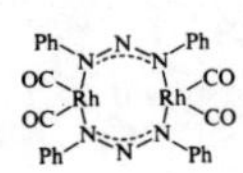

Rh-00269

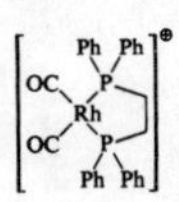

Rh-00270

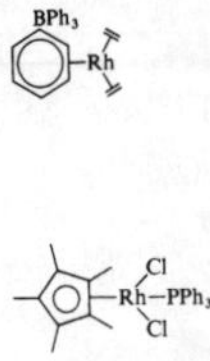

Rh-00271

Rh-00272

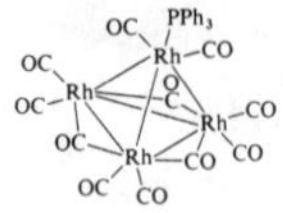

Rh-00273

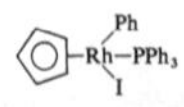

Rh-00274

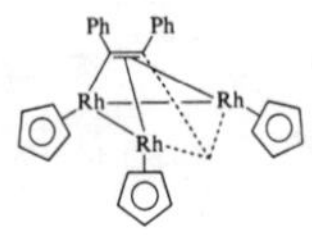

Rh-00275

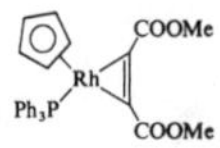

Rh-00276

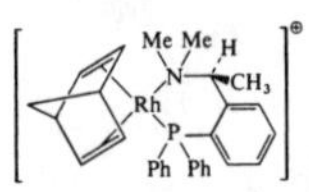

Rh-00277

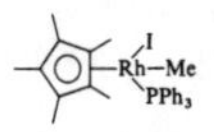

Rh-00278

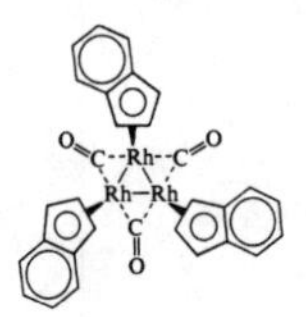

Rh-00279

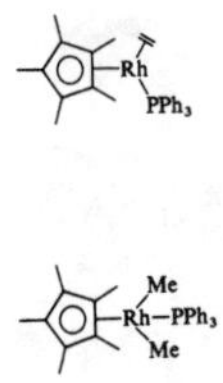

Rh-00280

Rh-00281

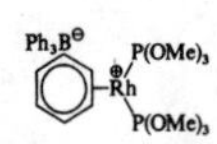

Rh-00282

As Rh-00111 with
R = –CH(CH$_3$)$_2$

Rh-00283

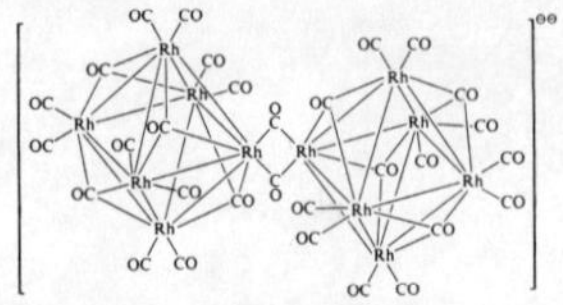

Rh-00284

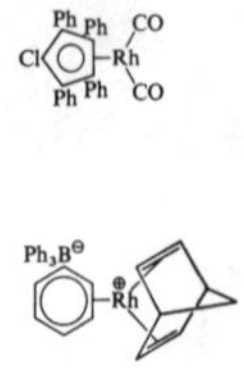

Rh-00285

Rh-00286

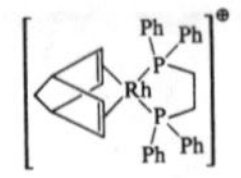

Rh-00287

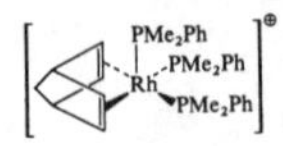

Rh-00288

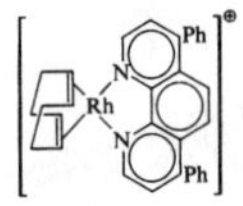

Rh-00289

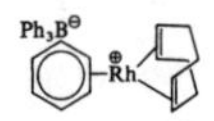

Rh-00290

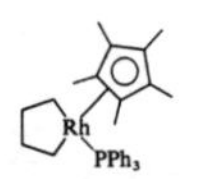

Rh-00291

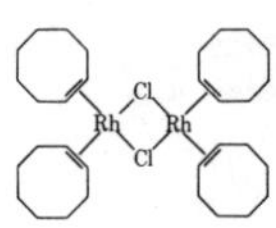

cis-form

Rh-00292

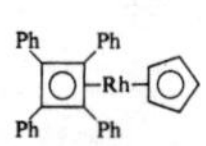

Rh-00293

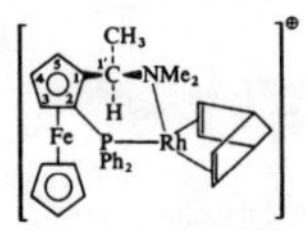

Rh-00294

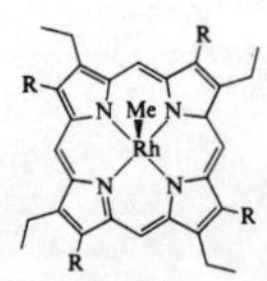

R = CH$_3$

Rh-00295

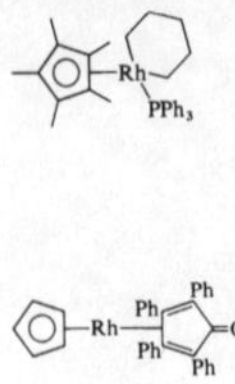

Rh-00296

Rh-00297

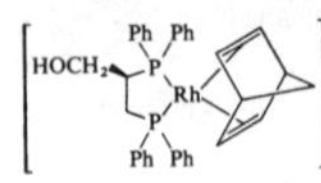

(*R*)-*form*

Rh-00298

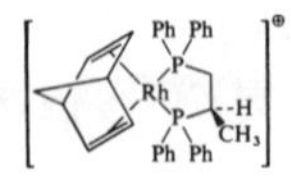

Rh-00299

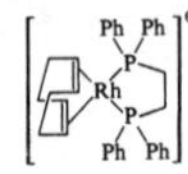

Rh-00300

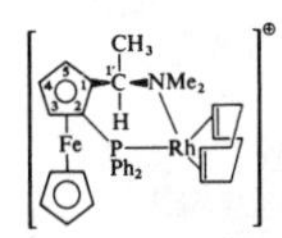

(1*S*,1'*R*)-*form*

Rh-00301

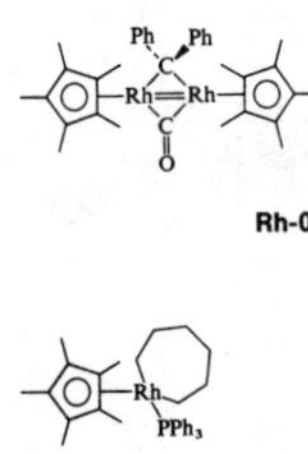

Rh-00302

Rh-00303

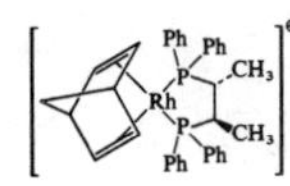

(2*R*,3*R*)-*form*

Rh-00304

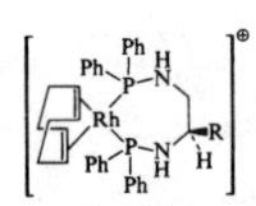

R = CH$_3$

Rh-00305

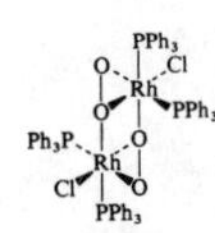

Dimer

Rh-00306

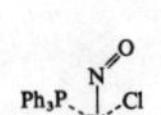

Rh-00307

$(Ph_3P)_2RhI_2(NO)$

Rh-00308

$(Ph_3P)_2Rh(NO)_2^{\oplus}$

Rh-00309

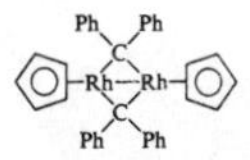

Rh-00310

$RhClH_2(PPh_3)_2$

Rh-00311

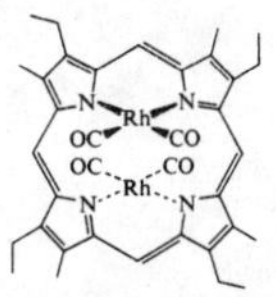

Rh-00312

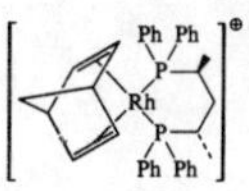

Rh-00313

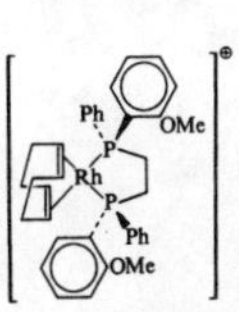

Rh-00314

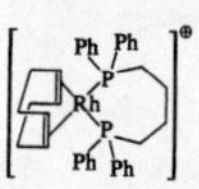

Rh-00315

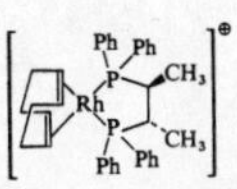

Rh-00316

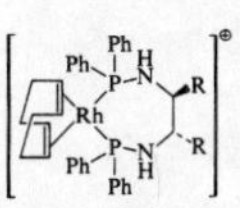

(S,S)-form

R = CH_3

Rh-00317

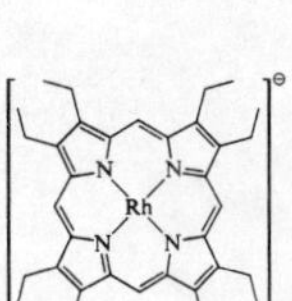

Rh-00318

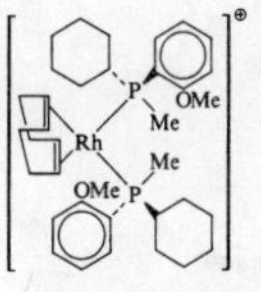

Rh-00319

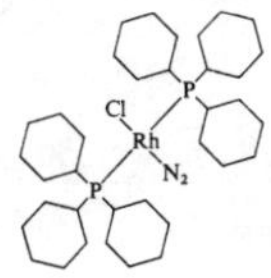

Rh-00320

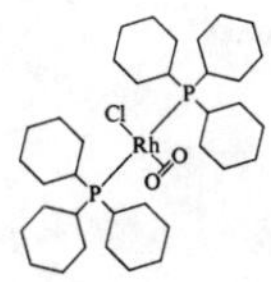

Rh-00321

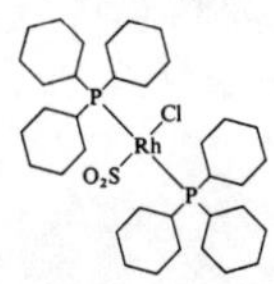

Rh-00322

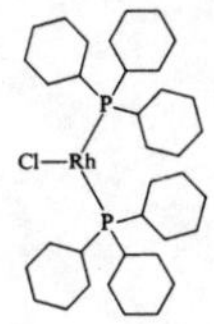

Rh-00323

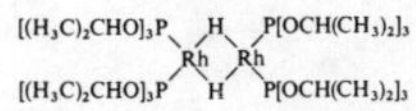

Rh-00324

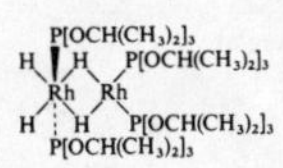

Rh-00325

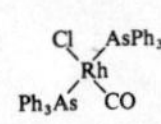

trans-form

Rh-00326

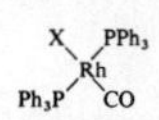

X = Br

Rh-00327

$Rh(PPh_3)_2(CO)Cl$

Rh-00328

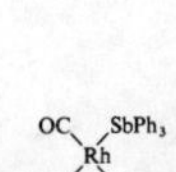

Rh-00329

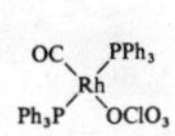

Rh-00330

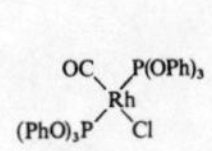

Rh-00331

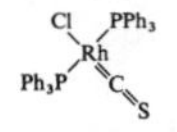

Rh-00332

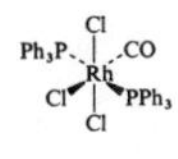

Rh-00333

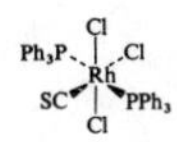

Rh-00334

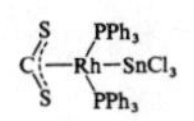

Rh-00335

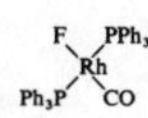

trans-form

Rh-00336

As Rh-00327 with
X = I

Rh-00337

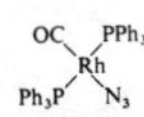

Rh-00338

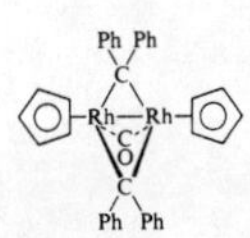

Rh-00339

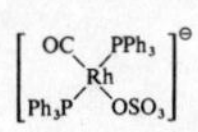

Rh-00340

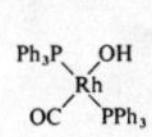

Rh-00341

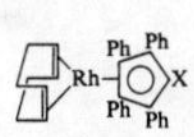

X = Br

Rh-00342

As Rh-00342 with
X = Cl

Rh-00343

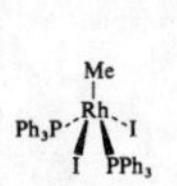

Rh-00344

$(Ph_3P)_2Rh(CO)(BH_4)$

Rh-00345

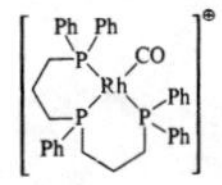

Rh-00346

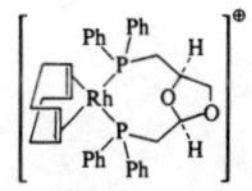

(2R,4R)-form

Rh-00347

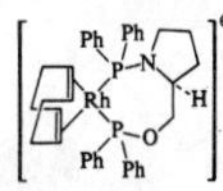

Rh-00348

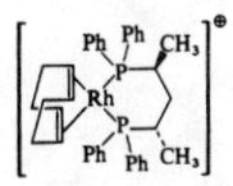

Rh-00349

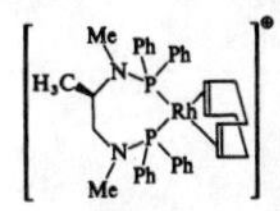

Rh-00350

As Rh-00295 with
R = CH_2CH_3

Rh-00351

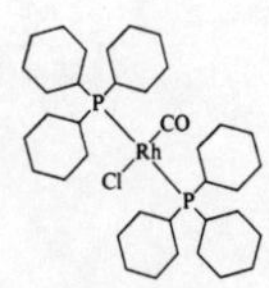

Rh-00352

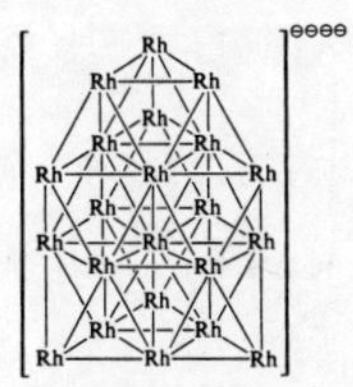

Rh skeleton only shown

Rh-00353

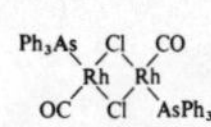

Rh-00354

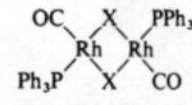

trans-form

X = Br

Rh-00355

As Rh-00355 with
X = Cl

Rh-00356

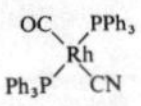

Rh-00357

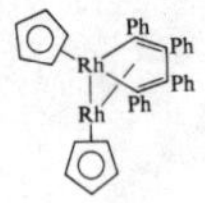

Rh-00358

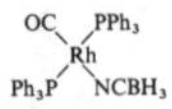

Rh-00359

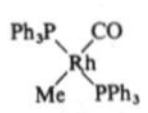

Rh-00360

Rh-00361

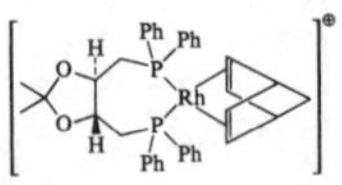

(R,R)-form

Rh-00362

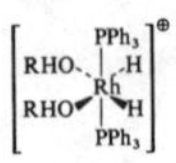

R = Me

Rh-00363

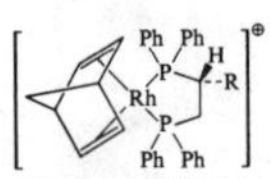

R = $-CH_2OC(CH_3)_3$

Rh-00364

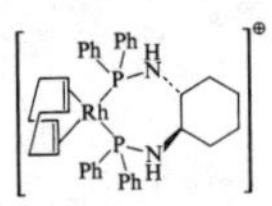

(R,R)-form

Rh-00365

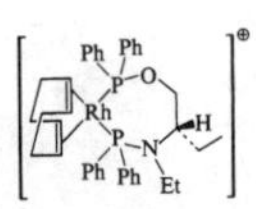

Rh-00366

$(Ph_3P)_2Rh(CO)(OOCCF_3)$

Rh-00367

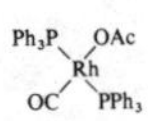

Rh-00368

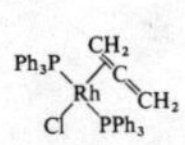

Rh-00369

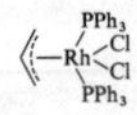

Rh-00370

$RhCl(PMePh_2)_3$

Rh-00371

As Rh-00364 with
R = Cyclohexyl

Rh-00372

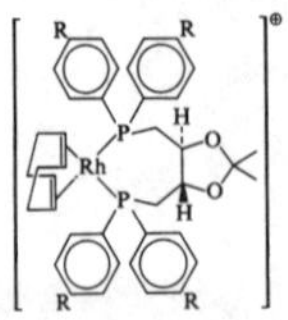

R = H

Rh-00373

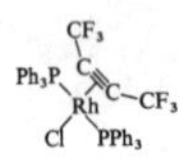

Rh-00374

$Rh(C{\equiv}CCH_3)(CO)(PPh_3)_2$

Rh-00375

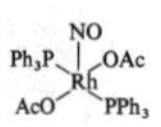

Rh-00376

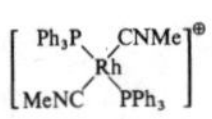

Rh-00377

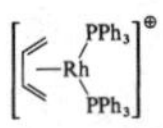

Rh-00378

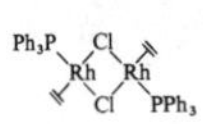

Rh-00379

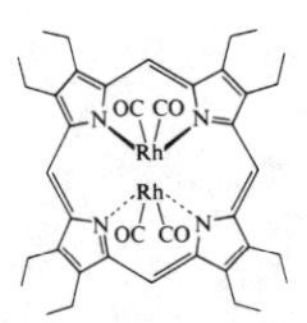

Rh-00380

As Rh-00382 with
n = 5

Rh-00381

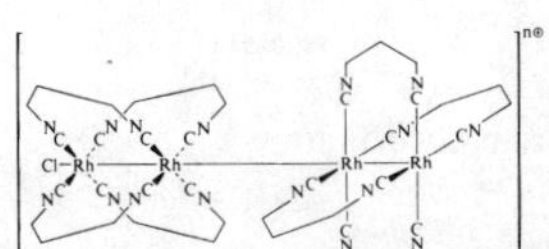

n = 5

Rh-00382

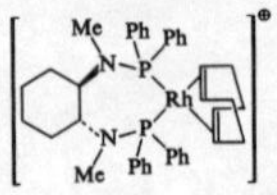

Rh-00383

As Rh-00382 with
n = 6

Rh-00384

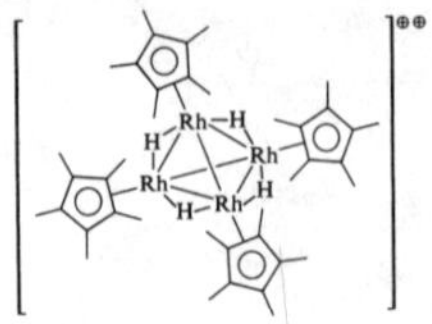

Rh-00385

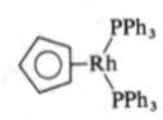

Rh-00386

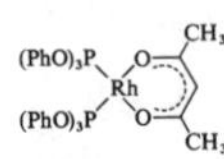

Rh-00387

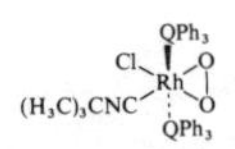

Q = As

Rh-00388

As Rh-00388 with
Q = P

Rh-00389

As Rh-00364 with
R = CH_2OCH_2Ph

Rh-00390

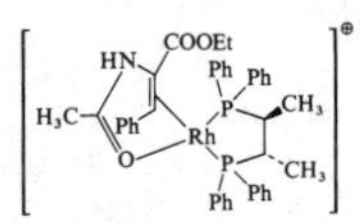

Rh-00391

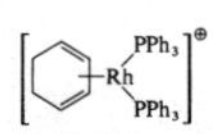

Rh-00392

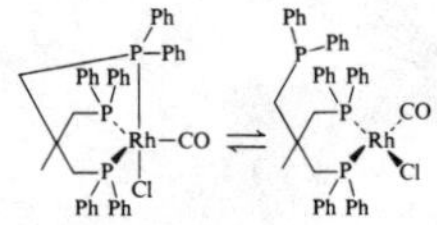

Rh-00393

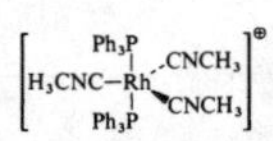

Rh-00394

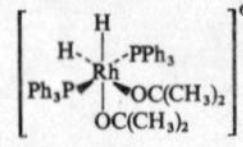

Rh-00395

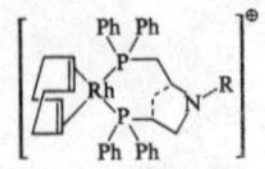

R = $COC(CH_3)_3$

Rh-00396

As Rh-00396 with
R = $-COOC(CH_3)_3$

Rh-00397

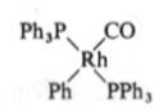

Rh-00398

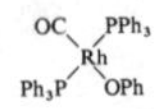

Rh-00399

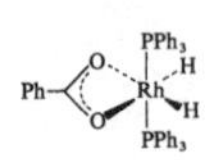

Rh-00400

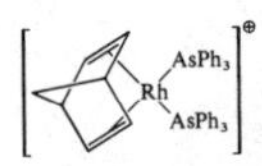

Rh-00401

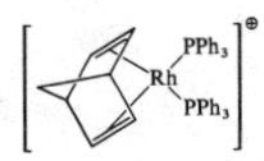

Rh-00402

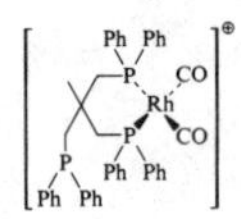

Rh-00403

As Rh-00373 with
R = CH_3

Rh-00404

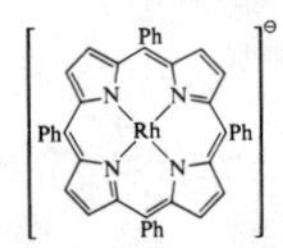

Rh-00405

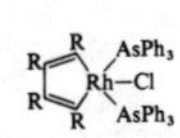

R = CF_3

Rh-00406

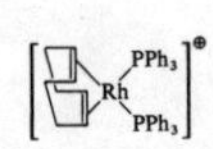

Rh-00407

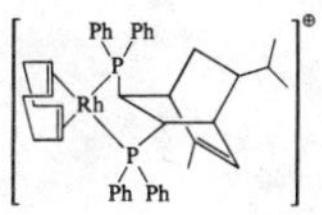

Rh-00408

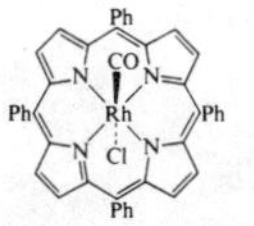

Rh-00409

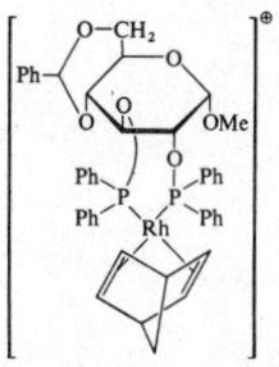

Rh-00410

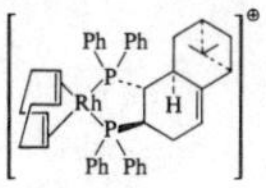

Rh-00411

As Rh-00317 with
R = Ph

Rh-00412

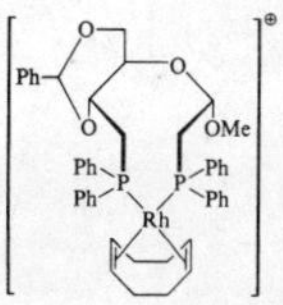

Rh-00413

As Rh-00406 with
R = COOMe

Rh-00414

HRhCO(PPh_3)[P(C_5H_4N)$_3$]$_2$

Rh-00415

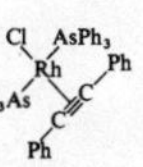

Rh-00416

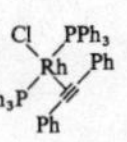

Rh-00417

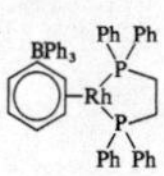

Rh-00418

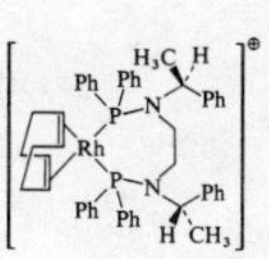

Rh-00419

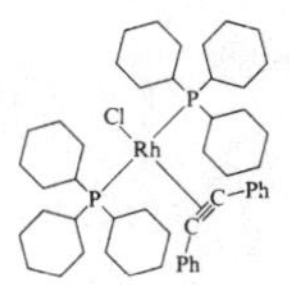

Rh-00420

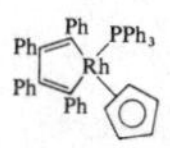

Rh-00421

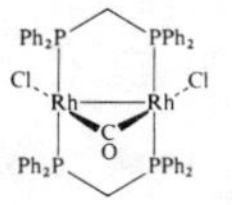

Rh-00422

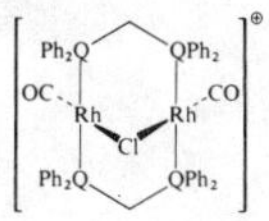

Q = As

Rh-00423

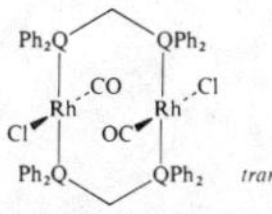

trans-form

Q = As

Rh-00424

As Rh-00423 with
Q = P

Rh-00425

As Rh-00424 with
Q = P

Rh-00426

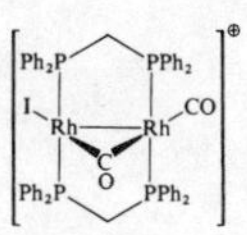

Rh-00427

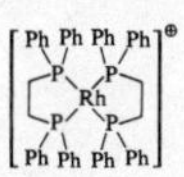

Rh-00428

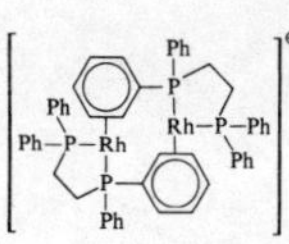

Rh-00429

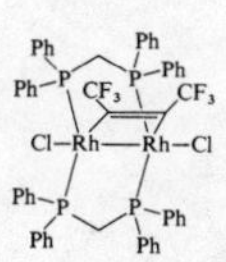

Rh-00430

Rh(PPh_3)$_3$Br

Rh-00431

(Ph_3P)$_3$RhCl

Rh-00432

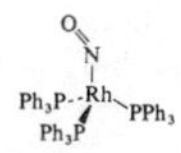

Rh-00433

(Ph_3P)$_3$RhHCl_2

Rh-00434

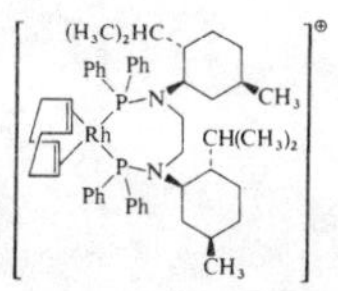

Rh-00435

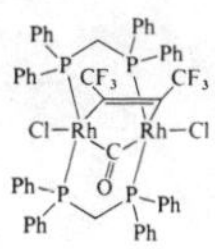

Rh-00436

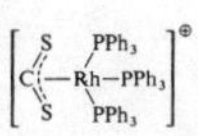

Rh-00437

HRh(PPh_3)$_3$(CO)

Rh-00438

(Ph_3P)$_3$RhMe

Rh-00439

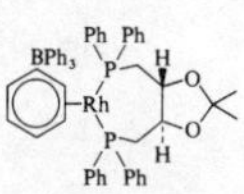

(R,R)-form

Rh-00440

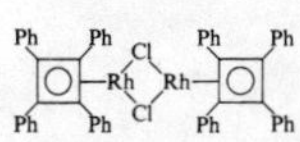

Rh-00441

Rh(OAc)(PPh_3)$_3$

Rh-00442

Rh_2(CO)$_2$(PPh_3)$_3$(CO_2)

Rh-00443

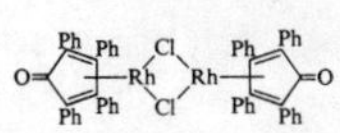

Rh-00444

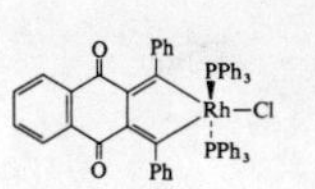

Rh-00445

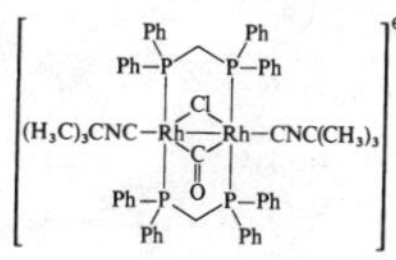

Rh-00446

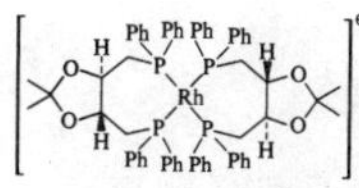

(All-R)-form

Rh-00447

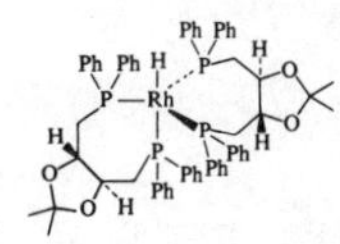

Rh-00448

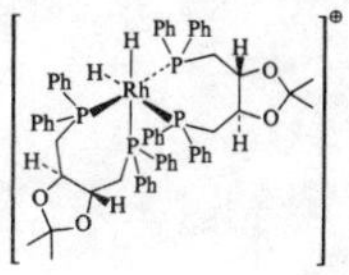

Rh-00449

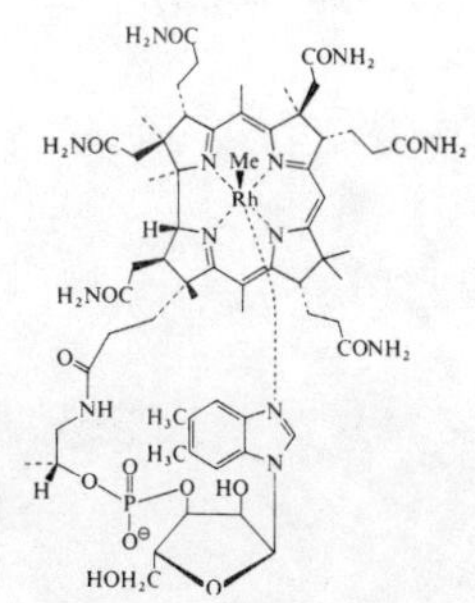

Rh-00450

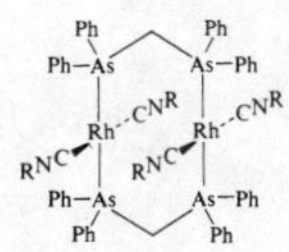

R = $(H_3C)_3C-$

Rh-00451

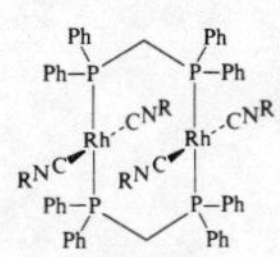

R = $(H_3C)_3C-$

Rh-00452

RhO_2($AsPh_3$)$_4$$^{\oplus}$

Rh-00453

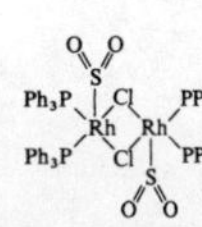

Rh-00454

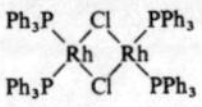

Rh-00455

$RhH[P(OPh)_3]_4$

Rh-00456

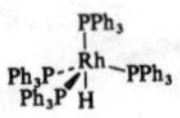

Rh-00457

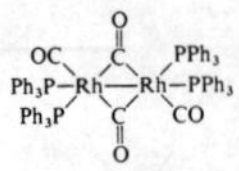

Rh-00458

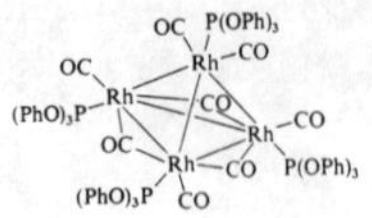

Rh-00459

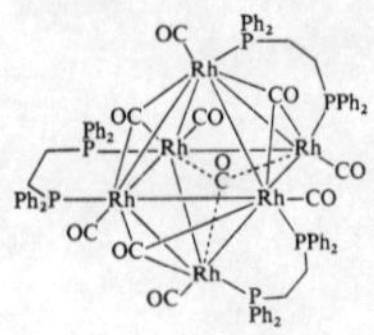

Rh-00460

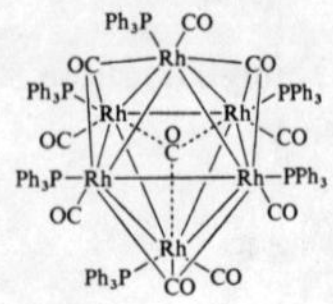

Rh-00461

$RhH(PF_3)_4$

Rh-00462

$(F_3P)_4RhRh(PF_3)_4$

Rh-00463

CBr_3ORh Rh-00001

Tribromocarbonylrhodium, 8CI

[27857-59-6]

$RhBr_3CO$

M 370.628

Dark-red cryst. ir ν_{CO} 2110 cm^{-1} (KBr).

Colton, R. *et al*, *Aust. J. Chem.*, 1970, **23**, 1351 (*synth, ir*)

$CBr_5ORh^{\ominus\ominus}$ Rh-00002

Pentabromocarbonylrhodate(2−), 10CI

[44628-04-8]

$RhBr_5CO^{\ominus\ominus}$ (O_h)

M 530.436 (ion)

Catalyst for the reduction of Fe^{III} to Fe^{II}.

Di-Cs salt: [22594-89-4].
CBr_5Cs_2ORh M 796.247
Orange-red cryst. ir ν_{CO} 2067 cm^{-1} (nujol).

Bis(tetraethylammonium) salt: [24596-67-6].
$C_{17}H_{40}Br_5N_2ORh$ M 790.941
Orange cryst. ir ν_{CO} 2070 cm^{-1} (nujol).

Cleare, M.J. *et al*, *J. Chem. Soc. (A)*, 1969, 372 (*synth, ir, raman*)
Forster, D., *Inorg. Chem.*, 1969, **8**, 2556 (*synth, ir*)
Colton, R. *et al*, *Aust. J. Chem.*, 1970, **23**, 1351 (*synth, ir*)
Dahm, D.J. *et al*, *Inorg. Nucl. Chem. Lett.*, 1970, **6**, 15 (*struct*)
Cleare, M.J. *et al*, *Spectrochim. Acta, Part A*, 1972, **28**, 2019 (*raman, ir*)
James, B.R. *et al*, *Can. J. Chem.*, 1977, **55**, 567 (*use*)

CCl_3ORh Rh-00003

Carbonyltrichlororhodium, 8CI

Trichlorocarbonylrhodium

[27857-62-1]

$RhCl_3CO$

M 237.275

Unstable when exposed to the atm. Characterised spectroscopically. ir ν_{CO} 2135 cm^{-1} (KBr).

Colton, R. *et al*, *Aust. J. Chem.*, 1970, **23**, 1351 (*synth, ir*)

$CCl_5ORh^{\ominus\ominus}$ Rh-00004

Carbonylpentachlororhodate(2−), 9CI

Pentachlorocarbonylrhodate(2−)

$RhCl_5CO^{\ominus\ominus}$ (O_h)

M 308.181 (ion)

Di-Cs salt: [22594-88-3].
CCl_5Cs_2ORh M 573.992
Orange cryst. ir ν_{CO} 2117 cm^{-1} (nujol).

Bis(tetramethylammonium) salt: [24596-66-5].
$C_9H_{24}Cl_5N_2ORh$ M 456.472
Orange solid. ir ν_{CO} ~2090br cm^{-1} (nujol).

Bis(tetraethylammonium) salt: [24596-65-4].
$C_{17}H_{40}Cl_5N_2ORh$ M 568.686
Orange cryst. ir ν_{CO} ~2070br cm^{-1} (nujol).

Cleare, M.J. *et al*, *J. Chem. Soc. (A)*, 1969, 372 (*synth, ir, raman*)
Forster, D., *Inorg. Chem.*, 1969, **8**, 2556 (*synth, ir*)
Stanko, J.A. *et al*, *J. Chem. Soc., Chem. Commun.*, 1969, 1100 (*synth, ir*)
Colton, R. *et al*, *Aust. J. Chem.*, 1970, **23**, 1351 (*synth, ir*)
Dahm, D.J. *et al*, *Inorg. Nucl. Chem. Lett.*, 1970, **6**, 15 (*struct*)
Cleare, M.J. *et al*, *Spectrochim. Acta, Part A*, 1972, **28**, 2019 (*ir, raman*)

CI_3ORh Rh-00005

Carbonyltriiodorhodium, 8CI

Triiodocarbonylrhodium

[27857-60-9]

RhI_3CO

M 511.629

Black cryst. ir ν_{CO} 2073 cm^{-1} (KBr).

Colton, R. *et al*, *Aust. J. Chem.*, 1970, **23**, 1351 (*synth, ir*)

$CI_5ORh^{\ominus\ominus}$ Rh-00006

Carbonylpentaiodorhodate(2−), 10CI

Pentaiodocarbonylrhodate(2−)

[38840-34-5]

$RhI_5CO^{\ominus\ominus}$ (O_h)

M 765.438 (ion)

Di-Cs salt: [29201-07-8].
CCs_2I_5ORh M 1031.249
Black cryst. ir ν_{CO} 2020vs cm^{-1} (nujol).

Bis(tetraethylammonium) salt:
$C_{17}H_{40}IN_2ORh$ M 518.326
Black cryst. Extremely insol. all solvs. ir ν_{CO} 2040 cm^{-1} (nujol).

Forster, D., *Inorg. Chem.*, 1969, **8**, 2556 (*synth, ir*)
Colton, R. *et al*, *Aust. J. Chem.*, 1970, **23**, 1351 (*synth, ir*)
Dahm, D.J. *et al*, *Inorg. Nucl. Chem. Lett.*, 1970, **6**, 15 (*struct*)
Griffith, W.P. *et al*, *J. Chem. Soc. (A)*, 1970, 2788 (*synth, ir, raman*)
Cleare, M.J. *et al*, *Spectrochim. Acta, Part A*, 1972, **28**, 2019 (*raman, ir*)

$C_2Br_2O_2Rh^{\ominus}$ Rh-00007

Dibromodicarbonylrhodate(1−), 10CI

[44516-32-7]

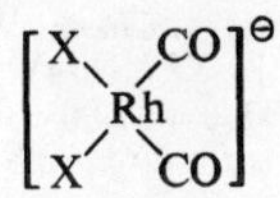

X = Br

M 318.734 (ion)

cis-form [44516-36-1]

Solns. become red when left in air owing to oxidn. to Rh(III) species.

Tetraphenylarsonium salt: [30193-20-5].
$C_{26}H_{20}AsBr_2O_2Rh$ M 702.078
Pale-yellow cryst. (Me_2CO/pet. ether). Mp 150° dec. ir ν_{CO} 2055, 1975 cm^{-1} (nujol).

Tetrapropylammonium salt: [66197-28-2].
$C_{14}H_{28}Br_2NO_2Rh$ M 505.094
Yellow cryst. (CH_2Cl_2/Et_2O). Mp 104-8° dec. ir ν_{CO} 2067, 1995 cm^{-1} (nujol).

Tetrabutylammonium salt: [30191-99-2].
$C_{18}H_{36}Br_2NO_2Rh$ M 561.201
Pale-yellow cryst. (Me_2CO/pet. ether). Mp 86° dec. ir ν_{CO} 2062, 1985 cm^{-1} (nujol).

Vallarino, L.M., *Inorg. Chem.*, 1965, **4**, 161 (*synth, ir*)
Kingston, J.F., *Inorg. Nucl. Chem. Lett.*, 1968, **4**, 65 (*synth, ir*)
Griffith, W.P. *et al*, *J. Chem. Soc. (A)*, 1970, 2788 (*synth, ir, raman*)
Browning, J. *et al*, *J. Chem. Soc., Dalton Trans.*, 1977, 2061 (*synth, ir, raman, cmr*)

Geoffroy, G.L. *et al*, *Inorg. Chem.*, 1977, **16**, 1950 (*uv, mcd*)
Brown, C. *et al*, *J. Organomet. Chem.*, 1980, **192**, 93 (*nmr*)

$C_2Br_4O_2Rh^{\ominus}$ Rh-00008

Tetrabromodicarbonylrhodate(1−), 9CI

$RhBr_4(CO)_2^{\ominus}$ (O_h)

M 478.542 (ion)

cis-form [24596-70-1]

Characterised spectroscopically. ir ν_{CO} 2146, 2115 cm^{-1} ($CHCl_3$).

trans-form [55659-62-6]

Characterised spectroscopically.

Forster, D., *Inorg. Chem.*, 1969, **8**, 2556 (*synth, ir*)
Daly, J.J. *et al*, *J. Am. Chem. Soc.*, 1975, **97**, 2551 (*synth*)

C_2ClO_2Rh Rh-00009

Dicarbonylchlororhodium

Dicarbonylrhodium chloride

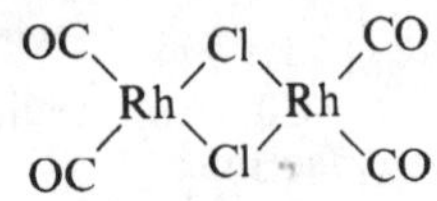

M 194.379

Dimeric.

Dimer: [14523-22-9]. *Tetracarbonyldi-μ-chlorodirhodium,* 9CI. *Bis(chlorodicarbonylrhodium).*

$C_4Cl_2O_4Rh_2$ M 388.759

Catalyst for strained hydrocarbon rearrangements and olefin hydroformylation. Starting material for Rh(I) carbonyl compds. Red needles. Insol. hydrocarbons. Mp 126.5-127°. Solns. dec. slowly in air. ν_{CO} 2115w, 2095s, 2045s (nujol).

Dahl, L.F. *et al*, *J. Am. Chem. Soc.*, 1961, **83**, 1761 (*struct*)
Clement, D.I. *et al*, *J. Chem. Soc., Chem. Commun.*, 1969, 1509 (*nmr*)
Colton, R. *et al*, *Aust. J. Chem.*, 1970, **23**, 1351 (*synth, ir*)
Gallay, J. *et al*, *J. Organomet. Chem.*, 1972, **38**, 179 (*synth, ir*)
Inorg. Synth., 1974, **15**, 14, 17 (*synth*)
Iqbal, M.Z., *CA*, 1980, **93**, 87637 (*synth*)
Fieser, M. *et al*, *Reagents for Organic Synthesis*, Wiley, 1967-83, **10**, 382.

$C_2Cl_2O_2Rh^{\ominus}$ Rh-00010

Dicarbonyldichlororhodate(1−), 10CI

[15550-00-2]

As Dibromodicarbonylrhodate(1−), Rh-00007 with

X = Cl

M 229.832 (ion)

Starting material for CO-containing Rh complexes. Catalyst for the reduction of NO by CO.

cis-form [44516-52-1]

Tetraphenylarsonium salt: [30193-19-2].

$C_{26}H_{20}AsCl_2O_2Rh$ M 613.176

Pale-yellow cryst. (Me_2CO/pet. ether). Mp 185-90° (181°) dec. ir ν_{CO} 2058, 1974 cm^{-1} (nujol).

Tetrabutylammonium salt: [17966-76-6].

$C_{18}H_{36}Cl_2NO_2Rh$ M 472.299

Pale-yellow cryst. (C_6H_6/hexane). Mp 81° dec. ir ν_{CO} 2060, 1975 cm^{-1} (nujol).

Vallarino, L.M., *Inorg. Chem.*, 1965, **4**, 161 (*synth, ir*)
Griffith, W.P. *et al*, *J. Chem. Soc. (A)*, 1970, 2788 (*synth, ir*)
Stanko, J.A. *et al*, *Inorg. Chem.*, 1971, **10**, 566 (*synth, struct*)
Varshavskii, Yu.S. *et al*, *J. Organomet. Chem.*, 1971, **31**, 119 (*use*)
Meyer, C.D. *et al*, *J. Am. Chem. Soc.*, 1976, **98**, 1364 (*use*)
Brown, C. *et al*, *J. Organomet. Chem.*, 1980, **192**, 93 (*nmr*)

$C_2Cl_4O_2Rh^{\ominus}$ Rh-00011

Dicarbonyltetrachlororhodate(1−)

Tetrachlorodicarbonylrhodate(1−)

$RhCl_4(CO)_2^{\ominus}$ (O_h)

M 300.738 (ion)

Characterised spectroscopically.

cis-form

ir ν_{CO} 2160s, 2125s cm^{-1} ($CHCl_3$).

Forster, D., *Inorg. Chem.*, 1969, **8**, 2556 (*synth, ir*)

$C_2H_3ClO_2NRh$ Rh-00012

Amminedicarbonylchlororhodium, 10CI

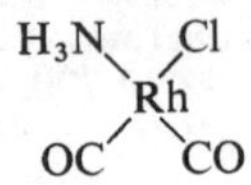

M 211.410

cis-form [72402-62-1]

Metallic-green cryst. ($CHCl_3$). Mp 145° dec. ir ν_{CO} 2100, 2070, 2020, 1990, 1968 cm^{-1} (KBr).

Vallarino, L.M. *et al*, *Inorg. Chim. Acta*, 1979, **36**, 243 (*synth, ir*)

$C_2I_2O_2Rh^{\ominus}$ Rh-00013

Dicarbonyldiiodorhodate(1−), 10CI

Diiododicarbonylrhodate(1−)

[38255-39-9]

As Dibromodicarbonylrhodate(1−), Rh-00007 with

X = I

M 412.735 (ion)

Catalyst for the carbonylation of alcohols, especially MeOH to AcOH.

cis-form [44516-57-6]

Tetraphenylarsonium salt: [13986-81-7].

$C_{26}H_{20}AsI_2O_2Rh$ M 796.079

Light-brown cryst. Mp 121° dec. ir ν_{CO} 2043, 1967 cm^{-1} (nujol). Cannot be recryst., loses CO rapidly in soln.

Tetrabutylammonium salt: [30192-00-8].

$C_{18}H_{36}I_2NO_2Rh$ M 655.202

Mustard-yellow cryst. Mp 78° dec. ir ν_{CO} 2052, 1985 cm^{-1} (nujol). Cannot be recryst., loses CO rapidly in soln.

Vallarino, L.M., *Inorg. Chem.*, 1965, **4**, 161 (*synth, ir*)
Griffith, W.P. *et al*, *J. Chem. Soc. (A)*, 1970, 2788 (*ir*)
Roth, J.F. *et al*, *Chem. Technol.*, 1971, 600 (*use*)
Forster, D., *J. Am. Chem. Soc.*, 1976, **98**, 846 (*use, ir*)
Hjortkjaer, J. *et al*, *J. Chem. Soc., Perkin Trans. 2*, 1978, 763 (*use*)

$C_2I_4O_2Rh^{\ominus}$ Rh-00014

Dicarbonyltetraiodorhodate(1−), 10CI

Tetraiododicarbonylrhodate(1−)

$RhI_4(CO)_2^{\ominus}$ (O_h)

M 666.544 (ion)

Catalyst for the water-gas shift reaction and for hydrogenolysis of MeOH.

cis-form

Isom. to the *trans*-form within a few minutes at r.t.

Tetraethylammonium salt: [24596-69-8].
$C_{10}H_{20}I_4NO_2Rh$ M 796.797
Characterised spectroscopically. ir ν_{CO} 2122, 2092 cm^{-1} ($CHCl_3$).

trans-form [44769-93-9]

Tetrapropylammonium salt: [55634-00-9].
$C_{14}H_{28}I_4NO_2Rh$ M 852.904
Black cryst. (CH_2Cl_2/CCl_4). ir ν_{CO} 2091 cm^{-1} (CH_2Cl_2).

Tetrabutylammonium salt: [55634-01-0].
$C_{18}H_{36}I_4NO_2Rh$ M 909.011
Black cryst.

Tetraphenylarsonium salt: [26004-02-4].
$C_{26}H_{20}AsI_4O_2Rh$ M 1049.888
Black cryst.

Forster, D., *Inorg. Chem.*, 1969, **8**, 2556 (*synth, ir*)
Daly, J.J. *et al*, *J. Am. Chem. Soc.*, 1975, **97**, 2551 (*synth, struct*)
Drury, D. *et al*, *J. Organomet. Chem.*, 1982, **236**, C23 (*use*)

C_3BrO_3Rh Rh-00015

Bromotricarbonylrhodium, 9CI
Bromotris(carbonyl)rhodium
[40784-46-1]

$Rh(CO)_3Br$

M 266.841
Unstable in the absence of a high CO pressure. ir ν_{CO} 2093, 2054 cm^{-1} (hexane).

Morris, D.E. *et al*, *J. Organomet. Chem.*, 1973, **49**, C53 (*synth, ir*)

C_3ClO_3Rh Rh-00016

Tricarbonylchlororhodium, 9CI
Chlorotris(carbonyl)rhodium
[37780-59-9]

$Rh(CO)_3Cl$

M 222.390
Catalyst for the polymerisation of propadiene. Unstable in the absence of a high CO pressure. ir ν_{CO} 2098, 2052 cm^{-1} (hexane).

Scholten, J.P. *et al*, *J. Polym. Sci., Part A-1*, 1972, **10**, 3067 (*synth, ir, use*)
Morris, D.E. *et al*, *J. Organomet. Chem.*, 1973, **49**, C53 (*synth, ir*)
Scholten, J.P. *et al*, *J. Polym. Sci., Polym. Chem. Ed.*, 1973, **11**, 3205 (*use*)
Rüger, C. *et al*, *J. Prakt. Chem.*, 1975, **317**, 867.

$C_3H_5Cl_2Rh$ Rh-00017

Dichloro(η^3-2-propenyl)rhodium, 10CI
π-Allyldichlororhodium, 8CI. *Dichloro(allyl)rhodium*
[12306-43-3]

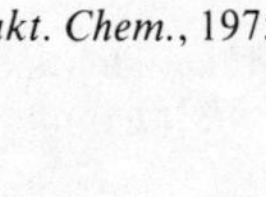

M 214.884
Polymeric. Red-brown amorph. solid.

Powell, J. *et al*, *J. Chem. Soc. (A)*, 1968, 583 (*ir, synth*)
Pruchnik, F. *et al*, *Rocz. Chem.*, 1977, **51**, 1581; *CA*, **88**, 49842g (*ir, uv*)

$C_3H_5F_9P_3Rh$ Rh-00018

Tris(phosphorous trifluoride)(η^3-2-propenyl)rhodium, 9CI
(η^3-Allyl)tris(trifluorophosphine)rhodium
[36462-60-9]

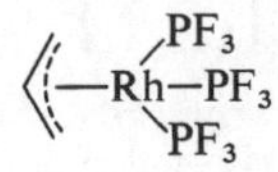

M 407.885
Pale-yellow volatile liq. Nmr indicates inter- and intramol. phosphine exchange processes.

Nixon, J.F. *et al*, *J. Fluorine Chem.*, 1973, **3**, 179 (*nmr*)
Nixon, J.F. *et al*, *J. Chem. Soc., Dalton Trans.*, 1974, 1993 (*synth, ir, ms, nmr*)

C_3IO_3Rh Rh-00019

Tricarbonyliodorhodium, 9CI
Iodotris(carbonyl)rhodium
[40784-47-2]

$Rh(CO)_3I$

M 313.841
Unstable in the absence of a high CO pressure. ir ν_{CO} 2084, 2054 cm^{-1} (hexane).

Morris, D.E. *et al*, *J. Organomet. Chem.*, 1973, **49**, C53 (*synth, ir*)

$C_4Br_2O_4Rh_2$ Rh-00020

Di-μ-bromotetracarbonyldirhodium, 10CI
Tetracarbonyldi-μ-bromodirhodium
[21475-96-7]

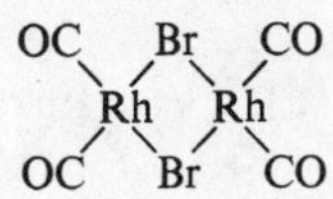

M 477.661
Bridged dimer. Red cryst. (hexane). Mp 118°. ir ν_{CO} 2040s, 2090s, 2115w cm^{-1} (nujol).

Monomer: Bromodicarbonylrhodium. Dicarbonylrhodium bromide.
C_2BrO_2Rh M 238.830
Unknown.

Hieber, W. *et al*, *Z. Anorg. Allg. Chem.*, 1943, **251**, 98 (*synth*)
Garland, C.W. *et al*, *J. Chem. Phys.*, 1962, **36**, 1094 (*ir*)
Lewis, J. *et al*, *Acc. Chem. Res.*, 1968, **1**, 245 (*ms*)
Colton, R. *et al*, *Aust. J. Chem.*, 1970, **23**, 1351 (*ir, synth*)
Brown, C. *et al*, *J. Organomet. Chem.*, 1980, **192**, 93 (*nmr, synth*)
Nixon, J.F. *et al*, *Inorg. Chim. Acta*, 1981, **47**, 147 (*synth, pe*)

$C_4H_4NO_4Rh$ Rh-00021

Dicarbonyl(glycinato-*N,O*)rhodium, 10CI
[62613-42-7]

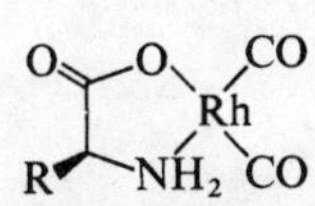

R = H

M 232.985
Yellow cryst. ir ν_{CO} 2089, 2015 cm^{-1} (CH_2Cl_2/MeOH 1:1).

Nagy-Magos, Z. *et al*, *Transition Met. Chem.*, 1980, **5**, 186 (*synth, ir*)

$C_4H_8Cl_2Rh^{\ominus}$ Rh-00022

Dichlorobis(ethylene)rhodate(1−)

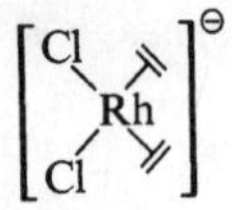

M 229.919 (ion)

Intermed. in the dimerisation of ethylene to linear butenes using $RhCl_3$.

Cramer, R., *J. Am. Chem. Soc.*, 1965, **87**, 4717 (*uv*)

$C_4N_2O_{10}Rh_2$ Rh-00023

Dinitratotetracarbonyldirhodium

Bis-μ-nitratobis[dicarbonylrhodium]

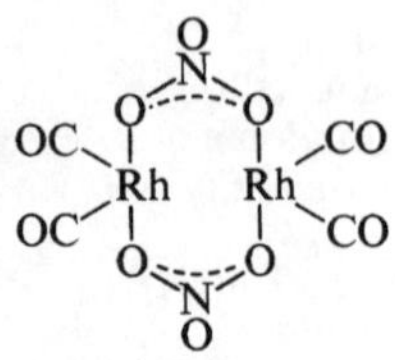

M 441.862

Moisture-sensitive, purple solid by subl. Mp 100-5° dec. Bp 80° subl. *in vacuo*. ir ν_{CO} 2123vs, 2081m, 2037vs, 2000s, 1992m cm^{-1} (nujol).

Lawson, D.N. *et al*, *J. Chem. Soc.*, 1965, 1900 (*synth, ir*)

$C_4O_8Rh_2S$ Rh-00024

Sulfatotetracarbonyldirhodium

μ-Sulfatobis[dicarbonylrhodium]

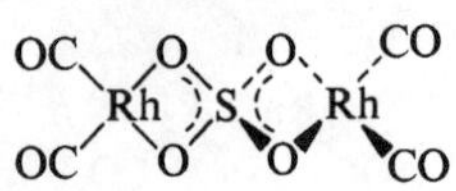

M 413.910

Green-purple solid (Me_2CO), v. sensitive to moisture. Mp ca. 50° dec. ir ν_{CO} 2093, 2035, 2014 cm^{-1} ($CHCl_3$).

Lawson, D.N. *et al*, *J. Chem. Soc.*, 1965, 1900 (*synth, ir*)

$C_5HN_5Rh^{\ominus\ominus\ominus}$ Rh-00025

Pentakis(cyano-*C*)hydrorhodate(3−), 10CI

Pentacyanohydrorhodate(3−), *8CI*. *Hydridopentakis(cyano)rhodate(3−)*

$RhH(CN)_5^{\ominus\ominus\ominus}$

M 234.002 (ion)

Tri-K salt: [19610-72-1].

$C_5HK_3N_5Rh$ M 351.297

White cryst. (MeOH). V. hygroscopic.

Krogmann, K. *et al*, *J. Organomet. Chem.*, 1968, **11**, P27 (*synth, ir, uv, pmr*)
Whitesides, G.M. *et al*, *J. Am. Chem. Soc.*, 1969, **91**, 4980 (*synth, pmr*)
Jewsbury, R.A. *et al*, *J. Chem. Soc. (A)*, 1971, 2847 (*synth, uv, ir, pmr, raman*)

$C_5H_5O_2Rh$ Rh-00026

Dicarbonyl[(2,3-η)-2-propenyl]rhodium, 9CI

π-Allyldicarbonylrhodium, *8CI*

[33292-30-7]

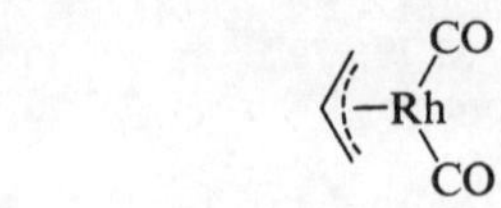

M 199.999

Yellow cryst. by subl. Mp 64° (55-60°) dec. ir ν_{CO} 2060, 1995 cm^{-1} ($CDCl_3$). Susceptible to atm. oxidn.

O'Brien, S., *J. Chem. Soc., Chem. Commun.*, 1968, 757 (*synth, ir, pmr*)
Abel, E.W. *et al*, *J. Chem. Soc., Dalton Trans.*, 1973, 1706 (*synth*)

$C_5H_6NO_2RhS_2$ Rh-00027

Dicarbonyl(dimethylcarbamodithiolato-*S,S'*)rhodium, 9CI

(Dimethyldithiocarbamato)dicarbonylrhodium

[39142-24-0]

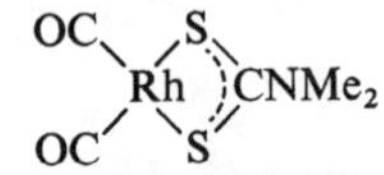

M 279.133

Orange-red cryst. Mp 130° dec. ir ν_{CO} 2071, 2005 cm^{-1}, ν_{CN} 1545 cm^{-1} (CH_2Cl_2).

Cotton, F.A. *et al*, *Inorg. Chem.*, 1964, **3**, 1398 (*synth, ir*)
Abel, E.W. *et al*, *J. Chem. Soc., Dalton Trans.*, 1973, 98 (*synth, pmr, ir, ms*)

$C_5H_6NO_4Rh$ Rh-00028

(Alaninato-*N,O*)dicarbonylrhodium, 10CI

As Dicarbonyl(glycinato-*N,O*)rhodium, Rh-00021 with

$R = CH_3$

M 247.012

(S)-form [62828-91-5]

Biologically active. Yellow cryst. ir ν_{CO} 2093, 2016 cm^{-1} (CH_2Cl_2).

HCl adduct:

$C_5H_7ClNO_4Rh$ M 283.473

Brilliant dark-red material. Elemental composition varies slightly.

Nagy-Magos, Z. *et al*, *Transition Met. Chem.*, 1980, **5**, 186 (*synth, ir, bibl*)

$C_5H_{12}ClORhS_2$ Rh-00029

Carbonylchlorobis(dimethyl sulfide)rhodium

Carbonylchlorobis[thiobis[methane]]rhodium, *10CI*, *9CI*

[56954-39-3]

$RhCl(CO)(SMe_2)_2$

M 290.628

Sterochem. not specified. Brown solid or cream-yellow cryst. ir ν_{CO} 2020, ν_{SO} 1122, 1112 cm^{-1} (nujol).

Kuyper, J. *et al*, *J. Organomet. Chem.*, 1975, **96**, 289 (*synth*)
Dutta, D.K. *et al*, *Transition Met. Chem.*, 1980, **5**, 244 (*synth, ir, uv*)

$C_5H_{12}ClO_3RhS_2$ Rh-00030

Carbonylchlorobis(dimethyl sulfoxide)rhodium

Carbonylchlorobis[sulfinylbis[methane]-S]rhodium, 10CI

[56954-39-3]

$(Me_2SO)_2RhCl(CO)$

M 322.627

Yellow solid. Sol. $CHCl_3$, Me_2CO. Mp 117-9° dec. ir ν_{CO} 2020, ν_{SO} 1122, 1112 cm^{-1} (nujol).

Lawson, D.N. *et al*, *J. Chem. Soc.*, 1965, 1900 (*synth, ir*)
Dutta, D.K. *et al*, *Transition. Met. Chem.*, 1980, **5**, 244 (*synth, ir, uv*)

$C_6H_2O_8Rh_2$ Rh-00031

Tetracarbonylbis[μ-(formato-O:O')]dirhodium, 10CI

Bis-μ-formatobis(dicarbonylrhodium)

[69041-48-1]

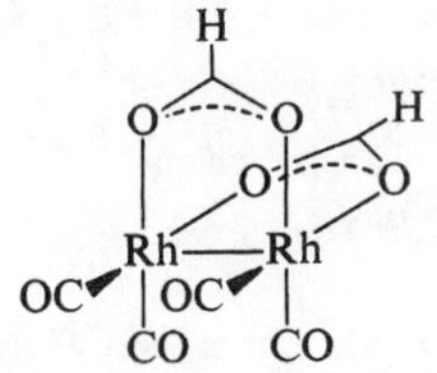

M 407.888

Golden-brown cryst. with metallic lustre, air-stable but moisture-sensitive. Sol. common org. solvs. Mp 115-9° dec. Colour changes to green-blue on grinding.

Kolomnikov, I.S. *et al*, *Koord. Chim.*, 1979, **5**, 1759.
Pruchnik, F. *et al*, *Inorg. Chim. Acta*, 1980, **40**, 203 (*synth, ir, ms, uv*)

$C_6H_5I_2ORh$ Rh-00032

Carbonyl(η^5-2,4-cyclopentadien-1-yl)diiodorhodium, 9CI

[12192-49-3]

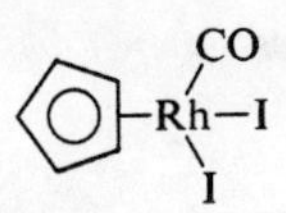

M 449.819

Brown-purple solid. Insol. Et_2O, pentane. ir ν_{CO} 2065 cm^{-1} (nujol).

King, R.B., *Inorg. Chem.*, 1966, **5**, 82 (*synth, ir, pmr*)
Gardner, S.A., *J. Organomet. Chem.*, 1974, **78**, 415.

$C_6H_6I_6O_4Rh_2^{\ominus\ominus}$ Rh-00033

Diacetyldicarbonyldi-μ-iodotetraiododirhodate(2−), 10CI, 9CI

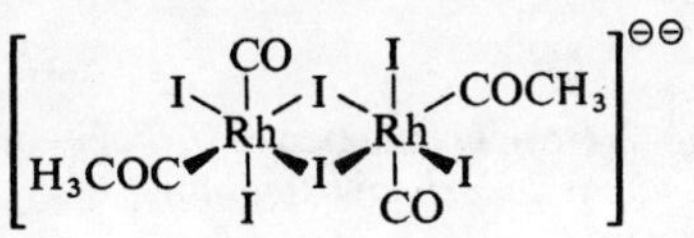

M 1109.349 (ion)

Intermed. in the rhodium-catalysed formation of acetic acid by carbonylation of MeOH. ir ν_{CO} 2062, 1711 cm^{-1} (MeOH/AcOH).

Trimethylphenylammonium salt: [53261-22-6].
$C_{24}H_{34}I_6N_2O_4Rh_2$ M 1381.782
Air-stable deep-red cryst.

Monomer: Acetylcarbonyltriiodorhodate(1−).
$C_3H_3I_3O_2Rh^{\ominus}$ M 554.675 (ion)
Unknown.

Forster, D. *et al*, *J. Organomet. Chem.*, 1974, **71**, C17 (*cryst struct*)
Forster, D. *et al*, *J. Am. Chem. Soc.*, 1976, **98**, 846 (*synth, ir, props*)

$C_6H_6N_2O_2Rh^{\oplus}$ Rh-00034

Bis(acetonitrile)dicarbonylrhodium(1+), 10CI

[33451-07-9]

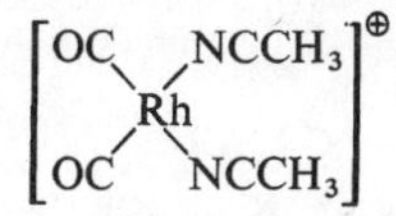

M 241.031 (ion)

Tetrafluoroborate: [53513-97-6].
$C_6H_6BF_4N_2O_2Rh$ M 327.835
Purple-blue cryst. (CH_2Cl_2/Et_2O).

Hexafluorophosphate: [82411-91-4].
$C_6H_6F_6N_2O_2PRh$ M 385.995
No phys. props. reported.

Osborn, J.A. *et al*, *J. Am. Chem. Soc.*, 1971, **93**, 3089.
Albano, V.C. *et al*, *J. Chem. Soc., Chem. Commun.*, 1974, 299 (*synth, ir*)
Connelly, N.G. *et al*, *J. Chem. Soc., Dalton Trans.*, 1977, 70.
Brown, C. *et al*, *J. Organomet. Chem.*, 1980, **192**, 93 (*nmr, synth*)
Valderramas, M. *et al*, *Can. J. Chem.*, 1982, **60**, 1044 (*ir*)

$C_6H_6O_4Rh_2S_2$ Rh-00035

Tetracarbonylbis[μ-(methanethiolato)]dirhodium, 10CI

Bis-μ-(methylthio)bis(dicarbonylrhodium)

[24818-83-5]

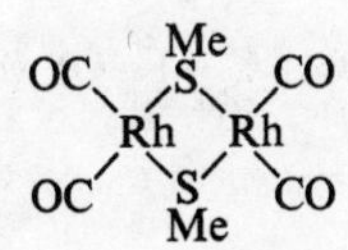

M 412.042

Bridged dimer. Dark-red cryst. (Et_2O/pet. ether). Mp 87.5°. ir ν_{CO} 1972vw, 2008s, 2047w, 2055s, 2074m cm^{-1} (CCl_4).

Monomer: Dicarbonyl(methanethiolato)rhodium. (Methylthio)dicarbonylrhodium.
$C_3H_3O_2RhS$ M 206.021
Unknown.

Bolton, E.S. *et al*, *J. Organomet. Chem.*, 1969, **18**, 153 (*synth, ms, ir, pmr*)
Hill, R. *et al*, *J. Chem. Soc., Chem. Commun.*, 1977, 434.

$C_6H_6O_6Rh_2$ Rh-00036

Tetracarbonyldi-μ-methoxydirhodium, 10CI

[49713-92-0]

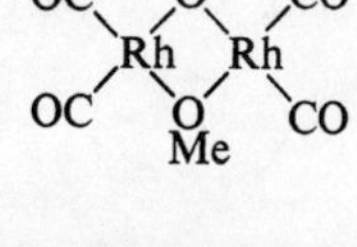

M 379.921

Yellow cryst. (heptane), stable in dry N_2 atm., extremely moisture-sensitive. Bp 50° subl. *in vacuo*.

Monomer: Dicarbonylmethoxyrhodium.
$C_3H_3O_3Rh$ M 189.960

Unknown.

Vizi-Orosz, A. *et al*, *J. Organomet. Chem.*, 1973, **57**, 379 (*ir, synth*)
Giordano, G. *et al*, *Gazz. Chim. Ital.*, 1975, **105**, 613 (*synth, ms, ir, pmr*)
Schumann, H. *et al*, *J. Organomet. Chem.*, 1978, **153**, 327.

$C_6H_8Cl_2O_2Rh_2$ Rh-00037

Dicarbonyldi-μ-chlorobis(η^2-ethene)dirhodium, 9CI
Dicarbonyldi-μ-chlorobis(ethylene)dirhodium, 8CI. *Di-μ-chlorodicarbonylbis(η^2-ethylene)dirhodium*
[12306-60-4]

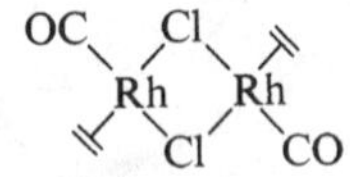

M 388.845
Bridged dimer. Useful synthetic intermed. Dark-red cryst. (toluene/hexane). Mp 113-4° dec. ir ν_{CO} 2010, 2000 cm^{-1} (C_6H_6).
Monomer: Carbonylchloroethylenerhodium.
C_3H_4ClORh M 194.423
Unknown.

Powell, J. *et al*, *J. Chem. Soc. (A)*, 1968, 211 (*synth, ir, pmr*)
Maisonnat, A. *et al*, *Inorg. Chem.*, 1974, **13**, 661 (*ir, pmr, use*)
Maisonnat, A. *et al*, *Inorg. Chem.*, 1974, **13**, 2996 (*use*)
Maisonnat, A. *et al*, *J. Organomet. Chem.*, 1974, **73**, C36 (*use*)

$C_6H_9I_3N_3Rh$ Rh-00038

Tris(isocyanomethane)triiodorhodium
Tris(methyl isocyanide)triiodorhodium

$RhI_3(CNMe)_3$

M 606.776
Red-brown cryst. (CH_2Cl_2/Et_2O). Darkens on heating, but does not melt <250°. Stereochem. not specified.

Balch, A.L. *et al*, *J. Organomet. Chem.*, 1971, **32**, 263 (*synth, ir*)

$C_7HF_6O_4Rh$ Rh-00039

Dicarbonyl(1,1,1,5,5,5-hexafluoro-2,4-pentanedionato-*O,O'*)rhodium, 10CI, 9CI
Dicarbonyl(hexafluoroacetylacetonato)rhodium
[18517-12-9]

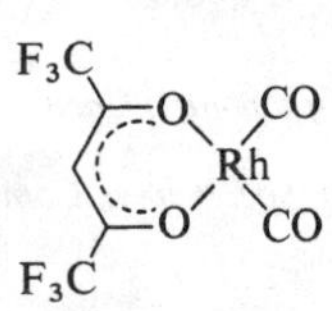

M 365.978
Dark-red needles by subl. Sol. Et_2O, hydrocarbons. Mp 71-4°.

Bonati, F. *et al*, *J. Chem. Soc.*, 1964, 3156 (*synth, ir, pmr, uv*)
Bailey, N.A. *et al*, *J. Chem. Soc., Chem. Commun.*, 1967, 1041 (*cryst struct, ir*)

$C_7H_4NO_4Rh$ Rh-00040

Dicarbonyl(η^5-1-nitro-2,4-cyclopentadien-1-yl)rhodium, 10CI
(η^5-Nitrocyclopentadienyl)dicarbonylrhodium
[75862-04-3]

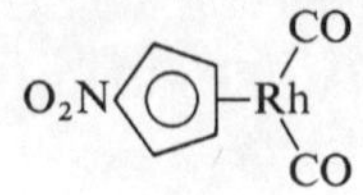

M 269.018
Red cryst. by subl. Mp 62-3°.

Rausch, M.D. *et al*, *J. Organomet. Chem.*, 1980, **197**, 225 (*synth, struct, ir, ms, pmr*)

$C_7H_5ClNO_2Rh$ Rh-00041

Dicarbonylchloro(pyridine)rhodium, 10CI, 9CI
Pyridinechlorodicarbonylrhodium
[14024-06-7]

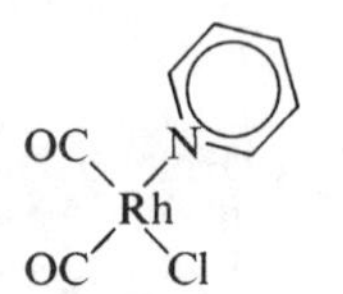

M 273.481
Catalyst for carbonylation of MeOH and hydroformylation of ethylene.
cis-form [61008-96-6]
Orange-red needles (pet. ether). Sol. common org. solvs. Mp 68° dec. ir ν_{CO} 2075s, 2060m, 2010vs, 1980m cm^{-1} (nujol).

Hieber, W. *et al*, *Chem. Ber.*, 1957, **90**, 2425.
Lawson, D.N. *et al*, *J. Chem. Soc.*, 1965, 1900 (*synth, ir*)
Brodzki, D. *et al*, *Bull. Soc. Chim. Fr.*, 1976, 61 (*use*)
King, R.B. *et al*, *J. Am. Chem. Soc.*, 1979, **101**, 4893 (*use*)

$C_7H_5F_3IORh$ Rh-00042

Carbonyl(η^5-2,4-cyclopentadien-1-yl)iodo(trifluoromethyl)-rhodium, 9CI
Iodocarbonylcyclopentadienyl(trifluoromethyl)rhodium
[12078-27-2]

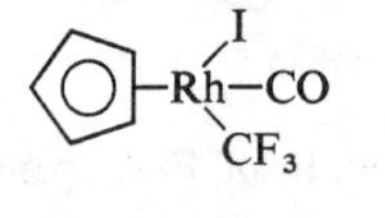

M 391.921
Dark-red needles (CH_2Cl_2/pet. ether). Sol. common org. solvs., spar. sol. CCl_4, CS_2. Mp 168-9°. $Bp_{0.1}$ 100° subl. ir ν_{CO} 2098 cm^{-1} (CS_2).

McCleverty, J.A. *et al*, *J. Chem. Soc.*, 1964, 4200 (*synth, nmr, ir*)
Gardner, S.A. *et al*, *Inorg. Chem.*, 1974, **13**, 997 (*synth, props*)

$C_7H_5O_2Rh$ Rh-00043

Dicarbonyl(η^5-2,4-cyclopentadien-1-yl)rhodium, 10CI, 9CI
Cyclopentadienyldicarbonylrhodium
[12192-97-1]

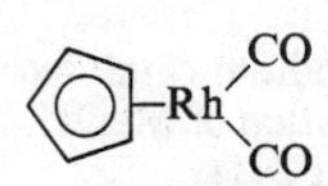

M 224.021
Catalyses hydroformylation and hydrogenation of alkenes, and the cyclotrimerization of alkynes. Orange liq. Mp −11°. Bp 240°.

Fischer, E.O. *et al*, *Z. Naturforsch.*, *B*, 1961, **16**, 225 (*synth, ir*)
Dickson, R.S. *et al*, *Aust. J. Chem.*, 1970, **23**, 1531 (*synth, ir, pmr*)
Knight, J. *et al*, *J. Chem. Soc.* (*A*), 1970, 654 (*synth*)
Axelson, D.E. *et al*, *Inorg. Nucl. Chem. Lett.*, 1973, **9**, 885 (*cmr*)
Gardner, S.A. *et al*, *Inorg. Chem.*, 1973, **12**, 2396 (*use*)
Dykstra, R.W. *et al*, *Rev. Sci. Instrum.*, 1981, **52**, 1690 (*nmr*)

$C_7H_7O_4Rh$ Rh-00044

Dicarbonyl(2,4-pentanedionato-*O,O'*)rhodium, 10CI, 9CI

Dicarbonylacetylacetonatorhodium. Rhodium dicarbonyl acetylacetonate

[14874-82-9]

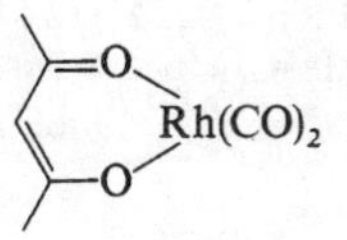

M 258.035

Catalyst for hydroformylation and hydrosilylation of alkenes. Also catalyses Fischer-Tropsch type reacns. Red-green dichroic cryst. (pet. ether). Mp 155°. $Bp_{0.1}$ 90° subl. Exhibits semiconducting props. ν_{CO} 2084(s), 2016(s) ($CHCl_3$).

Bonati, F. *et al*, *J. Chem. Soc.*, 1964, 3156 (*synth, pmr, ir*)
Bailey, N.A. *et al*, *J. Chem. Soc., Chem. Commun.*, 1967, 1041 (*cryst struct*)
Adams, D.M. *et al*, *J. Chem. Soc., Dalton Trans.*, 1974, 690 (*ir*)
Haegele, K.D. *et al*, *Inorg. Chem.*, 1974, **13**, 1960 (*ms*)
Huq, F. *et al*, *J. Cryst. Mol. Struct.*, 1974, **4**, 411 (*cryst struct*)
Jesse, A.C. *et al*, *Inorg. Chim. Acta*, 1978, **26**, 129 (*synth, ir, raman, cmr*)
Brown, C. *et al*, *J. Organomet. Chem.*, 1980, **192**, 93 (*cmr, nmr*)

$C_7H_{18}ClOP_2Rh$ Rh-00045

Carbonylchlorobis(trimethylphosphine)rhodium

[36713-95-8]

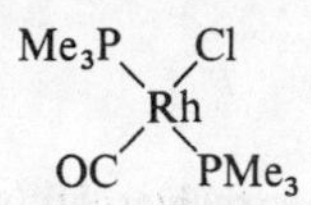

M 318.525

trans-form

Pale-yellow cryst. (pet. ether). Sol. CH_2Cl_2, $MeNO_2$. Mp 108° dec. ir ν_{CO} 1965 cm^{-1} (hexane).

Browning, J. *et al*, *J. Chem. Soc., Dalton Trans.*, 1977, 2061 (*synth, ir, nmr*)

$C_7H_{18}Cl_3OP_2Rh$ Rh-00046

Carbonyltrichlorobis(trimethylphosphine)rhodium

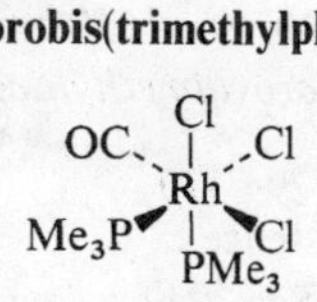

fac-form

M 389.431

Fac-form

cis-*form*

Formed when Carbonylchlorobis(trimethylphosphine)-rhodium, Rh-00045 is chlorinated in CH_2Cl_2. Yellow cryst. Sl. sol. CH_2Cl_2, insol. Et_2O. Mp 130° dec. ir ν_{CO} 2094 cm^{-1} (nujol).

Mer-form

trans-*form*

Formed in soln. when Carbonylchlorobis(trimethylphosphine)rhodium, Rh-00045 is chlorinated in C_6H_6. Partly isomerizes to *fac*-form when isolation is attempted. ir ν_{CO} 2077 cm^{-1} (C_6H_6).

Browning, J. *et al*, *J. Chem. Soc., Dalton Trans.*, 1977, 2061 (*synth, nmr, ir*)

$C_8F_6O_8Rh_2$ Rh-00047

Tetracarbonylbis[μ-(trifluoroacetato-*O*:*O'*)]dirhodium, 10CI

Bis(trifluoroacetato)tetracarbonyldirhodium. Bis-μ-trifluoroacetatobis[dicarbonylrhodium]

[14522-94-2]

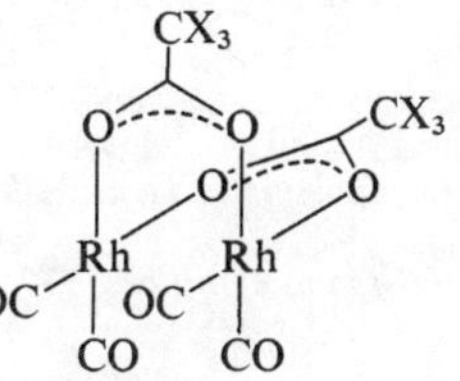

X = F

M 543.885

Dark-blue solid (Me_2CO). Mp 100-20° dec. ir ν_{CO} 2101m, 2081s, 2031m cm^{-1} (nujol).

Lawson, D.N. *et al*, *J. Chem. Soc.*, 1965, 1900 (*ir, synth*)
Balch, A.L. *et al*, *Inorg. Chem.*, 1977, **16**, 2840 (*uv*)

$C_8H_6O_8Rh_2$ Rh-00048

Bis[μ-acetato-*O*:*O'*]tetracarbonyldirhodium, 10CI

Bis-μ-acetatobis[dicarbonylrhodium]

[14057-86-4]

As Tetracarbonylbis[μ-(trifluoroacetato-*O*:*O'*)]-dirhodium, Rh-00047 with

X = H

M 435.942

Catalyst for the hydroformylation of ethylene. Green-red cryst. (pentane). Mp 160° dec. Bp ca. 100° subl. *in vacuo*. ir ν_{CO} 2083vs, 2067vs, 2027vs, 1980s cm^{-1} (nujol).

Lawson, D.N. *et al*, *J. Chem. Soc.*, 1965, 1900 (*synth, ir*)
Johnson, B.F.G. *et al*, *J. Chem. Soc.* (*A*), 1970, 1100 (*synth, ir*)
Balch, A.L. *et al*, *Inorg. Chem.*, 1977, **16**, 2840 (*uv*)
King, R.B. *et al*, *J. Am. Chem. Soc.*, 1979, **101**, 4893 (*use*)
Pruchnik, F. *et al*, *Inorg. Chim. Acta*, 1980, **40**, 203 (*ir, uv*)

$C_8H_7ClNO_2Rh$ Rh-00049

(Benzenamine)dicarbonylchlororhodium, 10CI

(Aniline)dicarbonylchlororhodium

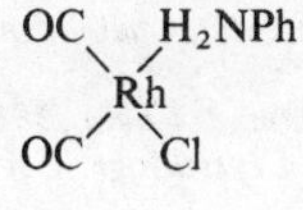

M 287.507

cis-form [72402-60-9]

Dichroic, fibre-like cryst. (CH_2Cl_2/hexane), violet by transmission, golden-green by reflectance. Sol. common org. solvs. ir ν_{CO} 2092vs, 2078m, 2020vs, 2010sh cm^{-1} (nujol). Solns. are pale yellow. Dec. at 168°.

Vallarino, L.M. *et al*, *Inorg. Chim. Acta*, 1979, **36**, 243 (*synth, ir, pmr, uv*)

$C_8H_8BN_4O_2Rh$ Rh-00050

Dicarbonyl[dihydrobis(1*H*-pyrazolato-*N'*)borato(1−)-N^3,$N^{3'}$]rhodium, 10CI

Dicarbonyl[bis(pyrazolyl)borato]rhodium

[33058-09-2]

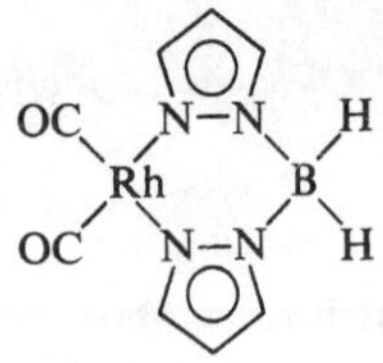

M 305.892

Catalyst for the hydroformylation of alkenes. Yellow cryst. (hexane). Mp 71° (66°). ir ν_{CO} 2088s, 2022s cm^{-1} (hexane).

King, R.B. *et al*, *J. Organomet. Chem.*, 1974, **73**, 115 (*synth, ir, pmr*)
Bonati, F. *et al*, *J. Organomet. Chem.*, 1975, **87**, 365 (*synth, pmr, ir, uv, ms*)
Cocivera, M. *et al*, *Organometallics*, 1982, **1**, 1125 (*ir, synth*)

$C_8H_{10}ClRh$ Rh-00051

Chloro(η^5-2,4-cyclopentadien-1-yl)(η^3-2-propenyl)rhodium, 9CI

π-Allylchloro-π-cyclopentadienylrhodium, 8CI

[12306-82-0]

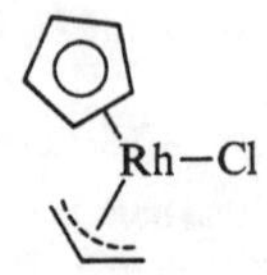

M 244.526

Orange-red microprisms. Mp 175-85° dec.

Powell, J. *et al*, *J. Chem. Soc. (A)*, 1968, 583 (*synth, ir, pmr*)
Nesmeyanov, A.N. *et al*, *J. Organomet. Chem.*, 1978, **164**, 259 (*synth, ir*)

$C_8H_{12}ClRh$ Rh-00052

Bis(η^4-1,3-butadiene)chlororhodium, 9CI

Bis(1,3-butadiene)rhodium chloride. Chlorobis(butadiene)rhodium

[12107-55-0]

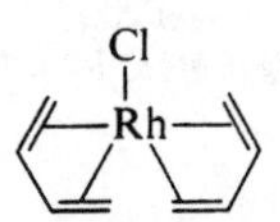

M 246.541

Yellow cryst. solid (EtOH), air-sensitive. Mp 38-40°.

Porri, L. *et al*, *J. Chem. Soc., Chem. Commun.*, 1965, 336 (*synth, ir, struct*)
Porri, L. *et al*, *J. Organomet. Chem.*, 1966, **6**, 422 (*ir*)
Immirzi, A. *et al*, *Acta Crystallogr., Sect. B*, 1969, **25**, 120 (*struct*)
Nelson, S.M. *et al*, *J. Chem. Soc., Dalton Trans.*, 1973, 2195.
Varshawskii, Yu.S. *et al*, *J. Organomet. Chem.*, 1979, **170**, 81.

$C_8H_{12}Cl_2N_4Rh^{\oplus}$ Rh-00053

Dichlorotetrakis(isocyanomethane)rhodium(1+), 10CI

Dichlorotetrakis(methyl isocyanide)rhodium(1+). Dichlorotetrakis(methylisonitrile)rhodium(1+)

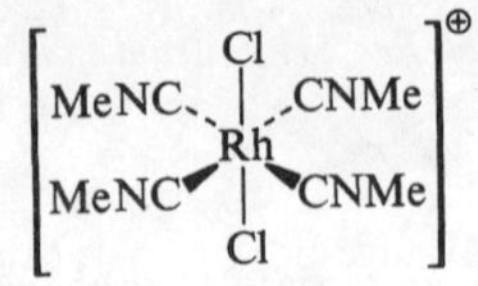

M 338.021 (ion)

***trans*-form**

Hexafluorophosphate: [51139-79-8].
$C_8H_{12}Cl_2F_6N_4PRh$ M 482.985
Yellow cryst. (Me_2CO/pet. ether).

Dart, J.W. *et al*, *J. Chem. Soc., Dalton Trans.*, 1973, 2039 (*synth, ir*)

$C_8H_{12}N_4Rh^{\oplus}$ Rh-00054

Tetrakis(isocyanomethane)rhodium(1+), 10CI

Tetrakis(methyl isocyanide)rhodium(1+). Tetrakis(methylisonitrile) rhodium(1+)

$[Rh(CNMe)_4]^{\oplus}$ (D_{4h})

M 267.115 (ion)

Hexafluorophosphate: [51139-63-0].
$C_8H_{12}F_6N_4PRh$ M 412.079
Violet cryst. (CH_2Cl_2/MeOH).

Tetraphenylborate: [34742-53-5].
$C_{32}H_{32}BN_4Rh$ M 586.347
Yellow cryst. (CH_2Cl_2/propanol). Mp 194-202° dec.

Balch, A.L. *et al*, *J. Organomet. Chem.*, 1971, **32**, 263 (*synth, ir*)
Dart, J.W. *et al*, *J. Chem. Soc., Dalton Trans.*, 1973, 2039 (*synth, ir*)
Kaneshima, T. *et al*, *Inorg. Chem.*, 1974, **13**, 2198 (*ir, pmr*)
Balch, A.L., *J. Am. Chem. Soc.*, 1976, **98**, 8049 (*synth*)
McCleverty, J.A. *et al*, *Transition Met. Chem.*, 1978, **3**, 205 (*synth, ir, pmr*)

$C_8H_{14}Cl_6N_2O_2Rh_2$ Rh-00055

Dicarbonyldi-μ-chlorotetrachlorobis[(dimethylamino)methylene]dirhodium, 9CI

[54438-28-7]

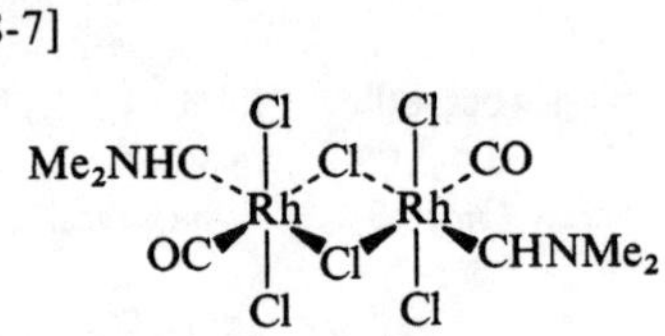

M 588.740

Bridged dimer. Yellow cryst. Insol. common org. solvs. Mp 300° dec. ir ν_{CO} 2140, ν_{CN} 1620 cm^{-1} (nujol).

Monomer: Carbonyltrichloro(dimethylaminomethylene)-rhodium.
$C_4H_7Cl_3NORh$ M 294.370
Has no independent existence.

Cetinkaya, B. *et al*, *J. Chem. Soc., Dalton Trans.*, 1974, 1591 (*synth, ir*)
Hartshorn, A.J. *et al*, *J. Chem. Soc., Dalton Trans.*, 1978, 348 (*props*)

$C_8H_{16}Br_2Rh_2$ **Rh-00056**

Di-μ-bromotetrakis(η^2-ethylene)dirhodium, 8CI
[12275-70-6]

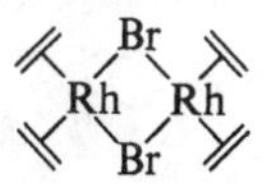

M 477.833
Bridged dimer. Orange-brown cryst. Mp 85-100° dec.
Monomer: Bromobis(ethylene)rhodium.
C_4H_8BrRh M 238.917
Unknown.

Adams, D.M. *et al*, *J. Chem. Soc. (A)*, 1969, 588 (*synth*, *ir*)

$C_8H_{16}Cl_2Rh_2$ **Rh-00057**

Di-μ-chlorotetrakis(η^2-ethene)dirhodium, 10CI
Tetrakis(ethylene)di-μ-chlorodirhodium
[12081-16-2]

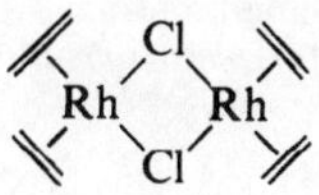

M 388.931
Bridged dimer. Catalyst for ethylene dimerisation, for cooligomerisation of ethylene and butadiene to give 1,3-hexadienes, and for alkene isomerisation. Useful catalyst precursor for asymmetric hydrogenation of olefins in the presence of chiral phosphine. Orange-red solid. Spar. sol. all solvs. Dec. at 115°. Stable at 5° but darkens slowly at r.t.
Monomer: Chlorobis(η^2-ethene)rhodium. Bis(ethylene)-rhodium chloride.
C_4H_8ClRh M 194.466
Unknown.

Cramer, R., *Inorg. Chem.*, 1962, **1**, 722 (*synth*, *struct*)
Cramer, R., *J. Am. Chem. Soc.*, 1964, **86**, 217 (*ir*)
Cramer, R., *J. Am. Chem. Soc.*, 1965, **87**, 4717 (*use*)
Inorg. Synth., 1974, **15**, 14 (*synth*, *ir*)
Fieser, M. *et al*, *Reagents for Organic Synthesis*, Wiley, 1967-83, **5**, 203 (*use*)

$C_8O_8Rh_2$ **Rh-00058**

Di-μ-carbonylhexacarbonyldirhodium, 10CI
Octacarbonyldirhodium, 9CI. Rhodium carbonyl, 8CI. Dirhodium octacarbonyl
[29658-60-4]

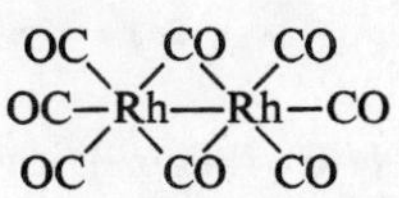

M 429.894
V. unstable. Disproportionates to $Rh_4(CO)_{12}$ at r.t. ir ν_{CO} 2060s, 2043s, 2038s, 1852w, 1830m cm^{-1}.

Whyman, R., *J. Chem. Soc., Dalton Trans.*, 1972, 1375 (*ir*)
Hanlan, L.A. *et al*, *J. Am. Chem. Soc.*, 1974, **96**, 6324 (*ir*)
Hanlan, A.J.L. *et al*, *Inorg. Chem.*, 1979, **18**, 1790 (*uv*)
Vidal, J.L. *et al*, *Inorg. Chem.*, 1981, **20**, 249 (*ir*)

$C_9H_5F_6RhS_2$ **Rh-00059**

π-Cyclopentadienyl[1,1,1,4,4,4-hexafluoro-2-butene-2,3-dithiolato(2−)]rhodium, 8CI
[12169-96-9]

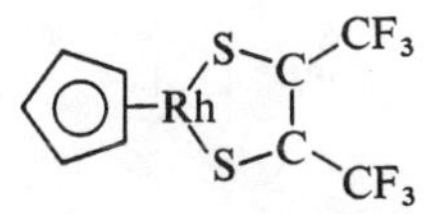

M 394.154
Red cryst. solid (CH_2Cl_2/hexane). Mp 179-80°.

Dessy, R.E. *et al*, *J. Am. Chem. Soc.*, 1966, **88**, 471.
King, R.B. *et al*, *Inorg. Chem.*, 1967, **6**, 469 (*synth*, *ir*, *nmr*, *uv*)

$C_9H_9F_6O_2Rh$ **Rh-00060**

Bis(η^2-ethene)(1,1,1,5,5,5-hexafluoro-2,4-pentanedionato-*O,O'*)rhodium, 9CI
Bis(η^2-ethylene)(1,1,1,5,5,5-hexafluoroacetylacetone)-rhodium
[55188-59-5]

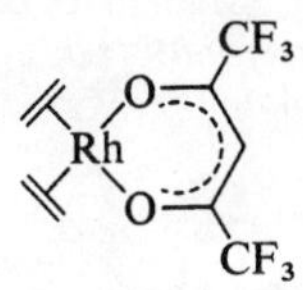

M 366.065
Red prisms. Mp 49-50°. $Bp_{0.1}$ 30° subl.

Ingrosso, G. *et al*, *J. Organomet. Chem.*, 1975, **84**, 75 (*synth*, *pmr*)
Van Dam, H. *et al*, *Inorg. Chem.*, 1980, **19**, 3448 (*pe*)

C_9H_9Rh **Rh-00061**

(η^4-1,3-Cyclobutadiene)(η^5-2,4-cyclopentadiene-1-yl)rhodium, 9CI
π-Cyclopentadienyl-π-cyclobutadienerhodium
[39475-05-3]

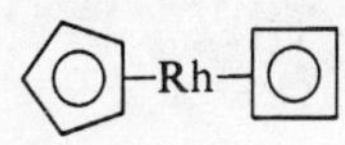

M 220.076
White cryst. by subl. Mp 129.5-130°.

Gardner, S.A. *et al*, *J. Organomet. Chem.*, 1973, **56**, 365 (*synth*, *ir*, *nmr*, *ms*)

$C_9H_{11}Rh$ **Rh-00062**

(η^4-1,3-Butadiene)(η^5-2,4-cyclopentadien-1-yl)rhodium, 9CI
[51232-43-0]

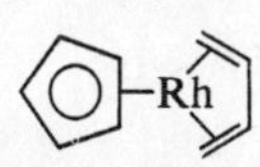

M 222.091
Yellow solid by subl. Mp 83-4°.

Nelson, S.M. *et al*, *J. Chem. Soc., Dalton Trans.*, 1973, 2195 (*pmr*, *synth*, *ir*, *ms*)
Arthurs, M. *et al*, *J. Chem. Soc., Dalton Trans.*, 1977, 779 (*pmr*)
Powell, P. *et al*, *J. Chem. Res. (S)*, 1978, 283.

$C_9H_{13}Rh$ **Rh-00063**

(η^5-2,4-Cyclopentadien-1-yl)bis(η^2-ethene)rhodium, 9CI

π-Cyclopentadienylbis(ethylene)rhodium

[12211-95-9]

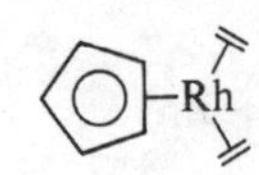

M 224.107

Yellow volatile cryst. (EtOH). Mp 72-3°.

King, R.B., *Inorg. Chem.*, 1963, **2**, 528 (*synth*)
Cramer, R. *et al*, *J. Am. Chem. Soc.*, 1964, **86**, 217 (*pmr, ir*)
King, R.B., *J. Organomet. Chem.*, 1968, **14**, P19 (*ms*)
Cramer, R. *et al*, *J. Am. Chem. Soc.*, 1969, **91**, 2519 (*pmr*)
Bodner, G.M. *et al*, *J. Chem. Soc., Chem. Commun.*, 1970, 1530 (*cmr*)
Howard, J. *et al*, *J. Chem. Soc., Faraday Trans. 2*, 1976, **72**, 513 (*ir, raman*)

$C_9H_{15}O_2Rh$ **Rh-00064**

Bis(η^2-ethene)(2,4-pentanedionato-*O,O'*)rhodium, 10CI, 9CI

Bis(ethylene)acetylacetonatorhodium. Bis(ethylene)-rhodium acetylacetonate

[12082-47-2]

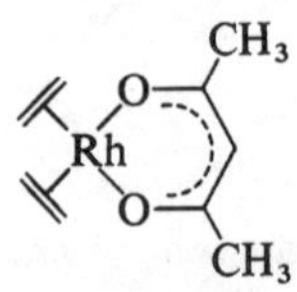

M 258.122

Catalyst for polymerisation of olefins. Orange-yellow platelets (Et_2O). Mp 144-6° dec. Best stored at 0°.

Cramer, R., *J. Am. Chem. Soc.*, 1964, **86**, 217 (*synth, ir, pmr*)
Russell, D.R. *et al*, *J. Chem. Soc., Chem. Commun.*, 1971, 197 (*cryst struct*)
Bouchal, K. *et al*, *Collect. Czech. Chem. Commun.*, 1972, **37**, 3081 (*synth*)
Haegele, K.D. *et al*, *Inorg. Chem.*, 1974, **13**, 1960 (*ms*)
Inorg. Synth., 1974, **15**, 14 (*synth, ir*)
Jesse, A.C. *et al*, *Inorg. Chim. Acta*, 1978, **26**, 129 (*ir, raman, cmr*)

$C_9H_{15}Rh$ **Rh-00065**

Tris(η^3-2-propenyl)rhodium, 10CI, 9CI

Tri(π-allyl)rhodium, 8CI

[12082-48-3]

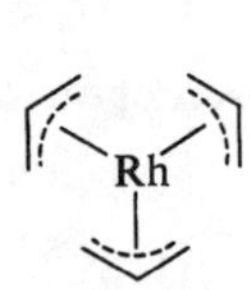

M 226.123

Catalyst for the polymerisation of butadiene and vinyl compds. Yellow prisms. Mp 80-5°. $Bp_{0.01}$ 40° subl. Dec. at 130°.

Becconsall, J.K. *et al*, *J. Chem. Soc., Chem. Commun.*, 1966, 720 (*pmr, ms*)
Powell, J. *et al*, *J. Chem. Soc. (A)*, 1968, 583 (*synth, pmr*)
Andrews, D.C. *et al*, *J. Organomet. Chem.*, 1973, **55**, 383 (*ir, raman*)
Chenskaye, T.B. *et al*, *J. Organomet. Chem.*, 1978, **148**, 85 (*ir, raman*)

$C_{10}H_6N_4O_4Rh_2$ **Rh-00066**

Tetracarbonylbis[μ-(1*H*-pyrazolato-N^1:N^2)]dirhodium, 10CI

[77244-11-2]

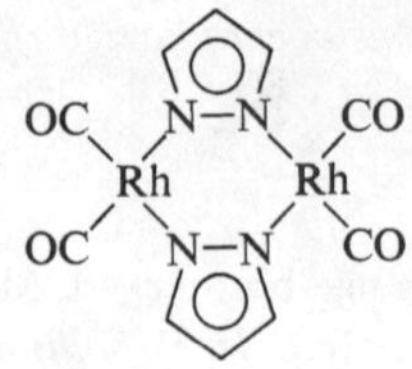

M 451.993

Air-stable yellow solid. ir ν_{CO} 2022, 2079, 2090 cm^{-1} (cyclohexane).

Borkett, N.F. *et al*, *J. Organomet. Chem.*, 1974, **65**, C51.
Usòn, R. *et al*, *J. Organomet. Chem.*, 1981, **205**, 247 (*synth, ir*)

$C_{10}H_{10}BCl_2N_6ORh$ **Rh-00067**

Carbonyldichloro[hydrotris(1*H*-pyrazolato-*N'*)borato(1–)]-rhodium

Carbonyldichloro[tris(pyrazolyl)borato]rhodium

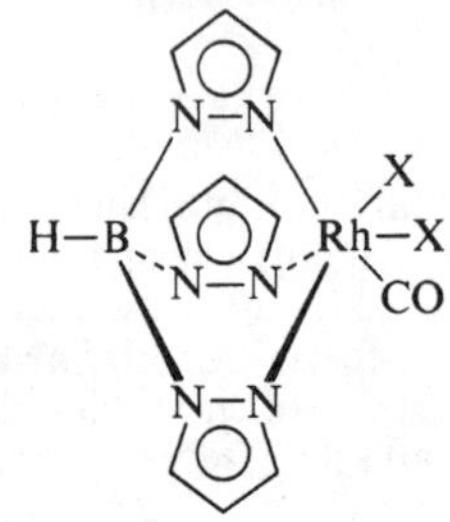

X = Cl

M 414.850

Struct. by analogy with Carbonyl[hydrotris(1*H*-pyrazolato-*N'*)borato(1–)]diiodorhodium, Rh-00068 . Loses CO slowly in solid state and rapidly on attempted recryst. ir ν_{CO} 2129 cm^{-1} (KBr).

Cocivera, M. *et al*, *Organometallics*, 1982, **1**, 1125 (*synth, ir*)

$C_{10}H_{10}BI_2N_6ORh$ **Rh-00068**

Carbonyl[hydrotris(1*H*-pyrazolato-*N'*)borato(1–)]diiodo-rhodium, 9CI

Carbonyldiiodo[tris(pyrazolyl)borato]rhodium

[52504-76-4]

As Carbonyldichloro[hydrotris(1*H*-pyrazolato-*N'*)-borato(1–)]rhodium, Rh-00067 with

X = I

M 597.753

Deep-red cryst. (CH_2Cl_2/hexane). ir ν_{CO} 2095 cm^{-1} (KBr).

Cocivera, M. *et al*, *Organometallics*, 1982, **1**, 1125 (*synth, pmr, ir, struct*)

$C_{10}H_{10}Cl_4Rh_2$ **Rh-00069**

Di-μ-chlorodichlorobis(η^5-2,4-cyclopentadien-1-yl)dirhodium, 9CI

Di-μ-chlorodichlorodi-π-cyclopentadienyldirhodium, 8CI. Tetrachlorodicyclopentadienyldirhodium

[12307-16-3]

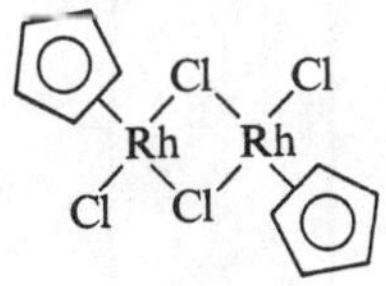

M 477.812

Bridged dimer. Brick-brown insol. solid. Insol. common org. solvs., H_2O. Mp >310° dec.

Monomer: Dichloro(cyclopentadienyl)rhodium.

$C_5H_5Cl_2Rh$ M 238.906

Unknown.

Powell, J. *et al*, *J. Chem. Soc. (A)*, 1968, 583 (*ir*)
Kang, J.W. *et al*, *J. Am. Chem. Soc.*, 1969, **91**, 5970 (*synth*)
Faraone, F. *et al*, *J. Organomet. Chem.*, 1978, **152**, 337.

$C_{10}H_{10}N_2O_2Rh_2$ **Rh-00070**

Bis(η^5-2,4-cyclopentadien-1-yl)di-μ-nitrosyldirhodium, 10CI

Bis[(η^5-cyclopentadienyl)nitrosylrhodium]

[67426-08-8]

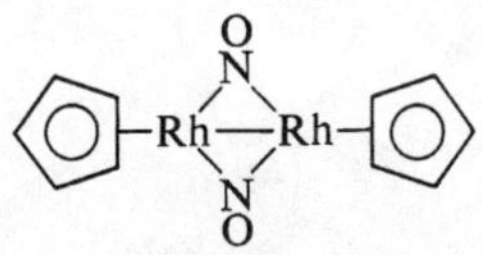

M 396.012

Brown cryst. (C_6H_6/pentane). Sol. $CHCl_3$. Struct. assigned by analogy with $Cp_2Co_2(NO)_2$.

Monomer: (η^5-2,4-Cyclopentadien-1-yl)-nitrosylrhodium.

C_5H_5NORh M 198.006

Unknown.

Dimas, P.A. *et al*, *Inorg. Chem.*, 1981, **20**, 281 (*synth, pmr, ir, props*)

$C_{10}H_{10}Rh^{\oplus}$ **Rh-00071**

Rhodocenium, 10CI, 9CI

Bis(cyclopentadienyl)rhodium

[37205-11-1]

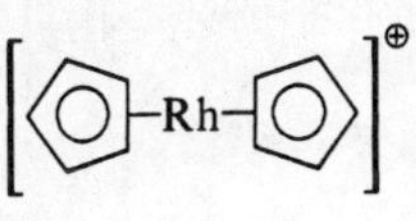

M 233.095 (ion)

May be reduced to rhodocene.

Tribromide:

$C_{10}H_{10}Br_3Rh$ M 472.807

Golden-yellow powder.

Hexafluorophosphate:

$C_{10}H_{10}F_6PRh$ M 378.059

Colourless.

Cotton, F.A. *et al*, *J. Am. Chem. Soc.*, 1953, **75**, 3586 (*synth, uv*)
Fischer, E.O. *et al*, *Chem. Ber.*, 1959, **92**, 1624 (*synth*)
Angelici, R.J. *et al*, *J. Am. Chem. Soc.*, 1963, **85**, 3733 (*synth*)
Fischer, E.O. *et al*, *J. Organomet. Chem.*, 1966, **5**, 559 (*synth, ir, uv*)

$C_{10}H_{11}Rh$ **Rh-00072**

[(1,2,3,4-η)-1,3-Cyclopentadiene](η^5-2,4-cyclopentadien-1-yl)rhodium, 9CI

[1293-77-2]

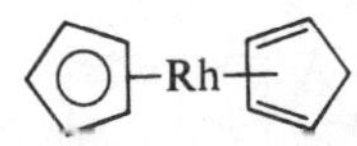

M 234.102

Orange-yellow cryst. solid (pentane). Mp 121-2° (118-20°). Sublimes. Dec. in air (~50% recoverable after 2 days).

Heinz, P.F., *Chem. Ber.*, 1959, **92**, 780 (*ir*)
Fischer, E.O. *et al*, *Chem. Ber.*, 1959, **92**, 1624 (*synth*)
Green, M.L.H. *et al*, *J. Chem. Soc.*, 1959, 3753 (*synth, uv, ir, pmr*)
Müller, J. *et al*, *Chem. Ber.*, 1975, **108**, 1074 (*ms*)

$C_{10}H_{13}Rh$ **Rh-00073**

(η^5-2,4-Cyclopentadien-1-yl)[(1,2,3,4-η)-1,3-pentadiene]rhodium

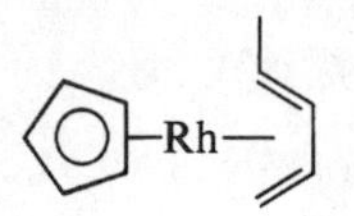

(*E*)-*form*

M 236.118

(***E***)-***form*** [51267-94-8]

Yellow oil.

(***Z***)-***form*** [51232-46-3]

Yellow oil. Can be isomerised to (*E*)-*form* by heating in inert solvent.

Nelson, S.M. *et al*, *J. Chem. Soc., Dalton Trans.*, 1973, 2195 (*synth, ir, ms, pmr*)
Arthurs, M. *et al*, *J. Chem. Soc., Dalton Trans.*, 1975, 1794 (*synth*)

$C_{10}H_{13}Rh$ **Rh-00074**

(η^5-2,4-Cyclopentadien-1-yl)[(1,2,4,5-η)-1,4-pentadiene]rhodium, 9CI

[57683-42-8]

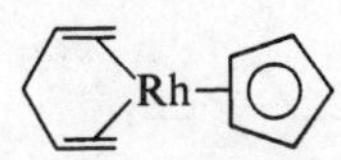

M 236.118

Red oil. Can be isomerised to (η^5-2,4-Cyclopentadien-1-yl)[(1,2,3,4-η)-1,3-pentadiene]rhodium, Rh-00073 .

Arthurs, M. *et al*, *J. Chem. Soc., Dalton Trans.*, 1975, 1794 (*synth, pmr, ms*)
Arthurs, M. *et al*, *J. Chem. Soc., Dalton Trans.*, 1977, 779 (*nmr*)

$C_{10}H_{14}Rh^{\oplus}$ **Rh-00075**

(η^6-Benzene)bis(η^2-ethene)rhodium(1+), 9CI

Benzenebisethylenerhodium(1+)

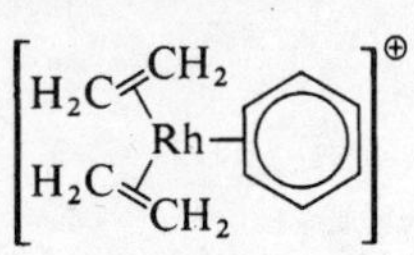

M 237.126 (ion)

Tetrafluoroborate: [36644-72-1].

$C_{10}H_{14}BF_4Rh$ M 323.930

Yellow cryst. (CH_2Cl_2/Et_2O). Mp 133-4° dec.

Green, M. *et al*, *J. Chem. Soc., Dalton Trans.*, 1972, 832 (*synth, pmr*)

$C_{10}H_{15}Cl_2Rh$ **Rh-00076**

Dichloropentamethylcyclopentadienylrhodium

Di-μ-chlorodichlorobis[(1,2,3,4,5-η)-1,2,3,4,5-pentamethyl-2,4-cyclopentadien-1-yl]dirhodium, 10CI, 9CI

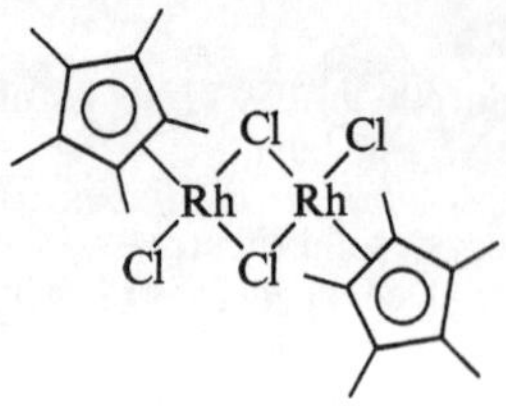

M 309.040

Dimeric.

Dimer: [12354-85-7].

$C_{20}H_{30}Cl_4Rh_2$ M 618.080

V. active hydrogenation catalyst for arenes and alkenes and homogeneous catalyst for acetaldehyde disproportionation. Starting material for pentamethylcyclopentadienyl-rhodium compds. Red cryst. solid. Mp >300°.

Booth, B.L. *et al*, *J. Chem. Soc. (A)*, 1969, 1299 (*synth, ms, uv, ir, pmr*)
Kang, J.W. *et al*, *J. Am. Chem. Soc.*, 1969, **91**, 5970 (*synth, pmr*)
Churchill, M.R. *et al*, *Inorg. Chem.*, 1977, **16**, 1137 (*struct*)
Russell, M.J. *et al*, *J. Chem. Soc., Chem. Commun.*, 1977, 427 (*use*)
Maitis, P.M., *Acc. Chem. Res.*, 1978, **11**, 301 (*rev*)
Gill, D.S. *et al*, *J. Magn. Reson.*, 1979, **35**, 459 (*nmr*)
Cook, J. *et al*, *J. Chem. Soc., Chem. Commun.*, 1980, 144 (*use*)
Fieser, M. *et al*, *Reagents for Organic Synthesis*, Wiley, 1967-83, **8**, 152 (*use*)

$C_{10}H_{15}F_6P_2Rh$ **Rh-00077**

[(1,2,3,4,5-η)-1,2,3,4,5-Pentamethyl-2,4-cyclopentadien-1-yl]bis(phosphorus trifluoride)rhodium, 9CI

(Pentamethylcyclopentadienyl)bis(trifluorophosphine)-rhodium

[34822-29-2]

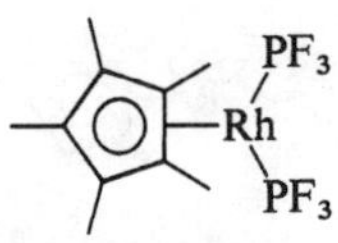

M 414.072

Deep-orange cryst. by subl. Mp 88-9°. $Bp_{0.01}$ 30° subl. Readily undergoes oxidative addition e.g. CF_3I gives $Rh(C_5Me_5)(PF_3)(CF_3)I$.

King, R.B. *et al*, *J. Am. Chem. Soc.*, 1972, **94**, 3768 (*synth, nmr, ir, ms*)
King, R.B. *et al*, *J. Organomet. Chem.*, 1972, **36**, 371 (*props*)

$C_{10}H_{18}N_2Rh^{\oplus}$ **Rh-00078**

Bis(acetonitrile)tris(η^2-ethene)rhodium(1+), 9CI

Bis(acetonitrile)tris(ethylene)rhodium(1+)

[37080-23-2]

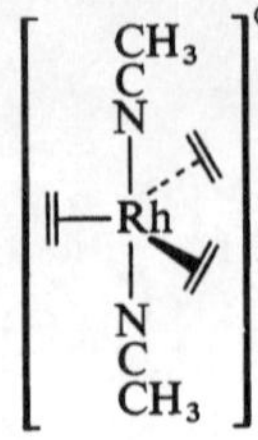

M 269.171 (ion)

Tetrafluoroborate:

$C_{10}H_{18}BF_4N_2Rh$ M 355.975

Yellow needles. Sol. with rapid dec. in most polar solvs. Stable in C_2H_4 atm. but loses C_2H_4 under Ar.

Maspero, F. *et al*, *J. Organomet. Chem.*, 1972, **38**, C43 (*synth, ir, pmr, struct*)
Del Piero, G. *et al*, *Cryst. Struct. Commun.*, 1974, **3**, 15 (*struct*)

$C_{10}H_{18}PRh$ **Rh-00079**

(η^5-2,4-Cyclopentadien-1-yl)(η^2-ethene)(trimethylphosphine)-rhodium, 10CI

Cyclopentadienyl(ethylene)trimethylphosphinerhodium

[69178-16-1]

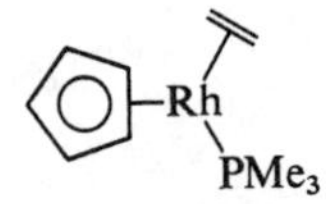

M 272.131

Yellow-brown cryst. (pentane). Mp 45° dec.

Werner, H. *et al*, *J. Organomet. Chem.*, 1982, **232**, 351 (*pmr, ms, synth*)
Werner, H. *et al*, *J. Organomet. Chem.*, 1982, **236**, C71.

$C_{10}H_{24}N_3Rh^{\oplus\oplus}$ **Rh-00080**

Triammine[(1,2,3,4,5-η)-1,2,3,4,5-pentamethyl-2,4-cyclopentadien-1-yl]rhodium(2+), 10CI

Tris(ammonia)-pentamethylcyclopentadienylrhodium(2+)

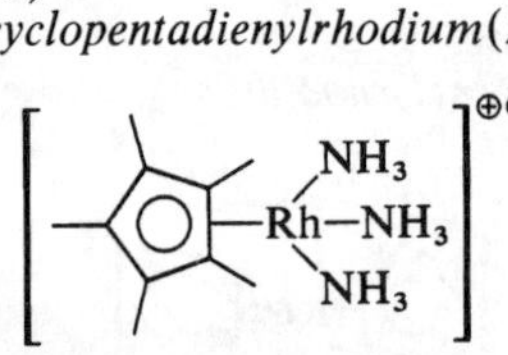

M 289.225 (ion)

Dichloride: [71412-47-0].

$C_{10}H_{24}Cl_2N_3Rh$ M 360.131

Pale-yellow, hygroscopic powder. Insol. Me_2CO, $CHCl_3$, sol. liq. SO_2.

Bis(hexafluorophosphate): [71412-49-2].

$C_{10}H_{24}F_{12}N_3P_2Rh$ M 579.154

Bright-yellow cryst. (Me_2CO/diisopropyl ether).

Rigby, W. *et al*, *J. Chem. Soc., Dalton Trans.*, 1979, 382 (*synth, pmr*)

$C_{11}H_6NO_3Rh$ **Rh-00081**

Dicarbonyl(8-quinolinolato-N^1,O^8)rhodium, 10CI, 9CI, 8CI

[20956-26-7]

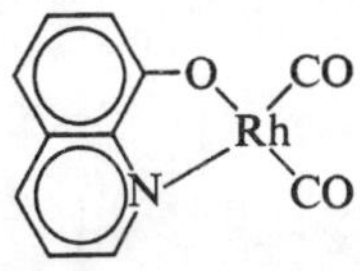

M 303.079

Biologically active. Catalyst for C=C hydrogenation, hydroformylation, hydrosilylation. Dichroic black-violet cryst., yellow-orange in $CHCl_3$. Mp 210°. ir ν_{CO} 2073vs, 2006s cm^{-1} ($CHCl_3$).

Ugo, R. *et al*, *J. Organomet. Chem.*, 1968, **11**, 159 (*synth*, *ir*, *props*)
Bresler, L.S. *et al*, *J. Organomet. Chem.*, 1979, **171**, 229 (*cmr*, *ir*)
Usòn, R. *et al*, *J. Organomet. Chem.*, 1981, **205**, 259.

$C_{11}H_7O_2Rh$ **Rh-00082**

Dicarbonyl(1,2,3,3*a*,7*a*-η)-1*H*-inden-1-yl)rhodium, 9CI

[12153-10-5]

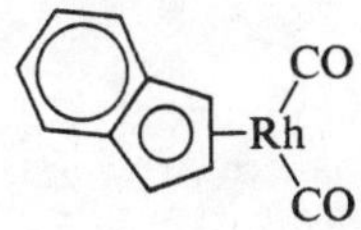

M 274.081

Cryst. Mp 54-55.5°.

Abel, E.W. *et al*, *J. Chem. Soc., Dalton Trans.*, 1973, 1706 (*synth*)
Caddy, P. *et al*, *J. Chem. Soc., Chem. Commun.*, 1978, 839 (*synth*)
Eshtigh-Hosseini, H. *et al*, *J. Less-Common Met.*, 1978, **61**, 107 (*synth*, *ir*, *nmr*)

$C_{11}H_8ClN_2ORh$ **Rh-00083**

(2,2′-Bipyridine-*N*,*N*′)carbonylchlororhodium

(2,2′-Bipyridyl)carbonylchlororhodium

[52659-24-2]

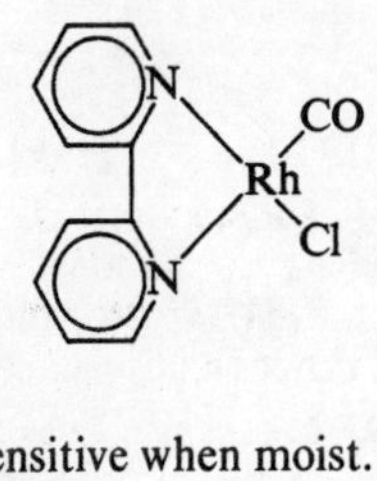

M 322.556

Black solid, air-sensitive when moist. Sol. $CHCl_3$. ir ν_{CO} 1958 cm^{-1} (nujol).

Mestroni, G. *et al*, *J. Organomet. Chem.*, 1974, **65**, 119 (*synth*, *ir*)
Drago, R.S. *et al*, *Inorg. Chem.*, 1981, **20**, 2461 (*ir*)

$C_{11}H_{11}Rh^{\oplus\oplus}$ **Rh-00084**

η⁶-Benzene(η⁵-2,4-cyclopentadien-1-yl)rhodium(2+), 9CI

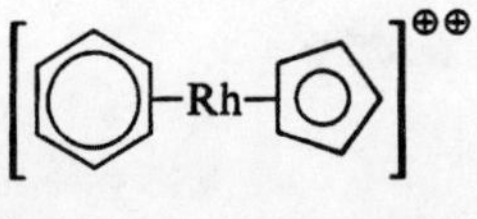

M 246.113 (ion)

Bis(tetrafluoroborate): [38959-26-1].
$C_{11}H_{11}B_2F_8Rh_2$ M 522.626
White solid.

Johnson, B.F.G. *et al*, *J. Chem. Soc., Dalton Trans.*, 1972, 2084 (*synth*)

$C_{11}H_{13}BCl_2N_7Rh$ **Rh-00085**

(Acetonitrile)dichloro[hydrotris(1*H*-pyrazolato-N^1)borato(1−)-N^2,$N^{2'}$,$N^{2''}$]rhodium, 10CI

(Hydrotrispyrazolylborato)dichloro(acetonitrile)rhodium

[73137-32-3]

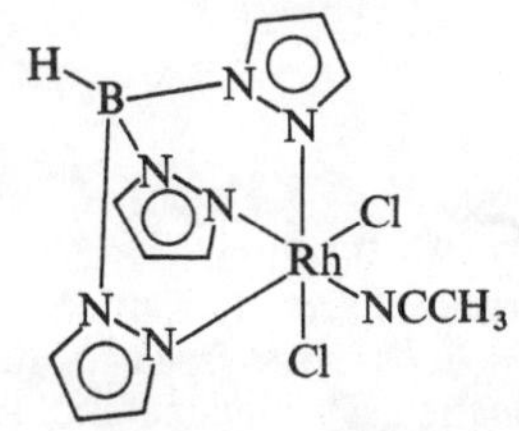

M 427.892

Starting material for hydrotris(1-pyrazolyl)boratorhodium complexes. Dull-gold prisms (MeCN). Mp >360°.

May, S. *et al*, *Inorg. Chem.*, 1980, **19**, 1582 (*synth*, *pmr*)

$C_{11}H_{13}Rh$ **Rh-00086**

[(1,2,3,4-η)-1,3-Cyclohexadiene](η⁵-2,4-cyclopentadien-1-yl-)rhodium, 9CI

[37685-50-0]

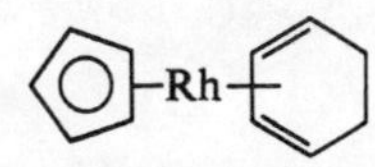

M 248.129

Yellow cryst. Mp 71-2°.

Johnson, B.F.G. *et al*, *J. Chem. Soc., Dalton Trans.*, 1972, 2084 (*pmr*)
Arthurs, M. *et al*, *J. Chem. Soc., Dalton Trans.*, 1975, 1794 (*synth*, *pmr*)
Arthurs, M. *et al*, *J. Chem. Soc., Dalton Trans.*, 1977, 779 (*nmr*)

$C_{11}H_{14}N_2Rh^{\oplus}$ **Rh-00087**

Bis(acetonitrile)[(2,3,5,6-η)-bicyclo[2.2.1]hepta-2,5-diene]rhodium(1+), 10CI

Bis(acetonitrile)(norbornadiene)rhodium(1+), 8CI

[46240-59-9]

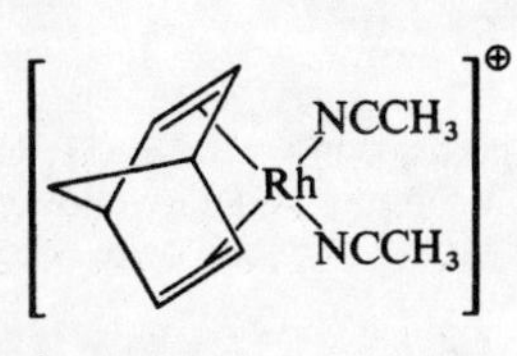

M 277.151 (ion)

Tetrafluoroborate: [34664-33-0].
$C_{11}H_{14}BF_4N_2Rh$ M 363.954
Air-stable yellow cryst. (CH_2Cl_2/Et_2O). Mp 111°.

Denise, B. *et al*, *J. Organomet. Chem.*, 1978, **161**, 171.
Green, M. *et al*, *J. Chem. Soc. (A)*, 1971, 2334 (*synth*, *ir*, *pmr*, *uv*)

$C_{11}H_{15}Rh$ **Rh-00088**

(η^5-2,4-Cyclopentadien-1-yl)-2-propenyl(η^3-2-propenyl)rhodium, 9CI

Allyl-π-allyl-π-cyclopentadienylrhodium, 8CI. Bis(allyl)-η-cyclopentadienylrhodium. Diallyl-η⁵-cyclopentadienylrhodium. Cyclopentadienylbis(2-propenyl)-rhodium

[12307-51-6]

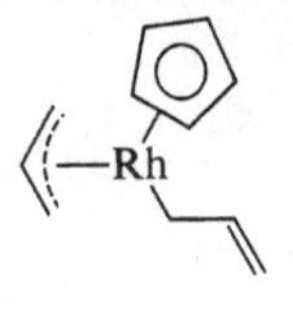

M 250.145

Yellow oil. Fp −20°. $Bp_{0.01}$ 30°.

Powell, J. *et al*, *J. Chem. Soc. (A)*, 1968, 583 (*synth, ir, pmr*)
Rubezhov, A.Z. *et al*, *Izv. Akad. Nauk SSSR, Ser. Khim.*, 1973, 951; *CA*, **79**, 42656e (*synth*)

$C_{11}H_{17}O_2Rh$ **Rh-00089**

(2,4-Pentanedionato-*O,O'*)bis(η^3-2-propenyl)rhodium, 10CI

Di-π-allyl(2,4-pentanedionato)rhodium, 8CI. Bis(η^3-allyl)(2,4-pentanedionato-O,O')rhodium. Rhodium bisallylacetylacetonate. Acetylacetonatodiallylrhodium

[12307-52-7]

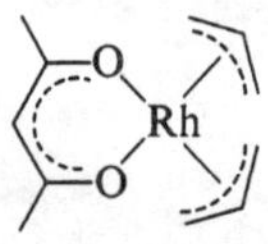

M 284.160

Catalyst for the hydrosilylation of unsaturated hydrocarbons and for C═C cleavage by molecular oxygen. Yellow cryst. Mp 56-9°. $Bp_{0.01}$ 35° subl.

Powell, J. *et al*, *J. Chem. Soc. (A)*, 1968, 583 (*synth, pmr*)
Boennemann, H. *et al*, *Helv. Chim. Acta*, 1983, **66**, 177 (*use*)

$C_{11}H_{17}Rh$ **Rh-00090**

[(1,2,5,6-η)-1,5-Cyclooctadiene](η^3-2-propenyl)rhodium, 10CI

π-Allyl(1,5-cyclooctadiene)rhodium, 8CI

[12176-45-3]

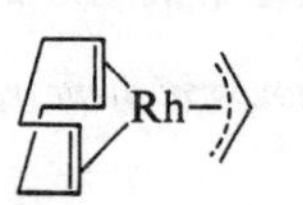

M 252.161

Starting material for synth. of η^3-$C_3H_5RhL_2$ and η^3-$C_3H_5RhL_3$ complexes. Pale-yellow cryst. by subl. Mp 30.5-31°, 108-12° dec. Slowly dec. at r.t., subl. *in vacuo* at r.t.

Kasahara, A. *et al*, *Bull. Chem. Soc. Jpn.*, 1966, **39**, 634 (*synth, pmr*)
Sivak, A.J. *et al*, *J. Am. Chem. Soc.*, 1979, **101**, 4878 (*synth, ms, pmr, use*)

$C_{11}H_{18}ClN_2ORh$ **Rh-00091**

Carbonylchlorobis(2-isocyano-2-methylpropane)rhodium, 10CI

Bis(tert-butyl isocyanide)carbonylchlororhodium

[72547-64-9]

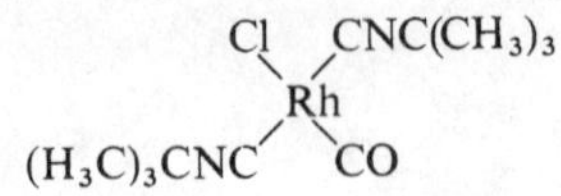

M 332.635

Blue-black solid. Sol. $CHCl_3$. Not fully characterized. ir ν_{CO} 2000 cm^{-1} (nujol).

Deeming, A.J., *J. Organomet. Chem.*, 1979, **175**, 105 (*synth, ir*)

$C_{11}H_{23}P_2Rh$ **Rh-00092**

(η^5-2,4-Cyclopentadien-1-yl)bis(trimethylphosphine)rhodium, 10CI

[69178-15-0]

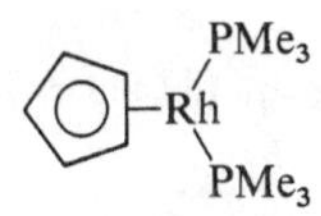

M 320.156

Red-brown needles (pentane at −78°). Mp 85°. Reacts readily with cationic electrophiles to form stable cationic rhodium complexes.

Werner, H. *et al*, *Chem. Ber.*, 1979, **112**, 834 (*synth, props*)

$C_{11}O_{11}Rh_4^{\ominus\ominus}$ **Rh-00093**

Hepta-μ-carbonyltetracarbonyltetrarhodate(2−), 9CI

Undecacarbonyltetrarhodate(2−)

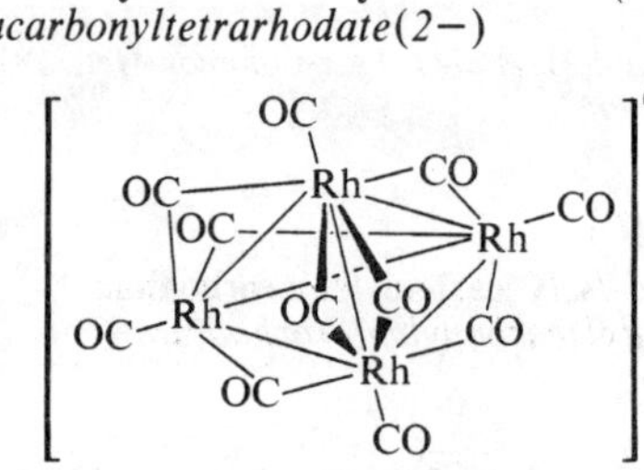

M 719.736 (ion)

ir ν_{CO} 1975, 1925, 1810 cm^{-1} (THF).

Bis[bis(triphenylphosphine)iminium]salt: [60815-43-2]. $C_{83}H_{60}N_2O_{11}P_4Rh_4$ M 1796.911
Orange-red cryst. (THF/2-propanol). Highly reactive towards CO and other rhodiumcarbonyl clusters.

Martinengo, S. *et al*, *J. Organomet. Chem.*, 1976, **116**, 333 (*synth, ir*)
Albano, V.G. *et al*, *J. Organomet. Chem.*, 1976, **116**, 343 (*cryst struct*)
Chini, P. *et al*, *Adv. Organomet. Chem.*, 1976, **14**, 285 (*ir*)

$C_{12}Co_2O_{12}Rh_2$ Rh-00094

Tri-μ-carbonyltetracarbonyl(pentacarbonyldicobalt)dirhodium, 9CI

Dodecacarbonyldicobaltdirhodium, 8CI

[50696-78-1]

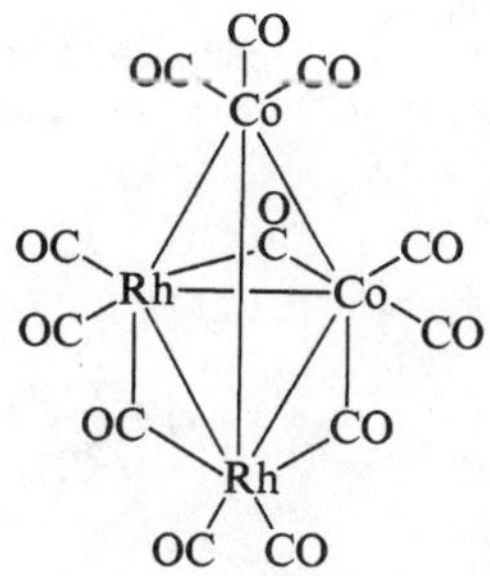

M 659.802

Proposed struct. shown. Hydrogenation and hydroformylation catalyst. Brown cryst. (pentane). Sol. hydrocarbons. Dec. at 120°.

Martinengo, S. *et al*, *J. Organomet. Chem.*, 1973, **59**, 379 (*synth, ir*)

Michelin-Lausarot, P. *et al*, *Inorg. Chim. Acta*, 1979, **35**, 227 (*ms*)

$C_{12}Co_3O_{12}Rh$ Rh-00095

Di-μ-carbonyldicarbonyl(μ-carbonylheptacarbonyltricobalt)rhodium, 9CI

Dodecacarbonyltricobaltrhodium, 8CI

[50696-79-2]

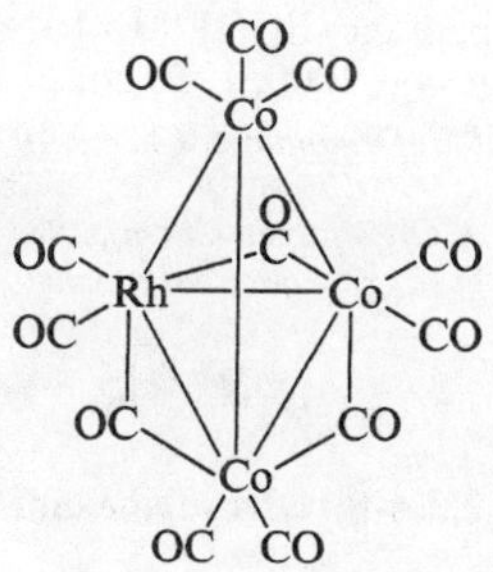

M 615.830

Fluxional at r.t. with CO exchange. Hydrogenation and hydroformylation catalyst. Brown cryst. (CH_2Cl_2/cyclohexane). Sol. hydrocarbons. Dec. at 130°.

Martinengo, S. *et al*, *J. Organomet. Chem.*, 1973, **59**, 379 (*synth, ir*)

Johnson, B.F.G. *et al*, *J. Chem. Soc., Chem. Commun.*, 1974, 441 (*cmr, struct*)

Labroue, D. *et al*, *J. Mol. Catal.*, 1977, **2**, 329 (*use*)

Evans, J. *et al*, *J. Chem. Soc., Dalton Trans.*, 1978, 626 (*cmr*)

Ichikawa, M., *J. Catal.*, 1979, **59**, 67 (*use*)

Michelin-Lausarot, P. *et al*, *Inorg. Chim. Acta*, 1979, **35**, 227 (*ms*)

Having problems with locating a compound? Have you checked the indexes?

$C_{12}H_7FeO_5Rh$ Rh-00097

Tricarbonyl[μ-[(1,2,3-η:4,5,6,7-η)-2,4,6-cycloheptatrien-1-yl]](dicarbonylrhodium)iron, 9CI

μ-[1,2,3-η:4,5,6,7-η-Cycloheptatrienyl]tricarbonylirondicarbonylrhodium

[60767-44-4]

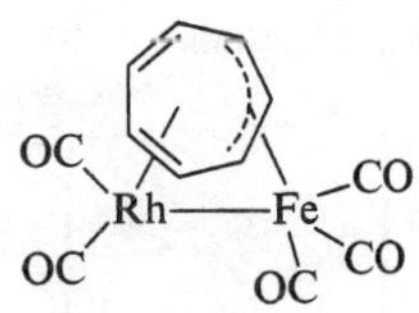

M 389.937

Fluxional molecule. Red-brown cryst. (pentane). Mp 96-8°.

Bennett, M.J. *et al*, *J. Am. Chem. Soc.*, 1976, **98**, 4810 (*synth, struct, pmr, cmr, ms, ir*)

$C_{12}H_8Cl_2O_4Rh_2$ Rh-00098

Di-μ-chlorobis[(2,3,5,6-η)-2,5-cyclohexadiene-1,4-dione]dirhodium, 9CI

*Di-μ-chlorobis(*p*-benzoquinone)dirhodium*

[41619-54-9]

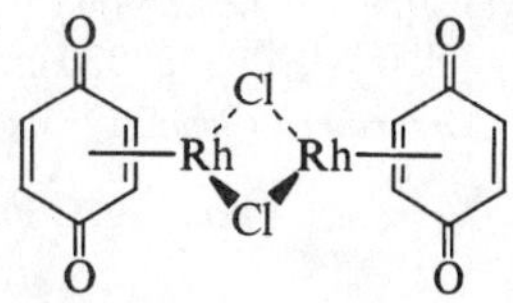

M 492.910

Red solid.

Monomer: Chloro(2,5-cyclohexadiene-1,4-dione)rhodium. p*-Benzoquinonechlororhodium.* p*-Benzoquinonerhodium chloride.*

$C_6H_4ClO_2Rh$ M 246.455

Unknown.

Partenheimer, W. *et al*, *J. Am. Chem. Soc.*, 1973, **95**, 2840 (*synth, ir*)

$C_{12}H_8N_2O_2Rh^{\oplus}$ Rh-00099

(2,2′-Bipyridine-*N,N′*)dicarbonylrhodium(1+), 10CI, 9CI

(2,2′-Bipyridyl)dicarbonylrhodium(1+)

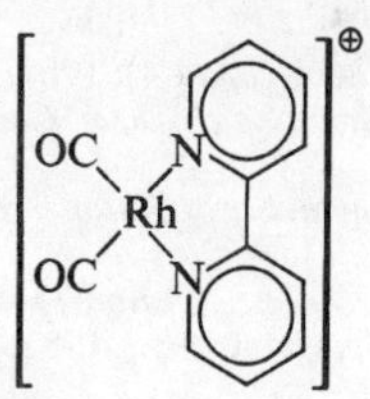

M 315.113 (ion)

Perchlorate: [25300-52-1].

$C_{12}H_8ClN_2O_6Rh$ M 414.564

Yellow solid. Mp 200-5°.

Tetrafluoroborate: [52699-87-3].

$C_{12}H_8BF_4N_2O_2Rh$ M 401.917

No phys. props. reported.

cis-*Dicarbonyldibromorhodate:* [69192-89-8].

$C_{14}H_8Br_2N_2O_4Rh_2$ M 633.847

Dark-brown cryst. with metallic lustre. Mod. sol. polar org. solvs. Mp 165-8° dec.

cis-*Dicarbonyldichlororhodate:* [55914-11-9].

$C_{14}H_8Cl_2N_2O_4Rh_2$ M 544.945

Brown cryst. with metallic lustre, blue-green when powdered. Mp 120-3° dec.

Cocevar, C. *et al*, *J. Organomet. Chem.*, 1972, **35**, 389 (*synth*)

Reddy, G.K.N. *et al*, *J. Organomet. Chem.*, 1974, **67**, 443 (*synth*)
Pruchnik, F. *et al*, *J. Organomet. Chem.*, 1979, **164**, 71 (*synth*, *ir*, *uv*)

$C_{12}H_{10}N_2O_2Rh^{\oplus}$ Rh-00100

Dicarbonylbis(pyridine)rhodium(1+), 9CI

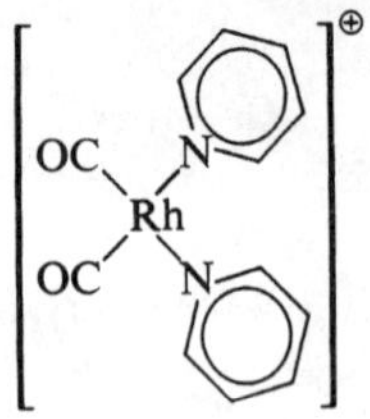

M 317.129 (ion)
Lit. refs. in disagreement over ir spectrum and struct.

cis-form

Perchlorate: [59091-29-1].
$C_{12}H_{10}ClN_2O_6Rh$ M 416.579
Solid. Mp 120° dec.
Hexafluorophosphate: [51455-33-5].
$C_{12}H_{10}F_6N_2O_2PRh$ M 462.093
Pale-yellow solid. ir ν_{CO} 2095, 2020 cm^{-1} ($CHCl_3$).

Denise, B. *et al*, *J. Organomet. Chem.*, 1973, **63**, 423 (*synth*, *ir*, *pmr*)
Uson, R. *et al*, *J. Organomet. Chem.*, 1976, **105**, 365 (*synth*, *ir*)

$C_{12}H_{13}Rh$ Rh-00101

[(2,3,5,6-η)Bicyclo[2.2.1]hepta-2,5-diene](η⁵-2,4-cyclopentadien-1-yl)rhodium, 10CI
π-Cyclopentadienyl(2,5-norbornadiene)rhodium, 8CI
[12295-14-6]

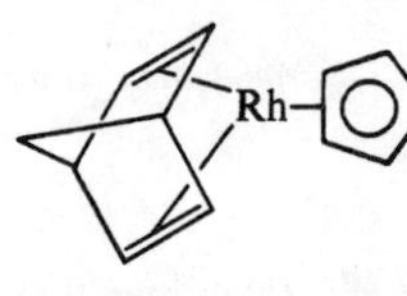

M 260.140
Pale-yellow cryst. Mp 71-2°. $Bp_{0.1}$ 75° subl.

Cook, D.J. *et al*, *J. Chem. Soc. (A)*, 1967, 1547 (*synth*)
Todd, L.J. *et al*, *J. Chem. Soc., Chem. Commun.*, 1970, 1530 (*cmr*)
Green, M. *et al*, *J. Chem. Soc., Dalton Trans.*, 1972, 832 (*synth*, *pmr*)
Evans, J. *et al*, *J. Chem. Soc., Dalton Trans.*, 1977, 510 (*synth*)
Maurer, E. *et al*, *Helv. Chim. Acta*, 1982, **65**, 26 (*nmr*)

$C_{12}H_{14}Rh^{\oplus}$ Rh-00102

(η⁶-Benzene)[(1,2,3,4-η)-1,3-cyclohexadiene]rhodium(1+), 10CI
1,3-Cyclohexadiene(benzene)rhodium(1+)

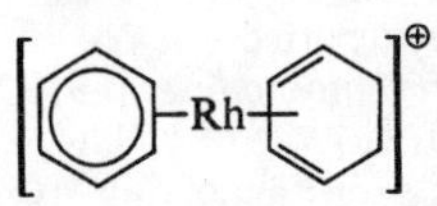

M 261.148 (ion)
Tetrafluoroborate: [64330-98-9].
$C_{12}H_{14}BF_4Rh$ M 347.952
Solid.

Schrock, R.R. *et al*, *J. Am. Chem. Soc.*, 1971, **93**, 3089 (*synth*)

Draggett, P.T. *et al*, *J. Organomet. Chem.*, 1977, **135**, C60 (*synth*)

$C_{12}H_{15}O_2Rh$ Rh-00103

[(2,3,5,6-η)-Bicyclo[2.2.1]hepta-2,5-diene](2,4-pentanedionato-*O*,*O'*)rhodium, 9CI
(2,5-Norbornadiene)(2,4-pentanedionato)rhodium, 8CI. *Acetylacetonato(η⁴-norbornadiene)rhodium*
[32354-50-0]

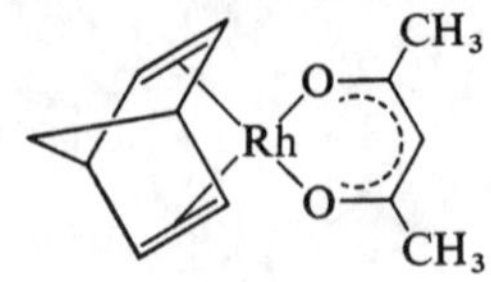

M 294.155
Yellow needles. Mp 176-7°.

Bonati, F. *et al*, *J. Chem. Soc.*, 1964, 3156 (*synth*, *ir*, *pmr*)
North, B. *et al*, *J. Organomet. Chem.*, 1970, **21**, 445 (*synth*)

$C_{12}H_{15}O_2Rh$ Rh-00104

Dicarbonyl[(1,2,3,4,5-η)-1,2,3,4,5-pentamethyl-2,4-cyclopentadien-1-yl]rhodium, 10CI
[32627-01-3]

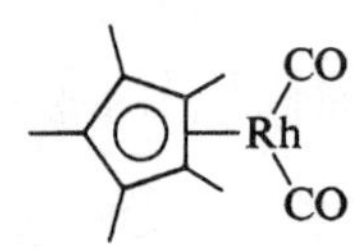

M 294.155
Orange cryst. (pentane/Et_2O). Mp 101-7° dec. $Bp_{0.001}$ 30-40° subl. ir ν_{CO} 2000, 1950 cm^{-1} (KBr).

Maitlis, P.M. *et al*, *J. Organomet. Chem.*, 1971, **26**, 393 (*ir*, *synth*, *pmr*)
Werner, H. *et al*, *J. Organomet. Chem.*, 1981, **218**, 395 (*synth*)
Herrmann, W.A. *et al*, *Z. Anorg. Allg. Chem.*, 1982, **487**, 85 (*ir*, *nmr*, *synth*)

$C_{12}H_{16}Cl_2Rh_2$ Rh-00105

Di-μ-chlorobis[(1,2,3,4-η)-1,3-cyclohexadiene]dirhodium, 10CI
[64429-21-6]

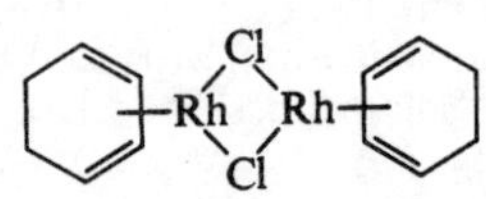

M 436.975
Yellow cryst. (pet. ether/CH_2Cl_2), unstable in air and soln. Mp >115° dec.
Monomer: Chloro-1,3-cyclohexadienerhodium. 1,3-Cyclohexadienerhodium chloride.
C_6H_8ClRh M 218.488
Has no independent existence.

Winkhaus, G. *et al*, *Chem. Ber.*, 1966, **99**, 3602 (*synth*, *pmr*, *ir*)
Draggett, P.T. *et al*, *J. Organomet. Chem.*, 1977, **135**, C60.

$C_{12}H_{18}N_2Rh^{\oplus}$ Rh-00106

Bis(acetonitrile)[(1,2,5,6-η)-1,5-cyclooctadiene]rhodium(1+), 10CI

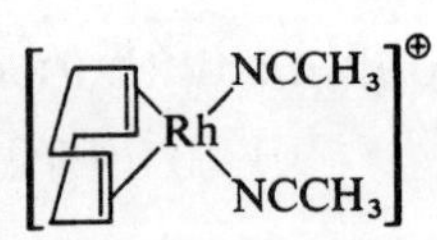

M 293.193 (ion)

Alkene hydrogenation catalyst.
Perchlorate: [72353-79-8].
$C_{12}H_{18}ClN_2O_4Rh$ M 392.644
No phys. props. reported.
Tetrafluoroborate: [32679-02-0].
$C_{12}H_{18}BF_4N_2Rh$ M 379.997
Air-stable bright-yellow cryst. Mp 188-90°.
Hexafluorophosphate: [53734-56-8].
$C_{12}H_{18}F_6N_2PRh$ M 438.157
No phys. props. reported.

Green, M. *et al*, *J. Chem. Soc. (A)*, 1971, 2334 (*synth, ir, pmr*)
Denise, B. *et al*, *J. Organomet. Chem.*, 1978, **161**, 171.
Uson, R. *et al*, *J. Organomet. Chem.*, 1979, **179**, 65 (*synth, ir, use*)

$C_{12}H_{20}Cl_2Rh_2$ Rh-00107

Di-μ-chlorotetrakis(η³-2-propenyl)dirhodium, 10CI, 9CI
Dichlorotetra-π-allyldirhodium, 8CI. *Tetrakis(η³-allyl)-di-μ-chlorodirhodium*
[12090-11-8]

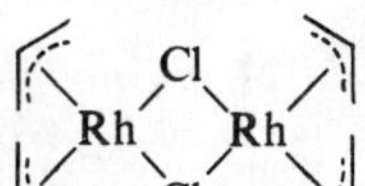

M 441.007
Bridged dimer. Catalyses hydrogenation of alkenes and alkynes. Yellow prisms (CH_2Cl_2/MeOH). Mp 180-5° dec.
Monomer: Chlorobis(η³-2-propenyl)rhodium. Bis(η³-allyl)rhodium chloride. Diallylchlororhodium.
$C_6H_{10}ClRh$ M 220.504
Unknown.

Powell, J. *et al*, *J. Chem. Soc., Chem. Commun.*, 1966, 236 (*synth*)
Mason, R. *et al*, *J. Chem. Soc., Chem. Commun.*, 1967, 16 (*struct*)
Powell, J. *et al*, *J. Chem. Soc. (A)*, 1968, 583 (*synth, ir, pmr*)
Pruchnik, F. *et al*, *Rocz. Chem.*, 1977, **51**, 1813; *CA*, **88**, 43314 (*uv*)

$C_{12}H_{20}Cl_4Rh_2$ Rh-00108

[μ-[(1,2-η:3,4-η)-1,3-Butadiene]]bis[(1,2,3-η)-2-butenyl]di-μ-chlorodichlorodirhodium, 10CI
(μ-1,3-Butadiene)di-μ-chlorodichlorobis(1-methyl-π-allyl)dirhodium, 8CI. *Tetrachloro(1,3-butadiene)-bis(1-methyl-π-allyl)dirhodium*
[12307-75-4]

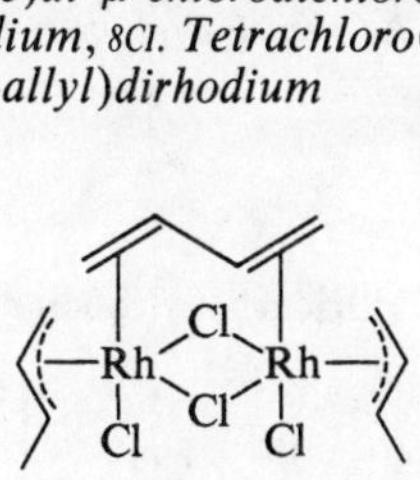

M 511.913
Catalyst for butadiene dimerisation and co-dimerisation of ethylene and butadiene. Golden-yellow or bronze plates. Mp 165-70°.

Cramer, R. *et al*, *J. Am. Chem. Soc.*, 1967, **89**, 1633 (*synth, ir, pmr*)
Powell, J. *et al*, *J. Chem. Soc. (A)*, 1968, 597 (*synth, pmr*)

$C_{12}H_{22}O_2Rh_2$ Rh-00109

Di-μ-hydroxytetrakis(η³-2-propenyl)dirhodium, 10CI
Di-μ-hydroxytetrakis(η³-allyl)dirhodium. Tetraallyldihydroxydirhodium
[73456-53-8]

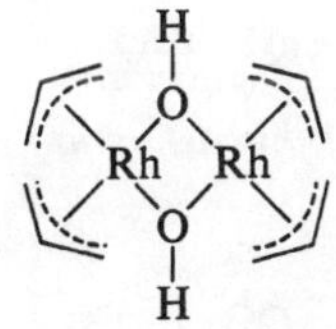

M 404.116
Bridged dimer. Yellow cryst. (DME). Mp 129-38° dec.
Monomer: Hydroxybis(η³-2-propenyl)rhodium. Diallylhydroxyrhodium.
$C_6H_{11}ORh$ M 202.058
Unknown.

Sakurai, F. *et al*, *J. Am. Chem. Soc.*, 1980, **102**, 1749 (*synth, pmr, ir*)

$C_{12}H_{23}OPb_2Rh$ Rh-00110

Carbonyl(η⁵-2,4-cyclopentadien-1-yl)bis(trimethylplumbyl)-rhodium, 9CI
Bis(trimethylplumbyl)(cyclopentadienyl)(carbonyl)rhodium
[55519-53-4]

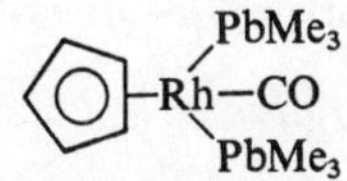

M 700.619
Pale-yellow solid. $Bp_{0.02}$ 25° subl. V. air-sensitive.

Hill, R. *et al*, *J. Chem. Soc., Dalton Trans.*, 1975, 2622.

$C_{12}H_{32}O_9P_3Rh$ Rh-00111

(η³-2-Propenyl)tris(trimethyl phosphite-*P*)rhodium, 10CI
Allyltris(trimethyl phosphite)rhodium
[59448-04-3]

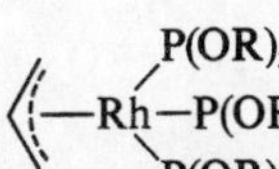

R = Me

M 516.206
Air-sensitive yellow cryst.

Green, M. *et al*, *J. Organomet. Chem.*, 1976, **111**, C6 (*synth, nmr*)
Sivak, A.J. *et al*, *J. Am. Chem. Soc.*, 1979, **101**, 4878 (*synth, pmr*)

$C_{12}H_{36}P_3Rh$ Rh-00112

Trimethyltris(trimethylphosphine)rhodium, 10CI
[64216-74-6]

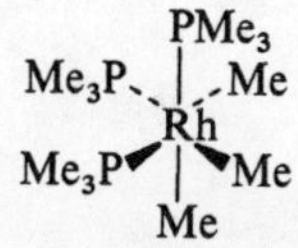

M 376.243

fac-form
Yellow needles. Mp 106-8°.

Andersen, R.A. *et al*, *J. Chem. Soc., Dalton Trans.*, 1978, 446 (*synth, ir, nmr*)

$C_{12}IrO_{12}Rh_3$ Rh-00113

Tricarbonyl(tri-μ-carbonylhexacarbonyltrirhodium)iridium, 9CI

Dodecacarbonyliridiumtrirhodium

[50864-90-9]

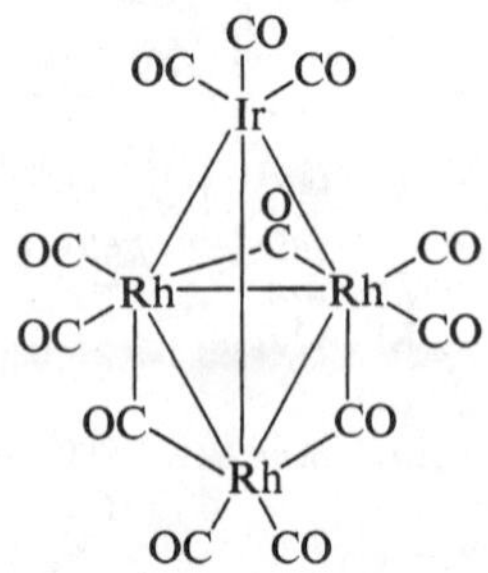

M 837.061

Proposed struct. shown. Orange cryst. Sol. hydrocarbons. Dec. at 130°.

Martinengo, S. *et al*, *J. Organomet. Chem.*, 1973, **59**, 379 (*synth, ir, ms*)

$C_{12}Ir_2O_{12}Rh_2$ Rh-00114

Di-μ-carbonylpentacarbonyl(μ-carbonyltetracarbonyldirhodium)diiridium, 9CI

Dodecacarbonyldiiridiumdirhodium

[50696-81-6]

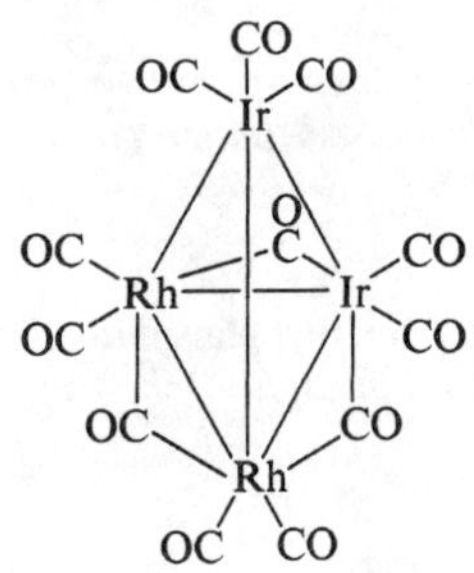

M 926.376

Proposed struct. shown. Orange cryst. Sol. hydrocarbons. Dec. ca. 150°.

Martinengo, S. *et al*, *J. Organomet. Chem.*, 1973, **59**, 379 (*synth, ir, uv, ms*)

$C_{12}O_{12}Rh_4$ Rh-00115

Tri-μ-carbonylnonacarbonyltetrarhodium, 10CI, 9CI

Tetrarhodium dodecacarbonyl

[19584-30-6]

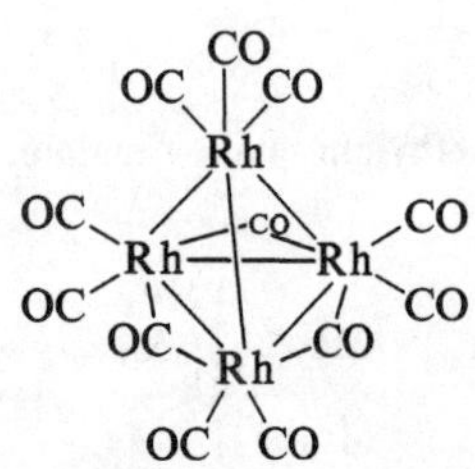

M 747.747

Catalyst for hydrogenation, hydroformylation and hydrosilylation of alkenes. Also catalyses the water gas shift reaction and Fischer-Tropsch type reactions. Dark-red cryst. Mp ca. 150° dec. Dec. above 130-40° → $Rh_6(CO)_{16}$.

Booth, B.L. *et al*, *J. Organomet. Chem.*, 1968, **14**, 417 (*synth*)
Chini, P. *et al*, *Inorg. Chim. Acta*, 1969, **3**, 315 (*synth*)
Wei, C.H., *Inorg. Chem.*, 1969, **8**, 2384 (*struct*)
Evans, J. *et al*, *J. Chem. Soc., Dalton Trans.*, 1978, 626 (*cmr*)
Michelin-Lausarot, P. *et al*, *Inorg. Chim. Acta*, 1979, **35**, 227 (*ms*)
Inorg. Synth., 1980, **20**, 209 (*synth*)
Creighton, J.A. *et al*, *J. Chem. Soc., Dalton Trans.*, 1981, 1498 (*ir, raman*)

$C_{13}H_3O_{13}Rh_4^{\ominus}$ Rh-00116

Undecacarbonyl(methoxycarbonyl)tetrarhodate(1–), 9CI

$[(OC)_{11}Rh_4(COOMe)]^{\ominus}$

M 778.781 (ion)

Struct. not reported. Orange, v. water-sensitive. Several salts obt.

Na salt: [60650-16-0].
$C_{13}H_3NaO_{13}Rh_4$ M 801.771
Red soln. in THF.

Bis(triphenylphosphine)iminium salt: [60866-37-7].
$C_{49}H_{33}NO_{13}P_2Rh_4$ M 1317.368
Dark-red cryst. stored under CO. Sol. C_6H_6, toluene, THF. Stable for several days at −15°; rapidly dec. by air or H_2O.

Martinengo, S. *et al*, *J. Organomet. Chem.*, 1976, **116**, 333 (*synth*)

$C_{13}H_8ClN_2ORh$ Rh-00117

Carbonylchloro(1,10-phenanthroline-N^1,N^{10})rhodium

[52614-24-1]

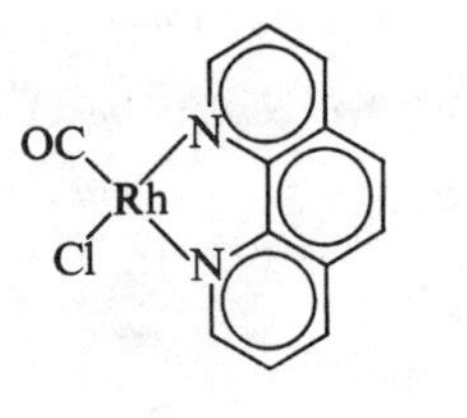

M 346.578

Red-violet cryst., air-sensitive in soln.

Mestroni, G. *et al*, *J. Organomet. Chem.*, 1974, **65**, 119 (*synth, ir*)
Gillard, R.D. *et al*, *J. Chem. Soc., Dalton Trans.*, 1975, 133 (*synth, ir*)

$C_{13}H_8N_2O_3Rh^{\oplus}$ Rh-00118

(2,2′-Bipyridine-*N*,*N*′)tricarbonylrhodium(1+), 10CI

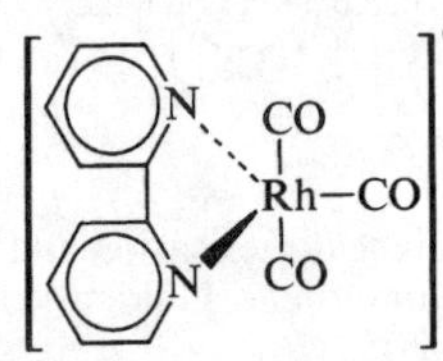

M 343.123 (ion)

ir ν_{CO} 2101, 2040 cm^{-1} (MeCN).

Hexafluorophosphate: [52614-08-1].
$C_{13}H_8F_6N_2O_3PRh$ M 488.087
Green solid. Sol. MeCN, insol. MeOH, Et_2O. Loses 1 mol. CO reversibly in MeCN soln.

Perchlorate: [72248-31-8].
$C_{13}H_8ClN_2O_7Rh$ M 442.574
Green solid.

Mestroni, G. *et al*, *J. Organomet. Chem.*, 1974, **65**, 119 (*synth, ir*)
Seth, P.A. *et al*, *J. Chem. Soc., Dalton Trans.*, 1979, 1569 (*synth, uv*)

$C_{13}H_{10}O_3Rh_2$ Rh-00119

μ-Carbonyldicarbonylbis(η⁵-2,4-cyclopentadien-1-yl)dirhodium, 9CI
μ-Carbonyldicarbonyldi-π-cyclopentadienyldirhodium, 8CI
[12318-88-6]

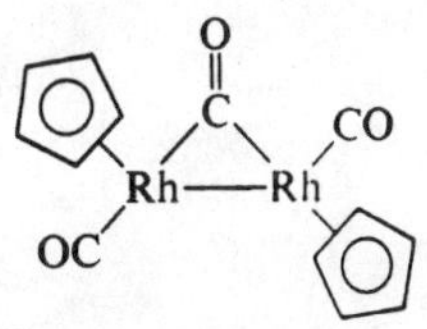

M 420.031
Red cryst. (pentane/Et_2O). Mp 124-5°, 139°. ir ν_{CO} 1963, 1816 cm^{-1} (KBr).

Mills, O.S. *et al*, *J. Organomet. Chem.*, 1967, **10**, 337 (*synth, struct, ms*)
Evans, J. *et al*, *J. Chem. Soc., Chem. Commun.*, 1973, 79 (*synth, ir, cmr*)
Hill, R. *et al*, *J. Chem. Soc., Dalton Trans.*, 1975, 2622 (*synth, ir, pmr*)
Herrmann, W.A. *et al*, *J. Organomet. Chem.*, 1977, **140**, 73 (*synth, ir*)
Evans, J. *et al*, *J. Chem. Soc., Dalton Trans.*, 1978, 626 (*nmr, ir*)
Lawson, R.J. *et al*, *Inorg. Chem.*, 1978, **17**, 2963 (*pmr*)

$C_{13}H_{11}O_3Rh_2^{\oplus}$ Rh-00120

μ-Carbonyldicarbonylbis(η⁵-2,4-cyclopentadien-1-yl)-μ-hydrodirhodium(1+), 10CI

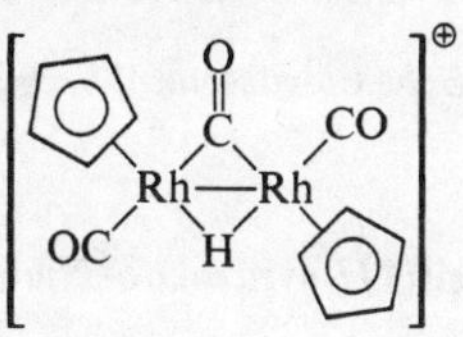

M 421.039 (ion)
ir ν_{CO} 1970, 1818 cm^{-1} (KBr).
Tetrafluoroborate: [76045-10-8].
$C_{13}H_{11}BF_4O_3Rh_2$ M 507.843
Prepd. by action of HBr_4 on (μ—CO)[(C_5H_5)-Rh(CO)]$_2$. Light-red powder. V. air- and moisture-sensitive. Dec. slowly at r.t. even in an inert atm. Dec. in common solvs. May be handled at −20° in $PhNO_2$.

Herrmann, W.A. *et al*, *J. Organomet. Chem.*, 1980, **190**, C47 (*synth, props, pmr, ir*)
Herrmann, W.A. *et al*, *J. Am. Chem. Soc.*, 1981, **103**, 63 (*synth, props, pmr, cmr*)

$C_{13}H_{12}Cl_5Rh$ Rh-00121

[(1,2,5,6-η)-1,5-Cyclooctadiene](η⁵-1,2,3,4,5-pentachloro-2,4-cyclopentadien-1-yl)rhodium, 9CI
η⁵-Pentachlorocyclopentadienyl(1,5-cyclooctadiene)rhodium
[56282-20-3]

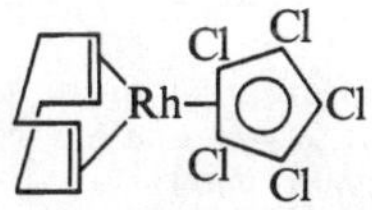

M 448.408
Cryst. struct. suggests that the cyclopentadienyl ring is bonded to the metal *via* two localised π bonds and one σ bond. Orange cryst. (hexane). Mp 92-4° dec. Somewhat thermally unstable, turns black at r.t. after several days.

Day, V.M. *et al*, *J. Chem. Soc., Chem. Commun.*, 1975, 403 (*synth, cryst struct*)
Reimer, K.J. *et al*, *Inorg. Chem.*, 1975, **14**, 2707 (*synth, struct, ir, cmr, nqr*)
Inorg. Synth., 1980, **20**, 188 (*synth*)

$C_{13}H_{12}O_2Rh_2$ Rh-00122

Dicarbonylbis(η⁵-2,4-cyclopentadien-1-yl)-μ-methylenedirhodium, 10CI
μ-Methylenedicarbonylbis(η-cyclopentadienyl)dirhodium.
μ-Methylenebis(carbonyl-η⁵-cyclopentadienylrhodium)
[62154-28-3]

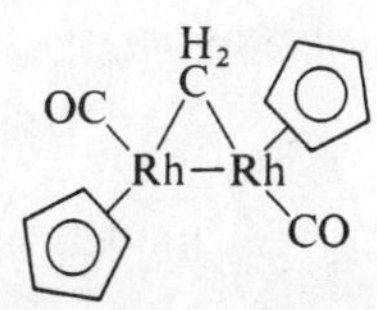

M 406.048
Orange-red cryst. (pentane). Mp 94°. ir ν_{CO} 1984 cm^{-1} (pentane).

Herrmann, W.A. *et al*, *Angew. Chem., Int. Ed. Engl.*, 1977, **16**, 334 (*synth, pmr, ir, ms, struct*)
Herrmann, W.A. *et al*, *J. Organomet. Chem.*, 1977, **140**, 73 (*synth, ir, pmr, ms, struct*)
Lawson, R.J. *et al*, *Inorg. Chem.*, 1978, **17**, 2963 (*nmr*)
Mayer, K.K. *et al*, *J. Organomet. Chem.*, 1979, **182**, 361 (*ms*)
Herrmann, W.A. *et al*, *J. Am. Chem. Soc.*, 1981, **103**, 63.
Takasagawa, F. *et al*, *Inorg. Chem.*, 1981, **20**, 3060 (*struct*)

$C_{13}H_{13}O_2Rh_2^{\oplus}$ Rh-00123

Dicarbonyl(η⁵-2,4-cyclopentadien-1-yl)-μ-hydro-μ-methylenedirhodium(1+), 10CI
μ-Hydrido-μ-methylenedicarbonylbis(η-cyclopentadienyl)dirhodium(1+)

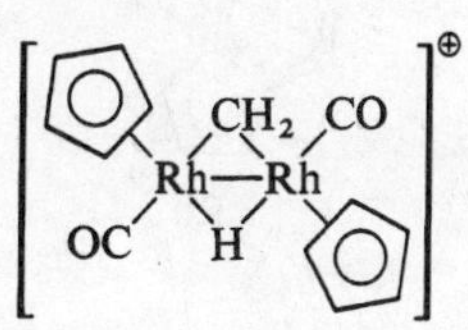

M 407.056 (ion)
Thermally unstable, air- and moisture-sensitive.
Bromide: [74666-50-5].
$C_{13}H_{13}BrO_2Rh_2$ M 486.960
Yellow solid.
Chloride: [74666-49-2].
$C_{13}H_{13}ClO_2Rh_2$ M 442.509
Yellow solid.

Fluorosulfonate: [74543-06-9].
$C_{13}H_{13}FO_5Rh_2S$ M 506.112
Yellow solid.
Trifluoromethanesulfonate: [74543-05-8].
$C_{14}H_{13}F_3O_5Rh_2S$ M 556.120
Yellow solid.
Tetrafluoroborate: [74543-04-7].
$C_{13}H_{13}BF_4O_2Rh_2$ M 493.859
Yellow solid.

Herrmann, W.A. *et al*, *J. Am. Chem. Soc.*, 1980, **102**, 5906; 1981, **103**, 63 (*synth, ir, nmr*)

$C_{13}H_{13}Rh$ Rh-00124

[(1,2,5,6-η)-1,3,5,7-Cyclooctatetraene](η⁵-2,4-cyclopentadien-1-yl)rhodium, 9CI
[55480-71-2]

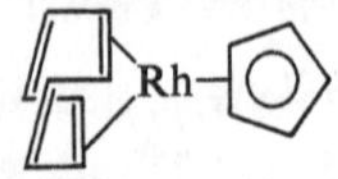

M 272.151
Air-stable yellow cryst. Mp 106°.

Davison, A. *et al*, *J. Chem. Soc.*, 1962, 4821 (*synth, ir, nmr*)
Evans, J. *et al*, *J. Chem. Soc., Dalton Trans.*, 1974, 2375 (*pmr*)

$C_{13}H_{14}Rh^{\oplus}$ Rh-00125

(η^6-Benzene)[(2,3,5,6-η)-bicyclo[2.2.1]hepta-2,5-diene]rhodium(1+), 9CI
Benzenenorbornadienerhodium

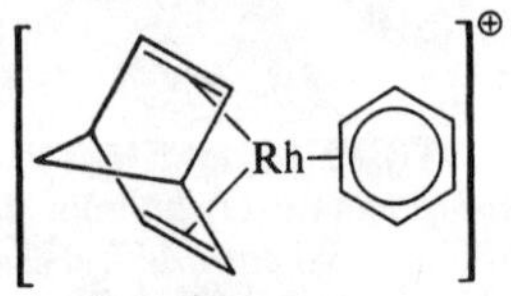

M 273.159 (ion)
Tetrafluoroborate: [36682-23-2].
$C_{13}H_{14}BF_4Rh$ M 359.963
Yellow cryst. (CH_2Cl_2/Et_2O). Mp 166-9° dec.

Green, M. *et al*, *J. Chem. Soc., Dalton Trans.*, 1972, 832 (*synth, pmr*)

$C_{13}H_{15}Rh$ Rh-00126

Bis(η^2-ethene)[(1,2,3,3a,7a-η)-1*H*-inden-1-yl]rhodium, 10CI
Bis(ethylene)(η^5-indenyl)rhodium. Indenylbis(ethylene)rhodium
[63428-46-6]

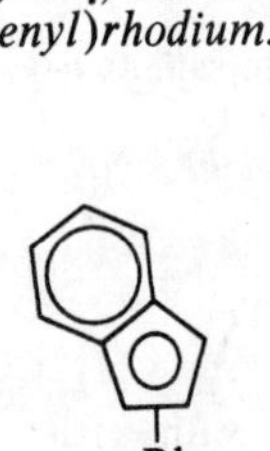

M 274.167
Yellow cryst. (pentane).

Caddy, P. *et al*, *J. Chem. Soc., Dalton Trans.*, 1980, 962 (*synth, ir, nmr, ms*)
Jeffrey, J.C. *et al*, *Organometallics*, 1982, **1**, 1597.

$C_{13}H_{17}ClNRh$ Rh-00127

Chloro[(1,2,5,6-η)-1,5-cyclooctadiene]pyridinerhodium, 9CI
[41812-78-6]

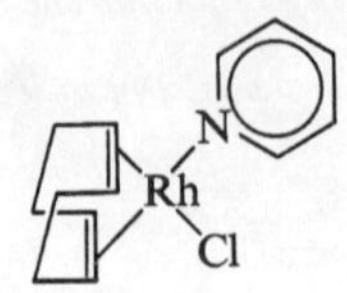

M 325.643
Hydroformylation catalyst. Yellow cryst. (hexane or pet. ether). Mp 165-6°.

Takesada, M. *et al*, *Nippon Kagaku Zasshi*, 1968, **89**, 1121; *CA*, **70**, 87940y (*synth, pmr*)
Fougeroux, P. *et al*, *J. Organomet. Chem.*, 1973, **60**, 375 (*synth, ir, pmr, use*)
Brodzki, D. *et al*, *J. Organomet. Chem.*, 1973, **63**, 431 (*ir*)
King, R.B. *et al*, *J. Am. Chem. Soc.*, 1979, **101**, 4893 (*synth, use*)

$C_{13}H_{17}Rh$ Rh-00128

(1,2,5,6-η)-1,5-Cyclooctadiene(η⁵-2,4-cyclopentadien-1-yl)rhodium, 10CI
Cyclopentadienylcyclooctadienerhodium
[32610-45-0]

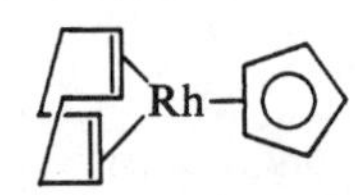

M 276.183
Orange-yellow cryst. (CH_2Cl_2). Sol. Et_2O. Mp 108-108.5°.

Chatt, J. *et al*, *J. Chem. Soc.*, 1957, 4735 (*synth*)
King, R.B., *J. Organomet. Chem.*, 1968, **14**, P19 (*ms*)
Kang, J.W. *et al*, *J. Am. Chem. Soc.*, 1969, **91**, 5970 (*synth, pmr*)
Todd, L.J. *et al*, *J. Chem. Soc., Chem. Commun.*, 1970, 1530 (*cmr*)
Axelson, D.E. *et al*, *Inorg. Nucl. Chem. Lett.*, 1973, **9**, 885 (*cmr*)
Arthurs, M. *et al*, *J. Chem. Soc., Dalton Trans.*, 1977, 779 (*pmr*)
Maurer, E. *et al*, *Helv. Chim. Acta*, 1982, **65**, 26 (*nmr*)

$C_{13}H_{18}BN_6Rh$ Rh-00129

Bis(ethene)[hydrotris(1*H*-pyrazolato-*N*)borato(1−)-N^2,$N^{2'}$]-rhodium, 9CI
Bis(ethylene)[tris(pyrazolyl)borato]rhodium
[52668-39-0]

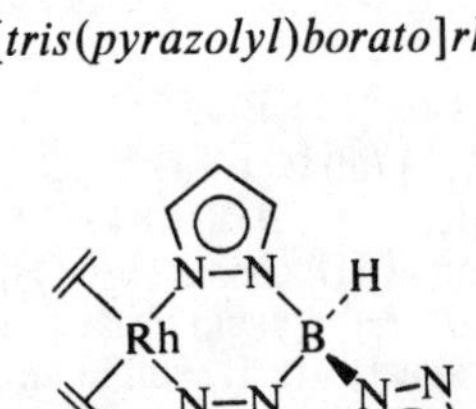

M 372.041
Pyrazolyl groups are fluxional and magnetically equivalent. Ethylene groups are also fluxional. Yellow, air-stable cryst. Dec. >140°, or on attempted subl.

Trofimenko, S. *et al*, *J. Am. Chem. Soc.*, 1969, **91**, 588 (*synth, pmr*)
O'Sullivan, D. *et al*, *J. Organomet. Chem.*, 1974, **65**, C47 (*struct*)

$C_{13}H_{19}O_2Rh$ **Rh-00130**

(1,2,5,6-η-1,5-Cyclooctadiene)(2,4-pentanedionato-*O,O'*)rhodium, 9CI

Acetylacetonato(1,5-cyclooctadiene)rhodium

[12245-39-5]

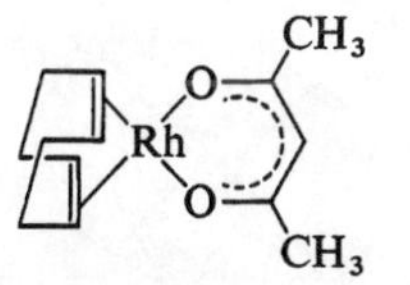

M 310.197

Yellow cryst. Mp 125-8° dec.

Bonati, F. *et al*, *J. Chem. Soc.*, 1964, 3156 (*synth, ir, pmr*)
Todd, L.J. *et al*, *J. Chem. Soc., Chem. Commun.*, 1970, 1530 (*nmr*)
Tucker, P.A. *et al*, *Acta Crystallogr., Sect. B*, 1975, 592 (*cryst struct*)
Bonnaire, R. *et al*, *J. Organomet. Chem.*, 1976, **104**, 107 (*cmr*)
Sinou, D. *et al*, *J. Organomet. Chem.*, 1976, **114**, 325 (*synth, pmr*)

$C_{13}H_{21}BCl_2N_3ORh$ **Rh-00131**

Dichloro(*N,N*-dimethylformamide-*O*)bis(pyridine)[tetrahydroborato(1−)-*H*]rhodium, 10CI

Dichloro(N,N-dimethylformamide)bis(pyridine)rhodium(1+) tetrahydroborate, 8CI

[72930-45-1]

$Rh(HCONMe_2)Py_2Cl_2(BH_4)$

M 419.950

Struct. not reported. Hydrogenation and hydroformylation catalyst for olefins. Dark-red cryst. ($CHCl_3$). ir ν_{CO} 1610 cm^{-1}. 1:1 Electrolyte in DMF with proposed dissoc. of $Cl^{\ominus}$ rather than $BH_4^{\ominus}$.

Abley, P. *et al*, *J. Chem. Soc. (C)*, 1971, 840 (*synth, ir, pmr, mv, ms*)
U.S.P., 4 178 312, (*1979*); *CA*, **92**, 110538p (*use*)

$C_{14}H_8N_2O_2Rh^{\oplus}$ **Rh-00132**

Dicarbonyl(1,10-phenanthroline-*N,N'*)rhodium(1+), 10CI

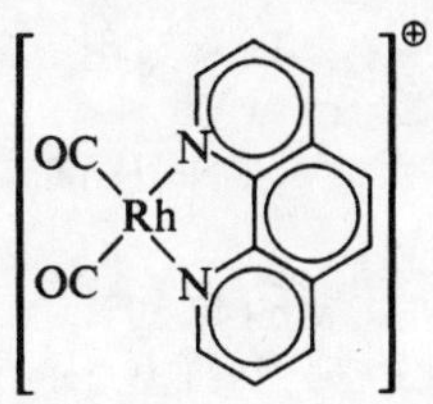

M 339.135 (ion)

Chloride: [65720-28-7].
$C_{14}H_8ClN_2O_2Rh$ M 374.588
Pinkish-yellow fine cryst. Insol. Me_2CO. Not ionic except when dissolved in H_2O.

Perchlorate: [25300-53-2].
$C_{14}H_8ClN_2O_6Rh$ M 438.586
Phototropic cryst., yellow when formed in dark, reversibly becoming green on exp. to light. Sl. sol. H_2O, common org. solvs.

Dibromodicarbonylrhodate:
$C_{16}H_8Br_2N_2O_4Rh_2$ M 657.869
Red-brown cryst. with metallic lustre, blue-green when powdered. Mp 160-3° dec.

Dicarbonyldichlororhodate: [55914-10-8].
$C_{16}H_8Cl_2N_2O_4Rh_2$ M 568.967
Red-brown cryst. with metallic lustre, blue-green when powdered. Mp 120-3° dec. Loses CO when powdered.

Reddy, G.K. *et al*, *J. Organomet. Chem.*, 1974, **67**, 443 (*synth*)
Gillard, R.D. *et al*, *J. Chem. Soc., Dalton Trans.*, 1975, 133 (*synth, ir, props*)
Pruchnik, F. *et al*, *J. Organomet. Chem.*, 1979, **164**, 71 (*synth, ir, uv, props*)

$C_{14}H_{14}F_3O_4Rh$ **Rh-00133**

Dicarbonyl[1,7,7-trimethyl-3-(trifluoroacetyl)bicyclo[2.2.1]-heptan-2-onato-*O,O'*]rhodium, 10CI

Dicarbonyl(3-trifluoroacetylcamphorato)rhodium

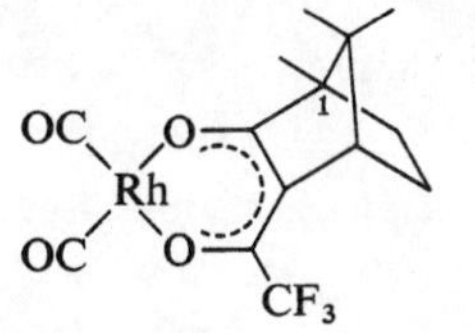

(1*R*)-*form*

M 406.163

(1R)-form [36654-16-7]

Stationary phase for the separation of monoolefins and resolution of chiral olefins by glc. Lemon-yellow cryst. (hexane). Mp 134° (131°). $[\alpha]_D^{23}$ +106° (c, 0.4 in $CHCl_3$).

(1S)-form [65450-85-3]

Lemon-yellow cryst. (hexane). Mp 134°.

(±)-form

Red-green dichroic cryst. (hexane). Mp 130.5°.

Gil-Av, E. *et al*, *Anal. Chem.*, 1971, **43**, 2030 (*use*)
Schurig, V., *Inorg. Chem.*, 1972, **11**, 736 (*synth*)
Haegele, K.D. *et al*, *Inorg. Chem.*, 1974, **13**, 1960 (*ms*)
Schurig, V., *Angew. Chem., Int. Ed. Engl.*, 1976, **15**, 304; 1977, **16**, 110 (*use*)
Schurig, V., *Angew. Chem., Int. Ed. Engl.*, 1981, **20**, 807 (*uv*)

$C_{14}H_{15}F_6O_4Rh$ **Rh-00134**

Aqua[(1,2,3,4,5-η)-1,2,3,4,5-pentamethyl-2,4-cyclopentadien-1-yl]bis(trifluoroacetato-*O*)rhodium

*[(1,2,3,4,5-η)-1,2,3,4,5-Pentamethyl-2,4-cyclopentadien-1-yl]bis(trifluoroacetato-*O*)rhodium*, 9CI. *Bis(trifluoroacetato)(pentamethylcyclopentadienyl)rhodium hydrate*

[34738-21-1]

$[Rh(\eta^5\text{-}C_5(CH_3)_5)(OOCCF_3)_2H_2O]_n$

M 464.166

Polymeric hydrate. Olefin hydrogenation catalyst. Orange-red, air-stable cryst. (Et_2O).

Kang, J.W. *et al*, *J. Organomet. Chem.*, 1971, **30**, 127 (*synth, nmr*)
White, C. *et al*, *J. Chem. Soc., Dalton Trans.*, 1973, 1901 (*use, props*)

$C_{14}H_{15}N_2RhS_2$ **Rh-00135**

[2,3-Dimercapto-2-butanedinitrilato(2−)-*S,S'*][(1,2,3,4,5-η)-1,2,3,4,5-pentamethyl-2,4-cyclopentadien-1-yl]rhodium, 10CI

(1,2-Dicyanoethylene-1,2-dithiolato)(η^5-pentamethylcyclopentadienyl)rhodium

[68447-94-9]

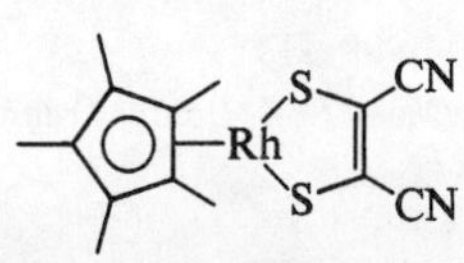

M 378.311

Red cryst. (CH_2Cl_2/pet. ether).

Russell, M.J.H. *et al*, *J. Chem. Soc., Dalton Trans.*, 1978, 849 (*synth, pmr, cmr*)

$C_{14}H_{15}Rh$ **Rh-00136**

[(1,2,3,3*a*,7*a*-η)-1*H*-Inden-1-yl][(1,2,3,4-η)-2-methyl-1,3-butadiene]rhodium, 10CI

η^5-Indenyl(η^4-isoprene)rhodium

[63428-37-5]

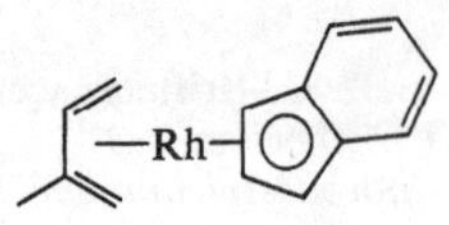

M 286.178

Air-stable yellow cryst. (hexane).

Caddy, P. *et al*, *J. Chem. Soc., Dalton Trans.*, 1980, 962 (*synth, ir, ms, nmr*)

$C_{14}H_{16}Cl_2Rh_2$ **Rh-00137**

Bis[(2,3,5,6-η)-bicyclo[2.2.1]hepta-2,5-diene]di-μ-chlorodirhodium, 10CI, 9CI

Di-μ-chlorobis(2,5-norbornadiene)dirhodium, 8CI. *Bis(norbornadiene-μ-chlororhodium)*

[12257-42-0]

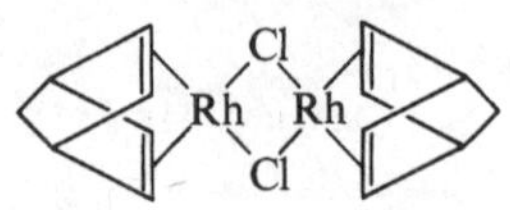

M 460.997

Useful precursor to rhodium hydrogenation catalysts. Yellow cryst. (CH_2Cl_2/MeOH). Mp 240° dec.

Monomer: [Bicyclo[2.2.1]hepta-2,5-diene]chlororhodium. Norbornadienerhodium chloride. Chloro(norbornadiene)rhodium.

C_7H_8ClRh M 230.499

Unknown.

Abel, E.W. *et al*, *J. Chem. Soc.*, 1959, 3178 (*synth, ir*)
Bennett, M.A. *et al*, *Inorg. Chem.*, 1967, **6**, 1647 (*ir*)
Volger, H.C. *et al*, *Inorg. Chim. Acta*, 1969, **3**, 145 (*synth, ir, pmr*)
Schrock, R.R. *et al*, *J. Am. Chem. Soc.*, 1971, **93**, 2397 (*synth*)
Denise, B. *et al*, *Bull. Soc. Chim. Fr.*, 1975, 1034 (*ir, uv, pmr, struct*)
Schrock, R.R. *et al*, *J. Am. Chem. Soc.*, 1976, **98**, 2134 (*use*)

$C_{14}H_{16}Rh^{\oplus}$ **Rh-00138**

[(2,3,5,6-η)-Bicyclo[2.2.1]hepta-2,5-diene][(1,2,3,4,5,6-η)-1,3,5-cycloheptatriene]rhodium(1+), 9CI

(1,3,5-Cycloheptatriene)(2,5-norbornadiene)rhodium(1+), 8CI. *(η^6-Cycloheptatriene)(η^4-norbornadiene)rhodium(1+)*

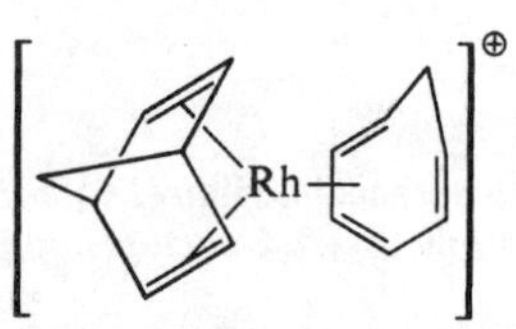

M 287.186 (ion)

Tetrafluoroborate: [36644-73-2].

$C_{14}H_{16}BF_4Rh$ M 373.990

Yellow cryst. Mp 209-11°.

Green, M. *et al*, *J. Chem. Soc., Dalton Trans.*, 1972, 832 (*synth, ir, pmr*)

$C_{14}H_{16}Rh^{\oplus}$ **Rh-00139**

Bis[(2,3,5,6-η)bicyclo[2.2.1]hepta-2,4-diene]rhodium(1+), 10CI

Bis(2,5-norbornadiene)rhodium(1+), 8CI

[35015-46-4]

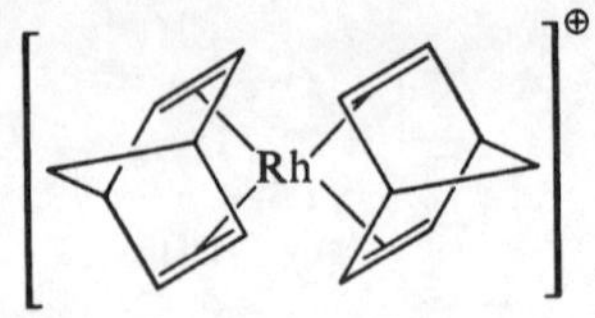

M 287.186 (ion)

Catalyst for olefin hydrogenation and disproportionation of 1,3-cyclohexadiene to cyclohexene and C_6H_6. Precursor for a range of cationic diene compds. containing group V donor ligands and of η^6-arene compds.

Tetrafluoroborate: [36620-11-8].

$C_{14}H_{16}BF_4Rh$ M 373.990

Red cryst. (CH_2Cl_2/Et_2O). Mp 157-9°.

Perchlorate: [60576-58-1].

$C_{14}H_{16}ClO_4Rh$ M 386.637

Rust-brown cryst. (CH_2Cl_2/Et_2O).

Green, M. *et al*, *J. Chem. Soc. (A)*, 1971, 2334 (*synth, ir, pmr*)
Osborn, J.A. *et al*, *J. Am. Chem. Soc.*, 1971, **93**, 3089 (*synth, use*)
Green, M. *et al*, *J. Chem. Soc., Dalton Trans.*, 1972, 832 (*use*)
Uson, R. *et al*, *Transition Met. Chem.*, 1979, **4**, 55 (*synth, pmr, use*)

$C_{14}H_{19}BCl_2N_7Rh$ **Rh-00140**

Dichloro[hydrotris(1*H*-pyrazolato-*N*[1])borato(1−)-*N*[2],*N*[2′],*N*[2″]](1-isocyanobutane)rhodium, 10CI

*(Butyl isonitrile)dichloro[hydrotris(1*H*-pyrazolato-N[1]-borato]rhodium*

[73117-45-0]

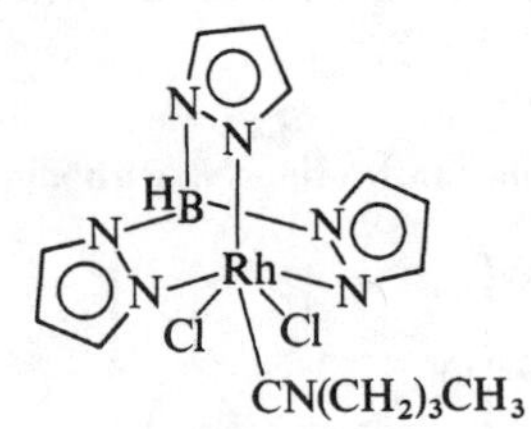

M 469.973

Bright yellow prisms. Mp 148-50°.

May, S. *et al*, *Inorg. Chem.*, 1980, **19**, 1582 (*synth, ir, pmr*)

$C_{14}H_{19}NRh^{\oplus}$ **Rh-00141**

[(1,2,3,4,5-η-1,2,3,4,5-Pentamethyl-2,4-cyclopentadien-1-yl](η[5]-1*H*-pyrrol-1-yl)rhodium(1+), 9CI

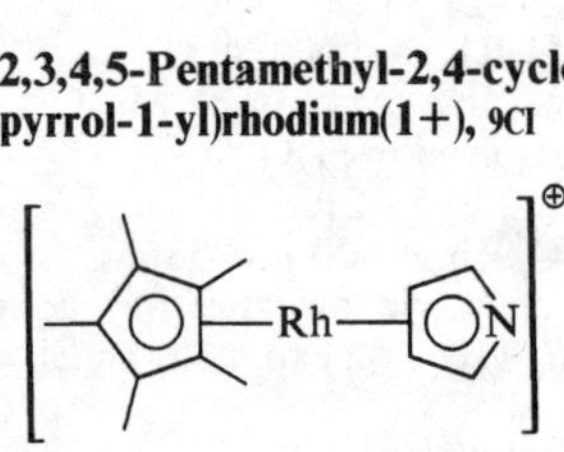

M 304.216 (ion)

Hexafluorophosphate: [35917-11-4].

$C_{14}H_{19}F_6NPRh$ M 449.180

Cryst. solid (CH_2Cl_2/Et_2O). Mp 249-51° dec.

White, C. *et al*, *J. Chem. Soc. (A)*, 1971, 3322 (*synth, pmr*)

$C_{14}H_{23}O_5Rh$ **Rh-00142**

Bis(acetato-*O*)aqua[(1,2,3,4,5-η)-1,2,3,4,5-pentamethyl-2,4-cyclopentadien-1-yl]rhodium, 10CI

Aquabis(acetato)(pentamethyl-π-cyclopentadienyl)rhodium, 8CI. *Diacetato(pentamethylcyclopentadienyl)rhodium hydrate*

[67849-36-9]

$$[Rh(\eta^5\text{-}C_5(CH_3)_5)(OAc)_2H_2O]_n$$

M 374.238

Polymeric hydrate. Olefin hydrogenation catalyst. Red-orange, air-stable cryst. (Et_2O/hexane). Mp 143-5°.

Kang, J.W. *et al*, *J. Am. Chem. Soc.*, 1969, **91**, 5970 (*synth, pmr*)
Gill, D.S. *et al*, *J. Chem. Soc., Dalton Trans.*, 1978, 617 (*use*)
Gill, D.S. *et al*, *J. Magn. Reson.*, 1979, **35**, 459 (*nmr*)

$C_{14}IO_{14}Rh_5^{\ominus\ominus}$ **Rh-00143**

Tetradecacarbonyliodopentarhodate(2–)

Hexa-μ-carbonyloctacarbonyliodopentarhodate(2–), 10CI

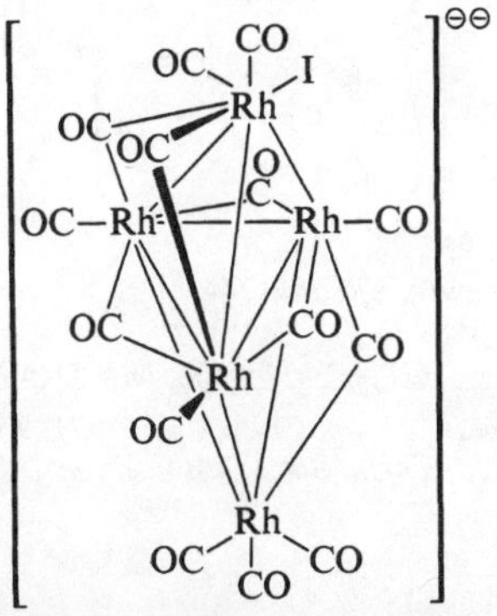

M 1033.578 (ion)

Stable only under an atm. of CO.

Bis(tetrabutylammonium) salt: [73711-40-7].
$C_{46}H_{72}IN_2O_{14}Rh_5$ M 1518.512
Red cryst. (THF/2-propanol).

Bis(tetraphenylarsonium) salt: [73711-41-8].
$C_{62}H_{40}As_2IO_{14}Rh_5$ M 1800.265
No phys. props. recorded.

Martinengo, S. *et al*, *J. Chem. Soc., Chem. Commun.*, 1979, 1059 (*synth, ir, cryst struct*)

$C_{14}O_{14}Rh_6^{\ominus\ominus\ominus\ominus}$ **Rh-00144**

Tetradecacarbonylhexarhodate(4–), 8CI

Octa-μ_3-carbonylhexacarbonylhexarhodate(4–)

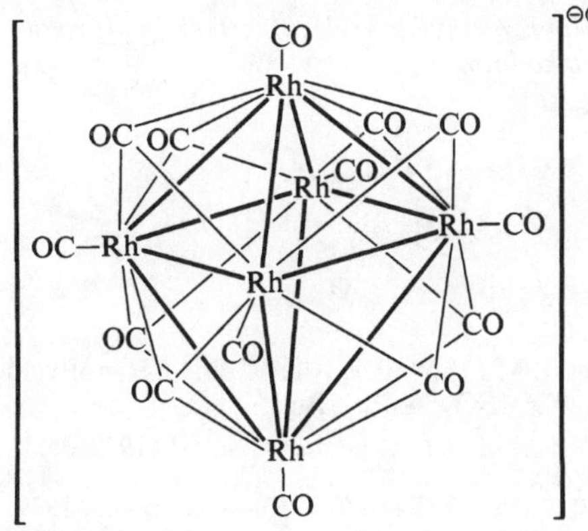

M 1009.579 (ion)

Struct. by analogy with Co analogue.

Tetra-K salt:
$C_{14}O_{14}K_4Rh_6$ M 1165.972
Dark-red cryst., extremely air-sensitive.

Chini, P. *et al*, *J. Chem. Soc., Chem. Commun.*, 1969, 1092 (*synth, ir, struct*)

$C_{15}ClO_{15}Rh_6^{\ominus}$ **Rh-00145**

Tetra-μ_3-carbonylundecacarbonylchlorohexarhodate(1–), 9CI

Pentadecacarbonylchlorohexarhodate(1–)

As Pentadecacarbonylhydrohexarhodate(1–), Rh-00146 with

X = Cl

M 1073.042 (ion)

Struct. by analogy with iodo deriv.

Tetrabutylammonium salt:
$C_{31}H_{36}ClNO_{15}Rh_6$ M 1315.509
Brown cryst. (Me_2CO aq.).

Benzyltrimethylammonium salt:
$C_{25}H_{16}ClNO_{15}Rh_6$ M 1223.285
Insol. H_2O.

Chini, P. *et al*, *Gazz. Chim. Ital.*, 1972, **102**, 330 (*ir, synth*)

$C_{15}HO_{15}Rh_6^{\ominus}$ **Rh-00146**

Pentadecacarbonylhydrohexarhodate(1–)

Hydridopentadecacarbonylhexarhodate(1–)

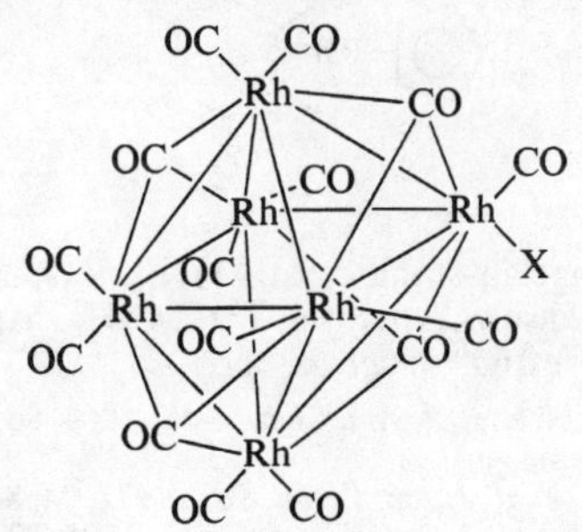

X = H

M 1038.597 (ion)

Formed initially when $[Rh_{12}(CO)_{30}]^{2-}$ reacts with H_2 in 2-propanol. Assumed to be isostructural with Tetra-μ_3-carbonylundecacarbonylchlorohexarhodate(1–), Rh-00145 .

Albano, V.G. *et al*, *J. Chem. Soc., Chem. Commun.*, 1975, 859.

$C_{15}H_{10}F_6ORh_2$ Rh-00147

μ-Carbonylbis(η⁵-2,4-cyclopentadien-1-yl)[μ-[(2,3-η:2,3-η)-1,1,1,4,4,4-hexafluoro-2-butyne]]dirhodium, 10CI

[Bis(trifluoromethyl)acetylene]carbonylbis(cyclopentadienyl)dirhodium

[71844-53-6]

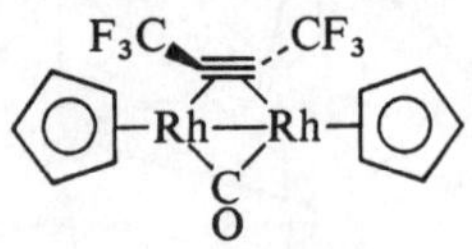

M 526.045

Green cryst. Mp 136°. ir ν_{CO} 1841 cm^{-1}. Readily adds 1 mole of CO, SO_2, $AsPh_3$, Bu^tNC etc.

Dickson, R.S. *et al*, *Acta Crystallogr., Sect. B*, 1979, **35**, 2321 (*cryst struct*)
Dickson, R.S. *et al*, *J. Chem. Soc., Chem. Commun.*, 1979, 277 (*synth, ir, nmr*)

$C_{15}H_{16}ClF_6O_2Rh$ Rh-00148

Chloro(1,1,1,5,5,5-hexafluoro-2,4-pentanedionato-*O,O'*)-[(1,2,3,4,5-η)-1,2,3,4,5-pentamethyl-2,4-cyclopentadien-1-yl]rhodium, 10CI

[71395-92-1]

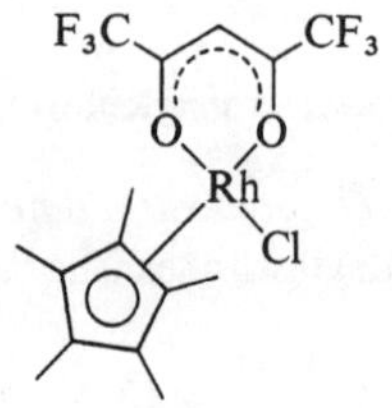

M 480.639

Orange-red cryst. solid (Me_2CO/pet. ether).

Rigby, W. *et al*, *J. Chem. Soc., Dalton Trans.*, 1979, 387 (*synth, nmr*)

$C_{15}H_{17}O_2Rh$ Rh-00149

(η⁵-2,4-Cyclopentadien-1-yl)[(2,3,5,6-η)-2,3,5,6-tetramethyl-2,5-cyclohexadiene-1,4-dione]rhodium, 9CI

*(Cyclopentadienyl)(tetramethyl-*p*-benzoquinone)rhodium. Cyclopentadienyl(duroquinone)rhodium*

[31854-81-6]

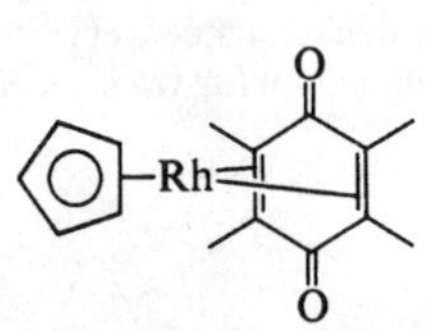

M 332.204

Yellow-orange air-stable cryst. (THF). V. sol. H_2O, chlorinated solvs., spar. sol. THF, C_6H_6. Mp 195° dec. Subl. above 160° *in vacuo*.

Schrauzer, G.N. *et al*, *J. Am. Chem. Soc.*, 1964, **86**, 3265 (*synth, ir, pmr, uv*)
Slocum, D.W. *et al*, *J. Am. Chem. Soc.*, 1972, **94**, 8596.
Bodner, G.M. *et al*, *J. Organomet. Chem.*, 1975, **88**, 391 (*cmr*)

$C_{15}H_{20}Rh^{\oplus}$ Rh-00150

[(2,3,5,6-η)Bicyclo[2.2.1]hepta-2,5-diene][(1,2,5,6-η)-1,5-cyclooctadiene]rhodium(1+), 9CI

(1,5-Cyclooctadiene)(2,5-norbornadiene)rhodium(1+), 8CI

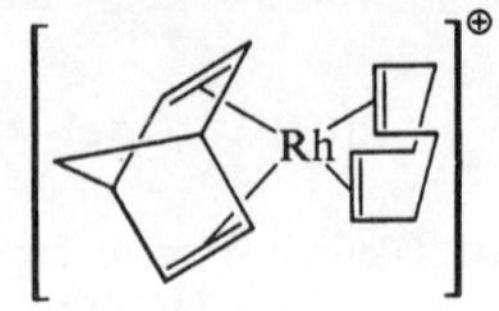

M 303.229 (ion)

Tetrafluoroborate: [38704-67-5].
$C_{15}H_{20}BF_4Rh$ M 390.032
Orange cryst.

Schrock, R.R. *et al*, *J. Am. Chem. Soc.*, 1971, **93**, 3089 (*synth*)

$C_{15}H_{20}Rh^{\oplus}$ Rh-00151

[(1,2,3,4,5,6-η)-1,3,5-Cycloheptatriene][(1,2,5,6-η)-1,5-cyclooctadiene]rhodium(1+), 9CI

1,5-Cyclooctadiene(cycloheptatriene)rhodium(1+)

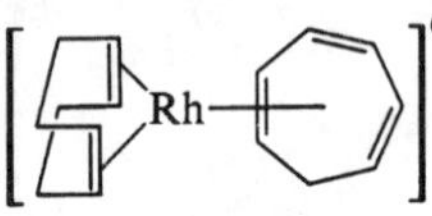

M 303.229 (ion)

Tetrafluoroborate: [36644-74-3].
$C_{15}H_{20}BF_4Rh$ M 390.032
Yellow cryst. (CH_2Cl_2/Et_2O). Mp 210-2°.

Schrock, R.R. *et al*, *J. Am. Chem. Soc.*, 1971, **93**, 3089 (*synth*)
Green, M. *et al*, *J. Chem. Soc., Dalton Trans.*, 1972, 832 (*synth, pmr*)

$C_{15}H_{20}Rh^{\oplus}$ Rh-00152

(η⁵-2,4-Cyclopentadien-1-yl)[(1,2,3,4,5-η)-1,2,3,4,5-pentamethyl-2,4-cyclopentadien-1-yl]rhodium(1+), 10CI

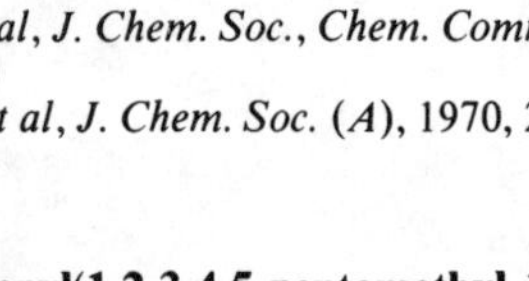

M 303.229 (ion)

Hexafluorophosphate: [33010-40-1].
$C_{15}H_{20}F_6PRh$ M 448.193
Cream cryst. ($CHCl_3/C_6H_6$). Mp >270°.

Moseley, K. *et al*, *J. Chem. Soc., Chem. Commun.*, 1969, 616 (*synth, pmr*)
Maitlis, P.M. *et al*, *J. Chem. Soc. (A)*, 1970, 2875 (*pmr, synth*)

$C_{15}H_{21}Rh$ Rh-00153

π-Cyclopentadienyl(1,2,3,4,5-pentamethyl-1,3-cyclopentadiene)rhodium, 8CI

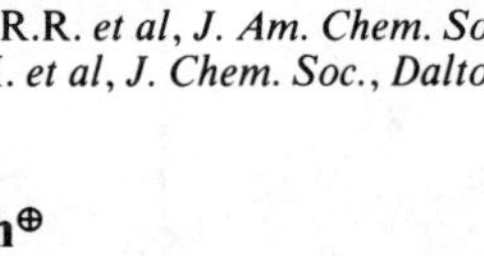

endo-form

M 304.236

endo-form [32698-19-4]
Yellow cryst. (pet. ether). Mp 80-2°.

exo-form [33503-63-8]
Yellow cryst. (Et_2O). Mp 106-7° dec.

Moseley, K. *et al*, *J. Chem. Soc., Dalton Trans.*, 1969, 616 (*synth, ir, ms, pmr*)

$C_{15}H_{22}ClO_2Rh$ **Rh-00154**

Chloro[(1,2,3,4,5-η)-1,2,3,4,5-pentamethyl-2,4-cyclopentadien-1-yl](2,4-pentanedionato-*O,O'*)rhodium, 10CI

*Chloro(2,4-pentanedionato-*O,O'*)(η^5-pentamethylcyclopentadienyl)rhodium. (Acetylacetonato)chloro(pentamethylcyclopentadienyl)rhodium*

[71395-90-9]

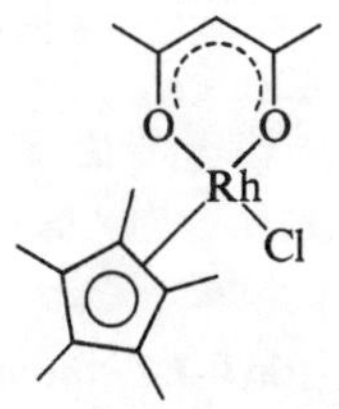

M 372.696

Orange-red cryst. solid (CH_2Cl_2/diisopropyl ether).

Rigby, W. *et al*, *J. Chem. Soc.*, *Dalton Trans.*, 1979, 387 (*synth*, *nmr*)

$C_{15}H_{25}N_2O_2Rh$ **Rh-00155**

Bis(2-isocyano-2-methylpropane)(2,4-pentanedionato-*O,O'*-)rhodium, 9CI

*(Acetylacetonato)bis(*tert*-butyl isocyanide)rhodium*

[57816-38-3]

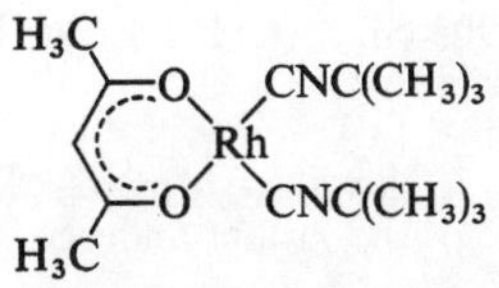

M 368.280

Yellow cryst. + 1/4 CH_2Cl_2 (CH_2Cl_2/pet. ether), unstable in soln. Mp 127-9°. Forms 1:1 adduct with tetracyanoethylene.

Kaneshima, T. *et al*, *Inorg. Chim. Acta*, 1975, **15**, 161 (*synth*, *ir*, *pmr*)

$C_{15}H_{25}P_2Rh$ **Rh-00156**

[(1,2,3,3*a*,7*a*-η)-1*H*-inden-1-yl]bis(trimethylphosphine)rhodium, 10CI

[75094-84-7]

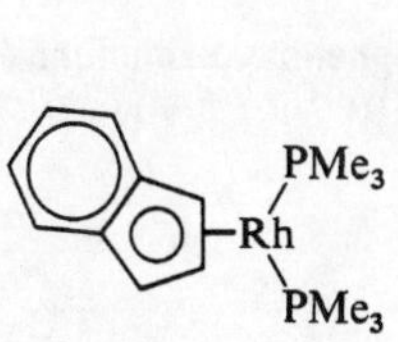

M 370.216

Air-sensitive red cryst. Mp 116°.

Werner, H. *et al*, *Z. Naturforsch.*, *B*, 1980, **35**, 689 (*synth*, *ms*, *pmr*)

$C_{15}H_{28}ClN_4ORh$ **Rh-00157**

Carbonylchlorobis(1,3-diethyl-2-imidazolidinylidene)rhodium, 9CI

[52409-41-3]

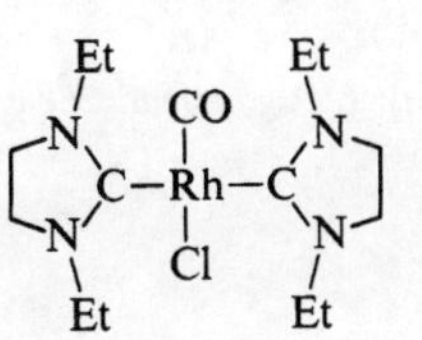

M 418.771

Dicarbene complex. Mp 184-6° dec.

Cardin, D.J. *et al*, *J. Organomet. Chem.*, 1974, **65**, C13 (*pmr*)
Doyle, M.J. *et al*, *J. Chem. Soc.*, *Chem. Commun.*, 1974, 679 (*pmr*)

$C_{15}H_{32}P_3Rh$ **Rh-00158**

Phenyltris(trimethylphosphine)rhodium

[71110-93-5]

$PhRh(PMe_3)_3$

M 408.245

Wrongly indexed in 10CI. Orange-yellow cryst. (pet. ether at −20°). Mp 85-90° dec. Reacts with H_2 to liberate C_6H_6. Dec. only slowly in air.

Jones, R.A. *et al*, *J. Chem. Soc.*, *Dalton Trans.*, 1979, 472 (*synth*, *ir*, *nmr*)

$C_{15}H_{45}O_{15}P_5Rh^{\oplus}$ **Rh-00159**

Pentakis(trimethyl phosphite-*P*)rhodium(1+), 10CI

Pentakis(phosphorous acid)rhodium(1+) pentadecamethyl ester, 8CI

[48077-64-1]

$Rh[P(OMe)_3]_5^{\oplus}$

M 723.286 (ion)

Tetraphenylborate: [33336-87-7].
$C_{39}H_{65}BO_{15}P_5Rh$ M 1042.518
White cryst. (CH_2Cl_2/MeOH). Mp 92°. Conductivity 73 ohm^{-1} cm^2 $mol.^{-1}$.

Hexafluorophosphate: [33336-88-8].
$C_{15}H_{45}F_6O_{15}P_6Rh$ M 868.250
White cryst. (MeOH aq.). Mp 94-6°. Conductivity 122 ohm^{-1} cm^2 $mol.^{-1}$.

Couch, D.A. *et al*, *J. Chem. Soc.*, *Chem. Commun.*, 1971, 1508 (*synth*)
Haines, L.M., *Inorg. Chem.*, 1971, **10**, 1685 (*synth*, *pmr*)
Meakin, P. *et al*, *J. Am. Chem. Soc.*, 1973, **95**, 7272; 1974, **96**, 5751 (*nmr*, *synth*)
English, A.D. *et al*, *J. Am. Chem. Soc.*, 1976, **98**, 7590.

$C_{15}NO_{15}Rh_6^{\ominus}$ **Rh-00160**

Nona-μ-carbonylhexacarbonyl-μ_6-nitridohexarhodate(1−), 10CI

Pentadecacarbonylnitridohexarhodate(1−)

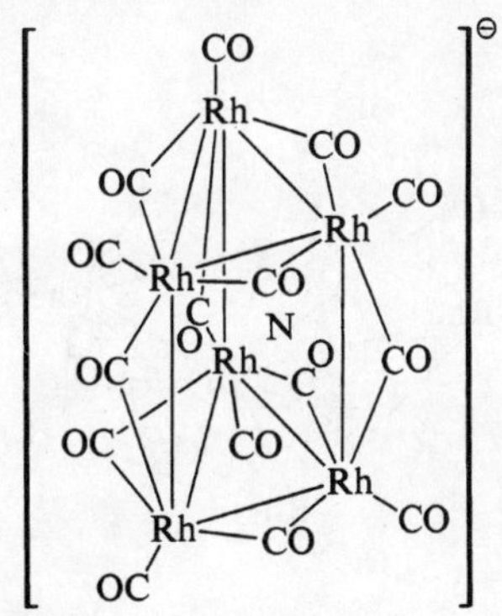

M 1051.596 (ion)

K salt: [72669-25-1].
$C_{15}KNO_{15}Rh_6$ M 1090.694
Yellow.

Bis(triphenylphosphine)iminium salt:
$C_{51}H_{30}N_2O_{15}P_2Rh_6$ M 1590.183
By metathesis of the K salt.

Tetraphenylphosphonium salt:

$C_{39}H_{20}NO_{15}PRh_6$ M 1390.991
By metathesis of the K salt.

Martinengo, S. *et al*, *J. Am. Chem. Soc.*, 1979, **101**, 7095 (*synth*, *cryst struct*, *nmr*, *ir*)

$C_{15}O_{15}Rh_5^{\ominus}$ Rh-00161

Pentadecacarbonylpentarhodate(1–)

Penta-μ-carbonyldecacarbonylpentarhodate(1–), 10CI

[72826-87-0]

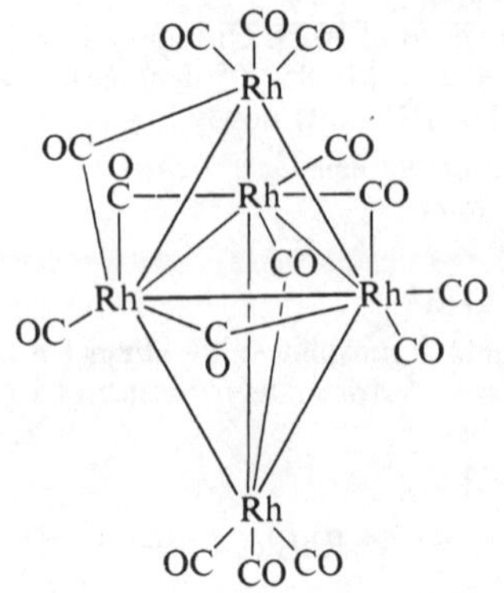

M 934.684 (ion)

Possible intermed. in the Rh—CO cluster catalysed synthesis of ethylene glycol from synthesis gas. Extremely reactive. Disproportionates to other Rh clusters in soln.

Bis(triphenylphosphine)iminium salt: [73700-43-3].
$C_{51}H_{30}NO_{15}P_2Rh_5$ M 1473.271
Red-brown cryst. (THF/2-propanol/CO). Sol. THF.

Fumagalli, A. *et al*, *J. Am. Chem. Soc.*, 1980, **102**, 1740 (*synth*, *cryst struct*, *cmr*, *ir*)
Vidal, J. *et al*, *Inorg. Chem.*, 1980, **19**, 896 (*use*)
Heaton, B.T. *et al*, *J. Chem. Soc., Chem. Commun.*, 1981, 331 (*cmr*)

$C_{15}O_{15}Rh_6^{\ominus\ominus}$ Rh-00162

Pentadecacarbonylhexarhodate(2–)

Tri-μ-carbonyltri-μ₃-carbonylnonacarbonylhexarhodate(2–), 10CI

[56731-55-6]

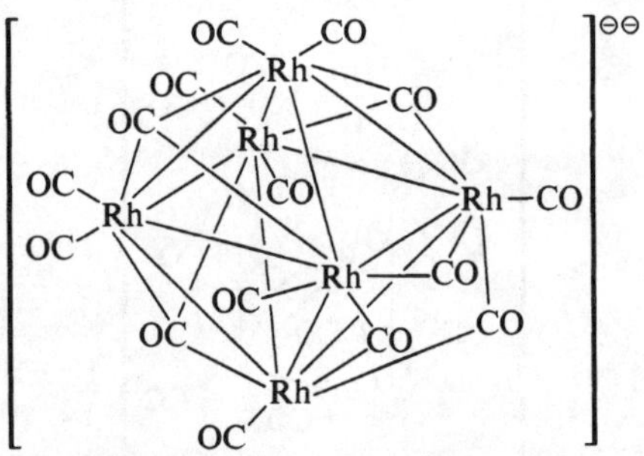

M 1037.589 (ion)

Struct. by analogy with the corresponding Co anion. All derivs. highly air-sensitive. Readily undergoes intramolecular carbonyl exchange at –70°.

Bis(tetraethylammonium) salt: [79306-13-1].
$C_{31}H_{40}N_2O_{15}Rh_6$ M 1298.094
V. air-sensitive.

Bis(tetrabutylammonium) salt:
$C_{47}H_{72}N_2O_{15}Rh_6$ M 1522.523
Green cryst. Insol. H_2O, sol. MeCN.

Bis(tetraphenylphosphonium) salt: [57472-89-6].
$C_{63}H_{40}O_{15}P_2Rh_6$ M 1716.381
Cryst. Sol. Me_2CO.

Bis[bis(triphenylphosphine)iminium] salt: [76330-61-5].
$C_{87}H_{60}N_2O_{15}P_4Rh_6$ M 2114.763
V. air-sensitive.

Martinengo, S. *et al*, *Gazz. Chim. Ital.*, 1972, **102**, 344 (*synth*)
Heaton, B.T. *et al*, *J. Chem. Soc., Chem. Commun.*, 1975, 523 (*cmr*)

$C_{16}H_{10}F_6O_2Rh_2$ Rh-00163

[μ-[1,2-Bis(trifluoromethyl)-1,2-ethenediyl]dicarbonylbis(η⁵-2,4-cyclopentadien-1-yl)dirhodium, 10CI

Dicarbonylbis(η⁵-cyclopentadienyl)-μ-(2,3-η-1,1,1,4,4,4-hexafluoro-2-butenyl)dirhodium

[57872-13-6]

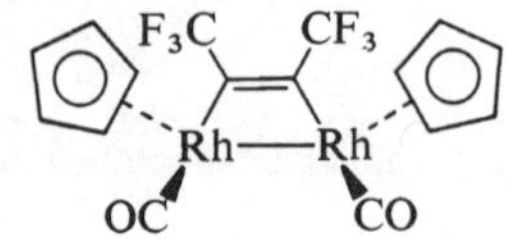

cis-form

M 554.055

cis-form [69881-14-7]
ir ν_{CO} 2040, 1993 cm^{-1} ($CHCl_3$). Slowly isomerises to *trans* isomer in soln.

trans-form [63950-13-0]
Orange-red cryst. (hexane). Mp 162-3°. ir ν_{CO} 1992 cm^{-1} ($CHCl_3$). CO groups undergo scrambling in soln. at r.t.

Dickson, R.S. *et al*, *Aust. J. Chem.*, 1972, **25**, 2535 (*synth*, *ir*, *pmr*)
Todd, L.J. *et al*, *J. Organomet. Chem.*, 1975, **101**, 133 (*nmr*)
Dickson, R.S. *et al*, *J. Organomet. Chem.*, 1975, **101**, C48 (*cryst struct*)
Dickson, R.S. *et al*, *Acta Crystallogr., Sect. B*, 1977, **33**, 2057 (*cryst struct*)
Dickson, R.S. *et al*, *J. Organomet. Chem.*, 1979, **166**, 385 (*synth*, *ir*, *nmr*)

$C_{16}H_{10}O_6Rh_2$ Rh-00164

Tetracarbonyldi-μ-phenoxydirhodium, 9CI

Bis(μ-phenolato)tetracarbonyldirhodium

[49713-88-4]

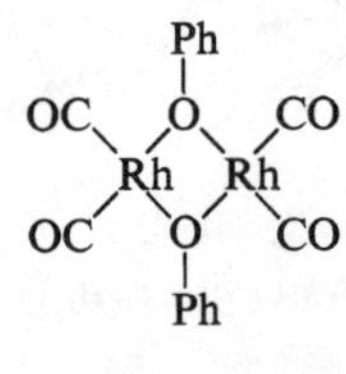

M 504.062

Yellow cryst. (hexane). Dec. on drying or prolonged storage in soln. even under N_2 or Ar. ir ν_{CO} 2095, 2081, 2078, 2018 cm^{-1} (hexane).

Monomer: Dicarbonylphenoxyrhodium.
$C_8H_5O_3Rh$ M 252.031
Unknown.

Vizi-Orosz, A. *et al*, *J. Organomet. Chem.*, 1973, **57**, 379 (*synth*, *ir*, *pmr*)

$C_{16}H_{16}Cl_2Rh_2$ **Rh-00165**

Di-μ-chlorobis[(1,2,5,6-η)-1,3,5,7-cyclooctatetraene]dirhodium, 10CI

[12308-49-5]

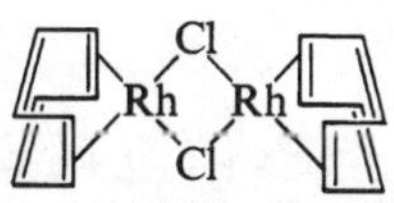

M 485.019

Orange cryst. Mp 140-5° dec.

Monomer: Chloro(cyclooctatetraene)rhodium. Cyclooctatetraenerhodium chloride.

C_8H_8ClRh M 242.510

Unknown.

Abel, E.W. *et al*, *J. Chem. Soc.*, 1959, 3178 (*synth*)
Bennett, M.A. *et al*, *Inorg. Chem.*, 1968, **7**, 321 (*synth, pmr, ir*)
Volger, H.L. *et al*, *Inorg. Chim. Acta*, 1969, **3**, 145 (*synth, pmr, ir*)

$C_{16}H_{20}Cl_2Rh_2$ **Rh-00167**

Di-μ-chlorobis[(1,2,5,6-η)-1,3,5-cyclooctatriene]dirhodium

[41371-55-5]

M 489.051

Yellow cryst.

Monomer: Chloro[(1,2,5,6-η)-1,3,5-cyclooctatriene]rhodium. 1,3,5-Cyclooctatrienerhodium chloride.

$C_8H_{10}ClRh$ M 244.526

Unknown.

Evans, J. *et al*, *J. Chem. Soc., Dalton Trans.*, 1972, 2668 (*synth, pmr, ir*)
Draggett, P.T. *et al*, *J. Organomet. Chem.*, 1977, **135**, C60.

$C_{16}H_{19}Rh$ **Rh-00166**

(1,5-Cyclooctadiene)(4-hydro-π-pentalenyl)rhodium

Hydropentalenyl-1,5-cyclooctadienerhodium

[12112-51-5]

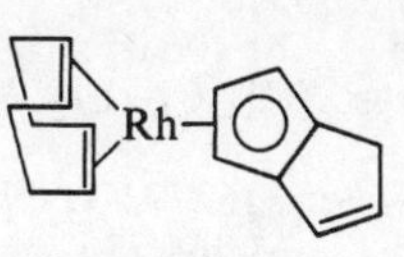

M 314.232

Incorrectly named in 8CA. Yellow cryst. by subl. Mp 85-6°. $Bp_{0.005}$ 85° subl.

Katz, T.J. *et al*, *J. Am. Chem. Soc.*, 1967, **89**, 1105 (*synth, ms, pmr, ir*)

$C_{16}H_{20}ORh^{\oplus}$ **Rh-00168**

[(1,2,3,4,5-η)-6-Oxo-2,4-cyclohexadien-1-yl][(1,2,3,4,5-η)-1,2,3,4,5-pentamethyl-2,4-cyclopentadien-1-yl]rhodium(1+), 10CI

[(1,2,3,4,5-η)-1,2,3,4,5-Pentamethyl-2,4-cyclopentadien-1-yl](η⁶-phenolato)rhodium(1+)

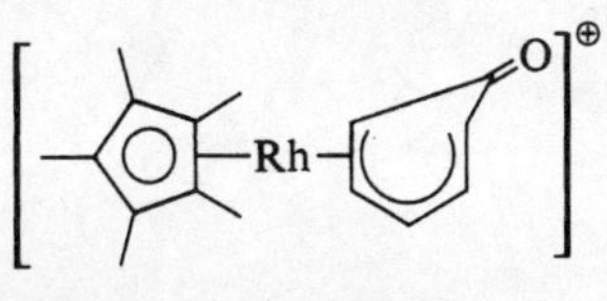

M 331.239 (ion)

Hexafluorophosphate: [62837-12-1].

$C_{16}H_{20}F_6OPRh$ M 476.203

Yellow solid (Me_2CO/Et_2O).

White, C. *et al*, *J. Organomet. Chem.*, 1977, **127**, 415 (*synth, ir, nmr*)

$C_{16}H_{20}Rh^{\oplus}$ **Rh-00169**

[(1,2,3,4,5,6-η)-1,3,5-Cyclooctatriene][(1,2,5,6-η)-1,3,5-cyclooctatriene]rhodium(1+), 10CI

Bis(1,3,5-Cyclooctatriene)rhodium(1+)

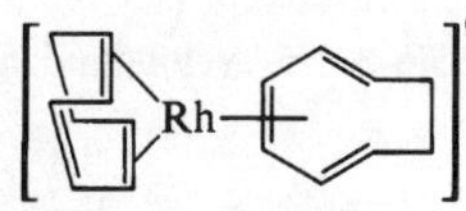

M 315.240 (ion)

Olefin hydrogenation catalyst.

Tetrafluoroborate: [64331-06-2].

$C_{16}H_{20}BF_4Rh$ M 402.043

No phys. props. given.

Draggett, P.J. *et al*, *J. Organomet. Chem.*, 1977, **135**, C60 (*synth, use*)

$C_{16}H_{21}Rh^{\oplus\oplus}$ **Rh-00170**

η⁶-Benzene[(1,2,3,4,5-η)-1,2,3,4,5-pentamethyl-2,4-cyclopentadien-1-yl]rhodium(2+)

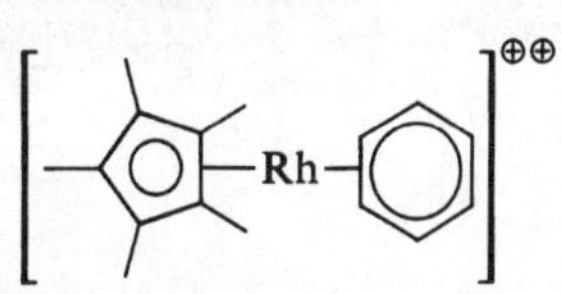

M 316.247 (ion)

Benzene ligand easily displaced by coordinating solvs.

Bis(hexafluorophosphate): [65074-04-6].

$C_{16}H_{21}F_{12}P_2Rh$ M 606.176

White solid (Me_2CO/Et_2O).

White, C. *et al*, *J. Chem. Soc., Dalton Trans.*, 1977, 1654 (*synth, pmr, cmr, props*)

$C_{16}H_{22}NRh^{\oplus\oplus}$ **Rh-00171**

(η⁶-Aniline)(pentamethylcyclopentadienyl)rhodium(2+)

(η⁶-Benzenamine)[(1,2,3,4,5-η)-1,2,3,4,5-pentamethyl-2,4-cyclopentadien-1-yl]rhodium(2+), 10CI

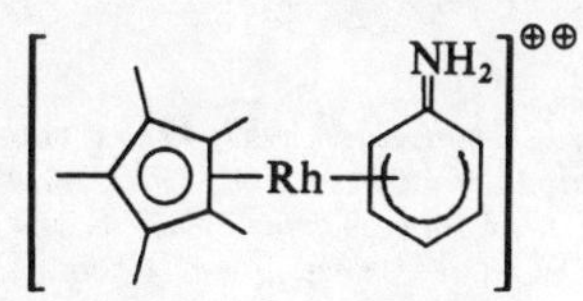

M 331.262 (ion)

Bis(hexafluorophosphate): [75583-64-1].

$C_{16}H_{22}F_6NPRh$ M 476.226

Yellow cryst. (Me_2CO/Et_2O).

Espinet, P. *et al*, *J. Chem. Soc., Dalton Trans.*, 1980, 1048 (*synth, nmr*)

$C_{16}H_{23}Rh$ **Rh-00172**

(1,3-Cyclohexadiene)(pentamethyl-π-cyclopentadienyl)rhodium, 8CI

[33519-75-4]

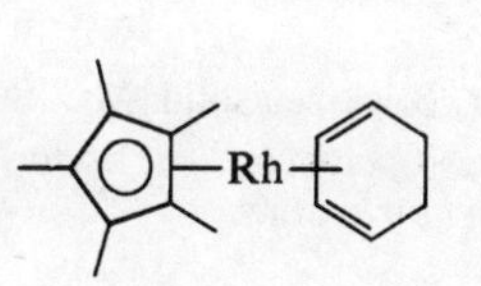

M 318.263

Catalyst for the disproportionation of 1,3-cyclohexadiene to cyclohexane and benzene. Somewhat air-sensitive yellow cryst. Mp 47-8°. $Bp_{0.001}$ 25° subl.

Moseley, K. *et al*, *J. Chem. Soc.* (*A*), 1970, 2884 (*synth*, *pmr*, *ms*, *use*)

$C_{16}H_{24}Br_2Rh_2$ Rh-00173

Di-μ-bromobis(1,2,5,6-η-1,5-cyclooctadiene)dirhodium, 9CI

[12092-45-4]

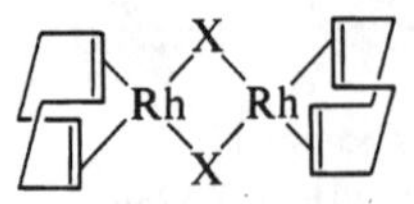

X = Br

M 581.985

Bridged dimer. Yellow cryst. Mp 207-12°, 222-8 dec.

Monomer: Bromo(1,5-cyclooctadiene)rhodium.

$C_8H_{12}BrRh$ M 290.992

Unknown.

Chatt, J. *et al*, *J. Chem. Soc.*, 1957, 4735 (*synth*)
Adams, D.M. *et al*, *J. Chem. Soc.* (*A*), 1969, 588 (*ir*)
Kingston, J.V. *et al*, *J. Chem. Soc.* (*A*), 1971, 3399 (*synth*, *ir*)
Tayim, H.A. *et al*, *J. Organomet. Chem.*, 1975, **92**, 107 (*synth*)

$C_{16}H_{24}Cl_2Rh_2$ Rh-00174

Di-μ-chlorobis[(1,2,5,6-η)-1,5-cyclooctadiene]dirhodium, 10CI, 9CI

Bis(1,5-cyclooctadiene-μ-chlororhodium)

[12092-47-6]

As Di-μ-bromobis(1,2,5,6-η-1,5-cyclooctadiene)-dirhodium, Rh-00173 with

X = Cl

M 493.083

Bridged dimer. Useful catalyst precursor for asymmetric hydrogenation of olefins (in the presence of a chiral phosphine). Orange cryst. (AcOH). Mp 256°.

Monomer: (1,5-Cyclooctadiene)chlororhodium.

$C_8H_{12}ClRh$ M 246.541

Unknown.

Chatt, J. *et al*, *J. Chem. Soc.*, 1957, 4735 (*synth*)
Ibers, J.A. *et al*, *J. Am. Chem. Soc.*, 1962, **84**, 495 (*struct*)
Bennett, M.A. *et al*, *Inorg. Chem.*, 1967, **6**, 1647 (*ir*)
v. Leeuwen, P.W. *et al*, *J. Organomet. Chem.*, 1969, **19**, 187 (*pmr*)
Axelson, D.E. *et al*, *Inorg. Nucl. Chem. Lett.*, 1973, **9**, 885 (*cmr*)
Inorg. Synth., 1979, **19**, 218 (*synth*, *ir*, *pmr*)

$C_{16}H_{24}I_2Rh_2$ Rh-00175

Bis[(1,2,5,6-η)-1,5-cyclooctadiene]di-μ-iododirhodium, 9CI

[12092-49-8]

As Di-μ-bromobis(1,2,5,6-η-1,5-cyclooctadiene)dirhodium, Rh-00173 with

X = I

M 675.986

Bridged dimer. Brick-red solid. Mp 180-200° dec.

Monomer: (1,5-Cyclooctadiene)iodorhodium. Iodo(1,5-cyclooctadiene)rhodium.

$C_8H_{12}IRh$ M 337.993

Unknown

Chatt, J. *et al*, *J. Chem. Soc.*, 1957, 4753 (*synth*)
Adams, D.M. *et al*, *J. Chem. Soc.* (*A*), 1969, 588 (*ir*)

$C_{16}H_{24}N_3Rh^{\oplus\oplus}$ Rh-00176

Tris(acetonitrile)[(1,2,3,4,5-η)-1,2,3,4,5-pentamethyl-2,4-cyclopentadien-1-yl]rhodium(2+), 10CI

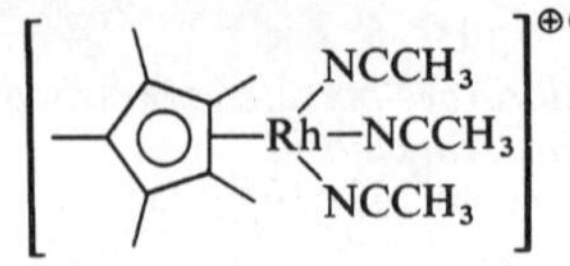

M 361.291 (ion)

Useful precursor to pentamethylcyclopentadienylrhodium(III) complexes. MeCN ligands readily displaced by *P*-donor ligands, olefins, etc.

Bis(hexafluorophosphate): [59738-28-2].

$C_{16}H_{24}F_{12}N_3P_2Rh$ M 651.220

Yellow, air-stable cryst. ($MeCN/Et_2O$).

White, C. *et al*, *J. Chem. Soc.*, *Dalton Trans.*, 1977, 1654 (*synth*, *pmr*, *ir*)
White, C. *et al*, *J. Chem. Soc.*, *Dalton Trans.*, 1978, 1305 (*props*)

$C_{16}H_{24}Rh^{\oplus}$ Rh-00177

Bis[(1,2,5,6-η)-1,5-cyclooctadiene]rhodium(1+), 10CI

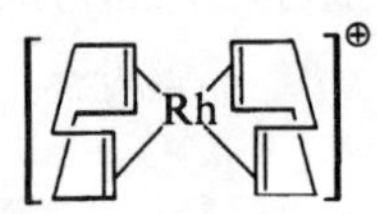

M 319.271 (ion)

Catalyst for hydrogenation and hydroformylation of olefins. Used to prepare $[Rh(cod)L_2]^{\oplus}$ complexes.

Tetrafluoroborate: [35138-22-8].

$C_{16}H_{24}BF_4Rh$ M 406.075

Orange-red cryst. (CH_2Cl_2/Et_2O). Mp 206-8° (200-1°).

Hexafluorophosphate: [62793-31-1].

$C_{16}H_{24}F_6PRh$ M 464.235

Red cryst.

Perchlorate: [59420-05-2].

$C_{16}H_{24}ClO_4Rh$ M 418.722

Red cryst.

Green, M. *et al*, *J. Chem. Soc.* (*A*), 1971, 2334 (*synth*, *ir*, *pmr*)
Schrock, R.R. *et al*, *J. Am. Chem. Soc.*, 1971, **93**, 3089 (*use*)
Uson, R. *et al*, *J. Organomet. Chem.*, 1976, **105**, 365 (*use*)

$C_{16}H_{26}BCl_2N_6ORh$ Rh-00178

Dichloro(methanol)[tris(3,5-dimethyl-1*H*-pyrazolato-*N*[1])-hydroborato(1–)-*N*[2],*N*[2'],*N*[2'']]rhodium

Hydrotris(3,5-dimethylpyrazolyl)borato]dichloro-(methanol)rhodium

[73117-95-0]

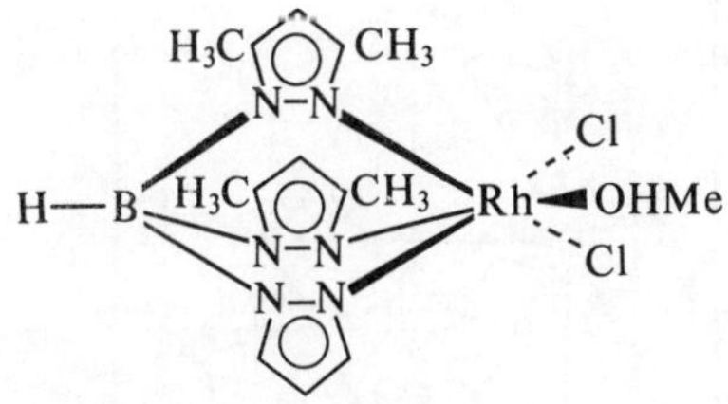

M 503.043

Incorrectly named in 10CI. Starting material for hydrotris(3,5-dimethylpyrazolyl)borato-rhodium complexes. Gold prisms (MeOH). Mp 296-8°.

May, S. *et al*, *Inorg. Chem.*, 1980, **19**, 1582 (*synth, pmr*)

$C_{16}H_{26}O_4Rh_2$ Rh-00179

Bis(μ-acetato)tetrakis(η³-2-propenyl)dirhodium

Bis(μ-acetato)tetra-π-allyldirhodium, 8CI

[12308-58-6]

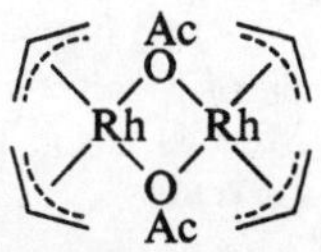

M 488.190

Yellow microprisms (CH_2Cl_2/pet. ether). Mp 110-5° dec.

Monomer: (Acetato)bis(2-propenyl)rhodium. Acetatodiallylrhodium.

$C_8H_{13}O_2Rh$ M 244.095

Unknown.

Powell, J. *et al*, *J. Chem. Soc. (A)*, 1968, 583 (*synth*)

$C_{16}H_{26}Rh^{\oplus}$ Rh-00180

Bis(η²-ethene)[(1,2,3,4,5,6-η)hexamethylbenzene]rhodium(1+), 10CI

Hexamethylbenzenebis(ethylene)rhodium(1+)

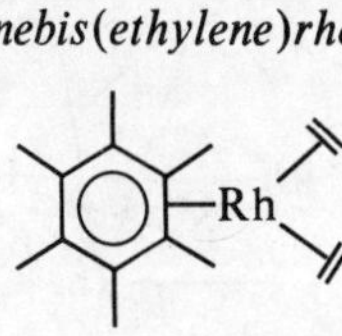

M 321.287 (ion)

Hexafluorophosphate: [76706-18-8].

$C_{16}H_{26}F_6PRh$ M 466.251

Cream cryst. (CH_2Cl_2/Et_2O). Mp 180° dec.

Tetrafluoroborate: [36644-70-9].

$C_{16}H_{26}BF_4Rh$ M 408.091

Cream cryst. (CH_2Cl_2/Et_2O). Mp 200-1° dec.

Green, M. *et al*, *J. Chem. Soc., Dalton Trans.*, 1972, 832 (*synth, pmr*)

Bianchi, F. *et al*, *J. Organomet. Chem.*, 1980, **202**, 99 (*synth, pmr*)

$C_{16}H_{28}FRh$ Rh-00181

Bis[(1,2-η)-cyclooctene]fluororhodium, 10CI

[63693-77-6]

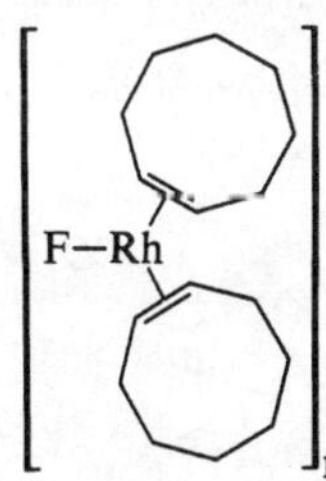

M 342.301

Prob. dimeric, n = 2. Useful starting compd. for fluororhodium(I) complexes. Yellow solid (C_6H_6/hexane). Can be stored in air at 25° for over a month.

Van Gaal, H.L.M. *et al*, *J. Organomet. Chem.*, 1976, **114**, C35 (*synth*)

Van Gaal, H.L.M. *et al*, *J. Organomet. Chem.*, 1977, **134**, 237 (*use*)

$C_{16}IO_{16}Rh_7^{\ominus\ominus}$ Rh-00182

Di-μ-carbonyltetra-μ₃-carbonyldecacarbonyl-μ-iodoheptarhodate(2–), 9CI

Hexadecacarbonyliodoheptarhodate(2–)

[57063-69-1]

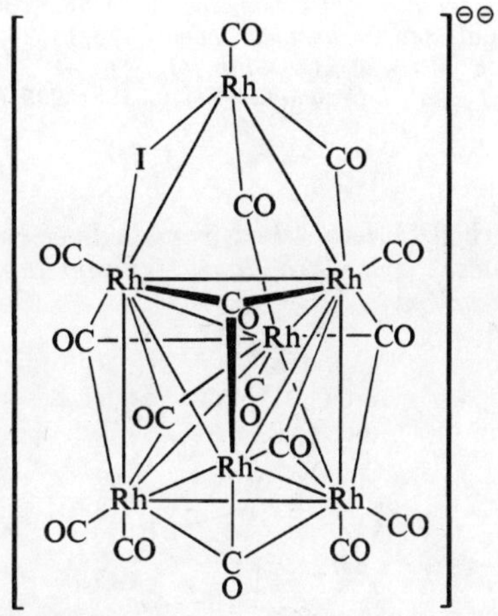

M 1295.409 (ion)

ir ν_{CO} 2058, 2018, 1996, 1777, 1763 cm^{-1} (MeCN).

Bis(tetramethylammonium) salt:

$C_{24}H_{24}IN_2O_{16}Rh_7$ M 1443.700

Sol. Me_2CO, MeCN, sl. sol. THF, insol. H_2O.

Bis(tetraethylammonium) salt: [57108-41-5].

$C_{32}H_{40}IN_2O_{16}Rh_7$ M 1555.915

Dark-brown cryst. (Me_2CO/2-propanol). Sol. Me_2CO, MeCN, sl. sol. THF, insol. H_2O. Begins to dec. >150°.

Martinengo, S. *et al*, *J. Organomet. Chem.*, 1975, **88**, 375 (*synth, ir*)

Albano, V.G. *et al*, *J. Organomet. Chem.*, 1975, **88**, 381 (*cryst struct*)

$C_{16}O_{15}Rh_6^{\ominus\ominus}$ Rh-00183

[μ_6-(Carbon)]nona-μ-carbonylhexacarbonylhexarhodate(2−), 10CI

μ_6-Carbidonona-μ-carbonylhexacarbonylhexarhodate(2−).

μ_6-(Carbon)pentadecacarbonylhexarhodate(2−)

[54065-66-6]

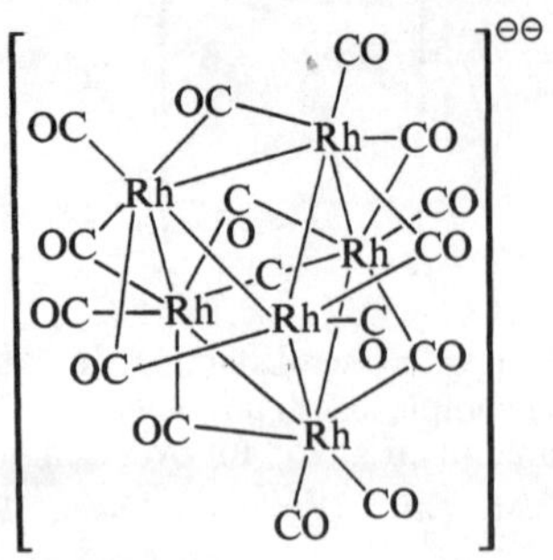

M 1049.600 (ion)

Di-K salt: [53468-95-4].

$C_{16}K_2O_{15}Rh_6$ M 1127.797

Yellow powder. Sol. H_2O. Air-sensitive.

Bis(benzyltrimethylammonium) salt: [39357-18-1].

$C_{36}H_{32}N_2O_{15}Rh_6$ M 1350.086

Yellow cryst. (Me_2CO/2-propanol).

Albano, V.G. *et al*, *J. Chem. Soc., Dalton Trans.*, 1973, 651 (*synth, struct*)
Albano, V.G. *et al*, *J. Am. Chem. Soc.*, 1974, **96**, 8106 (*cmr*)
Chini, P. *et al*, *Adv. Organomet. Chem.*, 1976, **14**, 285 (*rev*)
Inorg. Synth., 1980, **20**, 212 (*synth, ir*)
Heaton, B.T. *et al*, *J. Organomet. Chem.*, 1981, **215**, 415 (*nmr*)

$C_{16}O_{16}Rh_6$ Rh-00184

Tetra-μ_3-carbonyldodecacarbonylhexarhodium, 10CI, 9CI

Hexadecacarbonylhexarhodium. Hexarhodium hexadecacarbonyl

[28407-51-4]

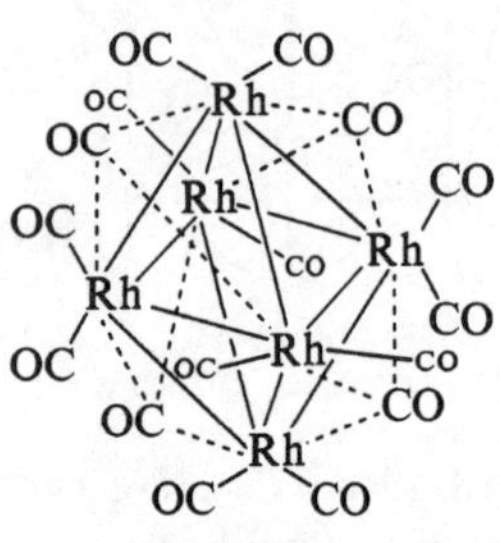

M 1065.599

Catalyses hydroformylation, water gas shift reaction, Fischer-Tropsch reaction, cyclopropanation of alkenes with ethyl diazoacetate, and the reduction of aromatic nitro compds. to arenes by CO + H_2O. Violet-brown to black cryst. ($CHCl_3$). Mp 238-42°. Dec. >200°. ν_{CO} (CCl_4) 2075 (s), 2044 (m), 1811 (m).

Corey, E.R. *et al*, *J. Am. Chem. Soc.*, 1963, **85**, 1202 (*struct*)
Cariati, F. *et al*, *Gazz. Chim. Ital.*, 1969, **99**, 1327 (*ir*)
Chini, P. *et al*, *Inorg. Chim. Acta*, 1969, **3**, 315 (*synth*)
Heaton, B.T. *et al*, *J. Chem. Soc., Chem. Commun.*, 1975, 523 (*cmr*)
Conrad, H. *et al*, *Chem. Phys. Lett.*, 1976, **42**, 115 (*uv*)
Inorg. Synth., 1976, **16**, 49 (*synth, ir*)
Laine, R.M. *et al*, *Ann. N.Y. Acad. Sci.*, 1980, **333**, 124 (*use*)
Fieser, M. *et al*, *Reagents for Organic Synthesis*, Wiley, 1967-83, **10**, 382 (*use*)

$C_{16}O_{16}Rh_7^{\ominus\ominus\ominus}$ Rh-00185

Hexadecacarbonylheptarhodate(3−)

Hexa-μ-carbonyltri-μ_3-carbonylheptacarbonylheptarhodate(3−), 10CI

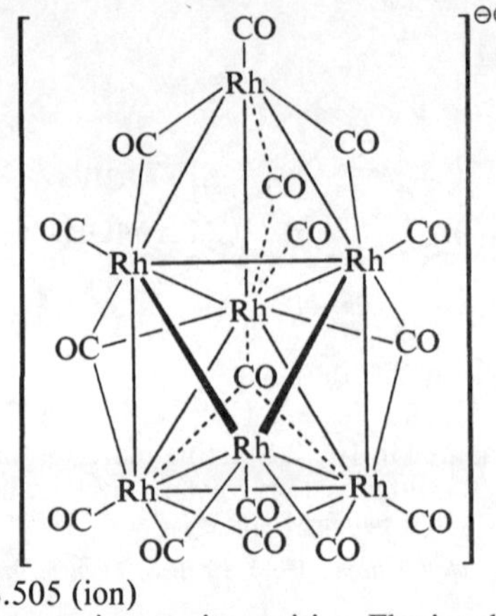

M 1168.505 (ion)

Dark-green anion, v. air-sensitive. Fluxional in soln. at r.t.

K salt: [12568-17-1].

$C_{16}K_3O_{17}Rh_7$ M 1301.799

V. sol. H_2O.

Tris(tetramethylammonium) salt:

$C_{28}H_{36}N_3O_{16}Rh_7$ M 1390.941

Prepd. by metathesis of K salt. Green cryst. (Me_2CO aq.). Sol. MeCN, Me_2CO, insol. THF, H_2O. Gradual dec. >200° in N_2 current.

Albano, V.G. *et al*, *J. Chem. Soc., Chem. Commun.*, 1969, 1024 (*cryst struct*)
Martinengo, S. *et al*, *Gazz. Chim. Ital.*, 1972, **102**, 344 (*synth*)
Heaton, B.T. *et al*, *J. Organomet. Chem.*, 1979, **169**, 309 (*nmr*)

$C_{17}H_3O_{17}Rh_6^{\ominus}$ Rh-00186

Pentadecacarbonyl(methoxycarbonyl)hexarhodate(1−), 10CI

Tetra-μ_3-carbonylundecacarbonyl(methoxycarbonyl)hexarhodate(1−)

[60650-14-8]

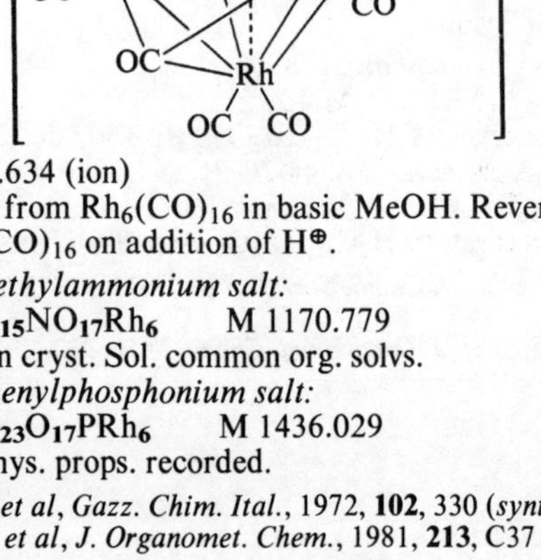

M 1096.634 (ion)

Formed from $Rh_6(CO)_{16}$ in basic MeOH. Reverts to $Rh_6(CO)_{16}$ on addition of $H^{\oplus}$.

Tetramethylammonium salt:

$C_{21}H_{15}NO_{17}Rh_6$ M 1170.779

Brown cryst. Sol. common org. solvs.

Tetraphenylphosphonium salt:

$C_{41}H_{23}O_{17}PRh_6$ M 1436.029

No phys. props. recorded.

Chini, P. *et al*, *Gazz. Chim. Ital.*, 1972, **102**, 330 (*synth, ir*)
Ciani, G. *et al*, *J. Organomet. Chem.*, 1981, **213**, C37 (*cryst struct*)

$C_{17}H_{16}N_2Rh^{\oplus}$ **Rh-00187**

[(2,3,5,6-η)-Bicyclo[2.2.1]hepta-2,5-diene](2,2′-bipyridine-*N*,*N′*)rhodium(1+), 9CI

(2,2′-Bipyridine)(norbornadiene)rhodium(1+)

[52445-03-1]

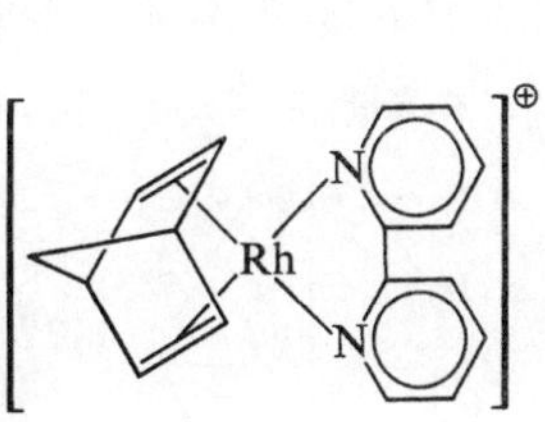

M 351.232 (ion)

Chloride: [56678-49-0].
$C_{17}H_{16}ClN_2Rh$ M 386.685
Red-violet cryst.

Hexafluorophosphate: [52445-05-3].
$C_{17}H_{16}F_6N_2PRh$ M 496.196
Orange cryst.

Perchlorate: [52445-04-2].
$C_{17}H_{16}ClN_2O_4Rh$ M 450.683
Orange or orange-red cryst. Mp >140°.

Tetraphenylborate: [66349-79-9].
$C_{41}H_{36}BN_2Rh$ M 670.464
Orange-red cryst.

Mestroni, G. *et al*, *J. Organomet. Chem.*, 1974, **65**, 119 (*synth*)
Zassinovich, G. *et al*, *J. Organomet. Chem.*, 1975, **91**, 379 (*synth*)
Uson, R. *et al*, *J. Organomet. Chem.*, 1976, **116**, C35 (*synth*)
Oro, L.A. *et al*, *J. Organomet. Chem.*, 1978, **148**, 81 (*synth*, *ir*)
Fordyce, L.A. *et al*, *Inorg. Chem.*, 1982, **21**, 1023 (*uv*)

$C_{17}H_{18}N_2Rh^{\oplus}$ **Rh-00188**

[(2,3,5,6-η)-Bicyclo[2.2.1]hepta-2,5-diene]bis(pyridine)rhodium(1+)

(Norbornadiene)bis(pyridine)rhodium(1+)

[60576-51-4]

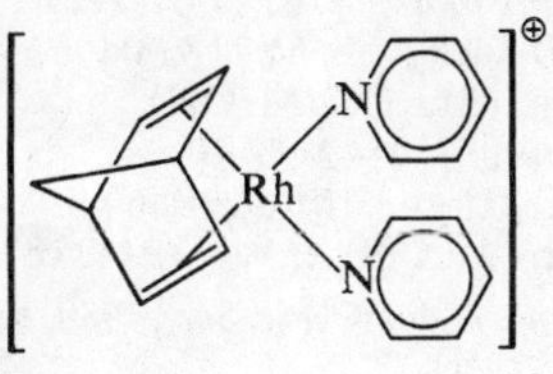

M 353.248 (ion)

Perchlorate: [60576-52-5].
$C_{17}H_{18}ClN_2O_4Rh$ M 452.699
Yellow cryst. Mp >140°.

Tetraphenylborate: [66349-85-7].
$C_{41}H_{38}BN_2Rh$ M 672.480
Yellow cryst.

Usòn, R. *et al*, *J. Organomet. Chem.*, 1976, **116**, C35 (*synth*)
Denise, B. *et al*, *J. Organomet. Chem.*, 1978, **161**, 171 (*synth*, *pmr*)
Oro, L.A. *et al*, *J. Organomet. Chem.*, 1978, **148**, 81 (*synth*, *ir*)

$C_{17}H_{22}BN_6Rh$ **Rh-00189**

[(1,2,5,6-η)-1,5-Cyclooctadiene][hydrotris(1*H*-pyrazolato-*N*[1])borato(1−)-*N*[2],*N*[2′]]rhodium, 9CI

[52668-35-6]

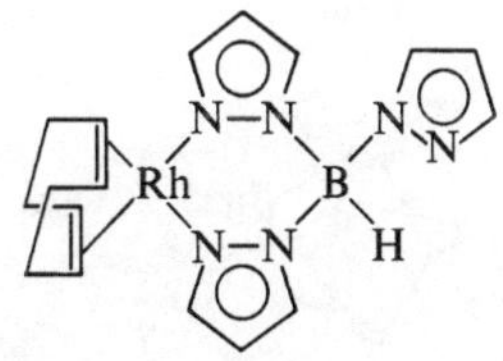

M 424.117

Yellow cryst. (CH_2Cl_2/Et_2O). Mp 167°. Stereochem. non-rigid at r.t.

O'Sullivan, D.J. *et al*, *J. Organomet. Chem.*, 1974, **65**, C47 (*synth*)
King, R.B. *et al*, *J. Organomet. Chem.*, 1974, **73**, 115 (*synth*, *nmr*)

$C_{17}H_{22}ClO_2P_2Rh$ **Rh-00190**

[(*C*,*O*-η)Carbon dioxide]chlorobis(dimethylphenylphosphine)rhodium, 10CI

[64596-40-3]

$RhCl(PMe_2Ph)_2(CO_2)$

M 458.666

Struct. not determined. Red. Loses CO_2 when heated above 80°. O_2 transfer from CO_2 to phosphine ligand occurs in polar solvs. eg. Me_2CO, CH_2Cl_2, EtOH.

Aresta, M. *et al*, *Inorg. Chim. Acta*, 1977, **24**, L49 (*synth*, *ir*, *props*)

$C_{17}H_{23}Rh^{\oplus\oplus}$ **Rh-00191**

η^6-Benzene[(1,2,3,4,5-η)-1-ethyl-2,3,4,5-tetramethyl-2,4-cyclopentadien-1-yl]rhodium(2+)

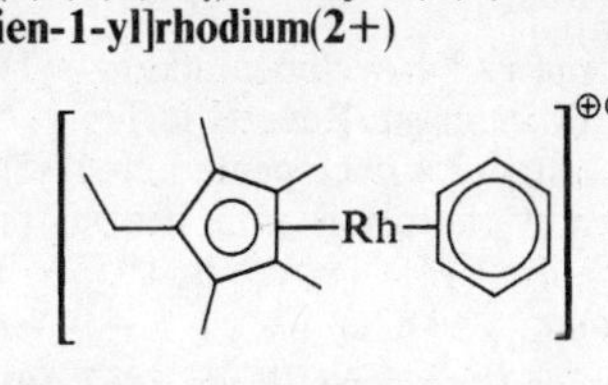

M 330.274 (ion)

Catalyst for the intramolecular cyclisation of substituted fluorophenylpropanols to chromans. Benzene easily displaced by strongly coordinating solvs.

Bis(hexafluorophosphate): [69762-18-1].
$C_{17}H_{23}F_{12}P_2Rh$ M 620.203
White, air-stable cryst. ($MeNO_2/Et_2O$).

Bis(tetrafluoroborate): [74776-06-0].
$C_{17}H_{23}B_2F_8Rh$ M 503.881
White, air-stable cryst. ($MeNO_2/Et_2O$).

Bailey, N.A. *et al*, *J. Chem. Soc., Dalton Trans.*, 1980, 829 (*synth*, *pmr*, *cmr*, *props*)
Houghton, R.P. *et al*, *J. Chem. Soc., Chem. Commun.*, 1980, 884 (*use*)

$C_{17}H_{23}Rh$ Rh-00192

Bicyclo[2.2.1]heptadiene(pentamethylcyclopentadienyl)rhodium

(2,5-Norbornadiene)(pentamethyl-π-cyclopentadienyl-)rhodium, 8CI

[32678-99-2]

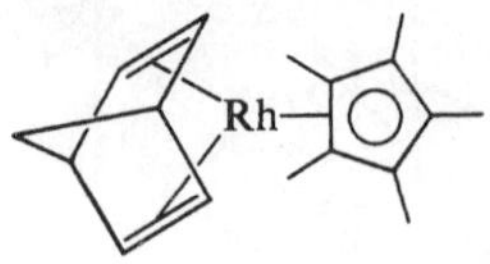

M 330.274

Yellow cryst. (pet. ether). Mp 120-8° dec. $Bp_{0.001}$ 45° subl.

Maitlis, P.M. *et al*, *J. Chem. Soc. (A)*, 1970, 2875 (*synth, pmr, ms*)

Lee, H.B. *et al*, *J. Chem. Soc., Dalton Trans.*, 1975, 2322.

$C_{18}H_{15}O_3Rh_3$ Rh-00193

Di-μ-carbonylcarbonyltris(η^5-2,4-cyclopentadien-1-yl)trirhodium, 9CI

Tricarbonyltris(η^5-2,4-cyclopentadien-1-yl)trirhodium-(isomer B). Tris(π-cyclopentadienylrhodiumcarbonyl)-(isomer B). Tris(cyclopentadienylcarbonylrhodium)-(isomer B)

[12305-50-9]

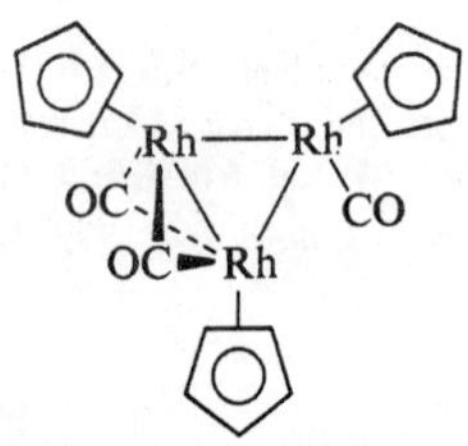

M 588.031

Isomer of Tri-μ-carbonyltris(η^5-2,4-cyclopentadien-1-yl)trirhodium, Rh-00194 . Black cryst. Reverts to Tri-μ-carbonyltris(η^5-2,4-cyclopentadien-1-yl)trirhodium, Rh-00194 on uv photolysis or on heating (within ca. 30 min. at 80°). ir ν_{CO} 1973s, 1967s, 1851w, 1797m, 1748s (KBr).

Paulus, E.F. *et al*, *J. Organomet. Chem.*, 1967, **10**, P3 (*struct, ir*)

Paulus, E.F. *et al*, *Acta Crystallogr., Sect. B*, 1969, **25**, 2206 (*struct*)

Lawson, R.J. *et al*, *Inorg. Chem.*, 1978, **17**, 772 (*synth, nmr, ms, ir*)

Pinhas, A.R. *et al*, *Helv. Chim. Acta*, 1980, **63**, 29.

$C_{18}H_{15}O_3Rh_3$ Rh-00194

Tri-μ-carbonyltris(η^5-2,4-cyclopentadien-1-yl)trirhodium, 9CI

Tri-μ-carbonyltri-π-cyclopentadienyltrirhodium, 8CI. *Tris(π-cyclopentadienylrhodiumcarbonyl)(isomer A). Tris(cyclopentadienylcarbonylrhodium)(isomer A)*

[12148-54-8]

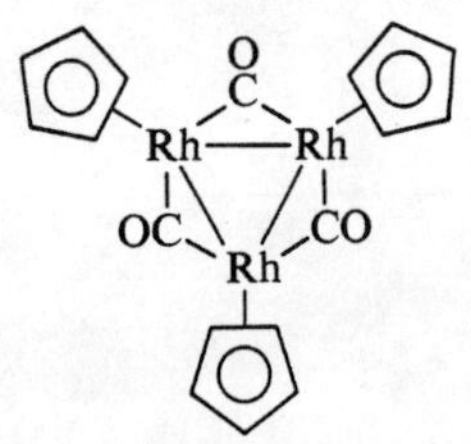

M 588.031

Isomer of Di-μ-carbonylcarbonyltris(η^5-2,4-cyclopentadien-1-yl)trirhodium Rh-00193 . Black cryst. (C_6H_6/hexane). Mp 265-70° dec. ir ν_{CO} 1827s, 1783m, 1766s cm^{-1} (KBr).

Mills, O.S. *et al*, *J. Chem. Soc., Chem. Commun.*, 1966, 815 (*struct*)

Mills, O.S. *et al*, *J. Organomet. Chem.*, 1967, **10**, 331 (*struct, synth*)

Lawson, R.J. *et al*, *J. Am. Chem. Soc.*, 1976, **98**, 7433 (*synth, ir, nmr*)

Lawson, R.J. *et al*, *Inorg. Chem.*, 1978, **17**, 772 (*cmr*)

Rausch, M.D. *et al*, *J. Organomet. Chem.*, 1981, **212**, 247 (*synth, pmr*)

$C_{18}H_{16}O_2Rh_3^{\oplus}$ Rh-00195

Di-μ-carbonyltris(η^5-2,4-cyclopentadien-1-yl)-μ_3-methylidynetrirhodium(1+), 10CI

Di-μ-carbonyl-μ_3-methynetris(η-cyclopentadienyl)trirhodium(1+)

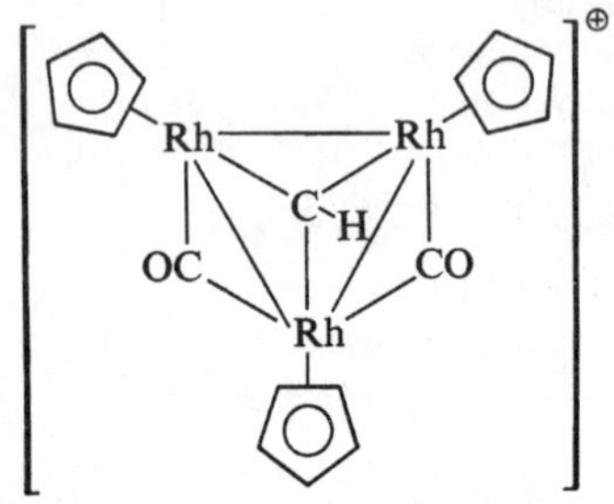

M 573.040 (ion)

Trifluoromethanesulfonate: [74543-02-5].

$C_{19}H_{16}F_3O_5Rh_3S$ M 722.104

Black cryst. (THF). ir ν_{CO} 1932, 1895 cm^{-1} (KBr).

Tetrafluoroborate: [74542-98-6].

$C_{18}H_{16}BF_4O_2Rh_3$ M 659.843

Black cryst. (THF). Sol. MeCN, Me_2CO, MeOH, DMF. Mp >280°. ir ν_{CO} 1929, 1884 cm^{-1} (KBr).

Hexafluorophosphate: [76114-01-7].

$C_{18}H_{16}F_6O_2PRh_3$ M 718.004

Purple cryst. (Me_2CO/MeOH).

Trifluoroacetate: [74542-99-7].

$C_{20}H_{16}F_3O_4Rh_3$ M 686.056

Black cryst. (MeCN). ir ν_{CO} 1918, 1870 cm^{-1} (KBr).

Dimas, P.A. *et al*, *J. Am. Chem. Soc.*, 1980, **102**, 7787 (*synth, ms, ir, nmr, struct*)

Herrmann, W.A. *et al*, *Angew. Chem., Int. Ed. Engl.*, 1980, **19**, 651 (*synth, nmr, ir, struct*)

Herrmann, W.A. *et al*, *J. Am. Chem. Soc.*, 1981, **103**, 63 (*synth, struct, nmr, ir*)

$C_{18}H_{16}O_3Rh_3^{\oplus}$ Rh-00196

Tri-μ-carbonyltris(η^5-2,4-cyclopentadien-1-yl)-μ_3-hydrotrirhodium(1+), 10CI

*Tri-μ-carbonyl-μ_3-hydrido-*cyclo*-tris[(η^5-cyclopentadienyl)rhodium](1+)*

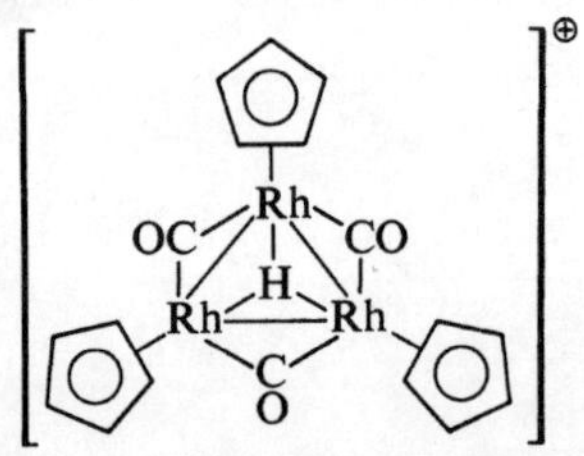

M 589.039 (ion)

Trifluoromethanesulfonate: [76082-92-3].

$C_{19}H_{16}F_3O_6Rh_3S$ M 738.104

Dark-brown greenish shiny powder or microcryst. Sol. $MeNO_2$ without deprotonation. Mp 177-9° dec. Air-stable for at least a few days. Slow dec. above ca. 135°, sinters at 167-70°. Deprotonation occurs in donor solvs. ir ν_{CO} 1889, 1835, 1795, 1767 cm^{-1} (KBr).

Herrmann, W.A. *et al*, *J. Am. Chem. Soc.*, 1981, **103**, 63 (*synth, ir*)

$C_{18}H_{20}N_2Rh^{\oplus}$ **Rh-00197**

(2,2′-Bypridine-*N,N′*)[(1,2,5,6-η)-1,5-cyclooctadiene]rhodium(1+), 9CI

η⁴-1,5-Cyclooctadiene(2,2′-bipyridyl)rhodium(1+)

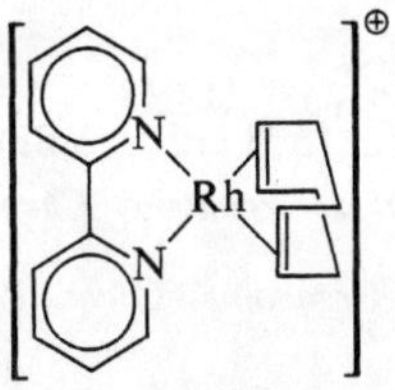

M 367.275 (ion)

Catalyst for carbonylation of MeOH to form AcOH.

Hexafluorophosphate: [37726-78-6].
$C_{18}H_{20}F_6N_2PRh$ M 512.239
Red cryst. (MeOH).

Tetraphenylborate: [37685-02-2].
$C_{42}H_{40}BN_2Rh$ M 686.507
Red cryst. (MeOH).

Perchlorate: [37685-01-1].
$C_{18}H_{20}ClN_2O_4Rh$ M 466.726
Red cryst. (MeOH).

Cocevar, C. *et al*, *J. Organomet. Chem.*, 1972, **35**, 389 (*synth*)
Brodzki, D. *et al*, *Bull. Soc. Chim. Fr.*, 1976, 61 (*use*)
Fosdyce, W.A. *et al*, *Inorg. Chem.*, 1982, **21**, 1023 (*uv*)
Fosdyce, W.A. *et al*, *Inorg. Chem.*, 1982, **21**, 1027 (*esr*)

$C_{18}H_{22}NRh^{\oplus\oplus}$ **Rh-00198**

[(3*a*,4,5,6,7,7*a*-η)-1*H*-Indole][(1,2,3,4,5-η)-1,2,3,4,5-pentamethyl-2,4-cyclopentadien-1-yl]rhodium(2+)

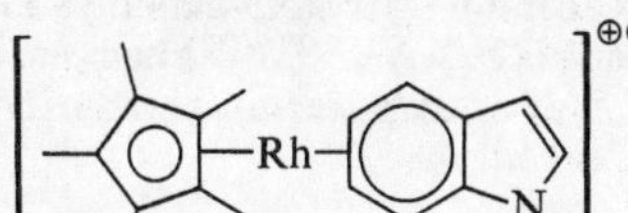

M 355.284 (ion)

Bis(hexafluorophosphate): [60937-00-0].
$C_{18}H_{22}F_{12}NP_2Rh$ M 645.212
Pale-yellow solid (Me_2CO/Et_2O).

White, C. *et al*, *J. Chem. Soc., Dalton Trans.*, 1977, 1654 (*synth, pmr, cmr, props*)

Suggestions for new Entries are welcomed. Please write to the Editor, Dictionary of Organometallic Compounds, Chapman and Hall Ltd, 11 New Fetter Lane, London EC4P 4EE

$C_{18}H_{22}N_2Rh^{\oplus}$ **Rh-00199**

[(1,2,5,6-η)-1,5-Cyclooctadiene]bis(pyridine)rhodium(1+), 9CI

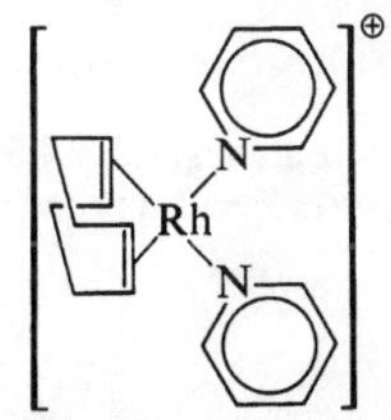

M 369.291 (ion)

Hexafluorophosphate: [51509-00-3].
$C_{18}H_{22}F_6N_2PRh$ M 514.255
Yellow solid (EtOH).

Perchlorate: [59328-05-1].
$C_{18}H_{22}ClN_2O_4Rh$ M 468.741
Yellow solid (CH_2Cl_2/pentane). Mp 135° dec.

Brodzki, D. *et al*, *J. Organomet. Chem.*, 1973, **63**, 431 (*ir*)
Denise, B. *et al*, *J. Organomet. Chem.*, 1973, **63**, 423 (*synth, ir, pmr*)
Uson, R. *et al*, *J. Organomet. Chem.*, 1976, **105**, 365 (*synth*)

$C_{18}H_{22}O_4Rh_2$ **Rh-00200**

Bis[μ-(acetato-*O*:*O′*)]bis[(2,3,5,6-η)-bicyclo[2.2.1]hepta-3,5-diene]dirhodium, 9CI

Di-μ-acetatobis(η⁴-norbornadiene)dirhodium

[36570-53-3]

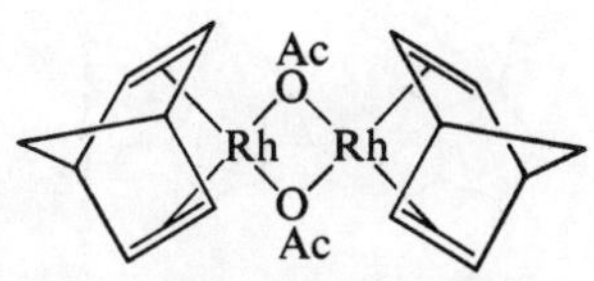

M 508.180

Catalyses valence isomerisation of quadricyclane to norbornadiene. Orange-red cryst. (EtOAc). Mp 215-8° dec.

Monomer: Acetato(norbornadiene)rhodium.
$C_9H_{11}O_2Rh$ M 254.090
Unknown.

Haszeldine, R.N. *et al*, *J. Chem. Soc. (A)*, 1971, 3696 (*synth, ir, pmr*)
Green, M. *et al*, *J. Chem. Soc., Dalton Trans.*, 1972, 832 (*synth, ir, pmr*)
Chen, M.J. *et al*, *Inorg. Chem.*, 1979, **18**, 1859 (*synth*)
Reis, A.H. *et al*, *Inorg. Chem.*, 1979, **18**, 1864 (*cryst struct*)

$C_{18}H_{23}Rh$ **Rh-00201**

Cyclooctatetraene(pentamethylcyclopentadienyl)rhodium

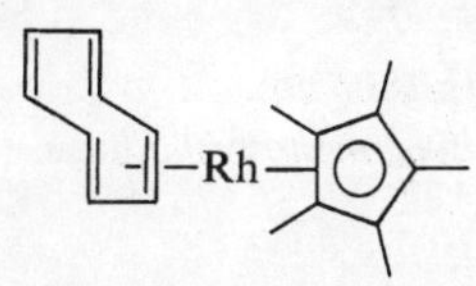

(1,2,3,4-η)-*form*

M 342.285

(1,2,3,4-η)-form [61721-82-2]
[(1,2,3,4-η)-1,3,5,7-Cyclooctatraene][(1,2,3,4,5-η)-1,2,3,4,5-pentamethyl-2,4-cyclopentadien-1-yl]rhodium, 10CI
Orange solid.

(1,2,5,6-η)-form [62206-80-8]
[(1,2,5,6-η)-1,3,5,7-Cyclooctatetraene][(1,2,3,4,5-η)-1,2,3,4,5-pentamethyl-2,4-cyclopentadien-1-yl]rhodium, 10CI
Yellow solid. Mp 127-30° dec. $Bp_{0.1}$ 100° subl.

Smith, A.K. *et al*, *J. Chem. Soc., Dalton Trans.*, 1976, 1773 (*synth, nmr, ir*)

$C_{18}H_{27}Rh$ Rh-00202

[(1,2,5,6-η)-1,5-Cyclooctadiene][(1,2,3,4,5-η)-1,2,3,4,5-pentamethyl-2,4-cyclopentadien-1-yl]rhodium, 9CI

[55787-63-8]

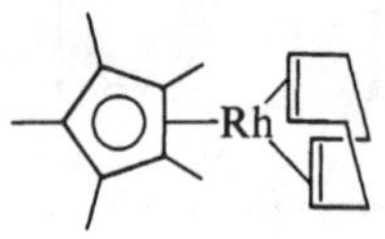

M 346.317

Yellow cryst. (Et_2O). Mp 160-3° (159°). $Bp_{0.001}$ 65° subl.

Kang, J.W. *et al*, *J. Am. Chem. Soc.*, 1968, **90**, 3259 (*synth, pmr, ms*)
Moseley, K. *et al*, *J. Chem. Soc., Chem. Commun.*, 1969, 1155 (*synth*)
Lee, H.B. *et al*, *J. Chem. Soc., Dalton Trans.*, 1975, 2322 (*synth*)
Müller, J. *et al*, *Chem. Ber.*, 1975, **108**, 1074 (*synth, pmr*)

$C_{18}H_{27}RhS^{\oplus\oplus}$ Rh-00203

[(1,2,3,4,5-η)-1,2,3,4,5-Pentamethyl-2,4-cyclopentadien-1-yl][(1,2,3,4,5-η)-tetramethylthiophene]rhodium(2+), 10CI

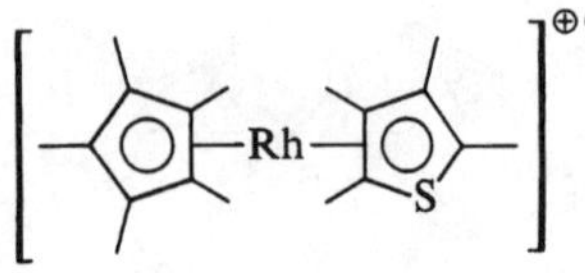

M 378.377 (ion)

Catalyst for the hydrogenation of cyclohexene.

Bis(hexafluorophosphate): [68246-36-6].
$C_{18}H_{27}F_{12}P_2RhS$ M 668.305
Cream cryst. (Me_2CO/Et_2O).

Russell, M.J.H. *et al*, *J. Chem. Soc., Dalton Trans.*, 1978, 857 (*synth, nmr, use*)

$C_{18}H_{30}O_2Rh_2$ Rh-00204

Bis[(1,2,5,6-η)-1,5-cyclooctadiene]dimethoxydirhodium, 9CI

[12148-72-0]
M 484.245

As Di-μ-bromobis(1,2,5,6-η)-1,5-cyclooctadiene)-dirhodium, Rh-00173 with

X = OMe

Yellow cryst. (MeOH). Mp 180° dec.

Monomer: 1,5-Cyclooctadienemethoxyrhodium.
$C_9H_{15}ORh$ M 242.122
Unknown.

Chatt, J. *et al*, *J. Chem. Soc.*, 1957, 4735 (*synth, ir*)
Robinson, S.D. *et al*, *J. Chem. Soc.*, 1965, 4997 (*pmr*)
Brodzki, D. *et al*, *Bull. Soc. Chim. Fr.*, 1976, 61.
Brodzki, D. *et al*, *J. Mol. Catal.*, 1977, **2**, 149.
Sakuno, F. *et al*, *J. Am. Chem. Soc.*, 1980, **102**, 1749.

$C_{19}H_{20}O_6Rh_2$ Rh-00205

Dicarbonylbis(η^5-2,4-cyclopentadien-1-yl)μ-[2-ethoxy-1-(ethoxycarbonyl)-2-oxoethylidene]]dirhodium, 10CI

[μ-Bis(ethoxycarbonyl)methylene]dicarbonylbis(η-cyclopentadienyl)dirhodium. [μ-Bis(ethoxycarbonyl)-methylene]bis(carbonyl-η^5-cyclopentadienylrhodium)

[65097-93-0]

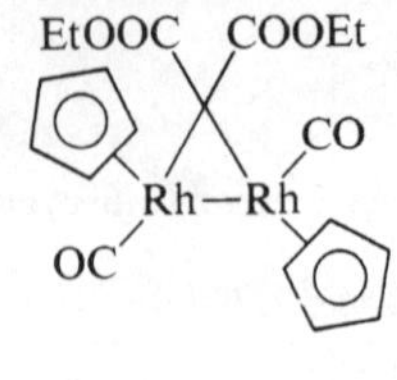

M 550.174

Light-red air-stable cryst. (Et_2O/CH_2Cl_2). Mp 197° dec. ir ν_{CO} 1980, 1972, 1678 cm^{-1} (KBr).

Herrmann, W.A. *et al*, *J. Organomet. Chem.*, 1977, **140**, 73 (*synth, ir, pmr*)
Mayer, K.K. *et al*, *J. Organomet. Chem.*, 1979, **182**, 361 (*ms*)

$C_{19}H_{22}Rh^{\oplus}$ Rh-00206

[(1,2,3,3a,7a-η)-1*H*-inden-1-yl][(1,2,3,4,5-η)-1,2,3,4,5-pentamethyl-2,4-cyclopentadien-1-yl]rhodium(1+), 10CI

Pentamethylcyclopentadienyl(indenyl)rhodium(1+)

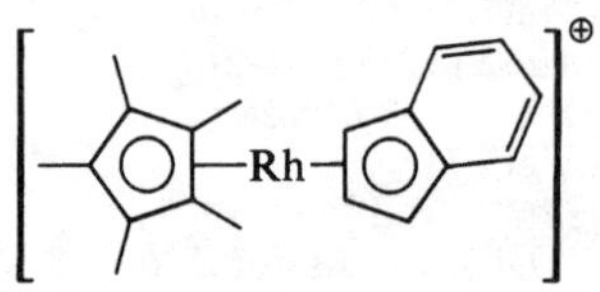

M 353.288 (ion)

Hexafluorophosphate: [60937-02-2].
$C_{19}H_{22}F_6PRh$ M 498.252
Orange solid (Me_2CO/Et_2O).

White, C. *et al*, *J. Chem. Soc., Dalton Trans.*, 1977, 1654 (*synth, nmr*)

$C_{19}H_{24}BF_6N_6O_5Rh$ Rh-00207

Aquabis(trifluoroacetato-*O*)[tris(3,5-dimethyl-1*H*-pyrazolato-N^1)hydroborato(1−)-N^2,$N^{2'}$,$N^{2''}$]rhodium, 10CI

[Hydrotris(3,5-dimethylpyrazolyl)borato]bis(trifluoroacetato)aquorhodium

[73117-64-3]

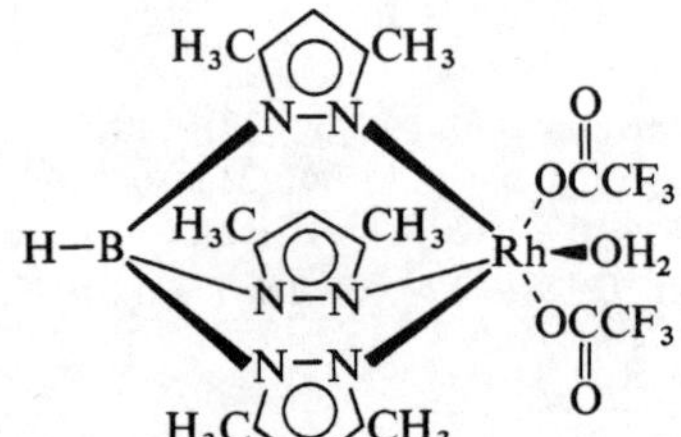

M 644.142

Pale-yellow plates (CH_2Cl_2/hexane). Mp 231-3°.

May, S. *et al*, *Inorg. Chem.*, 1980, **19**, 1582 (*synth, pmr*)

$C_{19}H_{25}BN_6Rh^{\oplus}$ **Rh-00208**

[Hydrotris(1*H*-pyrazolato-*N*[1])borato(1–)-*N*[2],*N*[2'],*N*[2''](*η*[5]-pentamethylcyclopentadienyl)rhodium(1+)

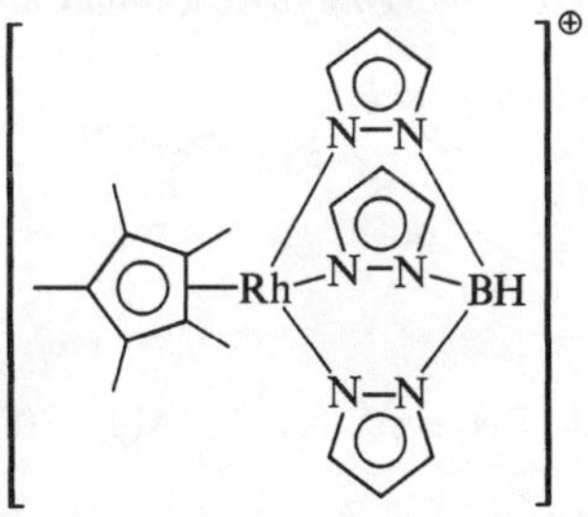

M 451.162 (ion)

Hexafluorophosphate: [52015-82-4].

$C_{19}H_{25}BF_6N_6PRh$ M 596.126

Yellow, air-stable cryst. (CH_2Cl_2/Et_2O). Mp 220-7°.

Restivo, R.J. *et al*, *Inorg. Chem.*, 1975, **14**, 3046 (*synth, ir, pmr, cryst struct*)

$C_{19}H_{25}N_2Rh$ **Rh-00209**

[(1,2,3,3*a*,7*a*-*η*)-1*H*-Inden-1-yl)]bis(2-isocyano-2-methylpropane)rhodium, 10CI

Bis(tert-*butyl isocyanide*)*indenylrhodium*

[63466-37-5]

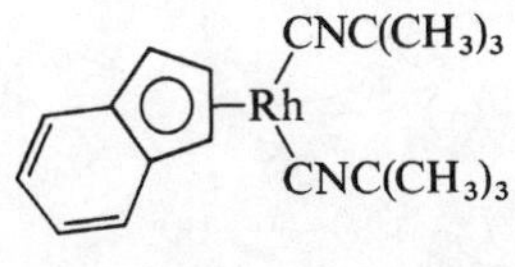

M 384.325

No phys. props. reported. In soln. slowly converts to μ-(CNBu[t])[Rh(η^5-C_9H_7)CNBu[t]]$_2$.

Caddy, P. *et al*, *Angew. Chem., Int. Ed. Engl.*, 1977, **16**, 648 (*synth, ir*)

$C_{19}H_{26}Rh^{\oplus}$ **Rh-00210**

[(2,3,5,6-*η*)Bicyclo[2.2.1]hepta-2,5-diene][(1,2,3,4,5,6-*η*)hexamethylbenzene]rhodium(1+), 9CI

(*Hexamethylbenzene*)(*2,5-norbornadiene*)*rhodium*(*1*+)-, *8CI*

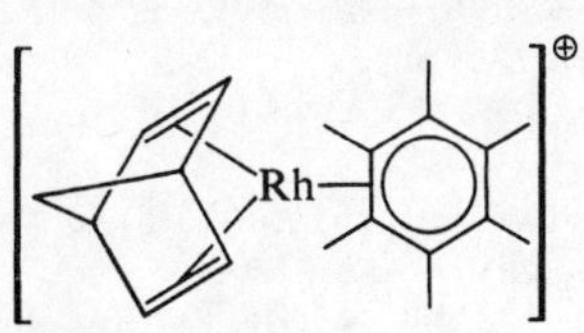

M 357.320 (ion)

Tetrafluoroborate: [36620-13-0].

$C_{19}H_{26}BF_4Rh$ M 444.124

White cryst. (CH_2Cl_2/Et_2O). Mp >270°.

Osborn, J.A. *et al*, *J. Am. Chem. Soc.*, 1971, **93**, 3089 (*synth*)
Green, M. *et al*, *J. Chem. Soc., Dalton Trans.*, 1972, 832 (*synth, pmr*)

$C_{19}H_{33}O_3Rh^{\oplus\oplus}$ **Rh-00211**

[(1,2,3,4,5-*η*)-1,2,3,4,5-Pentamethyl-2,4-cyclopentadien-1-yl]tris(2-propanone)rhodium(2+), 10CI

Tris(*acetone*)(*pentamethylcyclopentadienyl*)-*rhodium*(*2*+)

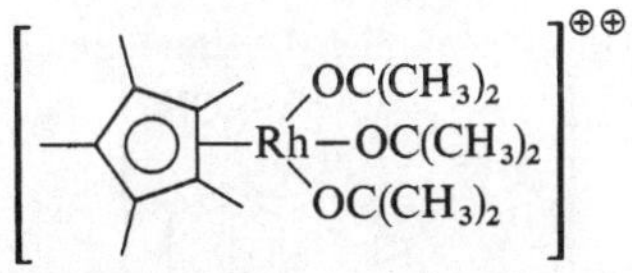

M 412.373 (ion)

Useful precursor to pentamethylcyclopentadienylrhodium(III) complexes. Slowly rearranges in soln. Me_2CO ligands readily displaced by arenes, *P*-donor ligands, olefins, etc.

Bis(*hexafluorophosphate*): [60936-90-5].

$C_{19}H_{33}F_{12}O_3P_2Rh$ M 702.302

Prepd. in Me_2CO soln. from $AgPF_6$ and [Rh(C_5Me_5)-Cl_2]$_2$. Not isol.

Thompson, S.J. *et al*, *Angew. Chem., Int. Ed. Engl.*, 1976, **15**, 490 (*props*)
White, C. *et al*, *J. Chem. Soc., Dalton Trans.*, 1977, 1654; 1978, 1305 (*synth, pmr, use*)
White, C. *et al*, *J. Organomet. Chem.*, 1977, **134**, 319 (*props*)

$C_{20}H_{20}N_2Rh^{\oplus}$ **Rh-00212**

[(1,2,5,6-*η*)-1,5-Cyclooctadiene](1,10-phenanthroline-*N*[1],*N*[10])rhodium(1+)

[47248-45-3]

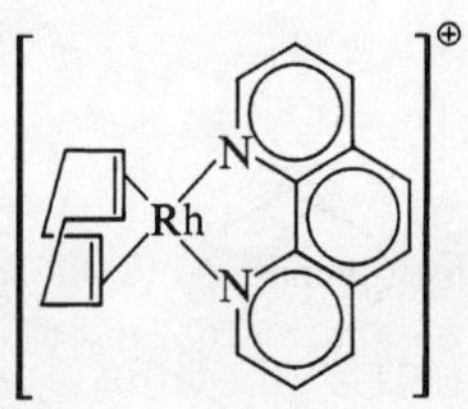

M 391.297 (ion)

Chloride: [54324-73-1].

$C_{20}H_{20}ClN_2Rh$ M 426.750

Red cryst. Mp 285-90° dec.

Hexafluorophosphate: [52513-60-7].

$C_{20}H_{20}F_6N_2PRh$ M 536.261

Cryst.

Tetraphenylborate: [52513-61-8].

$C_{44}H_{40}BN_2Rh$ M 710.529

Cryst.

Perchlorate: [37685-03-3].

$C_{20}H_{20}ClN_2O_4Rh$ M 490.748

Cryst.

Chatt, J. *et al*, *J. Chem. Soc.*, 1957, 4735 (*synth*)
Cocevar, C. *et al*, *J. Organomet. Chem.*, 1972, **35**, 389 (*synth, ir*)
Gillard, R.D. *et al*, *J. Chem. Soc., Dalton Trans.*, 1972, 133 (*synth, ir, pmr*)
Mestroni, G. *et al*, *J. Organomet. Chem.*, 1974, **65**, 119 (*synth, ir, pmr*)
Zassinovich, G. *et al*, *J. Mol. Catal.*, 1977, **2**, 63 (*synth*)

$C_{20}H_{22}N_2Rh^{\oplus}$ **Rh-00213**

(4,7-Dimethyl-1,10-phenanthroline-N^1,N^{10})[(1,2,5,6-η)-1,5-hexadiene]rhodium(1+), 10CI

1,5-Hexadiene(4,7-dimethylphenanthroline)-rhodium(1+)

[62630-29-9]

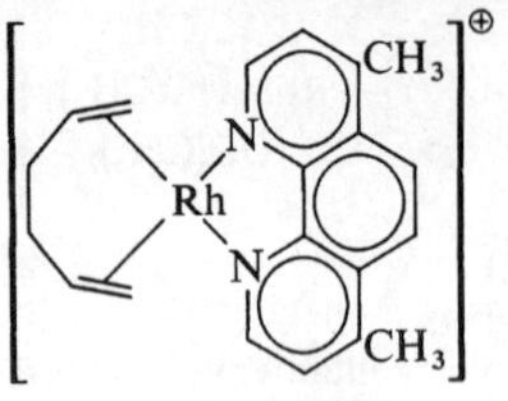

M 393.313 (ion)

Active catalyst for the selective reduction of C=O by H_2 or 2-propanol in the presence of base.

Hexafluorophosphate: [62630-30-2].
$C_{20}H_{22}F_6N_2PRh$ M 538.277
Orange-red cryst.

Tetraphenylborate: [62662-82-2].
$C_{44}H_{42}BN_2Rh$ M 712.545
No phys. props. reported.

Mestroni, G. *et al*, *J. Organomet. Chem.*, 1977, **140**, 63 (*synth, props, use*)

Zassinovich, G. *et al*, *J. Organomet. Chem.*, 1977, **168**, C37 (*use*)

$C_{20}H_{24}Cl_2N_8Rh_2^{\oplus\oplus}$ **Rh-00214**

Dichlorotetrakis[μ-(1,3-diisocyanopropane)]dirhodium(2+), 10CI

[68079-82-3]

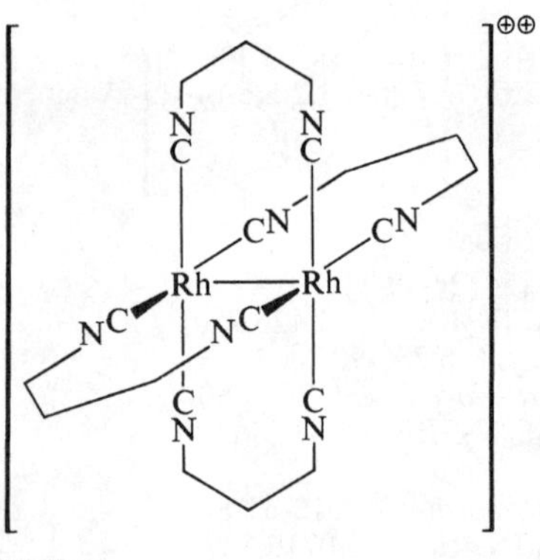

M 653.180 (ion)

Obt. by photolysis of Tetrakis[μ-(1,3-diisocyanopropane)]dirhodium(2+), Rh-00216 in conc. HCl.

Dichloride: [70982-68-2].
$C_{20}H_{24}Cl_4N_8Rh_2$ M 724.086
Yellow cryst. + $8H_2O$ (HCl aq.).

Mann, K.R. *et al*, *Inorg. Chem.*, 1979, **18**, 2671 (*cryst struct*)

Sigal, I.S. *et al*, *J. Am. Chem. Soc.*, 1980, **102**, 7252.

$C_{20}H_{24}Cl_2O_4Rh_2$ **Rh-00215**

Dichlorobis(2,3,5,6-tetramethyl-2,5-cyclohexadiene-1,4-dione)dirhodium

Dichlorobis(tetramethyl-p-benzoquinone)dirhodium, 8CI

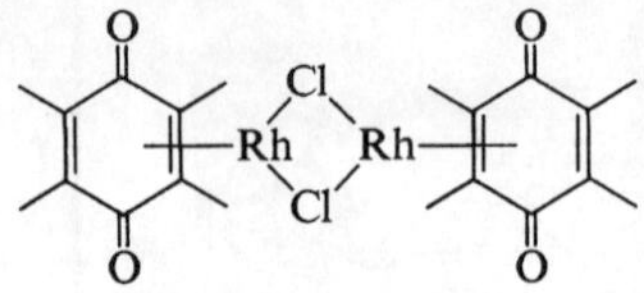

M 605.124

Bridged dimer. Red cryst. (C_6H_6). Mp >300° (dec. from 200°).

Monomer: Chloro(tetramethylbenzoquinone)rhodium. Duroquinone rhodium chloride.
$C_{10}H_{12}ClO_2Rh$ M 302.562
Unknown.

McVey, S. *et al*, *Can. J. Chem.*, 1966, **44**, 2429 (*synth, ir, pmr*)

Kang, J.W. *et al*, *Can. J. Chem.*, 1968, **46**, 3189 (*synth*)

$C_{20}H_{24}N_8Rh_2^{\oplus\oplus}$ **Rh-00216**

Tetrakis[μ-(1,3-diisocyanopropane)]dirhodium(2+)

[71375-02-5]

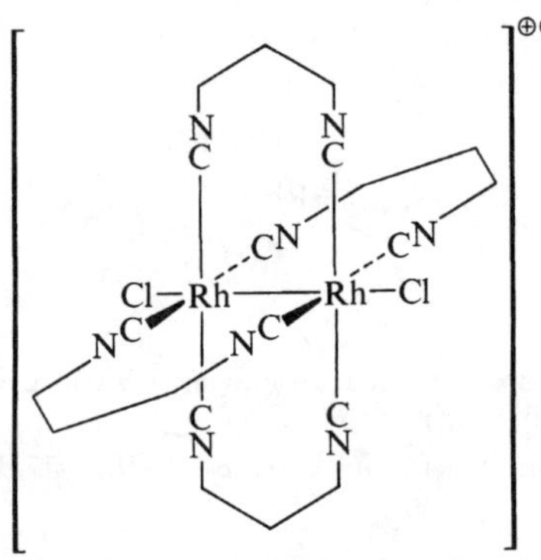

M 582.274 (ion)

Possible use in solar energy storage involving photocatalytic conversion of H_2O to $H_2 + O_2$. Photosensitive. Produces H_2 and Dichlorotetrakis[μ-(1,3-diisocyanopropane)]dirhodium(2+), Rh-00214 on treatment with conc. HCl and light.

Dichloride: [61156-13-6].
$C_{20}H_{24}Cl_2N_8Rh_2$ M 653.180
Blue powder.

Bis(tetraphenylborate): [61156-15-8].
$C_{68}H_{64}B_2N_8Rh_2$ M 1220.738
Purple cryst. + 2MeCN (MeCN).

Lewis, N.S. *et al*, *J. Am. Chem. Soc.*, 1976, **98**, 7461 (*synth, props*)

Gray, H.B. *et al*, *Adv. Chem. Ser.*, 1978, **168**, 44 (*use*)

Miskowski, V.M. *et al*, *J. Am. Chem. Soc.*, 1978, **100**, 485 (*synth, ir*)

Yaneff, P.V. *et al*, *J. Organomet. Chem.*, 1979, **179**, 101 (*synth, uv*)

Mann, K.R. *et al*, *J. Am. Chem. Soc.*, 1980, **102**, 3965 (*cryst struct*)

Sigal, I.S. *et al*, *J. Am. Chem. Soc.*, 1980, **102**, 7252 (*props, use*)

Fukuzumi, S. *et al*, *Bull. Chem. Soc. Jpn.*, 1982, **55**, 2892 (*props*)

$C_{20}H_{25}N_6Rh^{\oplus\oplus}$ **Rh-00217**

[1,1′,1″-Methylidynetris[1*H*-pyrazole]-*N*², *N*²′, *N*²″]-[(1,2,3,4,5-η)-1,2,3,4,5-pentamethyl-2,4-cyclopentadien-1-yl]rhodium(2+), 9CI

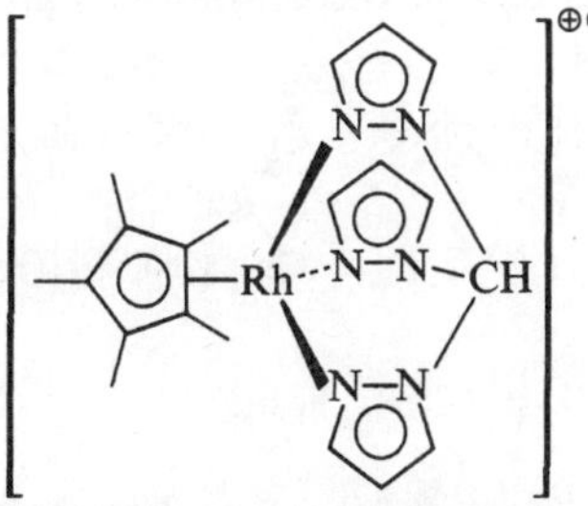

M 452.363 (ion)

Bis(hexafluorophosphate): [52165-51-2].
$C_{20}H_{25}F_{12}N_6P_2Rh$ M 742.292
Orange, air-stable cryst.

O'Sullivan, D.J. *et al*, *J. Organomet. Chem.*, 1973, **57**, C58 (*synth*)

$C_{20}H_{30}Br_4Rh_2$ **Rh-00218**

Di-μ-bromodibromobis[(1,2,3,4,5-η)-1,2,3,4,5-pentamethyl-2,4-cyclopentadien-1-yl]dirhodium, 10CI

Tetrabromobis(pentamethylcyclopentadienyl)dirhodium

[36484-11-4]

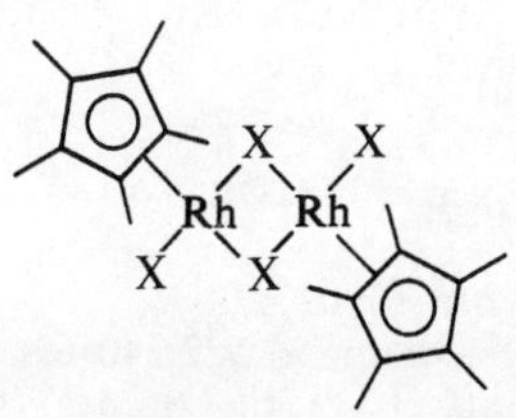

M 795.884

Hydrogenation catalyst. Dark-brown cryst. ($CHCl_3$/C_6H_6). Mp >260°.

Monomer: Dibromopentamethylcyclopentadienylrhodium.
$C_{10}H_{15}Br_2Rh$ M 397.942
Has no independent existence.

Gill, D.S. *et al*, *J. Organomet. Chem.*, 1975, **87**, 359 (*synth*, *pmr*)
Russell, M.J. *et al*, *J. Chem. Soc., Chem. Commun.*, 1977, 427 (*use*)
Churchill, M.R. *et al*, *Inorg. Chem.*, 1978, **17**, 3011 (*synth*, *struct*)
Gill, D.S. *et al*, *J. Chem. Soc., Dalton Trans.*, 1978, 617 (*use*)
Millan, A. *et al*, *J. Chem. Soc., Dalton Trans.*, 1982, 73 (*ir*)

$C_{20}H_{30}Cl_3Rh_2^{\oplus}$ **Rh-00219**

Tri-μ-chlorobis[(1,2,3,4,5-η)-1,2,3,4,5-pentamethyl-2,4-cyclopentadien-1-yl]dirhodium(1+), 10CI

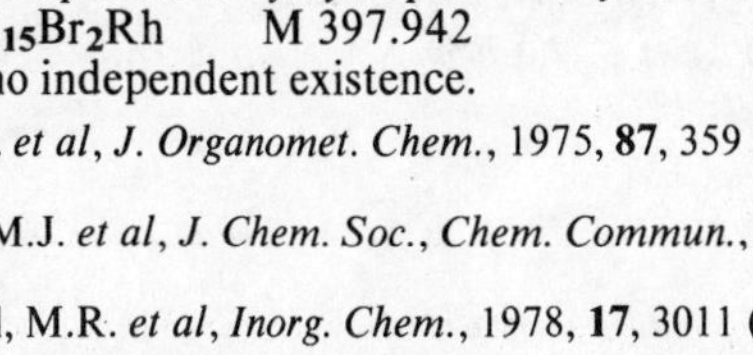

M 582.627 (ion)

Hexafluorophosphate: [71412-51-6].
$C_{20}H_{30}Cl_3F_6PRh_2$ M 727.591
No phys. props. reported.

Tetraphenylborate:
$C_{54}H_{50}BCl_3PRh_2$ M 1052.943
Red cryst.

Tetrafluoroborate: [71412-50-5].
$C_{20}H_{30}BCl_3F_4Rh_2$ M 669.431
No phys. props. reported.

Kang, J.W. *et al*, *J. Organomet. Chem.*, 1971, **30**, 127 (*pmr*, *ir*, *synth*)
Rigby, W. *et al*, *J. Chem. Soc., Dalton Trans.*, 1979, 382 (*synth*, *nmr*)

$C_{20}H_{30}I_4Rh_2$ **Rh-00220**

Di-μ-iododiiodobis[(1-5-η)-1,2,3,4,5-pentamethyl-2,4-cyclopentadien-1-yl]dirhodium, 10CI

Tetraiodobis(pentamethylcyclopentadienyl)rhodium

[67841-74-1]

M 983.886

Hydrogenation catalyst. Deep-red cryst. ($CHCl_3$/EtOH). Mp >230°.

As Di-μ-bromodibromobis[(1,2,3,4,5-η)-1,2,3,4,5-pentamethyl-2,4,-cyclopentadien-1-yl]dirhodium, Rh-00218 with

X = I

Monomer: Diiodopentamethylcyclopentadienylrhodium. Pentamethylcyclopentadienylrhodium diiodide.
$C_{10}H_{15}I_2Rh$ M 491.943
Unknown.

Booth, B.L. *et al*, *J. Chem. Soc. (A)*, 1969, 1299 (*synth*, *pmr*, *ir*)
Kang, J.W. *et al*, *J. Am. Chem. Soc.*, 1969, **91**, 5970 (*synth*, *pmr*)
Gill, D.S. *et al*, *J. Chem. Soc., Dalton Trans.*, 1978, 617 (*use*)
Rigby, W. *et al*, *J. Chem. Soc., Dalton Trans.*, 1979, 382.

$C_{20}H_{30}O_4Rh_2$ **Rh-00221**

Bis[μ-(acetato-*O*:*O*′)]bis[(1,2,5,6-η)-1,5-cyclooctadiene)dirhodium, 9CI

Di-μ-acetatobis(1,5-cyclooctadiene)dirhodium

[12097-36-8]

As Di-μ-bromobis(1,2,5,6-η-1,5-cyclooctadiene)dirhodium, Rh-00173 with

X = OAc

M 540.266

Bridged dimer. Butadiene polymn. catalyst. Orange cryst. Mp 197-8°.

Monomer: (Acetato)(1,5-cyclooctadiene)rhodium. (1,5-Cyclooctadiene)rhodium acetate.
$C_{10}H_{15}O_2Rh$ M 270.133
Unknown.

Chatt, J. *et al*, *J. Chem. Soc.*, 1957, 4753 (*synth*, *ir*)
Kormer, V.A. *et al*, *J. Polym. Sci., Part C*, 1965, **16**, 4351 (*use*)
Robinson, S.D. *et al*, *J. Chem. Soc.*, 1965, 4997 (*pmr*)
Sakurai, F. *et al*, *J. Am. Chem. Soc.*, 1980, **102**, 1749 (*synth*)

$C_{20}H_{30}Rh^{\oplus}$ **Rh-00222**

[(1,2,5,6-η)-1,5-cyclooctadiene][(1,2,3,4,5,6-η)hexamethylbenzene]rhodium(1+), 9CI

Hexamethylbenzene-1,5-cyclooctadienerhodium(1+)

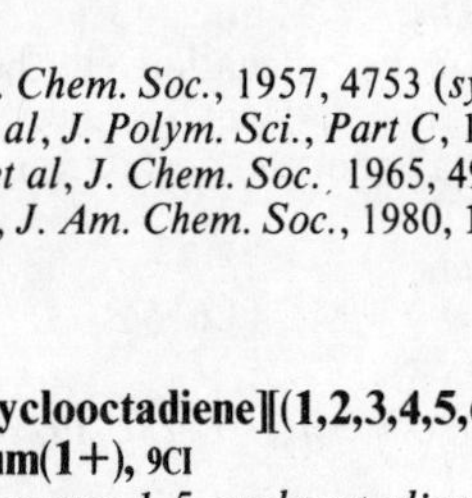

M 373.363 (ion)
Tetrafluoroborate: [36644-49-2].
$C_{20}H_{30}BF_4Rh$ M 460.166
White cryst. (CH_2Cl_2/Et_2O). Mp >270°.

Green, M. *et al, J. Chem. Soc., Dalton Trans.*, 1972, 832 (*synth, pmr*)

$C_{20}H_{31}Cl_3Rh_2$ **Rh-00223**

μ-Chlorodichloro-μ-hydrobis[(1,2,3,4,5-η)-1,2,3,4,5-pentamethyl-2,4-cyclopentadien-1-yl]dirhodium, 10CI
μ-Chloro-μ-hydridodichlorobis(pentamethylcyclopentadienyl)dirhodium
[41380-04-5]

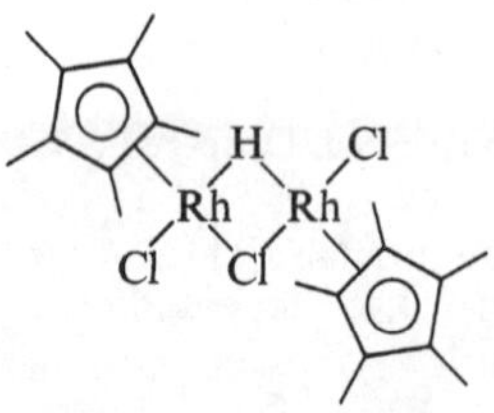

M 583.635
Catalyst for the hydrogenation of olefins. Purple cryst. (C_6H_6/cyclohexane). Mp >155° dec.

Maitlis, P.M. *et al, J. Chem. Soc., Chem. Commun.*, 1971, 734 (*synth, pmr, ir, use*)
Churchill, M.R. *et al, J. Am. Chem. Soc.*, 1973, **95**, 2150 (*struct*)
White, C. *et al, J. Chem. Soc., Dalton Trans.*, 1973, 1901 (*synth, pmr*)
Gill, D.S. *et al, J. Chem. Soc., Dalton Trans.*, 1978, 617 (*use*)

$C_{20}H_{33}O_3Rh_2^{\oplus}$ **Rh-00224**

Tri-μ-hydroxybis[(1,2,3,4,5-η)-1,2,3,4,5-pentamethyl-2,4-cyclopentadien-1-yl]dirhodium, 9CI
Tri-μ-hydroxobis(pentamethylcyclopentadienyl)-dirhodium

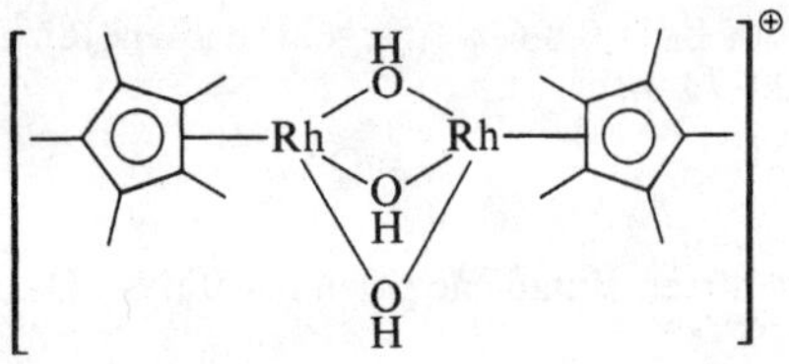

M 527.290 (ion)
Chloride:
$C_{20}H_{33}ClO_3Rh_2$ M 562.743
Catalyses olefin hydrogenation and the disproportionation of aldehydes in water to give corresponding alcohol and acid. Orange cryst. + $4H_2O$ (H_2O).
Tetraphenylborate:
$C_{44}H_{53}BO_3Rh$ M 743.616
Orange-red cryst.

Kang, J.W. *et al, J. Organomet. Chem.*, 1971, **30**, 127 (*synth, pmr*)
Hirai, K. *et al, J. Mol. Catal.*, 1981, **10**, 203 (*use*)
Nutton, A. *et al, J. Organomet. Chem.*, 1981, **213**, 313.
Cook, J. *et al, J. Chem. Soc., Dalton Trans.*, 1981, 2342 (*use*)

$C_{20}H_{36}Cl_2N_4Rh_2$ **Rh-00225**

Di-μ-chlorotetrakis(2-isocyano-2-methylpropane)dirhodium, 9CI
*Di-μ-chlorotetrakis(*tert*-butyl isocyanide)dirhodium. Bis[chlorobis(*tert*-butylisonitrile)rhodium]*
[37017-32-6]

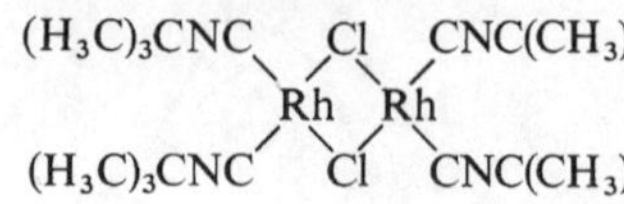

M 609.248
Pale-yellow cryst. (C_6H_6/hexane at 0°). Sol. CH_2Cl_2. Mp 91-3°.
*Monomer: Chlorobis(2-isocyano-2-methylpropane)rhodium. Chlorobis(*tert*-butylisocyanide)rhodium. Chlorobis(*tert*-butylisonitrile)rhodium.*
$C_{10}H_{18}ClN_2Rh$ M 304.624
Unknown.

Branson, P.R. *et al, J. Chem. Soc., Dalton Trans.*, 1972, 1303 (*synth, ir, pmr*)

$C_{20}H_{36}N_4Rh^{\oplus}$ **Rh-00226**

Tetrakis(2-isocyano-2-methylpropane)rhodium(1+), 10CI
*Tetrakis(*tert*-butylisocyanide)rhodium(1+),* 8CI.
*Tetrakis(*tert*-butylisonitrile)rhodium(1+)*
[34195-61-4]

$Rh[CNC(CH_3)_3]_4^{\oplus}$ (D_{4h})

M 435.437 (ion)
Chloride: [51139-75-4].
$C_{20}H_{36}ClN_4Rh$ M 470.890
Yellow cryst.
Tetrafluoroborate: [37017-30-4].
$C_{20}H_{36}BF_4N_4Rh$ M 522.240
Yellow cryst. (CH_2Cl_2/Et_2O). Mp 170° dec.
Perchlorate: [70443-05-9].
$C_{20}H_{36}ClN_4O_4Rh$ M 534.887
Yellow cryst. (CH_2Cl_2/ligroin).
Hexafluorophosphate: [42744-96-7].
$C_{20}H_{36}F_6N_4PRh$ M 580.401
Yellow cryst. (CH_2Cl_2/EtOH).

Branson, P.R. *et al, J. Chem. Soc., Dalton Trans.*, 1972, 1303 (*synth, ir, pmr*)
Dart, J.W. *et al, J. Chem. Soc., Dalton Trans.*, 1973, 2039 (*synth, ir*)
Branson, P.R. *et al, J. Chem. Soc., Chem. Commun.*, 1974, 364 (*cmr*)
Mann, K.R. *et al, Inorg. Chem.*, 1978, **17**, 828 (*synth, ir, uv*)
Kowae, R. *et al, Bull. Chem. Soc. Jpn.*, 1979, **52**, 1067 (*synth, ir, uv*)

$C_{20}O_{19}Rh_8$ Rh-00227

[μ_6-(Carbon)]hexa-μ-carbonyldi-μ_3-carbonyldecacarbonyloctarhodium, 9CI

(Carbon)nonadecacarbonyloctarhodium

[53321-99-6]

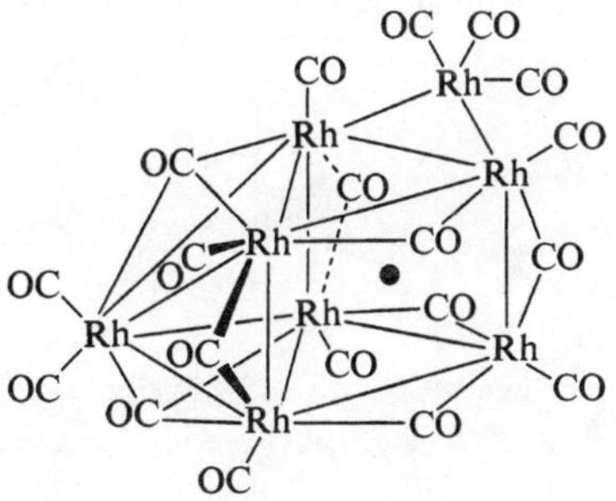

• = C

M 1367.453

Monocapped trigonal prism, plus one edge-bridging Rh. Black cryst.

Albano, V.G. *et al*, *J. Chem. Soc., Chem. Commun.*, 1974, 299 (*synth, struct*)

Albano, V.G. *et al*, *J. Chem. Soc., Dalton Trans.*, 1975, 305 (*cryst struct*)

$C_{21}H_8O_{14}Rh_6$ Rh-00228

[(2,3,5,6-η)-Bicyclo[2.2.1]hepta-2,5-diene]tetra-μ_3-carbonyldecacarbonylhexarhodium, 9CI

Tetradecacarbonyl(norbornadiene)hexarhodium

[51914-08-0]

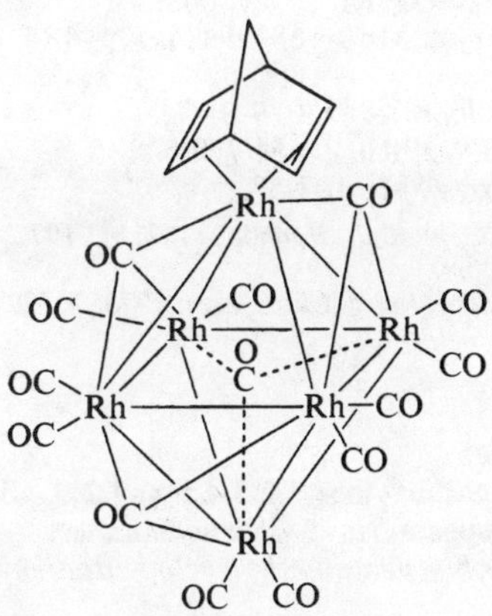

M 1101.719

Proposed struct. shown. Air-stable reddish-brown cryst. (hexane). Mp ~145° dec.

Kitamura, T. *et al*, *J. Organomet. Chem.*, 1974, **65**, 235 (*synth, ir*)

$C_{21}H_{20}ClNOPRh$ Rh-00229

Carbonylchloro[2-(diphenylphosphino)-*N*,*N*-dimethylaniline]rhodium

*Carbonylchloro[2-(diphenylphosphino)-*N,N*-dimethylbenzenamine-*N,P*]rhodium, 10CI*

[52594-71-5]

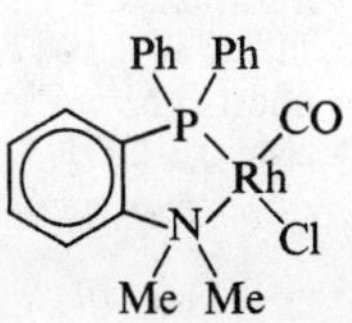

M 471.727

Yellow cryst. (EtOH at 0°). Mp 256-60° dec. (sealed tube). ir ν_{CO} 2005 cm^{-1} (CH_2Cl_2).

Rauchfuss, T.B. *et al*, *J. Am. Chem. Soc.*, 1974, **96**, 3098 (*synth, ir, props*)

Roundhill, D.M. *et al*, *Inorg. Chem.*, 1980, **19**, 284 (*nmr, stereochem*)

$C_{22}H_{20}O_2Rh_4$ Rh-00230

Di-μ_3-carbonyltetra(η^5-2,4-cyclopentadien-1-yl)tetrarhodium, 10CI

[61156-22-7]

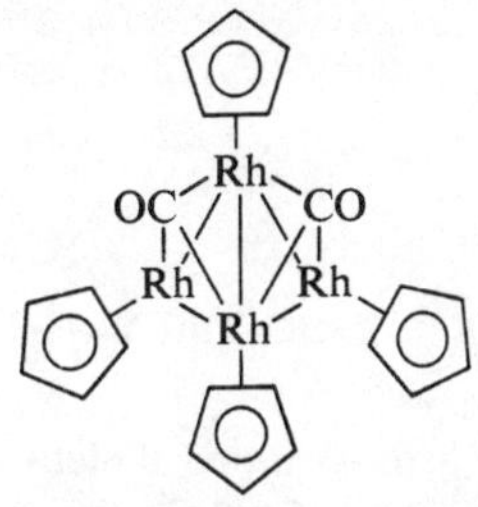

M 728.021

Black solid.

Lawson, R.J. *et al*, *J. Am. Chem. Soc.*, 1976, **98**, 7433 (*synth, ir, pmr, ms*)

Dickson, R.S. *et al*, *J. Chem. Soc., Chem. Commun.*, 1979, 277.

$C_{22}H_{30}N_4Rh_2$ Rh-00231

Bis[(1,2,5,6-η)-1,5-cyclooctadiene]bis[μ-(1*H*-pyrazolato-N^1:N^2)dirhodium, 10CI

[77244-03-2]

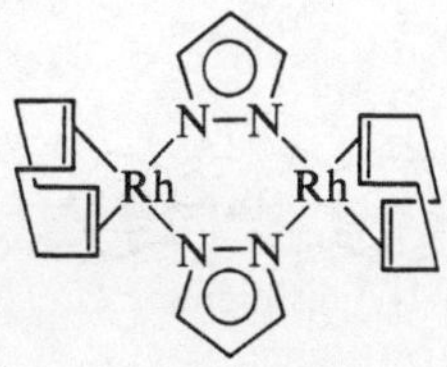

M 556.317

Catalyst precursor for the hydroformylation of olefins. Yellow solid. Sol. $CHCl_3$, insol. MeOH. Mp >210°.

Usòn, R. *et al*, *J. Organomet. Chem.*, 1981, **205**, 247 (*synth, props*)

Usòn, R. *et al*, *J. Mol. Catal.*, 1982, **14**, 375 (*use*)

$C_{22}H_{30}O_2Rh_2$ Rh-00232

Di-μ-carbonylbis(1,2,3,4,5-η)-1,2,3,4,5-pentamethyl-2,4-cyclopentadien-1-yldirhodium, 9CI

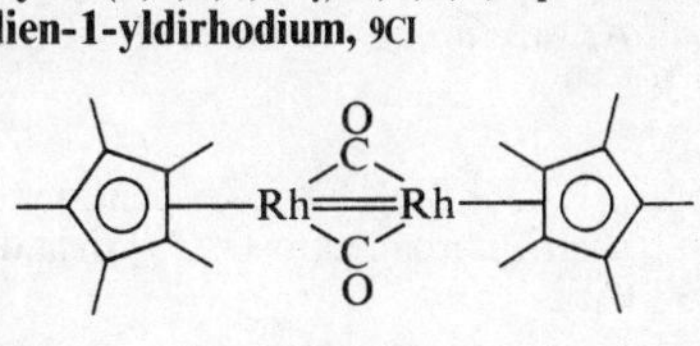

M 532.289

Dark-blue cryst. by subl., stable in air for 24 hr. ir ν_{CO} 1732 cm^{-1}.

Monomer: [69728-34-3]. *Carbonyl(pentamethylcyclopentadienyl)rhodium. Pentamethylcyclopentadienylrhodium carbonyl.*

$C_{11}H_{15}ORh$ M 266.144

Unknown.

Nutton, A. *et al*, *J. Organomet. Chem.*, 1979, **166**, C21 (*synth, nmr, ms, ir*)

Boag, N.M. *et al*, *J. Chem. Soc., Chem. Commun.*, 1980, 1171 (*synth*)

Clauss, A.D. *et al*, *J. Organomet. Chem.*, 1980, **201**, C31.

Herrmann, W.A. *et al*, *Angew. Chem., Int. Ed. Engl.*, 1981, **20**, 193 (*synth*)
Herrmann, W.A. *et al*, *Chem. Ber.*, 1982, **115**, 878 (*props*)

$C_{22}H_{32}F_3O_2Rh_2^{\oplus}$ Rh-00233

Di-μ-hydrobis[(1,2,3,4,5-η)-1,2,3,4,5-pentamethyl-2,4-cyclopentadien-1-yl][μ-(trifluoroacetato-O:O')]dirhodium(1+), 9CI

Di-μ-hydrido-μ-trifluoroacetatobis(pentamethylcyclopentadienylrhodium)(1+)

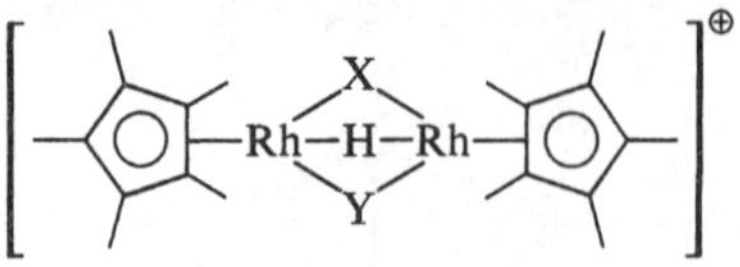

X = —OOCCF$_3$, Y = H

M 591.300 (ion)

Catalyst for the hydrogenation of olefins.

Hexafluorophosphate: [51472-01-6].
$C_{22}H_{32}F_9O_2PRh_2$ M 736.264
Red cryst. (CH_2Cl_2/Et_2O).

White, C. *et al*, *J. Chem. Soc., Dalton Trans.*, 1973, 1901 (*synth, pmr, use*)

$C_{22}H_{33}Rh^{\oplus\oplus}$ Rh-00234

[(1,2,3,4,5,6-η)-Hexamethylbenzene][(1,2,3,4,5-η-)-1,2,3,4,5-pentamethyl-2,4-cyclopentadien-1-yl]rhodium(2+), 9CI

Pentamethylcyclopentadienylhexamethylbenzenerhodium(2+)

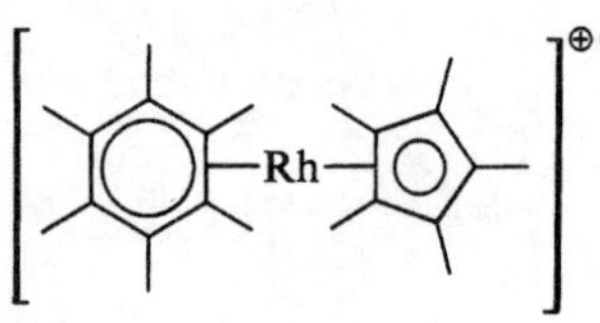

M 400.408 (ion)

Bis(hexafluorophosphate): [12715-84-3].
$C_{22}H_{33}F_{12}P_2Rh$ M 690.337
White cryst. (Me_2CO). Mp >250°.

White, C. *et al*, *J. Chem. Soc. (A)*, 1971, 3322 (*pmr, synth*)

$C_{22}H_{35}O_2Rh_2^{\oplus}$ Rh-00235

(μ-Acetato-O:O')di-μ-hydrobis[(1,2,3,4,5-η)-1,2,3,4,5-pentamethyl-2,4-cyclopentadien-1-yl]dirhodium(1+), 9CI

Di-μ-hydrido-μ-acetatobis(pentamethylcyclopentadienylrhodium)(1+)

As Di-μ-hydrobis[(1,2,3,4,5-η)-1,2,3,4,5-pentamethyl-2,4-cyclopenta dien-1-yl][μ-(trifluoroacetato-*O:O'*)]dirhodium(1+), Rh-00233 with

X = OAc, Y = H

M 537.328 (ion)

Catalyst for the hydrogenation of olefins and for disproportionation of acetaldehyde into AcOH and EtOH.

Hexafluorophosphate: [51471-99-9].
$C_{22}H_{35}F_6O_2PRh_2$ M 682.292
Dark-red cryst. (CH_2Cl_2/Et_2O).

White, C. *et al*, *J. Chem. Soc., Dalton Trans.*, 1973, 1901 (*synth, pmr*)
Cook, J. *et al*, *J. Chem. Soc., Dalton Trans.*, 1981, 2342 (*synth, use*)

Nutton, A. *et al*, *J. Organomet. Chem.*, 1981, **213**, 313 (*struct, synth, pmr, ir*)

$C_{23}H_{20}I_2PRh$ Rh-00236

(η⁵-2,4-Cyclopentadien-1-yl)diiodo(triphenylphosphine)rhodium, 9CI

[12131-90-7]

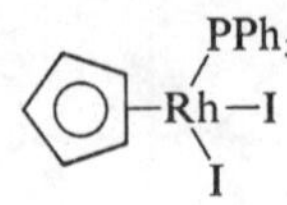

M 684.099

Air-stable brown cryst. Mp 245-50° dec.

Kasahara, A. *et al*, *Bull. Chem. Soc. Jpn.*, 1967, **40**, 699 (*synth, props*)
Gardner, S.A. *et al*, *J. Organomet. Chem.*, 1974, **78**, 415.

$C_{23}H_{20}NOPRh^{\oplus}$ Rh-00237

(η⁵-Cyclopentadienyl)nitrosyl(triphenylphosphine)rhodium(1+)

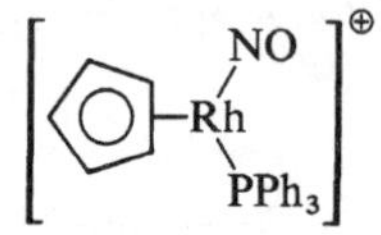

M 460.296 (ion)

Hexafluorophosphate: [37684-97-2].
$C_{23}H_{20}F_6NOP_2Rh$ M 605.261
Yellow cryst. Mp >168° dec. ir ν_{NO} 1847 cm^{-1} (nujol).

Tetraphenylborate: [53079-91-7].
$C_{47}H_{40}BNOPRh$ M 779.528
Yellow cryst. Mp 103-5°.

Connelly, N.G. *et al*, *J. Organomet. Chem.*, 1972, **38**, 385 (*synth, ir, pmr*)
Faraone, F. *et al*, *Inorg. Chim. Acta*, 1973, **7**, 729 (*synth, ir, pmr*)

$C_{23}H_{30}O_3Rh_2$ Rh-00238

μ-Carbonyldicarbonylbis[(1,2,3,4,5-η)-1,2,3,4,5-pentamethyl-2,4-cyclopentadien-1-yl]dirhodium, 10CI

Tricarbonylbis(pentamethylcyclopentadienyl)dirhodium

[75533-62-9]

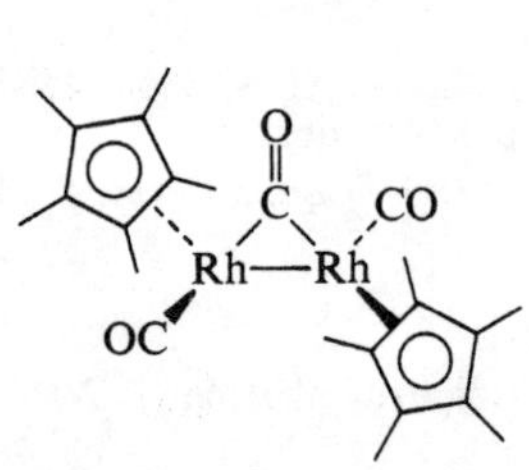

M 560.299

Violet-red, light-sensitive needles (THF/Et_2O at −25°). Stable in air for several days at r.t. in the dark. Sol. THF, CH_2Cl_2, C_6H_6. ir ν_{CO} 1947, 1794 cm^{-1} (KBr). Dec. at 141-3° or photochemically to give Rh(C_5Me_5)-(CO)$_2$ and [Rh(C_5Me_5)μ-CO]$_2$.

Plank, J. *et al*, *Angew. Chem., Int. Ed. Engl.*, 1980, **19**, 937 (*synth, ir, pmr, ms*)
Herrmann, W.A. *et al*, *Z. Anorg. Allg. Chem.*, 1982, **487**, 85 (*synth, cryst struct, props*)

$C_{23}H_{32}O_2Rh_2$ **Rh-00239**

Dicarbonyl-μ-methylenebis[(1,2,3,4,5-η)-1,2,3,4,5-pentamethyl-2,4-cyclopentadien-1-yl]dirhodium, 10CI

μ-Methylenedicarbonylbis(η-pentamethylcyclopentadienyl)dirhodium

[76545-02-3]

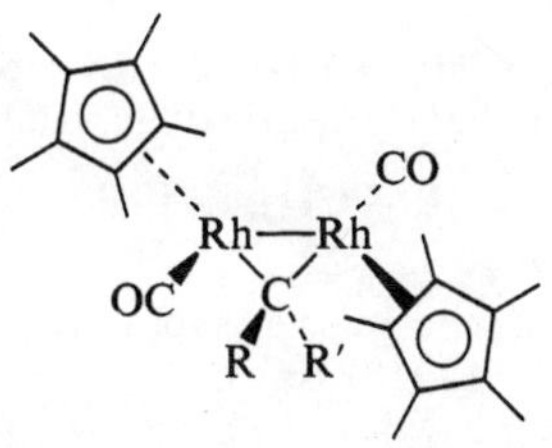

R = R′ = H

M 546.316

Dark-red cryst. (pentane or by subl.). Mp 169° dec. ir ν_{CO} 1945 cm^{-1} (hexane).

Boag, N.M. *et al*, *J. Chem. Soc., Chem. Commun.*, 1980, 1171 (*synth, ir, nmr*)

Claus, A.D. *et al*, *J. Organomet. Chem.*, 1980, **201**, C31 (*synth, pmr, ir*)

Herrmann, W.A. *et al*, *Angew. Chem., Int. Ed. Engl.*, 1981, **20**, 193 (*synth, ir, pmr*)

Herrmann, W.A. *et al*, *Chem. Ber.*, 1982, **115**, 878 (*ir, cmr*)

$C_{24}HO_{24}Rh_{13}^{\ominus\ominus\ominus\ominus}$ **Rh-00240**

Tetracosacarbonylhydrotridecarhodate(4−), 10CI

Dodeca-μ-carbonyldodecacarbonyltrihydrotridecarhodate(4−)

[72855-44-8]

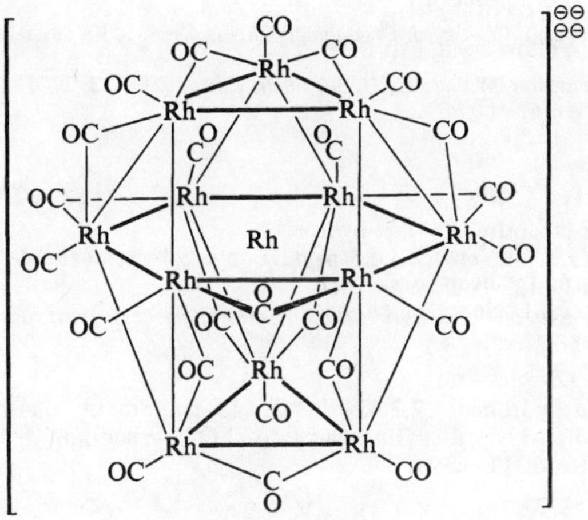

H not shown

M 2011.029 (ion)

H atom not shown, proposed to reside in cavity near a square RhRhRhRh face.

Tetrakis(tetramethylammonium) salt: [77001-70-8].
$C_{40}H_{49}N_4O_{24}Rh_{13}$ M 2307.611
Cryst. + 3 mol. 1-methyl-2-pyrrolidone (2-propanol/1-methyl-2-pyrrolidone).

Tetrakis(tetraethylammonium) salt: [74539-17-6].
$C_{56}H_{81}N_4O_{24}Rh_{13}$ M 2532.040
No phys. props. descr.

Ciani, G. *et al*, *J. Chem. Soc., Dalton Trans.*, 1981, 519 (*cryst struct, ir, pmr*)

$C_{24}H_2O_{24}Rh_{13}^{\ominus\ominus\ominus}$ **Rh-00241**

Tetracosacarbonyldihydrotridecarhodate(3−), 10CI

Dodeca-μ-carbonyldodecacarbonyltrihydrotridecarhodate(3−)

[58500-22-4]

M 2012.037 (ion)

Same struct. as
Tetracosacarbonylhydrotridecarhodate(4−), Rh-00240
Hcp array of Rh in cryst. H atoms not shown. Brown. Easily converted to
Tetracosacarbonyltrihydrotridecarhodate(2−), Rh-00242 with acid. Fluxional in Me_2CO soln.

Tris(tetrabutylammonium) salt: [71899-98-4].
$C_{72}H_{110}N_3O_{24}Rh_{13}$ M 2739.438
Phys. props. not descr.

Tris(benzyltriphenylphosphonium) salt: [71702-03-9].
$C_{99}H_{68}O_{24}P_3Rh_{13}$ M 3072.305
Phys. props. not descr. ir ν_{CO} 1995, 1810 cm^{-1} (MeCN).

Tris[bis(triphenylphosphine)iminium] salt: [79738-35-5].
$C_{132}H_{92}N_3O_{24}P_6Rh_{13}$ M 3627.799
Dark cryst. (Me_2CO/2-propanol).

Albano, V.G. *et al*, *J. Chem. Soc., Chem. Commun.*, 1975, 859 (*synth, ir, pmr*)

Martinengo, S. *et al*, *J. Chem. Soc., Chem. Commun.*, 1977, 39 (*pmr, cmr*)

Albano, V.G. *et al*, *J. Chem. Soc., Dalton Trans.*, 1979, 978 (*cryst struct*)

Vidal, J.L. *et al*, *Fundam. Res. Homogeneous Catal.*, 1979, **3**, 499 (*use, bibl*)

Vidal, J.L. *et al*, *J. Organomet. Chem.*, 1981, **218**, 217 (*synth, cryst struct, bibl*)

$C_{24}H_3O_{24}Rh_{13}^{\ominus\ominus}$ **Rh-00242**

Tetracosacarbonyltrihydrotridecarhodate(2−)

Dodeca-μ-carbonyldodecacarbonyltrihydrotridecarhodate(2−), 10CI

[58543-08-1]

M 2013.045 (ion)

Same struct. as
Tetracosacarbonylhydrotridecarhodate(4−), Rh-00240
Hcp array of Rh in cryst. H atoms not shown. Brown. Readily converted to
Tetracosacarbonyldihydrotridecarhodate(3−), Rh-00241 with base. Fluxional in Me_2CO soln.

Di-Na salt: [76723-33-6].
$C_{24}H_3Na_2O_{24}Rh_{13}$ M 2059.024
Phys. props. not descr.

Bis(tetrabutylammonium) salt: [71899-97-3].
$C_{56}H_{75}N_2O_{24}Rh_{13}$ M 2497.979
Phys. props. not descr.

Martinengo, S. *et al*, *J. Chem. Soc., Chem. Commun.*, 1975, 859 (*cryst struct, synth*)

Martinengo, S. *et al*, *J. Chem. Soc., Chem. Commun.*, 1977, 39 (*pmr, cmr*)

Vidal, J.L. *et al*, *Fundam. Res. Homogeneous. Catal.*, 1979, **3**, 499 (*use, bibl*)

$C_{24}H_{20}OPRh$ **Rh-00243**

Carbonyl(η^5-2,4-cyclopentadien-1-yl)(triphenylphosphine)rhodium

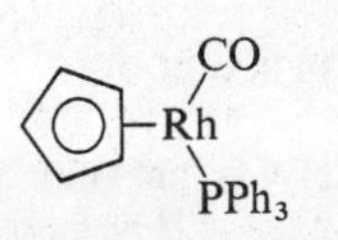

M 458.301

Orange cryst. (CH_2Cl_2/hexane). Sol. CH_2Cl_2. Mp 153-6° dec. ir ν_{CO} 1957 cm^{-1} (hexane). Reacts with MeI to form Acetyl(η^5-2,4-cyclopentadien-1-yl)iodo(triphenylphosphine)rhodium, Rh-00254 .

Hart-Davis, A.J. *et al*, *Inorg. Chem.*, 1970, **9**, 2658 (*synth, ir, props*)

$C_{24}H_{20}PRhS_2$ Rh-00244

(*C,S*-η)-Carbon disulfide(η^5-2,4-cyclopentadien-1-yl)(triphenylphosphine)rhodium, 9CI

[52367-57-4]

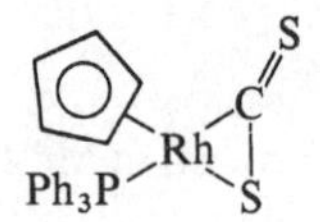

M 506.421

CA refers to a nonexistent (carbon disulfide-*S*) isomer. Dark-red cryst. Mp 138-40° dec.

Wakatsuki, Y. *et al*, *J. Organomet. Chem.*, 1974, **64**, 393 (*synth, ir*)

$C_{24}H_{22}O_3PRh$ Rh-00245

Carbonyl(2,4-pentanedionato-*O,O'*)triphenylphosphinerhodium

Acetylacetonatocarbonyl(triphenylphosphine)rhodium

[25470-96-6]

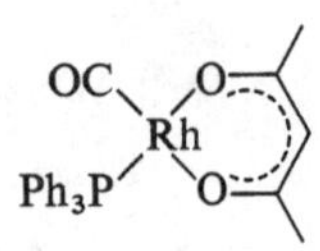

M 492.315

Olefin hydrogenation and hydroformylation catalyst. Yellow cryst. Mp ~175° dec. ir ν_{CO} 1981 cm^{-1} (nujol).

Bonati, F. *et al*, *J. Chem. Soc.*, 1964, 3156 (*synth, ir*)
Varshavskii, Yu.S. *et al*, *J. Organomet. Chem.*, 1974, **77**, 107 (*synth, ir*)
Leipoldt, J.G. *et al*, *Inorg. Chim. Acta*, 1978, **26**, L35 (*cryst struct*)
Brester, L.S. *et al*, *J. Organomet. Chem.*, 1979, **171**, 229 (*cmr*)

$C_{24}H_{31}F_6O_4Rh_2^{\oplus}$ Rh-00246

μ-Hydrobis[(1,2,3,4,5-η)-1,2,3,4,5-pentamethyl-2,4-cyclopentadien-1-yl]bis[μ-(trifluoroacetato-*O:O'*)]dirhodium(1+), 9CI

Di-μ-trifluoroacetato-μ-hydridobis(pentamethylcyclopentadienylrhodium)(1+)

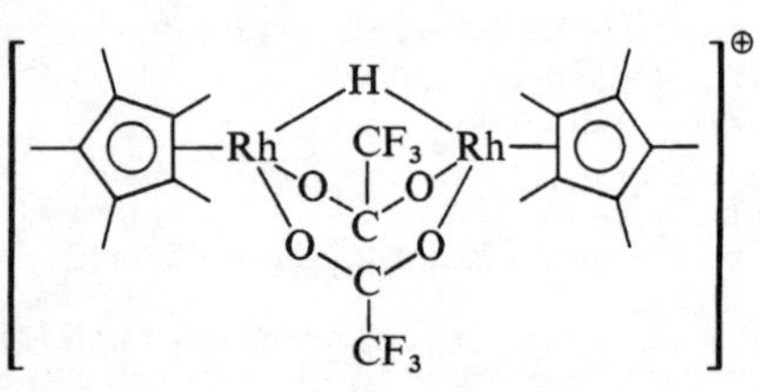

M 703.308 (ion)

Catalyst for the hydrogenation of olefins.

Hexafluorophosphate: [39385-29-0].
$C_{24}H_{31}F_{12}O_4PRh_2$ M 848.272
Orange cryst. (CH_2Cl_2/Et_2O). Mp 223°.

Trifluoroacetate: [39385-28-9].
$C_{26}H_{31}F_9O_6Rh_2$ M 816.324
Orange cryst. (CH_2Cl_2/hexane). Mp 149-50°.

Can. Pat., 915 697, (*1973*); *CA*, **78**, 76413t (*use*)
White, C. *et al*, *J. Chem. Soc., Dalton Trans.*, 1973, 1901 (*nmr, synth*)

$C_{24}H_{34}O_2Rh_2$ Rh-00247

Dicarbonyl-μ-ethylidenebis[(1,2,3,4,5-η)-1,2,3,4,5-pentamethyl-2,4-cyclopentadien-1-yl]dirhodium, 10CI

μ-Ethylidenedicarbonylbis(η-pentamethylcyclopentadienyl)dirhodium

[76582-87-1]

As Dicarbonyl-μ-methylenebis[(1,2,3,4,5-η)-1,2,3,4,5-pentamethyl-2,4-cyclopentadien-1-yl]dirhodium, Rh-00239 with

$$R = CH_3, R' = H$$

M 560.342

Deep-red cryst. (pentane or by subl.). Mp 162-3° dec. ir ν_{CO} 1937 cm^{-1} (THF).

Claus, A.D. *et al*, *J. Organomet. Chem.*, 1980, **201**, C31 (*synth, ir, pmr*)
Herrmann, W.A. *et al*, *Angew. Chem., Int. Ed. Engl.*, 1981, **20**, 193 (*synth, ir, pmr*)
Herrmann, W.A. *et al*, *Chem. Ber.*, 1982, **115**, 878 (*ir, pmr, synth*)

$C_{24}H_{36}Rh^{\oplus\oplus}$ Rh-00248

Bis[(1,2,3,4,5,6-η)hexamethylbenzene]rhodium(2+), 10CI

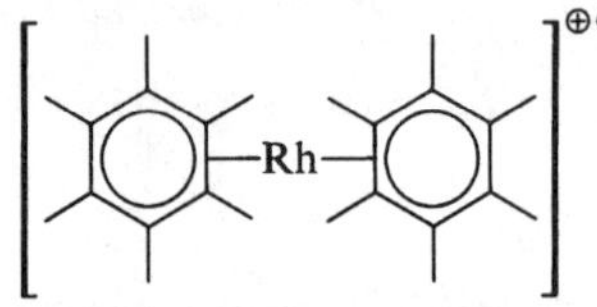

M 427.454 (ion)

Bis(hexafluorophosphate): [74276-80-5].
$C_{24}H_{36}F_{12}P_2Rh$ M 717.382
Red powder. μ_{eff} = 1.32 BM.

Fischer, E.O. *et al*, *J. Organomet. Chem.*, 1964, **1**, 307 (*synth, ir*)
Thompson, M.R. *et al*, *J. Am. Chem. Soc.*, 1980, **102**, 2979 (*synth*)

$C_{24}H_{37}O_4Rh_2^{\oplus}$ Rh-00249

Bis[μ-(acetato-*O:O'*)]-μ-hydrobis[(1,2,3,4,5-η)-1,2,3,4,5-pentamethyl-2,4-cyclopentadien-1-yl]dirhodium(1+), 9CI

Di-μ-acetato-μ-hydridobis(pentamethylcyclopentadienylrhodium)(1+)

As Di-μ-hydrobis[(1,2,3,4,5-η)-1,2,3,4,5-pentamethyl-2,4-cyclopentadien-1-yl][μ-(trifluoroacetato-*O:O'*)]dirhodium(1+), Rh-00233 with

$$X = Y = OAc$$

M 595.365 (ion)

Catalyst for hydrogenation of olefins, dehydrogenation of alcohols and disproportionation of CH_3CHO to AcOH and EtOH.

Hexafluorophosphate: [51472-05-0].
$C_{24}H_{37}F_6O_4PRh_2$ M 740.329
Orange-red cryst. (CH_2Cl_2/Et_2O).

Acetate, AcOH complex: [51614-72-3].
$C_{28}H_{44}O_8Rh_2$ M 714.462
Orange solid.

White, C. *et al*, *J. Chem. Soc., Dalton Trans.*, 1973, 1901 (*synth, pmr*)
Cook, J. *et al*, *J. Chem. Soc., Dalton Trans.*, 1981, 2342 (*synth, use*)

$C_{24}H_{40}Rh_2$ Rh-00250

Dimethyldi-μ-methylenebis[(1,2,3,4,5-η)-1,2,3,4,5-pentamethyl-2,4-cyclopentadien-1-yl]dirhodium, 10CI
Bis(μ-methylene)dimethylbis(pentamethylcyclopentadienyl)dirhodium

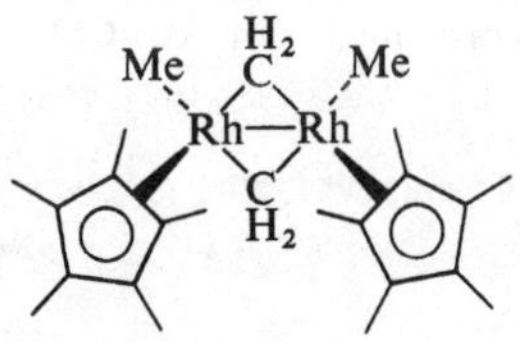

cis-form

M 534.391
cis-form [80410-45-3]
Red-brown cryst. (pentane).
trans-form [80446-42-0]
Red-brown cryst. (pentane).

Isobe, K. *et al*, *J. Chem. Soc., Chem. Commun.*, 1981, 809 (*synth, ms, nmr*)
Vázquez de Miguel, A. *et al*, *J. Chem. Soc., Chem. Commun.*, 1982, 758 (*synth*)

$C_{24}H_{48}Cl_2Rh_2$ Rh-00251

Di-μ-chlorotetrakis(η^3-2,3-dimethyl-2-butene)dirhodium
[69997-66-6]

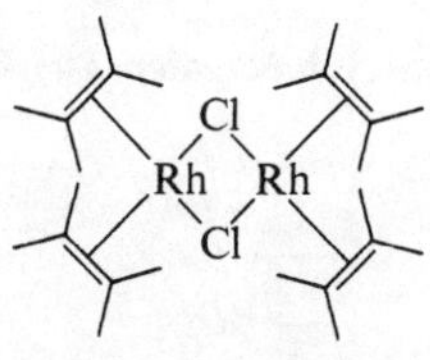

M 613.360
Bridged dimer. Orange cryst. solid (CH_2Cl_2/Et_2O).
Monomer: Chlorobis(2,3-dimethyl-2-butene)rhodium. Bis(2,3-dimethyl-2-butene)rhodium chloride.
$C_{12}H_{24}ClRh$ M 306.680
Unknown.

Inorg. Synth., 1979, **19**, 218 (*synth*)

$C_{24}H_{72}O_{24}P_8Rh_2$ Rh-00252

Octakis(trimethyl phosphite-*P*)dirhodium, 9CI
[52721-32-1]

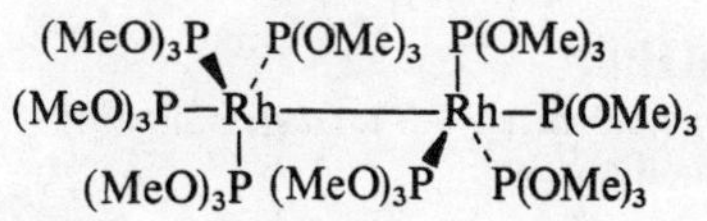

M 1198.419
Non-fluxional bicapped trigonal antiprismatic struct. at r.t. Pale-yellow cryst. Mp 75° dec.

Mathieu, R. *et al*, *J. Chem. Soc., Chem. Commun.*, 1974, 147 (*synth, nmr*)

$C_{25}H_{23}ClPRh$ Rh-00253

[(2,3,5,6-η)-Bicyclo[2.2.1]hepta-2,5-diene]chloro(triphenylphosphine)rhodium, 9CI
Chloro(2,5-norbornadiene)(triphenylphosphine)rhodium, 8CI
[12278-58-9]

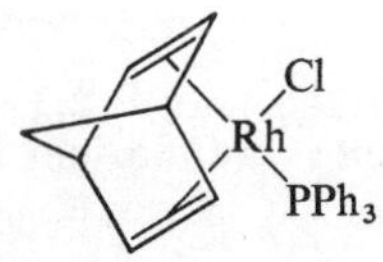

M 492.789
Yellow cryst. (EtOH or by subl.). Mp 224.5-225.5°.

Bennett, M.A. *et al*, *J. Chem. Soc.*, 1961, 1418 (*synth*)
Bennett, M.A. *et al*, *Inorg. Chem.*, 1967, **6**, 1647 (*ir*)
Dawes, J.L. *et al*, *J. Chem. Soc. (A)*, 1968, 2093 (*ir*)
Vrieze, K. *et al*, *J. Organomet. Chem.*, 1968, **14**, 185; **15**, 195 (*nmr, struct*)
Acton, N. *et al*, *J. Am. Chem. Soc.*, 1972, **94**, 5446 (*synth*)

$C_{25}H_{23}IOPRh$ Rh-00254

Acetyl(η^5-2,4-cyclopentadien-1-yl)iodo(triphenylphosphine)rhodium, 9CI

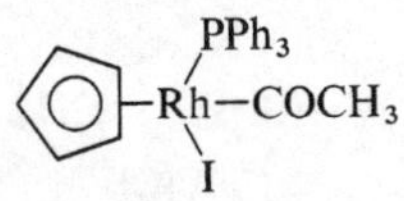

M 600.240
Formed by reaction of MeI with Carbonyl(η^5-2,4-cyclopentadien-1-yl)(triphenylphosphine)rhodium, Rh-00243 . Red-brown cryst. (CH_2Cl_2/hexane). Sol. CH_2Cl_2. Mp 170-5° dec. ir ν_{CO} 1666, 1643 (sh) cm^{-1} (CH_2Cl_2).

Hart-Davis, A.J. *et al*, *Inorg. Chem.*, 1970, **9**, 2658; 1971, **10**, 1653 (*synth, ir, pmr*)

$C_{25}H_{24}PRh$ Rh-00255

(η^5-2,4-Cyclopentadien-1-yl)(η^2-ethene)triphenylphosphinerhodium
π-Cyclopentadienyl(ethylene)(triphenylphosphine)rhodium, 8CI
[34922-78-6]

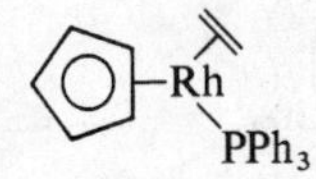

M 458.344
Orange-red cryst. Mp ~150° dec.

Oliver, J.A. *et al*, *Inorg. Chem.*, 1971, **10**, 1165 (*synth, pmr*)

$C_{25}H_{26}PRh$ Rh-00256

(η^5-2,4-Cyclopentadien-1-yl)bis(methyl)triphenylphosphinerhodium

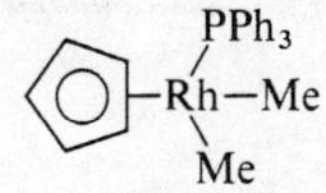

M 460.360
Synth. from (η^5-2,4-Cyclopentadien-1-yl)diiodo(triphenylphosphine)rhodium, Rh-00236 and MeMgI. Air-stable pale-yellow cryst. Sol. most org. solvs. Mp 115-9°.

Kasahara, A. *et al*, *Bull. Chem. Soc. Jpn.*, 1967, **40**, 699 (*synth, pmr*)

$C_{25}H_{30}F_6O_2Rh_2$ **Rh-00257**

Dicarbonylbis[(1,2,3,4,5-η)-1,2,3,4,5-pentamethyl-2,4-cyclopentadien-1-yl][μ-[2,2,2-trifluoro-1-(trifluoromethyl)ethylidene]]dirhodium, 10CI

μ-Bis(trifluoromethyl)methylenedicarbonylbis(η-pentamethylcyclopentadienyl)dirhodium

[76680-93-8]

As Dicarbonyl-μ-methylenebis[(1,2,3,4,5-η)-1,2,3,4,5-pentamethyl-2,4-cyclopentadien-1-yl]dirhodium, Rh-00239 with

R = R′ = CF_3

M 682.312

Red cryst. (pet. ether). ir ν_{CO} 1972 cm^{-1} (hexane).

Green, M. *et al*, *J. Chem. Soc., Dalton Trans.*, 1982, 1309 (*synth, ir, nmr*)

$C_{25}H_{42}N_3Rh^{\oplus\oplus}$ **Rh-00258**

Tris(2-isocyano-2-methylpropane)[(1,2,3,4,5-η)-1,2,3,4,5-pentamethyl-2,4-cyclopentadien-1-yl]rhodium(2+), 10CI

Pentamethylcyclopentadienyltris(tert-*butylisocyanide*)-*rhodium*(*2*+)

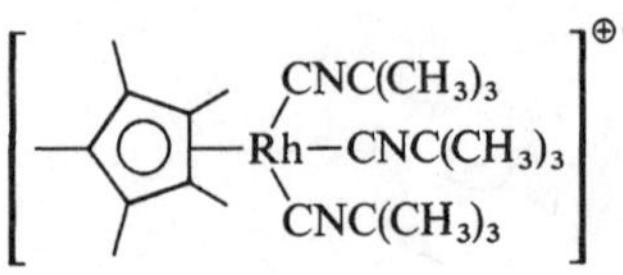

M 487.532 (ion)

Bis(hexafluorophosphate): [67933-70-4].

$C_{25}H_{42}F_{12}N_3P_2Rh$ M 777.461

White cryst. (Me_2CO/diisopropyl ether).

McCleverty, J.A. *et al*, *Transition Met. Chem.*, 1978, **3**, 201 (*synth, ir, pmr*)

$C_{25}O_{25}Rh_{14}^{\ominus\ominus\ominus\ominus}$ **Rh-00259**

Pentacosacarbonyltetradecarhodate(4−)

Hexadeca-μ-carbonylnonacarbonyltetradecarhodate(4−), *10CI*

[68714-69-2]

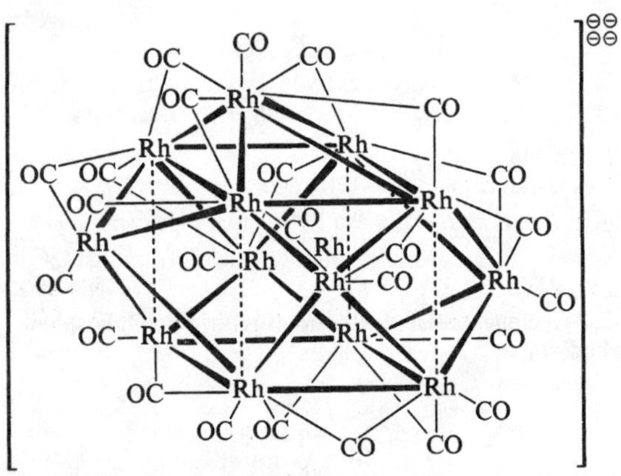

M 2140.937 (ion)

Bcc type struct. in cryst., with 9 terminal and 16 edge-bridging carbonyl groups.

Tetra-Cs salt: [75558-11-1].

$C_{25}Cs_4O_{25}Rh_{14}$ M 2672.559

No phys. props. reported.

Tetrakis(tetramethylammonium) salt: [75548-81-1].

$C_{41}H_{48}N_4O_{25}Rh_{14}$ M 2437.519

Cryst. (Me_2CO/2-propanol).

Tetrakis(tetraethylammonium) salt: [68714-70-5].

$C_{57}H_{80}N_4O_{25}Rh_{14}$ M 2661.948

Cryst. (Me_2CO/2-propanol).

Tetrakis(benzyltriethylammonium) salt: [75558-89-3].

$C_{77}H_{88}N_4O_{25}Rh_{14}$ M 2910.231

Cryst. (Me_2CO/2-propanol). Sol. Me_2CO.

Martinengo, S. *et al*, *J. Am. Chem. Soc.*, 1978, **100**, 7097 (*synth, cryst struct, ir*)

Vidal, J.L. *et al*, *Inorg. Chem.*, 1981, **20**, 265 (*synth, ir, cmr, cryst struct*)

$C_{26}H_{27}ClO_2PRhS$ **Rh-00260**

Chloro(1,5-cyclooctadiene)(sulfur dioxide)(triphenylphosphine)rhodium, 8CI

[32613-66-4]

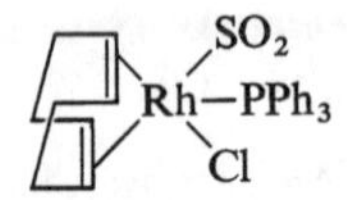

M 572.890

Air-stable red solid. ir ν_{SO} 1218, 1050 cm^{-1} (nujol). Loses SO_2 >80°.

Palazzi, A. *et al*, *J. Organomet. Chem.*, 1970, **25**, 249 (*synth, ir*)

$C_{26}H_{27}ClPRh$ **Rh-00261**

Chloro[(1,2,5,6-η)-1,5-cyclooctadiene](triphenylphosphine)rhodium, 9CI

[31781-57-4]

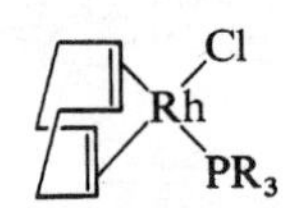

R = Ph

M 508.832

Yellow cryst. (EtOH). Mp 146°.

Chatt, J. *et al*, *J. Chem. Soc.*, 1957, 4735 (*synth*)

Vrieze, K. *et al*, *J. Organomet. Chem.*, 1968, **14**, 185 (*nmr, struct*)

Grim, S.O. *et al*, *Inorg. Chim. Acta*, 1970, **4**, 277 (*nmr, uv*)

Crociani, B. *et al*, *J. Organomet. Chem.*, 1979, **177**, 385 (*synth, ir, pmr*)

Naaktgeboren, A.J. *et al*, *J. Am. Chem. Soc.*, 1980, **102**, 3350 (*nmr*)

$C_{26}H_{29}ClN_2OPRh$ **Rh-00262**

Carbonylchloro(1,3-diethyl-2-imidazolidinylidene)triphenylphosphinerhodium, 9CI

[54686-50-9]

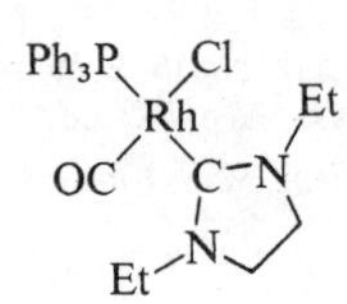

M 554.860

Carbene complex.

Doyle, M.J. *et al*, *J. Chem. Soc., Chem. Commun.*, 1974, 679 (*pmr*)

$C_{26}H_{45}ClPRh$ **Rh-00263**

Chloro[(1,2,5,6-η)-1,5-cyclooctadiene](tricyclohexylphosphine)rhodium, 10CI

[65394-49-2]

As Chloro[(1,2,5,6-η)-1,5-cyclooctadiene](triphenylphosphine)rhodium, Rh-00261 with

R = cyclohexyl

M 526.974
Cryst.

Crabtree, R.H. *et al*, *J. Organomet. Chem.*, 1977, **141**, 113 (*synth, pmr*)
DeWaal, D.J.A. *et al*, *Inorg. Chim. Acta*, 1978, **26**, 91 (*synth, uv*)
James, B.R. *et al*, *Adv. Chem. Ser.*, 1982, **196**, 145 (*synth, ir*)

$C_{27}H_{24}ClOP_2Rh$ Rh-00264

Carbonylchloro[1,2-ethanediylbis[diphenylphosphine]-*P,P'*]-rhodium, 10CI

Carbonylchloro[ethylenebis[diphenylphosphine]]rhodium, 8CI. *Carbonylchloro[bis(1,2-diphenylphosphino)-ethane]rhodium*

[15024-81-4]

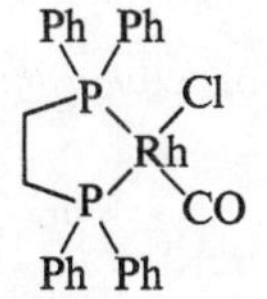

M 564.792
Catalyst for hydroformylation of 1-hexene and carbonylation of MeOH. Yellow cryst. (C_6H_6). Mp 135° dec. ir ν_{CO} 2010 cm^{-1} (CH_2Cl_2).

Hieber, W. *et al*, *Chem. Ber.*, 1967, **100**, 148 (*synth, ir*)
Sanger, A.R., *J. Chem. Soc., Dalton Trans.*, 1977, 120 (*synth, ir, nmr*)
Grimblot, J. *et al*, *Inorg. Chim. Acta*, 1979, **34**, 29 (*ir, use*)

$C_{27}H_{29}ORh$ Rh-00265

[(1,2,4,5-η)-1,5-Diphenyl-1,4-pentadien-3-one][(1,2,3,4,5-η)-1,2,3,4,5-pentamethyl-2,4-pentadien-1-yl]rhodium, 9CI

Dibenzylideneacetone(pentamethylcyclopentadienyl)-rhodium

[51177-80-1]

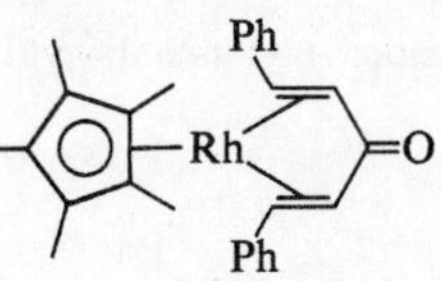

M 472.431
Red cryst.

Lee, H.B. *et al*, *J. Organomet. Chem.*, 1973, **57**, C87 (*synth, pmr, ir*)
Ibers, J.A., *J. Organomet. Chem.*, 1974, **73**, 389 (*struct*)

$C_{27}H_{64}P_3Rh$ Rh-00266

Hydrotris(triisopropylphosphine)rhodium

Hydrotris[tris(1-methylethyl)phosphine]rhodium

[71357-03-4]

$HRh[P[CH(CH_3)_2]_3]_3$

M 584.629
Fluxional in soln., planar in solid state. Catalyst precursor for water-gas shift reaction, and for hydrogenations of ketones. Yellow cryst. (pentane). Sol. Py, Me_2CO.

Yoshida, T. *et al*, *J. Organomet. Chem.*, 1979, **181**, 183 (*synth, ir, pmr*)
Yoshida, T. *et al*, *J. Am. Chem. Soc.*, 1980, **102**, 6451 (*cryst struct, nmr*)
Yoshida, T. *et al*, *J. Am. Chem. Soc.*, 1981, **103**, 3411 (*use*)
Tani, K. *et al*, *Polyhedron*, 1982, **1**, 839 (*use*)

$C_{27}O_{27}Rh_{15}^{\ominus\ominus\ominus}$ Rh-00267

Heptacosacarbonylpentadecarhodate(3−)

Tetradeca-μ-carbonyltridecacarbonylpentadecarhodate(3−), 10CI

[68517-96-4]

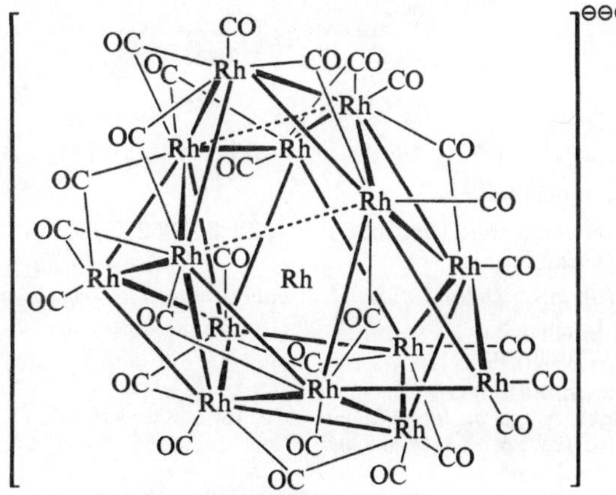

M 2299.863 (ion)
Struct. intermed. between that of Pentacosacarbonyltetradecarhodate(4−), Rh-00259 and Tetracosacarbonyldihydrotridecarhodate(3−), Rh-00241 .

Tris(tetramethylammonium) salt: [68679-12-9].
$C_{39}H_{36}N_3O_{27}Rh_{15}$ M 2522.300
No phys. props. recorded.

Tris(tetraethylammonium) salt: [68679-13-0].
$C_{51}H_{60}N_3O_{27}Rh_{15}$ M 2690.621
No phys. props. recorded.

Martinengo, S. *et al*, *J. Am. Chem. Soc.*, 1978, **100**, 7096 (*synth, cryst struct, ir*)
Vidal, J.L. *et al*, *J. Organomet. Chem.*, 1981, **215**, C11 (*synth, props*)

$C_{28}H_{20}N_4Rh^{\oplus}$ Rh-00268

Tetrakis(isocyanobenzene)rhodium(1+)

Octakis(isocyanobenzene)dirhodium. Tetrakis(phenyl isocyanide)rhodium. Octakis(phenyl isocyanide)-dirhodium

$[(PhNC)_4Rh—Rh(CNPh)_4]^{\oplus\oplus}$ (D_{4d})

M 515.398 (ion)
Dimeric in the solid state and in concentrated MeCN soln. Dil. MeCN solns. are yellow, conc. solns. blue.

Chloride:
$C_{56}H_{40}Cl_2N_8Rh_2$ M 1101.703
Dark-green powder. Sol. H_2O, insol. C_6H_6, Et_2O.

Tetrafluoroborate: [65405-04-1].
$C_{56}H_{40}B_2F_8N_8Rh_2$ M 1204.404
Brown cryst. (MeCN). Sol. Me_3CN, insol. H_2O.

Hexafluorophosphate: [56192-49-5].
$C_{56}H_{40}F_{12}N_8P_2Rh_2$ M 1320.725
Yellow cryst. (MeCN). Sol. Me_3CN, insol. H_2O.

Tetraphenylborate: [65405-08-5].
$C_{104}H_{80}B_2N_8Rh_2$ M 1669.261
Purple needles (MeCN). Sol. MeCN, insol. H_2O.

Malatesta, L. *et al*, *J. Chem. Soc.*, 1956, 1867 (*synth*)
Malatesta, L. *et al*, *Isocyanide Complexes of Metals*, 1969, Wiley-Interscience, London, 140.

Mann, K.R. *et al*, *J. Am. Chem. Soc.*, 1975, **97**, 3553 (*uv*)
Mann, K.R. *et al*, *Inorg. Chem.*, 1978, **17**, 828 (*cryst struct*, *uv*)

$C_{28}H_{20}N_6O_4Rh_2$ Rh-00269

Tetracarbonylbis[μ-(1,3-diphenyl-1-triazenato-N^1:N^3)]dirhodium, 10CI
Bis[1,3-diphenyltriazenidodicarbonylrhodium]
[36126-17-7]

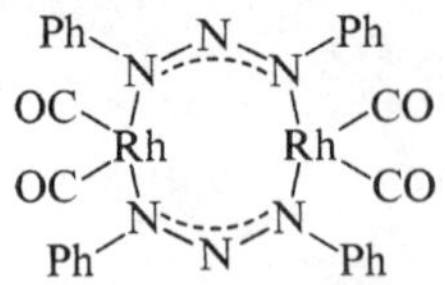

M 710.315
Red cryst. solid (hexane). ir ν_{CO} 2088, 2062, 2021 cm^{-1} (hexane).
Monomer: Dicarbonyl(1,3-diphenyltriazenato)rhodium.
$C_{14}H_{10}N_3O_2Rh$ M 355.157
Unknown.

Knoth, W.H., *Inorg. Chem.*, 1973, **12**, 38 (*synth*, *ir*)
Balch, A.L. *et al*, *Inorg. Chem.*, 1977, **16**, 2840 (*synth*, *ir*, *uv*)
Abel, E.W. *et al*, *J. Organomet. Chem.*, 1978, **155**, 299 (*synth*, *ms*, *ir*, *pmr*)
Abel, E.W. *et al*, *J. Organomet. Chem.*, 1980, **193**, 389.

$C_{28}H_{24}O_2P_2Rh^{\oplus}$ Rh-00270

Dicarbonyl[1,2-ethanediylbis[diphenylphosphine]-*P*,*P'*]rhodium, 10CI
Dicarbonyl[ethylenebis(diphenylphosphine)]-rhodium(1+), 8CI. *[1,2-Bis(diphenylphosphino)ethane]-dicarbonylrhodium(1+)*

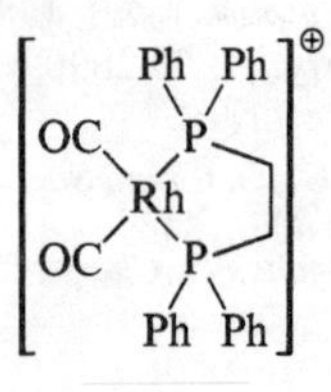

M 557.349 (ion)
Chloride: [26004-12-6].
$C_{28}H_{24}ClO_2P_2Rh$ M 592.802
Yellow solid (CH_2Cl_2/pentane). Λ_M 26.8 ohm^{-1} cm^2 mol^{-1} (CH_2Cl_2, 25°), ir ν_{CO} 2062, 1982 cm^{-1} (nujol).
Tetraphenylborate: [74404-48-1].
$C_{52}H_{44}BO_2P_2Rh$ M 876.581
Yellow solid (CH_2Cl_2/pentane at −20°). ir ν_{CO} 2100, 2055 cm^{-1} (CH_2Cl_2).

Uguagliati, P. *et al*, *Inorg. Chem.*, 1969, **8**, 1625 (*synth*, *ir*)
Albano, P. *et al*, *J. Organomet. Chem.*, 1980, **190**, 243 (*synth*, *ir*)

$C_{28}H_{28}BRh$ Rh-00271

Bis(ethylene)[tetraphenylborato(1−)]rhodium, 8CI
Bis(η^2-ethene)(η^6-tetraphenylborato)rhodium

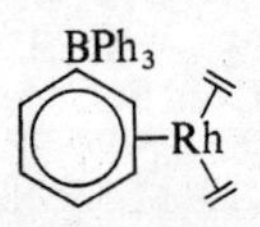

M 478.245
Yellow cryst. (CH_2Cl_2/MeOH).

Schrock, R.R. *et al*, *Inorg. Chem.*, 1970, **9**, 2339 (*synth*, *ir*)

$C_{28}H_{30}Cl_2PRh$ Rh-00272

Dichloro[(1,2,3,4,5-η)-1,2,3,4,5-pentamethyl-2,4-cyclopentadien-1-yl](triphenylphosphine)rhodium
[63179-49-7]

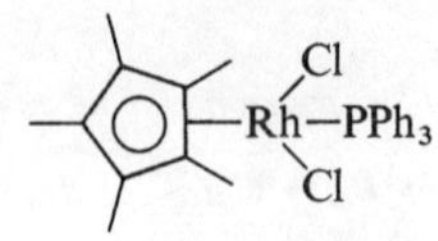

M 571.330
Reddish-brown, air-stable cryst. Mp >230°. Readily dialkylated with Grignard reagents.

Kang, J.W. *et al*, *J. Am. Chem. Soc.*, 1969, **91**, 5970 (*synth*, *pmr*)
Diversi, P. *et al*, *Inorg. Chem.*, 1980, **19**, 3590 (*props*)

$C_{29}H_{15}O_{11}PRh_4$ Rh-00273

Tri-μ-carbonyloctacarbonyl(triphenylphosphine)tetrarhodium, 10CI
Undecacarbonyl(triphenylphosphine)tetrarhodium, 8CI
[37451-27-7]

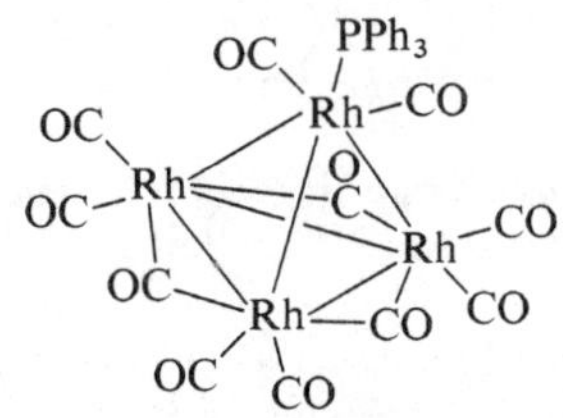

M 982.027
Red cryst. Air-stable. Sol. $CHCl_3$, toluene, insol. pentane. Fluxional in soln., when more than one isomer may be formed.

Evans, J. *et al*, *J. Chem. Soc., Dalton Trans.*, 1978, 626 (*cmr*, *struct*)
Heaton, B.T. *et al*, *J. Organomet. Chem.*, 1981, **213**, 333 (*synth*, *ir*, *nmr*)

$C_{29}H_{25}IPRh$ Rh-00274

(η^5-2,4-Cyclopentadien-1-yl)iodophenyl(triphenylphosphine)-rhodium, 9CI
[50276-32-9]

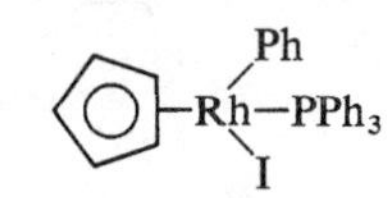

M 634.300
Red solid (C_6H_6/hexane). Mp 154-5°.

Gardner, S.A. *et al*, *Inorg. Chem.*, 1974, **13**, 997 (*synth*, *ir*, *pmr*)

$C_{29}H_{25}Rh_3$ **Rh-00275**

(Diphenylacetylene)tris(cyclopentadienyl)trirhodium

Tris(η^5-2,4-cyclopentadien-1-yl)[μ_3-[1,1′-(1,2-ethynediyl)bis[benzene]]]trirhodium, 9CI

[39346-08-2]

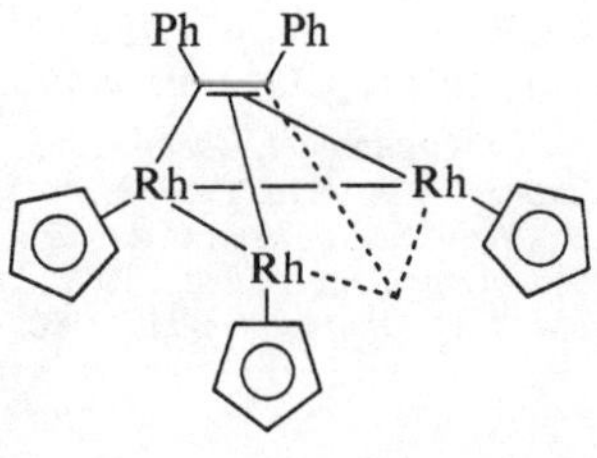

M 682.233

Formed during the cyclotrimerisation of PhC≡CPh by $(C_5H_5)Rh(CO)_2$. Orange-red solid (C_6H_6/hexane).

Gardner, S.A. *et al*, Inorg. Chem., 1973, **12**, 2396 (*synth, ir, pmr, ms*)

$C_{29}H_{26}O_4PRh$ **Rh-00276**

(η^5-2,4-Cyclopentadien-1-yl)[(2,3-η)-dimethyl-2-butynedioate](triphenylphosphine)rhodium, 10CI

Cyclopentadienyl(dimethylacetylenedicarboxylate)(triphenylphosphine)rhodium

[50277-80-0]

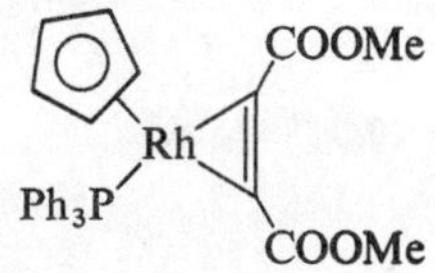

M 572.401

Yellow cryst. Mp 163-8° dec.

Wakatsuki, Y. *et al*, J. Am. Chem. Soc., 1973, **95**, 5781 (*synth*)
Wakatsuki, Y. *et al*, J. Organomet. Chem., 1974, **64**, 393 (*synth*)

$C_{29}H_{32}NPRh^{\oplus}$ **Rh-00277**

[(2,3,5,6-η)-Bicyclo[2.2.1]hepta-2,5-diene][2-(diphenylphosphino)-*N,N,α*-trimethylbenzenemethanamine-*N,P*]rhodium(1+)

[N,N-*Dimethyl-1-*[O-*(diphenylphosphino)phenyl*]*ethylamine*](*norbornadiene*)*rhodium*(*1*+)

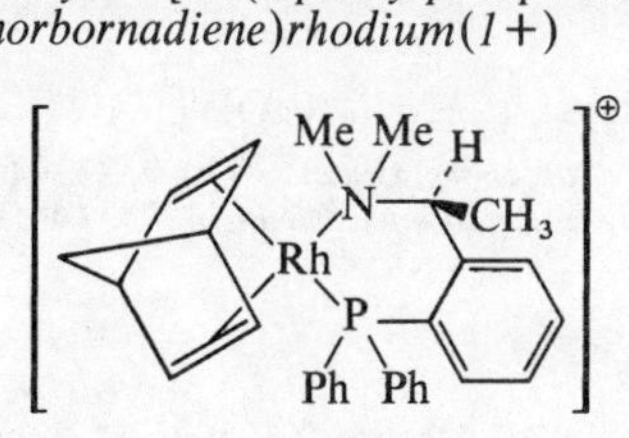

M 528.458 (ion)

(*S*)-*form*

Catalyst precursor for the asymmetric hydrosilylation of ketones.

Perchlorate: [79215-49-9].
$C_{29}H_{32}ClNO_4PRh$ M 627.908
Cryst. (CH_2Cl_2/Et_2O). Mp 170-80° dec.

Payne, N.C. *et al*, Inorg. Chem., 1982, **21**, 182 (*synth, nmr, use*)

$C_{29}H_{33}IPRh$ **Rh-00278**

Iodomethyl[(1,2,3,4,5-η)-1,2,3,4,5-pentamethyl-2,4-cyclopentadien-1-yl]triphenylphosphinerhodium

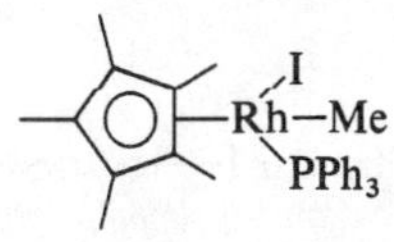

M 642.363

Reddish-brown cryst. (Et_2O). Mp 110-5° dec.

Kang, J.W. *et al*, J. Am. Chem. Soc., 1969, **91**, 5970 (*synth, pmr*)

$C_{30}H_{21}O_3Rh_3$ **Rh-00279**

Tri-μ-carbonyltris[(1,2,3,3*a*,7*a*-η)-1*H*-inden-1-yl]trirhodium, 10CI

[63466-35-3]

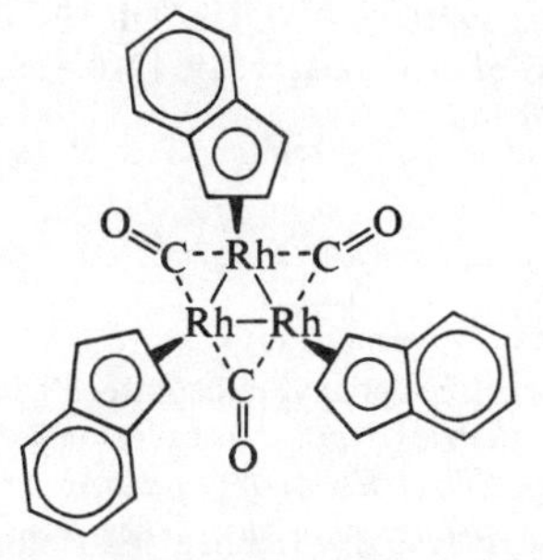

M 738.211

Dark-green solid. ir ν_{CO} 1852, 1805 cm^{-1} (CH_2Cl_2).

Caddy, P. *et al*, Angew. Chem., Int. Ed. Engl., 1977, **16**, 648 (*synth, ir*)
Al-Obaidi, Y.N. *et al*, J. Chem. Soc., Dalton Trans., 1981, 2321.

$C_{30}H_{34}PRh$ **Rh-00280**

(η^2-Ethene)[(1,2,3,4,5-η)-1,2,3,4,5-pentamethyl-2,4-cyclopentadien-1-yl](triphenylphosphine)rhodium, 10CI

Ethylene(pentamethylcyclopentadienyl)triphenylphosphinerhodium

[63162-05-0]

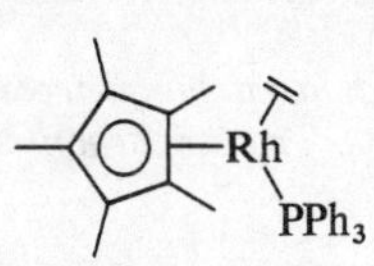

M 528.478

Orange-red needles (pentane at −30°). Mp 150-5° dec.

Porzio, W. *et al*, J. Am. Chem. Soc., 1978, **100**, 2048 (*cryst struct*)
Diversi, P. *et al*, J. Organomet. Chem., 1979, **165**, 253 (*synth, ms*)

$C_{30}H_{36}PRh$ **Rh-00281**

Dimethyl(η^5-pentamethylcyclopentadienyl)triphenylphosphinerhodium

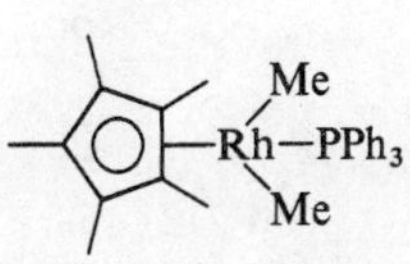

M 530.494

Yellow cryst. (Et_2O). Mp 184-6° dec.

Kang, J.W. *et al*, *J. Am. Chem. Soc.*, 1969, **91**, 5970 (*synth*, *pmr*)

$C_{30}H_{38}BO_6P_2Rh$ Rh-00282

[(η^6-Phenyl)triphenylborato(1−)]bis(trimethyl phosphite-*P*)-rhodium, 9CI

Bis(phosphorous acid)[tetraphenylborato(1−)]rhodium, hexamethyl ester, 8CI. *(Tetraphenylboron)bis(trimethyl phosphite)rhodium*

[33362-35-5]

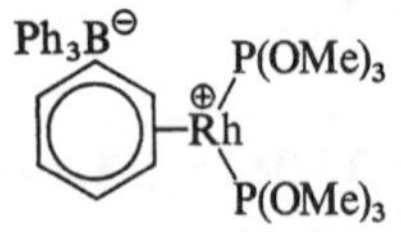

M 670.290

Yellow cryst. (CH_2Cl_2/MeOH). Mp 163-5°.

Haines, L.M. *et al*, *Inorg. Chem.*, 1971, **10**, 1685 (*synth*, *ir*, *pmr*, *cryst struct*)
Nolte, M.J. *et al*, *Acta Crystallogr., Sect. B*, 1974, **30**, 738 (*cryst struct*)

$C_{30}H_{68}O_9P_3Rh$ Rh-00283

η^3-2-Propenyltris(triisopropyl phosphite-*P*)rhodium

(η^3-2-Propenyl)tris[tris(1-methylethyl)phosphite-P]-rhodium, 10CI. *Tris(triisopropyl phosphite)allylrhodium. Allyltris(triisopropyl phosphite)rhodium*

[71729-51-6]

As (η^3-2-Propenyl)tris(trimethyl phosphite-*P*)rhodium, Rh-00111 with

$$R = -CH(CH_3)_2$$

M 768.689

Yellow cryst. Mp 71.5-72°. Allyl-rhodium bond cleaved by H_2 at 20° to give a rhodium hydride.

Sivak, A.J. *et al*, *J. Am. Chem. Soc.*, 1979, **101**, 4878 (*synth*, *pmr*)

$C_{30}O_{30}Rh_{12}^{\ominus\ominus}$ Rh-00284

Di-μ-carbonylocta-μ_3-carbonyleicosacarbonyldodecarhodate(2−), 10CI

Octa-μ_3-carbonyldi-μ-carbonyleicosacarbonyldodecarhodate(2−), 10CI. *Tricosacarbonyldodecarhodate(2−)*

[12775-50-7]

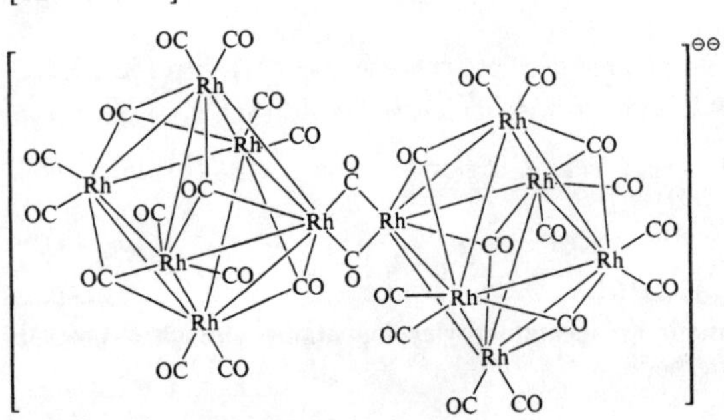

M 2075.178 (ion)

Di-Na salt: [12576-08-8].

$C_{30}Na_2O_{30}Rh_{12}$ M 2121.158

Violet cryst. Sol. H_2O. Mp 150° dec.

Bis(tetramethylammonium) salt:

$C_{38}H_{24}N_2O_{30}Rh_{12}$ M 2223.469

Dark-violet cryst. (Me_2CO/2-propanol). Mp 150° dec.

Bis(tetrabutylammonium) salt: [12576-05-5].

$C_{62}H_{72}N_2O_{30}Rh_{12}$ M 2560.112

Violet cryst. (Me_2CO/2-propanol).

Bis(benzyltrimethylammonium) salt:

$C_{58}H_{32}N_2O_{30}Rh_{12}$ M 2471.752

Dark-violet cryst. (Me_2CO/2-propanol). Mp 135° dec.

Albano, V.G. *et al*, *J. Organomet. Chem.*, 1969, **19**, 405 (*struct*)
Chini, P. *et al*, *Inorg. Chim. Acta*, 1969, **3**, 299 (*synth*, *ir*)
Chini, P. *et al*, *J. Chem. Soc., Chem. Commun.*, 1974, 310 (*cmr*)
Chini, P. *et al*, *Adv. Organomet. Chem.*, 1976, **14**, 285 (*rev*)
Heaton, B.T. *et al*, *J. Organomet. Chem.*, 1980, **194**, C29 (*nmr*)

$C_{31}H_{20}ClO_2Rh$ Rh-00285

Dicarbonyl[(1,2,3,4,5-η)-1-chloro-2,3,4,5-tetraphenyl-2,4-cyclopentadien-1-yl]rhodium, 9CI

[52394-68-0]

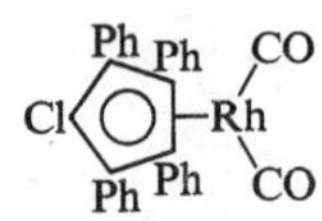

M 562.856

Red cryst. (hexane). Mp 78-81°. ir ν_{CO} 2048, 1987 cm^{-1} (cyclohexane).

Day, V.M. *et al*, *J. Am. Chem. Soc.*, 1974, **96**, 1227 (*synth*)
Reimer, K.J. *et al*, *J. Organomet. Chem.*, 1975, **93**, 239 (*synth*, *ir*, *pmr*)
Inorg. Synth., 1980, **20**, 188 (*synth*)

$C_{31}H_{28}BRh$ Rh-00286

[(2,3,5,6-η)-Bicyclo[2.2.1]hepta-2,5-diene][(η^6-phenyl)triphenylborato(1−)]rhodium, 10CI

(2,5-Norbornadiene)[tetraphenylborato(1−)]rhodium, 8CI

[31884-67-0]

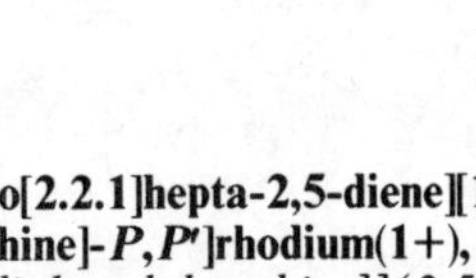

M 514.278

Pale-yellow cryst. (CH_2Cl_2/EtOH).

Schrock, R.R. *et al*, *Inorg. Chem.*, 1970, **9**, 2339 (*synth*, *ir*, *pmr*)
Oro, L.A. *et al*, *J. Organomet. Chem.*, 1978, **148**, 81 (*synth*, *ir*)

$C_{31}H_{28}P_2Rh^{\oplus}$ Rh-00287

[(2,3,5,6-η)Bicyclo[2.2.1]hepta-2,5-diene][1,2-ethanediylbis-[diphenylphosphine]-*P*,*P'*]rhodium(1+), 10CI

[Ethylenebis[diphenylphosphine]](2,5-norbornadiene)-rhodium(1+), 8CI. *[Bis(1,2-diphenylphosphino)ethane]-(norbornadiene)rhodium(1+)*

[47768-38-7]

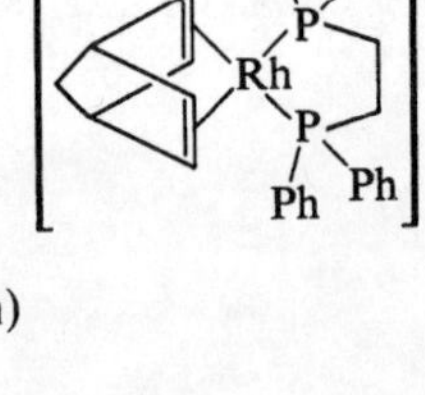

M 565.415 (ion)

Catalyst for selective hydrogenation of alkynes to *cis*-olefins and of dienes to monoenes and for hydrogen transfer from alcohols to ketones.

Perchlorate: [32799-34-1].
$C_{31}H_{28}ClO_4P_2Rh$ M 664.866
Orange cryst. (CH_2Cl_2/Et_2O). Wrongly named in 10CI.

Hexafluorophosphate: [60470-22-6].
$C_{31}H_{28}F_6P_3Rh$ M 710.379
Orange cryst. (CH_2Cl_2/Et_2O).

Tetrafluoroborate: [60430-43-5].
$C_{31}H_{28}BF_4P_2Rh$ M 652.219
Orange cryst. (CH_2Cl_2/Et_2O).

Osborn, J.A. *et al*, *J. Am. Chem. Soc.*, 1971, **93**, 2397 (*synth, pmr*)
Schrock, R.R. *et al*, *J. Am. Chem. Soc.*, 1976, **98**, 2143, 4450 (*use*)
Slack, D.A. *et al*, *Inorg. Chem.*, 1979, **18**, 3125 (*nmr*)
Spogliarich, R. *et al*, *J. Organomet. Chem.*, 1979, **179**, C45 (*use*)
Brown, J.M. *et al*, *J. Organomet. Chem.*, 1981, **216**, 263 (*synth, nmr, use*)

$C_{31}H_{41}P_3Rh^{\oplus}$ Rh-00288

[(2,3,5,6-η)Bicyclo[2.2.1]hepta-2,5-diene]tris(dimethylphenylphosphine)rhodium(1+), 10CI

Tris(dimethylphenylphosphine)(2,5-norbornadiene)rhodium(1+), 8CI. *Tris(dimethylphenylphosphine)-norbornadienerhodium(1+)*

[47752-21-6]

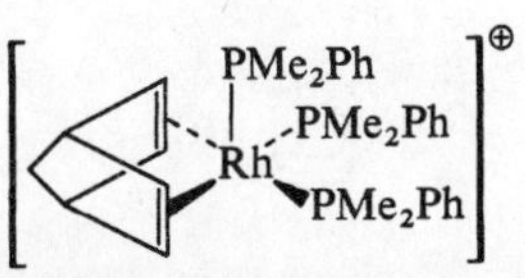

M 609.492 (ion)

Catalyst for selective hydrogenation of alkynes to *cis*-olefins and for aminomethylation of olefins.

Hexafluorophosphate: [32761-50-5].
$C_{31}H_{41}F_6P_4Rh$ M 754.456
Yellow cryst. (CH_2Cl_2/Et_2O).

Perchlorate: [32878-39-0].
$C_{31}H_{41}ClO_4P_3Rh$ M 708.942
Yellow cryst. (CH_2Cl_2/Et_2O). Slowly dec. over several weeks in air.

Osborn, J.A. *et al*, *J. Am. Chem. Soc.*, 1971, **93**, 2397 (*synth, nmr*)
Schrock, R.R. *et al*, *J. Am. Chem. Soc.*, 1976, **98**, 2134 (*use*)
Denise, B. *et al*, *J. Organomet. Chem.*, 1978, **148**, 155.
Jachimowicz, F. *et al*, *J. Org. Chem.*, 1982, **47**, 445 (*use*)

$C_{32}H_{28}N_2Rh^{\oplus}$ Rh-00289

[(1,2,5,6-η)-1,5-Cyclooctadiene](4,7-diphenyl-1,10-phenanthroline-N^1,N^{10})rhodium(1+), 10CI

[67675-70-1]

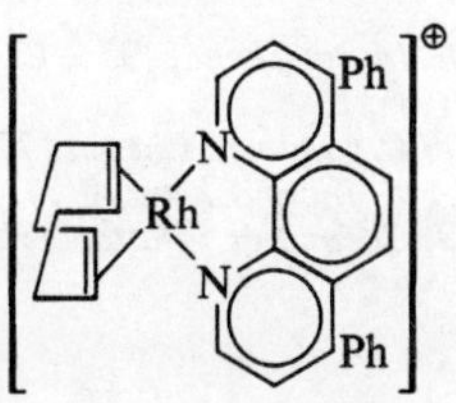

M 543.492 (ion)

Catalyst precursor for the hydroformylation of alkenes.

Perchlorate: [80263-16-7].
$C_{32}H_{28}ClN_2O_4Rh$ M 642.943
Solid (CH_2Cl_2/Et_2O). Sol. Me_2CO.

Claver, C. *et al*, *Transition. Met. Chem.*, 1982, **7**, 246 (*synth, use*)
Fordyce, W.A. *et al*, *Inorg. Chem.*, 1982, **21**, 1023, 1027 (*uv, props*)

$C_{32}H_{32}BRh$ Rh-00290

[(1,2,5,6-η)-1,5-Cyclooctadiene][(η⁶-phenyl)triphenylborato(1−)]rhodium, 9CI

(1,5-Cyclooctadiene)[(tetraphenylborato(1−)]rhodium, 8CI

[31974-01-3]

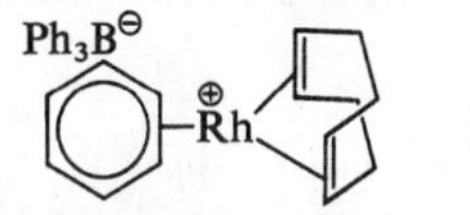

M 530.320

Pale-yellow cryst. Spar. sol. org. solvs.

Schrock, R.R. *et al*, *Inorg. Chem.*, 1970, **9**, 2339 (*synth, ir*)

$C_{32}H_{38}PRh$ Rh-00291

1,4-Butanediyl[(1,2,3,4,5-η)-1,2,3,4,5-pentamethyl-2,4-cyclopentadien-1-yl](triphenylphosphine)rhodium, 10CI

(Pentamethylcyclopentadienyl)(triphenylphosphine)-(1,4-butanediyl)rhodium

[63162-06-1]

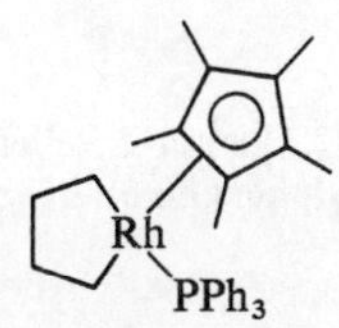

M 556.531

Yellow needles (pentane). Mp 115° dec.

Diversi, P. *et al*, *J. Organomet. Chem.*, 1979, **165**, 253 (*synth, ms*)
Diversi, P. *et al*, *Inorg. Chem.*, 1980, **19**, 3590 (*synth, pmr, cryst struct*)

$C_{32}H_{56}Cl_2Rh_2$ Rh-00292

Di-μ-chlorotetrakis(1,2-η)cyclooctenedirhodium, 9CI

M 717.511

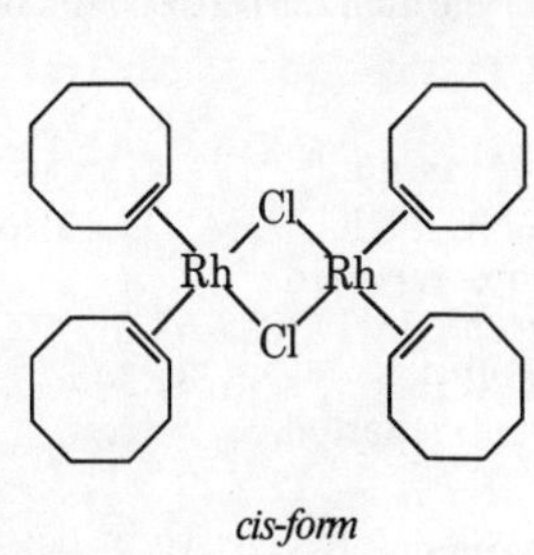

cis-form

***cis*-cyclooctene-form** [12279-09-3]
Starting material for synth. of Rh(I) complexes. Reddish-brown cryst. solid (toluene/MeOH). Dec. >150°.

***trans*-cyclooctene-form** [63902-12-5]
Orange cryst. (toluene/MeOH).

Monomer: [*μ-Chlorobis*(*1,2-η*)*cyclooctene*]*rhodium*, *9CI*. *Dicyclooctenerhodium chloride. Bis*(*cyclooctene*)-*rhodium chloride.*

$C_{16}H_{28}ClRh$ M 358.756

Unknown.

Winkhaus, G. *et al*, *Chem. Ber.*, 1966, **99**, 3602 (*ir*)
Bennett, M.A. *et al*, *Inorg. Chem.*, 1967, **6**, 1647 (*ir*)
James, B.R. *et al*, *J. Chem. Soc., Chem. Commun.*, 1970, 908.
Inorg. Synth., 1973, **14**, 92 (*synth*)
Epstein, R.A. *et al*, *Inorg. Chem.*, 1979, **18**, 478 (*uv, mcd*)
Saito, K., *ACS Symp. Ser.*, 1980 No. 119 (*trans form*)

$C_{33}H_{25}Rh$ Rh-00293

η⁵-Cyclopentadienyl-η⁴-tetraphenylcyclobutadienerhodium

[*1,1′,1″,1‴-*[*η⁴-1,3-Cyclobutadiene-1,2,3,4-tetrayl*)*tetrakis*(*benzene*)]](*η⁵-2,4-cyclopentadien-1-yl*)*rhodium*, *9CI*

[41572-80-9]

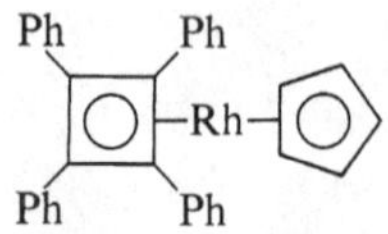

M 524.466

Yellow cryst. (C_6H_6/pet. ether). Mp 230-1° (227-8°). Dec. when exposed to compds. other than hydrocarbons. Rapidly degraded by CCl_4.

Cash, G.G. *et al*, *J. Organomet. Chem.*, 1973, **50**, 277 (*struct, synth, pmr, ms, ir*)
Kooti, M. *et al*, *Inorg. Nucl. Chem. Lett.*, 1973, **9**, 1031 (*synth*)
Nixon, J.F. *et al*, *J. Organomet. Chem.*, 1976, **104**, 231 (*synth, ir, pmr, ms*)

$C_{33}H_{36}FeNPRh^{\oplus}$ Rh-00294

[(2,3,5,6-η)-Bicyclo[2.2.1]hepta-2,5-diene][1-[1-(dimethylamino)ethyl]-2-diphenylphosphino)ferrocene-*N,P*]rhodium(1+), 10CI

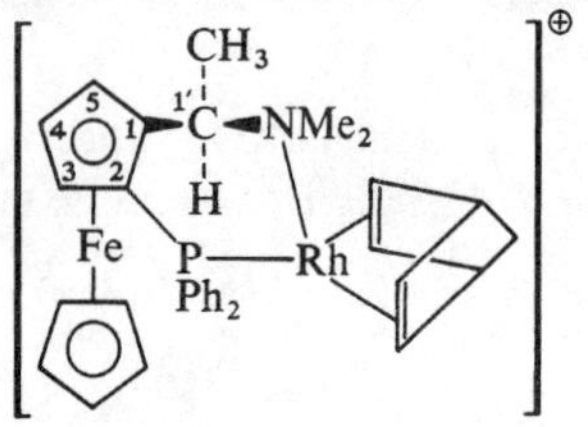

M 636.380 (ion)

Catalyst precursor for the asymmetric hydrogenation of acetamidoacrylic acid derivatives, itaconic acid and styrene.

(1R,1′S)-form

Perchlorate: [73135-89-4].

$C_{33}H_{36}ClFeNO_4PRh$ M 735.831

No phys. props. reported.

Hexafluorophosphate: [73089-67-5].

$C_{33}H_{36}F_6FeNP_2Rh$ M 781.345

No phys. props. reported.

(1S,1′R)-form

Hexafluorophosphate: [65107-19-9]. No phys. props. reported.

(1RS,1′RS)-form

Perchlorate: Orange-brown solid. Mp 190° dec.

Hexafluorophosphate: Red solid (CH_2Cl_2/EtOH/Et_2O). Sol. $CHCl_3$. Mp 192° dec.

Cullen, W.R. *et al*, *J. Am. Chem. Soc.*, 1980, **102**, 988 (*synth, cryst struct, use, pmr*)
Cullen, W.R. *et al*, *Can. J. Chem.*, 1982, **60**, 1793 (*use*)

$C_{33}H_{39}N_4Rh_2$ Rh-00295

Methyl[2,7,12,17-tetraethyl-3,8,13,18-tetramethyl-21*H*,23*H*-porphinato(2−)-*N*²¹,*N*²²,*N*²³,*N*²⁴]rhodium, 10CI

Methyl(*etioporphyrinato*)*rhodium*

[59539-66-1]

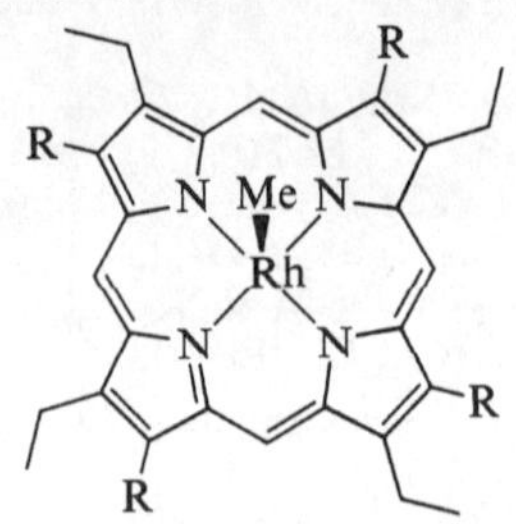

R = CH_3

M 697.509

Formed by reaction of MeI with (*μ*-Etioporphyrinato)-tetracarbonyldirhodium, Rh-00312 . Red plates ($CHCl_3$). Mp >300°.

Abeysekera, A.M. *et al*, *J. Chem. Soc., Perkin Trans. 1*, 1977, 36 (*synth, pmr*)

$C_{33}H_{40}PRh$ Rh-00296

[(1,2,3,4,5-η)-1,2,3,4,5-Pentamethyl-2,4-cyclopentadien-1-yl]-1,5-pentanediyl(triphenylphosphine)rhodium, 10CI

[63162-03-8]

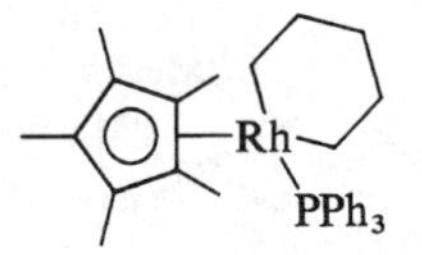

M 570.558

Orange cryst. (Me_2CO). Sol. pentane. Mp 109° dec.

Diversi, P. *et al*, *Inorg. Chem.*, 1980, **19**, 3590 (*synth, ms, pmr*)
Cuccuru, A. *et al*, *J. Organomet. Chem.*, 1981, **204**, 123.

$C_{34}H_{25}ORh$ Rh-00297

(η⁵-2,4-Cyclopentadien-1-yl)[(2,3,4,5-η)-2,3,4,5-tetraphenyl-2,4-cyclopentadien-1-one]rhodium, 9CI

π-Tetraphenylcyclopentadienone-π-cyclopentadienylrhodium. (*Cyclopentadienyl*)(*tetracyclone*)*rhodium*

[31851-08-8]

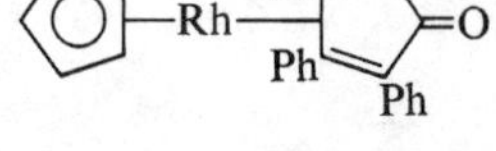

M 552.476

Yellow cryst. ($CHCl_3$/hexane). Mp 334-6° (331-2°). pK_a 2.54 (50% dioxane aq.).

Dickson, R.S. *et al*, *Aust. J. Chem.*, 1970, **23**, 1531 (*synth, ir, ms, pmr, uv*)
Gardner, S.A. *et al*, *Inorg. Chem.*, 1973, **12**, 2396 (*synth, ir, pmr*)
Sheats, J.E. *et al*, *J. Organomet. Chem.*, 1975, **96**, 115.
Van Willigen, H. *et al*, *Inorg. Chem.*, 1977, **16**, 581.
Baker, P.K. *et al*, *J. Chem. Soc., Dalton Trans.*, 1980, 1710.

$C_{34}H_{34}OP_2Rh^{\oplus}$ **Rh-00298**

[(2,3,5,6-η)Bicyclo[2.2.1]hepta-2,5-diene][2,3-bis(diphenylphosphino)-1-propanol-*P*,*P′*]rhodium(1+), 10CI

[1,2-Bis(diphenylphosphino)-3-propanol](norbornadiene)rhodium(1+)

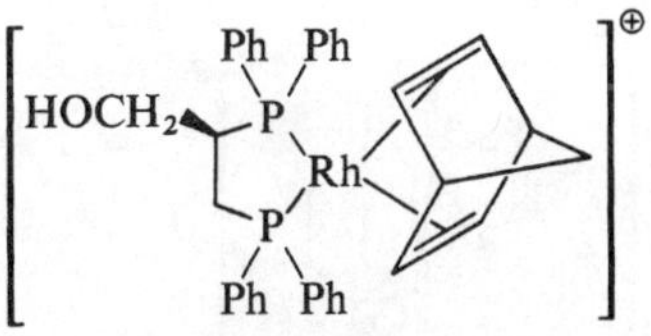

(*R*)-*form*

M 623.495 (ion)

Catalyst for the asymmetric hydrogenation of dehydroamino acid derivs.

(*R*)-*form*

Tetrafluoroborate: [80145-70-6].

$C_{34}H_{34}BF_4OP_2Rh$ M 710.299

Yellow solid.

Padmakumari Amma, J. *et al*, *J. Org. Chem.*, 1982, **47**, 468 (*synth*, *use*)

$C_{34}H_{34}P_2Rh^{\oplus}$ **Rh-00299**

[(2,3,5,6-η)Bicyclo[2.2.1]hepta-2,5-diene][(1-methyl-1,2-ethanediyl)bis[diphenylphosphine]-*P*,*P′*]rhodium(1+)

[1,2-Bis(diphenylphosphino)propane]norbornadienerhodium(1+). Norbornadiene[PROPHOS]-rhodium(1+)

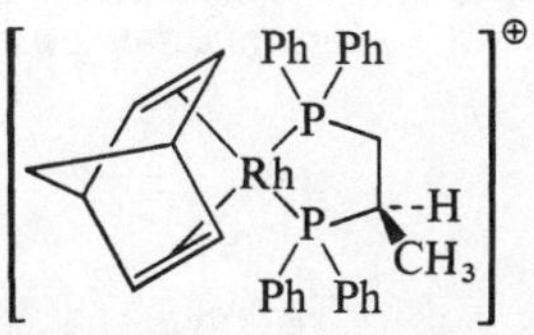

M 607.496 (ion)

(*R*)-*form* [67881-58-7]

Catalytic precursor for the asymmetric hydrogenation of *N*-acylaminoacrylic acids. Complex can be used to asymmetrically hydrogenate $H_2C{=}C(OAc)COOEt$ to give the chiral starting material for the (*R*)-PROPHOS ligand.

Perchlorate: [67881-59-8].

$C_{34}H_{34}ClO_4P_2Rh$ M 706.946

Orange-red blocks + $0.5CH_2Cl_2$ (CH_2Cl_2/THF/hexane). Sol. CH_2Cl_2, spar. sol. THF.

Fryzuk, M.D. *et al*, *J. Am. Chem. Soc.*, 1978, **100**, 5491 (*synth*, *use*)
Slack, D.A. *et al*, *Inorg. Chem.*, 1979, **18**, 3125 (*nmr*, *props*)

$C_{34}H_{36}P_2Rh^{\oplus}$ **Rh-00300**

[(1,2,5,6-η)-1,5-Cyclooctadiene][1,2-ethanediylbis[diphenylphosphine]-*P*,*P′*]rhodium(1+), 9CI

(1,5-Cyclooctadiene)[ethylenebis(diphenylphosphine)]-rhodium(1+). [1,2-Bis(diphenylphosphino)ethane]-1,5-cyclooctadienerhodium(1+)

[47778-85-8]

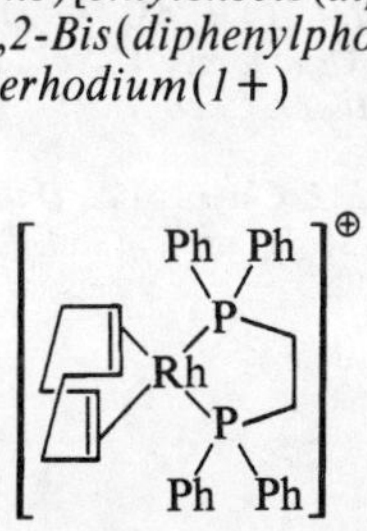

M 609.511 (ion)

Hydrogen transfer catalyst.

Perchlorate: [32799-70-5].

$C_{34}H_{36}ClO_4P_2Rh$ M 708.962

Cryst.

Tetrafluoroborate: [34664-29-4].

$C_{34}H_{36}BF_4P_2Rh$ M 696.315

Orange-yellow cryst. Mp 24°.

Hexafluoroantimonate: [72138-24-0].

$C_{34}H_{36}F_6P_2RhSb$ M 845.252

Cryst.

Green, M. *et al*, *J. Chem. Soc. (A)*, 1971, 2334 (*synth*, *pmr*)
Schrock, R.R. *et al*, *J. Am. Chem. Soc.*, 1971, **93**, 2397 (*synth*, *nmr*)
Burckett-St. Laurent, J.C.T.R. *et al*, *Inorg. Chem.*, 1980, **19**, 577 (*synth*, *use*)
Maciel, G.E. *et al*, *Adv. Chem. Ser.*, 1982, **196**, 389 (*nmr*)

$C_{34}H_{40}FeNPRh^{\oplus}$ **Rh-00301**

[(1,2,5,6-η)-1,5-Cyclooctadiene][1-[1-(dimethylamino)ethyl]-2-(diphenylphosphino)ferrocene-*N*,*P*]rhodium(1+), 10CI

(1,5-Cyclooctadiene)[(2-diphenylphosphinoferrocenyl)-ethyldimethylamine]rhodium(1+). Cyclooctadiene(PPFA)rhodium(1+)

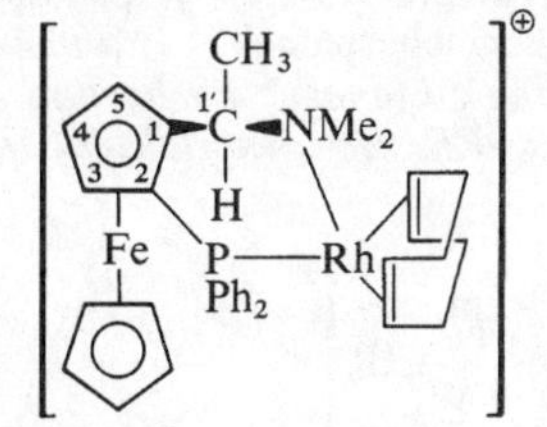

(1*S*,1′*R*)-*form*

M 652.423 (ion)

Catalyst precursor for the asymmetric hydrogenation of acylaminocinnamic acid and acylaminoacrylic acids. Opt. active forms known.

Perchlorate:

$C_{34}H_{40}ClFeNO_4PRh$ M 751.874

Orange-red cryst. (EtOH). Mp 185° dec. Conductivity Λ 73.6 Ω^{-1} cm^{-1} M^{-1}.

Tetrafluoroborate:

$C_{34}H_{40}BF_4FeNPRh$ M 739.227

Orange-red solid (EtOH). Mp 190° dec. Conductivity Λ 73.9 Ω^{-1} cm^{-1} M^{-1}.

Tetraphenylborate:

$C_{58}H_{60}BFeNPRh$ M 971.655

Orange-yellow solid. Mp 150-2° dec. Conductivity Λ 50.13 Ω^{-1} cm^{-1} M^{-1}.

Cullen, W.R. *et al*, *J. Am. Chem. Soc.*, 1980, **102**, 988 (*synth*, *use*)

$C_{34}H_{40}ORh_2$ **Rh-00302**

μ-Carbonyl[μ-(diphenylmethylene)]bis[(1,2,3,4,5-η)-1,2,3,4,5-pentamethyl-2,4-cyclopentadien-1-yl]dirhodium, 10CI

[76545-06-7]

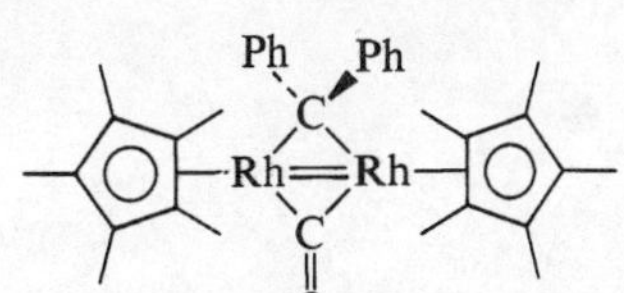

M 670.500

Dark-green, air-stable, rhombic cryst. (CH_2Cl_2/Et_2O). V. sol. C_6H_6, CH_2Cl_2, THF. Mp >250°. ir ν_{co} 1751 cm^{-1} (cyclohexane). Solns. air-stable at r.t.

Clauss, A.D. *et al*, *J. Organomet. Chem.*, 1980, **201**, C31 (*synth*, *ir*, *pmr*)

Herrmann, W.A. *et al*, *Chem. Ber.*, 1982, **115**, 878 (*synth, props, ir, pmr, ms*)

$C_{34}H_{42}PRh$ Rh-00303

1,6-Hexanediyl[(1,2,3,4,5-η)-1,2,3,4,5-pentamethyl-2,4-cyclopentadien-1-yl](triphenylphosphine)rhodium, 10CI

[63162-04-9]

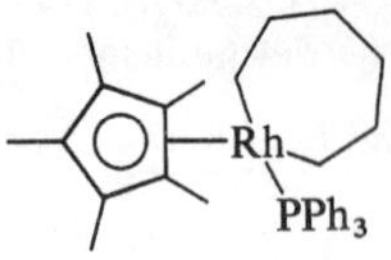

M 584.585

Orange cryst. (Me_2CO). Sol. pentane. Mp 107.2°.

Diversi, P. *et al*, *Inorg. Chem.*, 1980, **19**, 3590 (*synth, ms, pmr*)
Cuccuru, A. *et al*, *J. Organomet. Chem.*, 1981, **204**, 123.

$C_{35}H_{36}P_2Rh^{\oplus}$ Rh-00304

[(2,3,5,6-η)-Bicyclo[2.2.1]hepta-2,5-diene][(1,2-dimethyl-1,2-ethanediyl)bis[diphenylphosphine]-*P,P'*]rhodium(1+)

Bis(diphenylphosphino)butane(η⁴-norbornadiene)-rhodium(1+). (CHIRAPHOS)(η⁴-norbornadiene)-rhodium(1+)

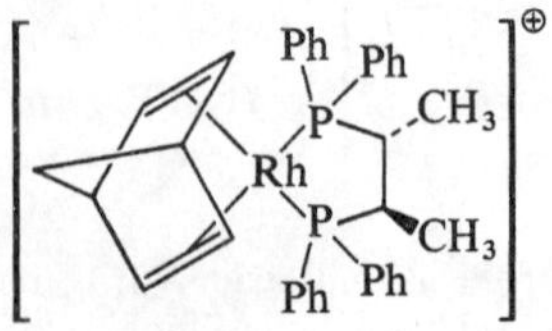

(2*R*,3*R*)-*form*

M 621.522 (ion)

Wrongly named in 10CI.

(2*R*,3*R*)-*form*

Catalyst for asymmetric hydrogenations, e.g. of α-*N*-acylaminoacrylic acids.

Perchlorate: [74892-62-9].
$C_{35}H_{36}ClO_4P_2Rh$ M 720.973
Orange-red cryst. (CH_2Cl_2/hexane).

(2*S*,3*S*)-*form* [65012-73-9]

Catalyst for asymmetric hydrogenations.

Perchlorate: [65012-74-0]. Orange-red cryst. Loses catalytic activity if stored in air.

Tetrafluoroborate: [79790-89-9].
$C_{35}H_{36}BF_4P_2Rh$ M 708.326
Orange-red cryst. (CH_2Cl_2/Et_2O).

Fryzuk, M.D. *et al*, *J. Am. Chem. Soc.*, 1977, **99**, 6262 (*pmr, use, synth*)
Slack, D.A. *et al*, *Inorg. Chem.*, 1979, **18**, 3125 (*nmr, use*)
Brown, J.M. *et al*, *J. Am. Chem. Soc.*, 1980, **102**, 3040 (*synth, use, nmr*)
Koettner, J. *et al*, *Chem. Ber.*, 1980, **113**, 2323 (*synth, use*)
Brown, J.M. *et al*, *J. Organomet. Chem.*, 1981, **216**, 263 (*synth, nmr, use*)

$C_{35}H_{40}N_2P_2Rh^{\oplus}$ Rh-00305

[(1,2,5,6-η)-1,5-Cyclooctadiene][*N,N'*-(1-methyl-1,2-ethanediyl)bis[*P,P*-diphenylphosphinous amide]-*P,P'*]rhodium(1+), 10CI

(1,5-Cyclooctadiene)[[1,2-bis(diphenylphosphino)amino]propane]rhodium(1+)

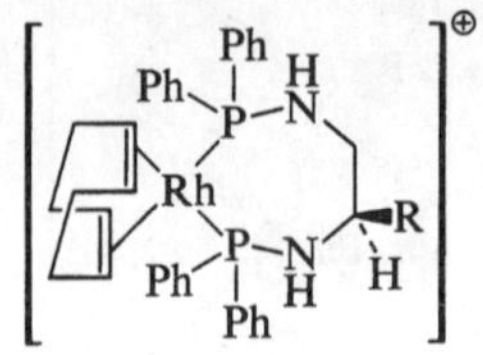

R = CH_3

M 653.567 (ion)

(*R*)-*form*

Asymmetric hydrogenation catalyst for prochiral olefins.

Hexafluorophosphate: [75730-44-8].
$C_{35}H_{40}F_6N_2P_3Rh$ M 798.532
Orange blocks + 1$CHCl_3$ ($CHCl_3/Et_2O$). $[\alpha]_D^{25}$ −0.5° (c, 0.20 in $CHCl_3$).

Kashiwabara, K. *et al*, *Bull. Chem. Soc. Jpn.*, 1980, **53**, 2275 (*synth, pmr, cd, use*)

$C_{36}H_{30}ClO_2P_2Rh$ Rh-00306

Chloro(dioxygen)bis(triphenylphosphine)rhodium, 10CI

Chloroperoxybis(triphenylphosphine)rhodium

[59561-97-6]

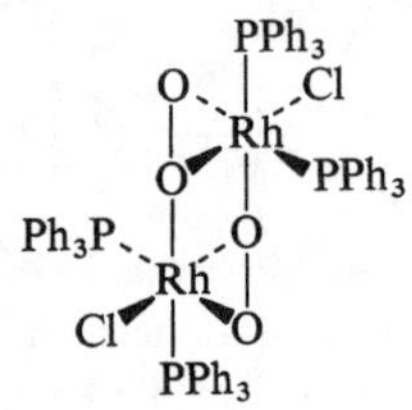

Dimer

M 694.938

Dimeric.

Dimer: [33461-21-1]. *Dichlorobis[μ-(peroxy-O:O')]tetrakis(triphenylphosphine)dirhodium. Dichlorodi-μ-peroxytetrakis(triphenylphosphine)dirhodium.*
$C_{72}H_{60}Cl_2O_4P_4Rh_2$ M 1389.876
Catalyst for alkene oxidn. and ring-opening of bicycloalkenes. Brown cryst. (CH_2Cl_2). Diamagnetic. Slowly oxid. in air to phosphine oxide-dioxygen species.

CH_2Cl_2 complex: [62623-64-7].
$C_{74}H_{64}Cl_6O_4P_4Rh_2$ M 1559.741
Red-brown cryst. (CH_2Cl_2). Mp 133-4° dec.

Bennett, M.J. *et al*, *J. Am. Chem. Soc.*, 1971, **93**, 3307 (*synth, ir, struct*)
Muto, S. *et al*, *J. Catal.*, 1976, **41**, 148 (*use*)
Beach, D.L. *et al*, *J. Organomet. Chem.*, 1977, **142**, 211 (*use*)
Bennett, M.J. *et al*, *Inorg. Chem.*, 1977, **16**, 1585 (*synth, ir, struct*)
Augustine, R.L. *et al*, *J. Chem. Soc., Dalton Trans.*, 1979, 832.
Atlay, M.T. *et al*, *J. Mol. Catal.*, 1980, **7**, 31.

$C_{36}H_{30}Cl_2NOP_2Rh$ Rh-00307

Dichloronitrosylbis(triphenylphosphine)rhodium, 10CI
[20097-11-4]

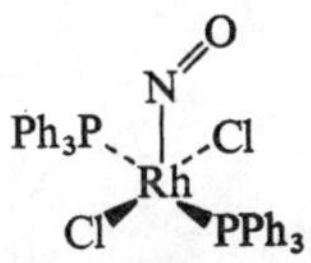

M 728.398
Brown cryst. ($CHCl_3$/EtOH). Mp 207-10°, 230° dec. ir ν_{NO} 1630 cm^{-1} (nujol).

Crooks, G.R. *et al*, *J. Chem. Soc. (A)*, 1970, 1662 (*synth, ir*)
Robinson, S.D. *et al*, *J. Chem. Soc. (A)*, 1970, 2947; 1971, 1254 (*synth, ir*)
Caulton, K.G., *Inorg. Chem.*, 1974, **13**, 1774 (*nmr*)
Goldberg, S.Z. *et al*, *Inorg. Chem.*, 1975, **14**, 1650 (*struct, synth, nmr*)

$C_{36}H_{30}I_2NOP_2Rh$ Rh-00308

Diiodo(nitrosyl)bis(triphenylphosphine)rhodium, 9CI
[40241-25-6]

$(Ph_3P)_2RhI_2(NO)$

M 911.301
Dark-red microcryst. or chocolate-brown cryst. (CH_2Cl_2/hexane). ir ν_{NO} 1626 cm^{-1} (nujol).

Dolcetti, G. *et al*, *Inorg. Chim. Acta*, 1972, **6**, 531 (*synth*)
Ghedini, M. *et al*, *Inorg. Chem.*, 1976, **15**, 2385 (*ir*)
Connelly, N.G. *et al*, *J. Chem. Soc., Dalton Trans.*, 1977, 70 (*synth, ir*)

$C_{36}H_{30}N_2O_2P_2Rh^{\oplus}$ Rh-00309

Dinitrosylbis(triphenylphosphine)rhodium(1+), 9CI
[47841-81-6]

$(Ph_3P)_2Rh(NO)_2^{\oplus}$

M 687.498 (ion)
Catalyses conversion of mixts. of NO and CO to N_2O and CO_2.
Hexafluorophosphate: [36462-94-9].
$C_{36}H_{30}F_6N_2O_2P_3Rh$ M 832.462
Black cryst. (CH_2Cl_2/MeOH). ir ν_{NO} 1840, 1785 cm^{-1} (nujol).
Perchlorate: [57650-15-4].
$C_{36}H_{30}ClN_2O_6P_2Rh$ M 786.949
Black cryst. (CH_2Cl_2/EtOH).
Tetrafluoroborate: [41749-95-5].
$C_{36}H_{30}BF_4N_2O_2P_2Rh$ M 774.302
Black cryst. Mp 156-8°.

Dolcetti, G. *et al*, *Inorg. Chim. Acta*, 1972, **6**, 531 (*synth, ir*)
Finn, P. *et al*, *Inorg. Chem.*, 1972, **11**, 893 (*ir, pe*)
Beck, W. *et al*, *Chem. Ber.*, 1973, **106**, 868 (*synth, ir*)
Hudak, J.A. *et al*, *Inorg. Chem.*, 1975, **14**, 3070 (*synth, ir, cryst struct*)
Bhaduri, S. *et al*, *J. Chem. Soc., Dalton Trans.*, 1977, 2085 (*synth, ir*)
Sacco, A. *et al*, *Inorg. Chim. Acta*, 1979, **32**, 171 (*synth, ir*)

$C_{36}H_{30}Rh_2$ Rh-00310

Bis(η^5-2,4-cyclopentadien-1-yl)bis[μ-(diphenylmethylene)]-dirhodium, 9CI
Bis(μ-diphenylmethylene)bis(η-cyclopentadienyl)dirhodium
[54032-12-1]

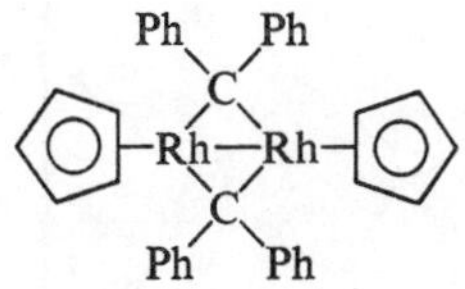

M 668.444

Yamamoto, T. *et al*, *J. Chem. Soc., Chem. Commun.*, 1974, 354 (*cmr, synth*)

$C_{36}H_{32}ClP_2Rh$ Rh-00311

Chlorodihydrobis(triphenylphosphine)rhodium, 10CI
[12119-41-4]

$RhClH_2(PPh_3)_2$

M 664.955
Hydrogenation catalyst for allenes and the selective redn. of mono- and disubstituted double bonds. White cryst.

Osborn, J.A. *et al*, *J. Chem. Soc. (A)*, 1966, 1711 (*synth, ir, pmr*)
Biellmann, J.F. *et al*, *J. Am. Chem. Soc.*, 1968, **90**, 1673 (*use*)
Rousseau, C. *et al*, *J. Mol. Catal.*, 1978, **3**, 309 (*use*)

$C_{36}H_{36}N_4O_4Rh_2$ Rh-00312

(μ-Etioporphyrinato)tetracarbonyldirhodium
Tetracarbonyl[μ-2,7,12,17-tetraethyl-3,8,13,18-tetramethyl-21H,23H-porphinato(2−)]dirhodium, 10CI. *(μ-Aetioporphyrinato)tetracarbonyldirhodium*
[59539-70-7]

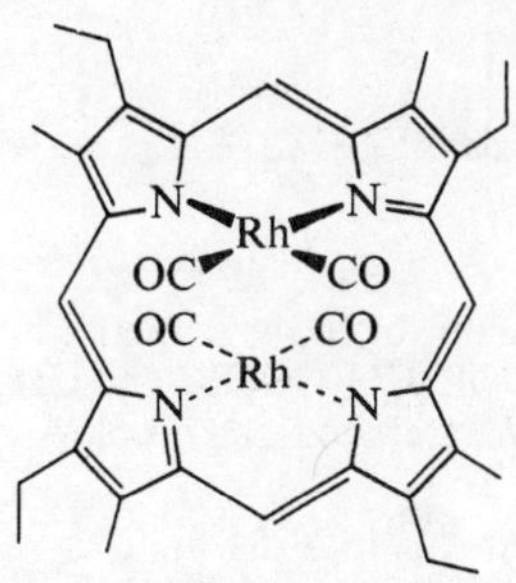

M 794.516
Dark-brown air-stable needles ($CHCl_3$/pet. ether). Sol. $CHCl_3$. Mp >300°. Undergoes a range of oxidative-addition reactions including reaction with MeI to form Methyl[2,7,12,17-tetraethyl-3,8,13,18-tetramethyl-21*H*,23*H*-porphinato(2−)-N^{21},N^{22},N^{23},N^{24}]rhodium, Rh-00295 .

Abeysekera, A.M. *et al*, *J. Chem. Soc., Perkin Trans. 1*, 1977, 36 (*synth, pmr, ir, props*)

$C_{36}H_{38}P_2Rh^{\oplus}$ **Rh-00313**

[(2,3,5,6-η)Bicyclo[2.2.1]hepta-2,5-diene][(1,3-dimethyl-1,3-propanediyl)bis[diphenylphosphine]-*P,P'*]rhodium(1+), 10CI

[2,4-Bis(diphenylphosphino)pentane]norbornadienerhodium(1+). Norbornadiene(SKEWPHOS)rhodium(1+)

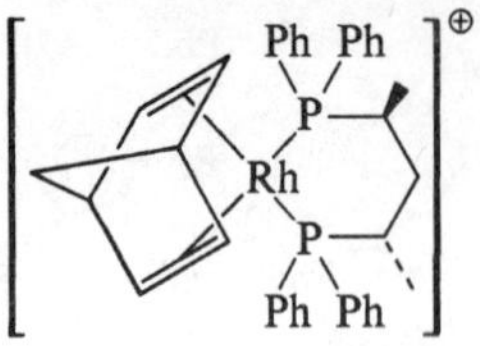

M 635.549 (ion)

(S,S)-form

Catalytic precursor for the asymmetric hydrogenation of (*N*-acylamido)acrylic acids.

Perchlorate: [77864-85-8].

$C_{36}H_{38}ClO_4P_2Rh$ M 735.000

Orange, air-sensitive prisms. Sol. MeOH, CH_2Cl_2, spar. sol. THF. Stored under N_2 at 5°.

MacNeil, P.A. *et al*, *J. Am. Chem. Soc.*, 1981, **103**, 2273 (*synth*)

$C_{36}H_{40}O_2P_2Rh^{\oplus}$ **Rh-00314**

[(1,2,5,6-η)-1,5-Cyclooctadiene][1,2-ethanediylbis[(2-methoxyphenyl)phenylphosphine]-*P,P'*]rhodium(1+), 10CI

η⁴-1,5-Cyclooctadiene(DIPAMP)rhodium(1+)

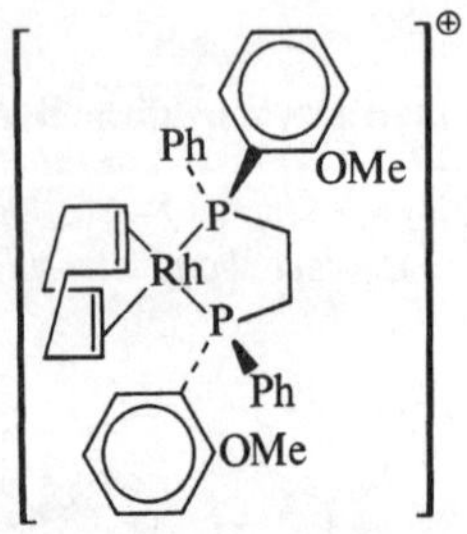

M 669.564 (ion)

(R,R)-form

Catalytic precursor for the asymmetric hydrogenation of $H_2C{=}C(COOR)CH_2COOR'$ and $CH_2{=}C(OAc)R$. Chirality *R,R* refers to free phosphine. Config. in complex is *S,S*.

Tetrafluoroborate: [71423-54-6].

$C_{36}H_{40}BF_4O_2P_2Rh$ M 756.367

Red-orange cryst. (EtOH).

Perchlorate: [81445-13-8].

$C_{36}H_{40}ClO_6P_2Rh$ M 769.014

Characterised spectroscopically, not isol. Sol. Me_2CO.

Vineyard, B.D. *et al*, *J. Am. Chem. Soc.*, 1977, **99**, 5946 (*synth, cryst struct*)
Christopfel, W.C. *et al*, *J. Am. Chem. Soc.*, 1979, **101**, 4406 (*use*)
Koenig, K.E. *et al*, *J. Org. Chem.*, 1980, **45**, 2362 (*synth, use*)
Stults, B.R. *et al*, *J. Am. Chem. Soc.*, 1981, **103**, 3235 (*struct*)
Maciel, G.E. *et al*, *Adv. Chem. Ser.*, 1982, **196**, 389 (*nmr, synth*)

$C_{36}H_{40}P_2Rh^{\oplus}$ **Rh-00315**

[1,4-Butanediylbis[diphenylphosphine]-*P,P'*][(1,2,5,6-η)-1,5-cyclooctadiene]rhodium(1+), 10CI

[75752-92-0]

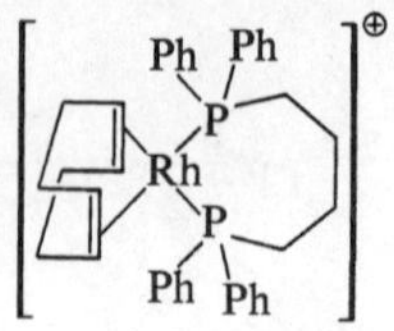

M 637.565 (ion)

Catalyst for the hydrogenation of ketones and olefins.

Tetrafluoroborate: [79255-71-3].

$C_{36}H_{40}BF_4P_2Rh$ M 724.369

Orange cryst. (CH_2Cl_2/Et_2O). Mp 215-20° dec.

Spogliarich, R. *et al*, *J. Organomet. Chem.*, 1980, **198**, 81 (*use*)
Anderson, M.P. *et al*, *Inorg. Chem.*, 1981, **20**, 4101 (*synth, nmr, struct*)
Brown, J.M. *et al*, *J. Organomet. Chem.*, 1981, **216**, 263 (*synth, use*)

$C_{36}H_{40}P_2Rh^{\oplus}$ **Rh-00316**

[(1,2,5,6-η)-1,5-Cyclooctadiene][1,2-dimethyl-1,2-ethanediylbis[diphenylphosphine]-*P,P'*]rhodium(1+), 10CI

η⁴-1,5-Cyclooctadiene[2,3-bis(diphenylphosphino)butane]rhodium(1+). CHIRAPHOS(η⁴-1,5-Cyclooctadiene)rhodium(1+)

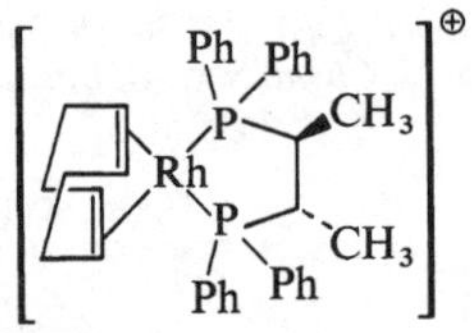

M 637.565 (ion)

(S,S)-form

Reagent for asymmetric hydrogenation of α-*N*-acylaminoacrylic acids and geminal substituted vinyl acetates.

Perchlorate: [61886-02-0].

$C_{36}H_{40}ClO_4P_2Rh$ M 737.016

Orange cryst. (THF).

Perchlorate, THF complex: [61886-03-1].

$C_{40}H_{48}ClO_5P_2Rh$ M 809.122

Orange cryst. (THF).

2,4-Pentanedione salt: [73173-81-6]. No phys. props. recorded.

Ball, R.G. *et al*, *Inorg. Chem.*, 1977, **16**, 1187 (*struct*)
Fryzuk, M.N. *et al*, *J. Am. Chem. Soc.*, 1977, **99**, 6262 (*pmr, synth, use*)
Koenig, K.E. *et al*, *J. Org. Chem.*, 1980, **45**, 2362 (*synth, use*)
Ojima, I. *et al*, *J. Org. Chem.*, 1980, **45**, 4728 (*nmr*)

$C_{36}H_{42}N_2P_2Rh^{\oplus}$ Rh-00317

[(1,2,5,6-η)-1,5-Cyclooctadiene][*N,N'*-(1,2-dimethyl-1,2-ethanediyl)bis[*P,P*-diphenylphosphinous amide]-*P,P''*]rhodium(1+), 10CI

(1,5-Cyclooctadiene)[2,3-bis(diphenylphosphinamino)-butane]rhodium(1+)

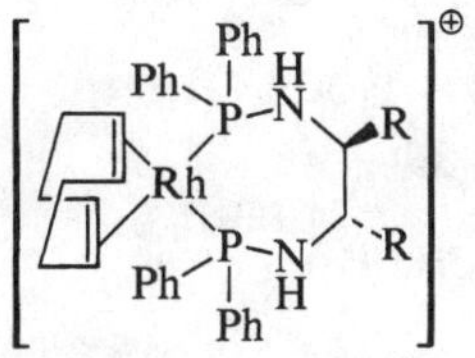

(S,S)-form

R = CH_3

M 667.594 (ion)

***(S,S)*-form** [72087-58-2]

Catalyst for the asymmetric hydrogenation of prochiral alkenes.

Perchlorate: [72087-59-3]. Cryst. + $\frac{1}{2}CH_2Cl_2$. Mp 143-4° dec.

Onuma, K. *et al*, *Bull. Chem. Soc. Jpn.*, 1980, **53**, 2012, 2016 (*synth, use*)

$C_{36}H_{44}N_4Rh^{\ominus}$ Rh-00318

[2,3,7,8,12,13,17,18-Octaethyl-21*H*,23*H*-porphinato(2−)-N^{21},N^{22},N^{23},N^{24}]rhodate(1−), 9CI

[57307-09-2]

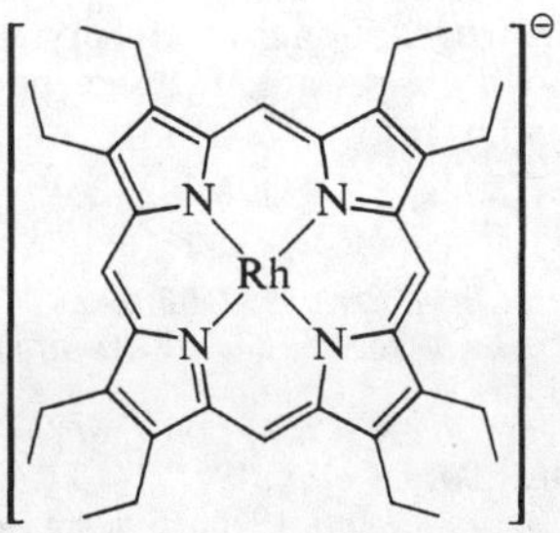

M 635.676 (ion)

Obt. and used in soln. λ_{max} 384, 420, 521 nm.

Ogoshi, H. *et al*, *J. Chem. Soc., Chem. Commun.*, 1975, 572 (*synth, props*)

$C_{36}H_{54}O_2P_2Rh^{\oplus}$ Rh-00319

Bis[cyclohexyl(2-methoxyphenyl)methylphosphine-*P*]-[(1,2,5,6-η)-1,5-cyclooctadiene]rhodium(1+)

Bis(o-anisylcyclohexylmethylphosphine)-η^4-1,5-cyclooctadienerhodium(1+). (1,5-Cyclooctadiene)bis[cyclohexyl(o-methoxyphenyl)methylphosphine]-rhodium(1+)

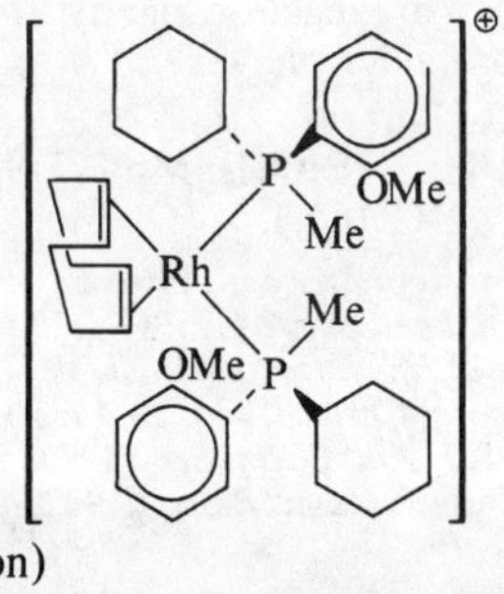

M 683.674 (ion)

***(R,R)*-form**

Catalyst for asymmetric hydrogenations and hydroformylation of prochiral alkenes.

Tetrafluoroborate: [65375-70-4].

$C_{36}H_{54}BF_4O_2P_2Rh$ M 770.478

Orange cryst. (EtOH). Mp 170-5° dec.

Tetraphenylborate: [78962-52-4].

$C_{60}H_{74}BO_2P_2Rh$ M 1002.906

No phys. props. reported.

Solodar, J., *J. Org. Chem.*, 1978, **43**, 1787 (*use*)
Koenig, K.E. *et al*, *J. Org. Chem.*, 1980, **45**, 2362 (*synth, use*)
Stults, B.R. *et al*, *J. Am. Chem. Soc.*, 1981, **103**, 3235.

$C_{36}H_{66}ClN_2P_2Rh$ Rh-00320

Chloro(dinitrogen)bis(tricyclohexylphosphine)rhodium, 9CI

[52637-52-2]

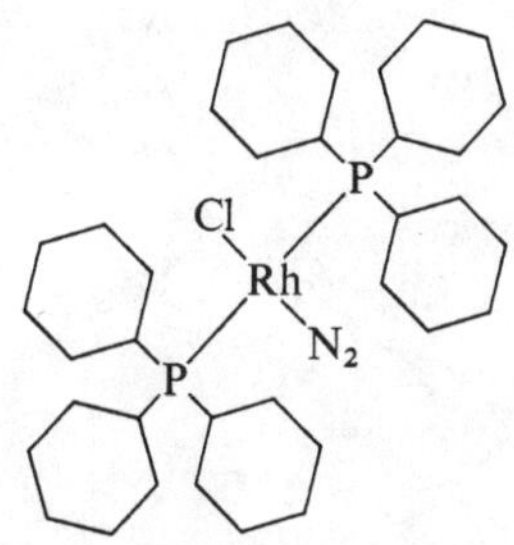

M 727.237

Pale-green solid. ir ν_{N_2} 2100 cm^{-1} (C_6H_6). N_2 displaced by O_2 or CO. Stable at r.t. under N_2.

Van Gaal, H.L.M. *et al*, *J. Organomet. Chem.*, 1977, **134**, 237 (*synth, ir, nmr*)
Dilworth, J.R. *et al*, *Inorg. Chim. Acta*, 1981, **53**, L161 (*nmr*)

$C_{36}H_{66}ClO_2P_2Rh$ Rh-00321

Chloro(dioxygen)bis(tricyclohexylphosphine)rhodium, 9CI

Chloroperoxybis(tricyclohexylphosphine)rhodium

[52637-53-3]

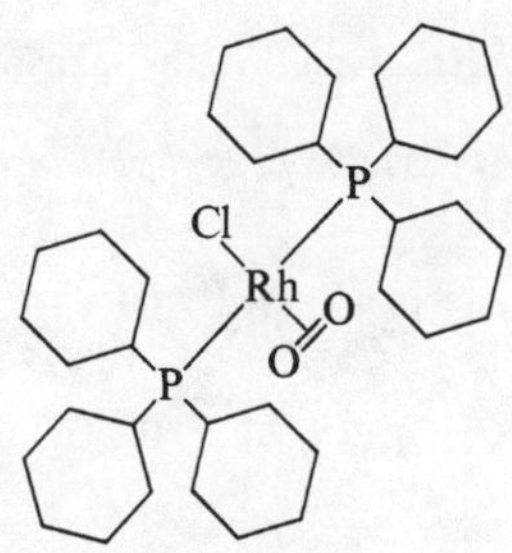

M 731.222

Blue-green solid. ir ν_{O_2} 993 cm^{-1} (nujol).

Van Gaal, H.L.M. *et al*, *J. Organomet. Chem.*, 1977, **134**, 237 (*synth, ir, nmr*)

$C_{36}H_{66}ClO_2P_2RhS$ Rh-00322

Chloro(sulfur dioxide-*S*)bis(tricyclohexylphosphine)rhodium, 10CI

[63742-63-2]

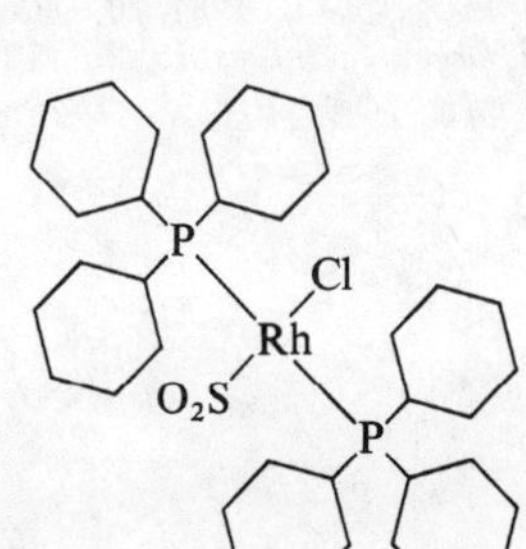

M 763.282

Brown cryst. Mp 192° dec.

Van Gaal, H.L.M. *et al*, *J. Organomet. Chem.*, 1977, **133**, 93 (*synth, ir, pmr*)
Meij, R. *et al*, *J. Organomet. Chem.*, 1978, **164**, 353 (*ir*)
Kubas, G.J. *et al*, *Inorg. Chem.*, 1979, **18**, 182 (*ir*)
Kubas, G.J. *et al*, *Inorg. Chim. Acta*, 1981, **47**, 131 (*cryst struct*)

$C_{36}H_{66}ClP_2Rh$ Rh-00323

Chlorobis(tricyclohexylphosphine)rhodium, 10CI

[63683-89-6]

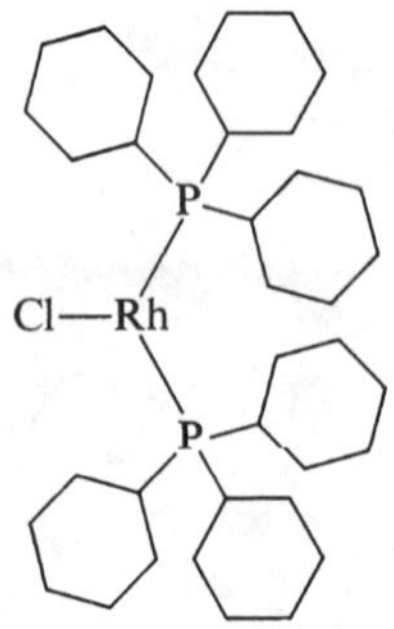

M 699.223

Partially associated in soln. Lilac solid. Sol. C_6H_6. Dec. instantly in air and slowly under an inert atm. Forms adducts with O_2, N_2, CO. Forms paramagnetic Rh(II) complexes with Cl_2 and MeCl.

Van Gaal, H.L.M. *et al*, *Inorg. Chim. Acta*, 1977, **23**, 43 (*props*)
Van Gaal, H.L.M. *et al*, *J. Organomet. Chem.*, 1977, **134**, 237 (*synth, nmr*)

$C_{36}H_{86}O_{12}P_4Rh_2$ Rh-00324

Di-μ-hydrotetrakis[tris(1-methylethyl)phosphite-*P*]dirhodium, 10CI

*Di-μ-hydrotetrakis(triisopropyl phosphite-*P*)dirhodium*

[70727-44-5]

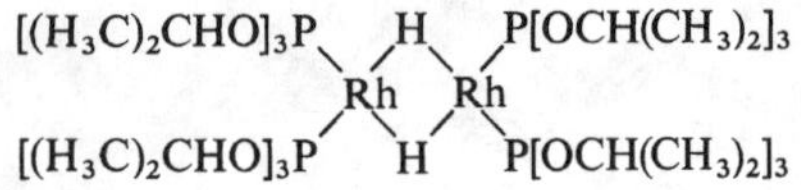

M 1040.774

Catalyst for the hydrogenation of olefins and alkynes. Green cryst. (pentane). Mp 123° dec.

Monomer: Hydridobis(triisopropyl phosphite)rhodium. $C_{18}H_{43}O_6P_2Rh$ M 520.387
Has no independent existence.

Day, V.W. *et al*, *J. Am. Chem. Soc.*, 1977, **99**, 8091 (*nmr*)
Sivak, A.J. *et al*, *J. Am. Chem. Soc.*, 1979, **101**, 4878 (*synth, nmr, use*)
Burch, R.R. *et al*, *J. Am. Chem. Soc.*, 1981, **103**, 5517.
Teller, R.G. *et al*, *Inorg. Chem.*, 1981, **20**, 1806 (*struct*)
Burch, R.R. *et al*, *Organometallics*, 1982, **1**, 188 (*synth*)

$C_{36}H_{88}O_{12}P_4Rh_2$ Rh-00325

Di-μ-hydrodihydrotetrakis[tris(1-methylethyl)phosphite-*P*-]dirhodium, 10CI

*Di-μ-hydridodihydridotetrakis(triisopropyl phosphite-*P*)dirhodium*

[78557-77-4]

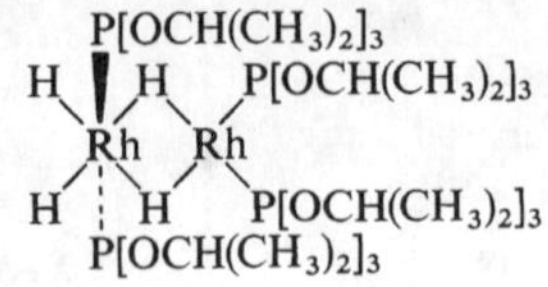

M 1042.790

Catalytic intermed. in the hydrogenation of alkenes and alkynes using Di-μ-hydrotetrakis[tris(1-methylethyl)phosphite-*P*]dirhodium Rh-00324 .

Sivak, A.J. *et al*, *J. Am. Chem. Soc.*, 1979, **101**, 4878.
Burch, R.R. *et al*, *J. Am. Chem. Soc.*, 1983, **105**, 3546.

$C_{37}H_{30}As_2ClORh$ Rh-00326

Chlorocarbonylbis(triphenylarsine)rhodium, 10CI

[14877-90-8]

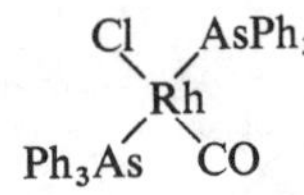

trans-form

M 778.845

Catalyst for hydrogenation, hydroformylation and hydrosilylation of unsaturated hydrocarbons.

trans-form [16970-35-7]

Yellow cryst. (C_6H_6/EtOH). Mp 242-4°. ir ν_{CO} 1963 cm^{-1} (KBr). $\mu = 0.52\ \Omega^{-1}\ cm^2\ mol.^{-1}$.

Hieber, W. *et al*, *Chem. Ber.*, 1957, **90**, 2425 (*ir*)
Vallerino, L., *J. Chem. Soc.*, 1957, 2287 (*synth*)
Inorg. Synth., 1966, **8**, 214 (*synth*)
Bennett, M.A. *et al*, *Inorg. Chem.*, 1967, **6**, 1647 (*ir*)
Rony, P.R. *et al*, *J. Mol. Catal.*, 1975, **1**, 13 (*use*)
Brady, R. *et al*, *Inorg. Chem.*, 1976, **15**, 1485 (*uv*)

$C_{37}H_{30}BrOP_2Rh$ Rh-00327

Bromocarbonylbis(triphenylphosphine)rhodium, 10CI

Carbonylbis(triphenylphosphine)rhodium bromide

[14056-79-2]

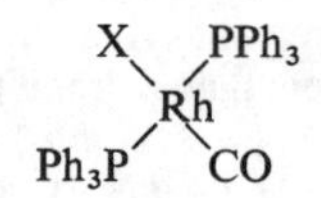

X = Br

M 735.400

Catalyst for the hydrogenation and hydroformylation of olefins and formaldehyde.

trans-form [17070-17-6]

Yellow cryst. ($CHCl_3$/EtOH). Mp 177-81° (173-5°). ir ν_{CO} 1979 cm^{-1} ($CHCl_3$).

Bennett, M.A. *et al*, *Inorg. Chem.*, 1967, **6**, 1647 (*synth, ir*)
Evans, D. *et al*, *J. Chem. Soc. (A)*, 1968, 3133 (*use*)
Strohmeier, W. *et al*, *Z. Naturforsch., B*, 1971, **26**, 61 (*use*)
Brady, R. *et al*, *Inorg. Chem.*, 1976, **15**, 1485 (*uv*)
Bartlett, P.D. *et al*, *J. Am. Chem. Soc.*, 1977, **99**, 5334 (*synth*)
Colquhoun, I.J. *et al*, *J. Magn. Reson.*, 1982, **46**, 525 (*nmr*)

$C_{37}H_{30}ClOP_2Rh$ Rh-00328

Carbonylchlorobis(triphenylphosphine)rhodium, 9CI

Bis(triphenylphosphine)rhodium carbonyl chloride. Bis(triphenylphosphine)carbonylrhodium chloride

[13938-94-8]

Rh$(PPh_3)_2$(CO)Cl

M 690.949

Catalyst for rearrangements, alkylations, carboxylations, H exchange, redox reactions and for decarbonylation of aldehydes and of acyl and aroyl halides. Hydroformylating agent. Used in a synth. of ketones from acid chlorides. Yellow cryst. (C_6H_6). Mp 195-7°. As normally prepared is predominately the *trans*-isomer.

***cis*-form** [16353-77-8]
Mp 204-5°. ν_{CO} 1978 cm^{-1}.

***trans*-form** [15318-33-9]
Mp 209-10°. ν_{CO} 1965 cm^{-1}.

Inorg. Synth., 1966, **8**, 214 (*synth*)
Blum, J. *et al*, *J. Am. Chem. Soc.*, 1967, **89**, 2338 (*use*)
Inorg. Synth., 1968, **11**, 99 (*synth*)
Winzer, A. *et al*, *Z. Chem.*, 1972, **12**, 181 (*synth*)
Rueger, C., *Z. Chem.*, 1974, **14**, 196 (*nmr*)
Svoboda, P. *et al*, *Collect. Czech. Chem. Commun.*, 1975, **40**, 1746 (*synth*)
Fieser, M. *et al*, *Reagents for Organic Synthesis*, Wiley, 1967-83, **6**, 105 (*use*)

$C_{37}H_{30}ClORhSb_2$ Rh-00329

Carbonylchlorobis(triphenylstibine)rhodium, 10CI, 9CI

[17967-63-4]

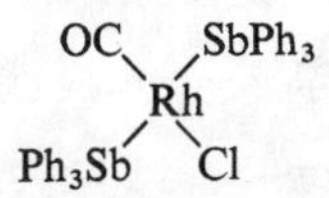

M 872.502

***trans*-form**
Red-purple prisms (CH_2Cl_2/diisopropyl ether). Sol. chlorinated solvs., C_6H_6, $PhNO_2$, sl. sol. Me_2CO, Et_2O, insol. MeOH, hexane. Dec. at 151-3°.

Vallanino, L., *J. Chem. Soc.*, 1957, 2287 (*synth*)
Brady, R. *et al*, *Inorg. Chem.*, 1976, **15**, 1485 (*uv*)

$C_{37}H_{30}ClO_5P_2Rh$ Rh-00330

Carbonyl(perchlorato-*O*)bis(triphenylphosphine)rhodium, 9CI

[36593-52-9]

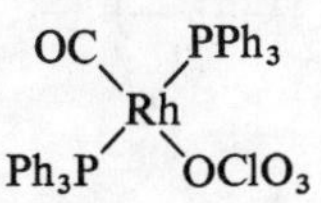

M 754.947

***trans*-form** [32354-26-0]
Yellow cryst. (C_6H_6/hexane). ir ν_{CO} 1990 cm^{-1} ($CHCl_3$).

Peone, J. *et al*, *Angew. Chem., Int. Ed. Engl.*, 1971, **10**, 511 (*synth, ir*)
Vaska, L. *et al*, *J. Chem. Soc., Chem. Commun.*, 1971, 418 (*synth, ir*)
Inorg. Synth., 1974, **15**, 68 (*synth, ir, uv*)
Brady, R. *et al*, *Inorg. Chem.*, 1976, **15**, 1485 (*uv*)

$C_{37}H_{30}ClO_7P_2Rh$ Rh-00331

Carbonylchlorobis(triphenyl phosphite-*P*)rhodium, 10CI

[25780-71-6]

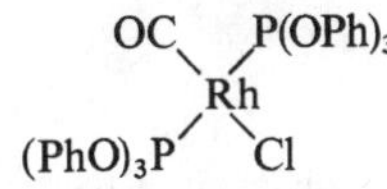

M 786.946

***trans*-form**
Yellow cryst. ($CHCl_3$/EtOH). Sol. CH_2Cl_2, $CHCl_3$, C_6H_6, $PhNO_2$, mod. sol. Me_2CO, Et_2O, insol. MeOH, hexane. Dec. at 160-70°.

Vallarino, L., *J. Chem. Soc.*, 1957, 2473 (*synth*)
Dutta, D.K. *et al*, *Transition Met. Chem.*, 1980, **5**, 244 (*synth, ir, uv*)

$C_{37}H_{30}ClP_2RhS$ Rh-00332

(Carbonothioyl)chlorobis(triphenylphosphine)rhodium, 10CI

Chloro(thiocarbonyl)bis(triphenylphosphine)rhodium, 8CI

[26500-10-7]

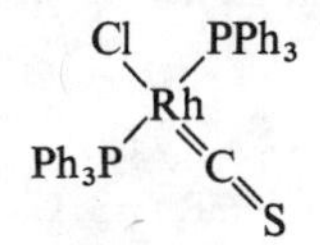

M 707.010

***trans*-form** [59349-68-7]
Orange cryst. (CH_2Cl_2/Et_2O). Mp 250-2°.

de Boer, J.L. *et al*, *J. Chem. Soc., Chem. Commun.*, 1966, 756 (*struct*)
Baird, M.C. *et al*, *J. Chem. Soc. (A)*, 1967, 2037 (*synth, ir*)
Thewissen, D.H.M.W., *J. Organomet. Chem.*, 1980, **188**, 211 (*synth, ir*)
Feltham, R.D. *et al*, *J. Am. Chem. Soc.*, 1982, **104**, 641.

$C_{37}H_{30}Cl_3OP_2Rh$ Rh-00333

Carbonyltrichlorobis(triphenylphosphine)rhodium, 10CI

[16969-79-2]

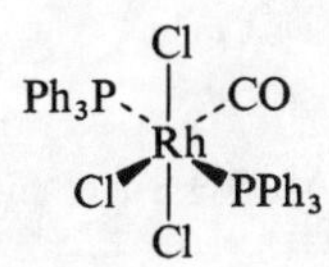

M 761.855

***trans*-form** [24322-80-3]
Yellow cryst. Dec. at 190-5°.

Vallarino, L.M., *J. Inorg. Nucl. Chem.*, 1958, **8**, 288 (*synth*)
Bennett, M.A. *et al*, *Inorg. Chem.*, 1967, **6**, 1647 (*ir*)
Slack, D.A. *et al*, *J. Organomet. Chem.*, 1978, **146**, 71 (*nmr*)
Bresler, L.S. *et al*, *J. Organomet. Chem.*, 1979, **171**, 229 (*cmr*)

$C_{37}H_{30}Cl_3P_2RhS$ Rh-00334

(Carbothioyl)trichlorobis(triphenylphosphine)rhodium, 9CI

Trichloro(thiocarbonyl)bis(triphenylphosphine)rhodium, 8CI

[15638-83-2]

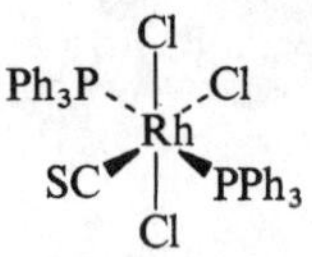

M 777.916

Orange cryst. (MeOH/C_6H_6). Mp 141°, 189-92° dec. ir ν_{CS} 1362 cm^{-1} (C_6H_6).

Baird, M.C. *et al*, *J. Chem. Soc. (A)*, 1967, 2037 (*synth, ir*)
Kubota, M. *et al*, *Inorg. Chem.*, 1974, **13**, 2277 (*synth, ir*)

$C_{37}H_{30}Cl_3P_2RhS_2Sn$ Rh-00335

Dithiocarboxylato(trichlorostannyl)bis(triphenylphosphine)-rhodium, 8CI

(π-Carbon disulfide)-trichlorostannatobis(triphenylphosphine)rhodium

[22143-03-9]

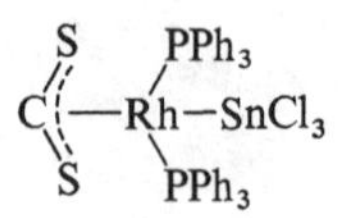

M 928.666

Green cryst. Unstable in air.

Yagupsky, M.P. *et al*, *J. Chem. Soc. (A)*, 1968, 2813 (*synth, ir*)

$C_{37}H_{30}FOP_2Rh$ Rh-00336

Carbonylfluorobis(triphenylphosphine)rhodium, 10CI

Fluorocarbonylbis(triphenylphosphine)rhodium

[58167-05-8]

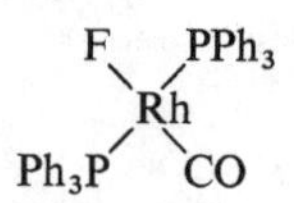

trans-form

M 674.495

Sol. C_6H_6, $CHCl_3$, MeOH, spar. sol. EtOH, Et_2O. ir ν_{CO} 1971 cm^{-1} ($CHCl_3$).

trans-form [22481-17-0]

Catalyst for the hydroformylation of HCHO. Used in prepn. of related Rh complexes. Yellow cryst.

Vaska, L. *et al*, *J. Chem. Soc., Chem. Commun.*, 1971, 418 (*synth, ir, use*)
Inorg. Synth., 1974, **15**, 64 (*synth, ir, uv*)
Brady, R. *et al*, *Inorg. Chem.*, 1976, **15**, 1485 (*uv*)
Spencer, A. *et al*, *J. Organomet. Chem.*, 1980, **194**, 113 (*use*)
Colquhoun, I.J. *et al*, *J. Magn. Reson.*, 1982, **46**, 525 (*nmr*)

$C_{37}H_{30}IOP_2Rh$ Rh-00337

Carbonyliodobis(triphenylphosphine)rhodium, 10CI

Iodocarbonylbis(triphenylphosphine)rhodium. Carbonylbis(triphenylphosphine)rhodium iodide

[15094-66-3]

As Bromocarbonylbis(triphenylphosphine)rhodium, Rh-00327 with

X = I

M 782.401

Catalyst for the hydrogenation and hydroformylation of olefins and formaldehyde. Catalyst for the decarbonylation of benzoyl bromide and phenacyl chloride.

trans-form [21006-49-5]

Yellow cryst. (C_6H_6/hexane), unstable in soln. Mp 130° dec. ir ν_{CO} 1970 cm^{-1} (nujol).

Vallarino, L., *J. Chem. Soc.*, 1957, 2287 (*synth*)
Lawson, D.N. *et al*, *J. Chem. Soc. (A)*, 1966, 1733 (*synth, ir*)
Evans, D. *et al*, *J. Chem. Soc. (A)*, 1968, 3133 (*use*)
Kingston, J.W. *et al*, *J. Inorg. Nucl. Chem.*, 1969, **31**, 2557 (*synth, ir*)
Brady, R. *et al*, *Inorg. Chem.*, 1976, **15**, 1485 (*uv*)
Colquhoun, I.J. *et al*, *J. Magn. Reson.*, 1982, **46**, 525 (*nmr*)

$C_{37}H_{30}N_3OP_2Rh$ Rh-00338

Azidocarbonylbis(triphenylphosphine)rhodium, 9CI

[32354-30-6]

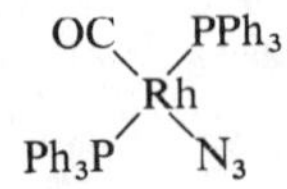

M 697.517

ir ν_{CO} 1980 cm^{-1} ($CHCl_3$).

trans-form [32345-30-6]

Yellow cryst. (CH_2Cl_2/C_6H_6). Mp 167-71° dec.

Busetto, L. *et al*, *Inorg. Chem.*, 1970, **9**, 2792 (*synth*)
Vaska, L. *et al*, *J. Chem. Soc., Chem. Commun.*, 1971, 418 (*synth, ir*)
Beck, W. *et al*, *Chem. Ber.*, 1972, **105**, 3209 (*ir, nmr*)
Beck, W. *et al*, *Chem. Ber.*, 1973, **106**, 868.
Brady, R. *et al*, *Inorg. Chem.*, 1976, **15**, 1485 (*uv*)

$C_{37}H_{30}ORh_2$ Rh-00339

μ-Carbonylbis(η^5-2,4-cyclopentadien-1-yl)bis(μ-diphenylmethylene)dirhodium, 9CI

μ-Carbonylbis(μ-diphenylmethylene)bis(η-cyclopentadienyl)dirhodium

[54183-84-5]

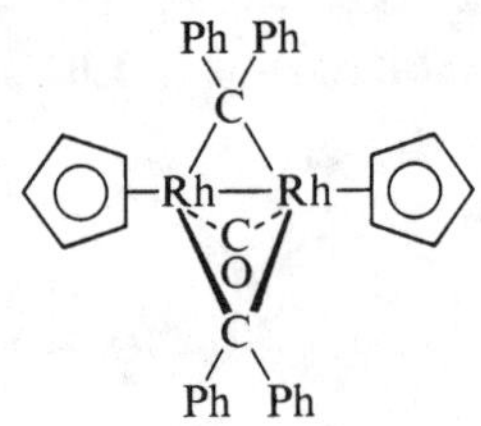

M 696.454

Green cryst. Mp 179-82° dec. ir ν_{CO} 1840 cm^{-1}. Slowly loses CO on heating or if stored in soln.

Hong, P. *et al*, *J. Chem. Soc., Chem. Commun.*, 1972, 993 (*synth, ir, ms*)
Yamamoto, T. *et al*, *J. Chem. Soc., Chem. Commun.*, 1974, 354 (*cmr*)
Ueda, H. *et al*, *Bull. Chem. Soc. Jpn.*, 1977, **50**, 2250 (*struct*)

$C_{37}H_{30}O_5P_2RhS^{\ominus}$ Rh-00340

Carbonyl[sulfato(2−)-*O*]bis(triphenylphosphine)rhodate(1−), 9CI

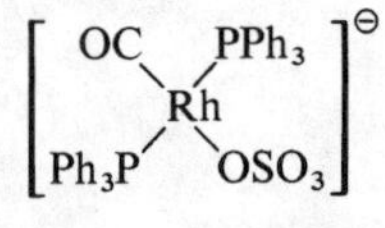

M 751.554 (ion)

Na salt: [42756-34-3].
$C_{37}H_{30}NaO_5P_2RhS$ M 774.544
No phys. props. reported.

Flynn, B.B. *et al*, *J. Am. Chem. Soc.*, 1973, **95**, 5081 (*synth*)

$C_{37}H_{31}O_2P_2Rh$ Rh-00341

Carbonylhydroxybis(triphenylphosphine)rhodium, 10CI, 9CI, 8CI

Carbonylhydroxobis(triphenylphosphine)rhodium

[24444-24-4]

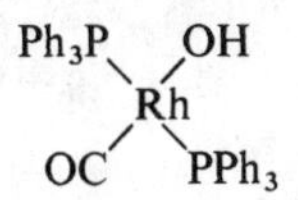

M 672.504

Trans-form [32354-36-2]

Catalyst for the hydration of nitriles to carboxamides. Yellow cryst. (C_6H_6/hexane). Mp 176-8°. ir ν_{CO} 1948cm^{-1} (nujol).

Gregorio, G. *et al*, *Inorg. Chim. Acta*, 1969, **3**, 89 (*synth, ir*)
Vaska, L. *et al*, *J. Chem. Soc., Chem. Commun.*, 1971, 418 (*synth, ir*)
Bennett, M.A. *et al*, *J. Am. Chem. Soc.*, 1973, **95**, 3030 (*use*)
Brady, R. *et al*, *Inorg. Chem.*, 1976, **15**, 1485 (*uv*)
Jewsbury, R.A., *Inorg. Chim. Acta*, 1981, **49**, 141 (*ir, nmr*)

$C_{37}H_{32}BrRh$ Rh-00342

[(1,2,3,4,5-η)-1-Bromo-2,3,4,5-tetraphenyl-2,4-cyclopentadien-1-yl][(1,2,5,6-η)-1,5-cyclooctadiene]rhodium, 9CI

[52394-65-7]

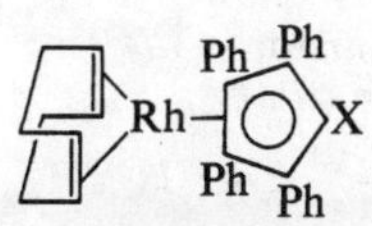

X = Br

M 659.469

Yellow cryst. (pentane). Mp 202-4° dec.

Day, V.M. *et al*, *J. Am. Chem. Soc.*, 1974, **96**, 1227 (*synth*)
Reimer, K.J. *et al*, *J. Organomet. Chem.*, 1975, **93**, 239 (*synth*)
Inorg. Synth., 1980, **20**, 188 (*synth*)

$C_{37}H_{32}ClRh$ Rh-00343

[(1,2,3,4,5-η)-1-Chloro-2,3,4,5-tetraphenyl-2,4-cyclopentadien-1-yl][(1,2,5,6-η)-1,5-cyclooctadiene]rhodium, 9CI

[52410-07-8]

As [(1,2,3,4,5-η)-1-Bromo-2,3,4,5-tetraphenyl-2,4-cyclopentadien-1-yl][(1,2,5,6-η)-1,5-cyclooctadiene]rhodium, Rh-00342 with

X = Cl

M 615.018

Yellow cryst. (pentane). Mp 208-10°.

Day, V.M. *et al*, *J. Am. Chem. Soc.*, 1974, **96**, 1227 (*synth*)
Reimer, K.J. *et al*, *J. Organomet. Chem.*, 1975, **93**, 239 (*synth, pmr*)
Inorg. Synth., 1980, **20**, 188 (*synth*)

$C_{37}H_{33}I_2P_2Rh$ Rh-00344

Diiodomethylbis(triphenylphosphine)rhodium, 10CI

[47829-28-7]

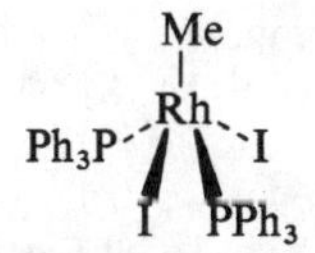

M 896.330

Initially incorrectly formulated as RhIClMe(PPh$_3$)$_2$(MeI). Methylates alkenyl-, alkynyl- and aryl-mercurials. Green cryst. + 1C_6H_6 (C_6H_6). Spar. sol. C_6H_6, $CHCl_3$, $MeNO_2$, insol. Me_2CO, pet. ether. Mp 170° dec.

Lawson, D.N. *et al*, *J. Chem. Soc. (A)*, 1966, 1733 (*synth, pmr, ir, uv*)
Troughton, P.G.H. *et al*, *J. Chem. Soc., Chem. Commun.*, 1968, 575 (*cryst struct*)
Larock, R.C. *et al*, *Tetrahedron Lett.*, 1981, **22**, 2443 (*use*)
Fieser, M. *et al*, *Reagents for Organic Synthesis*, Wiley, 1967-83, **10**, 149 (*use*)

$C_{37}H_{34}BOP_2Rh$ Rh-00345

Carbonyl[tetrahydroborato(1−)-*H,H'*]bis(triphenylphosphine)rhodium, 9CI

[36606-35-6]

$(Ph_3P)_2Rh(CO)(BH_4)$

M 670.338

Thought to contain doubly bridging BH_4 moiety. Yellow cryst. ir ν_{BH} 2452, 2416, ν_{CO} 1955 cm^{-1} (nujol). Stable in air as solid but dec. rapidly in soln.

Flynn, B.R. *et al*, *J. Chem. Soc., Chem. Commun.*, 1971, 1615 (*synth, ir, pmr*)

$C_{37}H_{37}OP_3Rh^{\oplus}$ Rh-00346

[Bis[3-(diphenylphosphino)propyl]phenylphosphine-*P,P'*,*P''*]carbonylrhodium(1+), 10CI

[47829-31-2]

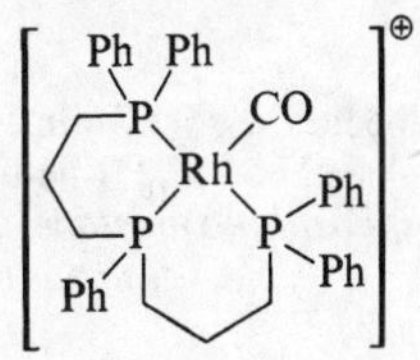

M 693.525 (ion)

Hexafluorophosphate: [36480-96-3].
$C_{34}H_{37}F_6OP_4Rh$ M 802.457
No phys. props. recorded.

Christoph, G.C. *et al*, *Inorg. Chem.*, 1979, **18**, 894 (*nmr*)

$C_{37}H_{40}O_2P_2Rh^{\oplus}$ Rh-00347

1,5-Cyclooctadiene[2,4-bis(diphenylphosphinomethyl)-1,3-dioxolane]rhodium(1+)

1,5-Cyclooctadiene(DIOXOP)rhodium(1+)

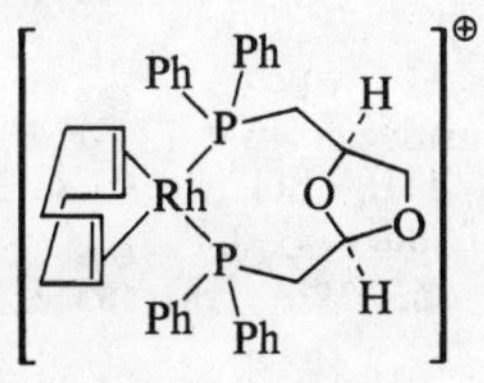

(2*R*,4*R*)-*form*

M 681.575 (ion)

(2R,4R)-form [67063-37-0]

Catalyst for the asymmetric hydrogenation of α-acylaminoacrylic acids and α-acylaminocinnamic acids and esters.

Perchlorate: [67063-38-1].

$C_{37}H_{40}ClO_6P_2Rh$ M 781.025

$[\alpha]_D^{20}$ −112° (c, 1.1 in CH_2Cl_2).

Descotes, G. *et al*, *J. Organomet. Chem.*, 1978, **150**, C14 (*use*)
Brown, J.M. *et al*, *J. Chem. Soc., Chem. Commun.*, 1979, 611 (*nmr*)
Lafont, D. *et al*, *J. Organomet. Chem.*, 1979, **169**, 87 (*synth, use*)
Lafont, D. *et al*, *J. Mol. Catal.*, 1981, **10**, 305 (*use*)

$C_{37}H_{41}NOP_2Rh^{\oplus}$ Rh-00348

[(1,2,5,6-η)-1,5-Cyclooctadiene][[1-(diphenylphosphino)-2-pyrrolidinyl]methyldiphenylphosphinite-*P,P'*]rhodium(1+), 10CI

(1,5-Cyclooctadiene)[N-*(diphenylphosphino)-2-diphenylphosphinoxymethylpyrrolidine*]*rhodium(1+)*. *(1,5-Cyclooctadiene)*[*PROLOPHOS*]*rhodium(1+)*

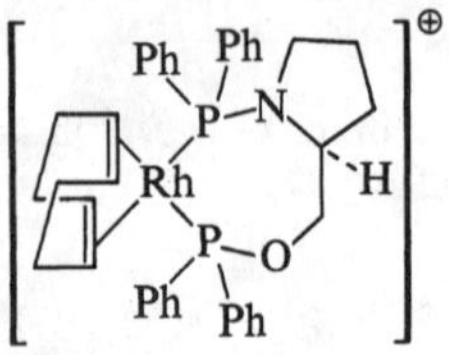

M 680.590 (ion)

(S)-form

Catalyst precursor for the asymmetric hydrogenation of olefins e.g. of subst. acrylic acids.

Perchlorate: [83714-95-8].

$C_{37}H_{41}ClNO_5P_2Rh$ M 780.041

Yellow-orange cryst. + 1THF or + $1CH_2Cl_2$ (CH_2Cl_2/THF/Et_2O).

Cesarotti, E. *et al*, *Tetrahedron Lett.*, 1982, **23**, 2995 (*use*)
Cesarotti, E. *et al*, *J. Organomet. Chem.*, 1983, **251**, 79 (*synth, cryst struct, use*)

$C_{37}H_{42}P_2Rh^{\oplus}$ Rh-00349

[(1,2,5,6-η)-1,5-Cyclooctadiene][(1,3-dimethyl-1,3-propanediyl)bis[diphenylphosphine]-*P,P'*]rhodium(1+), 10CI

[*2,4-Bis(diphenylphosphino)pentane*]*(1,5-cyclooctadiene)rhodium(1+)*. *(1,5-Cyclooctadiene)(SKEWPHOS)rhodium(1+)*

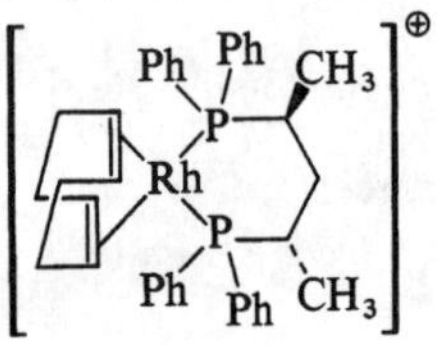

M 651.592 (ion)

(S,S)-form

Catalyst precursor for the asymmetric hydrogenation of *N*-acylamidoacrylic acids.

Perchlorate: [77864-87-0].

$C_{37}H_{42}ClO_4P_2Rh$ M 751.042

Yellow-orange, air-sensitive cryst. + 1THF + $0.25CH_2Cl_2$ (CH_2Cl_2/THF). Sol. MeOH, CH_2Cl_2, spar. sol. THF. Stored under N_2 at 5°.

MacNeil, P.A. *et al*, *J. Am. Chem. Soc.*, 1981, **103**, 2273 (*synth, use*)

$C_{37}H_{44}N_2P_2Rh^{\oplus}$ Rh-00350

[(1,2,5,6-η)-1,5-Cyclooctadiene][*N,N'*-(1-methyl-1,2-ethanediyl)bis[*N*-methyl-*P,P*-diphenylphosphinous amide]-*P,P''*]rhodium(1+), 10CI

(1,5-Cyclooctadiene)[*1,2-bis*[N-*methyl(diphenylphosphino)amino*]*propane*]*rhodium(1+)*

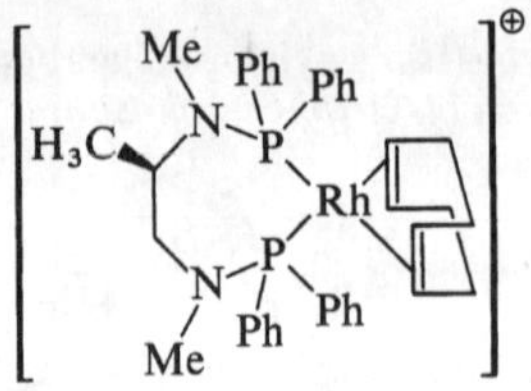

M 681.621 (ion)

(R)-form

Asymmetric hydrogenation catalyst for prochiral olefins.

Hexafluorophosphate: [75730-42-6].

$C_{37}H_{44}F_6N_2P_3Rh$ M 826.585

Orange needles ($CHCl_3$/Et_2O). $[\alpha]_D^{25}$ −97.6° (*c* 0.17 in $CHCl_3$).

Kashiwabara, K. *et al*, *Bull. Chem. Soc. Jpn.*, 1980, **53**, 2275 (*synth, pmr, cd, use*)

$C_{37}H_{47}N_4Rh$ Rh-00351

Methyl[2,3,7,8,12,13,17,18-octaethyl-21*H*,23*H*-porphinato(2−)-N^{21},N^{22},N^{23},N^{24}]rhodium

Methyl(octaethylporphyrinato)rhodium

[36643-85-3]

As Methyl[2,7,12,17-tetraethyl-3,8,13,18-tetramethyl-21*H*,23*H*-porphinato(2−)-N^{21},N^{22},N^{23},N^{24}]rhodium, Rh-00295 with

R = CH_2CH_3

M 650.711

Red cryst. (1,2-dichloroethane). Sol. $CHCl_3$, THF.

Ogoshi, H. *et al*, *J. Am. Chem. Soc.*, 1975, **97**, 6461 (*synth, pmr, uv*)
Setsune, J. *et al*, *J. Chem. Soc., Perkin Trans. 1*, 1980, 1641 (*props*)

$C_{37}H_{66}ClOP_2Rh$ Rh-00352

Carbonylchlorobis(tricyclohexylphosphine)rhodium

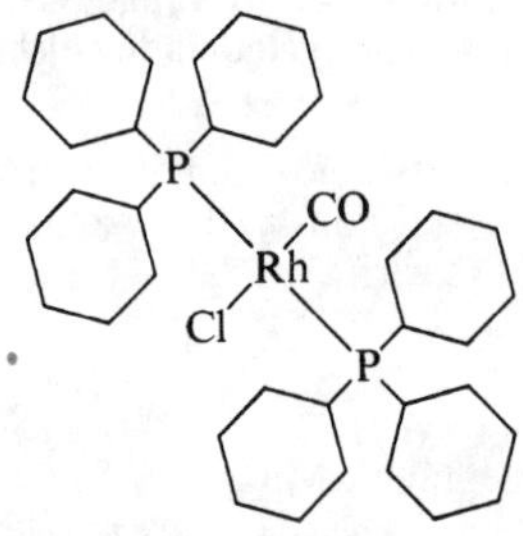

M 727.234

trans-form [26534-14-5]

Air-stable yellow needles (toluene/$CHCl_3$). Insol. Et_2O, pentane. ir ν_{CO} 1941 cm^{-1} (C_6H_6).

Moers, F.G. *et al*, *J. Inorg. Nucl. Chem.*, 1973, **35**, 1915 (*synth, ir*)
Uson, R. *et al*, *Transition Met. Chem.*, 1980, **5**, 327 (*synth, props*)

$C_{37}O_{37}Rh_{22}^{\ominus\ominus\ominus\ominus}$ **Rh-00353**

Octadeca-μ-carbonylhepta-μ$_3$-carbonyldodecacarbonyldocosarhodate(4−), 10CI

Heptatricosacarbonyldocosarhodate(4−)

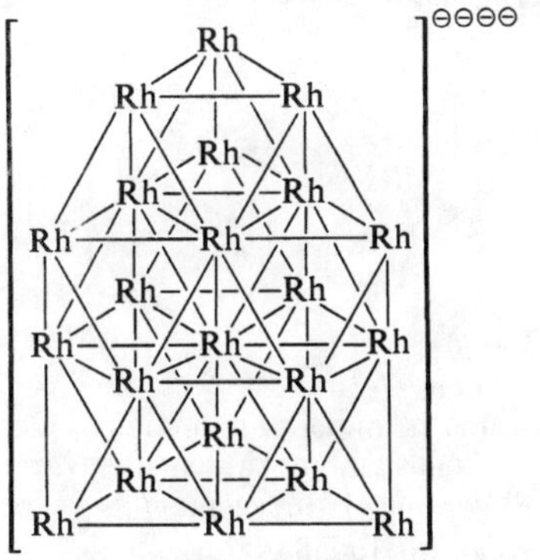

Rh skeleton only shown

M 3300.306 (ion)

Struct. consists of a mixed hcp/ccp lattice with seven face-bridging CO, eighteen edge bridging CO and twelve terminal CO groups.

Tetra-Na salt: [75718-89-7].
$C_{37}Na_4O_{37}Rh_{22}$ M 3392.265
Brown microcryst. with a metallic appearance.

Tetrakis(triethylpropylammonium) salt: [75899-66-0].
$C_{73}H_{88}N_4O_{37}Rh_{22}$ M 3877.424
Brown cryst.

Martinengo, S. *et al*, *J. Am. Chem. Soc.*, 1980, **102**, 7564 (*synth*, *ir*, *cryst struct*)

$C_{38}H_{30}As_2Cl_2O_2Rh_2$ **Rh-00354**

Dicarbonyldi-μ-chlorobis(triphenylarsine)dirhodium, 9CI

Bis[μ-chlorocarbonyl(triphenylarsine)rhodium]

[38651-41-1]

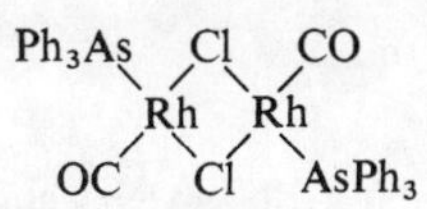

M 945.214

Rust-orange cryst. Mp 180° dec. ir ν_{CO} 1980 cm^{-1} ($CHCl_3$).

Monomer: Chlorocarbonyl(triphenylphosphine)-rhodium.
$C_{19}H_{15}AsClORh$ M 472.607
Unknown.

Barlex, D.M. *et al*, *J. Organomet. Chem.*, 1972, **43**, 425 (*synth*, *ir*)

$C_{38}H_{30}Br_2O_2P_2Rh_2$ **Rh-00355**

Di-μ-bromodicarbonylbis(triphenylphosphine)dirhodium, 9CI

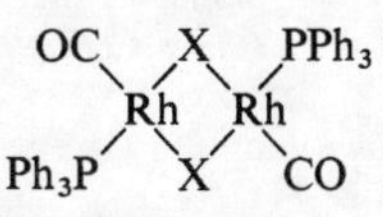

trans-form

X = Br

M 946.220

trans-form

Orange-brown powder. Mp 153-6°, 185° dec. ir ν_{CO} 1982 cm^{-1}.

Monomer: Bromocarbonyl(triphenylphosphine)-rhodium.
$C_{19}H_{15}BrOPRh$ M 473.110
Unknown.

Barlex, D.M. *et al*, *J. Organomet. Chem.*, 1972, **43**, 425 (*synth*, *ir*, *props*)
Steele, D.G. *et al*, *J. Chem. Soc., Dalton Trans.*, 1972, 2161 (*synth*, *ir*, *props*)

$C_{38}H_{30}Cl_2O_2P_2Rh_2$ **Rh-00356**

Dicarbonyldi-μ-chlorobis(triphenylphosphine)dirhodium, 9CI

As Di-μ-bromodicarbonylbis(triphenylphosphine)dirhodium, Rh-00355 with

X = Cl

M 857.318

Originally thought to be $Rh(CO)_2Cl(PPh_3)$.

cis-form

ir ν_{CO} 2090, 2023 cm^{-1}. Isom. to *trans*-form in C_6H_6.

trans-form

Air-stable golden-yellow cryst. (C_6H_6). Mp 200-4° dec. ir ν_{CO} 1980 cm^{-1}.

Monomer: Carbonylchloro(triphenylphosphine)-rhodium.
$C_{19}H_{15}ClOPRh$ M 428.659
Unknown.

Steele, D.F. *et al*, *Inorg. Nucl. Chem. Lett.*, 1971, **7**, 877 (*props*)
Barlex, D.M. *et al*, *J. Organomet. Chem.*, 1972, **43**, 425 (*synth*, *ir*, *props*)
Steele, D.F. *et al*, *J. Chem. Soc., Dalton Trans.*, 1972, 2161 (*synth*, *ir*, *props*)

$C_{38}H_{30}NOP_2Rh$ **Rh-00357**

Carbonyl(cyano-*C*)bis(triphenylphosphine)rhodium, 9CI

[32109-05-0]

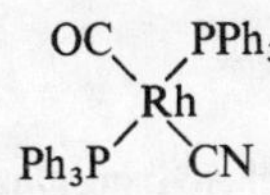

M 681.514

trans-form

Pale-yellow cryst. (CH_2Cl_2/pentane). ir ν_{CN} 2117, ν_{CO} 2003 cm^{-1} (nujol).

von Werner, K. *et al*, *Chem. Ber.*, 1972, **105**, 3947 (*synth*, *ir*, *nmr*)
Brady, R. *et al*, *Inorg. Chem.*, 1976, **15**, 1485 (*uv*)

$C_{38}H_{30}Rh_2$ **Rh-00358**

Bis(η^5-2,4-cyclopentadien-1-yl)[(1,2,3,4-η:1,4-η)-1,2,3,4-tetraphenyl-1,3-butadiene-1,4-diyl]dirhodium, 9CI

[39363-58-1]

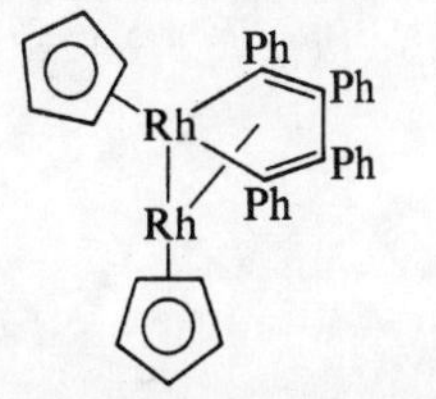

M 692.466

Formed during the catalytic cyclotrimerisation of PhC≡CPh by $(C_5H_5)Rh(CO)_2$. Deep-red cryst. (C_6H_6/hexane).

Gardner, S.A. *et al*, *Inorg. Chem.*, 1973, **12**, 2396 (*synth*, *ir*, *pmr*, *ms*)

$C_{38}H_{33}BNOP_2Rh$ Rh-00359

Carbonyl[(cyano-*C*)trihydroborato(1−)-*N*)]bis(triphenylphosphine)rhodium, 9CI

Carbonyl(cyanotrihydridoborato)bis(triphenylphosphine)rhodium

[36606-39-0]

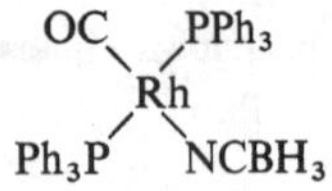

M 695.348

Air-stable yellow solid. Sol. $CHCl_3$, C_6H_6, insol. hexane, alcohols. ir ν_{BH} 2360, ν_{CN} 2192, ν_{CO} 1997 cm^{-1} (nujol).

Vaska, L. *et al*, *J. Chem. Soc., Chem. Commun.*, 1971, 1615 (*ir*, *synth*)
Inorg. Synth., 1974, **15**, 72 (*synth*, *ir*)
Brady, R. *et al*, *Inorg. Chem.*, 1976, **15**, 1485 (*uv*)

$C_{38}H_{33}OP_2Rh$ Rh-00360

Carbonylmethylbis(triphenylphosphine)rhodium, 10CI, 9CI

[82190-13-4]

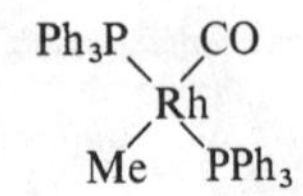

M 670.531

***trans*-form**

Yellow cryst. Insol. EtOH.

Dahlenburg, D. *et al*, *Z. Naturforsch.*, *B*, 1982, **37**, 310 (*synth*, *ir*, *nmr*)

$C_{38}H_{34}ClP_2Rh$ Rh-00361

Chloro(η^2-ethene)bis(triphenylphosphine)rhodium, 9CI

Chloro(ethylene)bis(triphenylphosphine)rhodium, 8CI. *Bis(triphenylphosphine)ethylenechlororhodium. Ethylenebis(triphenylphosphine)rhodium chloride*

[12120-14-8]

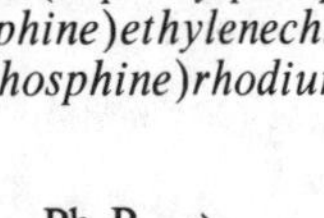

M 690.993

Hydrogenation catalyst for alkenes and alkynes. Yellow cryst. Loses ethylene rapidly.

Osborn, J.A. *et al*, *J. Chem. Soc. (A)*, 1966, 1711 (*synth*)
Candlin, J.P. *et al*, *Disc. Faraday Soc.*, 1968, 60 (*use*)
Mason, R. *et al*, *J. Chem. Soc., Dalton Trans.*, 1972, 1729 (*uv*)
Tolman, C.A. *et al*, *J. Am. Chem. Soc.*, 1974, **96**, 2762 (*synth*, *nmr*, *ir*, *uv*)

$C_{38}H_{40}O_2P_2Rh^{\oplus}$ Rh-00362

[(2,3,5,6-η)-Bicyclo[2.2.1]hepta-2,5-diene][[(2,2-dimethyl-1,3-dioxolane-4,5-diyl)bis(methylene)]bis(diphenylphosphine)-*P*,*P'*]rhodium(1+), 10CI

(DIOP)norbornadienerhodium(1+)

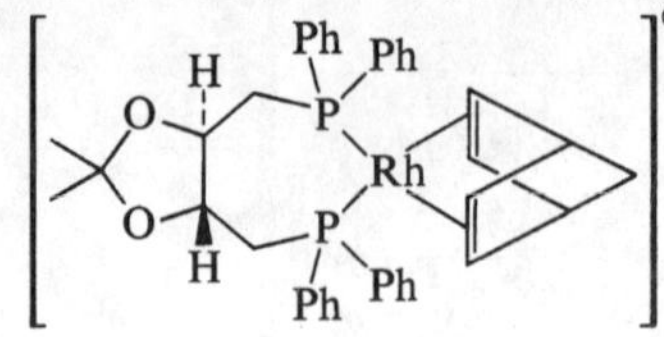

(R,R)-form

M 693.586 (ion)

***(R,R)*-form** [55663-97-3]

Catalyst for the asymmetric reduction of C=N, C=O and C=C bonds and for the asymmetric hydroformylation of vinyl acetate.

Tetrafluoroborate: [60584-05-6].
$C_{38}H_{40}BF_4O_2P_2Rh$ M 780.389
Orange cryst. (MeOH).

Tetraphenylborate: [67110-67-2].
$C_{62}H_{60}BO_2P_2Rh$ M 1012.818
No phys. props. reported.

***(S,S)*-form**

Catalyst for asymmetric hydrogenations and hydroformylations.

Perchlorate: [68629-11-8].
$C_{38}H_{40}ClO_6P_2Rh$ M 793.036
No phys. props. reported.

Tetraphenylborate: [79548-77-9].
$C_{62}H_{60}BO_2P_2Rh$ M 1012.818
No phys. props. reported.

Levi, A. *et al*, *J. Chem. Soc., Chem. Commun.*, 1975, 6 (*synth*, *use*)
Toros, S. *et al*, *J. Organomet. Chem.*, 1978, **159**, 401 (*use*)
Slack, D.A. *et al*, *Inorg. Chem.*, 1979, **18**, 3125 (*nmr*, *use*)
Hobbs, C.F. *et al*, *J. Org. Chem.*, 1981, **46**, 4422 (*use*)
Stults, B.R. *et al*, *J. Am. Chem. Soc.*, 1981, **103**, 3235 (*struct*)
Brown, J.M. *et al*, *J. Chem. Soc., Perkin Trans. 2*, 1982, 711 (*use*, *synth*)

$C_{38}H_{40}O_2P_2Rh^{\oplus}$ Rh-00363

Dihydrobis(methanol)bis(triphenylphosphine)rhodium(1+)

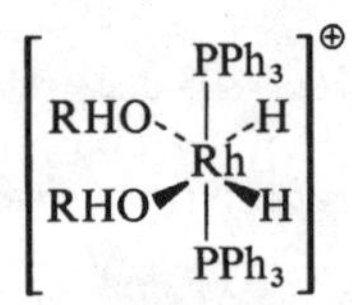

R = Me

M 693.586 (ion)

Intermed. in hydrogenations carried out in MeOH using $[Rh(PPh_3)_2(diene)]^{\oplus}$ complexes.

Tetrafluoroborate: [68811-63-2].
$C_{38}H_{40}BF_4O_2P_2Rh$ M 780.389
Known only in soln.

Brown, J.M. *et al*, *J. Chem. Soc., Chem. Commun.*, 1978, 646 (*synth*, *nmr*)

$C_{38}H_{42}OP_2Rh^{\oplus}$ Rh-00364

[(2,3,5,6-η)-Bicyclo[2.2.1]hepta-2,5-diene][[1-[(1,1-dimethylethoxy)methyl]-1,2-ethanediyl]bis[diphenylphosphine]-*P,P'*]rhodium(1+), 10CI

*[1,2-Bis(diphenylphosphino)-3-*tert*-butoxypropane]-norbornadienerhodium(1+)*

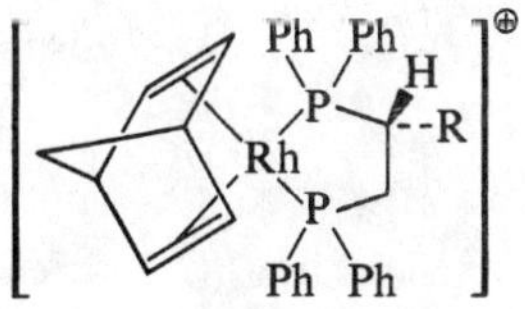

R = $-CH_2OC(CH_3)_3$

M 679.602 (ion)

***(R)*-form**

Catalyst for the asymmetric hydrogenation of dehydroamino acids.

Tetrafluoroborate: [80145-72-8].
$C_{38}H_{42}BF_4OP_2Rh$ M 766.406
No phys. props. reported.

Padmakumari Amma, J. *et al, J. Org. Chem.*, 1982, **47**, 468 (*synth, use*)

$C_{38}H_{44}N_2P_2Rh^{\oplus}$ Rh-00365

[*N,N'*-1,2-Cyclohexanediylbis[*P,P*-diphenylphosphinous amide]-*P,P'*][(1,2,5,6-η)-1,5-cyclooctadiene]rhodium(1+), 10CI

[1,2-Bis(diphenylphosphinamino)cyclohexane](1,5-cyclooctadiene)rhodium(1+)

[77699-33-3]

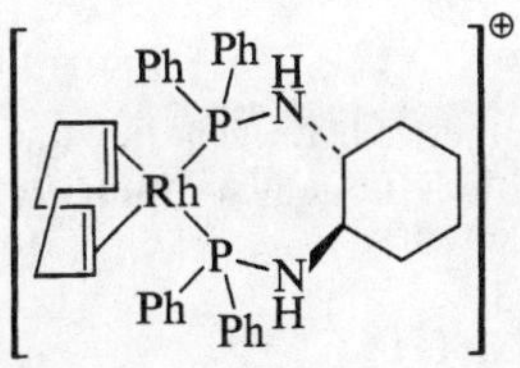

(R,R)-form

M 693.632 (ion)

Asymmetric hydrogenation catalyst for prochiral olefins.

***(R,R)*-form** [70901-06-3]

Hexafluorophosphate: [75766-16-4].
$C_{38}H_{44}F_6N_2P_3Rh$ M 838.596
Orange blocks + $1CHCl_3$ ($CHCl_3/Et_2O$). $[\alpha]_D^{25}$ −94.8° (c, 0.13 in $CHCl_3$).

Perchlorate: [70901-07-4].
$C_{38}H_{44}ClN_2O_4P_2Rh$ M 793.083
Cryst. + $\frac{1}{2}CH_2Cl_2$ (Me_2CO/CH_2Cl_2). Mp 142-5° dec.

***(S,S)*-form** [72121-10-9]

Perchlorate: [72121-11-0]. Wine-red cryst. + $\frac{1}{2}CH_2Cl_2$ (Me_2CO/CH_2Cl_2). Mp 142-5° dec.

Kashiwabara, K. *et al, Bull. Chem. Soc. Jpn.*, 1980, **53**, 2275 (*synth, pmr, cd, use*)
Onuma, K. *et al, Bull. Chem. Soc. Jpn.*, 1980, **53**, 2012, 2016 (*synth, use*)
Onuma, K. *et al, Bull. Chem. Soc. Jpn.*, 1981, **54**, 761 (*cryst struct*)

$C_{38}H_{45}NOP_2Rh^{\oplus}$ Rh-00366

[(1,2,5,6-η)-1,5-Cyclooctadiene][2-[(diphenylphosphino)ethylamino]butyldiphenylphosphinite-*P,P'*]rhodium(1+), 10CI

(1,5-Cyclooctadiene)(1-diphenylphosphinoxy-2-N-ethyl-N-diphenylphosphinoaminobutane)rhodium(1+).
(BUTAPHOS)(1,5-Cyclooctadiene)rhodium(1+)

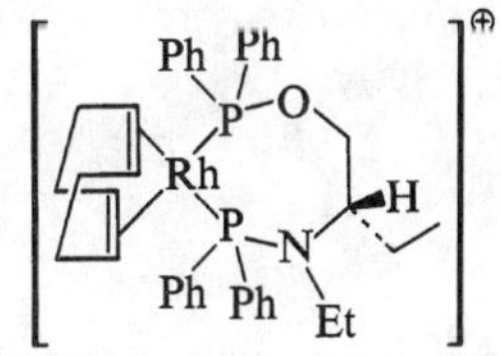

M 696.633 (ion)

***(S)*-form**

Perchlorate: [83714-97-0].
$C_{38}H_{45}ClNO_5P_2Rh$ M 796.083
Solid + $1CH_2Cl_2$ ($CH_2Cl_2/THF/Et_2O$).

Cesarotti, E. *et al, Tetrahedron Lett.*, 1982, **23**, 2995 (*use*)
Cesarotti, E. *et al, J. Organomet. Chem.*, 1983, **251**, 79 (*synth*)

$C_{39}H_{30}F_3O_3P_2Rh$ Rh-00367

Carbonyl(trifluoroacetato-*O*)bis(triphenylphosphine)rhodium, 10CI

[41509-48-2]

$(Ph_3P)_2Rh(CO)(OOCCF_3)$

M 768.512

Orange solid. ir ν_{CO} 1989 cm^{-1} (nujol).

Van Vliet, P.I. *et al, J. Organomet. Chem.*, 1977, **139**, 337 (*synth, ir, nmr*)

$C_{39}H_{33}O_3P_2Rh$ Rh-00368

(Acetato-*O*)carbonylbis(triphenylphosphine)rhodium, 10CI

[65165-69-7]

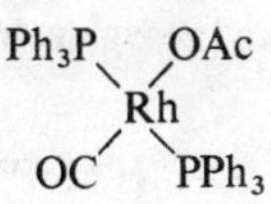

M 714.541

Catalyst for the hydroformylation of cyclic dienes and formaldehyde.

***trans*-form** [32354-34-0]

Yellow cryst. (C_6H_6/Et_2O). Mp 150°, 171-3° dec. ν_{CO} 1972 cm^{-1} ($CHCl_3$).

Mitchell, R.W. *et al, J. Chem. Soc. (A)*, 1971, 3224 (*synth, ir*)
Robinson, S.D. *et al, J. Chem. Soc., Dalton Trans.*, 1973, 1912 (*synth, ir, pmr*)
Spencer, A., *J. Organomet. Chem.*, 1977, **124**, 85; 1980, **194**, 113 (*use, synth*)
Colquhoun, I-J. *et al, J. Magn. Reson.*, 1982, **46**, 525 (*nmr*)
Bird, C. *et al, J. Chem. Soc., Dalton Trans.*, 1982, 517 (*synth*)

$C_{39}H_{34}ClP_2Rh$ **Rh-00369**

Chloropropadienebis(triphenylphosphine)rhodium

Allenechlorobis(triphenylphosphine)rhodium, 8CI. *Chloro(methyleneethylene)bis(triphenylphosphine)rhodium*, 8CI

[22391-90-8]

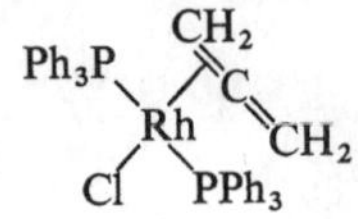

M 703.004

Catalyses tetramerisation and polymerisation of allene. Air-stable yellow cryst. (C_6H_6). Mp 140-2°.

Otsuka, S. *et al*, *J. Polym. Sci., Part B*, 1967, **5**, 973 (*synth, use*)
Otsuka, S. *et al*, *J. Chem. Soc. (A)*, 1969, 1404 (*synth, ir, nmr, uv*)

$C_{39}H_{35}Cl_2P_2Rh$ **Rh-00370**

Dichloro(η^3-2-propenyl)bis(triphenylphosphine)rhodium

π-Allyldichlorobis(triphenylphosphine)rhodium, 8CI

[12120-31-9]

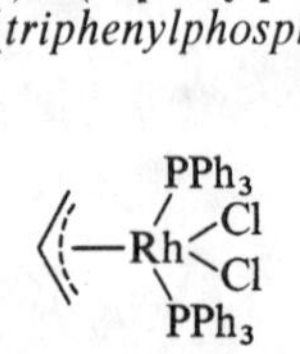

M 739.465

Orange-red or yellow cryst. Mp 144-6°, 169-73°.

Lawson, D.N. *et al*, *J. Chem. Soc. (A)*, 1966, 1733 (*synth, ir, pmr*)
Volger, H.C. *et al*, *J. Organomet. Chem.*, 1967, **9**, 527 (*synth*)

$C_{39}H_{39}ClP_3Rh$ **Rh-00371**

Chlorotris(methyldiphenylphosphine)rhodium

[19531-77-2]

$RhCl(PMePh_2)_3$

M 739.017

Catalyst for decarbonylation of aldehydo sugars and for hydrosilylation of olefins. Yellow cryst. (toluene). Mp 153° dec.

Dewhirst, K.C. *et al*, *Inorg. Chem.*, 1968, **7**, 546 (*synth, nmr*)
Ward, D.I. *et al*, *Chem. Ind. (London)*, 1976, 162 (*use*)
Fieser, M. *et al*, *Reagents for Organic Synthesis*, Wiley, 1967-83, **7**, 67.

$C_{39}H_{42}P_2Rh^{\oplus}$ **Rh-00372**

[(2,3,5,6-η)-Bicyclo[2.2.1]hepta-2,5-diene][(1-cyclohexyl-1,2-ethanediyl)bis[diphenylphosphine]-*P,P'*]rhodium(1+)

[1,2-Bis(diphenylphosphino)-1-cyclohexylethane]norbornadienerhodium(1+). CYCPHOS(norbornadiene)-rhodium(1+)

As [(2,3,5,6-η)-Bicyclo[2.2.1]hepta-2,5-diene][[1-[(1,1-dimethylethoxy)methyl]-1,2-ethanediyl]bis[diphenylphosphine]-*P,P'*]rhodium(1+), Rh-00364 with

R = Cyclohexyl

M 675.614 (ion)

(R)-form

Catalyst for the asymmetric hydrogenation of prochiral olefins.

Hexafluorophosphate: [75421-76-0].
$C_{39}H_{42}F_6P_3Rh$ M 820.578
Orange solid. V. sol. CH_2Cl_2, EtOAc.

Riley, D.P. *et al*, *J. Org. Chem.*, 1980, **45**, 5187 (*synth, use*)
Riley, D.P. *et al*, *J. Organomet. Chem.*, 1982, **234**, 85 (*nmr*)

$C_{39}H_{44}O_2P_2Rh^{\oplus}$ **Rh-00373**

[(1,2,5,6-η)-1,5-Cyclooctadiene][[(2,2-dimethyl-1,3-dioxolane-4,5-diyl)bis[methylene]]bis(diphenylphosphine)-*P,P'*]-rhodium(1+), 9CI

(1,5-Cyclooctadiene)[isopropylidene-2,3-dihydroxy-1,4-bis(diphenylphosphino)butane]rhodium(1+). 1,5-Cyclooctadiene(DIOP)rhodium(1+). [4,5-Bis(diphenylphosphinomethyl)-2,2-dimethyl-1,3-dioxolan](1,5-cyclooctadiene)rhodium(1+)

[60104-67-8]

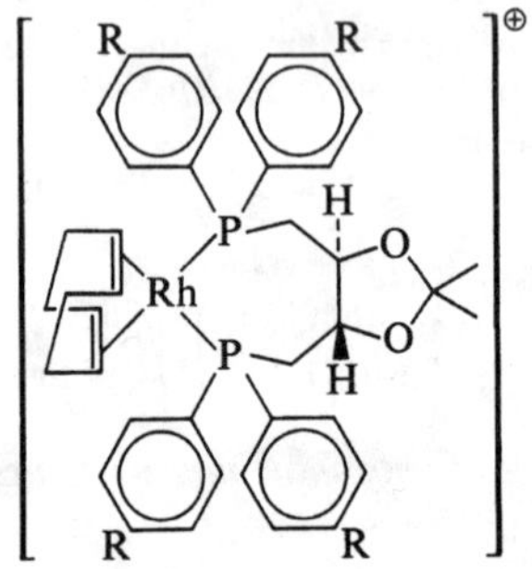

R = H

M 709.628 (ion)

Catalyst for the asymmetric hydrogenation and hydroformylation of olefins, and asymmetric hydrosilylation of ketones.

(R,R)-form [79547-74-3]

Chloride: [70832-59-6].
$C_{39}H_{44}ClO_2P_2Rh$ M 745.081
Brown solid (Me_2CO/Et_2O). $[\alpha]_D^{25}$ +10.2° (Me_2CO).

Tetrafluoroborate: [60104-68-9].
$C_{39}H_{44}BF_4O_2P_2Rh$ M 796.432
Yellow-orange, air-sensitive solid. Sol. MeOH, EtOH.

Perchlorate: [70832-57-4].
$C_{39}H_{44}ClO_6P_2Rh$ M 809.079
Sol. MeOH, C_6H_6. λ_{max} 443 nm (ϵ 2450) (C_6H_6).

Tetraphenylborate: [67068-51-3].
$C_{63}H_{64}BO_2P_2Rh$ M 1028.860
No phys. props. reported.

(S,S)-form [60594-32-3]

Perchlorate: [60594-33-4]. Yellow-orange, air-sensitive powder. $[\alpha]_D^{25}$ −9.2° (c, 1).

Glaser, R. *et al*, *J. Organomet. Chem.*, 1976, **112**, 355 (*synth, use*)
Sinou, D. *et al*, *J. Organomet. Chem.*, 1976, **114**, 325 (*synth, pmr, use*)
Kolb, I. *et al*, *Collect. Czech. Chem. Commun.*, 1980, **45**, 2224 (*synth, pmr, use, uv*)
Spogliarich, R. *et al*, *J. Organomet. Chem.*, 1980, **198**, 81 (*use*)
Hobbs, C.F. *et al*, *J. Org. Chem.*, 1981, **46**, 4422 (*use*)
Uehara, A. *et al*, *J. Organomet. Chem.*, 1982, **239**, 11 (*synth, use*)

$C_{40}H_{30}ClF_6P_2Rh$ Rh-00374

Chloro[(2,3-η)-1,1,1,4,4,4-hexafluoro-2-butyne]bis(triphenylphosphine)rhodium, 10CI

[Bis(trifluoromethyl)acetylene]chlorobis(triphenylphosphine)rhodium

[12194-63-7]

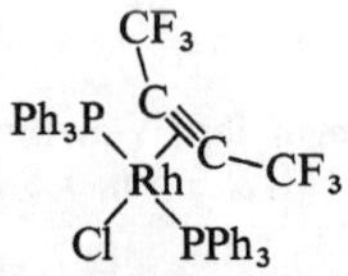

M 824.973

Orange cryst. (Et_2O/hexane). Mp 141° dec.

Mays, M.J. *et al*, *J. Chem. Soc.*, 1965, 6629 (*synth, ir, nmr*)
Van Gaal, H.L.M. *et al*, *J. Organomet. Chem.*, 1977, **131**, 453 (*ir*)

$C_{40}H_{33}OP_2Rh$ Rh-00375

Carbonyl-1-propynylbis(triphenylphosphine) rhodium, 9CI

Carbonyl(methylethynyl)bis(triphenylphosphine)rhodium. Propynylcarbonylbis(triphenylphosphine)-rhodium

[34720-47-3]

$Rh(C{\equiv}CCH_3)(CO)(PPh_3)_2$

M 694.553

trans-form

Unstable yellow cryst. (toluene/EtOH). ir ν_{CO} 1968 cm^{-1} (nujol).

Brown, C.K. *et al*, *J. Chem. Soc. (A)*, 1971, 3120 (*ir, pmr, synth*)

$C_{40}H_{36}NO_5P_2Rh$ Rh-00376

Bis(acetato-*O*)nitrosylbis(triphenylphosphine)rhodium, 9CI

Diacetoxynitrosylbis(triphenylphosphine)rhodium

[50661-88-6]

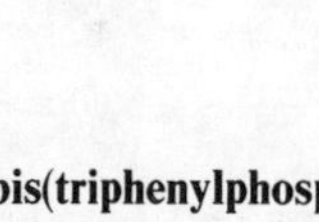

M 775.581

Green cryst. Mp 210-11°. ir ν_{NO} 1614 cm^{-1} (nujol).

Robinson, S.D. *et al*, *J. Chem. Soc., Dalton Trans.*, 1973, 1912 (*synth, ir, pmr*)
Inorg. Synth., 1977, **17**, 124 (*synth, ir*)

$C_{40}H_{36}N_2P_2Rh^{\oplus}$ Rh-00377

Bis(isocyanomethane)bis(triphenylphosphine)rhodium(1+), 9CI

Bis(methylisocyanide)bis(triphenylphosphine)-rhodium(1+)

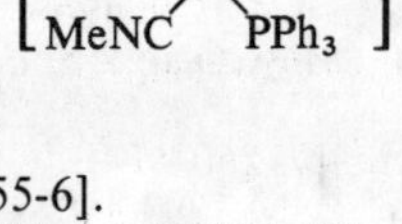

M 709.591 (ion)

Iodide: [51743-55-6].
$C_{40}H_{36}IN_2P_2Rh$ M 836.495
Cryst.

Hexafluorophosphate: [51147-77-4].
$C_{40}H_{36}F_6N_2P_3Rh$ M 854.555
Yellow cryst. (CH_2Cl_2/MeOH).

Perchlorate: [51703-33-4].
$C_{40}H_{36}ClN_2O_4P_2Rh$ M 809.041
Cryst.

Dart, J.W. *et al*, *J. Chem. Soc., Dalton Trans.*, 1973, 2046 (*synth, ir, pmr*)
Kaneshima, T. *et al*, *Inorg. Chem.*, 1974, **13**, 2198 (*synth, ir, pmr*)
Balch, A. *et al*, *J. Am. Chem. Soc.*, 1976, **98**, 8049 (*synth, ir, pmr, uv*)

$C_{40}H_{36}P_2Rh^{\oplus}$ Rh-00378

(1,3-Butadiene)bis(triphenylphosphine)rhodium(1+), 8CI

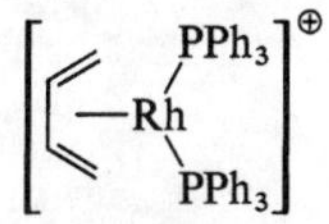

M 681.577 (ion)

Hexafluorophosphate: [32757-44-1].
$C_{40}H_{36}F_6P_3Rh$ M 826.542
Red solid. Disproportionates if recryst. is attempted.

Osborn, J.A. *et al*, *J. Am. Chem. Soc.*, 1971, **93**, 2397 (*synth, pmr*)

$C_{40}H_{38}Cl_2P_2Rh_2$ Rh-00379

Di-μ-chlorobis(η^2-ethene)bis(triphenylphosphine)dirhodium

Di-μ-chlorobis(η^2-ethylene)bis(triphenylphosphine)dirhodium, 8CI

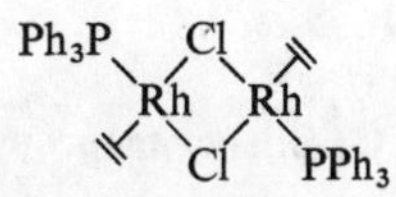

M 857.405

Bridged dimer. Allene polymn. catalyst.

Monomer: Chloro(ethylene)(triphenylphosphine)-rhodium.
$C_{20}H_{19}ClPRh$ M 428.702
Unknown.

Otsuka, S. *et al*, *Kogyo Kagaku Zasshi*, 1967, **70**, 2007; *CA*, **69**, 10750h (*synth, use*)

$C_{40}H_{44}N_4O_4Rh_2$ Rh-00380

Tetracarbonyl[μ-[2,3,7,8,12,13,17,18-octaethyl-21*H*,23*H*-porphinato(2–)-N^{21},N^{22}:N^{23},N^{24}]]dirhodium, 9CI

μ-Octaethylporphyrinbis[dicarbonylrhodium]

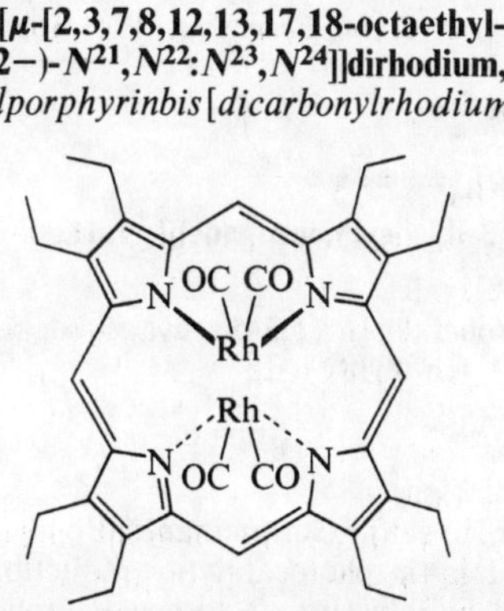

M 850.623

Green-blue needles ($CHCl_3/C_6H_6$). Mp >300°.

Takenaka, A. *et al*, *J. Chem. Soc., Chem. Commun.*, 1973, 792 (*cryst struct*)

Ogoshi, H. *et al*, *J. Am. Chem. Soc.*, 1975, **97**, 6461 (*synth, pmr, uv*)
Takenaka, A. *et al*, *Acta Crystallogr., Sect. B*, 1975, **31**, 1 (*cryst struct*)
Grigg, R. *et al*, *J. Chem. Soc., Perkin Trans. 1*, 1977, 36 (*synth*)
Sigal, I.S. *et al*, *J. Am. Chem. Soc.*, 1980, **102**, 7252 (*uv, props, synth*)
Mann, K.R. *et al*, *Inorg. Chem.*, 1981, **20**, 1921 (*synth, uv, props*)
Sigal, I.S. *et al*, *J. Am. Chem. Soc.*, 1981, **103**, 2220 (*synth, props*)

$C_{40}H_{48}Cl_2N_{16}Rh_4$⊕⊕⊕⊕ Rh-00382

Dichlorooctakis[μ-(1,3-diisocyanopropane)]tetrarhodium(4+)

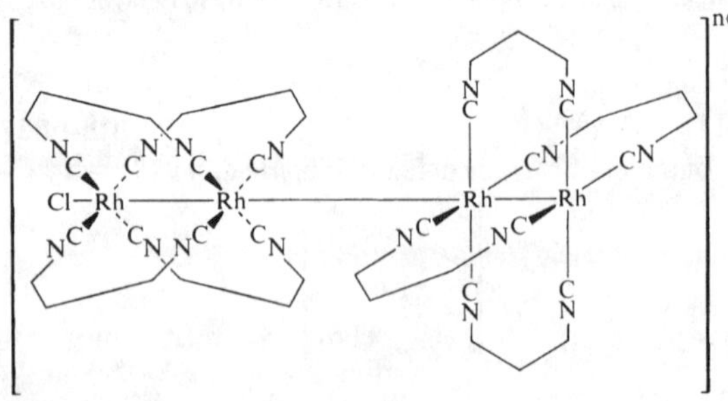

n = 4

M 1235.454 (ion)

Principal species present when Octakis[μ-(1,3-diisocyanopropane)]tetrarhodium(6+), Rh-00384 is dissolved in HCl with $[Cl^-]<0.01M$.

Gray, H.B. *et al*, *Fundam. Res. Homogeneous Catal.*, 1979, **3**, 819 (*props*)
Miskowski, V.M. *et al*, *J. Am. Chem. Soc.*, 1979, **101**, 4383 (*props*)
Sigal, I.S. *et al*, *J. Am. Chem. Soc.*, 1980, **102**, 7252 (*synth, props*)

$C_{40}H_{48}ClN_{16}Rh_4$⊕⊕⊕⊕⊕ Rh-00381

Dichlorooctakis[μ-(1,3-diisocyanopropane)]tetrarhodium(5+)

As Dichlorooctakis[μ-(1,3-disocyanopropane)]tetrarhodium(4+), Rh-00382 with

n = 5

M 1200.001 (ion)

Principal species present when Octakis[μ-(1,3-diisocyanopropane)]tetrarhodium(6+), Rh-00384 is dissolved in HCl with $[Cl^-]<0.01M$.

Trihydrogentetrakis(tetrachlorocobaltate):
$C_{40}H_{51}Cl_{17}Co_4N_{16}Rh_4$ M 2006.006
Dark-green, extremely photosensitive cryst. + $6H_2O$.

Mann, K.R. *et al*, *J. Am. Chem. Soc.*, 1980, **102**, 3965 (*cryst struct*)

$C_{40}H_{48}N_{16}Rh_4$⊕⊕⊕⊕⊕⊕ Rh-00384

Octakis[μ-(1,3-diisocyanopropane)]tetrarhodium(6+), 10CI

[76900-21-5]

As Dichlorooctakis[μ-(1,3-disocyanopropane)]tetrarhodium(4+), Rh-00382 with

n = 6

M 1164.548 (ion)

Purple soln. in sulfuric or methanesulfonic acid. Reactive intermed. in the photocatalytic production of H_2 from HCl and Tetrakis[μ-(1,3-diisocyanopropane)]dirhodium(2+), Rh-00216.

Hexakis(tetrafluoroborate): [77285-27-9].
$C_{40}H_{48}B_6F_{24}N_{16}Rh_4$ M 1685.370
Black photosensitive powder. ir ν_{CN} 2218 cm^{-1} (KBr).

$C_{40}H_{48}N_2P_2Rh$⊕ Rh-00383

[*N,N′*-1,2-Cyclohexanediylbis[*N*-methyl-*P,P*-diphenylphosphinous amide]-*P,P′*][(1,2,5,6-η)-1,5-cyclooctadiene]rhodium(1+), 10CI

*[1,2-Bis[*N*-methyl(diphenylphosphino)amino]cyclohexane](1,5-cyclooctadiene)rhodium(1+)*

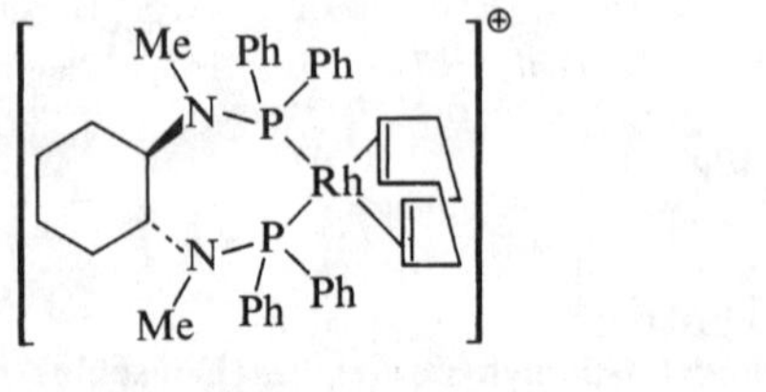

M 721.686 (ion)

(R,R)-form [70718-95-5]

Asymmetric hydrogenation catalyst for prochiral olefins.

Hexafluorophosphate: [75766-15-3].
$C_{40}H_{48}F_6N_2P_3Rh$ M 866.650
Orange cryst. ($CHCl_3/Et_2O$). $[\alpha]_D^{25}$ −79.1° (*c* 0.18 in $CHCl_3$).

Perchlorate: [70718-96-6].
$C_{40}H_{48}ClN_2O_4P_2Rh$ M 821.136
No phys. props. reported.

Fiorini, M. *et al*, *J. Mol. Catal.*, 1979, **5**, 303 (*use*)
Kashiwabara, K. *et al*, *Bull. Chem. Soc. Jpn.*, 1980, **53**, 2275 (*synth, pmr, cd, use*)

$C_{40}H_{64}Rh_4$⊕⊕ Rh-00385

Tetra-μ-hydrotetrakis[(1,2,3,4,5-η)-1,2,3,4,5-pentamethyl-2,4-cyclopentadien-1-yl]tetrarhodium(2+)

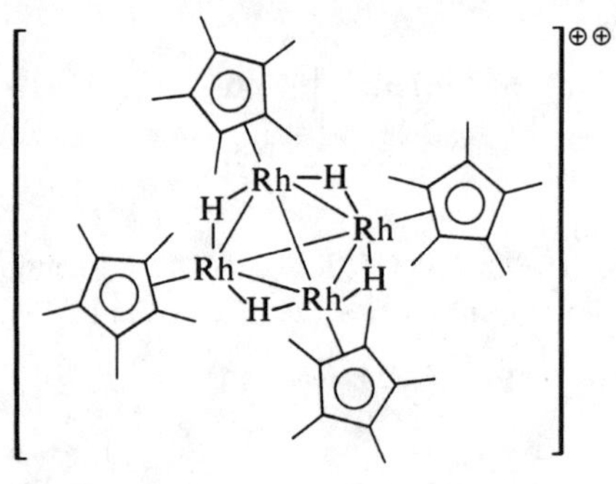

M 956.568 (ion)

Dichloride: [71204-91-6]. Glassy-blackish solid + $8H_2O$. Not obt. pure.

Bis(tetrafluoroborate): [71250-46-9].
$C_{40}H_{64}B_2F_8Rh_4$ M 1130.175
Black cryst. (H_2O). Mod. sol. H_2O, sol. Me_2CO.

Bis(hexafluorophosphate): [71250-45-8].
$C_{40}H_{64}F_{12}P_2Rh_4$ M 1246.496
Dark-green cryst. (Me_2CO aq.). Sol. Me_2CO, insol. H_2O.

Tetrachloroplatinate: [71250-47-0].
$C_{40}H_{64}Cl_4PtRh_4$ M 1293.460
Dark-green cryst. Insol.

Espinet, P., *Inorg. Chem.*, 1979, **18**, 2706 (*synth, cryst struct, nmr, ir, ms*)
Nutton, A. *et al*, *J. Organomet. Chem.*, 1981, **213**, 313 (*synth*)

$C_{41}H_{35}P_2Rh$ Rh-00386

(η^5-2,4-Cyclopentadien-1-yl)bis(triphenylphosphine)rhodium, 10CI, 9CI

[33677-50-8]

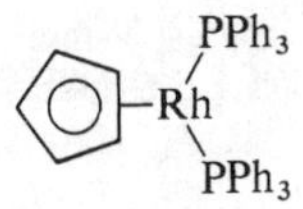

M 692.581

Pale-brown cryst.

C_6H_6 solvate:

$C_{47}H_{41}P_2Rh$ M 770.694

Dark-brown cryst. (C_6H_6/hexane). Mp 190° dec. Air-stable in the solid state.

Yamazaki, H. *et al*, *Bull. Chem. Soc. Jpn.*, 1971, **44**, 2260 (*synth*)
Watatsuki, Y. *et al*, *J. Organomet. Chem.*, 1974, **64**, 393 (*synth, props*)

$C_{41}H_{37}O_8P_2Rh$ Rh-00387

(2,4-Pentanedionato-*O,O'*)bis(triphenyl phosphite-*P*)rhodium, 10CI

Acetylacetonatobis(triphenyl phosphite)rhodium

[25966-19-2]

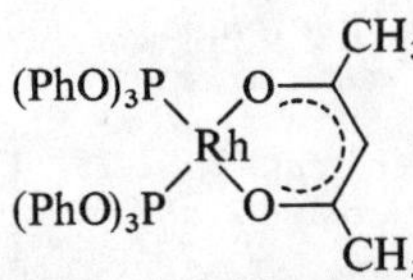

M 822.592

Catalyst for the hydrogenation of arenes and olefins. Yellow solid.

Trzeciak, A.M. *et al*, *Inorg. Chim. Acta Lett.*, 1982, **64**, L267 (*synth, ir, uv, nmr*)
Pieta, D. *et al*, *J. Mol. Catal.*, 1983, **18**, 193 (*use*)

$C_{41}H_{39}As_2ClNO_2Rh$ Rh-00388

Chloro(dioxygen)(2-isocyano-2-methylpropane)bis(triphenylarsine)rhodium, 9CI

(tert-*Butylisocyanide)chloro(dioxygen)bis(triphenylarsine)rhodium*

[39176-38-0]

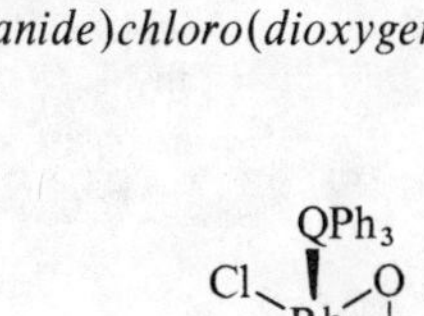

Q = As

M 865.966

Possibly polymeric in solid state. Green cryst. Insol. org. solvs. Mp 180° dec. ir ν_{CN} 2145, $\nu_{O\text{-}O}$ 884 cm^{-1}.

Nakamura, A. *et al*, *Inorg. Chem.*, 1972, **11**, 2058 (*synth, ir, uv*)

$C_{41}H_{39}ClNO_2P_2Rh$ Rh-00389

Chloro(dioxygen)(2-isocyano-2-methylpropane)bis(triphenylphosphine)rhodium, 9CI

Chloro(2-isocyano-2-methylpropane)peroxybis(triphenylphosphine)rhodium, 9CI. (tert-*Butylisocyanide)chloro(dioxygen)bis(triphenylphosphine)rhodium*

[39047-07-9]

As Chloro(dioxygen)(2-isocyano-2-methylpropane)bis(triphenylarsine)rhodium, Rh-00388 with

Q = P

M 778.071

Possibly polymeric in solid state. Green powder. Insol. org. solvs. Mp 147-50° dec. ir ν_{CN} 2149, $\nu_{O\text{-}O}$ 892 cm^{-1}. Loses O_2 in refluxing C_6H_6 suspension.

Nakamura, A. *et al*, *Inorg. Chem.*, 1972, **11**, 2058 (*synth, ir, uv*)

$C_{41}H_{40}OP_2Rh^{\oplus}$ Rh-00390

[(2,3,5,6-η)-Bicyclo[2.2.1]hepta-2,5-diene][[1-(phenylmethoxy)methyl]-1,2-ethanediyl]bis[diphenylphosphine]-*P,P'*]rhodium(1+), 10CI

[1,2-Bis(diphenylphosphino)-3-(benzyloxy)propane]-norbornadienerhodium(1+)

As [(2,3,5,6-η)-Bicyclo[2.2.1]hepta-2,5-diene][[1-[(1,1-dimethylethoxy)methyl]-1,2-ethanediyl]bis[diphenylphosphine]-*P,P'*]rhodium(1+), Rh-00364 with

R = CH_2OCH_2Ph

M 713.619 (ion)

***(R)*-form**

Catalyst for the asymmetric hydrogenation of dehydroamino acids.

Tetrafluoroborate: [80145-74-0].

$C_{41}H_{40}BF_4OP_2Rh$ M 800.423

Yellow solid.

Padmakumari Amma, J. *et al*, *J. Org. Chem.*, 1982, **47**, 468 (*synth, use, nmr*)

$C_{41}H_{43}NO_3P_2Rh^{\oplus}$ Rh-00391

[(1,2-Dimethyl-1,2-ethanediyl)bis[diphenylphosphine]-*P,P'*-][(2,3-η)-ethyl-2-(acetylamino)-3-phenyl-2-propenoate-*O*-[2]]rhodium(1+), 10CI

[(1,2-Dimethyl-1,2-ethanediyl)bis[diphenylphosphine]-][ethyl α-acetamidocinnamate]rhodium(1+). (CHIRAPHOS)(Ethyl α-acetamidocinnamate)rhodium(1+)

[74858-55-2]

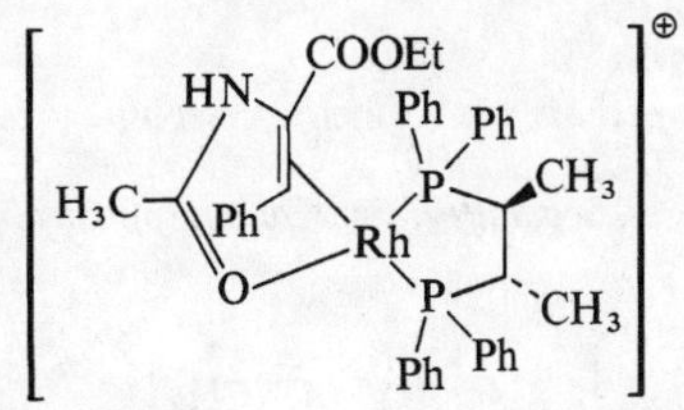

M 762.649 (ion)

Intermed. in Rh(I)-chiraphos-catalysed asymmetric hydrogenation of ethyl α-acetamidocinnamate. Rh preferentially binds to olefinic face shown but the opposite diastereoisomer adds H_2 more rapidly.

Perchlorate: [74892-35-6].

$C_{41}H_{43}ClNO_7P_2Rh$ M 862.099

Sol. MeOH.

Chan, A.S.C. *et al*, *J. Am. Chem. Soc.*, 1980, **102**, 5952 (*synth, cryst struct, props*)
Chua, P.S. *et al*, *J. Chem. Soc., Chem. Commun.*, 1981, 1278 (*cd*)

$C_{42}H_{38}P_2Rh^{\oplus}$ **Rh-00392**

[(1,2,3,4-η)-1,3-Cyclohexadiene]bis(triphenylphosphine)rhodium(1+)

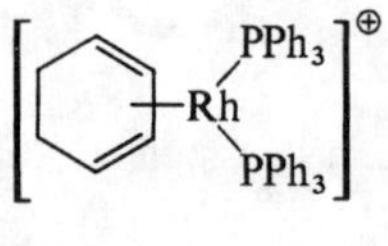

M 707.615 (ion)

Catalyses hydrogenation of cyclohexadienes to cyclohexene.

Perchlorate: [32799-73-8].
$C_{42}H_{38}ClO_4P_2Rh$ M 807.066
Red-orange cryst.

Schrock, R.R. *et al*, *J. Am. Chem. Soc.*, 1971, **93**, 2397 (*synth, pmr*)
Schrock, R.R. *et al*, *J. Am. Chem. Soc.*, 1976, **98**, 4450 (*synth*)

$C_{42}H_{39}ClOP_3Rh$ **Rh-00393**

Carbonylchloro[1,1,1-tris(diphenylphosphinomethyl)ethane]-rhodium

[2,2-Bis[(diphenylphosphinomethyl)propyl]diphenylphosphine]carbonylchlororhodium, 8CI.
Carbonylchloro(TRIPOD)rhodium

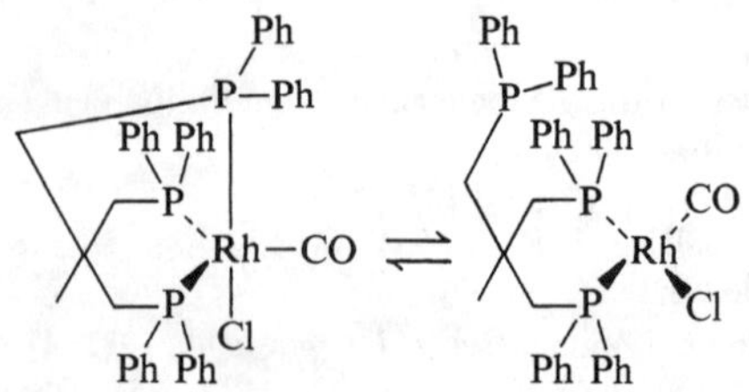

M 791.049

Thought to exist as equilibrium mixt. of 4- and 5-coordinate forms. Air-sensitive red-brown powder. Sol. CH_2Cl_2, Py, sl. sol. C_6H_6, MeCN, MeOH, insol. toluene, Me_2CO. ir ν_{CO} 2005, 1915 cm^{-1} (KBr). Reacts readily with CO to form Dicarbonyl[1,1,1-tris(diphenylphosphinomethyl)ethane]rhodium(1+), Rh-00403 .

Siegl, W.O. *et al*, *Inorg. Chem.*, 1971, **10**, 2158 (*synth, ir, props*)

$C_{42}H_{39}N_3P_2Rh^{\oplus}$ **Rh-00394**

Tris(isocyanomethane)bis(triphenylphosphine)rhodium(1+), 10CI

Tris(methyl isocyanide)bis(triphenylphosphine)-rhodium(1+)

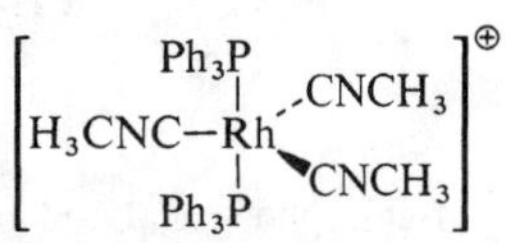

M 750.643 (ion)
ir $\nu_{N\equiv C}$ 2174 cm^{-1}.

Hexafluorophosphate: [67802-86-2].
$C_{42}H_{39}F_6N_3P_3Rh$ M 895.607
Yellow cryst. (cold CH_2Cl_2/diisopropyl ether). Sol. CH_2Cl_2, Me_2CO.

McCleverty, J.A. *et al*, *Transition Met. Chem.*, 1978, **3**, 205 (*synth, ir, pmr*)

$C_{42}H_{44}O_2P_2Rh^{\oplus}$ **Rh-00395**

Dihydrobis(2-propanone)bis(triphenylphosphine)rhodium(1+), 10CI

Bis(acetone)dihydridobis(triphenylphosphine)-rhodium(1+)
[24419-59-8]

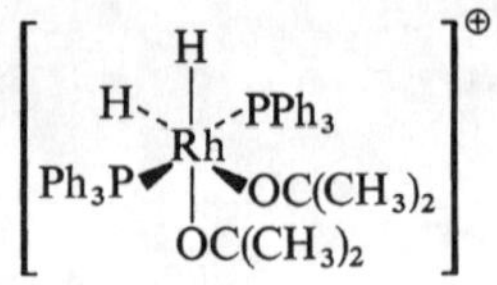

M 745.661 (ion)

Olefin hydrogenation catalyst. Tends to lose hydrogen thermally, esp. in soln.

Hexafluorophosphate: [79027-52-4].
$C_{42}H_{44}F_6O_2P_3Rh$ M 890.626
No phys. props. reported.

Shapley, J.R. *et al*, *J. Am. Chem. Soc.*, 1969, **91**, 2816 (*synth, use*)
Howarth, O.W. *et al*, *J. Chem. Soc., Dalton Trans.*, 1981, 1481; 1982, 541 (*pmr, props*)

$C_{42}H_{49}NOP_2Rh^{\oplus}$ **Rh-00396**

[(1,2,5,6-η)-1,5-Cyclooctadiene][1-(2,2-dimethyl-1-oxopropyl)-4-(diphenylphosphino)-2-[(diphenylphosphino)methyl]-pyrrolidine-*P,P'*]rhodium(1+), 10CI

*(1,5-Cyclooctadiene)[*N*-pivaloyl-4-(diphenylphosphino)-2-[(diphenylphosphino)methyl]pyrrolidine]rhodium(1+)*. *(1,5-Cyclooctadiene)(PPPM)rhodium(1+)*

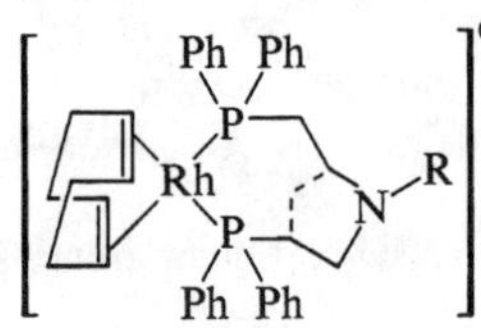

R = $COC(CH_3)_3$

M 748.708 (ion)

(2S,4S)-form

Catalyst precursor for asymmetric hydrogenations, e.g. of $H_2C{=}C(COOH)CH_2COOH$.

Perchlorate: [67322-51-4].
$C_{42}H_{49}ClNO_5P_2Rh$ M 848.159
Orange cryst. Sol. $CHCl_3$, insol. THF.

Achiwa, K., *Chem. Lett.*, 1978, 561 (*synth, use*)
Ojima, I. *et al*, *J. Org. Chem.*, 1980, **45**, 4728 (*synth, cryst struct, nmr*)

$C_{42}H_{49}NO_2P_2Rh^{\oplus}$ **Rh-00397**

[(1,2,5,6-η)-1,5-Cyclooctadiene][1,1-dimethylethyl 2-(diphenylphosphino)-5-[(diphenylphosphino)methyl]-1-pyrrolidinecarboxylate-*P,P'*]rhodium(+), 10CI

*[(1,2,5,6-η)-1,5-Cyclooctadiene][*N*-butoxycarbonyl-4-diphenylphosphino-2-(diphenylphosphino)methylpyrrolidine]rhodium(1+)*. *(BPPM)(1,5-Cyclooctadiene)-rhodium(1+)*

As [(1,2,5,6-η)-1,5-Cyclooctadiene][1-(2,2-dimethyl-1-oxopropyl)-4-(diphenylphosphino)-2-[(diphenylphosphino)methyl]-pyrrolidine-*P,P'*]rhodium(1+), Rh-00396 with

R = $-COOC(CH_3)_3$

M 764.708 (ion)

***(2S,4S)*-form**
Catalyst percursor for the asymmetric hydrogenation of α-keto esters and olefins.
Perchlorate: [67322-49-0].
$C_{42}H_{49}ClNO_6P_2Rh$ M 864.158
Orange solid + $1Et_2O$ (THF/Et_2O). Sol. THF, $CHCl_3$.

Koenig, K.E. *et al*, *J. Org. Chem.*, 1980, **45**, 2362 (*synth*, *use*)
Ojima, I. *et al*, *J. Org. Chem.*, 1980, **45**, 4728 (*synth*, *nmr*, *use*)
Ojima, I. *et al*, *J. Organomet. Chem.*, 1980, **195**, 239 (*synth*, *use*)
Maciel, G.E. *et al*, *Adv. Chem. Ser.*, 1982, **196**, 389 (*synth*, *nmr*)

$C_{43}H_{35}OP_2Rh$ Rh-00398
Carbonylphenylbis(triphenylphosphine)rhodium, 10CI, 9CI

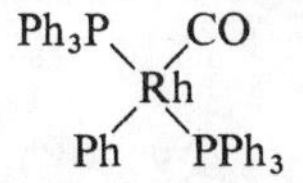

M 732.602
***trans*-form** [82190-10-1]
Reagent for phenylation of acyl halides. Yellow needles. Insol. EtOH. Mp 170-80° (dec.). ir ν_{CO} 1960 cm^{-1} (nujol). Reacts with CH_2Cl_2 to form red $Rh_2(\mu\text{—}CO)_2(PPh_3)_4(CH_2Cl_2)_2$.

Keim, W., *J. Organomet. Chem.*, 1969, **19**, 161 (*synth*, *ir*)
Hegedus, L.S. *et al*, *J. Am. Chem. Soc.*, 1973, **95**, 3040 (*synth*, *use*)
Dahlenburg, D. *et al*, *Z. Naturforsch.*, *B*, 1982, **37**, 310 (*synth*, *ir*, *nmr*)
Fieser, M. *et al*, *Reagents for Organic Synthesis*, Wiley, 1967-83, **5**, 45 (*use*)

$C_{43}H_{35}O_2P_2Rh$ Rh-00399
Carbonylphenoxybis(triphenylphosphine)rhodium, 9CI
Phenolatobis(triphenylphosphine)carbonylrhodium
[49713-89-5]

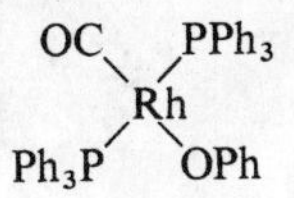

M 748.601
Yellow cryst. (C_6H_6). ir ν_{CO} 1970 cm^{-1} ($CHCl_3$).
***trans*-form**
Yellow cryst. (C_6H_6).

Vaska, L. *et al*, *J. Chem. Soc., Chem. Commun.*, 1971, 418 (*synth*, *ir*)
Vizi-Orosz, A. *et al*, *J. Organomet. Chem.*, 1973, **57**, 379 (*synth*)
Schiavon, G. *et al*, *J. Inorg. Nucl. Chem.*, 1977, **39**, 115 (*props*)

$C_{43}H_{37}O_2P_2Rh$ Rh-00400
(Benzoato-*O,O'*)dihydrobis(triphenylphosphine)rhodium
[62866-34-6]

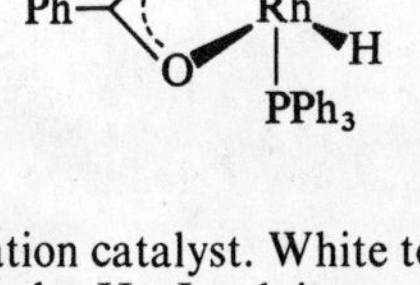

M 750.617
Olefin hydrogenation catalyst. White to pale-yellow cryst., stored under H_2. Insol. inert org. solvs.

Nagy-Magos, Z. *et al*, *Transition Met. Chem.*, 1976, **1**, 215 (*synth*, *ir*, *use*)

$C_{43}H_{38}As_2Rh^{\oplus}$ Rh-00401
[(2,3,5,6-η)-Bicyclo[2.2.1]hepta-2,5-diene]bis(triphenylarsine)rhodium(1+), 10CI
Norbornadienebis(triphenylarsine)rhodium(1+)

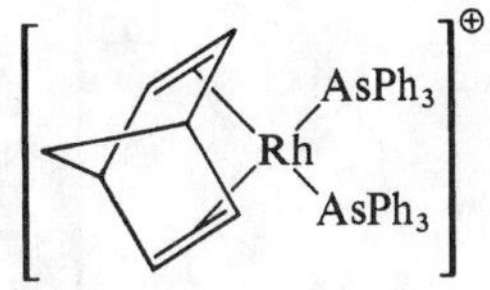

M 807.522 (ion)
Perchlorate: [32802-71-4].
$C_{43}H_{38}As_2ClO_4Rh$ M 906.973
Air-stable orange cryst. + $\frac{1}{2}CH_2Cl_2$ ($CH_2Cl_2/EtOH$). Sol. Me_2CO, insol. Et_2O.

Schrock, R.A. *et al*, *J. Am. Chem. Soc.*, 1971, **93**, 2397 (*synth*, *pmr*)
Schrock, R.A. *et al*, *J. Am. Chem. Soc.*, 1976, **98**, 2134 (*use*)
Valderrama, M. *et al*, *Synth. React. Inorg. Metal-Org. Chem.*, 1979, **9**, 577.

$C_{43}H_{38}P_2Rh^{\oplus}$ Rh-00402
[(2,3,5,6-η)-Bicyclo[2.2.1]hepta-2,5-diene]bis(triphenylphosphine)rhodium(1+), 10CI
(2,5-Norbornadiene)bis(triphenylphosphine)-rhodium(1+), 8CI
[47859-60-9]

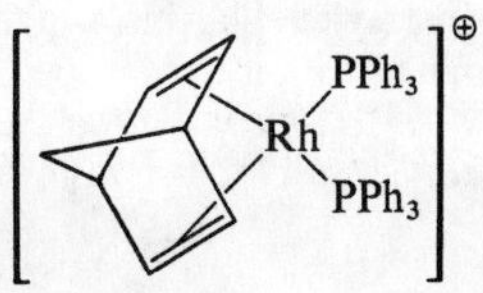

M 719.626 (ion)
Catalyst for the hydrogenation of olefins, alkynes and ketones.
Hexafluorophosphate: [32799-32-9].
$C_{43}H_{38}F_6P_3Rh$ M 864.590
Orange cryst. ($CH_2Cl_2/EtOH$ or Et_2O).
Perchlorate: [32799-31-8].
$C_{43}H_{38}ClO_4P_2Rh$ M 819.077
Orange cryst. ($CH_2Cl_2/EtOH$ or Et_2O).
Tetrafluoroborate: [34664-30-7].
$C_{43}H_{38}BF_4P_2Rh$ M 806.430
Orange cryst. (CH_2Cl_2/Et_2O). Mp 124-6°.
Tetraphenylborate: [31666-38-3].
$C_{67}H_{58}BP_2Rh$ M 1038.858
Red cryst. ($CH_2Cl_2/MeOH$).

Haines, L.M., *Inorg. Chem.*, 1970, **9**, 1517 (*pmr*, *synth*)
Green, M. *et al*, *J. Chem. Soc. (A)*, 1971, 2334 (*synth*)
Osborn, J.A. *et al*, *J. Am. Chem. Soc.*, 1971, **93**, 2397 (*synth*)
Schrock, R.R. *et al*, *J. Am. Chem. Soc.*, 1976, **98**, 2134, 2143 (*use*)
Brown, J.M. *et al*, *J. Chem. Soc., Chem. Commun.*, 1978, 646 (*nmr*)
Uson, R. *et al*, *Transition Met. Chem.*, 1979, **4**, 55 (*synth*, *use*)

$C_{43}H_{39}O_2P_3Rh^{\oplus}$ **Rh-00403**

Dicarbonyl[1,1,1-tris(diphenylphosphinomethyl)ethane]rhodium(1+)

[2,2-Bis[(diphenylphosphinomethyl)propyl]diphenylphosphine]dicarbonylrhodium(1+), 8CI.
Dicarbonyl(TRIPOD)rhodium(1+)

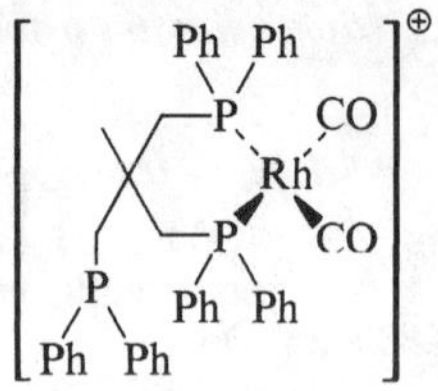

M 783.607 (ion)
In soln. believed to exist as a mixt. of 4- and 5- coordinate complexes.
Chloride:
$C_{43}H_{39}ClO_2P_3Rh$ M 819.060
Formed by bubbling CO through a CH_2Cl_2 sol. of Carbonylchloro[1,1,1-tris(diphenylphosphinomethyl)ethane]rhodium, Rh-00393 . 1:1 Electrolyte in $MeNO_2$.
Tetraphenylborate:
$C_{63}H_{59}BO_2P_3Rh$ M 1054.795
Formed from the chloride by metathesis. Yellow-orange powder. ir ν_{CO} 2048, 1975 cm^{-1} (KBr).

Siegl, W.O. *et al*, *Inorg. Chem.*, 1971, **10**, 2158 (*synth, ir, props*)

$C_{43}H_{52}O_2P_2Rh^{\oplus}$ **Rh-00404**

[(1,2,5,6-η)-1,5-Cyclooctadiene][[(2,2-dimethyl-1,3-dioxolane-4,5-diyl)bis[methylene]]bis[bis(4-methylphenyl)phosphine]-*P,P'*]rhodium(1+), 10CI

(1,5-Cyclooctadiene)[isopropylidine-2,3-dihydroxy-1,4-bis[di(4-methylphenyl)phosphino]butane]-rhodium(1+)

As [(1,2,5,6-η)-1,5-Cyclooctadiene][[(2,2-dimethyl-1,3-dioxolane-4,5-diyl)bis[methylene]]bis(diphenylphosphine)-*P,P'*-]rhodium(1+), Rh-00373 with

R = CH_3

M 765.736 (ion)

(S,S)-form

Catalyst precursor for the asymmetric hydrogenation of olefins e.g. $H_3CCH{=}C(NHAc)Ph$.
Perchlorate: [60552-11-6].
$C_{43}H_{52}ClO_6P_2Rh$ M 865.186
Yellow solid. Sol. $CHCl_3$. $[\alpha]_D^{25}$ −14.2° (c, 1).

Sinou, D. *et al*, *J. Organomet. Chem.*, 1976, **114**, 325 (*synth, pmr, use*)

$C_{44}H_{28}N_4Rh$ **Rh-00405**

[5,10,15,20-Tetraphenyl-21*H*,23*H*-porphinato(2−)-N^{21},N^{22},N^{23},N^{24}]rhodium

[38856-19-8]

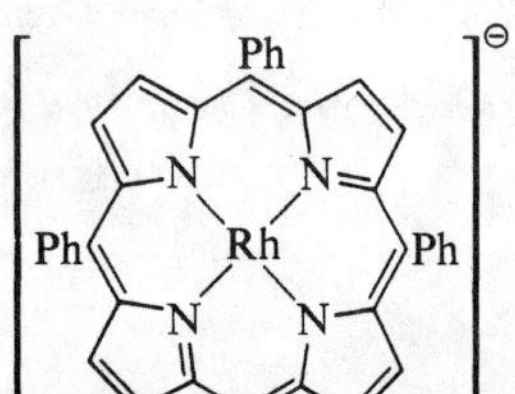

M 715.638
Catalyses the hydrogenation of acetylene and the conversion of hydrogen-oxygen mixtures to water. Reduced by H_2 to Rh(I); the reverse reaction occurs on exposure of the Rh(I) solution to air.
H deriv: [39153-29-2]. *Hydro(tetraphenylporphyrinato)-rhodium.*
$C_{44}H_{29}N_4Rh$ M 716.645
Extremely air-sensitive.

James, B.R. *et al*, *J. Am. Chem. Soc.*, 1972, **94**, 6225 (*synth, esr, uv, props*)

$C_{44}H_{30}As_2ClF_{12}Rh$ **Rh-00406**

Chloro[1,2,3,4-tetrakis(trifluoromethyl)-1,3-butadiene-1,4-diyl]bis(triphenylarsine)rhodium, 9CI

Chlorobis(triphenylarsine)tetrakis(trifluoromethyl)rhodacyclopentadiene

[52612-92-7]

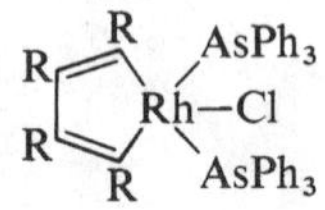

R = CF_3

M 1074.904
Bright-yellow cryst. (1,1-dichloroethane), orange-yellow prisms (Et_2O/C_6H_6). Mp 195-7°.

Mague, J.T. *et al*, *Inorg. Chem.*, 1968, **7**, 542 (*synth, ir*)
Mague, J.T. *et al*, *J. Chem. Soc., Dalton Trans.*, 1973, 2578 (*synth, ir, nmr*)

$C_{44}H_{42}P_2Rh^{\oplus}$ **Rh-00407**

[(1,2,5,6-η)-1,5-Cyclooctadiene]bis(triphenylphosphine)rhodium(1+), 10CI

[47864-84-6]

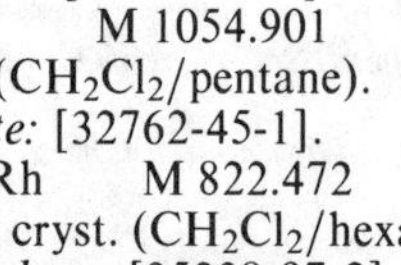

M 735.669 (ion)
Catalyst for hydrogenation and hydroformylation of olefins.
Tetraphenylborate: [31761-02-1].
$C_{68}H_{62}BP_2Rh$ M 1054.901
Orange cryst. (CH_2Cl_2/pentane).
Tetrafluoroborate: [32762-45-1].
$C_{44}H_{42}BF_4P_2Rh$ M 822.472
Yellow-orange cryst. (CH_2Cl_2/hexane). Mp 192-4°.
Hexafluorophosphate: [35238-97-2].
$C_{44}H_{42}F_6P_3Rh$ M 880.633
Orange cryst. (CH_2Cl_2/EtOH or Et_2O).
Perchlorate: [32799-30-7].
$C_{44}H_{42}ClO_4P_2Rh$ M 835.119
Orange cryst. (CH_2Cl_2/EtOH or Et_2O).

Haines, L.M., *Inorg. Chem.*, 1970, **9**, 1517 (*synth, pmr*)
Osborn, J.A. *et al*, *J. Am. Chem. Soc.*, 1971, **93**, 2397 (*synth, pmr*)
Green, M. *et al*, *J. Chem. Soc. (A)*, 1971, 2334 (*synth, pmr*)
Lewis, J. *et al*, *J. Chem. Soc. (A)*, 1971, 2699 (*synth, pmr*)
Crabtree, R.H. *et al*, *J. Organomet. Chem.*, 1977, **141**, 113 (*use*)
Crabtree, R.H. *et al*, *J. Mol. Catal.*, 1979, **5**, 75 (*use*)

$C_{44}H_{50}P_2Rh^{\oplus}$ **Rh-00408**

[(1,2,5,6-η)-1,5-Cyclooctadiene[5-methyl-7-(1-methylethyl-)bicyclo[2.2.2]oct-5-ene-2,3-diyl]bis[diphenylphosphine]]-rhodium(1+), 10CI

1,5-Cyclooctadiene(PHELLANPHOS)rhodium(1+)

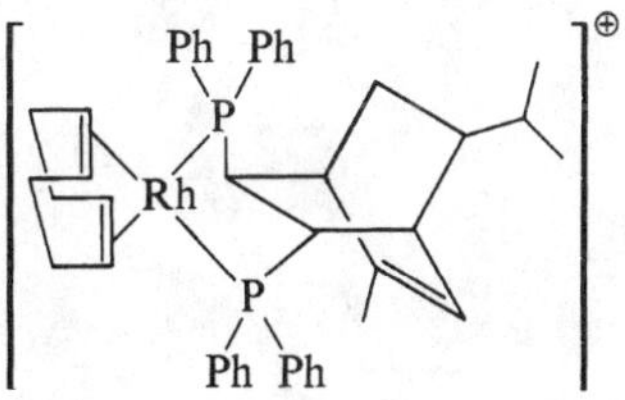

M 743.732 (ion)

(S,S)-form

Catalyst precursor for the asymmetric hydrogenation of *N*-acetyl α-aminoacrylic acids and other prochiral olefins.

Hexafluorophosphate: [72034-28-7].
$C_{44}H_{50}F_6P_3Rh$ M 888.696
Red or yellow air-stable solid. Sol. EtOH, $CHCl_3$. $[\alpha]_D^{20}$ +106° (c, 0.2 in $CHCl_3$). In the presence of oxygen a less active, less stereospecific brown complex obt.

Lauer, M. *et al*, *J. Organomet. Chem.*, 1979, **177**, 309 (*synth, use*)
Samuel, O. *et al*, *Nouv. J. Chim.*, 1981, **5**, 15 (*synth, use, nmr*)

$C_{45}H_{28}ClN_4ORh$ **Rh-00409**

Carbonylchloro[5,10,15,20-tetraphenyl-21*H*,23*H*-porphinato(2−)-N^{21},N^{22},N^{23},N^{24}]rhodium, 9CI

Carbonylchloro(tetraphenylporphyrin)rhodium

[42892-91-1]

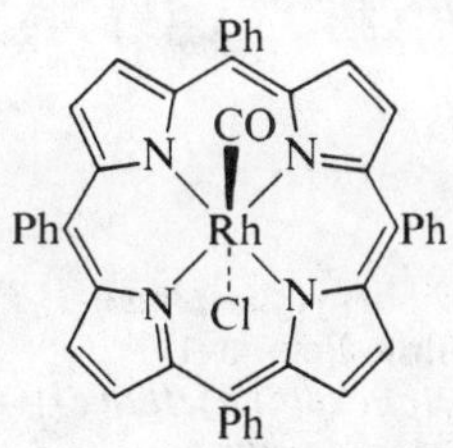

M 779.101

Cryst. Diamagnetic.

Fleischer, E.B. *et al*, *J. Chem. Soc., Chem. Commun.*, 1969, 475 (*synth, ir*)
Cohen, I.A. *et al*, *Inorg. Chem.*, 1974, **13**, 488 (*props*)

$C_{45}H_{44}O_6P_2Rh^{\oplus}$ **Rh-00410**

[(2,3,5,6-η)Bicyclo[2.2.1]hepta-2,5-diene][methyl-4,6-*O*--(phenylmethylene)glucopyranosidebis(diphenylphosphinite)-*P*,*P*′]rhodium(1+), 10CI

[Methyl-2,3-bis(O-diphenylphosphino)-4,6-O-benzylideneglucopyranoside]norbornadienerhodium(1+)

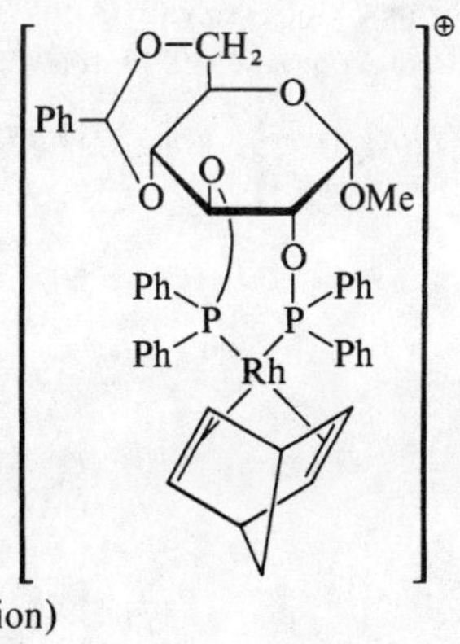

M 845.692 (ion)

α-D-form

Hexafluorophosphate: [68089-18-9].
$C_{45}H_{44}F_6O_6P_3Rh$ M 990.656
Catalyst for the asymmetric hydrogenation of α-acetamidoacrylic acids and esters. No phys. props. reported.

Cullen, W.R. *et al*, *Tetrahedron Lett.*, 1978, 1635 (*synth, use*)

$C_{45}H_{50}P_2Rh^{\oplus}$ **Rh-00411**

[(1,2,5,6-η)-1,5-Cyclooctadiene][(1,2,3,4,4*a*,5,6,7-octahydro-2,2-dimethyl-1,3-methanonaphthalene-5,6-diyl)bis[diphenylphosphine]rhodium(1+), 10CI

1,5-Cyclooctadiene(NOPAPHOS)rhodium(1+)

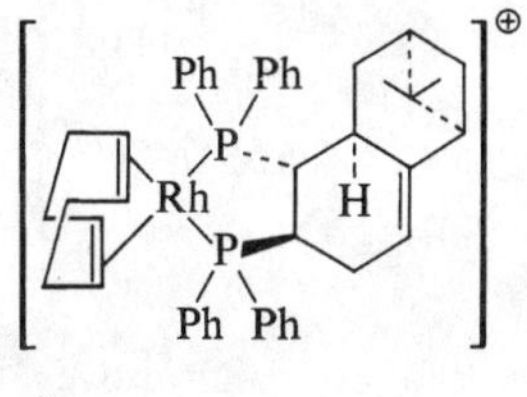

M 755.743 (ion)

(R,R)-form

Catalyst precursor for the asymmetric hydrogenation of *N*-acetyl α-aminoacrylic acids and other prochiral olefins.

Hexafluorophosphate: [73172-45-9].
$C_{45}H_{50}F_6P_3Rh$ M 900.707
Yellow air-stable solid. Sol. $CHCl_3$, EtOH. $[\alpha]_D^{20}$ −23° (c, 0.2 in $CHCl_3$).

Samuel, O. *et al*, *Nouv. J. Chim.*, 1981, **5**, 15 (*synth, nmr, use*)

$C_{46}H_{46}N_2P_2Rh^{\oplus}$ **Rh-00412**

[(1,2,5,6-η)-1,5-Cyclooctadiene][*N*,*N*′-(1,2-diphenyl-1,2-ethanediyl)bis[*P*,*P*-diphenylphosphinous amide]-*P*,*P*′]rhodium(1+), 10CI

(1,5-Cyclooctadiene)[N,N′-bis(diphenylphosphino)-1,2-diphenylethylenediamine]rhodium(1+)

As [(1,2,5,6-η)-1,5-Cyclooctadiene][*N*,*N*′-(1,2-dimethyl-1,2-ethanediy)bis[*P*,*P*-diphenylphosphinous amide]-*P*,*P*″]-rhodium(1+), Rh-00317 with

R = Ph

M 791.736 (ion)

(S,S)-form

Perchlorate:
$C_{46}H_{46}ClN_2O_4P_2Rh$ M 891.186
No phys. props. recorded.

Fiorini, M. *et al*, *J. Mol. Catal.*, 1979, **5**, 303 (*synth, use*)
Fiorini, M. *et al*, *J. Mol. Catal.*, 1980, **7**, 411 (*use*)

The first digit of the Entry number defines the Supplement in which the Entry is found. 0 indicates the Main Work

$C_{46}H_{50}O_4P_2Rh^{\oplus}$ **Rh-00413**

[(1,2,5,6-η)-1,5-Cyclooctadiene][[[5-[2-(diphenylphosphino)-1-methoxyethoxy]-2-phenyl-1,3-dioxan-4-yl]methyl]-diphenylphosphine-*P*,*P'*]rhodium(1+), 10CI

[1,5-Cyclooctadiene][O-benzylidene-2,4-O-(diphenylphosphino-2'-methoxy-1-ethyl)-3-desoxyl-1-diphenylphosphino-1-D-erythritol]rhodium(1+). (1,5-Cyclooctadiene)(ERYTHRITOP)rhodium(1+)

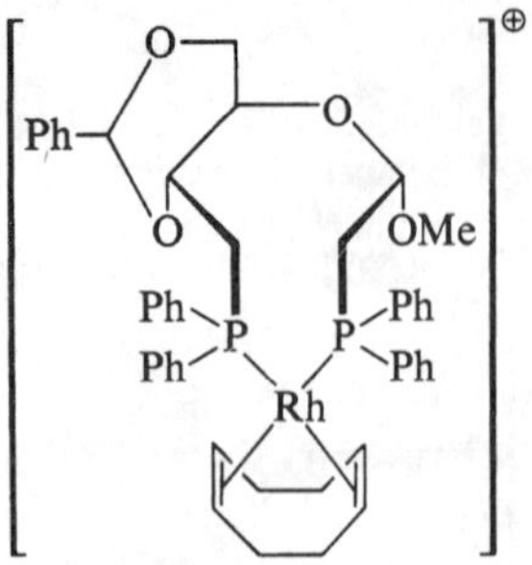

M 831.752 (ion)

Catalyst precursor for asymm. hydrogenations.

Perchlorate: [70163-66-5].
$C_{46}H_{50}ClO_8P_2Rh$ M 931.202
Solid. $[\alpha]^{20}_{D}$ −52° (c, 0.52 in CH_2Cl_2).

Lafont, D. *et al*, *J. Organomet. Chem.*, 1979, **169**, 87 (*synth, use*)

$C_{48}H_{42}As_2ClO_8Rh$ **Rh-00414**

Chloro[1,2,3,4-tetrakis(methoxycarbonyl)-1,3-butadiene-1,4-diyl]bis(triphenylarsine)rhodium, 9CI

Chloro(1,2,3,4-tetracarboxy-1,3-butadienylene)bis(triphenylarsine)rhodium tetramethyl ester, 8CI. Chloro[tetra(methoxycarbonyl)-1,3-butadienylene]bis(triphenylarsine)rhodium

[18989-29-2]

As Chloro[1,2,3,4-tetrakis(trifluoromethyl)-1,3-butadiene-1,4-diyl]bis(triphenylarsine)rhodium, Rh-00406 with

R = COOMe

M 1035.057

Orange cryst. ($CHCl_3$/MeOH). Mp 238-40°.

Collman, J.P. *et al*, *Inorg. Chem.*, 1968, **7**, 1298 (*synth, ir, nmr*)

$C_{49}H_{40}N_6OP_3Rh$ **Rh-00415**

Carbonylhydrobis[2,2',2''-phosphinidynetris(pyridine)-*P*](triphenylphosphine)rhodium, 10CI

Carbonylhydrobis[tris(2-pyridyl)phosphine]triphenylphosphinerhodium

[73413-34-0]

$HRhCO(PPh_3)[P(C_5H_4N)_3]_2$

M 924.721

Catalyst for the hydroformylation of 1-hexene in H_2O. Yellow solid. Sol. H_2O, most org. solvs., insol. hydrocarbons. Struct. not fully descr.

Kurtev, K. *et al*, *J. Chem. Soc., Dalton Trans.*, 1980, 55 (*ir, synth, pmr*)

$C_{50}H_{40}As_2ClRh$ **Rh-00416**

Chloro(η^2-diphenylacetylene)bis(triphenylarsine)rhodium, 8CI

[12124-05-9]

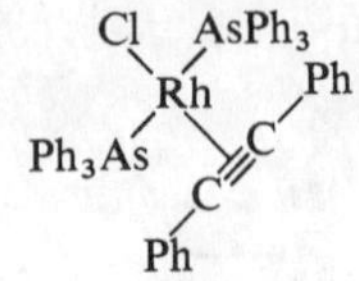

M 929.068

Orange cryst. Mp 218-20° dec.

Mague, J.T. *et al*, *J. Chem. Soc. (A)*, 1966, 1736 (*synth, ir*)

$C_{50}H_{40}ClP_2Rh$ **Rh-00417**

Chloro[1,1'-(η^2-1,2-ethynediyl)bis[benzene]]bis(triphenylphosphine)rhodium, 10CI

Chlorodiphenylacetylenebis(triphenylphosphine)rhodium, 8CI. Diphenylacetylenebis(triphenylphosphine)-rhodium chloride

[12124-06-0]

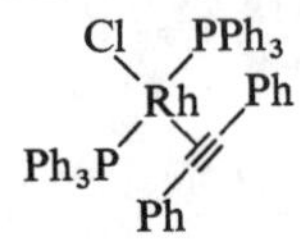

M 841.172

trans-form

Yellow cryst. Mp 157-63° dec. Dissociates in soln. and cannot be recryst.

Mague, J.T. *et al*, *J. Chem. Soc. (A)*, 1966, 1736 (*synth, ir*)
Van Gaal, H.L.M. *et al*, *J. Organomet. Chem.*, 1977, **131**, 453 (*ir*)

$C_{50}H_{44}BP_2Rh$ **Rh-00418**

[1,2-Ethanediylbis(diphenylphosphine)-*P*,*P'*][(η^6-phenyl)triphenylborato(1−)]rhodium, 10CI

[1,2-Bis(diphenylphosphino)ethane][(η^6-phenyl)triphenylborato(1−)]rhodium

[72332-27-5]

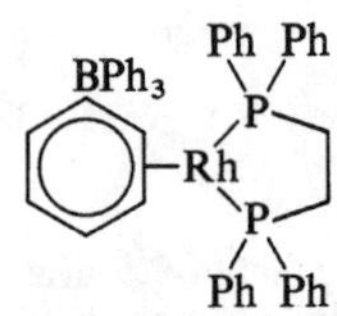

M 820.561

Effective catalyst for methylacetylene oligomerisation and allene polymerisation. Orange, air-stable cryst. Mp 230° dec. Forms CH_2Cl_2 solvate. Adds CO_2 reversibly.

Albano, P. *et al*, *Inorg. Chem.*, 1980, **19**, 1069 (*synth, cryst struct*)
Albano, P. *et al*, *J. Organomet. Chem.*, 1980, **190**, 243 (*use*)

$C_{50}H_{54}N_2P_2Rh^{\oplus}$ Rh-00419

[(1,2,5,6-η)-1,5-Cyclooctadiene][*N,N'*-1,2-ethanediylbis[*P,P*-diphenyl-*N*-(1-phenylethyl)phosphinous amide]-*N,N'*]rhodium(1+), 10CI

1,5-Cyclooctadiene[N,N'-*bis(α-methylbenzyl)*-N,N'-*bis(diphenylphosphino)ethylene*]*rhodium(1+)*. (*1,5-Cyclooctadiene*)[N,N'-*bis(diphenylphosphino)*-N,N'-*bis(α-methylbenzyl)ethylenediamine*]*rhodium(1+)*

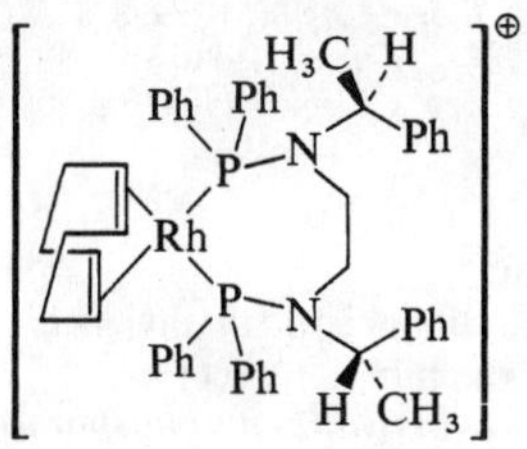

M 847.843 (ion)

Catalyst for the asymmetric hydrogenation of ketones and alkenes.

***(S,S)*-form**

Perchlorate: [66010-33-1].
$C_{50}H_{54}ClN_2O_4P_2Rh$ M 947.294
Red-orange solid (THF/Et_2O).

Fiorini, M. *et al*, *J. Mol. Catal.*, 1977, **3**, 385 (*use*)
Fiorini, M. *et al*, *J. Mol. Catal.*, 1978, **4**, 125 (*synth, pmr, use*)

$C_{50}H_{76}ClP_2Rh$ Rh-00420

Chloro(η^2-diphenylacetylene)bis(tricyclohexylphosphine)rhodium

Chloro[1,1'-(η^2-1,2-ethynediyl)bis[benzene]]bis(tricyclohexylphosphine)rhodium, 10CI

[63443-47-0]

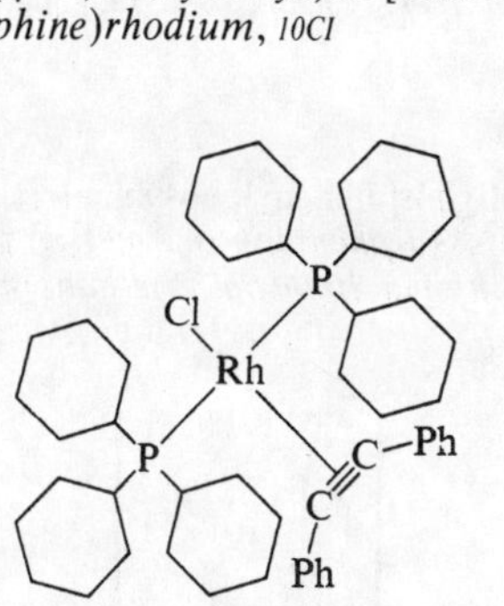

M 877.456

Yellow cryst.

Van Gaal, H.L.M. *et al*, *J. Organomet. Chem.*, 1977, **131**, 453; **133**, 93; **134**, 237 (*synth, ir, pmr*)

$C_{51}H_{40}PRh$ Rh-00421

(η^5-2,4-Cyclopentadien-1-yl)(1,2,3,4-tetraphenyl-1,3-butadiene-1,4-diyl)(triphenylphosphine)rhodium, 9CI

1-(π-Cyclopentadienyl)-1-(triphenylphosphine)-2,3,4,5-tetraphenylrhodole

[54913-14-3]

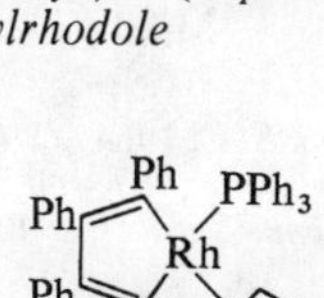

M 786.756

Orange-brown cryst. (C_6H_6/hexane). Mp 275-7°.

Gardner, S.A. *et al*, *J. Organomet. Chem.*, 1974, **78**, 415 (*synth, ir, ms, pmr*)

$C_{51}H_{44}Cl_2OP_4Rh_2$ Rh-00422

μ-Carbonyldichlorobis[μ-[methylenebis[diphenylphosphine]-*P:P'*]]dirhodium, 10CI

μ-Carbonylbis[μ-bis(diphenylphosphino)methane]dichlorodirhodium

[73680-37-2]

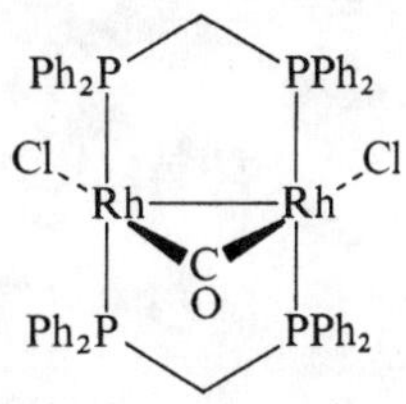

M 1073.520

Obt. from Dicarbonyldichlorobis[μ-[methylenebis[diphenylphosphine]-*P:P'*]]dirhodium, Rh-00426 by heating. Red cryst. (CH_2Cl_2/Et_2O). ir ν_{CO} 1745 cm⁻ (nujol).

Cowie, M. *et al*, *Inorg. Chem.*, 1980, **19**, 2500 (*synth, ir, nmr*)

$C_{52}H_{44}As_4ClO_2Rh_2^{\oplus}$ Rh-00423

Dicarbonyl-μ-chlorobis[μ-[methylenebis[diphenylarsine]-*As:As'*]]dirhodium(1+), 10CI

μ-Chlorobis[μ-bis(diphenylarsino)methane]dicarbonyldirhodium(1+)

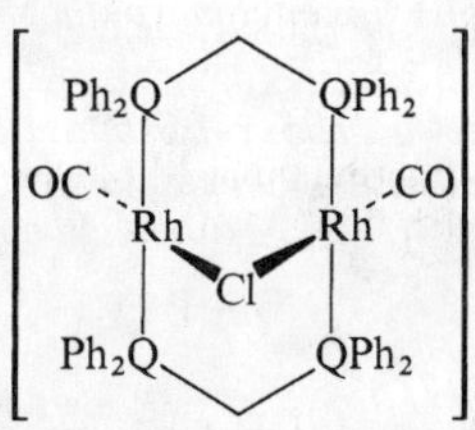

Q = As

M 1241.869 (ion)

ir ν_{CO} 2003, 1992 cm^{-1} (CH_2Cl_2).

Dicarbonyldichlororhodate: [72260-34-5].
$C_{54}H_{44}As_4Cl_3O_4Rh_3$ M 1471.701
Yellow-orange cryst. (CH_2Cl_2/Et_2O or C_6H_6).

Hexafluorophosphate: [74965-10-9].
$C_{52}H_{44}As_4ClF_6O_2PRh_2$ M 1386.833
Yellow-orange cryst. (Me_2CO/Et_2O). Mp 265° dec.

Tetraphenylborate: [70561-97-6].
$C_{76}H_{64}As_4BClO_2Rh_2$ M 1561.101
Orange cryst. Mp 154° dec.

Mague, J.T. *et al*, *Inorg. Chem.*, 1979, **18**, 2060 (*synth, ir, nmr*)
Olmstead, M.M. *et al*, *J. Organomet. Chem.*, 1979, **179**, 289 (*synth*)
Mague, J.T. *et al*, *Inorg. Chem.*, 1980, **19**, 3743.

$C_{52}H_{44}As_4Cl_2O_2Rh_2$ Rh-00424

Dicarbonyldichlorobis[μ-[methylenebis[diphenylarsine]-*As:As'*]]dirhodium, 9CI

Bis[[μ-bis(diphenylarsino)methane]carbonylchlororhodium]. Bis[μ-bis(diphenylarsino)methane]-dicarbonyldichlorodirhodium

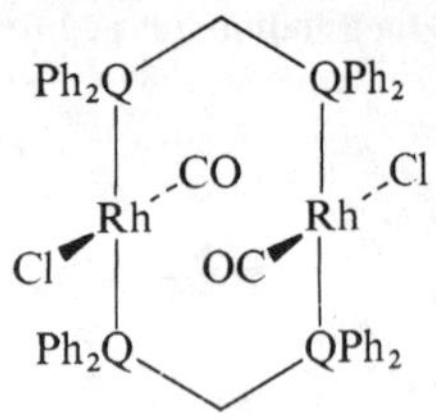

trans-form

Q = As

M 1277.322

Orange cryst. (CH_2Cl_2/Et_2O). Mp 223° dec. ir ν_{CO} 1973 cm^{-1} ($CHCl_3$).

CH_2Cl_2 solvate: [76529-81-2].

$C_{53}H_{46}As_4Cl_4O_2Rh_4$ M 1568.066

Dark-red prisms (CS_2 or CH_2Cl_2/Et_2O).

Mague, J.T. *et al*, *Inorg. Chem.*, 1969, **8**, 119 (*synth, ir*)
Mague, J.T., *Inorg. Chem.*, 1969, **8**, 1975 (*cryst struct*)
Balch, A.L., *J. Am. Chem. Soc.*, 1976, **98**, 8049 (*ir, uv, props*)
Cowie, M. *et al*, *Inorg. Chem.*, 1981, **20**, 1534 (*cryst struct*)

$C_{52}H_{44}ClO_2P_4Rh_2^{\oplus}$ Rh-00425

Dicarbonyl-μ-chlorobis[μ-[methylenebis(diphenylphosphine)-*P:P'*]]dirhodium(1+), 9CI

μ-Chlorobis[μ-bis(diphenylphosphino)methane]dicarbonyldirhodium(1+)

As Dicarbonyl-μ-chlorobis[μ-[methylenebis[diphenylarsine]-*As:As'*]]dirhodium(1+), Rh-00423 with

Q = P

M 1066.077 (ion)

ir ν_{CO} 2006, 1992 cm^{-1} (CH_2Cl_2).

Tetraphenylborate: [67202-35-1].

$C_{76}H_{64}BClO_2P_4Rh_2$ M 1385.309

Orange cryst. Mp 170° dec. Adds carbon monoxide reversibly.

Dicarbonyldichlororhodate: [72260-35-6].

$C_{54}H_{44}Cl_3O_4P_4Rh_3$ M 1295.910

Orange cryst. (CH_2Cl_2/Et_2O).

Chloride: [71647-00-2].

$C_{52}H_{44}Cl_2O_2P_4Rh_2$ M 1101.530

Not fully characterised.

Tetrafluoroborate:

$C_{52}H_{44}BClF_4O_2P_4Rh_2$ M 1152.881

Orange cryst. (Me_2CO/Et_2O).

Mague, J.T. *et al*, *Inorg. Chem.*, 1979, **18**, 2060 (*synth, ir, nmr*)
Cowie, M. *et al*, *Inorg. Chem.*, 1979, **18**, 2700 (*cryst struct*)
Olmstead, M.M. *et al*, *J. Organomet. Chem.*, 1979, **179**, 289.
Cowie, M. *et al*, *Inorg. Chem.*, 1980, **19**, 209 (*props*)

$C_{52}H_{44}Cl_2O_2P_4Rh_2$ Rh-00426

Dicarbonyldichlorobis[μ-[methylenebis[diphenylphosphine]-*P:P'*]]dirhodium, 9CI

Bis[[μ-bis(diphenylphosphino)methane]carbonylchlororhodium]

As Dicarbonyldichlorobis[μ-[methylenebis[diphenylarsine]-*As:As'*]]dirhodium, Rh-00424 with

Q = P

M 1101.530

cis-form [69063-77-0]

Yellow-orange cryst. Sol. CH_2Cl_2, Me_2CO. ν_{CO} 1971 cm^{-1} (nujol); formulation as *cis*-isomer based on chemical reactivity.

trans-form [22427-58-3]

Orange cryst. Insol. CH_2Cl_2, Me_2CO. Mp 221° dec. ir ν_{CO} 1976, 1937 cm^{-1} (nujol).

Mague, J.T. *et al*, *Inorg. Chem.*, 1969, **8**, 119 (*synth, ir*)
Cowie, M. *et al*, *Inorg. Chem.*, 1980, **19**, 209 (*nmr, ir*)
Cowie, M. *et al*, *Inorg. Chem.*, 1980, **19**, 2500 (*cryst struct*)

$C_{52}H_{44}IO_2P_4Rh_2^{\oplus}$ Rh-00427

μ-Carbonylcarbonyliodobis[μ-[methylenebis[diphenylphosphine]-*P:P'*]]dirhodium(1+), 9CI

μ-Carbonylbis[μ-bis(diphenylphosphino)methane]carbonyliododirhodium(1+)

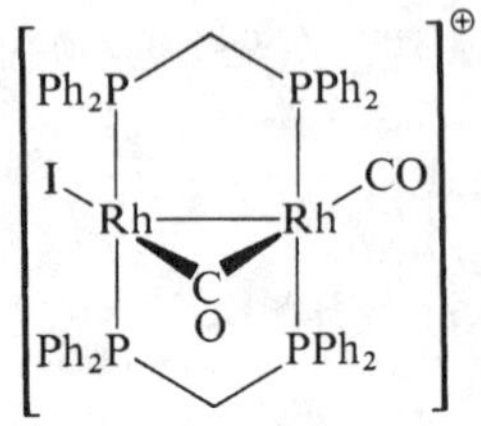

M 1157.529 (ion)

ir ν_{CO} 1955, 1810 cm^{-1} (nujol).

Iodide: [73680-38-3].

$C_{52}H_{44}I_2O_2P_4Rh_2$ M 1284.433

Stable in soln. under nitrogen without loss of CO.

Cowie, M. *et al*, *Inorg. Chem.*, 1980, **19**, 2500 (*synth, ir, nmr*)

$C_{52}H_{48}P_4Rh^{\oplus}$ Rh-00428

Bis[1,2-ethanediylbis[diphenylphosphine]-*P,P'*]rhodium(1+)

Bis[ethylenebis[diphenylphosphine]]rhodium(1+), 8CI. *Bis[1,2-bis(diphenylphosphino)ethane]rhodium(1+)*

[47895-57-8]

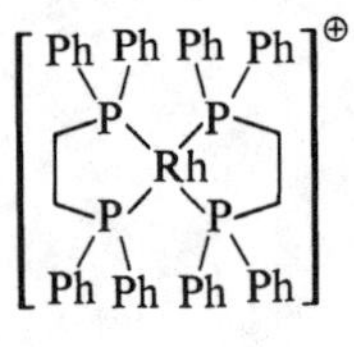

M 899.752 (ion)

Chloride: [15043-47-7].

$C_{52}H_{48}ClP_4Rh$ M 935.205

Yellow needles (CH_2Cl_2/C_6H_6). Sol. MeOH, CH_2Cl_2, sl. sol. THF, Me_2CO. Dec. ca. 240°.

Hexafluorophosphate: [17374-65-1].

$C_{52}H_{48}F_6P_5Rh$ M 1044.716

Yolk-yellow cryst. (MeOH). Mp 265°.

Perchlorate: [30513-14-5].

$C_{52}H_{48}ClO_4P_4Rh$ M 999.202

Yellow plates.

Sacco, A. *et al*, *J. Chem. Soc.*, 1964, 3274 (*synth*)
Hieber, W. *et al*, *Chem. Ber.*, 1967, **100**, 148 (*synth, ir*)
Hall, M.C. *et al*, *J. Chem. Soc. (A)*, 1970, 2539 (*cryst struct*)

$C_{52}H_{48}P_4Rh_2^{\oplus\oplus}$ Rh-00429

Bis[μ-[[2-(diphenylphosphino)ethyl]phenyl(η⁶-phenyl)phosphine-*P,P'*]dirhodium(2+), 10CI

Bis[1,2-bis(diphenylphosphino)ethane]dirhodium(2+)

[64867-97-6]

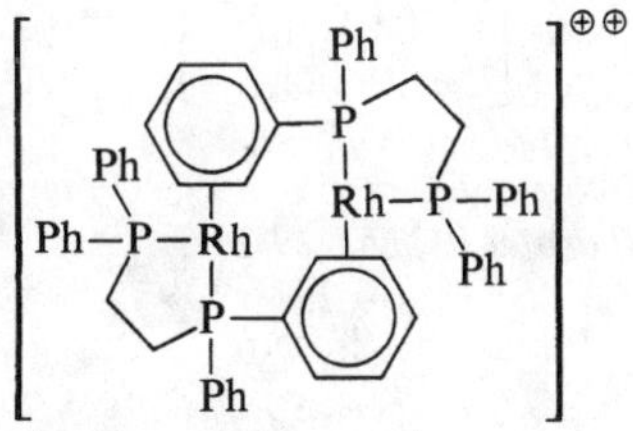

M 1002.657 (ion)

Olefin hydrogenation catalyst.

Bis(tetrafluoroborate): [64867-98-7]. $C_{52}H_{48}B_2F_8P_4Rh_2$ M 1176.264 Cryst.

Halpern, J. *et al*, *J. Am. Chem. Soc.*, 1977, **99**, 8055 (*synth, uv, nmr, cryst struct*)
Halpern, J. *et al*, *Adv. Chem. Ser.*, 1979, **173**, 16 (*synth*)

$C_{54}H_{44}Cl_2F_6P_4Rh_2$ Rh-00430

Dichloro[μ-[1,2-bis(trifluoromethyl)-1,2-ethenediyl]]bis[μ-[methylenebis[diphenylphosphine]-*P:P'*]dirhodium, 10CI

Dichlorobis[μ-bis(diphenylphosphino)methane]-μ-hexafluorobutynedirhodium

[77590-00-2]

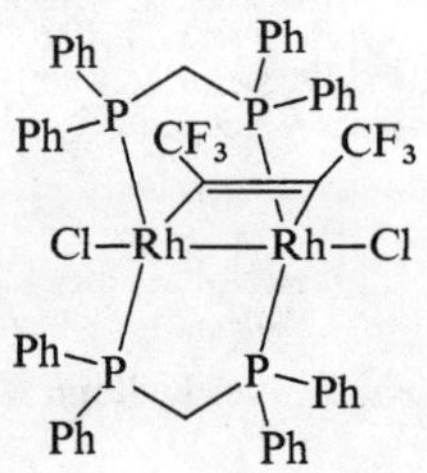

M 1207.544

Green solid.

Cowie, M. *et al*, *J. Organomet. Chem.*, 1980, **193**, C46 (*synth, ir, nmr*)
Cowie, M. *et al*, *Inorg. Chem.*, 1981, **20**, 2682 (*synth, ir, nmr, cryst struct*)
Cowie, M. *et al*, *Inorg. Chem.*, 1982, **21**, 246 (*synth, ir, nmr*)

$C_{54}H_{45}BrP_3Rh$ Rh-00431

Bromotris(triphenylphosphine)rhodium

[14973-89-8]

$Rh(PPh_3)_3Br$

M 969.680

Exists as two forms with identical spectroscopic and chemical props. Catalyst for the hydrogenation and hydrosilylation of olefins.

Orange-form

Orange cryst. Mp 133-4°.

Brown-form

Brown cryst. Mp 129-30°.

Bennett, M.A. *et al*, *Chem. Ind. (London)*, 1965, 846 (*synth*)
Osborn, J.A. *et al*, *J. Chem. Soc. (A)*, 1966, 1711 (*use, synth*)
Bennett, M.A. *et al*, *Inorg. Chem.*, 1967, **6**, 1647 (*ir*)
Inorg. Synth., 1967, **10**, 67 (*synth*)
Brown, J.H. *et al*, *J. Am. Chem. Soc.*, 1970, **92**, 2359 (*nmr, uv*)
Reed, J. *et al*, *J. Chem. Soc., Chem. Commun.*, 1977, 628 (*struct*)

$C_{54}H_{45}ClP_3Rh$ Rh-00432

Chlorotris(triphenylphosphine)rhodium, 10CI

Tris(triphenylphosphine)chlororhodium. Tris(triphenylphosphine)rhodium chloride. Wilkinson's catalyst

[14694-95-2]

$(Ph_3P)_3RhCl$

M 925.229

Active catalyst for the hydrogenation and hydroformylation of alkenes and alkynes, for the hydrosilylation of alkenes and carbonyl compds., for the selective deuteration of unsaturated compds. and the decarbonylation of acid halides and aldehydes. Also functions as an olefin isomerization catalyst. Red-purple cryst. (EtOH). Mp 157-8°. Also exists in a yellow form.

Osborn, J.A. *et al*, *J. Chem. Soc. (A)*, 1966, 1711 (*synth, props*)
Inorg. Synth., 1967, **10**, 67 (*synth*)
Brown, T. *et al*, *J. Am. Chem. Soc.*, 1970, **92**, 2359 (*nmr*)
Carlton, L. *et al*, *J. Chem. Soc., Perkin Trans. 1*, 1978, 1631 (*use*)
Catellani, M. *et al*, *Symp. Rhodium Homogeneous Catal.* [*Proc.*], 1978, 136; *CA*, **90**, 54427w (*use*)
Daub, J. *et al*, *Chem. Ber.*, 1978, **111**, 2877 (*use*)
Kampmeier, J.A. *et al*, *J. Org. Chem.*, 1980, **45**, 315 (*use*)
Halpern, J., *Inorg. Chim. Acta*, 1981, **50**, 11 (*props*)
Fieser, M. *et al*, *Reagents for Organic Synthesis*, Wiley, 1967-83, **10**, 98 (*use*)

$C_{54}H_{45}NOP_3Rh$ Rh-00433

Nitrosyltris(triphenylphosphine)rhodium, 10CI

[21558-94-1]

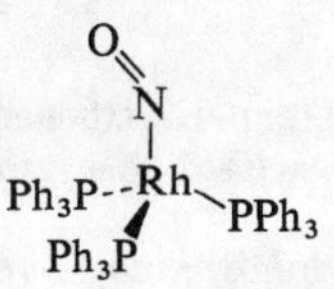

M 919.782

Catalyst for the hydrogenation of olefins. Red cryst. (CH_2Cl_2/EtOH). Mp 160° dec. (in air), 205-6° (under N_2). ir ν_{NO} 1610 cm^{-1} (nujol). Dec. if repeatedly recryst.

Collman, J.P. *et al*, *J. Am. Chem. Soc.*, 1969, **91**, 5659 (*synth, ir, use*)
Robinson, S.D. *et al*, *J. Chem. Soc. (A)*, 1970, 2947 (*synth, ir*)
Caulton, K.G., *Inorg. Chem.*, 1974, **13**, 1774 (*nmr*)
Inorg. Synth., 1974, **15**, 45; 1976, **16**, 32 (*synth, ir*)
Kaduk, J.A. *et al*, *Isr. J. Chem.*, 1977, **15**, 143 (*struct*)

$C_{54}H_{46}Cl_2P_3Rh$ Rh-00434

Dichlorohydrotris(triphenylphosphine)rhodium, 10CI

Dichlorohydridotris(triphenylphosphine)rhodium

[12124-48-0]

$(Ph_3P)_3RhHCl_2$

M 961.690

Catalyst for isomerisation of alkenes, hydrogenation of alkenes and alkynes and dec. of secondary carboxamides.

α-form

Yellow cryst. Mp 160° dec. ir ν_{RhH} 2220 cm^{-1} (nujol).

β-form

Pale-yellow cryst. Mp 100° dec. ν_{RhH} 2120 cm^{-1} (nujol). Loses HCl in EtOH soln.

Sacco, A. *et al*, *J. Chem. Soc. (A)*, 1966, 1670 (*synth, ir*)
Candlin, J.P. *et al*, *Disc. Faraday Soc.*, 1968, **46**, 60 (*synth, use*)
Hudson, B. *et al*, *Disc. Faraday Soc.*, 1968, **46**, 37 (*use*)
Blum, J. *et al*, *Tetrahedron*, 1973, **29**, 1073 (*use*)

$C_{54}H_{74}N_2P_2Rh^{\oplus}$ Rh-00435

[(1,2,5,6-η)-1,5-Cyclooctadiene][*N,N'*-1,2-ethanediylbis[*N*-[5-methyl-2-(1-methylethyl)cyclohexyl]-*P,P*-diphenylphosphinous amide]-*N,N'*]rhodium(1+), 10CI

1,5-Cyclooctadiene[N,N'-*bis(menthyl)*-N,N'-*bis(diphenylphosphino)ethylene*]*rhodium(1+)*

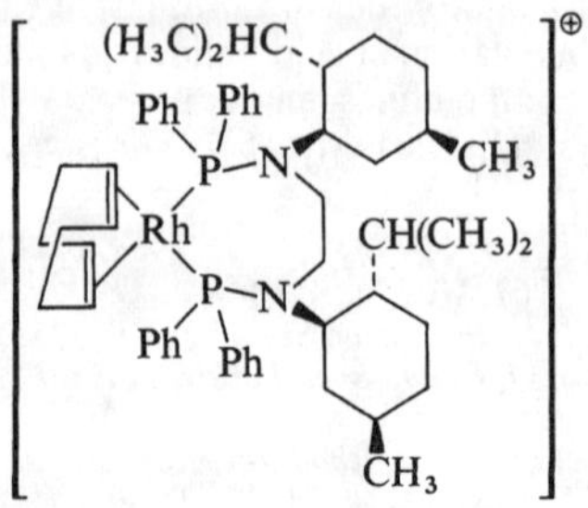

M 916.045 (ion)

(–)-form

Catalyst for the asymmetric hydrogenation of ketones and olefins.

Perchlorate: [66010-35-3].

$C_{54}H_{74}ClN_2O_4P_2Rh$ M 1015.496

Orange-yellow.

Fiorini, M. *et al*, *J. Mol. Catal.*, 1977, **3**, 385 (*use*)
Fiorini, M. *et al*, *J. Mol. Catal.*, 1978, **4**, 125 (*synth, use*)

$C_{55}H_{44}Cl_2F_6OP_4Rh_2$ Rh-00436

[μ-[1,2-Bis(trifluoromethyl)-1,2-ethenediyl]]-μ-carbonyldichlorobis[μ-[methylenebis[diphenylphosphine]-*P:P'*]]dirhodium, 10CI

[Bis-μ-bis(diphenylphosphino)methane]-μ-carbonyldichloro-μ-hexafluorobutynedirhodium

[75503-68-3]

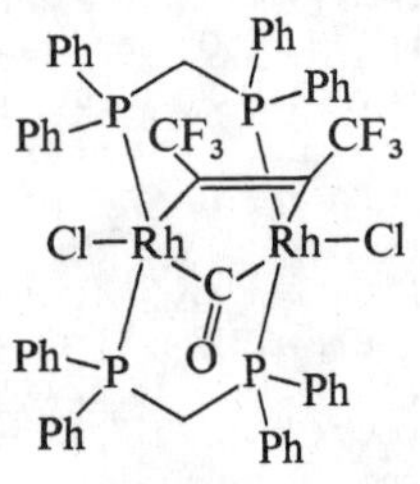

M 1235.554

Green solid. ir ν_{CO} 1705 cm^{-1} (nujol).

Cowie, M. *et al*, *J. Organomet. Chem.*, 1980, **193**, C46 (*synth, ir, nmr*)
Cowie, M. *et al*, *Inorg. Chem.*, 1981, **20**, 2682 (*synth*)
Cowie, M. *et al*, *Inorg. Chem.*, 1982, **21**, 246 (*synth, ir, nmr*)

$C_{55}H_{45}P_3RhS_2^{\oplus}$ Rh-00437

(Dithiocarboxylato)tris(triphenylphosphine)rhodium(1+), 8CI

(π-Carbon disulfide)tris(triphenylphosphine)rhodium

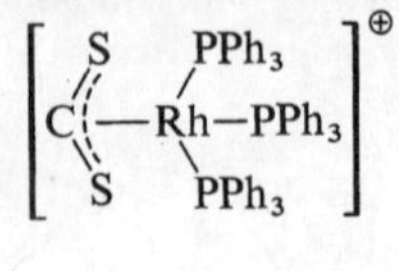

M 965.907 (ion)

Tetraphenylborate: [22257-89-2].

$C_{79}H_{65}BP_3RhS_2$ M 1285.139

Stable as solid but dec. in soln.

Yagupsky, M.P. *et al*, *J. Chem. Soc. (A)*, 1968, 2813 (*synth, ir*)

$C_{55}H_{46}OP_3Rh$ Rh-00438

Carbonylhydrotris(triphenylphosphine)rhodium, 10CI

Carbonylhydridotris(triphenylphosphine)rhodium. Tris(triphenylphosphine)carbonylhydridorhodium. Tris(triphenylphosphine)rhodium carbonyl hydride

[17185-29-4]

$$HRh(PPh_3)_3(CO)$$

M 918.795

Catalyst for hydroformylation and the selective hydrogenation and oxygenation of terminal olefins. Yellow cryst. Mp 120-2° (in air), 172-4° (under N_2). ν_{CO} 1923, ν_{RhH} 2040 cm^{-1} (nujol).

O'Connor, C. *et al*, *J. Chem. Soc. (A)*, 1968, 2665 (*use*)
Gregorio, G. *et al*, *Inorg. Chim. Acta*, 1969, **3**, 89 (*use*)
Kasai, N., *CA*, 1969, **70**, 45 (*pmr, cryst struct*)
Levison, J.J. *et al*, *J. Chem. Soc. (A)*, 1970, 2947 (*synth*)
Inorg. Synth., 1974, **15**, 59 (*synth, ir, pmr*)
Fieser, M. *et al*, *Reagents for Organic Synthesis*, Wiley, 1967-83, **7**, 53.

$C_{55}H_{48}P_3Rh$ Rh-00439

Methyltris(triphenylphosphine)rhodium, 9CI

[15320-81-7]

$$(Ph_3P)_3RhMe$$

M 904.811

Reacts with aryl and alkenyl halides to prod. methylarenes and methyl alkenes. Mp 120-40° dec. Very sensitive to air and moisture.

Keim, W., *J. Organomet. Chem.*, 1968, **14**, 179 (*synth*)
Fieser, M. *et al*, *Reagents for Organic Synthesis*, Wiley, 1967-83, **5**, 463.

$C_{55}H_{52}BO_2P_2Rh$ Rh-00440

[(2,2-Dimethyl-1,3-dioxolane-4,5-diyl)bis(methylene)]bis[diphenylphosphine]-*P,P'*][(η^6-phenyl)triphenylborato(1–)]-rhodium, 10CI

[4,5-Bis(diphenylphosphinomethyl)-2,2-dimethyldioxolan]rhodium η^6-tetraphenylborate. DIOP(η^6-Tetraphenylborato)rhodium

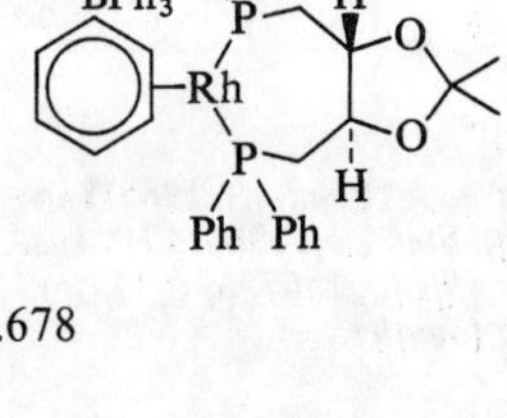

(R,R)-form

M 920.678

(R,R)-form [79790-95-7]

Catalyst precursor for asymmetric hydrogenations. Yellow solid, stable in air for several hours. Sol. CH_2Cl_2. Mp 129-30°. Forms solvates with alcohols.

Dang, T.P. *et al*, *J. Organomet. Chem.*, 1975, **91**, 105 (*synth, use*)
Chauvin, Y. *et al*, *J. Organomet. Chem.*, 1978, **146**, 311 (*synth*)
Brown, J.M. *et al*, *J. Organomet. Chem.*, 1981, **216**, 263 (*synth, nmr*)

$C_{56}H_{40}Cl_2Rh_2$ Rh-00441

Di-μ-chlorobis(tetraphenylcyclobutadiene)dirhodium

Di-μ-chlorobis[1,1′,1″,1‴-(η⁴-1,3-cyclobutadien-e-1,2,3,4-tetrayl)tetrakis[benzene]]dirhodium, 9CI

[50477-60-6]

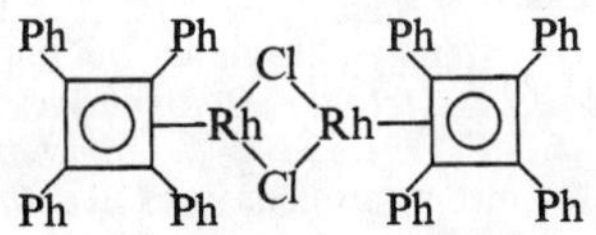

M 989.649

Red cryst. (C_6H_6/hexane). Mp 200° dec.

Monomer: Tetraphenylcyclobutadienerhodium chloride. Chloro(tetraphenylcyclobutadiene)rhodium.
$C_{28}H_{20}ClRh$ M 494.825
Unknown.

Kooti, M. *et al*, *Inorg. Nucl. Chem. Lett.*, 1973, **9**, 1031 (*synth*)
Nixon, J.F. *et al*, *J. Organomet. Chem.*, 1976, **104**, 231 (*synth, ms, ir, pmr*)

$C_{56}H_{48}O_2P_3Rh$ Rh-00442

(Acetato-*O*)tris(triphenylphosphine)rhodium, 10CI

Acetoxytris(triphenylphosphine)rhodium

[34731-03-8]

$Rh(OAc)(PPh_3)_3$

M 948.821

Hydrogenation catalyst for alkenes and alkynes. Orange cryst. (MeOH). Mp 134-6°, 208-9°.

Mitchell, R.W. *et al*, *J. Chem. Soc. (A)*, 1971, 3224 (*synth, use, ir*)
Robinson, S.D. *et al*, *J. Chem. Soc., Dalton Trans.*, 1973, 1912 (*synth, ir*)
Komiya, S. *et al*, *J. Organomet. Chem.*, 1975, **87**, 333 (*nmr, ir*)
Spencer, A. *et al*, *J. Organomet. Chem.*, 1975, **93**, 389 (*synth, use*)
Inorg. Synth., 1977, **17**, 124 (*synth*)

$C_{57}H_{45}O_4P_3Rh_2$ Rh-00443

(Carbon dioxide)dicarbonyltris(triphenylphosphine)dirhodium, 10CI, 9CI

[39448-22-1]

$Rh_2(CO)_2(PPh_3)_3(CO_2)$

M 1092.712

Mod. air-stable yellow needles $+1C_6H_6$. Sol. $CHCl_3$, CH_2Cl_2, C_6H_6. Dec. at 118° in presence of O_2 but stable >150° in an inert atm.

Iwashita, Y. *et al*, *J. Am. Chem. Soc.*, 1969, **91**, 2525 (*synth, ir*)
Lycyak, T.V. *et al*, *Koord. Khim.*, 1982, **8**, 183 (*synth, props*)

$C_{58}H_{40}Cl_2O_2Rh_2$ Rh-00444

Dichlorobis(tetraphenylcyclopentadienone)dirhodium

Bis(tetraphenylcyclopentadienone)dirhodium dichloride

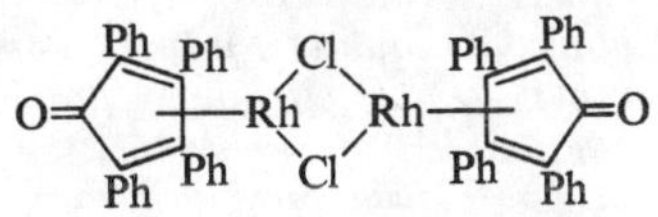

M 1045.670

Orange-red cryst. (cyclohexane). Mp 316-8°.

Monomer: Tetraphenylcyclopentadienonerhodium chloride. Chloro(tetraphenylcyclopentadienone)rhodium. Tetracyclonerhodium chloride.
$C_{29}H_{20}ClORh$ M 522.835
Unknown.

Maitlis, P.M. *et al*, *J. Organomet. Chem.*, 1965, **4**, 254 (*synth*)
McVey, S. *et al*, *Can. J. Chem.*, 1966, **44**, 2429 (*synth, ir*)

$C_{60}H_{44}ClO_2P_2Rh$ Rh-00445

Chloro[(1,4-dihydro-1,4-dioxo-2,3-naphthalenediylidene)bis(phenylmethylidyne)]bis(triphenylphosphine)rhodium, 9CI

[36463-91-9]

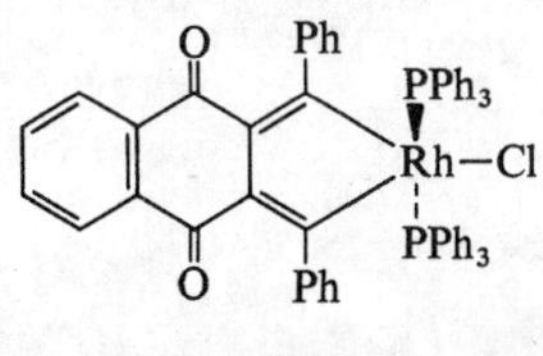

M 997.312

Reacts with alkynes to give substituted quinone derivs. Red-orange cryst. (EtOAc). Mp 244-8° dec.

Müller, E. *et al*, *Justus Liebigs Ann. Chem.*, 1971, **754**, 64 (*synth, use*)
Müller, E. *et al*, *Z. Naturforsch., B*, 1971, **26**, 305 (*struct*)
Müller, E. *et al*, *Justus Liebigs Ann. Chem.*, 1975, 1435 (*use*)
Müller, E. *et al*, *Synthesis*, 1975, 265.
Winter, W., *J. Organomet. Chem.*, 1975, **92**, 97 (*nmr*)

$C_{61}H_{62}ClN_2OP_4Rh_2^{\oplus}$ Rh-00446

μ-Carbonyl-μ-chlorobis(2-isocyano-2-methylpropane)bis[μ--[methylenebis[diphenylphosphine]-*P*:*P*′]]dirhodium(1+), 10CI

*Bis(*tert*-butylisocyanide)-μ-carbonyl-μ-chlorobis[bis(diphenylphosphino)methane]dirhodium(1+)*

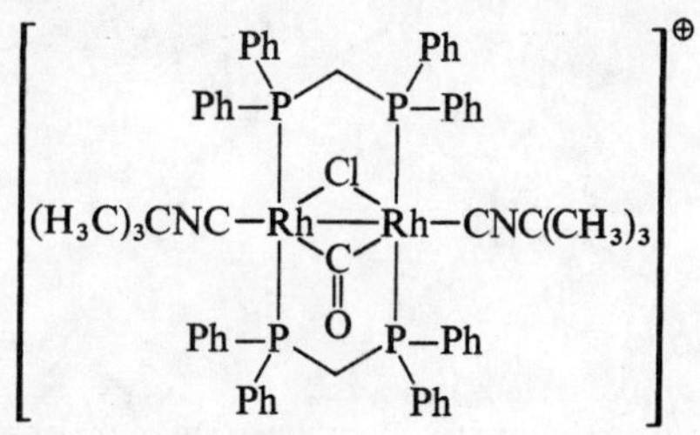

M 1204.333 (ion)

Tetraphenylborate: [74977-91-6].
$C_{85}H_{82}BClN_2OP_4Rh_2$ M 1523.565
Fluffy orange-yellow microcryst. Mp 211° dec. ir ν_{CN} 2155, ν_{CO} 1827 cm^{-1} (nujol).

Mague, J.F. *et al*, *Inorg. Chem.*, 1980, **19**, 3743 (*synth, ir, nmr, uv*)

$C_{62}H_{64}O_4P_4Rh^{\oplus}$ Rh-00447

Bis[[(2,2-dimethyl-1,3-dioxolane-4,5-diyl)bis[methylene]]bis-[diphenylphosphine]-*P,P'*]rhodium(1+), 10CI

Bis[isopropylidene-2,3-dihydroxy-1,4-bis(diphenylphosphino)butane]rhodium(1+). Bis[4,5-bis(diphenylphosphinomethyl)-2,2-dimethyldioxolan]rhodium(1+). Bis(DIOP)rhodium(1+)

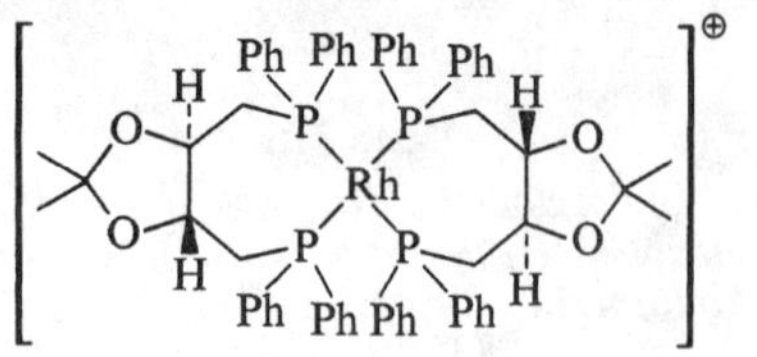

(All-*R*)-*form*

M 1099.986 (ion)

Catalyst for the asymmetric hydrogenation of olefins and the asymmetric hydrosilylation of ketones.

(all-R)-form

Perchlorate: [77447-76-8]. No phys. props. reported.

(all-S)-form

Perchlorate: [60606-76-0].
$C_{62}H_{64}ClO_8P_4Rh$ M 1199.436
Orange cryst. ($EtOH/Et_2O$). $[\alpha]_D^{25}$ −51°.

Tetrafluoroborate: [70196-24-6].
$C_{62}H_{64}BF_4O_4P_4Rh$ M 1186.789
Light-orange solid (CH_2Cl_2/Et_2O). Sol. MeOH, $CHCl_3$. Mp 174-7° dec.

Hexafluorophosphate: [60564-67-2].
$C_{62}H_{64}F_6O_4P_5Rh$ M 1244.950
Dark-red cryst. ($EtOH/Et_2O$). $[\alpha]_D^{25}$ −87°.

Sinou, D. *et al*, *J. Organomet. Chem.*, 1976, **114**, 325 (*synth, pmr*)
James, B.R. *et al*, *Can. J. Chem.*, 1979, **57**, 180 (*synth, nmr, uv, use*)
James, B.R. *et al*, *Can. J. Chem.*, 1980, **58**, 996 (*props*)
Kolb, I. *et al*, *Collect. Czech. Chem. Commun.*, 1980, **45**, 2808 (*synth, pmr, uv, use*)

$C_{62}H_{65}O_4P_4Rh$ Rh-00448

Bis[[(2,2-dimethyl-1,3-dioxolane-4,5-diyl)bis[methylene]]bis-[diphenylphosphine]-*P,P'*]hydrorhodium, 10CI

Hydridobis(DIOP)rhodium

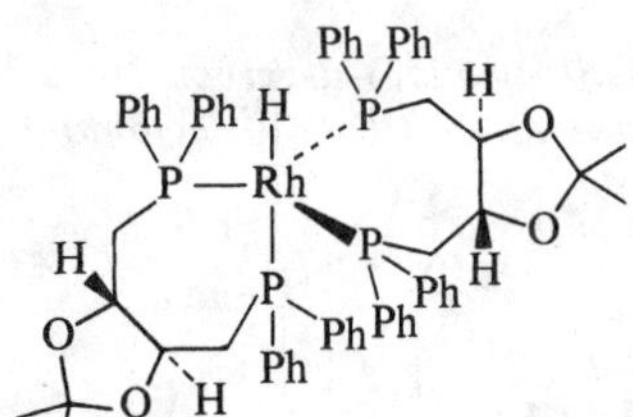

M 1100.994

(all-S)-form [65573-60-6]

Catalyst for asymmetric hydrogenation of olefins. Yellow, air-stable cryst. (CH_2Cl_2/pet. ether). Sol. C_6H_6, CH_2Cl_2, spar. sol. hexane, insol. EtOH. Mp 210° dec. $[\alpha]_D^{25}$ −3.4° (c, 2.5 in C_6H_6). Cryst. from CH_2Cl_2 gives 1:1 solvate Mp 185°. Solns. reversibly absorb 1 mol. CO.

Cullen, W.R. *et al*, *Inorg. Nucl. Chem. Lett.*, 1974, **10**, 167 (*synth, ir, pmr, use*)
Inorg. Synth., **17**, 81 (*synth, ir, pmr*)
Ball, R.G. *et al*, *Inorg. Chem.*, 1981, **20**, 254 (*cryst struct, nmr, use*)

$C_{62}H_{66}O_4P_4Rh^{\oplus}$ Rh-00449

Bis[[(2,2-dimethyl-1,3-dioxolane-4,5-diyl)bis[methylene]bis-[diphenylphosphine]-*P,P'*]dihydrorhodium(1+), 10CI

Dihydridobis[isopropylidene-2,3-dihydroxy-1,4-bis(diphenylphosphino)butane]rhodium(1+). Bis[4,5-bis(diphenylphosphinomethyl)-2,2-dimethyldioxolan]dihydrorhodium(1+). Bis(DIOP)dihydridorhodium(1+)

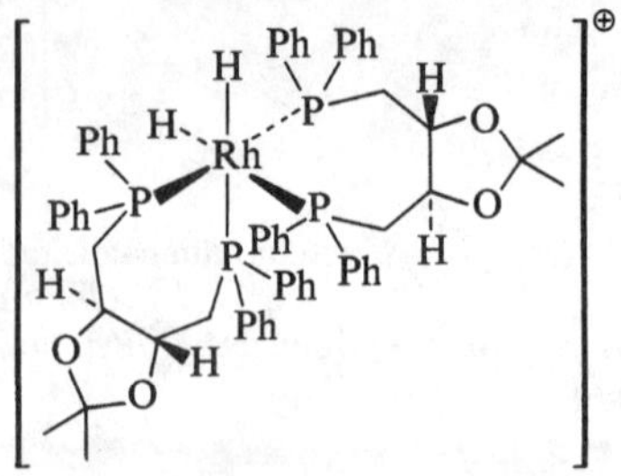

M 1102.002 (ion)

(all-R)-form

Catalyst for asymm. hydrogenations. Readily loses H_2 in soln.

Tetrafluoroborate: [74455-55-3].
$C_{62}H_{66}BF_4O_4P_4Rh$ M 1188.805
White cryst. Sol. CH_2Cl_2. Mp 153° dec. ir ν_{RhH} 2030, 2020 cm^{-1} (nujol).

Hexafluorophosphate: [60564-70-7].
$C_{62}H_{66}F_6O_4P_5Rh$ M 1246.966
White cryst. Sol. CH_2Cl_2. ir ν_{RhH} 2030 cm^{-1} (KBr).

Sinou, D. *et al*, *J. Organomet. Chem.*, 1976, **114**, 325 (*synth, pmr, ir*)
James, B.R. *et al*, *Can. J. Chem.*, 1980, **58**, 996 (*synth, nmr*)

$C_{63}H_{91}N_{13}O_{14}Rh$ Rh-00450

Methylrhodibalamin

Methylrhodibinamide hydroxide dihydrogen phosphate(ester) inner salt 3'-ester with 5,6-dimethyl-1-α-D-ribofuranosyl-1H-benzimidazole, 9CI

[53848-96-7]

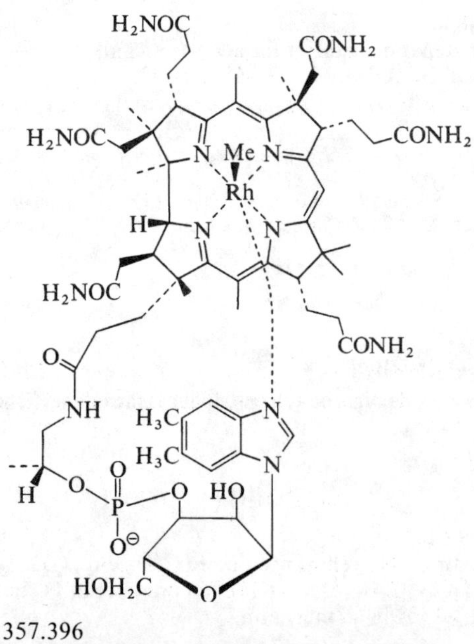

M 1357.396

Thick orange needles (Me_2CO aq.). Sol. H_2O. pmr δ −1.64 (Me, D_2O).

Koppenhagen, V.B. *et al*, *J. Biol. Chem.*, 1974, **249**, 6532 (*synth, uv, pmr, props*)

$C_{70}H_{80}As_4N_4Rh_2^{\oplus\oplus}$ **Rh-00451**

Tetrakis(2-isocyano-2-methylpropane)bis[μ-[methylenebis-[diphenylarsine]-*As*:*As'*]]dirhodium(2+), 10CI

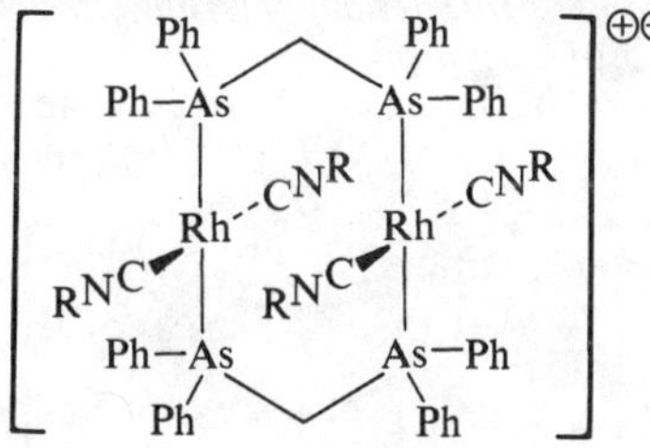

R = $(H_3C)_3C-$

M 1482.926 (ion)

Bis(tetraphenylborate): [69420-82-2].
$C_{118}H_{120}As_4B_2N_4Rh_2$ M 2121.390
Purple cryst. (Me_2CO/Et_2O). Mp 180° dec.

Balch, A.L. *et al*, *Inorg. Chem.*, 1979, **18**, 1224 (*synth, ir, uv*)
Mague, J.F. *et al*, *Inorg. Chem.*, 1980, **19**, 3743 (*synth, ir, nmr, uv*)

$C_{70}H_{80}N_4P_4Rh_2^{\oplus\oplus}$ **Rh-00452**

Tetrakis(2-isocyano-2-methylpropane)bis[μ-[methylenebis-[diphenylphosphine]-*P*:*P'*]]dirhodium(2+), 10CI

Tetrakis(tert-*butylisocyanidebis[bis(diphenylphosphino)methane]dirhodium(2+)*

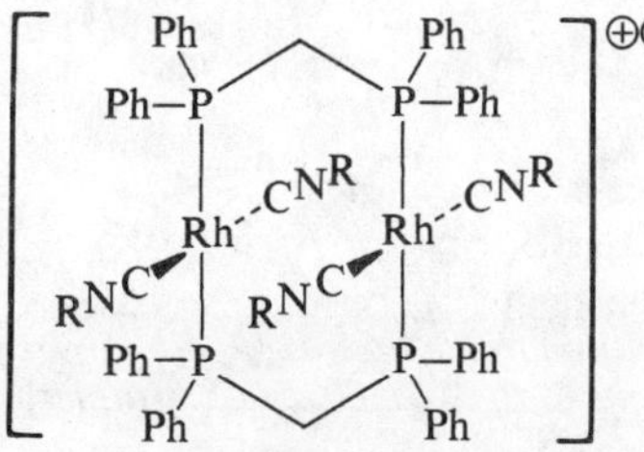

R = $(H_3C)_3C-$

M 1307.135 (ion)

Bis(hexafluorophosphate): [61160-77-8].
$C_{70}H_{80}F_{12}N_4P_6Rh_2$ M 1597.063
Red cryst.

Bis(tetraphenylborate): [74977-80-3].
$C_{118}H_{120}B_2N_4P_4Rh_2$ M 1945.599
No phys. props. reported.

Balch, A.L. *et al*, *J. Am. Chem. Soc.*, 1976, **98**, 8049 (*synth, ir, pmr, uv*)
Mague, J.T. *et al*, *Inorg. Chem.*, 1980, **19**, 3743 (*pmr*)

$C_{72}H_{60}As_4O_2Rh^{\oplus}$ **Rh-00453**

Peroxytetrakis(triphenylarsine)rhodium(1+), 10CI, 9CI

Dioxygentetrakis(triphenylarsine)rhodium

[68538-50-1]

$RhO_2(AsPh_3)_4^{\oplus}$

M 1359.857 (ion)

Reagent for selective oxidn. of terminal alkenes to ketones.

Hexafluorophosphate:
$C_{72}H_{60}As_4F_6O_2PRh$ M 1504.821
No phys. props. reported.

Haines, L.M., *Inorg. Chem.*, 1971, **10**, 1685 (*synth*)
Igersheim, F. *et al*, *J. Chem. Soc., Chem. Commun.*, 1978, 559 (*synth, use*)
Igersheim, F. *et al*, *Nouv. J. Chim.*, 1980, **4**, 161 (*use*)
Fieser, M. *et al*, *Reagents for Organic Synthesis*, Wiley, 1967-83, **8**, 206 (*use*)

$C_{72}H_{60}Cl_2O_4P_4Rh_2S_2$ **Rh-00454**

Di-μ-chlorobis(sulfur dioxide-*S*)tetrakis(triphenylphosphine)-dirhodium, 10CI

[63455-06-1]

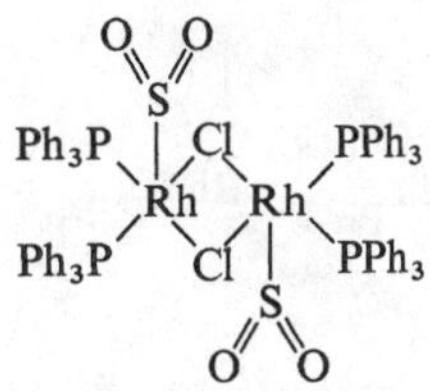

M 1453.996

Red-orange cryst. + $2CHCl_3$ ($CHCl_3/EtOH$). ir ν_{SO} 1166 and 1031 cm^{-1}. Loses $SO_2 + CHCl_3$ on heating.

Monomer: Chloro(sulfur dioxide-S)bis(triphenylphosphine)rhodium.
$C_{36}H_{30}ClO_2P_2RhS$ M 726.998
Unknown.

Kubas, G.J. *et al*, *Cryst. Struct. Commun.*, 1977, **6**, 295 (*cryst struct*)
Kubas, G.J., *Inorg. Chem.*, 1979, **18**, 182 (*synth, ir*)

$C_{72}H_{60}Cl_2P_4Rh_2$ **Rh-00455**

Di-μ-chlorotetrakis(triphenylphosphine)dirhodium, 10CI

[14653-50-0]

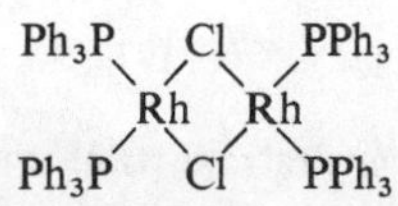

M 1325.878

Catalyst for the hydrogenation of unsaturated hydrocarbons. Red cryst. Almost insol. org. solvs. Mp 260°. Slowly takes up oxygen in air to give Ph_3PO + $Rh-Ph_3PO-O_2$ complexes.

Monomer: Chlorobis(triphenylphosphine)rhodium.
$C_{36}H_{30}ClP_2Rh$ M 662.939
Unknown.

EtOAc complex: [67351-11-5].
$C_{76}H_{68}Cl_2O_2P_4Rh$ M 1311.079
Red cryst. ($CH_2Cl_2/EtOAc$).

Bennett, M.A. *et al*, *Chem. Ind.* (*London*), 1965, 846 (*synth*)
Osborn, J.A. *et al*, *J. Chem. Soc.* (*A*), 1966, 1711 (*synth*)
Bennett, M.A. *et al*, *Inorg. Chem.*, 1967, **6**, 1647 (*ir*)
Tolman, C.A. *et al*, *J. Am. Chem. Soc.*, 1974, **96**, 2762 (*use, uv*)
Geoffroy, G.L. *et al*, *Inorg. Chem.*, 1977, **16**, 205 (*props*)
Curtis, M.D. *et al*, *Inorg. Chem.*, 1978, **17**, 2928 (*struct*)
Naaktgeboren, A.J. *et al*, *J. Am. Chem. Soc.*, 1980, **102**, 3350 (*nmr*)

$C_{72}H_{61}O_{12}P_4Rh$ **Rh-00456**

Hydridotetrakis(triphenyl phosphite)rhodium

Hydrotetrakis(phosphorous acid)rhodium dodecaphenyl ester, 8CI

[24651-65-8]

$RhH[P(OPh)_3]_4$

M 1345.067

Struct. not elucidated. Pale-yellow cryst. ($CH_2Cl_2/MeOH$), unstable in air.

Levison, J.J. *et al*, *J. Chem. Soc., Chem. Commun.*, 1968, 1405 (*synth, pmr*)
Parshall, G.W. *et al*, *J. Am. Chem. Soc.*, 1969, **91**, 4990 (*props*)
Inorg. Synth., 1971, **13**, 105 (*synth*)

$C_{72}H_{61}P_4Rh$ Rh-00457

Hydrotetrakis(triphenylphosphine)rhodium, 10CI
Hydridotetrakis(triphenylphosphine)rhodium
[18284-36-1]

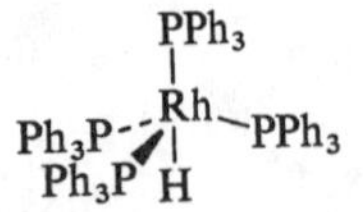

M 1153.074
PPh_3 ligands tetrahedrally arranged around Rh. Catalyst for the hydrogenation and hydrosilylation of olefins. Yellow cryst. (toluene). Mp 145-7°, 162-3°, 168° (154-5° under N_2). ir ν_{RhH} 2147 cm^{-1} (nujol).
C_6H_6 complex (2:1): [27102-75-6].
$C_{150}H_{128}P_8Rh_2$ M 2384.262
Orange cryst. (C_6H_6).

Yamamoto, A. *et al*, *J. Am. Chem. Soc.*, 1968, **90**, 1089 (*synth, ir*)
Dewhirst, K.C. *et al*, *Inorg. Chem.*, 1968, **7**, 546 (*synth, ir, nmr*)
Baker, R.W. *et al*, *J. Chem. Soc., Chem. Commun.*, 1969, 1495 (*cryst struct*)
Ito, T. *et al*, *J. Am. Chem. Soc.*, 1970, **92**, 3011 (*synth, ir*)
Robinson, S.D. *et al*, *J. Chem. Soc. (A)*, 1970, 2947 (*synth, ir*)
Ahmad, N. *et al*, *J. Chem. Soc., Dalton Trans.*, 1972, 843 (*synth*)

$C_{76}H_{60}O_4P_4Rh_2$ Rh-00458

Di-μ-carbonyldicarbonyltetrakis(triphenylphosphine)dirhodium
Bis[μ-carbonylcarbonylbis(triphenylphosphine)rhodium]
[24917-31-5]

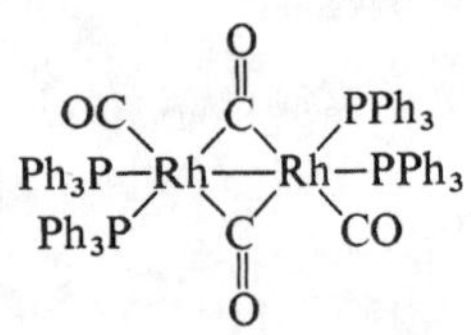

M 1367.014
Hydroformylation and hydrogenation catalyst. Unstable yellow cryst. Insol. hydrocarbons, sol. CH_2Cl_2, EtOH. ir ν_{CO} 2005, 1985, 1790, 1765 cm^{-1} (CH_2Cl_2). Can be handled only in an atm. of CO for short periods. Readily loses 2 mol. CO reversibly in CH_2Cl_2 soln. to form $[Rh(CO)(PPh_3)_2CH_2Cl_2]_2$.
Monomer: Dicarbonylbis(triphenylphosphine)rhodium.
$C_{36}H_{30}O_2P_2Rh$ M 659.485
Unknown.

Evans, D. *et al*, *J. Chem. Soc. (A)*, 1968, 2660 (*synth, ir, props*)
Booth, B.L. *et al*, *J. Organomet. Chem.*, 1971, **27**, 119 (*synth, ir, props*)

$C_{80}H_{60}O_{20}P_4Rh_4$ Rh-00459

Tri-μ-carbonylpentacarbonyltetrakis(triphenyl phosphite-*P*)tetrarhodium, 10CI
Octacarbonyltetra(triphenyl phosphite)tetrarhodium
[62767-14-0]

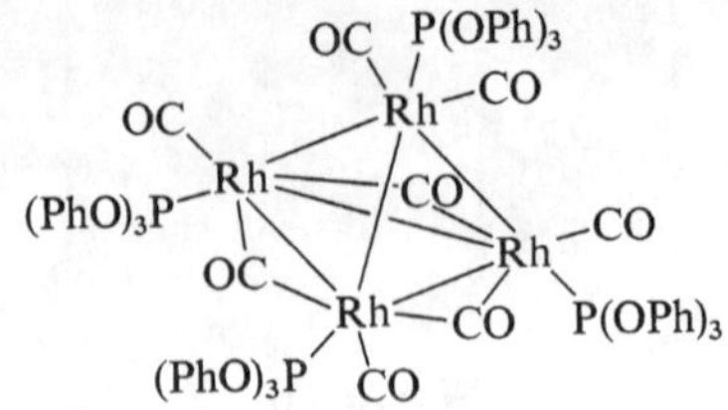

M 1876.859
Cryst. (THF/2-propanol).

Ciani, G. *et al*, *J. Organomet. Chem.*, 1977, **129**, C25 (*cryst struct*)
Heaton, B.T. *et al*, *J. Organomet. Chem.*, 1980, **192**, 431 (*cmr, synth*)

$C_{88}H_{72}O_{10}P_6Rh_6$ Rh-00460

Tetra-μ_3-carbonylhexacarbonyltris[1,2-ethanediylbis[diphenylphosphine]-*P,P'*]hexarhodium
Tetra-μ_3-carbonylhexacarbonyltris[μ-[ethylenebis[diphenylphosphine]]]hexarhodium, 8CI. *Tris[1,2-bis(diphenylphosphino)ethane]decacarbonylhexarhodium*
[12582-63-7]

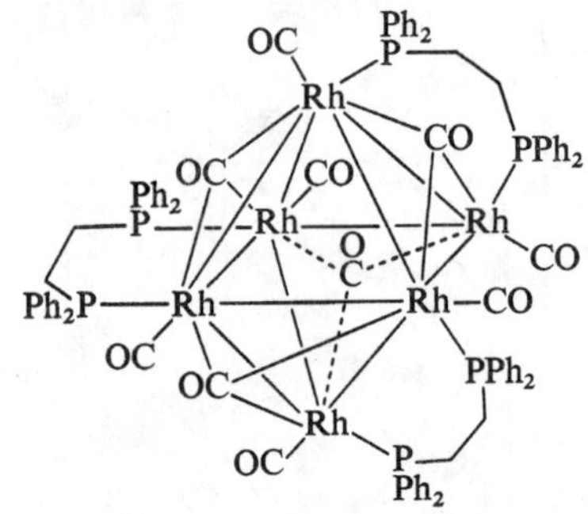

M 2092.806
Proposed struct. shown. Red cryst. ($CHCl_3/Et_2O$). Insol.

Johnson, B.F.G. *et al*, *J. Chem. Soc. (A)*, 1970, 1100 (*synth, ir*)

$C_{118}H_{90}O_{10}P_6Rh_6$ Rh-00461

Decacarbonylhexakis(triphenylphosphine)hexarhodium
Tetra-μ_3-carbonylhexacarbonylhexakis(triphenylphosphine)hexarhodium, 10CI
[11061-21-5]

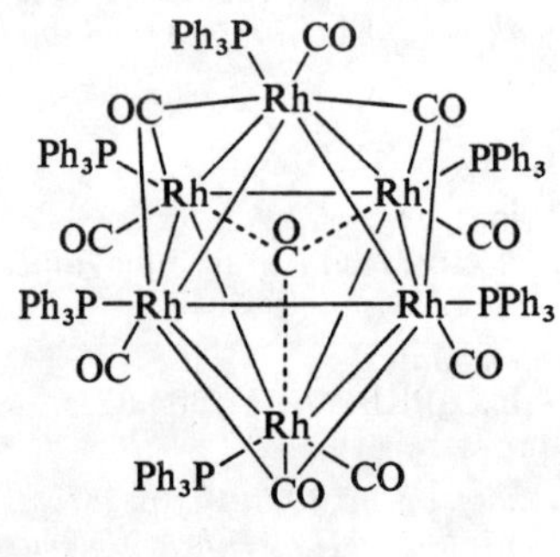

M 2471.279

Proposed struct. shown. Catalyst for hydrogenation of cyclohexene. Dark-red cryst. (CH_2Cl_2/hexane).

Booth, B.L. *et al*, *J. Organomet. Chem.*, 1971, **27**, 119 (*synth, ir*)
Smith, A.K. *et al*, *Fundam. Res. Homogeneous Catal.*, 1979, **3**, 621 (*use*)
Reimann, W. *et al*, *J. Mol. Catal.*, 1980, **9**, 349 (*use*)

$F_{12}HP_4Rh$ Rh-00462

Hydrotetrakis(phosphorous trifluoride)rhodium, 9CI
Hydrotetrakis(trifluorophosphine)rhodium. Tetrakis(trifluorophosphine)rhodium hydride
[16949-48-7]

$RhH(PF_3)_4$

M 455.789
Clear non-viscous liq. with an irritating odour. Mp −40°. Bp_{125} 89°. Stereochemically non-rigid in soln.

Kruck, T. *et al*, *Z. Naturforsch., B*, 1965, **20**, 705; 1966, **21**, 1020 (*synth, ir, nmr*)
Kruck, T. *et al*, *Angew. Chem., Int. Ed. Engl.*, 1967, **6**, 53 (*rev*)
Kruck, T. *et al*, *Chem. Ber.*, 1968, **101**, 3816 (*synth*)
Bennett, M.A. *et al*, *Inorg. Chem.*, 1971, **10**, 2387 (*ir*)
Meakin, P. *et al*, *J. Am. Chem. Soc.*, 1972, **94**, 5271 (*nmr*)
Head, R.A. *et al*, *J. Chem. Soc., Dalton Trans.*, 1975, 2054 (*pe*)
Bennett, M.A. *et al*, *Inorg. Chem.*, 1976, **15**, 2938 (*nmr, struct*)

$F_{24}P_8Rh_2$ Rh-00463

Octakis(phosphorus trifluoride)dirhodium, 9CI
Octakis(trifluorophosphine)dirhodium
[14876-96-1]

$(F_3P)_4RhRh(PF_3)_4$

M 909.563
Orange cryst. (pentane). Mp 100° dec. (92.5°). $Bp_{0.001}$ 20° subl.

Kruck, T. *et al*, *Z. Naturforsch., B*, 1966, **21**, 1020 (*synth, ir*)
Bennett, M.A. *et al*, *Inorg. Chem.*, 1971, **10**, 2387 (*synth, ir, nmr*)
Bennett, M.A. *et al*, *Inorg. Chem.*, 1976, **15**, 2938 (*nmr, struct*)

Name Index

Molecular Formula Index

$C_{33}H_{36}ClFeNO_4PRh$
[(2,3,5,6-η)-Bicyclo[2.2.1]hepta-2,5-diene][1-[1-(dimethylamino)ethyl]-2-diphenylphosphino)-ferrocene-*N*,*P*]rhodium(1+); Perchlorate, *in* Rh-00294

$C_{33}H_{36}F_6FeNP_2Rh$
[(2,3,5,6-η)-Bicyclo[2.2.1]hepta-2,5-diene][1-[1-(dimethylamino)ethyl]-2-diphenylphosphino)-ferrocene-*N*,*P*]rhodium(1+); Hexafluorophosphate, *in* Rh-00294

$C_{33}H_{36}FeNPRh^{\oplus}$
[(2,3,5,6-η)-Bicyclo[2.2.1]hepta-2,5-diene][1-[1-(dimethylamino)ethyl]-2-diphenylphosphino)-ferrocene-*N*,*P*]rhodium(1+), Rh-00294

$C_{33}H_{39}N_4Rh_2$
Methyl[2,7,12,17-tetraethyl-3,8,13,18-tetramethyl-21*H*,23*H*-porphinato(2−)-N^{21},N^{22},N^{23},N^{24}]rhodium, Rh-00295

$C_{33}H_{40}PRh$
[(1,2,3,4,5-η)-1,2,3,4,5-Pentamethyl-2,4-cyclopentadien-1-yl]-1,5-pentanediyl(triphenylphosphine)rhodium, Rh-00296

$C_{34}H_{25}ORh$
(η^5-2,4-Cyclopentadien-1-yl)[(2,3,4,5-η)-2,3,4,5-tetraphenyl-2,4-cyclopentadien-1-one]rhodium, Rh-00297

$C_{34}H_{26}BCoF_4O$
1-Hydroxy-2,3,4,5-tetraphenylcobaltocenium(1+); Tetrafluoroborate, *in* Co-00316

$C_{34}H_{26}BrCo$
η^6-Benzene(η^4-tetraphenylcyclobutadiene)cobalt(1+); Bromide, *in* Co-00315

$C_{34}H_{26}BrCoO$
1-Hydroxy-2,3,4,5-tetraphenylcobaltocenium(1+); Bromide, *in* Co-00316

$C_{34}H_{26}ClCoO_5$
1-Hydroxy-2,3,4,5-tetraphenylcobaltocenium(1+); Perchlorate, *in* Co-00316

$C_{34}H_{26}Co^{\oplus}$
η^6-Benzene(η^4-tetraphenylcyclobutadiene)cobalt(1+), Co-00315

$C_{34}H_{26}CoO^{\oplus}$
1-Hydroxy-2,3,4,5-tetraphenylcobaltocenium(1+), Co-00316

$C_{34}H_{26}F_6CoP$
η^6-Benzene(η^4-tetraphenylcyclobutadiene)cobalt(1+); Hexafluorophosphate, *in* Co-00315

$C_{34}H_{28}CoF_6P_3$
1,1′-Bis(diphenylphosphinyl)cobaltocenium(1+); Hexafluorophosphate, *in* Co-00317

$C_{34}H_{28}CoP_2^{\oplus}$
1,1′-Bis(diphenylphosphinyl)cobaltocenium(1+), Co-00317

$C_{34}H_{31}BCoN$
Aminocobaltocenium; Tetraphenylborate, *in* Co-00067

$C_{34}H_{32}BCoN_2$
1,1′-Diaminocobaltocenium; Tetraphenylborate, *in* Co-00068

$C_{34}H_{34}BF_4OP_2Rh$
[(2,3,5,6-η)Bicyclo[2.2.1]hepta-2,5-diene][2,3-bis(diphenylphosphino)-1-propanol-*P*,*P*′]-rhodium(1+); Tetrafluoroborate, *in* Rh-00298

$C_{34}H_{34}ClO_4P_2Rh$
[(2,3,5,6-η)Bicyclo[2.2.1]hepta-2,5-diene][(1-methyl-1,2-ethanediyl)bis[diphenylphosphine]-*P*,*P*′]-rhodium(1+); Perchlorate, *in* Rh-00299

$C_{34}H_{34}OP_2Rh^{\oplus}$
[(2,3,5,6-η)Bicyclo[2.2.1]hepta-2,5-diene][2,3-bis(diphenylphosphino)-1-propanol-*P*,*P*′]-rhodium(1+), Rh-00298

$C_{34}H_{34}P_2Rh^{\oplus}$
[(2,3,5,6-η)Bicyclo[2.2.1]hepta-2,5-diene][(1-methyl-1,2-ethanediyl)bis[diphenylphosphine]-*P*,*P*′]-rhodium(1+), Rh-00299

$C_{34}H_{35}B_{10}IrNOP$
Benzonitrilecarbonyl(triphenylphosphine)[7-phenyl-1,7-dicarba-*closo*-dodecaborane(12)]iridium, Ir-00154

$C_{34}H_{36}BF_4IrP_2$
[(1,2,5,6-η)-1,5-Cyclooctadiene][1,2-ethanediylbis[diphenylphosphine]-*P*,*P*′]iridium(1+); Tetrafluoroborate, *in* Ir-00155

$C_{34}H_{36}BF_4P_2Rh$
[(1,2,5,6-η)-1,5-Cyclooctadiene][1,2-ethanediylbis[diphenylphosphine]-*P*,*P*′]rhodium(1+); Tetrafluoroborate, *in* Rh-00300

$C_{34}H_{36}ClO_4P_2Rh$
[(1,2,5,6-η)-1,5-Cyclooctadiene][1,2-ethanediylbis[diphenylphosphine]-*P*,*P*′]rhodium(1+); Perchlorate, *in* Rh-00300

$C_{34}H_{36}F_6IrP_3$
[(1,2,5,6-η)-1,5-Cyclooctadiene][1,2-ethanediylbis[diphenylphosphine]-*P*,*P*′]iridium(1+); Hexafluorophosphate, *in* Ir-00155

$C_{34}H_{36}F_6P_2RhSb$
[(1,2,5,6-η)-1,5-Cyclooctadiene][1,2-ethanediylbis[diphenylphosphine]-*P*,*P*′]rhodium(1+); Hexafluoroantimonate, *in* Rh-00300

$C_{34}H_{36}IrP_2^{\oplus}$
[(1,2,5,6-η)-1,5-Cyclooctadiene][1,2-ethanediylbis[diphenylphosphine]-*P*,*P*′]iridium(1+), Ir-00155

$C_{34}H_{36}P_2Rh^{\oplus}$
[(1,2,5,6-η)-1,5-Cyclooctadiene][1,2-ethanediylbis[diphenylphosphine]-*P*,*P*′]rhodium(1+), Rh-00300

$C_{34}H_{37}F_6OP_4Rh$
[Bis[3-(diphenylphosphino)propyl]phenylphosphine-*P*,*P*′,*P*″]carbonylrhodium(1+); Hexafluorophosphate, *in* Rh-00346

$C_{34}H_{38}F_6IrP_3$
(1,2,5,6-η)-1,5-Cyclooctadienebis(methyldiphenylphosphine)-iridium(1+); Hexafluorophosphate, *in* Ir-00156

$C_{34}H_{38}IrP_2^{\oplus}$
(1,2,5,6-η)-1,5-Cyclooctadienebis(methyldiphenylphosphine)-iridium(1+), Ir-00156

$C_{34}H_{40}BF_4FeNPRh$
[(1,2,5,6-η)-1,5-Cyclooctadiene][1-[1-(dimethylamino)-ethyl]-2-(diphenylphosphino)ferrocene-*N*,*P*]-rhodium(1+); Tetrafluoroborate, *in* Rh-00301

$C_{34}H_{40}ClFeNO_4PRh$
[(1,2,5,6-η)-1,5-Cyclooctadiene][1-[1-(dimethylamino)-ethyl]-2-(diphenylphosphino)ferrocene-*N*,*P*]-rhodium(1+); Perchlorate, *in* Rh-00301

$C_{34}H_{40}Co_2Mo_2O_4$
Di-μ-carbonyl[μ_4-(η^2-carbonyl-*C:C:C*)][μ-carbonylbis[(1,2,3,4,5-η)-1,2,3,4,5-pentamethyl-2,4-cyclopentadien-1-yl]dicobalt]bis(η^5-2,4-cyclopentadien-1-yl)dimolybdenum, Co-00318

$C_{34}H_{40}FeNPRh^{\oplus}$
[(1,2,5,6-η)-1,5-Cyclooctadiene][1-[1-(dimethylamino)-ethyl]-2-(diphenylphosphino)ferrocene-*N*,*P*]-rhodium(1+), Rh-00301

$C_{34}H_{40}ORh_2$
μ-Carbonyl[μ-(diphenylmethylene)]bis[(1,2,3,4,5-η)-1,2,3,4,5-pentamethyl-2,4-cyclopentadien-1-yl]-dirhodium, Rh-00302

$C_{34}H_{42}PRh$
1,6-Hexanediyl[(1,2,3,4,5-η)-1,2,3,4,5-pentamethyl-2,4-cyclopentadien-1-yl](triphenylphosphine)rhodium, Rh-00303

$C_{35}H_{20}BrIr_4O_{11}P$
Bromotri-μ-carbonyloctacarbonyltetrairidate(1−); Tetraphenylphosphonium salt, *in* Ir-00045

$C_{40}H_{48}ClO_5P_2Rh$
[(1,2,5,6-η)-1,5-Cyclooctadiene][1,2-dimethyl-1,2-ethanediylbis[diphenylphosphine]-*P*,*P'*]rhodium(1+); Perchlorate, THF complex, *in* Rh-00316

$C_{40}H_{48}ClN_{16}Rh_4^{\oplus\oplus\oplus\oplus\oplus}$
Dichlorooctakis[μ-(1,3-disocyanopropane)]tetrarhodium(5+), Rh-00381

$C_{40}H_{48}Cl_2N_{16}Rh_4^{\oplus\oplus\oplus\oplus}$
Dichlorooctakis[μ-(1,3-disocyanopropane)]-tetrarhodium(4+), Rh-00382

$C_{40}H_{48}F_6N_2P_3Rh$
[*N*,*N'*-1,2-Cyclohexanediylbis[*N*-methyl-*P*,*P*-diphenylphosphinous amide]-*P*,*P'*][(1,2,5,6-η)-1,5-cyclooctadiene]rhodium(1+); Hexafluorophosphate, *in* Rh-00383

$C_{40}H_{48}N_2P_2Rh^{\oplus}$
[*N*,*N'*-1,2-Cyclohexanediylbis[*N*-methyl-*P*,*P*-diphenylphosphinous amide]-*P*,*P'*][(1,2,5,6-η)-1,5-cyclooctadiene]rhodium(1+), Rh-00383

$C_{40}H_{48}N_{16}Rh_4^{\oplus\oplus\oplus\oplus\oplus\oplus}$
Octakis[μ-(1,3-diisocyanopropane)]tetrarhodium(6+), Rh-00384

$C_{40}H_{49}N_4O_{24}Rh_{13}$
Tetracosacarbonylhydrotridecarhodate(4−); Tetrakis(tetramethylammonium) salt, *in* Rh-00240

$C_{40}H_{50}BCoN_4$
Dimethyl(2,3,9,10-tetramethyl-1,4,8,11-tetraazacyclotetradeca-1,3,8,10-tetraene-N^1,N^4,N^8,N^{11})cobalt(1+); Tetraphenylborate, *in* Co-00199

$C_{40}H_{51}Cl_{17}Co_4N_{16}Rh_4$
Chlorooctakis[μ-(1,3-disocyanopropane)]tetrarhodium(5+); Trihydrogentetrakis(tetrachlorocobaltate), *in* Rh-00381

$C_{40}H_{64}B_2F_8Rh_4$
Tetra-μ-hydrotetrakis[(1,2,3,4,5-η)-1,2,3,4,5-pentamethyl-2,4-cyclopentadien-1-yl]-tetrarhodium(2+); Bis(tetrafluoroborate), *in* Rh-00385

$C_{40}H_{64}Cl_4PtRh_4$
Tetra-μ-hydrotetrakis[(1,2,3,4,5-η)-1,2,3,4,5-pentamethyl-2,4-cyclopentadien-1-yl]-tetrarhodium(2+); Tetrachloroplatinate, *in* Rh-00385

$C_{40}H_{64}F_{12}P_2Rh_4$
Tetra-μ-hydrotetrakis[(1,2,3,4,5-η)-1,2,3,4,5-pentamethyl-2,4-cyclopentadien-1-yl]-tetrarhodium(2+); Bis(hexafluorophosphate), *in* Rh-00385

$C_{40}H_{64}Rh_4^{\oplus\oplus}$
Tetra-μ-hydrotetrakis[(1,2,3,4,5-η)-1,2,3,4,5-pentamethyl-2,4-cyclopentadien-1-yl]-tetrarhodium(2+), Rh-00385

$C_{40}H_{72}Co_2N_8$
Octakis(2-isocyano-2-methylpropane)dicobalt, Co-00335

$C_{41}H_{23}O_{17}PRh_6$
Pentadecacarbonyl(methoxycarbonyl)hexarhodate(1−); Tetraphenylphosphonium salt, *in* Rh-00186

$C_{41}H_{35}CoP_2$
η^5-2,4-Cyclopentadien-1-ylbis(triphenylphosphine)-cobalt, Co-00336

$C_{41}H_{35}P_2Rh$
(η^5-2,4-Cyclopentadien-1-yl)bis(triphenylphosphine)-rhodium, Rh-00386

$C_{41}H_{36}BN_2Rh$
[(2,3,5,6-η)-Bicyclo[2.2.1]hepta-2,5-diene](2,2'-bipyridine-*N*,*N'*)rhodium(1+); Tetraphenylborate, *in* Rh-00187

$C_{41}H_{37}IrOP_2$
[(1,2,3-η)-2-Butenyl]carbonylbis(triphenylphosphine)-iridium, Ir-00208
Carbonyl(1-methyl-1-propenyl)bis(triphenylphosphine)-iridium, Ir-00209
Carbonyl[(1,2,3-η)-2-methyl-2-propenyl]-bis(triphenylphosphine)iridium, Ir-00210

$C_{41}H_{37}O_8P_2Rh$
(2,4-Pentanedionato-*O*,*O'*)bis(triphenyl phosphite-*P*)rhodium, Rh-00387

$C_{41}H_{38}BN_2Rh$
[(2,3,5,6-η)-Bicyclo[2.2.1]hepta-2,5-diene]-bis(pyridine)rhodium(1+); Tetraphenylborate, *in* Rh-00188

$C_{41}H_{39}As_2ClNO_2Rh$
Chloro(dioxygen)(2-isocyano-2-methylpropane)-bis(triphenylarsine)rhodium, Rh-00388

$C_{41}H_{39}ClNO_2P_2Rh$
Chloro(dioxygen)(2-isocyano-2-methylpropane)-bis(triphenylphosphine)rhodium, Rh-00389

$C_{41}H_{40}BF_4OP_2Rh$
[(2,3,5,6-η)-Bicyclo[2.2.1]hepta-2,5-diene][[[1-(phenylmethoxy)methyl]-1,2-ethanediyl]-bis[diphenylphosphine]-*P*,*P'*]rhodium(1+); Tetrafluoroborate, *in* Rh-00390

$C_{41}H_{40}OP_2Rh^{\oplus}$
[(2,3,5,6-η)-Bicyclo[2.2.1]hepta-2,5-diene][[[1-(phenylmethoxy)methyl]-1,2-ethanediyl]-bis[diphenylphosphine]-*P*,*P'*]rhodium(1+), Rh-00390

$C_{41}H_{43}ClNO_7P_2Rh$
[(1,2-Dimethyl-1,2-ethanediyl)bis[diphenylphosphine]-*P*,*P'*][(2,3-η)-ethyl-2-(acetylamino)-3-phenyl-2-propenoate-O^2]rhodium(1+); Perchlorate, *in* Rh-00391

$C_{41}H_{43}IrO_2P_2Si_2$
Carbonylhydro(1,1,3,3-tetramethyl-1,3-disiloxanediyl)-bis(triphenylphosphine)iridium, Ir-00211

$C_{41}H_{43}NO_3P_2Rh^{\oplus}$
[(1,2-Dimethyl-1,2-ethanediyl)bis[diphenylphosphine]-*P*,*P'*][(2,3-η)-ethyl-2-(acetylamino)-3-phenyl-2-propenoate-O^2]rhodium(1+), Rh-00391

$C_{41}H_{48}BCoN_4$
(2,12-Dimethyl-3,7,11,17-tetraazabicyclo[11.3.1]-heptadeca-1(17),2,11,13,15-pentaene-N^3,N^7,N^{11},N^{17})dimethylcobalt(1+); Tetraphenylborate, *in* Co-00209

$C_{41}H_{48}N_4O_{25}Rh_{14}$
Pentacosacarbonyltetradecarhodate(4−); Tetrakis(tetramethylammonium) salt, *in* Rh-00259

$C_{42}H_{30}Cl_2Ir_2O_6P_2Sn$
Hexacarbonyl[μ-(dichlorostannylene)]-bis(triphenylphosphine)diiridium, Ir-00212

$C_{42}H_{30}Co_2O_6P_2$
Hexacarbonylbis(triphenylphosphine)dicobalt, Co-00337

$C_{42}H_{30}Co_2O_{12}P_2$
Hexacarbonylbis(triphenyl phosphite)dicobalt, Co-00338

$C_{42}H_{30}Ir_2O_6P_2$
Hexacarbonylbis(triphenylphosphine)diiridium, Ir-00213

$C_{42}H_{36}ClIrO_4P_2$
Chlorobis(triphenylphosphine)(dimethyl acetylenedicarboxylate)iridium, Ir-00214

$C_{42}H_{38}BIrN_2$
[(1,2,5,6-η)-1,5-Hexadiene](1,10-phenanthroline-N^1,N^{10})-iridium(1+); Tetraphenylborate, *in* Ir-00099

$C_{42}H_{38}ClO_4P_2Rh$
[(1,2,3,4-η)-1,3-Cyclohexadiene]bis(triphenylphosphine)-rhodium(1+); Perchlorate, *in* Rh-00392

$C_{42}H_{38}P_2Rh^{\oplus}$
[(1,2,3,4-η)-1,3-Cyclohexadiene]bis(triphenylphosphine)-rhodium(1+), Rh-00392

$C_{42}H_{39}ClOP_3Rh$
Carbonylchloro[1,1,1-tris(diphenylphosphinomethyl)-ethane]rhodium, Rh-00393

CAS Registry Number Index

43

15015